CELLULAR MECHANOTRANSDUCTION

"Mechanotransduction" is the term for the ability, first described by nineteenth-century anatomist Julius Wolff, of living tissues to sense mechanical stress and respond by tissue remodeling. More recently, the scope of mechanotransduction has been expanded to include the sensation of stress, its translation into a biochemical signal, and the sequence of biological responses it produces. This book looks at mechanotransduction in a more restricted sense, focusing on the process of stress sensing and transducing a mechanical force into a cascade of biochemical signals. This stress has become increasingly recognized as one of the primary and essential factors controlling biological functions, ultimately affecting the function of the cells, tissues, and organs. A primary goal of this broad book is also to help define the new field of mechanomics, which attempts to describe the complete mechanical state of a biological system.

Dr. Mohammad R. K. Mofrad is currently Assistant Professor of Bioengineering at the University of California, Berkeley, where he is also an affiliated faculty member of graduate programs in applied science and technology and biophysics. Dr. Mofrad received his B.A.Sc. degree from Sharif University of Technology. After earning M.A.Sc. and Ph.D. degrees from the Universities of Waterloo and Toronto, respectively, he spent two years at MIT and Harvard Medical School/Massachusetts General Hospital as a post-doctoral Fellow. Before joining the faculty at Berkeley, Dr. Mofrad was a Principal Research Scientist at MIT for nearly two years. At Berkeley, he has developed and taught several courses, including Cell Mechanics and Mechanotransduction and Molecular Cell Biomechanics. He is the founder of the Mechanotransduction Knowledgebase Web site, mechanotransduction.org.

Dr. Roger D. Kamm has long been interested in biomechanics, beginning with his work in vascular and pulmonary physiology and leading to his more recent work in cell and molecular mechanics in the context of cellular responses to mechanical stress. Dr. Kamm has been on the faculty at MIT since receiving his Ph.D. in 1977 and now holds a joint appointment in the Biological Engineering and Mechanical Engineering Departments. He is currently the Chair of the U.S. National Committee on Biomechanics and the World Council on Biomechanics, and he is Director of the Global Enterprise for MicroMechanics and Molecular Medicine. Kamm has a long-standing interest in bioengineering education, directs a National Institute of Health–funded biomechanics training program; co-chaired the committee to form MIT's new undergraduate major in biological engineering; and helped to develop MIT's course on molecular, cellular, and tissue biomechanics.

Cellular Mechanotransduction

DIVERSE PERSPECTIVES FROM MOLECULES TO TISSUES

Edited by
Mohammad R. K. Mofrad
University of California Berkeley
Roger D. Kamm
Massachusetts Institute of Technology

CAMBRIDGE
UNIVERSITY PRESS

32 Avenue of the Americas, New York NY 10013-2473, USA

Cambridge University Press is part of the University of Cambridge.

It furthers the University's mission by disseminating knowledge in the pursuit of education, learning and research at the highest international levels of excellence.

www.cambridge.org
Information on this title: www.cambridge.org/9781107682467

First published 2010
First paperback edition 2014

A catalogue record for this publication is available from the British Library

Library of Congress Cataloguing in Publication data

Mofrad, Mohammad R. K.
Cellular mechanotransduction : diverse perspectives from molecules to tissues / edited by Mohammad Mofrad, Roger Kamm.
p. cm.
Includes bibliographical references and index.
ISBN 978-0-521-89523-1 (hardback)
1. Cells–Mechanical properties. I. Kamm, Roger D. II. Title.

QH645.5.M64 2009
571.6–dc22 2008049985

ISBN 978-0-521-89523-1 Hardback
ISBN 978-1-107-68246-7 Paperback

Contents

Contributors *page* vii
Preface xi

1. **Introduction**
Roger D. Kamm and Mohammad R. K. Mofrad 1
2. **Endothelial Mechanotransduction**
Peter F. Davies and Brian P. Helmke 20
3. **Role of the Plasma Membrane in Endothelial Cell Mechanosensation of Shear Stress**
Peter J. Butler and Shu Chien 61
4. **Mechanotransduction by Membrane-Mediated Activation of G-Protein Coupled Receptors and G-Proteins**
Yan-Liang Zhang, John A. Frangos, and Mirianas Chachisvilis 89
5. **Cellular Mechanotransduction: Interactions with the Extracellular Matrix**
Andrew D. Doyle and Kenneth M. Yamada 120
6. **Role of Ion Channels in Cellular Mechanotransduction – Lessons from the Vascular Endothelium**
Abdul I. Barakat and Andrea Gojova 161
7. **Toward a Modular Analysis of Cell Mechanosensing and Mechanotransduction: A Manual for Cell Mechanics**
Benjamin J. Dubin-Thaler and Michael P. Sheetz 181
8. **Tensegrity as a Mechanism for Integrating Molecular and Cellular Mechanotransduction Mechanisms**
Donald E. Ingber 196
9. **Nuclear Mechanics and Mechanotransduction**
Shinji Deguchi and Masaaki Sato 220

10. **Microtubule Bending and Breaking in Cellular Mechanotransduction**
Andrew D. Bicek, Dominique Seetapun, and David J. Odde 234

11. **A Molecular Perspective on Mechanotransduction in Focal Adhesions**
Seung E. Lee, Roger D. Kamm, and Mohammad R. K. Mofrad. 250

12. **Protein Conformational Change: A Molecular Basis of Mechanotransduction**
Gang Bao 269

13. **Translating Mechanical Force into Discrete Biochemical Signal Changes: Multimodularity Imposes Unique Properties to Mechanotransductive Proteins**
Vesa P. Hytönen, Michael L. Smith, and Viola Vogel 286

14. **Mechanotransduction through Local Autocrine Signaling**
Nikola Kojic and Daniel J. Tschumperlin 339

15. **The Interaction between Fluid-Wall Shear Stress and Solid Circumferential Strain Affects Endothelial Cell Mechanobiology**
John M. Tarbell 360

16. **Micro- and Nanoscale Force Techniques for Mechanotransduction**
Nathan J. Sniadecki, Wesley R. Legant, and Christopher S. Chen. 377

17. **Mechanical Regulation of Stem Cells: Implications in Tissue Remodeling**
Kyle Kurpinski, Randall R. R. Janairo, Shu Chien, and Song Li 403

18. **Mechanotransduction: Role of Nuclear Pore Mechanics and Nucleocytoplasmic Transport**
Christopher B. Wolf and Mohammad R. K. Mofrad 417

19. **Summary and Outlook**
Mohammad R. K. Mofrad and Roger D. Kamm 438

Index 445

Color plates follow page 180

Contributors

Gang Bao
Department of Biomedical Engineering
Georgia Institute of Technology and
Emory University

Abdul I. Barakat
Department of Mechanical and Aeronautical Engineering
University of California, Davis

Andrew D. Bicek
Department of Biomedical Engineering
University of Minnesota

Peter J. Butler
Department of Bioengineering
The Pennsylvania State University

Mirianas Chachisvilis
La Jolla Bioengineering Institute
La Jolla, California

Christopher S. Chen
Department of Bioengineering
University of Pennsylvania

Shu Chien
Departments of Bioengineering and Medicine
University of California, San Diego

Peter F. Davies
Department of Pathology and Laboratory Medicine
Institute for Medicine and Engineering
University of Pennsylvania

Shinji Deguchi
Graduate School of Natural Science and Technology
Okayama University

Andrew D. Doyle
National Institute of Dental and Craniofacial Research
National Institutes of Health

Benjamin J. Dubin-Thaler
Cell Motion Laboratories, Inc.

John A. Frangos
La Jolla Bioengineering Institute
La Jolla, California

Andrea Gojova
Department of Mechanical and Aeronautical Engineering
University of California, Davis

Brian P. Helmke
Department of Biomedical Engineering
University of Virginia

Vesa P. Hytönen
Laboratory of Biologically Oriented Materials, Department of Materials
ETH Zurich

Donald E. Ingber
Departments of Pathology and Surgery
Harvard Medical School and Children's Hospital

Randall R. R. Janairo
Department of Bioengineering
University of California, Berkeley

Roger D. Kamm
Departments of Mechanical Engineering and Biological Engineering
Massachusetts Institute of Technology

Nikola Kojic
Harvard-MIT Division of Health Science and Technology
Boston, Massachusetts

Kyle Kurpinski
Department of Bioengineering
University of California, Berkeley

Seung E. Lee
Department of Bioengineering
University of California, Berkeley

Wesley R. Legant
Department of Bioengineering
University of Pennsylvania

Song Li
Department of Bioengineering
University of California, Berkeley

Mohammad R. K. Mofrad
Department of Bioengineering
University of California, Berkeley

David J. Odde
Department of Biomedical Engineering
University of Minnesota

Masaaki Sato
Department of Bioengineering and Robotics
Tohoku University

Dominique Seetapun
Department of Biomedical Engineering
University of Minnesota

Michael P. Sheetz
Department of Biological Sciences
Columbia University

Michael L. Smith
Laboratory of Biologically Oriented Materials, Department of Materials
ETH Zurich

Nathan J. Sniadecki
Department of Bioengineering
University of Pennsylvania

John M. Tarbell
Department of Biomedical Engineering
The City College of New York

Daniel J. Tschumperlin
Molecular and Integrative Physiological Sciences
Harvard School of Public Health

Viola Vogel
Laboratory of Biologically Oriented Materials, Department of Materials
ETH Zurich

Christopher B. Wolf
Department of Bioengineering
University of California, Berkeley

Kenneth M. Yamada
National Institute of Dental and Craniofacial Research
National Institutes of Health

Yan-Liang Zhang
La Jolla Bioengineering Institute
La Jolla, California

Preface

Many studies during the past two decades have shed light on a wide range of cellular responses to mechanical stimulation. It is now widely accepted that stresses experienced *in vivo* are instrumental in numerous pathologies. One of the first diseases found to be linked to cellular stress was atherosclerosis, where it was demonstrated that hemodynamic shear stress influences endothelial function, and that conditions of low or oscillatory shear stress are conducive to the formation and growth of atherosclerotic lesions. Even before then, the role of mechanical stress on bone growth and healing was widely recognized, and since then, many other stress-influenced cell functions have been identified.

Many have investigated the signaling cascades that become activated as a consequence of mechanical stress, and these are generally well characterized. The initiating process, however, by which cells convert the applied force into a biochemical signal, termed "mechanotransduction," is much more poorly understood, and only recently have researchers begun to unravel some of these fundamental mechanisms. Various processes and theories have been proposed to explain this phenomenon. The objective of this book is to bring together these different viewpoints to cellular mechanotransduction, ranging from the molecular basis of mechanotransduction phenomena to the tissue-specific events that lead to such processes. Our intent is to present in a single text the many and varied ways in which cellular mechanotransduction is viewed and, in doing so, spur on new experiments to test the theories, or the development of new theories themselves. We view this as an ongoing debate, where one of the leading proponents of each viewpoint could present his or her most compelling arguments in support of the model, so that members of the larger scientific community could form their own opinions. As such, this was intended to be a monograph that captured the current state of a rapidly evolving field. Since we began this project, however, it has been suggested that this book might meet the growing need for a text for courses taught specifically on cellular mechanotransduction. More broadly, it could be used to introduce concepts at the intersection of mechanics and biology, a field of study that has come to be termed "mechanobiology." Or, even more broadly, this collection might be useful as supplemental readings for a course that covers a range of topics in molecular, cellular, and tissue biomechanics. In any event, our hope is that this presentation might prove stimulating and educational to engineers, physicists, and biologists wishing to expand their understanding of the critical importance of mechanotransduction in cell function, and the various ways in which it might be understood.

In the end, we would like to extend our enormous thanks to all the contributing authors whose expert contributions made this text possible. Finally, we wish to express our deepest gratitude to Mr. Peter Gordon and his colleagues at Cambridge University Press, who provided us with the encouragement, technical assistance, and overall guidance that were essential to the ultimate success of this endeavor. In addition, we would like to acknowledge Ms. Elise Oranges, who steered us through the final stages of editing.

Mohammad R. K. Mofrad
University of California Berkeley

Roger D. Kamm
Massachusetts Institute of Technology

1

Introduction

Roger D. Kamm and Mohammad R. K. Mofrad

1.1 Mechanotransduction – Historical Development

Julius Wolff, a nineteenth-century anatomist, first observed that bone will adapt to the stresses it experiences and is capable of remodeling if the state of stress changes. This became known as Wolff's Law and stands today as perhaps the earliest recognized example of the ability of living tissues to sense mechanical stress and respond by tissue remodeling (see Chapter 17 for a detailed historical review). More recently, the term "mechanotransduction" has been introduced to represent this process, often including the sensation of stress, its transduction into a biochemical signal, and the sequence of biological responses it produces. Here we use mechanotransduction in a somewhat more restricted sense, and specifically use it for the process of stress sensing itself, transducing a mechanical force into a cascade of biochemical signals.

Since Wolff's early insight, the influence of mechanical force or stress has become increasingly recognized as one of the primary and essential factors controlling biological function. We now appreciate that the sensation of stress occurs at cellular or even subcellular scales, and that nearly every tissue and every cell type in the body is capable of sensing and responding to mechanical stimuli. Another manifestation of mechanotransduction is known as Murray's Law [1, 2], which states that the flow rate passing through a given artery scales with the third power of its radius. This has been widely recognized to be a response of the arterial endothelium and the smooth muscle cells to remodel the arterial wall to maintain a nearly constant level of hemodynamic shear stress (at $\sim$ 1 Pa), leading to the third power relationship.[1] One aspect of this response is the alignment of endothelial cells in the direction of stress, first observed in studies of arterial wall morphology [4], and later vividly demonstrated in controlled *in vitro* experiments [5]. Other biological factors, such as soft tissue remodeling [6], changes in the thickness of the arterial wall in response to circumferential stress [7], calcification in the heart valve tissue in response to pathological solid and fluid mechanical patterns, and bone loss in microgravity [8, 9], have all been found to be influenced by mechanical stress.

[1] Although accepted for years, recent evidence casts doubt on the validity of Murray's Law and suggests instead that flow rate varies as the vessel radius to the second power [3].

Mechanotransduction is also instrumental in our other senses, touch and hearing in particular [72]. Hearing, for example, is recognized to be mediated by the tension produced in a small filament, termed a tip-link, connecting adjacent stereocilia that project from the surface of the inner and outer hair cells in the form of a conical bundle of fine filaments and can be in direct contact with the tectorial membrane. Oscillations of the membrane cause the stereocilia to slide relative to one another, inducing tension in the tip-link. As a result of this tension, a channel is activated that leads to an increase in calcium ion concentration, initiating a signal transmitted to the brain and heard as sound.

Not only are cells exquisitely sensitive to *externally* imposed stress, but they also *generate* stresses internally, by actomyosin contractions, for example, that allow a cell to probe its mechanical environment, presumably through the response of the surrounding extracellular matrix to these internally generated forces. This is likely an important factor in biological development, guiding cells through a series of mechanical cues (in addition to the more widely studied biochemical ones), and influencing cellular differentiation (see Chapter 17). Stresses have recently been shown to guide the differentiation of stem cells (see Chapter 17); mesenchymal stem cells will differentiate into an osteogenic phenotype when subjected to low levels of strain [10], but into a cardiovascular lineage at higher strains [11]. Other types of cell behavior are also influenced by the stiffness of the matrix on which they are grown, and it is becoming clear that cells can sense their mechanical environment and respond accordingly [12]. Phenomena such as these give rise to the concept of mechanical signaling, both outside-in and inside-out, discussed again later in this chapter and this book.

1.2 Role of Mechanotransduction in Disease

Aside from its central role in a variety of normal, even essential, biological functions, mechanotransduction has a dark side, in that it has also been demonstrated to be a major factor in many pathological processes. We have known for many years that thickening and calcification of the arterial wall associated with atherosclerosis occurs predominantly at localized sites in the circulation of "disturbed flow" – regions prone to complex flow patterns, or low and possibly reversing hemodynamic shear stress. Studies over the past 30 years have led to an increasing appreciation of the central role played by the arterial endothelium, in the initial thickening of the arterial wall intima [13], to the recruitment and activation of circulating monocytes [14, 15], to the changes in endothelial permeability [16, 17], all of which contribute to disease progression. Other studies have demonstrated a link between mechanotransduction and arthritis [18], damage to articular cartilage [19], asthma [20, 21], other types of pulmonary diseases and lung injuries [22], and polycystic kidney disease [23]. (See [24] for a recent review.) These processes are mediated by a host of signaling cascades that are initiated by shear stress, and these are discussed in Chapter 2.

In this context, it is useful to discuss the magnitude of the mechanical stimuli that elicit a biological response. For example, in the vascular system, mean values of shear

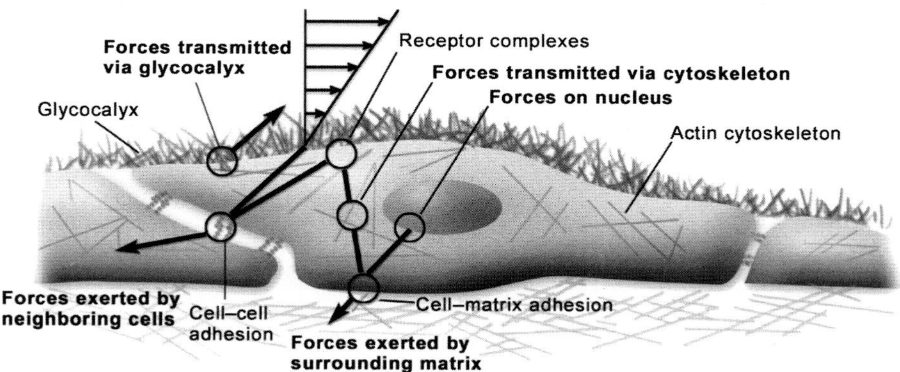

Figure 1.1. Forces experienced by endothelial cells lining a vessel wall (in bold) and the structures, both intra- and extracellular, that transmit these forces (nonbold).

stress associated with arterial blood flow can range from 0.1 to 10 Pa, with even wider variations observed instantaneously during the cardiac cycle; ample evidence exists to suggest that endothelial cells sense this level of stress and regulate their behavior accordingly. Similarly, changes in internal pressure give rise to circumferential strains in the arterial wall in the range of 2 to 18% [25] (see also Chapter 16). Strains of comparable magnitude occur in the lung, so all the cell types contained either in the lung or arterial wall are subjected to these levels of deformation. Airway epithelial cells have also been shown to be responsive to transepithelial pressures in the range of those induced by airway smooth muscle activation, about 1 to 4 kPa [26], and alveolar epithelium has been demonstrated to be stretch sensitive [22]. At the other extreme is bone, where the strains are much smaller, on the order of 1000 $\mu\varepsilon$ [73]. Even these minute strains, however, are known to be sensed by resident cells, and it has been suggested that bone possesses a special mechanism to enhance sensitivity [27].

1.3 *In Vitro* Tests of Mechanosensation

Its critical role in disease has led investigators to develop a wide variety of experimental tools to probe the effects of mechanotransduction, both *in vivo* and *in vitro*. And because they enable closer control of the various factors, *in vitro* experiments have proven to be particularly informative. These can be categorized in terms of the nature in which force is applied, as illustrated in Figure 1.1, and are discussed in more detail in Chapter 16.

Shear Stress. One of the first observed manifestations of force on cell function was the alignment of endothelial cells subjected to shear stress (Figure 1.2(a)), so it is not surprising that shear stress was one of the first methods used to elicit a response *in vitro*. Several geometries have been utilized including simple unidirectional flow chambers, where the cells, grown on one surface of a rectangular channel, are

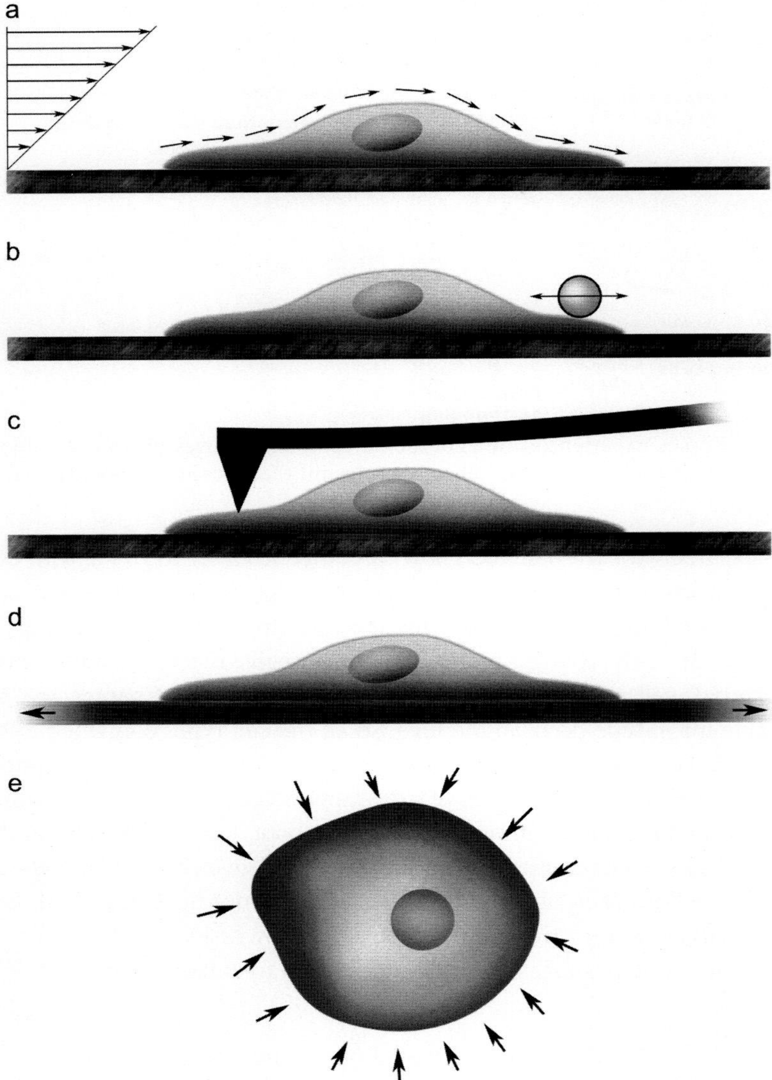

Figure 1.2. Various methods used to apply force to cells either *in vitro* or *in vivo*. (a) Fluid shear stress. (b) Forces applied to microbeads that are tethered to the cell via membrane receptors. (c) Indentation by an atomic force microscope probe. (d) Substrate stretch. (e) Hydrostatic pressure.

exposed to the shear associated with a fully developed flow of medium. Advantages of this technique are its simplicity and, provided an adequate entrance length is used to produce fully developed flow, a uniform shear stress that can be either steady or unsteady, depending only on the ability to produce time-dependent flow waveforms. An alternative system that also produces a well-defined, time-dependent, and spatially uniform shear stress distribution is the cone-and-plate rheometer [5]. By controlling the rotation of the cone, any time-varying shear stress existing in the circulation can be reproduced. One complicating factor, however, is that the shear stress acting on a given cell is nonuniform due to its uneven surface contour, and even the average shear stress can vary from cell to cell in a given monolayer, due to surface variations [28]. In addition, transmission of shear stress to the cell is known to be mediated by the glycocalyx, a surface glycoprotein layer that coats the apical surface of most endothelial monolayers (see Chapter 2). Not surprisingly, the presence or absence of a glycocalyx has been found to be a major determinant of the cell's response to shear stress [29].

Bead Forcing. At times, it can be useful to apply force in a more localized manner, and with improved force or displacement control or measurement accuracy. To meet these objectives, many have turned to the use of micron-sized beads or microspheres that can be tethered to the cell's membrane receptors by using coatings of an appropriate ligand and manipulated using either magnetic or optical traps (Figure 1.2(b)). In a magnetic trap, paramagnetic or ferromagnetic beads are used, and force is generated by an externally imposed magnetic field. Either linear force or rotational torque can be applied, while the bead's motion is monitored optically. With an optical trap or tweezers, forces are produced that draw the bead toward the center of a focused laser beam, and displacements are, once again, measured optically. In either of these cases, interpretation of the force-displacement data is subject to a number of uncertainties (e.g., strength of attachment to the cell, assumptions concerning the relative importance of membrane and cytoskeleton, active response of the cell to forcing), however, and these methods have been criticized on the basis that they are nonphysiological. Interestingly, endothelial cells exhibit a definitive response to bead forcing at a force level of about 1 nN, roughly corresponding to the shear stress of 1 Pa integrated over the surface area of a typical cell, suggesting that similar processes may be responsible for an endothelial cell's response to blood flow [30].

Atomic Force Microscopy (AFM). Similar experiments can be conducted using the tip of an AFM, either in its normal configuration or with a microbead attached to the cantilever in place of the pyramidal tip (Figure 1.2(c)). An advantage of the AFM is its excellent spatial resolution, but among its disadvantages are that it is difficult to simultaneously apply force to the cell and observe its response.

Substrate Stretch. Cells in the heart, the walls of arteries and veins, and in the lung are all subjected to cyclic strain or stretch, which also influences their behavior. Experimentalists have developed a variety of methods and devices to simulate these effects. In most, cells are grown on a flexible membrane and the membrane is stretched either by mechanical or hydraulic/pneumatic actuation (Figure 1.2(d)). Stretch can be uniaxial or biaxial, and either static or oscillatory. It can also act in a synergistic manner with shear stress, as discussed in Chapter 15. Cells in three-dimensional matrices or gels can also be subjected to strains, for example, by unconfined compression of the gel, which better mimics the environment of chondrocytes, for example.

One of the often ignored limitations of all of these experiments, and virtually all *in vitro* experiments, in fact, is that they almost universally are limited in duration to several days. Some *in vivo* studies have examined changes that occur over longer periods of time; these rarely look at the mechanism of mechanotranduction, but rather, the long-term remodeling that occurs in response to mechanical stimuli. These time-scales need to be compared to those of disease progression, where the changes generally occur over many years. For example, cartilage or joint damage from sports injuries often leads to degeneration and arthritis later in life. Apparently mechanical trauma initiates a sequence of biological events that ultimately lead to deterioration of the cartilage. While short-term experiments are obviously enormously useful and provide important insights into the longer term processes from studies of the initial event, some degree of speculation regarding the detailed connections between the two is always necessary, and represents an important area of ongoing research.

Often, the response to strain is complex. Of particular interest is that stem cells have been shown to change their differentiation pathway depending on the *level* of strain they experience, differentiating into an osteogenic lineage under low strains [10] but a vascular or muscle lineage at higher strains, greater than about 5% [11].

Hydrostatic Pressure. In most instances of mechanotransduction, there is a clear and measurable deformation that occurs in connection with the applied force, so most of the mechanisms described in the following are possible. With hydrostatic pressure (Figure 1.2(e)), however, especially at normal physiological loads, the amount of deformation experienced by the cell is minute, corresponding essentially to the compressibility of water, so the mechanisms of force sensation are less obvious. Numerous studies have been published, however, showing cellular responses to changes in hydrostatic pressure as small as 0.4 kPa [31–33], and it has been postulated that the response may be associated with a corresponding change in membrane-free volume and membrane fluidity, and consequently in the mobility of membrane-associated proteins [34]. Since the mechanism remains unclear, this continues to be an active area of investigation.

1.4 A Focus on Basic Mechanisms

Numerous reviews have been written addressing the signaling pathways that become activated by mechanical stress, the second messengers that convey these signals, and the changes in biological function that occur due to changes in gene expression, protein synthesis, and post-translational processing (see, e.g., [35–37]). In this collection of chapters, we focus instead on the fundamental mechanisms by which a cell senses and transduces mechanical force. That is, we address the factors that activate the various signaling pathways, ultimately leading to the observed biological response, and refer the reader to these other excellent sources for illumination of the detailed pathways that lie downstream of these initiating events. Each of these basic mechanisms is discussed in at least one chapter in this book, and we provide here just a brief summary of the concepts described in much more detail in later chapters.

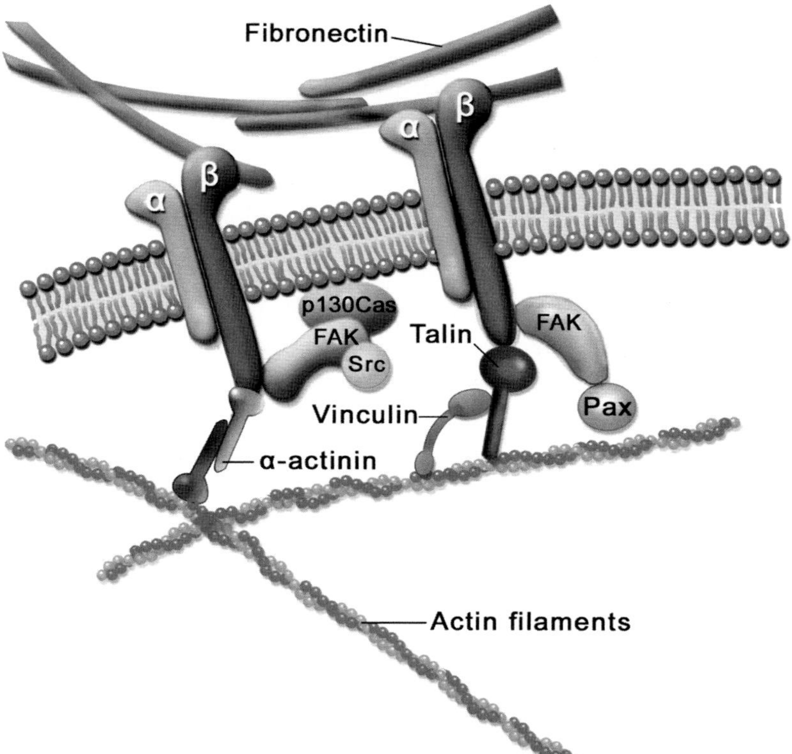

Figure 1.3. Some of the proteins within a focal adhesion that transmit forces from the extracellular matrix, across the cell membrane, to the cytoskeleton. Any of these force-transmitting proteins are candidates for mechanosensation through force-induced conformational change.

In the simplest of terms, mechanotransduction can be viewed as any force-induced process that initiates a biochemical response. That response could be as simple as changing the binding affinity of one protein to another or altering the phosphorylation state of a protein; or the response could be more complex, such as initiating a signaling pathway with a range of downstream consequences including changes in gene expression, protein synthesis, or change in cellular phenotype.

In the quest for mechanosensors, it seems logical to look first at those sites where forces might be amplified. Some of these are obvious, such as the sites where the cell anchors itself to its environment, either at focal adhesions (Figure 1.3) or cell–cell junctions. Others are less evident, since we know that the cell has the capability to focus forces at locations remote from the site of force application (e.g., [38]) and that forces such as shear stress, for example, tend to be distributed over large regions of the cell. In this section, we discuss evidence for a variety of transduction mechanisms, some requiring localized force and others for which the forces are spatially dispersed.

Stretch-Activated Channels. An enormous variety of ion and water channels have been identified and characterized, a small subset of which has been identified or proposed to be mechanosensitive (e.g., see Chapter 6). Some of these have been well characterized, and a few have even been modeled in a way that provides some insight into the transduction mechanism involved, two of which will be mentioned here. Even these, however, as well as others that are much less well understood, remain the subject of considerable debate.

One channel that has received considerable attention is found in the stereocilia of hair cells in the inner ear, which initiate the signals that are ultimately transmitted to the central nervous system, allowing us to hear. The mechanics of this system are fascinating, especially with regard to the mechanism of activation, which has largely been discovered. A cone-shaped collection of stereocilia is located on the top of a hair cell in the inner ear. These communicate at their tips with the tectorial membrane, which is in contact with the fluid of the inner ear and oscillates due to the propagation of waves in the chochlea [74]. As the membrane oscillates, the stereocilia move back and forth, and as they do, the tension in a small filament that connects the tip of one cilium with the side of its neighbor varies. Although the detailed arrangement is still being elucidated, a stretch-activated channel is known to be located near the attachment point of the filament on the lateral side of the cilium. As force is applied by the connecting filament, the channel's conductance changes, giving rise to a transient rise in Ca^{2+} concentration. Although the details are less clear, there also appears to be a mechanism to adjust the resting level of stress acting on the channel, and this provides a potential means of "tuning" its sensitivity (see Chapter 6).

A second channel, the mechanosensitive channel of large conductance (MscL), is found in bacteria but is notable because its crystal structure is known and it can therefore be studied by molecular dynamics. The MscL is known to be activated at levels of tension of $\sim$ 10 mN/m [39]. Two groups have investigated the change in conformation of MscL when membrane stress is applied [40, 41], showing that an initial conformational change can be induced by tensions of this magnitude. However, full activation was not achieved in this simulation, suggesting that this may be the initiating event in a sequence that ultimately leads to the observed change in conductance.

Shear stress or flow is also known to give rise to changes in the channel conductance of, for example, Na^+, K^+, or Ca^{2+}; however, the mechanisms are less well understood and could result from force interactions of the channel either with membrane lipids or the cortical matrix [42–44], and in the case of membrane interactions, either membrane tension or membrane curvature has been implicated [42, 44]. Whatever the mechanism, these channels are exquisitely sensitive, being affected by shear stresses as low as 0.01 Pa. Given the low energy levels required for shear activation (as low as $\sim$ 0.01 kT), it seems unlikely that flow is the direct effector of channel activation in the presence of thermal noise [45]. What role the cortical cytoskeleton or glycocalyx plays in this process remains a subject of debate [46].

Membrane Mechanotransduction. Other membrane-associated proteins have also been implicated in mechanotransduction, in particular G-proteins, G-protein coupled receptors, and various proteins that are found in focal adhesion complexes or the cytoskeleton. These mechanisms are discussed in detail in Chapter 4.

Experiments on G-proteins isolated in lipid vesicles and subjected to shear can become activated by shear [47], showing convincingly that G-proteins by themselves can act as mechanosensors, and that these effects appear to be associated with changes in membrane viscosity.

Changes in membrane fluidity have also been proposed as an initiating event. These ideas arose from early work by Butler [48], demonstrating that membrane fluidity increased in response to shear stress. They hypothesized that this could initiate a mechanoresponsive event through one of several mechanisms. First, increased membrane fluidity implies an increase in diffusivity of the transmembrane, or membrane-associated proteins, and lipids. If the reaction is diffusion limited, then an increase in fluidity would be expected to increase the likelihood of protein interaction [49]. G-proteins and G-protein complexes have been the focus of much attention, since G-protein hydrolysis is diffusion dependent and G-proteins are often implicated in mechanosensation.

Membrane stress, either in tension or in bending, can also directly influence the conformation of transmembrane proteins, or consequently could influence their tendency for activation or interaction with other proteins. This is especially relevant to structures such as calveoli [77], where a relatively minor increase in membrane tension could produce large changes in membrane curvature. This has recently been reported in connection with MAPK activation [50]. Other potential sites that might be influenced by membrane stress include lipid rafts, where G_i-proteins (Chapter 4), frequently implicated in mechanotransduction, are often found. Stress might also alter the thickness of the lipid bilayer, which, due to the complex hydrophobic interactions within the membrane and their influence on protein conformation, could also influence protein function. The primary mechanisms for these effects remain largely unknown.

Mechanotransduction in Focal Adhesion Complexes. A considerable amount of evidence has been reported regarding the role of focal adhesion proteins in transduction events (see Chapter 5). In one set of experiments, cells were first grown on a compliant substrate, and then the cell membranes were removed by application of a detergent, Triton X. The cells were then stretched in the presence of a variety of cytoskeletal proteins that contained a photocleavable botin tag [51]; by comparing the newly bound proteins to stretched and nonstretched cells, those proteins that preferentially bind to stretched cytoskelatal networks could be identified. From these experiments, binding of paxillin, focal adhesion kinase, p130Cas, and PKB/Akt were all found to be enhanced under 10% stretch, providing convincing evidence that mechanotransduction is not simply a membrane-mediated process, and that the focal adhesion complex contains a variety of proteins whose binding affinities are influenced by stretch.

Another vivid demonstration of this can be found in experiments by Wang et al. [52], who developed an assay for activation consisting of phosphorylation of a domain taken from a cSRC subtrate, p130Cas. Activation, in this case, produced a conformational change that could be observed by FRET and exhibited a wave of activation emanating from the site at which a tethered bead applied force to the cell.

In another recently developed technique, the fact that cysteines, which are generally buried in the protein core due to their hydrophobicity and inaccessible, become exposed by the application of force to a cell or single protein [53] is used as a means of detecting these changes in conformation. Introduction of a thiol-reactive fluorescent dye (IAEDANS) to the cell creates a fluorescent signal indicative of cysteine exposure and binding to the IAEDANS. This has been used to explore unfolding in spectrin, nonmuscle myosin IIA, and vimentin, but the method has the potential to be applied to a wide variety of proteins.

Role of the Glycocalyx in Mechanotransduction. Recently, it has been increasingly recognized, especially in the context of endothelial mechanotransduction, that the lipid bilayer is rarely subjected to fluid shear stress directly, but that stresses are instead transmitted via the glycocalyx, the glycoprotein layer that coats most endothelial cells *in vivo* and *in vitro*. Studies have convincingly demonstrated that many of the known responses of the endothelial cells to fluid shear are dramatically influenced by whether or not this layer is intact [54, 55]. This dependence seems a logical consequence of the fact that forces transmitted via the glycocalyx connect with different intracellular structures than forces applied directly to the bilayer. Studies are now attempting to determine which membrane or intracellular structures ultimately bear the load from the glycocalyx, and how these forces are subsequently distributed throughout the cell [29].

Cytoskeletal Transduction. Since the cytoskeleton is the primary pathway for force or stress transmission through the cell, some have suggested that one or more of its component proteins might serve as mechanosensors (see, e.g., Chapters 7, 8, and 10). Some direct evidence for this already exists. Several cytoskeletal proteins have been suggested as mechanosensors, including some actin cross-linking proteins [75] as well as microtubules [56]. Again, these seem likely candidates due to their role in force transmission. For example, forces sufficient to bend or possibly break a microtubule can influence the rate of filament growth or the binding of microtubule-associated proteins [56]. Similarly, forces acting through actin cross-linking proteins can rupture the bond or cause domain unfolding, either of which can lead to cytoskeletal remodeling or changes in the actin microstructure.

Direct Effects of Force on Gene Expression. It has been demonstrated in various ways, both *in vivo* and *in vitro*, that forces are transmitted to the nucleus via the surrounding cytoskeleton, causing changes in the nuclear shape [57, 58]. Just as forces acting on cytoskeletal proteins can change their conformation, DNA can be unwound under applied force to expose a transcription sequence. Pulling on a single strand of DNA can cause histone release and nucleosomal disruption [59], so it is not unreasonable to hypothesize that force could also influence gene expression and replication. Although it has been demonstrated that forces are, indeed, transmitted to the nucleus, evidence for direct control of gene expression by these transmitted forces has not yet been reported. Forces might also be generated internal to the nucleus, since it contains contractile proteins as well as their substrate filaments. These issues are discussed in more depth in Chapter 9. It is also conceivable that the external or internal forces might affect nucleocytoplasmic transport via the nuclear pore complexes, which can ultimately affect the regulation of cellular phenotype.

Protein Unfolding or Conformational Change. Many of the postulated mechanisms for mechanotransduction rely on protein conformational change or domain unfolding to initiate phosphorylation, altered enzymatic activity, altered binding, or other biochemical events that can lead to activation of one of the signaling pathways (see Chapters 11, 12, and 13). In the case of stretch-activated ion channels, it seems clear that channel activation involves a change in conformation of the channel that leads to an increase in pore dimension. Other cases are less clear. Attention has focused on proteins that are likely to be found along the force transmission pathway or where forces tend to be focused, for example, those located in focal adhesions, cell–cell adhesions, or in cytoskeleton-associated proteins. Some of the most recent insights have come from experiments, either on living cells or with isolated proteins. AFM studies have been conducted to pull on individual proteins and determine the force required for unfolding, which tend to fall in the range of 50–200 pN [76].

Computational methods have also shed light on protein unfolding, although the force levels needed to unravel a protein within the time scale of a fully explicit molecular dynamics calculation (in which each of the water molecules is included) tend to be much higher than are typical in AFM since high force is needed to reduce the computational times to reasonable levels, on the order of tens of nanoseconds. Simulations of, for example, fibronectin [60] and spectrin [61] employ forces in the range of 500–1000 pN. (See Chapter 12 for further details.)

Other recent simulation studies suggest that a more subtle conformational change might also lead to protein activation. With force levels on the order of 20 pN, it has been shown that the vinculin binding site on talin, normally buried within a five-helix bundle in the talin rod domain, can rotate out of the helix core when force is applied [62]. Once rotated, this domain becomes accessible for binding. In the case of talin, it seems likely that unfolding is not required, since the structure of the bound complex is consistent with a simple helix exchange from talin to vinculin, with the secondary structure of the binding domain remaining intact. An alternative mechanism has also been proposed [63] involving a breakdown of the rod domain in order to expose the vinculin bindng site.

Constrained Autocrine Signaling. A totally different mechanism of mechanotransduction is conceivable involving forces that act to produce changes at the cellular rather than the molecular scale (see Chapter 14). Studies had shown that stresses applied to a layer of airway epithelial cells grown on a porous membrane exhibit changes in gene expression [20], signaling [21], and ERK phosphorylation [26], but the mechanism remained a mystery for a number of years. It was finally discovered that the mechanism was related to the tendency for the cells to spread out when subjected to a transmembrane pressure, reducing the lateral intercellular space. The lateral confinement associated with a reduction in fluid volume leads to an elevation in the concentration of the EGF ligand within the lateral intercellular space (LIS). As the ligand concentration rises, activation of the EGF receptor increases, thus initiating EGF receptor signaling [26]. Some of the observed biological consequences of this are upregulation of the transforming growth factor β (TGF-β) and early growth response–1 (egr-1) [20]. Over the longer term, these could subsequently contribute to the wall thickening observed in asthma, as a consequence of repeated activation of the smooth muscle.

In reality, there are likely to be many mechanisms of force sensing, and certainly not just one. Cells have apparently evolved so that they can respond to different types of force, and do so by activation of numerous signaling pathways. Some of the responses are mediated by changes in gene expression, but many are not. This gives rise to the ability of a cell to initiate both immediate and long-term responses.

1.5 Force Levels That Are Sensed by Cells

In order to gauge the significance of mechanotransduction, it is useful to consider the levels of force (or energy) needed for mechanical activation of a cell or protein, and compare these to the forces (or thermal energy) that either occur at rest or that a cell might be expected to experience. One might consider the following two "design criteria" for effective mechanosensation: 1) The energy associated with activation should be above the threshold of thermal energy, since otherwise activation would be frequent and random, simply due to Brownian fluctuations. 2) The force levels should exceed those experienced by a cell under normal, resting conditions. One might add a further criterion: The activation level should not be too much greater than these threshold levels in order to achieve the greatest sensitivity. As we will see (Table 1.1), it appears that proteins exist in a state in which they are primed for mechanical activation, in that the forces or energies required to produce significant conformational changes lie just above the thermal noise floor, an exquisite balance between protein stability and conformational dynamics.

Resting levels of stress might arise due to a variety of factors, such as intracellular contraction due to the actomyosin machinery intrinsic to the cytoskeleton. As a result, even cells plated onto an adherent surface or within a three-dimensional matrix are under some level of baseline stress. Presumably, a cell would want to have the capability to sense changes from this baseline condition, as a signal to change its homeostatic state. Stresses at rest in the focal adhesions formed in various types of cells have been measured by a method called traction force microscopy, in which the tension exerted by a cell on a flexible substrate (e.g., compliant gels, thin elastic membranes, or flexible posts) is inferred from the deformations of the substrate along with a knowledge of its elastic properties [64–67]. By this method, the shear stress exerted by fibroblasts on a fibronectin-coated elastic membrane has been found to be about 5.5 kPa [66], and this figure seems to be relatively independent of the size of the adhesion complex. We can estimate the average tensile stress within the cell by taking this figure, estimating that about 1% of the cell–substrate interface

Table 1.1. *Level of force, stress, or strain known to elicit a response from cells.*

Stimulus type	Magnitude	Reference
Membrane tension	2 mN/m	[61]
Shear stress (endothelial cells)	1 Pa	[27]
Bead forcing	1 nN	[20]
Substrate strain (endothelial cells)	1–3%	[27]
Strain (bone)	100 με	[79]
Single protein forces	10 pN	(estimate)
Transepithelial pressure	0.5 kPa	[12]
Typical, resting values	**Magnitude**	**Reference**
Membrane tension	0.02–0.4 mN/m	[87]
Focal adhesion shear stress	5.5 kPa	[59]
Cytoskeletal prestress	>350 Pa	[86]
Molecular motor force (kinesin)	5–7 pN	[85]

contains focal adhesions, and taking a typical cell to be about 50 μm in lateral extent and 5 μ in height. A force balance through the cell midsection then gives an estimate of about 200 Pa as a measure of internal stress, at least in these cells. Interesting comparisons can be made to the fluid shear stress sufficient to elicit a cellular response of ~1 Pa and the shear modulus of cells, in the range of 100–1000 Pa. The first comparison suggests that the response to shear stress is not likely initiated by changes in cytoskeletal or focal adhesion stress, since the perturbation due to shear is so small compared to resting levels. It also hints at some sort of stress focusing mechanism as being essential for cells to sense such low levels of shear stress, as suggested in several recent studies [29, 38]. The second comparison suggests that the cytoskeletal matrix is under considerable prestress, and helps to explain the retraction observed when, for example, cells that are tethered to a surface are suddenly released. These effects have been extensively examined in the context of cellular tensegrity and are discussed further in Chapter 8.

Membrane tensions at rest are reported to be in the range of about 0.01 mN/m [68]. At the other extreme, bilayers rupture or lyse at strains of about 4%, corresponding to a tension of about 10 mN/m [69]. Potentially as a protective mechanism to prevent rupture, some cells contain stretch-activated channels such as the MscL discussed previously. From patch clamp experiments in which part of the cell membrane is aspirated into a micropipette, the threshold pressure for channel activation can be measured, and from that, membrane tension can be computed. Changes in channel conductance are typically observed to occur in the range of 3–4 kPa (Figure 1.4), corresponding to a membrane stress of ~ 2 mN/m, which is consistent with their role in preventing membrane rupture. We can also compute a crude estimate of the force acting on a single channel, by using this value for tension and the outside dimensions of the MscL (~ 5 nm), and estimate an activation force on the order

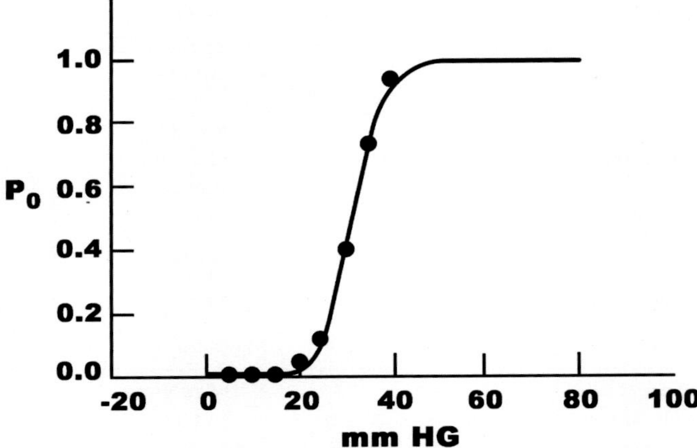

Figure 1.4. Response of the mechanosensitive channel of low conductance (MscL) to membrane tension. Activation (measured in terms of the probability P_0 of the channel opening) occurs at suction pressures (horizontal axis), applied by a micropipette, of about 30 mmHg or 4 kPa. (Adapted from *Physical Biology of the Cell*, Phillips, Kondev, Theriot, 2007.)

of about 10 pN. Consistent with this, the change in free energy required for activation falls in the range of about 10 kT. By comparison, a shear stress of 1 Pa (sufficient to induce a response in endothelial cells) integrated over a cell length of ~ 50 μm produces a maximum membrane tension of ~ 0.1 mN/m, about an order of magnitude higher than reported resting stress levels, but an order of magnitude lower than the tension needed to activate other known stretch-activated channels. These issues are discussed in greater depth in Chapter 6.

If mechanotransduction occurs at the single molecule scale, as in the case of force-induced conformational change or domain unfolding, then the range of critical forces can be estimated with a reasonable degree of confidence. Unfolding forces of a variety of proteins have been measured by AFM and generally fall in the range of 20–100 pN; rupturing of noncovalent bonds is often observed at forces only slightly higher than this. Conformational changes should be somewhat easier to achieve but are generally reported to occur in a similar range of force, possibly extending down to a scale of about 2 pN. In either unfolding or conformational change, protein extensions will likely be in the range of 1–10 nm. Using these estimates, the energy for activation of a single protein would be in the range of 2–1000 pN·nm, or about 0.5–250 kT. One further constraint needs to be imposed: The energy of activation should be in excess of the protein's thermal energy, otherwise the protein would often exist in the activated state even in the absence of force. Based on this line of reasoning, we would expect the force of activation to be ~ 10–50 pN with an activation energy of ~ 2–100 kT.

Unfortunately, these computed force levels fail to answer some of the most critical questions in mechanotransduction. For example, it is difficult to understand how shear stresses as low as 1 Pa or less can elicit a response; there must be some means of force amplification that we have yet to identify. Also, while much of the data in the literature can be explained by the activation of individual proteins or channels, at least one known mechanism, the activation of airway epithelial cells by compression (Chapter 14), occurs at the whole-cell level. There are likely to be other mechanisms, perhaps involving changes in transport as in this one case, that are yet to be identified.

1.6 The "Mechanome"

Over the past decade, the field of mechanobiology has flourished, and the pace of discoveries has even accelerated. Mechanotransduction represents one important aspect of this multifaceted discipline that bridges between chemistry, physics, bioengineering, and biology. But it constitutes one piece in a larger picture, representing the interface between forces transmitted through tissues, cells, and molecules, and the biochemical reactions and signaling pathways they trigger that ultimately affect the function of the cells, tissues, and organs. Other aspects of the broader field include the sensing by a cell of its mechanical environment, as illustrated by recent experiments showing that cell differentiation can be regulated by the stiffness of its surroundings [12], and the observation, with potentially critical clinical implications, that disease states such as malaria [70] and cancer [71] are reflected by changes in the mechanical stiffness of a cell. Therefore, recognizing the broader scope that mechanics plays, it is useful to cast our thoughts in terms of the overall mechanical state of a biological system, and what this implies in terms of its diverse range of functions. One term that captures the essence of this idea is the "mechanome," which we can define as the complete mechanical state of a biological system including the distribution of stress in the system, the mechanical properties of each of its subelements, as well as the collection of interactions between mechanics (force or stiffness) with ongoing biochemical processes. Force transmission might then be thought of as a mechanical signaling pathway, having direct parallels to, and numerous points of communication with, the biochemical signaling pathways that have been the subject of intense study for some time. One important purpose of this book, then, is to help define the field of mechanomics and all that it encompasses.

REFERENCES

[1] Kamiya, A. and T. Takahashi, Quantitative assessments of morphological and functional properties of biological trees based on their fractal nature. *J Appl Physiol*, 2007. **102**(6): 2315–23.

[2] Reneman, R. S. and A. P. Hoeks, Wall shear stress as measured in vivo: Consequences for the design of the arterial system. *Med Biol Eng Comput*, 2008. **46**(5): 499–507.

[3] Cheng, C., et al., Large variations in absolute wall shear stress levels within one species and between species. *Atherosclerosis*, 2007. **195**(2): 225–35.

[4] Nerem, R. M., M. J. Levesque, and J. F. Cornhill, Vascular endothelial morphology as an indicator of the pattern of blood flow. *J Biomech Eng*, 1981. **103**(3): 172–6.

[5] Dewey, C. F., Jr., et al., The dynamic response of vascular endothelial cells to fluid shear stress. *J Biomech Eng*, 1981. **103**(3): 177–85.

[6] Driessen, N. J., et al., Remodelling of continuously distributed collagen fibres in soft connective tissues. *J Biomech*, 2003. **36**(8): 1151–8.

[7] Price, R. J. and T. C. Skalak, Circumferential wall stress as a mechanism for arteriolar rarefaction and proliferation in a network model. *Microvasc Res*, 1994. **47**(2): 188–202.

[8] Yegorov, A. D., L. I. Kakurin, and Y. G. Nefyodov, Effects of an 18-day flight on the human body. *Life Sci Space Res*, 1972. **10**: 57–60.

[9] Neuman, W. F., Calcium metabolism in space flight. *Life Sci Space Res*, 1970. **8**: 309–15.

[10] Simmons, C. A., et al., Cyclic strain enhances matrix mineralization by adult human mesenchymal stem cells via the extracellular signal-regulated kinase (ERK1/2) signaling pathway. *J Biomech*, 2003. **36**(8): 1087–96.

[11] Schmelter, M., et al., Embryonic stem cells utilize reactive oxygen species as transducers of mechanical strain-induced cardiovascular differentiation. *Faseb J*, 2006. **20**(8): 1182–4.

[12] Discher, D. E., P. Janmey, and Y. L. Wang, Tissue cells feel and respond to the stiffness of their substrate. *Science*, 2005. **310**(5751): 1139–43.

[13] Hollander, W., Role of hypertension in atherosclerosis and cardiovascular disease. *Am J Cardiol*, 1976. **38**(6): 786–800.

[14] Nerem, R. M., Vascular fluid mechanics, the arterial wall, and atherosclerosis. *J Biomech Eng*, 1992. **114**(3): 274–82.

[15] Schwartz, C. J., et al., Pathophysiology of the atherogenic process. *Am J Cardiol*, 1989. **64**(13): 23G–30G.

[16] Jo, H., et al., Endothelial albumin permeability is shear dependent, time dependent, and reversible. *Am J Physiol*, 1991. **260**(6 Pt 2): H1992–6.

[17] Friedman, M. H. and D. L. Fry, Arterial permeability dynamics and vascular disease. *Atherosclerosis*, 1993. **104**(1–2): 189–94.

[18] Shieh, A. C. and K. A. Athanasiou, Principles of cell mechanics for cartilage tissue engineering. *Ann Biomed Eng*, 2003. **31**(1): 1–11.

[19] Kisiday, J. D., et al., Effects of dynamic compressive loading on chondrocyte biosynthesis in self-assembling peptide scaffolds. *J Biomech*, 2004. **37**(5): 595–604.

[20] Ressler, B., et al., Molecular responses of rat tracheal epithelial cells to transmembrane pressure. *Am J Physiol Lung Cell Mol Physiol*, 2000. **278**(6): L1264–72.

[21] Swartz, M. A., et al., Mechanical stress is communicated between different cell types to elicit matrix remodeling. *Proc Natl Acad Sci USA*, 2001. **98**(11): 6180–5.

[22] Waters, C. M., et al., Cellular biomechanics in the lung. *Am J Physiol Lung Cell Mol Physiol*, 2002. **283**(3): L503–9.

[23] Delmas, P., Polycystins: From mechanosensation to gene regulation. *Cell*, 2004. **118**(2): 145–8.

[24] Ingber, D. E., Mechanobiology and diseases of mechanotransduction. *Ann Med*, 2003. **35**(8): 564–77.

[25] Dobrin, P. B., Mechanical properties of arteries. *Physiol Rev*, 1978. **58**(2): 397–460.

[26] Tschumperlin, D. J., et al., Mechanotransduction through growth-factor shedding into the extracellular space. *Nature*, 2004. **429**(6987): 83–6.

[27] Wang, Y., et al., A model for the role of integrins in flow induced mechanotransduction in osteocytes. *Proc Natl Acad Sci USA*, 2007. **104**(40): 15941–6.

[28] Barbee, K. A., et al., Subcellular distribution of shear stress at the surface of flow-aligned and nonaligned endothelial monolayers. *Am J Physiol*, 1995. **268**(4 Pt 2): H1765–72.

[29] Weinbaum, S., et al., Mechanotransduction and flow across the endothelial glycocalyx. *Proc Natl Acad Sci USA*, 2003. **100**(13): 7988–95.
[30] Mack, P. J., et al., Force-induced focal adhesion translocation: Effects of force amplitude and frequency. *Am J Physiol Cell Physiol*, 2004. **287**(4): C954–62.
[31] Salwen, S. A., et al., Three-dimensional changes of the cytoskeleton of vascular endothelial cells exposed to sustained hydrostatic pressure. *Med Biol Eng Comput*, 1998. **36**(4): 520–7.
[32] Schwartz, E. A., et al., Exposure of human vascular endothelial cells to sustained hydrostatic pressure stimulates proliferation. Involvement of the alphaV integrins. *Circ Res*, 1999. **84**(3): 315–22.
[33] Hishikawa, K., et al., Pressure promotes DNA synthesis in rat cultured vascular smooth muscle cells. *J Clin Invest*, 1994. **93**(5): 1975–80.
[34] Muller, H. J. and H. J. Galla, Pressure variation of the lateral diffusion in lipid bilayer membranes. *Biochim Biophys Acta*, 1983. **733**(2): 291–4.
[35] Chien, S., Mechanotransduction and endothelial cell homeostasis: The wisdom of the cell. *Am J Physiol Heart Circ Physiol*, 2007. **292**(3): H1209–24.
[36] Lehoux, S., Y. Castier, and A. Tedgui, Molecular mechanisms of the vascular responses to haemodynamic forces. *J Intern Med*, 2006. **259**(4): 381–92.
[37] Liedert, A., et al., Signal transduction pathways involved in mechanotransduction in bone cells. *Biochem Biophys Res Commun*, 2006. **349**(1): 1–5.
[38] Hu, S. and N. Wang, Control of stress propagation in the cytoplasm by prestress and loading frequency. *Mol Cell Biomech*, 2006. **3**(2): 49–60.
[39] Moe, P. and P. Blount, Assessment of potential stimuli for mechano-dependent gating of MscL: Effects of pressure, tension, and lipid headgroups. *Biochemistry*, 2005. **44**(36): 12239–44.
[40] Gullingsrud, J., D. Kosztin, and K. Schulten, Structural determinants of MscL gating studied by molecular dynamics simulations. *Biophys J*, 2001. **80**(5): 2074–81.
[41] Bilston, L. E. and K. Mylvaganam, Molecular simulations of the large conductance mechanosensitive (MscL) channel under mechanical loading. *FEBS Lett*, 2002. **512**(1–3): 185–90.
[42] Kung, C., A possible unifying principle for mechanosensation. *Nature*, 2005. **436**(7051): 647–54.
[43] Martinac, B., Mechanosensitive ion channels: Molecules of mechanotransduction. *J Cell Sci*, 2004. **117**(Pt 12): 2449–60.
[44] Perozo, E., et al., Open channel structure of MscL and the gating mechanism of mechanosensitive channels. *Nature*, 2002. **418**(6901): 942–8.
[45] Barakat, A. I., D. K. Lieu, and A. Gojova, Secrets of the code: Do vascular endothelial cells use ion channels to decipher complex flow signals? *Biomaterials*, 2006. **27**(5): 671–8.
[46] Pahakis, M. Y., et al., The role of endothelial glycocalyx components in mechanotransduction of fluid shear stress. *Biochem Biophys Res Commun*, 2007. **355**(1): 228–33.
[47] Gudi, S., J. P. Nolan, and J. A. Frangos, Modulation of GTPase activity of G proteins by fluid shear stress and phospholipid composition. *Proc Natl Acad Sci USA*, 1998. **95**(5): 2515–9.
[48] Butler, P. J., et al., Shear stress induces a time- and position-dependent increase in endothelial cell membrane fluidity. *Am J Physiol Cell Physiol*, 2001. **280**(4): C962–9.
[49] Axelrod, D., Lateral motion of membrane proteins and biological function. *J Membr Biol*, 1983. **75**(1): 1–10.
[50] Czarny, M. and J. E. Schnitzer, Neutral sphingomyelinase inhibitor scyphostatin prevents and ceramide mimics mechanotransduction in vascular endothelium. *Am J Physiol Heart Circ Physiol*, 2004. **287**(3): H1344–52.

[51] Sawada, Y. and M. P. Sheetz, Force transduction by Triton cytoskeletons. *J Cell Biol*, 2002. **156**(4): 609–15.

[52] Wang, Y., et al., Visualizing the mechanical activation of Src. *Nature*, 2005. **434**(7036): 1040–5.

[53] Johnson, C. P., et al., Forced unfolding of proteins within cells. *Science*, 2007. **317**(5838): 663–6.

[54] Mochizuki, S., et al., Role of hyaluronic acid glycosaminoglycans in shear-induced endothelium-derived nitric oxide release. *Am J Physiol Heart Circ Physiol*, 2003. **285**(2): H722–6.

[55] Florian, J. A., et al., Heparan sulfate proteoglycan is a mechanosensor on endothelial cells. *Circ Res*, 2003. **93**(10): e136–42.

[56] Odde, D. J., et al., Microtubule bending and breaking in living fibroblast cells. *J Cell Sci*, 1999. **112**(Pt 19): 3283–8.

[57] Maniotis, A. J., C. S. Chen, and D. E. Ingber, Demonstration of mechanical connections between integrins, cytoskeletal filaments, and nucleoplasm that stabilize nuclear structure. *Proc Natl Acad Sci USA*, 1997. **94**(3): 849–54.

[58] Deguchi, S., et al., Flow-induced hardening of endothelial nucleus as an intracellular stress-bearing organelle. *J Biomech*, 2005. **38**(9): 1751–9.

[59] Brower-Toland, B. D., et al., Mechanical disruption of individual nucleosomes reveals a reversible multistage release of DNA. *Proc Natl Acad Sci USA*, 2002. **99**(4): 1960–5.

[60] Gao, M., et al., Identifying unfolding intermediates of FN-III(10) by steered molecular dynamics. *J Mol Biol*, 2002. **323**(5): 939–50.

[61] Ortiz, V., et al., Unfolding a linker between helical repeats. *J Mol Biol*, 2005. **349**(3): 638–47.

[62] Lee, S. E., R. D. Kamm, and M. R. Mofrad, Force-induced activation of talin and its possible role in focal adhesion mechanotransduction. *J Biomech*, 2007. **40**(9): 2096–106.

[63] Hytonen, V. P. and V. Vogel, How force might activate talin's vinculin binding sites: SMD reveals a structural mechanism. *PLoS Comput Biol*, 2008. **4**(2): e24.

[64] Munevar, S., Y. Wang, and M. Dembo, Traction force microscopy of migrating normal and H-ras transformed 3T3 fibroblasts. *Biophys J*, 2001. **80**(4): 1744–57.

[65] Butler, J. P., et al., Traction fields, moments, and strain energy that cells exert on their surroundings. *Am J Physiol Cell Physiol*, 2002. **282**(3): C595–605.

[66] Balaban, N. Q., et al., Force and focal adhesion assembly: A close relationship studied using elastic micropatterned substrates. *Nat Cell Biol*, 2001. **3**(5): 466–72.

[67] Tan, J. L., et al., Cells lying on a bed of microneedles: An approach to isolate mechanical force. *Proc Natl Acad Sci USA*, 2003. **100**(4): 1484–9.

[68] Sheetz, M. P., J. E. Sable, and H. G. Dobereiner, Continuous membrane-cytoskeleton adhesion requires continuous accommodation to lipid and cytoskeleton dynamics. *Ann Rev Biophys Biomol Struct*, 2006. **35**: 417–34.

[69] Dai, J., et al., Myosin I contributes to the generation of resting cortical tension. *Biophys J*, 1999. **77**(2): 1168–76.

[70] Mills, J. P., et al., Effect of plasmodial RESA protein on deformability of human red blood cells harboring Plasmodium falciparum. *Proc Natl Acad Sci USA*, 2007. **104**(22): 9213–7.

[71] Suresh, S., Biomechanics and biophysics of cancer cells. *Acta Biomater*, 2007. **3**(4): 413–38.

[72] Hudspeth, A. J., How the ear's works work: Mechanoelectrical transduction and amplification by hair cells. *C R Biol*, 2005. **328**(2): 155–62.

[73] You, L., S. C. Cowin, M. B. Schaffler, and S. Weinbaum, A model for strain amplification in the actin cytoskeleton of osteocytes due to fluid drag on pericellur matrix. *J Biomech*, 2001. **34**(11): 1375–86.

[74] Gummer, A. W., W. Hemmert, and H. P. Zenner, Resonant tectorial membrane motion in the inner ear: Its crucial role in frequency tuning. *Proc Natl Acad Sci USA*, 1996. **93**(16): 8727–32.

[75] Kolahi, K. S. and M. R. Mofrad, Molecular mechanics of filamin's rod domain. *Biophys J*, 2008. **94**(3): 1075–83.

[76] Carrion-Vazquez, M., A. F. Oberhauser, T. E. Fisher, P. E. Marszalek, H. Li, and J. M. Fernandez, Mechanical design of proteins studied by single-molecule force spectroscopy and protein engineering. *Prog Biophys Mol Biol*, 2000. **74**(1–2): 63–91. Review.

[77] Rizzo, V., A. Sung, P. Oh, and J. E. Schnitzer, Rapid mechanotransduction in situ at the luminal cell surface of vascular endothelium and its caveolae. *J Biol Chem*, 1998. **273**(41): 26323–9.

2

Endothelial Mechanotransduction

Peter F. Davies and Brian P. Helmke

2.1 Introduction

The endothelium is one of the most intensively studied tissues in cellular mechanotransduction. It is the interface between blood (or lymph) and the underlying vessel walls in arteries, microcirculatory beds, veins, and lymphatics. Mechanotransduction is of particular significance in high-pressure, high-flow arteries where considerable blood flow forces act on the endothelium lining the inner boundaries of the vessel walls. Consequently, endothelial mechanotransduction is studied principally as a *flow-mediated* mechanism.

Efficient vascular transport systems are central to the evolutionary success of all higher organisms. Throughout phylogeny there is a consistent pattern of structural relationships in branching vessels. For example, much of the mammalian arterial circulation obeys mathematical relationships of vessel geometry that ensure a continuum of flow characteristics (volumetric, velocity, flow profile, and shear relationships), where the major distributing arteries repeatedly branch to provide blood to the complex volume of widely dispersed tissues and organs throughout the body; Murray's Law [1, 3] and Zamir's Law [2] are examples. Similar relationships are found in fluid transport systems of primitive marine animals. General principles such as these reflect the interdependence of flow with vessel structure and function throughout evolution that ensures the efficient and successful distribution of fluid in primitive life forms and blood circulation in mammals.

Integral to the movement of blood through arteries are the accompanying hemodynamic forces that arise from the pressure differentials necessary for circulation. The local biomechanical environment of arterial cells arises from the interactions of blood with the vessel wall. This is particularly so for the endothelium, a polarized cell specialized in its functions. The luminal side is exposed to flowing blood, the cell edges are anchored to adjacent cells through junctional complexes, and the abluminal surface away from the flow is attached to a subcellular connective tissue matrix. In arteries, complex flow characteristics are resolvable into principal force vectors of pressure (normal to the wall) and shear stress, with the frictional

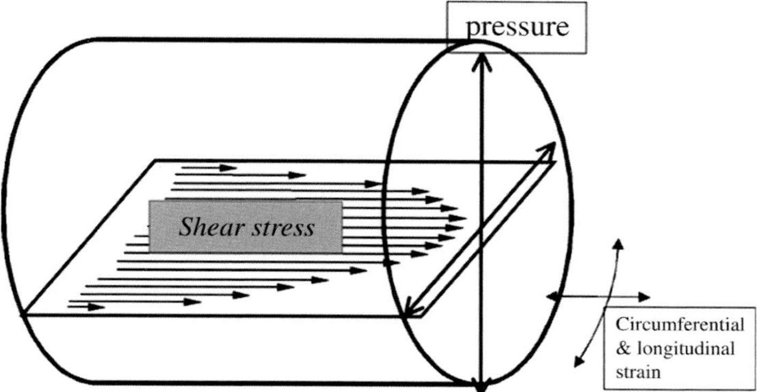

Figure 2.1. Resolution of hemodynamic forces into simplified principal vectors of shear stress parallel to the vessel wall and pressure perpendicular to the wall. These stress forces result in cell and tissue deformation (strain) in the endothelium and underlying vessel wall.

force at the endothelial luminal cell surface acting in the direction of the flow (Figure 2.1). These stresses create cyclic surface and intracellular strains in the endothelium during each cardiac cycle. A large literature demonstrates that hemodynamic forces regulate arterial physiology through endothelial signaling responses, confer patho-susceptibility through control of phenotype expression, and contribute to the development of atherosclerotic lesions through modulation of interactions with blood cells and plasma molecules [3–7]. There is considerable experimental evidence that the shear stress component exerts a dominant effect on cell signaling in the overall hemodynamic responses of the endothelium [8]. However, precise measurement of the complex flow dynamics present *in vivo* is challenging, and other hemodynamic elements undoubtedly contribute to endothelial responses that in turn are an important component of the integrated arterial vessel wall response. The interactions between hemodynamics and the endothelium are therefore one of the central determinants of vascular physiology and pathology in mammalian evolution, development, survival, and morbidity.

It is unclear whether the organization of mechanotransduction mechanisms to regulate such diverse and complex functions is hierarchical or parallel. The large number of different pathways activated by changes in the biomechanical environment of the endothelium suggests that they are unlikely to be downstream of a single "mechanotransducer." Nevertheless, the initial physical interaction between the fluid dynamics and the cell is limited to one surface of a polarized monolayer, and local mechanotransduction responses certainly occur at that surface. Examples include ion channel activation [9], G-protein activation [10], and increased membrane fluidity [11]. This chapter outlines the local and decentralized models of endothelial mechanotransduction with emphasis on a summary of the particular work of the authors in support of decentralized mechanisms. Since other chapters in this book report experimental detail pertaining to different mechanisms of mechanotransduction, much of it in the endothelium, we have tried to avoid duplication and refer the reader to those articles. Furthermore, dissection of endothelial mechanotransduction mechanisms *in vivo* is difficult because of the regulatory influences that are part of, or distinct from, flow-induced deformation and over which the investigator has little control. Consequently, mechanotransduction mechanisms are largely studied *in vitro* under controlled conditions of flow. Nevertheless, important mechanisms identified *in vitro* must be tested in the *in vivo* environment. Therefore we report recent attempts to bridge these approaches through genomics and proteomics mapping of endothelial phenotypes in regions where distinct flow characteristics are associated with susceptibility to or protection from atherogenesis.

2.2 Endothelial Functions

The principal functions of endothelium are the maintenance of anti-coagulant properties, the physiological control of lumen diameter (vasoregulation), the regulation of vascular permeability, and mediation of both pathological consequences and protective responses associated with acute and chronic inflammation, wound healing, and major cardiovascular disorders such as atherogenesis [12]. There is considerable local regulation of these functions by blood flow. Physiological mechanisms that regulate arterial tone act in large measure through the endothelium and are greatly influenced by shear stress [13]. Acute dilatation and constriction of arteries in response to changes of flow are controlled through endothelial-derived nitrovasodilators, prostaglandins, lipoxygenases, hyperpolarizing factors, and related molecules [14–16]. Sustained changes in the local hemodynamic conditions stimulate structural remodeling of the artery wall through chronic mechanisms that are also endothelium-dependent [17] and that involve a carefully orchestrated sequence of gene and protein expressions to facilitate major artery wall structural adaptation. Endothelial influences mediated by flow occur very early; recent reports show the importance of flow as a regulator of embryonic cardiovascular development [18]. Furthermore, hemodynamics play a critical role in the arterial circulation at locations where most vascular pathologies originate and develop their morbidity [19, 20]. The endothelium, as "gatekeeper" between the blood and artery wall, strongly influences the initiation and development of atherosclerosis [21]. Thus, endothelial biomechanics is a regulatory mechanism that influences the homeostatic, adaptive, and pathological mechanisms in major arteries.

2.3 Endothelial Mechanotransduction: Descriptors

Biological effects caused by the deformation of endothelial cells and their subcellular components resulting directly and indirectly from hemodynamic forces are generally described under the term "endothelial mechanotransduction." Strictly speaking, mechanotransduction is the conversion of a mechanical stimulus to chemical activity that generates biological consequence(s). There are, however, several elements in these descriptions that require further clarification.

First, "deformation" is a change of shape. In engineering terms it is the response to an applied force(s) that in the current context results from blood flow. This response may be a direct deformation of, for example, a cell surface feature(s) or of structures attached to the site(s) of direct deformation. However, change of position (displacement) of a structure in the cell may also occur passively and without deformation if the entire microvolume in which it is contained is displaced by the effects of adjacent, true deformation events. This phenomenon complicates efforts to spatially image and define the location of mechanically induced deformation leading to transduction. Separation of passive and active deformation events is desirable whenever possible in spatiotemporal studies of mechanotransduction (e.g., [22]).

Second, when flow dynamics change, the local transport characteristics of plasma cells and molecules within the fluid volume also change, a process distinct from structural deformation. The effects of changes of convective transport on endothelial responses are difficult to measure and are often overlooked in hemodynamic response studies. Examples include the altered retention time of plasma cells and molecules in regions of flow separation and disturbance [20] and the contributions to cell responses of boundary layer concentrations of short-lived agonists generated at the enzyme-rich endothelial surface or secreted from the cell (see also the chapter on autocrine signaling by Tschumperlin in this book) [23, 24]. Changes of blood flow affect the convective transport of molecules to and from the cell surface, where their effective concentration will influence agonist–receptor interactions [25].

Endothelial mechanotransduction can be further refined to include "mechano-*transmission*," the transfer of structural deformation through the continuity of dynamic structural elements within each cell as well as via the physical connections between cells. Thus, a local deformation of the luminal cell surface may be propagated near-instantaneously to cellular sites distant from the fluid-imposed stimulus via membrane proteins attached to the cytoskeleton. The distinction between transmission and transduction is useful because, as noted above, translational movement can be passive and disconnected from mechanotransduction, or deformation may remain below a threshold required to elicit a transduction response. Advances in cell and molecular imaging provide access to measurable subcellular deformation with high spatiotemporal resolution and simultaneous imaging of intracellular molecular responses. The combination provides a more detailed interpretation of the structural requirements for transduction.

2.4 Endothelial Mechanotransduction: A Decentralized Mechanism

The introduction of endothelial tissue culture in the 1970s paved the way for controlled studies of the effects of hemodynamic forces on endothelial cell biology. Prior to that period, observations were limited to *ex vivo* blood vessel organ cultures (where the endothelial integrity was rapidly lost) and observational studies of fixed blood vessels [26]. In an engineering-medicine collaborative effort, Dewey and colleagues [27] performed the first successful *in vitro* demonstrations of the morphological plasticity of endothelial cells in response to defined shear stresses. This was quickly followed by demonstrations of biochemical and cell biological responses [28–30]. The prevailing view throughout most of the following decade was that a mechanotransduction event occurred at the site of direct force interaction with the luminal cell surface. Measurements of molecular events at the cell surface, initially the identification of K^+ inward-rectifying ion channel responsiveness to shear stress [9], supported this view. Coincident with these developments, an integrated consideration of the spatial characteristics of the cytoskeleton and its membrane integrin anchorage sites in the context of the tensional integrity of cells was developing [31, 32] (see also Chapter 8 by Ingber in this book) which showed that a range of equilibrium states in the cytoskeletal tension of anchorage-dependent cells in general is required for the proper regulation of cell signaling. These findings inspired hemodynamics investigators to look into subcellular structures beyond the luminal endothelial surface in spatial studies of shear stress–induced signaling. Live-cell imaging of endothelial anchorage sites revealed the dynamic behavior of cell–substratum adhesion sites [33]. When directional flow was applied to the luminal surface of endothelial cells, the abluminal (attached) surface responded by reorganizing regions of the adhesion site alignment that related to the direction of flow of the luminal surface [34]. Subsequent biochemical measurements of focal adhesion-restricted signaling responses to shear stress [7, 35] have provided strong evidence for mechanoinitiated transduction at these locations, which are far removed from the initial deformation of the luminal cell surface. Other important work has identified rapid shear stress signaling responses at the cell junctions centered on the shear-induced phosphorylation of junctional molecules, particularly the involvement of platelet–endothelial cell adhesion molecule–1 (PECAM-1) in endothelium [36, 37]. These findings are discussed in greater detail in the following sections. A reasonable overall interpretation is (1) that mechanotransduction can occur at some distance (micrometers) from the initial deformation, and (2) that transmission though continuity of cell structure may transfer, amplify, modulate, filter, redirect, and/or transduce the force to elicit signaling at the luminal and abluminal surfaces, cell junctions, nuclear membrane, or, as is most likely, all of the above. The most pervasive cellular structures for mechanotransmission are the cytoskeletal actin filaments, intermediate filaments, and microtubules. Compelling evidence for a central role of cytoskeletal elements in endothelial flow responses was subsequently demonstrated by live endothelial cell imaging of intermediate filament displacement using fluorescent cytoskeleton reporter molecules [22, 38–40]. A decentralized model of force transmission-transduction was introduced in concept in 1993 [41] and expanded in detail in 1995 [3] and 2002 [42].

The essentials of the decentralized model are illustrated in Figure 2.2. The model proposes that although shear stress acts initially at the luminal (apical) cell surface with immediate biochemical responses located there, surface forces are simultaneously transmitted throughout the body of the cell such that multiple elements located away from the luminal surface may, independently or in concert, transduce the mechanical signals into biochemical activities. It accommodates many separate experimental findings regarding endothelial shear stress responses associated with disparate cellular locations. It essentially regards the entire cell as an assembly of mechanotransduction sites and mechanisms connected principally by

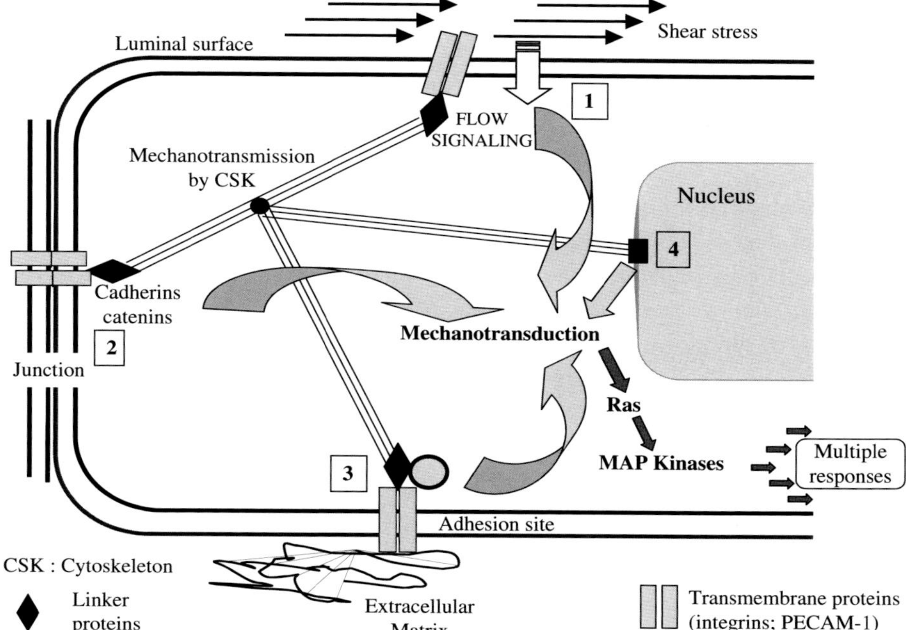

Figure 2.2. Decentralized model of endothelial mechanotransduction by shear stress. The cytoskeleton plays a central role in the mechanotransmission of tension changes throughout the cell. Examples of direct signaling at the luminal surface (1), junctional signaling (2), and adhesion site signaling (3) are shown. Nuclear deformation also likely results in mechanosignaling (4). The locations are based on direct or indirect experimental evidence, are not mutually exclusive, and are likely interconnected. At 1, deformation of the luminal cell surface, possibly via the glycocalyx. Examples include localized activation of K^+, Na^+, and Cl^- ion channels; phospholipase activity leading to calcium signaling; G-protein activation; and caveolar signaling. At 2, transmission of forces to the intercellular junction protein complexes via the cortical and/or filamentous cytoskeleton. VEGFR2 located at the luminal surface (site 1) or near the junction (site 2) may associate with VE-cadherin, β-catenin, and phosphatidylinositol 3-kinase to phosphorylate Akt and the primary transmembrane protein at this location, PECAM-1. At 3, cytoskeletal force transmission to adhesion sites. Transmembrane integrins bound to the extracellular matrix serve as a focus for deformation that results in the autophosphorylation of FAK, which binds the SH2 domain of c-Src, a kinase family that phosphorylates paxillin and p130 Cas, leading via Ras G-proteins to integrin-dependent activation of MAP kinases. In a variation of a similar assembly cascade, PYK2 found in multiple locations within the cell contributes to phosphorylation on translocation to adhesion sites, the cytoskeleton itself, or the nucleus. A second parallel integrin-mediated pathway involves the activation of Shc, which binds Src family kinases through SH2 domains. Shc phosphorylation leads to Ras-MAP kinase activity. A third integrin-mediated pathway is via RhoA activation, which profoundly influences actin assembly. At 4, nuclear deformation/displacement may result in mechanotransduction possibly via lamins in the nuclear membrane. Other possible direct effects are on macromolecular conformation relevant to gene regulation.

the cytoskeleton, while acknowledging that the structural elements are themselves dynamic. Thus mechanotransduction is decentralized and there is the potential for multiple "mechanosensors" remote from the blood interface as well as at the initial (luminal) surface of deformation. In effect the entire cell is a mechanotransducer. The degree to which mechanotransduction originating at one subcellular location may interact with that at another is not known.

2.5 General Overview of Spatiotemporal Flow Responses

The elements of endothelial mechanoresponses can be considered in temporal and spatial terms as *(1) the physical deformation of a living surface; (2) intra- and pericellular transmission of force/strain; (3) conversion of mechanical force to chemical activity (mechanotransduction); and (4) downstream signaling with feedback*. The separation of elements is somewhat arbitrary because the temporal relationships are not firmly established, and it is likely that complex interplay occurs between them. Furthermore, within each element are subtle differences and refinements that may vary from cell to cell. However, the conceptual template can be used to assemble existing data and can accommodate new findings.

1. *Physical deformation.* Displacement of one or more cellular elements is required for a mechanotransduction response. This is the most immediate event and is required to initiate any subsequent responses. Since shear stress acts at the luminal cell surface, structures associated with the surface participate in a mechanical-to-chemical-to-biological transduction. The best examples are direct measurement of ion channel activities [43], phospholipid activation [10], and membrane fluidity changes [11]. The distribution of forces acting on the luminal surface of the endothelial monolayer is determined by the microgeometry of the surface secondary to the bulk characteristics of the blood flow. Detailed mapping of the monolayer surface by atomic force microscopy (AFM) followed by computational modeling of flow [44, 45] identified considerable heterogeneity in the cell-to-cell and subcellular distributions of stress concentrations. As would be predicted, regions of the cell that extend furthest into the flow field are subjected to the highest shear stress forces. Until recently, the "tallest" structures were considered to be the cell surface extending over the nuclear region, where a typical thickness of 5–7 μm has been measured by AFM [44]. However, specialized structures, if present, may extend considerably further into the flow. Two in particular are under investigation and are discussed more fully later in this chapter: (1) the glycocalyx [46], a glycoprotein-rich extension of the cell surface that may serve as a surface mechanosensor, and (2) the "rediscovery" of a primary cilium on many endothelial cells that extends *through* the luminal cell surface and is connected to the cytoskeleton within the cell.
2. *Force transmission (mechanotransmission).* The deformation of a connected system under tension results in virtually instantaneous transmission of the forces throughout the connected elements. In endothelial cells, the transmission conduit is principally the cytoskeletal filaments distributed throughout the cell body and the submembranous spectrin-like cortical cytoskeleton at the periphery of

the cell. They are interconnected and also linked to membrane proteins throughout the cell. They provide rigidity, shape, and structure to the cell and are a dominant biomechanical determinant of cell structural dynamics. Interference with cytoskeletal assembly and dynamics inhibits flow responses in various experimental systems [47]. Demonstrations of flow-initiated intermediate filament displacement [22, 38, 39], actin filament deformation [40, 48], and directed motion of mitochondria attached to microtubules [49] support the view that forces at the luminal cell surface are transmitted to "remote" sites via cytoskeletal deformations and displacements that may be a function of the prestressed tension in the cytoskeleton [50]. Mechanotransmission in the decentralization model therefore facilitates mechanotransduction not only at the interface of fluid–cell interaction but also elsewhere throughout the cell. Furthermore, the deformation effects may extend to adjacent cells, communicated through cell junctional structures and possibly the extracellular matrix.

3. *Mechanical force conversion to chemical activity ("true" mechanotransduction).* It follows from step 2 that this critical event may occur at a number of locations simultaneously involving parallel, convergent, or divergent mechanisms and may be associated with a variety of different proteins and lipids, but all occurring secondarily to the structural deformation or displacement resulting from flow forces. Mechanotransduction mechanisms are a prime target for investigative and/or therapeutic intervention. A single mechanism appears unlikely considering the multiplicity of subcellular sites at which responses have been measured. Proposed mechanisms include mechanical induction of conformational changes in membrane proteins, physical effects on the lateral mobility of membrane molecules, direct force effects on ion channels, separation of assembled cell–cell oligomeric proteins at junctions, deformation of caveolar structures, and physical inhibition of integrin dynamics and clustering, especially at the basal sites of cell adhesion to extracellular matrices [3]. Physical deformation may also displace local cofactors engaged in homeostatic regulation, including soluble and bound forms.
4. *Immediate and downstream signaling responses.* The fastest responses are those most likely to be at or close to the force conversion event. Fast responses include luminal membrane ion channel activation, intracellular calcium release, membrane lipid cleavage, membrane fluidity changes, and protein phosphorylation events, all of which lead in to signaling pathways. These "immediate" responses occur at multiple locations, including the abluminal surface and the cell junctions. Many different slower responses occur with a time constant of minutes or longer. We speculate that multiple parallel responses allow the cells to efficiently react to acute flow-mediated physiological signaling events as well as to adapt to the local vascular environment in a structured process when changes in hemodynamics are sustained.

These dynamic elements engage a wide range of material properties that greatly influence transduction mechanisms.

2.6 Endothelial Cell Biomechanics

Of paramount importance to understanding mechanical sensing by the endothelium (or any organ) are the material properties of the cells comprising the tissue. The constitutive properties of the cell determine the spatial and temporal scales of force transmission and frictional dissipation at work in response to chemical or mechanical stimulation. A number of experimental and theoretical approaches have been employed to estimate viscoelastic moduli at a subcellular length scale and taking into account adhesive interactions with substrates of varying stiffness.

A constitutive equation describes a property of a material. In mechanical engineering, the relationship between the stress field in a material and the strain field can be represented mathematically by a constitutive equation. The most common models of organs, blood, and cells describe biological materials as *continua*, that is, the spatial distribution of mass is continuous in space, and the spatial gradient of mass density exists mathematically everywhere. The continuity assumption serves as a reasonable approximation if the characteristic length scales of stress and strain are large with respect to the molecular structure of the cell or tissue.

In early models, the stress–strain relationship for tissues and cells was fitted empirically to equations of linear viscoelasticity (Table 2.1). In these models, a spring represents an elastic element whose stress–strain relationship is represented by Hooke's law, and a dashpot element represents a Newtonian viscous fluid in which the stress in the material is proportional to the strain rate. Phenomenological linear viscoelasticity models combine these elements in series and/or parallel arrangements so that the resulting material properties are described by the linear superposition of the stress–strain relationships of individual elements. By recognizing that the stress must be identical in elements arranged in series and that the strain must be identical in elements arranged in parallel, constitutive equations governing the behavior of any combination of elements are easily derived. The specific solutions to the differential equations depend on the time history of the input stress or strain function. For example, creep functions are derived by solving for the strain $E(t)$ when the input stress is a Heaviside step function, that is,

$$T(t) = H(t) = \begin{cases} 0, & t<0 \\ 1, & t\geqslant 0 \end{cases}$$

Similarly, stress relaxation functions are derived by solving for $T(t)$ when the input is $E(t) = H(t)$. In general, the linear superposition of stress responses to an arbitrary strain input function of time results in the following convolution, known as a *hereditary integral*:

$$T(t) = \int_{-\infty}^{t} G(t-\tau)\dot{E}(\tau)d\tau$$

Table 2.1. *Linear viscoelastic constitutive models, equations, and complex moduli. For homogeneous materials, one-dimensional models represented by a linear combination of springs and dashpots predict the relationship among stress T, strain E, and strain rate $\dot{E}$. The spring constants k, k_p, and k_s and the dashpot viscosity μ are determined by empirically fitting experimental stress–strain measurements, often using an oscillatory test to fit the functional form of the complex modulus $G^*(\omega)$.*

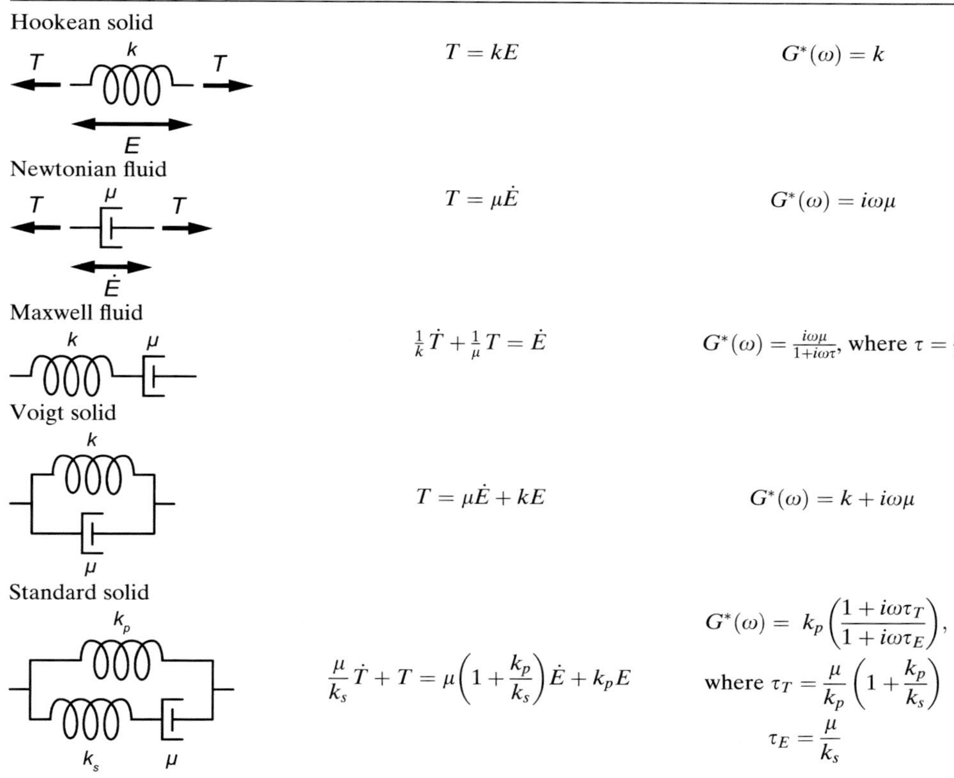

Model	Constitutive equation	Complex modulus
Hookean solid	$T = kE$	$G^*(\omega) = k$
Newtonian fluid	$T = \mu\dot{E}$	$G^*(\omega) = i\omega\mu$
Maxwell fluid	$\frac{1}{k}\dot{T} + \frac{1}{\mu}T = \dot{E}$	$G^*(\omega) = \frac{i\omega\mu}{1+i\omega\tau}$, where $\tau = \frac{\mu}{k}$
Voigt solid	$T = \mu\dot{E} + kE$	$G^*(\omega) = k + i\omega\mu$
Standard solid	$\frac{\mu}{k_s}\dot{T} + T = \mu\left(1 + \frac{k_p}{k_s}\right)\dot{E} + k_pE$	$G^*(\omega) = k_p\left(\frac{1 + i\omega\tau_T}{1 + i\omega\tau_E}\right)$, where $\tau_T = \frac{\mu}{k_p}\left(1 + \frac{k_p}{k_s}\right)$, $\tau_E = \frac{\mu}{k_s}$

where $\dot{E}(t)$ is the strain rate and $G(t)$ is a generalized stress relaxation function. By analogy to Hooke's Law, the *memory function G* represents a modulus that varies with the time history of deformation.

Sinusoidal input functions are useful for biological materials such as protein gels, since varying the frequency of the imposed stress or strain reveals the history dependence of the modulus. Since sinusoidal functions are easily represented as complex numbers, the one-dimensional modulus derived for a homogeneous material by solving the constitutive equation is often termed the *complex modulus* $G^*(\omega)$ and can be represented in two forms:

$$G^*(\omega) = G'(\omega) + iG''(\omega) = G_0(\omega)e^{i\delta(\omega)}$$

where $i = \sqrt{-1}$ and the sinusoid frequency ω is the independent variable. In the first form, the real part G' is the *elastic (storage) modulus* and the imaginary part G'' is the *viscous (loss) modulus*. Alternatively, G_0 gives a relative magnitude of stress and strain in the material, and the loss tangent δ (representing the phase shift between stress and strain) increases with frictional dissipation. For a Hookean solid, $G'' = 0$ and $\delta = 0$; for a Newtonian liquid, $G' = 0$ and $\delta = \pi/2$. Thus, the complex modulus values falling between these extremes are useful for predicting the relative amounts of energy stored as potential energy due to elastic deformation and dissipated as frictional heat.

One advantage of computing the complex modulus is that interpretation is not limited to constant-value estimates of parameters associated with the spring-dashpot models. A more general frequency response of the biological material reveals changes in material properties as a function of stress or strain history. For example, Janmey et al. [51] compared G' and creep functions for cytoskeletal protein polymer gels sheared in a cone and plate viscometer. The elastic modulus of an F-actin gel at 2 mg/ml was 283 Pa and remained constant as a function of both frequency and applied strain magnitude. Furthermore, the gel did not creep under a constant stress. Thus, F-actin gels *in vitro* can be represented to a first approximation as elastic. A microtubule gel at the same protein concentration has an elastic modulus of 34 Pa, and its modulus is also constant with frequency and strain magnitude. However, the microtubule gel exhibits slow creep with a shear rate of 0.0001 s^{-1}. Thus, microtubule networks appear to act as elastic gels at relatively fast shear rates but as viscous fluids at very low shear rates. Interestingly, vimentin gels exhibit both creep and strain hardening properties. At low stress, both the creep response and the elastic modulus of intermediate filament gels are similar to those of microtubule gels. Whereas microtubule gels flow freely with low elasticity at a strain magnitude of ~50%, intermediate filament networks continue to deform elastically without rupturing at strains approaching 80%. These measurements demonstrate that the individual polymer networks of the endothelial cytoskeleton contribute unique properties to the cell. Interestingly, the elastic modulus of a vimentin-actin copolymer gel is approximately one-third that of the pure actin gel [52], suggesting that the interactions among filament networks must also be relevant to predicting constitutive properties. Thus, measurements of stress–strain relationships must ideally be made in living cells *in situ*.

Viscoelastic parameters describing living cells were first estimated by micropipette aspiration of leukocytes and the derivation of a model consisting of a Maxwell fluid encased in an elastic cortical shell [53]. Using similar experimental techniques, order of magnitude estimates of Young's modulus for leukocytes or detached endothelial cells are typically in the 100-Pa range [54]. Sato and colleagues [55] represented detached endothelial cells as a standard linear solid. Pretreatment of cells either with cytochalasin B to disrupt assembly by capping actin filament barbed ends or with colchicine to promote microtubule disassembly decreased the values of elastic and viscous parameters (Table 2.2), suggesting that the cytoskeletal structure contributes most of the existing mechanical stiffness to material properties of living cells. After adaptation to steady unidirectional shear stress *in vitro* and in regions of low flow profile complexity *in vivo*, endothelial cells exhibit increased cytoskeletal

Table 2.2. *Fold change of viscoelastic parameter values relative to untreated endothelial cells (computed from values reported by Sato and colleagues [55, 58]).*

	Cytochalasin B	Colchicine	6-h Flow*	24-h Flow†
k_p	0.12	0.30	1.11	2.22
k_s	0.23	0.34	1.07	2.25
μ	0.18	0.32	1.32	2.79

* Parameter values are not significantly different from those in untreated cells.
† Cell shapes were significantly elongated, and stress fiber contact was increased.

filament bundling that suggests a reinforced structure [27, 56, 57]. The values of both elastic and viscous parameters in flow-adapted endothelial cells are significantly increased [58], indicating increased mechanical stiffness. Thus, cytoskeletal structure is likely a primary determinant of mechanotransmission through endothelial cells, but these phenomenological models cannot account for material structure.

Including an explicit structural dependence in continuum-based constitutive models has been proposed to explain aspects of small deformation behavior. For example, Satcher and Dewey [59] represented the "distributed cytoplasmic structural actin" (DCSA) network as a homogeneous linear elastic foam composed of cuboidal unit cells. Each unit cell in the open lattice contained edges represented as bendable beams connected rigidly at the corners. This model predicts that the presence of stress fibers significantly increases the effective Young's modulus of the cells, supporting the hypothesis that cytoskeletal structure contributes to force distribution through cytoplasm. This model predicts that a shear stress of 2 Pa acting at the cell surface would result in a shear strain of only 0.01%. However, the experimentally measured median displacement of cytoskeletal filaments near the apical surface of endothelial cells after onset of 1.5-Pa shear stress was 0.2 μm, and individual filament connections were displaced by as much as 1 μm [38]. Furthermore, a whole-cell finite element model derived from multimodal optical imaging of endothelial cells during the onset of shear stress predicts maximum strains near the basal cell surface in the range of 1–4% [60]. These direct experimental measurements indicate that structure-based models must be capable of accounting for both infinitesimal and finite deformations of living cells.

One such model is based on *tensegrity* (tensional integrity) structures, first described by Buckminster Fuller [61], which are stable equilibrium structures of elastic cables and rigid struts (see also Chapter 8). Wang and colleagues [62] were the first to use magnetic bead twisting cytometry to measure cell stiffness attributable to the cytoskeleton. Ferromagnetic beads were coated with ligands for cell-surface integrins, which interact physically with adhesion sites and cytoskeletons. Beads were adhered to the cell surface and magnetized in a directional field. The field was rotated 90°, and the angular strain of the beads was measured. The relationship between the stress imposed by rotating the magnetic field and the angular strain of the beads yields the effective stiffness of the cytoskeleton. Advantages of this technique are that it allows for input of an arbitrary stress function in magnitude and time and for finite deformation locally near the beads. These experiments were the first to demonstrate that cytoskeletal stiffness increased proportionally to the applied stress. The tensegrity model reproduces this relationship, but linear viscoelastic models cannot. Another important implication of the tensegrity model is that a nonzero continuous distribution of *prestress*, or mechanical load, exists in the structural cytoskeletal elements even in the absence of applied external loads. The distribution of prestress determines the cell shape and biophysical steady state on which tissue forces and hemodynamic shear stress are superimposed.

The existence of nonzero prestress does not necessarily require a tensegrity structure, and tensegrity cannot capture the ability of cells to remodel their cytoskeletal structures during processes such as spreading and migration. One way to reconcile this difficulty is to propose that fast deformation rates during these processes push the cytoskeletal polymer into a more liquid-like state described by an increased phase shift in the complex modulus. A new, adapted structure would then emerge that regains the elastic equilibrium properties of a tensed structure. In an interesting intersection of condensed soft matter physics and cell biomechanics, recent models have proposed that the cell cytoplasm exists physiologically near this sol–gel phase transition, similar to a soft glassy material [63]. Using magnetic bead twisting cytometry to apply an oscillating stress to the cytoskeleton, a theoretical framework has been developed that suggests a weak power-law relationship between modulus magnitude and oscillation frequency; that is, G' and G'' are proportional to f^{x-1}. In quiescent living cells, $x \approx 1.2$ and the loss tangent $\delta \approx 0.1$. This value of loss tangent and the weak power-law relationships are characteristic of soft glassy materials.

In these models, the value of x decreased when cells were treated with contractile agonists and increased when contractile mechanisms were relaxed or the actin cytoskeleton was disrupted [64]. These results suggest that intracellular contractility imposes increased structural order on the cytoplasm, decreasing the probability that external stresses will cause adaptation of structure and material properties. If the cell actively reorders its structure in response to an external stimulus, then adaptation should cause a decreased creep response as the material properties become more like those of an elastic solid. Indeed, the creep function magnitude decreases with increasing waiting time after an applied force [65]. This result is predicted for a soft glassy material that undergoes *aging*, the accumulation of local microscale reorganization events in material structures. The models demonstrate that characteristic time-scales of constitutive remodeling in living cells are consistent with soft glassy rheology theory, whether due to passive or active (ATP-dependent) processes. Whether these characteristics can be translated into biochemical signaling events remains to be seen.

The constitutive models each predict some but not all behaviors of endothelial cells as biomaterials. Viscoelastic models require a continuous distribution of mechanical energy across the plasma membrane. Prestressed structural network models provide a capability to predict the balance of internal forces with external forces acting at discrete locations. Soft glassy material models hold promise for deriving physical explanations of structural adaptation to external energy inputs.

A unifying principle of all of these models is the assumption of continuity at a cellular length scale. However, the decentralization hypothesis suggests that the subcellular structure and discrete locations of interaction with the extracellular matrix are crucial in triggering signaling networks in response to the local shear stress profile. Thus, a new focus in cell physiology has emerged to identify biochemical mechanisms that result from local molecular biophysical processes. In the following sections, we examine spatial compartments that are likely to redistribute or integrate forces at the subcellular level.

2.7 Shear Stress Acting at the Luminal Surface

Fluid shear stress acts directly at the luminal surface of the endothelium. The spatial and temporal profiles of shear stress are traditionally computed based on the bulk blood flow characteristics [66], but recognition of the role of local hemodynamic disturbances in vascular disease progression is increasing [67]. At a subcellular length scale, the spatial distribution of shear stress is influenced by the cell topography, and adaptation to shear stress serves to decrease both the mean shear stress and the spatial gradient magnitude [45]. Putative mechanosensitive cell surface–associated agents include ion channels, heterotrimeric G-proteins, cell surface receptors, and glycolipid microdomains. Whether shear forces act directly on cell surface–associated mechanosensitive molecules or indirectly through an intermediate surface structure remains unsolved.

Using tension field theory, Fung and Liu [68] proposed that the plasma membrane and its associated cortical cytoskeleton act as an elastic shell in which tension is transmitted continuously from the luminal plasma membrane to lateral junctions and the abluminal surface. Experimental measurements suggest that the onset of shear stress increases lipid bilayer fluidity [11, 69] in a manner consistent with the presence of increased lateral tension. The principal effect of a shear stress–mediated increase in the lateral tension in the plasma membrane is likely to be on ion channels. Inward-rectifying endothelial K^+ channels (Kir) sensitive to shear stress have been described [9], and of several Kir proteins in endothelium, Kir2.2 appears dominantly expressed [70]. However, the most well-studied stretch-activated channels are bacterial Msc channels involved in osmoregulation [71] and homologous degenerin/epithelial sodium channels (DEG/ENaC) implicated in hearing and touch [72]. Since the total free energy associated with a given conformation of the ion channel in the plasma membrane can be represented by the product of the channel cross-sectional area and the lateral tension acting on it, the change in channel cross-sectional area associated with opening may reflect a lower free energy state [71]. Alternatively, increased lateral tension may result in lipid bilayer thinning by 5%, and channel opening becomes favorable by changing protein conformation to reduce the hydrophobic mismatch at the new lipid layer thickness [73]. In vascular endothelial cells, a similar mechanism has been proposed for shear stress–induced activation of heterotrimeric G-proteins [74] and may also be at work in channels responsible for a shear stress–sensitive volume-regulated anion current (VRAC) [75].

Although changes in plasma membrane lateral tension are consistent with activation of ion channels or G-proteins, it is not clear whether mechanical energy is transferred directly to the plasma membrane at a sufficient magnitude to cause these events. For example, consider that the average wall shear stress acting on the endothelium surface is of order 1 Pa. Using the estimates of Sukharev et al. [71], the cross-sectional area of an open MscL channel is of order 10^{-17} m^2. Thus, the average shear force acting on a single ion channel is of order 10^{-17} N. By comparison, the average traction force per integrin molecule in adhesion complexes is five orders of magnitude larger [76]. Therefore it is likely that some other surface structure serves to amplify the fluid shear forces acting on integral membrane proteins involved in mechanotransduction.

Endothelial Glycocalyx. One candidate structure is the endothelial glycocalyx layer (for comprehensive reviews, see [46, 77]). In the vascular endothelial glycocalyx, proteoglycans serve as the primary structural components. The primary core proteins are syndecans, a family of four integral membrane proteins, and glypicans, a six-member family of glycosylphosphatidylinositol (GPI)-anchored proteins. Bound to the core proteins is a dense array of glycosaminoglycan (GAG) polymer side chains, including heparan sulfate, chondroitin sulfate, keratan sulfate, dermatan sulfate, and hyaluronan. Heparan sulfate constitutes most of the GAGs in the vascular endothelial glycocalyx, but the relative abundance of GAG types varies with vascular bed, shear stress, and biochemical stimulus (e.g., inflammatory cytokines). In addition to the proteoglycans, the glycocalyx also contains glycoproteins rich in carbohydrate side chains such as selectins, integrins, immunoglobulin receptors, and coagulation cascade-related proteins.

In vivo, bulk flow of blood plasma is excluded from the glycocalyx, as demonstrated by near-wall microparticle image velocimetry (μ-PIV) [78]. Instantaneous velocities as a function of radial coordinate in a mid-sagittal plane of mouse venules were measured by dual-flash strobe illumination of fluorescent microspheres administered intravenously. Extrapolation of the velocity profile near the wall (assuming a no-slip boundary condition at the glycocalyx surface) yields a hydrodynamic thickness of 0.30–0.35 μm. This distance is confirmed by degrading the glycocalyx using light-dye treatment and re-estimating the near-wall zero-velocity position. Thus, shear stress from the bulk flow may be transmitted to the glycocalyx rather than to the endothelial cell plasma membrane directly, and the mechanics of momentum transfer through the glycocalyx may determine the stress distribution acting on mechanosensitive elements in the cell cortex or elsewhere.

Although the widely accepted structure of the glycocalyx is that of a dense polymer mesh, direct measurements of material properties relevant to mechanotransmission and mechanotransduction remain elusive. Weinbaum and colleagues [79] have proposed that the material properties are determined primarily by the flexural rigidity of the core proteins, which they estimated from the recovery time of the glycocalyx thickness after "crushing" by a passing leukocyte in a capillary *in vivo*. Based on this estimate and a model glycocalyx consisting of tightly packed bush-like structures, a wall shear stress of 1 Pa is predicted to cause a 6-nm lateral displacement of actin filaments in the submembranous cortical web. This displacement magnitude is similar to that of the myosin motor step size, implying that the force transmitted from wall shear stress is of similar magnitude to individual actomyosin contraction events. More recently, this hypothesis has been extended to propose that the forces transmitted from shear stress cause sufficient deformation that the actin-dense peripheral bands in regions near cell–cell junctions interact mechanically like the rubber fenders on bumper cars [80]. In this model, the submembranous actin cortical web and the dense peripheral bands act as a functional mechanical unit, and force transmission from the cell surface to junctional regions occurs through the cytoskeleton in a manner consistent with the decentralization hypothesis. The implications of mechanotransmission through the cytoskeleton for signaling are discussed in detail in the following section.

A growing body of evidence suggests that the glycocalyx plays a physiologically relevant role in mechanotransduction in arterial endothelium. For example, when transgenic mice expressing a mutant human apolipoprotein E gene associated with familial dyslipidemia were fed an atherogenic diet, the glycocalyx was significantly thinned in common carotid arteries [81]. Intimal thickening in these regions was increased, consistent with early atherogenesis. Furthermore, when normogenic age-matched control mice were fed an atherogenic diet, the glycocalyx in the internal carotid sinus was thinned (and the intimal layer thickened) compared to that in the common carotid artery, suggesting that glycocalyx degradation correlates with atherosclerosis-prone regions that exhibit disturbed flow profiles. This result was somewhat surprising, since an atherogenic diet alone is not sufficient to cause human-like atherosclerotic lesions in mice without genetic modification. Increased oxidized low-density lipoprotein (ox-LDL) content in blood also degrades the glycocalyx and stimulates inflammatory cell adhesion to the endothelium [82], implying that the function of the glycocalyx is to modify early susceptibility to atherogenic stimuli whether or not full lesions eventually form.

A number of studies relate atherosclerosis risk factors to endothelial mechanotransduction more directly. A classical readout of flow-mediated endothelial mechanotransduction is nitric oxide (NO) production by the endothelium to induce smooth muscle cell relaxation and vasodilation, and NO production in inflamed endothelium overlying atherosclerotic lesions is impaired. In perfused rabbit arteries, neuraminidase treatment eliminated shear stress–mediated vasodilation [83] and NO production [84] without affecting shear stress–induced prostacyclin (PGI_2) production or acetylcholine-induced NO production. These observations have more recently been analyzed in well-controlled *ex vivo* studies. NO production in bovine aortic endothelial cell monolayers exposed to shear stress in a parallel plate flow chamber was reduced after neuraminidase treatment, but shear stress–mediated PGI_2 production remained unaffected [85]. Hyaluronidase or heparinase III [86] treatment also inhibited NO production specifically, but chondroitinase did not affect either NO or PGI_2 production in response to shear stress. Sustained exposure to shear stress increases hyaluronan expression in human umbilical vein endothelial cells [87], but changes in expression of other GAGs in response to shear stress profiles have not been measured. At longer time-scales, digestion of heparan sulfate with heparinase III inhibited shear stress–mediated formation of F-actin stress fibers and peripheral redistribution of vinculin-labeled adhesions [80]. In addition, shear stress–induced reorganization of zona occludens–1 (ZO-1) in tight junctions and connexin-43 in gap junctions was inhibited in heparinase-treated cells. Finally, heparinase treatment eliminated decreases in endothelial cell motility and proliferation rate that are induced by adaptation to shear stress [88]. Taken together, these data support the hypothesis that the glycocalyx serves to amplify force transmission either locally, to signal transduction complexes associated with the luminal surface, or remotely, at least to dense peripheral bands in the actin cytoskeleton. Both *in vitro* and *in vivo* analyses place the glycocalyx near the initiation of physiologically relevant endothelial mechanotransmission and mechanotransduction mechanisms.

Primary Cilium. A second candidate structure that could amplify shear forces directly is the primary cilium. It has been known for decades that mammalian cells including endothelium are capable of expressing a single primary cilium of undetermined function [89], but endothelial primary cilia have been studied much less than the glycocalyx. The structures are not obvious in cultured endothelium observed by light microscopy and have not been reported in AFM scans of living and fixed cells. However, they have been identified in corneal vessels by electron microscopy and in cultured human umbilical vein endothelial cells by antibody staining of acetyl-α-tubulin. Although the glycocalyx may extend the cell boundary by $<1\ \mu m$, its distribution generally follows the topography of the cell membrane. In contrast, erect cilia may extend several micrometers into the flow field, where higher velocities impose a greater drag force. Primary cilia may act as amplifiers to open transient receptor potential (TRP) channels or to transmit stress to the cortical (and hence cytoplasmic) cytoskeleton. Exposure of human umbilical vein endothelial cells to laminar shear stress resulted in primary cilium disassembly [90]. *In vivo*, endothelial cilia have been mapped to high shear locations in the embryonic chicken cardiovascular system [91]. Recent *in situ* immunostaining of large-vessel endothelium in adult mice has revealed expressions of endothelial cilia restricted to sites of flow disturbance [92]. This distribution and the induction of arterial endothelial cilia by the experimental creation of flow disturbance (low and disturbed shear stresses) have led to the proposal that primary cilia act as fluid shear stress sensors in endothelium. The mechanical properties of cilia, a critical consideration for flow mechanotransduction, are unstudied. Further mechanistic evidence of cilium-mediated mechanotransduction using genetic manipulation will help evaluate the role of these structures in arteries.

2.8 Integrated Mechanotransmission through the Cytoskeleton

As noted earlier, a unifying element in endothelial mechanotransduction is the cytoskeleton. As in all anchorage-dependent cells, the endothelial cytoskeleton exists in a state of intracellular tension (prestress) arising from the association of cytoskeletal elements with each other and with cellular membrane-associated proteins and organelles. When endothelium is subjected to flow (or changes of flow), the hemodynamic forces alter the existing balance between intracellular cytoskeletal tension and external reaction forces from adjacent cells and the basement membrane. Inherent to this hypothesis is the notion that the intracellular milieu exhibits spatially heterogeneous material properties, resulting in integration or amplification of mechanical energy density near spatial locations associated with the initiation of mechanotransduction.

Does intracellular spatial heterogeneity in mechanosignaling exist? Several direct measurements of signaling activation illustrate a spatially heterogeneous mechanotransduction response. One of the fastest responses to shear stress is the influx of calcium [93] by a mechanism that requires ATP stimulation of plasma membrane purinergic receptors [25]. ATP-induced Ca^{2+} waves initiate preferentially near cell surface caveolae located near peripheral edges and along actin stress fibers [94], and the initiation sites relocate along with caveolae to the trailing edges in cells stimulated by shear stress to migrate directionally [95]. A second example of local force-mediated signal activation involved direct visualization of Src kinase activation using Förster resonance energy transfer (FRET) [96]. A genetically encoded FRET donor fused to the Src homology type 2 (SH2) domain of c-Src was linked to a FRET acceptor-tagged substrate peptide. When cells expressing the FRET sensor were mechanically probed with fibronectin-coated beads in an optical trap, waves of activated Src were initiated near the site of force application and propagated distally at an average wavefront speed of 18 nm/s. Force-induced Src waves required an intact cytoskeleton. A final example of heterogeneity is spatial regulation of small GTPase activity in endothelial cells in response to shear stress [97]. Shear stress–mediated migration of endothelial cells requires directional polarization of the lamellipodium extension. A FRET sensor demonstrated that Rac1 is activated preferentially in lamellipodia and requires new integrin ligation by an extracellular matrix, implicating Rac in the establishment of planar cell polarity and cytoskeletal remodeling in response to shear stress. Taken together, these examples emphasize the importance of elucidating the role of the cytoskeleton in integrated mechanotransmission through the cytoplasm to coordinate the activation of diverse signaling networks.

A set of techniques to sample the spatial heterogeneity of complex modulus in living cells is based on tracking microinjected microparticles with high time resolution [98–100]. If the particle size is comparable to the porosity of the cytoskeleton, then mean squared displacement of the particles gives information about the dependence of the complex modulus on the local polymerization state of the cytoskeleton. The elastic modulus in lamellipodia is 82 Pa, which is consistent with the modulus values reported previously that were estimated using extracellular beads interacting with cytoskeleton-associated adhesions, whereas the modulus in perinuclear regions of the cell is 33 Pa. The spatial gradient of the modulus magnitude requires intact actin polymerization dynamics in the lamellipodium. These measurements demonstrate that local polymerization dynamics and cross-linking density can cause spatial gradients in material properties at a subcellular length scale.

Spatial gradients in an elastic modulus with the course-grained continuum assumption predict cytoskeleton-dependent heterogeneous force distribution in the cytoplasm, and structural models reinforce this hypothesis. For example, the actin cytoskeleton clearly contributes to intracellular prestress, as discussed previously. Disruption of microtubules in the tensegrity model results in the redistribution of stress to the cell–substrate interface [101]. Disruption or genetic depletion of vimentin intermediate filaments caused reduced cell stiffening under applied stress, as predicted by a tensegrity model that included intermediate filaments as slack elastic cables that became stretched with increased cell spreading or deformation [102]. This structure is consistent with the hypothesis that intermediate filaments serve to maintain the structural integrity of the elastic cytoskeleton during shear stress–induced remodeling of actin and microtubule networks [38, 51]. Thus, prestressed network models predict that the nature of interactions among cytoskeletal elements will affect the redistribution of intracellular tension during applied external stresses.

If internal prestress is spatially coordinated during changes in external fluid shear stress, then it should be reflected in the spatiotemporal distribution of displacement and strain in the cytoskeletal network. The first measurements of cytoskeletal displacement caused by shear stress were completed in endothelial cell monolayers expressing a green fluorescent protein (GFP)–vimentin fusion protein that is distributed to endogenous intermediate filaments [38, 39]. The intermediate filament network serves as a reliable indicator of cytoskeletal displacement, since the geometry of the filament mesh remains constant over time-scales that are long relative to those of mechanotransmission and early mechanotransduction. Within minutes after the onset of steady unidirectional shear stress at 1.2 Pa, individual GFP–vimentin filaments were displaced, bent, and rotated in three dimensions. Displacement magnitude approached 1 μm in the first 90 s after flow onset, in contrast to random small-magnitude displacements of individual intermediate filaments in the absence of external applied forces. Statistical analysis of the three-dimensional displacement field revealed a complex pattern induced by flow onset. On average, the displacement magnitude was larger near the apical than basal surface and larger in downstream than upstream regions. In confluent monolayers, the displacement magnitude at locations along edges of adjacent cells often correlated closely, suggesting mechanical communication between the cytoskeletal networks. The largest displacement magnitude occurred within 3 min after flow onset. However, displacement patterns in individual cells were spatially and temporally heterogeneous, indicating that geometric factors such as local cytoskeletal structure and cell shape influenced cytoskeletal displacement.

More relevant to mechanotransmission is the strain field computed from the relative displacement of intermediate filaments [22]. High-magnitude spatially localized *strain focusing* occurred at sites consistent with cytoskeletal connection

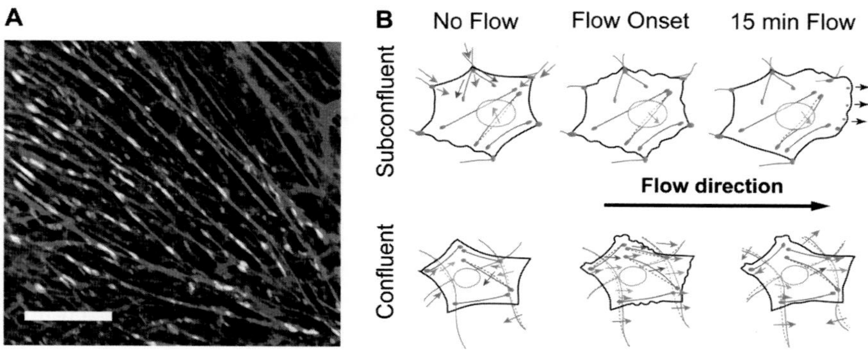

Figure 2.3. Live cell imaging demonstrates mechanical connectivity from the cytoskeleton to the extracellular matrix. (A) Relative displacement maps are generated by tracking adhesion sites containing GFP-vinculin (green), mRFP-actin cytoskeleton (red, a kind gift from E. Fuchs, Rockefeller University), and coumarin-labeled fibronectin fibrils in the extracellular matrix (blue). Scale bar, 10 μm. Adapted from [143]. (B) Patterns of structural dynamics in endothelial cells depend on cell density. In subconfluent cells, flow onset increases actin polymerization at cell edges, changes the direction of lateral stress fiber displacement (blue, dashed-blue), reduces the centripetal remodeling of actin stress fibers (blue) and adhesion sites (green), and reduces the displacement of the underlying fibronectin matrix (red). After 15 min, new actin polymerization becomes polarized in the downstream direction, and new adhesion sites coalesce and stabilize in downstream lamellipodia. In confluent cells, flow onset generates a rapid and transient burst of edge ruffling and a transient downstream displacement of the subcellular structures. The displacement patterns evolve to become more heterogeneous with time. Adapted from [40].

to adhesions and intercellular junctions but not near the apical surface. Strain focusing is only evident in high-resolution spatial maps of the Lagrangian strain tensor; the median strain computed over the whole cell is constant before and after flow onset. Finally, the principal strain direction near cell edges in confluent monolayers rotated from parallel to the edge (along the local filament axis) to perpendicular to the edge. Shear stress–induced mechanical stretch oriented across the cell boundary may be related to changes in junctional permeability and the onset of structural reorganization. Overall, these measurements in the intermediate filament cytoskeleton demonstrate that external stress is not only transmitted to but also amplified and focused near adhesions and junctions.

Fluid shear stress rapidly alters actin dynamics and stress fiber remodeling [40, 48]. The magnitude and time course of basal stress fiber displacement following flow onset closely parallel those of intermediate filaments, as measured in endothelial cells co-expressing GFP–vinculin and a monomeric red fluorescent protein (mRFP)–actin fusion protein (Figure 2.3(A); Plate 1). In subconfluent layers of endothelial cells, single endothelial cells exhibit downstream-directed mRFP-actin edge ruffles on a time-scale of 8–12 min, and centripetal stress fiber remodeling in the cell periphery is arrested (Figure 2.3(B); Plate 1). In confluent monolayers, shear stress induces increased edge activity even though adjacent cells are in close physical contact. Instead of a broad lamellipodium extension, actin-mediated blebbing in confluent monolayers consists primarily of smaller blebbing and a wavelike motion.

The spatially and temporally heterogeneous structural dynamics in the endothelial cytoskeleton reflect the redistribution of intracellular tension in response to the fluid shear stress profile. Strain focusing, actin edge ruffling, and stress fiber remodeling are associated with mechanotransmission to sites where mechanotransduction is initiated. The following sections outline the evidence supporting this hypothesis for adhesions and intercellular junctions.

2.9 Integrin-Mediated Mechanotransduction

The physical association of transmembrane integrins with the extracellular matrix on the outside of the cell and with signaling molecule complexes, linker proteins, and the cytoskeleton on the inside of the cell confer an important role for integrins in the regulation of mechanotransduction [103]. *In vivo*, disturbed flow profiles in inflammation-prone regions of the artery wall are associated with the remodeling of the structure and composition of the extracellular matrix. *In vitro*, time-lapse tandem scanning confocal microscopy images demonstrate that during adaptation to shear stress, adhesions coalesce, elongate, and align parallel to the direction of the shear stress on a time-scale similar to that of the cytoskeletal and matrix adaptation [34]. These observations were followed by demonstrations of shear-responsive phosphorylation of proteins recruited to adhesion sites, including focal adhesion kinase (FAK), paxillin, Shc, and other integrin-linked scaffold complexes [3, 104–106]. Furthermore, mechanotransduction requires dynamic ligation and detachment of integrins from the extracellular matrix, corresponding to temporal dynamics in the recruitment and activation of signaling molecules in adhesions [107]. Mechanotransmission from the cell surface to integrins in adhesion sites may therefore critically determine the physical interactions of cells with the extracellular matrix and specify the signaling networks involved in endothelial mechanotransduction.

Spatiotemporal patterns of displacement during the onset of shear stress reveal a close correlation among the cytoskeleton, adhesions, and extracellular matrix fibrils (Figure 2.3(B); Plate 1) [40], suggesting that mechanical continuity from inside to outside the cell is modulated by shear stress. In subconfluent cell layers, paxillin rapidly assembles into new adhesions in shear stress–induced lamellipodia, and paxillin in mature fibrillar focal contacts disassembles and translocates centripetally along stress fibers [108]. In contrast, centripetal remodeling of vinculin-containing adhesions and their associated stress fiber termini in the cell periphery is arrested after flow onset [40]. Reduced adhesion site displacement after flow onset also correlated with reduced displacement of fluorescently labeled fibronectin fragments located nearby (within micrometers). More relevant to *in vivo* endothelial physiology are the physical interactions within confluent monolayers (Figure 2.3(B); Plate 1). In cells expressing GFP–vinculin, the displacement rate of adhesions was not significantly different before compared to after flow onset. However, the displacement direction changed from a random distribution to a significant correlation with the flow direction within 1 min, and the directional displacement again became more heterogeneous after 15 min of exposure to shear stress. This temporal pattern of adhesion

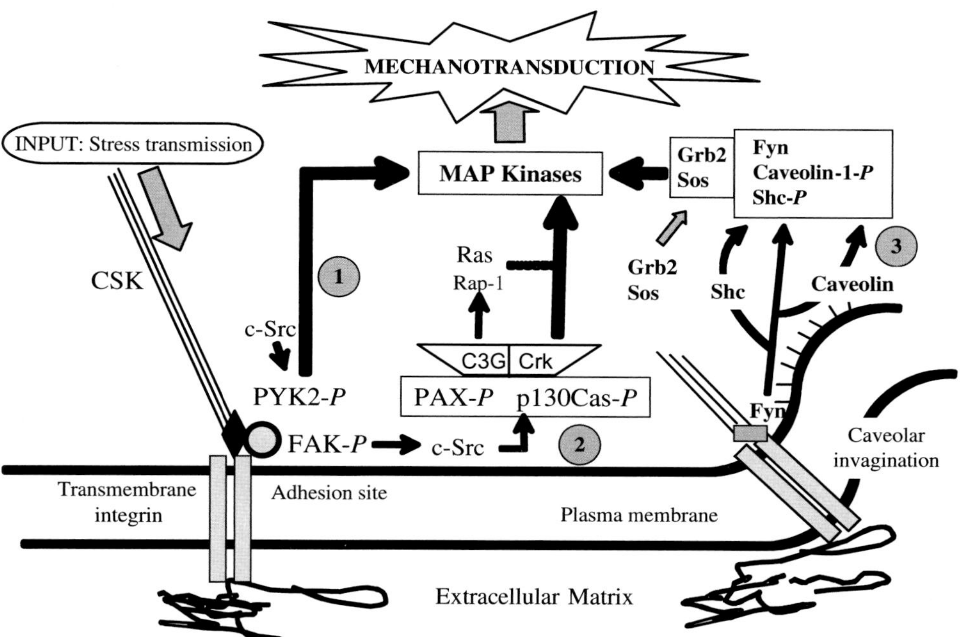

Figure 2.4. Integrin-mediated mechanotransduction. The physical association of transmembrane integrins with the extracellular matrix on the outside of the cell and with signaling molecule complexes, linker proteins, and the cytoskeleton on the inside of the cell confer an important role for integrins in the regulation of mechanotransduction. Three interacting mechanisms proposed to be initiated by stress transmission via the cytoskeleton (CSK) are illustrated. All three converge at the activation of MAP kinases that lead to multiple downstream flow responses. At 1, phosphorylation of proline-rich tyrosine kinase PYK2 by c-Src leads to the activation of MAPK possibly through coordination with a calcium-dependent reactive oxidant species mechanism [109]. In the other two mechanisms the ordered assembly of various complexes is required to activate MAPK. At 2, self-phosphorylation of integrin-associated focal adhesion kinase (FAK) stimulates Src to phosphorylate paxillin and p130Cas. These associate to form a scaffold to assemble a guanine nucleotide exchange factor C3G and an adapter protein Crk. C3G then activates Rap1, a small GTPase member of the Ras family, which helps coordinate the activation of MAPK by Crk. At 3, the caveolar protein caveolin-1 interacts with local integrins through the Src family kinase Fyn. Upon integrin activation, the adapter protein Shc and caveolin-1 are phosphorylated by Fyn to create a caveolin-*P*–Fyn–Shc-*P* complex that recruits growth factor receptor binding protein-2 (Grb2) and Son of sevenless (Sos), and this molecular assembly leads to the activation of the Ras-MAPK pathways. In these elements of the decentralized model of mechanotransduction, cytoskeletal mechanotransmission to the integrin-rich adhesion sites elicits principally cascades of protein phosphorylation, leading to the activation of MAPK.

displacement during flow onset was mirrored by fluorescently labeled fibronectin fibrils in the cell-assembled extracellular matrix. Interestingly, shear stress–mediated induction of directional matrix fibril displacement was not affected by pretreatment with low concentrations of latrunculin A to inhibit actin polymerization. However, the displacement magnitude was significantly reduced as a function of latrunculin A exposure time. Overall, these experiments suggest that shear stress at the luminal surface can regulate physical interactions through integrin-containing adhesion sites and that the resulting redistribution of prestress in the cytoskeleton and extracellular matrix triggers abluminal mechanotransduction.

Several mechanisms involving integrins in mechanotransduction have been proposed and are summarized in Figure 2.4. For example, Shyy and Chien [7] integrated the results of endothelial flow studies with data obtained from platelet activation and growth factor signaling (both are integrin-mediated). In their proposed mechanism, flow induces conformational activation of integrins to increase both affinity and avidity for extracellular matrix ligands. Competitive inhibition of integrin activation with Arg-Gly-Asp (RGD) peptides that occupy ligation sites or by monoclonal antibodies against β-integrin subunits inhibits typical downstream responses to shear stress such as mitogen-activated protein kinase (MAPK) phosphorylation and nuclear factor (NF)-κB activation.

On the cytosolic side of the plasma membrane, one mechanism by which integrins mediate shear stress mechanotransduction is proposed to be through the recruitment and phosphorylation of FAK. Phosphorylation of FAK on Tyr397 leads to its association with the SH2 domain of c-Src, which then phosphorylates paxillin and p130Cas. This complex acts as a scaffold to assemble C3G and the adapter protein Crk. C3G is a guanine nucleotide exchange factor (GEF) that activates Rap1, a small GTPase member of the Ras family, and Crk mediates the activation of MAPKs. A parallel pathway exists involving the tyrosine kinase Pyk2, another member of the FAK family [109]. Evidence exists to support recruitment and phosphorylation of both FAK and Pyk2 to sites of dynamic integrin ligation in endothelial cells subjected to shear stress, suggesting that the scaffold assembly is likely to be highly localized in the plasma membrane.

A similar spatial relationship may occur in a second integrin-dependent mechanism of mechanotransduction by shear stress. Flow conditioning of endothelial cells stimulates increased numbers of plasma membrane caveolae and recruits them into mechanotransduction pathways [110]. Caveolin-1, the cytoplasmic coat protein of caveolae, is constitutively associated with the Src family kinase Fyn. Activation of $\alpha_1\beta_1$, $\alpha_1\beta_5$, or $\alpha_V\beta_3$ integrins by shear stress results in the recruitment and phosphorylation of the adapter protein Shc. The mechanism may involve α-integrin interaction with phosphorylated caveolin-1 and subsequent activation of Fyn, which then binds the SH2 domain of Shc and phosphorylates it on Tyr317. The caveolin-Fyn-Shc complex recruits the growth factor receptor binding protein-2 (Grb2) and Son of sevenless (Sos), and this molecular assembly leads to the activation of the Ras-MAPK pathways that are prominent in endothelial shear stress responses.

Recent studies also implicate integrins in shear stress–mediated signaling that controls actin dynamics and remodeling. Caveolin-1 interacts not only with α-integrins but also with β_1-integrins to recruit and activate the Src family kinase inhibitor Csk, which leads to increased myosin light chain (MLC) phosphorylation [111]. The proposed mechanism places Csk as the switch between the activation of Src family kinases and small GTPases. Phosphorylated FAK and

Src have been implicated in downregulating RhoA activity transiently after flow onset [112]. In spreading cells, transient Rho downregulation occurs as a result of Src-mediated phosphorylation of the GTPase activating protein p190RhoGAP [113, 114]. Integrin-caveolin-Csk association in response to shear stress would therefore inhibit Src and p190RhoGAP phosphorylation, thereby maintaining Rho activity [111]. Activated Rho phosphorylates the serine/threonine kinase ROCK and promotes actin stress fiber contractility and structural stability by two mechanisms [115]. First, ROCK phosphorylates LIM-kinase (LIMK). LIMK-mediated phosphorylation of cofilin prevents actin depolymerization and thereby stabilizes stress fiber assembly. Second, ROCK inhibits myosin phosphatase, resulting in increased MLC phosphorylation by its kinase MLCK. This cascade serves to increase intracellular contractility and would be expected to increase prestress in the actin cytoskeleton. Many of these Rho/ROCK–mediated signaling events are down-regulated in the early response to shear stress, and emerging data demonstrate that inhibiting p190RhoGAP phosphorylation after flow onset prevents the arrest of adhesion and stress fiber remodeling in subconfluent cell layers and disrupts coordinated directional displacements in confluent endothelial monolayers (R. E. Mott and B. P. Helmke, unpublished data). Overall, one must conclude that control of small GTPase signaling by mechanotransduction through integrins represents a critical step in guiding actin dynamics and remodeling in response to shear stress.

Activation of integrin-mediated mechanotransduction pathways is specific to extracellular matrix composition. Shear stress triggers activation of pro-inflammatory NF-κB in endothelial cells on fibronectin or fibrinogen matrices but not in cells on collagen or laminin [116]. Shear stress induces conformational activation and ligation of $\alpha_5\beta_1$ and $\alpha_V\beta_3$ integrins but not $\alpha_2\beta_1$ in cells on fibronectin or fibrinogen, whereas $\alpha_2\beta_1$ but not $\alpha_5\beta_1$ or $\alpha_V\beta_3$ is activated in cells on collagen [117]. In cells on fibronectin, protein kinase C type α (PKCα) is activated by flow, and PKCα suppresses $\alpha_2\beta_1$ conformational activation. In contrast, flow activates protein kinase A (PKA) in cells on collagen, which leads to the suppression of flow-induced $\alpha_V\beta_3$ activation. Flow-induced $\alpha_2\beta_1$ ligation in cells on collagen prevents NF-κB activation by shear stress through a pathway that includes the activation of p38 MAPK locally near adhesion sites. Surprisingly, exogenous activation of p38 in cells on fibronectin prevents flow-mediated NF-κB activation. Finally, shear stress causes activation of p21-activated kinase (PAK) in cells on fibronectin but not on Matrigel (a tumor-derived basement membrane substitute) [118]. PAK is a downstream effector of the small GTPases Rac and Cdc42 and binds scaffold proteins such as Nck and Grb2. The flow-induced activation and relocalization of PAK to intercellular junctions is associated with endothelial permeability changes (discussed later). These matrix-specific and integrin-specific signaling networks may be involved in early atherogenesis, since focal deposition of fibronectin and fibrinogen into a provisional matrix *in vivo* occurs before other signs of atherosclerosis.

Flow-mediated mechanotransduction through integrins is intimately related to the binding of cytoskeletal elements at adhesion sites and cell junctions. Notably, integrins not located at the abluminal surface of the cell may be activated by shear stress without the engagement of the extracellular matrix [119]. Furthermore, mechanisms of mechanoactivation of the Ras pathway have been proposed that involve direct activation of heterotrimeric G-proteins in the plasma membrane by the deformation of membrane lipids without cytoskeletal or integrin involvement, as noted previously [10]. These events, which are spatially and temporally complex, undoubtedly involve levels of additional regulation yet to be discovered.

2.10 Primary Transduction of Shear Stress through Junctional Mechanosensors

In arteries, endothelial cells are in direct communication with each other mechanically through tight junctions and adherens junctions and electrochemically through gap junctions. Fujiwara and colleagues [36, 120] first demonstrated flow- and stretch-induced phosphorylation of PECAM-1 in endothelial junctions and suggested that PECAM-1 is a primary mechanosensor. Flow-mediated phosphorylation of ligated PECAM-1 serves to recruit Shp-2, a protein tyrosine phosphatase, and the adapter molecule Gab1 to junctions upstream of ERK activation. Furthermore, Shp-2 associates with PECAM-1 in the transactivation of Tie-2 [121], suggesting a role for a PECAM-1–Shp-2–Tie-2 pathway in flow-mediated signal transduction. Shay-Salit et al. [122] demonstrated flow-mediated activation of vascular endothelial growth factor receptor 2 (VEGFR2) and its assembly with junction-associated phosphorylated vascular endothelial cadherin (VE-cadherin), β-catenin, and PI3-K, which leads to the phosphorylation of Akt-1. Building on these findings, Tzima et al. [37] reported that this mechanosensory complex involving junctional proteins regulates a subset of mechanotransduction pathways through integrins. In transfection studies of nonendothelial cells, they showed that PECAM-1, VE-cadherin, and VEGFR2 are sufficient to transduce shear stress in the presence of an intact cytoskeleton and its associated proteins. Conversely, the selective inhibition or deletion of any of these molecules in endothelial cells inhibited flow-related transduction. Flow-induced activation of VEGFR2 did not occur in PECAM-1$^{-/-}$ and VE-cadherin$^{-/-}$ endothelial cells, nor was there activation of pathways that are dependent on integrin activation. In support of the mechanism *in vivo*, local regions of endothelial activation (NF-κB nuclear translocation) correlated with sites of disturbed hemodynamics in wild-type mice but were not activated in PECAM-1$^{-/-}$ mice. PECAM-1 is unlikely to be unique as a mechanosensor, since PECAM-1$^{-/-}$ endothelial cells *do* adapt to unidirectional flow by aligning parallel to the flow direction. Nevertheless, these data support a central role for PECAM-1 in flow-mediated mechanosensing.

PAK may play a central role in integrating mechanotransduction pathways at junctions with those at adhesions. PAK activation and localization to PECAM-1–labeled junctions are stimulated by several pro-atherogenic stimuli, including flow onset, disturbed flow waveforms, ox-LDL, and cytokines [118]. Activated PAK at junctions stimulates intercellular pore formation, whereas inhibiting PAK kinase activity inhibits flow-stimulated increases in endothelial permeability. Activated

PAK increases myosin phosphorylation and contractility, which may increase intracellular tension acting on junction proteins like PECAM-1 and VE-cadherin. How are mechanisms of PAK activation at junctions integrated with those at adhesion sites? Rac activation by flow is independent of extracellular matrix composition, but integrins may serve to modulate PAK activation by flow. For example, flow-mediated activation of $\alpha_2\beta_1$ integrins in cells on collagen leads to activation of PKA, and PAK phosphorylation by PKA inhibits its kinase activity. This mechanism may explain enhanced PAK activation by flow in cells on fibronectin relative to those on Matrigel. These observations support the idea that shear stress forces are transmitted to junctional regions and adhesion sites in a manner consistent with a decentralized model of mechanotransduction.

Do these mechanisms play a role in flow-mediated mechanotransduction *in vivo* during atherogenesis? Two hallmarks of early atherogenesis are increased endothelial permeability and deposition of a fibronectin-rich provisional matrix in place of the physiological subendothelial basement membrane comprised primarily of collagen IV and laminin. In an $ApoE^{-/-}$ model of atherosclerosis, arterial sites of PAK activation and increased fibronectin deposition correlate with atherosclerosis-prone regions such as the internal carotid sinus [118]. Furthermore, inhibiting PAK reduces endothelial permeability at those locations. Thus, recent progress in elucidating mechanotransduction mechanisms in reductionist models based on the decentralization hypothesis shows promise for a better understanding of mechanotransduction mechanisms relevant to vascular pathologies *in vivo*.

2.11 *In Vitro* to *In Vivo*: Hemodynamics and Arterial Endothelial Phenotypes *In Vivo*

Study of the biomechanics of endothelial cells in tissue culture is paralleled by *in vivo* (or *in situ*) investigations. Immunocytochemistry is useful to estimate the *in vivo* expression of candidate molecules identified *in vitro*, and despite limited control over the local mechanical and chemical environments *in vivo*, regional genomic analyses of endothelial phenotypes strongly support a biomechanical influence on endothelial regulation of arterial homeostasis *in vivo*.

A range of arterial geometries are associated with susceptibility to the formation of focal atherosclerotic plaques [19, 123]. In arterial regions where there are curves, branches, and bifurcations, an abrupt separation of the principal flow vectors creates complex transient vortices that expose the endothelial cells to temporary flow reversals together with multiple frequencies and magnitudes of pressure and shear stress, collectively referred to as "disturbed" flow (Figure 2.5). The endothelium is in a different biomechanical environment in a disturbed flow that leads to changes in its biology (phenotype) when compared to undisturbed flow regions. There is accumulating evidence that the altered phenotype may be linked to athero-*susceptibility* (or, conversely, that the endothelium in an undisturbed flow expresses a *protective* phenotype). Adding to the complexity, arterial geometry influences the characteristics of flow, both undisturbed laminar flow and disturbed flow, at different phases of

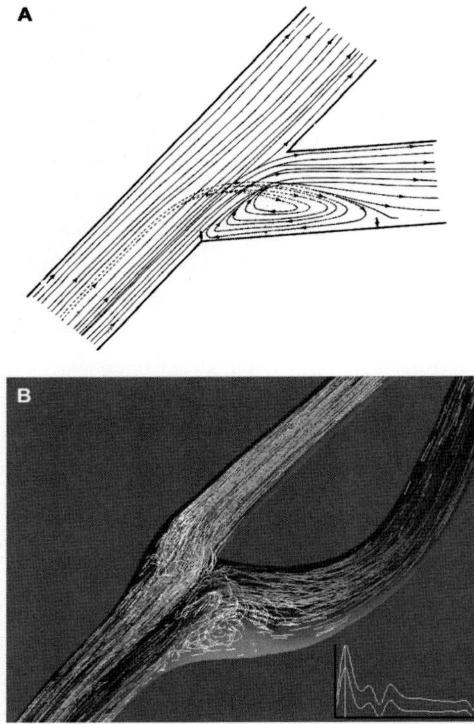

Figure 2.5. Hemodynamics in arteries can be experimentally modeled and measured *in vivo*. (A) Tracings of particle paths during steady flow in a glass tube. In this particular case, manipulation of the volumes passing through each branch resulted in flow separation in one of the tubes, creating a vortex in steady flow conditions. Particles entered the vortex from the mid-flow region of the parent tube. From velocity measurements inside and outside of the separated flow region, the peak wall shear stresses in the vortex were approximately an order of magnitude lower than at the flow divider and the parent tube. The shear stress was near zero at the stagnation regions near the boundaries of the vortex. Tube diameters 3 mm. Inflow Reynolds number 299. Adapted from [144]. (B) Flow characteristics in the normal male human carotid artery at the bifurcation into internal and external branches near peak systole showing spatial characteristics of the flow. Note the transition from unidirectional laminar flow (and shear stress) in the carotid artery sinus, a site of high susceptibility for atherosclerosis, where complex disturbed flow develops with oscillating flow reversal and the formation of transient vortices. Flow varies in velocity but is laminar and unidirectional elsewhere. This frame captures an instantaneous event at peak systole during unsteady arterial blood flow. As diastole develops, the flow velocities (indicated by bar-length) fall sharply in the flow separation region as predicted from the steady flow model of (A). Mean common carotid artery diameter in males is $\sim$ 6.3 mm. Frame from a combined medical imaging and computational fluid dynamics sequence of human blood flow by Professor David A. Steinman of the Department of Mechanical and Industrial Engineering, University of Toronto (Image gallery URL: http://www.mie.utoronto.ca/labs/bsl/). Kindly provided by Dr. Steinman as a high-resolution frame. Also see [145].

the cardiac cycle. A waveform for carotid artery undisturbed flow will always display forward motion, whereas in the nearby disturbed flow region of the carotid sinus near the carotid bifurcation, a flow reversal phase occurs in early diastole (Figure 2.5(B)) [124]. Thus the endothelium at both sites accommodates *temporal* changes of shear stress. However, the addition of transient flow reversal in a disturbed flow imposes steeper gradients of shear stress, a lower average magnitude of shear stress, and possibly of greatest importance, a change of direction in the deformation forces approximately every second. These differences create additional major *spatial* differences within the already spatially defined disturbed flow region. Low and high shear stress levels have been proposed to be both atheropermissive and atheroprotective [125, 126]. Some reconciliation of what appeared to be opposing views has resulted from the demonstration that temporal and spatial *gradients* of shear stress may be of importance to the endothelium; such gradients exist in regions of both high *and* low average shear stress, and both occur in disturbed flow locations. This was first demonstrated in an extreme model of disturbed flow, turbulence, where the characteristics of the flow and not the magnitude of shear stress elicited an important endothelial response (cell cycle initiation) [29]. This principal has recently been rediscovered in more controlled types of disturbed flows [127]. Furthermore, the concept of atheroprotection at the gene expression level by a nondisturbed flow as opposed to a disturbed flow or occasional turbulence suggests that the endothelial (and possibly the smooth muscle) cell is able to discern complex hemodynamic signaling patterns. As the spatial scale decreases, the challenge of mapping not only regional but also subregional endothelial phenotype heterogeneity increases. Efforts to date have successfully identified regional profiles both by global genomic analyses [128–130] and by more limited candidate gene (and protein) measurements [2, 131–134].

2.12 Regional Arterial Endothelial Heterogeneity Associated with Disturbed Flow

Profiling the expression of many genes (transcript profiling) within disturbed versus undisturbed flow locations of arteries and heart valves has recently provided insights into regional heterogeneity that also maps to athero-susceptibility. Passerini et al. [128] demonstrated that a delicate balance of pro- and anti-atherosclerotic mechanisms may exist simultaneously in the endothelium of lesion-prone sites to create a setting of vulnerability to atherogenesis. Following analysis of >13,000 genes, both pro-inflammatory and (protective) antioxidative mechanisms that converged at NF-κB were shown to coexist in regions of disturbed flow (atherosusceptible region). No lesions were evident and there was no indication of inflammatory protein expression at these sites. Thus in the normal animal, the endothelium of disturbed flow regions appeared to be primed for inflammation but was held in check by compensatory mechanisms that inhibited NF-κB activation.

Gene expression profiling can usefully guide the exploration of flow-related heterogeneous protein expression and protein function. For example, transcript profiles indicated differential mRNA expressions of protein kinase C (PKC) in disturbed flow versus undisturbed flow sites. PKCs are an important family of kinase enzymes that phosphorylate many regulatory proteins involved in key cellular pathways. Magid and Davies [131] identified site-specific differences of up to fivefold in the total enzyme activities of PKCs, a reflection of function; however, the expression levels of the protein isoforms α, β, ε, ι, λ, and ζ were similar in the different regions. They showed that the differences were attributable to a post-translational modification of one isoform, PKCζ, in which differential phosphorylation of Threonine 410 and Threonine 560 that determined PKCζ activation and degradation, respectively, accounted for the site-specific differences of enzyme activities. The experiments demonstrate that spatial endothelial phenotyping, including function-relevant measurements, can provide mechanistic insights *in vivo* that add significantly to reductionist *in vitro* flow experiments. A second example is the recent linking of the transcription factor Kruppel-like Factor-2 (KLF2) in endothelium to arterial regions protected from the development of atherosclerosis. Dekker et al. [135] demonstrated KLF2 to be the first endothelial transcription factor that is uniquely induced by flow and, on the basis of its *in situ* expression in atheroprotected regions, suggested that it might be at the molecular basis of the physiological, healthy, flow-exposed state of the endothelial cell. This group has subsequently characterized KLF2 mechanisms in endothelial cells and provided important evidence for a central role of KLF2 in endothelial pathological quiescence [136]. Much of this work has been confirmed and extended by others [137]. However, the biomechanical mechanism leading to the regulation of KLF2 is unknown.

An important example of a candidate gene experiment that addresses a flow-induced cause-and-effect hypothesis is that reported by Cheng et al. [138], who generated transgenic mice that express human eNOS in fusion with GFP as a reporter of eNOS protein expression. Gradations of shear stress (τ) and separations of flow were created by the placement of a tapered cast around the mid-portion of the left common carotid artery, resulting in a corresponding gradual narrowing of the artery lumen over a length of several millimeters. This is normally a region of pulsatile laminar flow without flow separation and where τ forces are unidirectional. It is also a site resistant to atherosclerotic lesion development. Since $\tau \propto 1/(\text{radius})^3$, shear stress increases rapidly throughout the length of the taper. Downstream of the cast the lumen widens to create a short region of oscillating separated flow similar to that recorded at atherosclerosis-susceptible locations elsewhere *in vivo*. Within 24 h following placement, the hemodynamics were spatially mapped to *en face* cell responses. Cheng et al. [138] reported that eNOS gene and protein expression was elevated as a function of τ within the tapered cast, consistent with shear stress experiments *in vitro*, and that the intracellular redistribution of eNOS, including its activated form (phosphorylated serine 1177), was significantly increased by both elevated τ and oscillatory flow. This particular study demonstrated that altered hemodynamics and the expression of an important vasoregulatory molecule were causally linked *in vivo*, confirming *in vitro* predictions and extending the correlative approach of other studies.

2.13 Spatial Heterogeneity of Endothelial Phenotypes and Side-Specific Vulnerability to Calcification in Normal Porcine Aortic Valves: A Role for Local Hemodynamics?

Calcific aortic valve sclerosis, characterized by thickening and calcification of the valve leaflets, is a common disease associated with significant morbidity due to

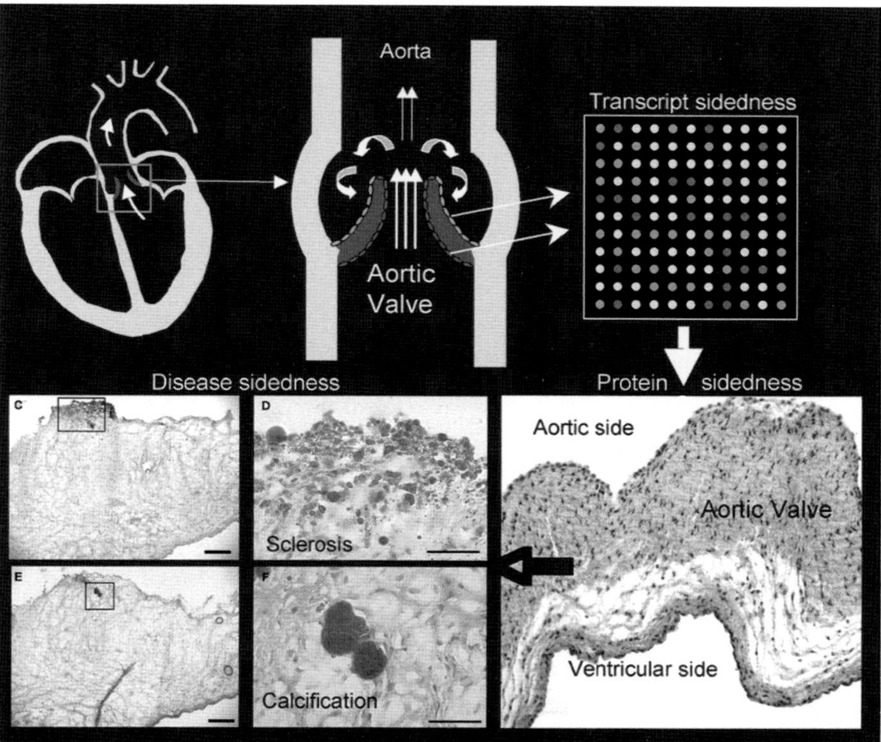

Figure 2.6. Flow characteristics around the aortic valve correlate with side-specific vulnerability to calcific sclerosis in endothelial phenotype profiles. The aortic valve is a site of complex blood flow in which the flow is highly disturbed on the downstream aortic side compared with the ventricular side. In the normal valve endothelium isolated from each side of the valve leaflets displays differential phenotypes, with the patho-susceptible aortic side showing pro-calcific gene expression. The aortic side is also more sensitive to the effects of hypercholesterolemia. The figure illustrates the approaches described in the main text to differentially profile the expression of a large number of swine endothelial genes on each side. It also shows the preferential development of inflammatory sclerosis and calcification on the aortic side when valve pathology develops.

cardiovascular causes [139]. Until recently, calcific aortic sclerosis was considered a passive degenerative process secondary to aging and the accumulation of mechanical damage to the valve matrix. However, recent studies demonstrate an association between clinical risk factors for atherosclerosis and the development of valvular disease [140], suggesting a more complex etiology than previously appreciated. Early degenerative lesions in human valves are characterized by increased cellularity, increased extracellular matrix deposition, accumulation of oxidized lipoproteins within the valve interstitium, and the presence of an inflammatory infiltrate containing nonfoam and foam cell macrophages and occasional T cells. These histological findings resemble early sclerotic lesions of the vasculature and, together with the shared risk factors, suggest that, as in atherosclerosis, the initiation of aortic valvular sclerosis involves chronic inflammatory processes potentiated by systemic factors. A further parallel with arterial pathologies is that calcific sclerosis occurs preferentially

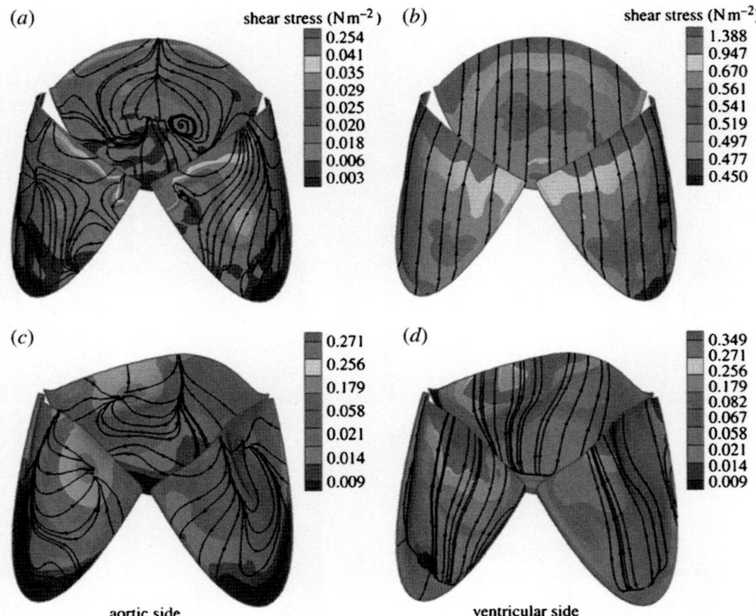

Figure 2.7. Aortic valve leaflets. Flow frictional streamlines and shear stress magnitude plots on the aortic (a, c) and ventricular (b, d) sides of fully open (a, b) and early closing (c, d) phases of the cardiac cycle. During the open phase, the streamlines show smooth and straight accelerating flow on the ventricular side. In contrast, the hemodynamics on the aortic side are influenced by the expanded sinus region to create complex flow separations. The magnitude of shear stress is much higher on the ventricular side (note range differences). As the leaflets commence closure, the magnitudes of shear stress become similar; but while the streamlines remain regular on the ventricular side, spiraling profiles persist on the aortic side. The hemodynamics on the patho-susceptible aortic side are therefore characterized by low shear stress magnitudes and complex flow patterns similar to atherosusceptible flow separation regions of large arteries. From [141]. Reproduced with permission.

on the aortic side of the valve leaflet (Figure 2.6; Plate 2), where the endothelium is subjected to hemodynamics characterized as a highly disturbed flow, in contrast to the protected ventricular side (Figure 2.7; Plate 3) [141]. Local biomechanical differences may contribute to the differences in endothelial phenotypes identified on each side [130]. In advanced calcific valvular lesions, prominent features include mineralized deposits composed of hydroxyapatite (the mineral phase found in bone tissue), several bone matrix proteins, and mature bone-forming osteoblasts and bone-resorbing osteoclasts. Thus, valvular calcification is not due to passive, unregulated precipitation of calcium phosphate but appears to be a highly regulated, active ossification process that, together with local inflammation, may be more closely associated with the local spatial biomechanical environment than previously appreciated.

Using RNA amplification and transcriptional profiling of aortic-side versus ventricular-side endothelium from normal porcine valves, Simmons et al. [130] identified globally distinct endothelial phenotypes on opposite sides of the aortic valve *in vivo*. The differential gene expression profiles suggest that the endothelium on the disease-prone aortic side of the valve is permissive to calcification but is protected in the normal valve against inflammation and lesion initiation by anti-oxidative mechanisms. The co-existence of susceptible and protective transcript profiles in susceptible valve endothelium is similar to that noted in arterial endothelium (see above), albeit with contextual differences appropriate to the prevalent pathological susceptibility (calcification, inflammation, etc.). The findings demonstrate that the endothelium may play a critical role in valve susceptibility to calcific sclerosis through a balance of inhibitory and stimulatory mechanisms. Within the demonstrated phenotype sidedness, there is regional heterogeneity of the endothelial expression of participating proteins within each side. Such spatial heterogeneity of valvular endothelial phenotypes may contribute to the focal susceptibility for lesion development. It should be possible to compare regional cell phenotype profiles to regions of predicted mechanical strain intensity to refine the correlations. In recent studies (M. Guerraty et al., manuscript submitted), the challenge of hypercholesterolemia for 2 weeks substantially altered the endothelial phenotype profile. There were significant changes in the expression of more than 800 endothelial genes on the susceptible aortic side of the valve leaflet; in contrast, only 87 endothelial genes were affected on the ventricular side. Thus the susceptible-side endothelium is more sensitive to hypercholesterolemia than is the protected side. Furthermore, an unexpected finding was that the endothelial phenotype profile on the susceptible aortic side was changed by hypercholesterolemia to a more protective profile of gene expression than was expressed in normocholesterolemia. The pro-calcific gene bone morphogenic protein 4 (BMP4) and osteopontin were down-regulated in these cells together with protective changes of expression of genes involved in cellular proliferation, molecular transport, and lipid trafficking. The data suggest that the phenotype "balance" is shifted to protect the tissue from modest pathological insult. It remains to be determined whether this phenotype change persists as significant valve pathology develops.

2.14 Scale, Hemodynamics, and Endothelial Heterogeneity

At the *single cell* level, endothelial responses to the hemodynamic environment are heterogeneous. Such variations are important for the spatial interpretation of vascular responses to flow and for an understanding of mechanotransduction mechanisms at the level of single or small groups of cells. Exposure of endothelium to a nominally uniform macro-flow field *in vivo* and *in vitro* frequently results in a heterogeneous distribution of individual cell responses. Extremes in response levels are often noted in neighboring cells. Examples include the expression of the vascular cell adhesion molecule–1 (VCAM-1) and the intercellular adhesion molecule–1 (ICAM-1) mRNA and protein, elevation of intracellular calcium ($[Ca^{++}]_i$), and the induction and nuclear localization of the transcription factors c-fos, egr-1, and NF-κB/IκB [142]. AFM measurements of cell-surface topography in living endothelium both *in vitro* and *in situ* combined with computational fluid dynamics demonstrated large cell-to-cell variations in the distribution of forces throughout the surface of individual cells of the monolayer [45]. The measurements show that the endothelial three-dimensional surface geometry defines the distribution of shear stresses and gradients at the single cell level and that there are large variations in force magnitude and distribution among neighboring cells. The measurements support the contribution of cell topology to differential endothelial responses to flow observed *in vivo* and *in vitro* [142].

A compelling conclusion arising from these studies is that the microscopic hemodynamic forces acting on individual endothelial cells vary considerably from cell to cell and within small regions of the endothelial monolayer. This presents a potential problem because nearly all biochemical measurements of flow responsiveness are averaged over many cells, and the heterogeneity, which may be minor or significant, goes unnoticed. A related consideration is that the physiological significance of heterogeneity may be that of a larger group of cells rather than one or a few, since the endothelium is a confluent monolayer and the cells interact both by direct contact and via secreted molecules. It simply is not known at present what weight to apply to the heterogeneity and its scale despite its potentially important relevance to the focal nature of vascular pathologies.

If differential responsiveness to *macro*scopic shear stresses occurs because of *micro*scopic heterogeneities of cell geometry that result in differential signaling and gene expression, it follows that pathological *permissiveness and protection are likely focally heterogeneous within regions.* It is reasonable to propose that in focal atherogenesis the phenotypes of only a limited number of cells within a region may be most relevant to the subsequent pathological changes (initiation and development of the lesion). Clarification is likely to come from making careful distinctions between pathological change and physiological remodeling mechanisms, from better real time imaging of endothelial cells in arteries and from the application of functional genomics and proteomics to endothelial regional and focal heterogeneity.

2.15 Acknowledgments

The authors thank R. E. Mott for assistance with the fluorescence images and acknowledge support by grants from the U.S. National Institutes of Health to PFD (HL36049 MERIT, HL62250, HL64388, HL70128) and BPH (HL071958, HL076499, HL080956), from the Whitaker Foundation (RG-02-0545 to BPH), from the UVa Biotechnology Training Program (NIH grant GM008715 to BPH), and from the National Space Biomedical Research Institute of NASA (to PFD).

REFERENCES

[1] LaBarbera M. Principles of design of fluid transport systems in zoology. *Science* 1990; 249:992–1000.

[2] Cheng C, Helderman F, Tempel D, Segers D, Hierck B, Poelmann R, et al. Large variations in absolute wall shear stress levels within one species and between species. *Atherosclerosis* 2007; 195:225–35.

[3] Davies PF. Flow-mediated endothelial mechanotransduction. *Physiol. Rev.* 1995; 75:519–60.
[4] Nerem RM. Hemodynamics and the vascular endothelium. *J. Biomech. Eng.* 1993; 115:510–14.
[5] Resnick N, Gimbrone MA, Jr. Hemodynamic forces are complex regulators of endothelial gene expression. *FASEB J.* 1995; 9:874–82.
[6] Davies PF, Polacek DC, Handen JS, Helmke BP, DePaola N. A spatial approach to transcriptional profiling: mechanotransduction and the focal origin of atherosclerosis. *Trends Biotechnol.* 1999; 17:347–51.
[7] Shyy JYJ, Chien S. Role of integrins in endothelial mechanosensing of shear stress. *Circ. Res.* 2002; 91:769–75.
[8] Koller A, Sun D, Kaley G. Role of shear stress and endothelial prostaglandins in flow- and viscosity-induced dilation of arterioles in vitro. *Circ. Res.* 1993; 72:1276–84.
[9] Olesen S-P, Clapham DE, Davies PF. Hemodynamic shear stress activates a K^+ current in vascular endothelial cells. *Nature* 1988; 331:168–70.
[10] Gudi SR, Nolan JP, Frangos JA. Modulation of GTPase activity of G proteins by fluid shear stress and phospholipid composition. *Proc. Natl. Acad. Sci. USA* 1998; 95:2515–19.
[11] Butler PJ, Norwich G, Weinbaum S, Chien S. Shear stress induces a time- and position-dependent increase in endothelial cell membrane fluidity. *Am. J. Physiol.* 2001; 280:C962–9.
[12] Aird WC, ed. *Endothelial Biomedicine.* New York: Cambridge University Press, 2007.
[13] Pohl U, Holtz J, Busse R, Bassenge E. Crucial role of endothelium in the vasodilator response to increased flow in vivo. *Hypertension* 1986; 8:37–44.
[14] Moncada S. Adventures in vascular biology: A tale of two mediators. *Philos. Trans. R. Soc. Lond. B Biol. Sci.* 2006; 361:735–59.
[15] Corson MA, James NL, Latta SE, Nerem RM, Berk BC, Harrison DG. Phosphorylation of endothelial nitric oxide synthase in response to fluid shear stress. *Circ. Res.* 1996; 79:984–91.
[16] Griffith TM. Endothelial control of vascular tone by nitric oxide and gap junctions: a haemodynamic perspective. *Biorheology* 2002; 39:307–18.
[17] Langille BL, O'Donnell F. Reductions in arterial diameter produced by chronic decreases in blood flow are endothelium-dependent. *Science* 1986; 231:405–7.
[18] Lee JS, Yu Q, Shin JT, Sebzda E, Bertozzi C, Chen M, et al. Klf2 is an essential regulator of vascular hemodynamic forces in vivo. *Dev. Cell* 2006; 11:845–57.
[19] Glagov S, Zarins CK, Giddens DP, Ku DN. Hemodynamics and atherosclerosis. Insights and perspectives gained from studies of human arteries. *Arch. Pathol. Lab. Med.* 1988; 112:1018–31.
[20] Ku DN, Giddens DP. Pulsatile flow in a model carotid bifurcation. *Arteriosclerosis* 1983; 3:31–9.
[21] Steinberg D. Atherogenesis in perspective: Hypercholesterolemia and inflammation as partners in crime. *Nat. Med.* 2002; 8:1211–17.
[22] Helmke BP, Rosen AB, Davies PF. Mapping mechanical strain of an endogenous cytoskeletal network in living endothelial cells. *Biophys. J.* 2003; 84:2691–9.
[23] Dull RO, Davies PF. Flow modulation of agonist (ATP)-response (Ca^{2+}) coupling in vascular endothelial cells. *Am. J. Physiol.* 1991; 261:H149–H54.
[24] Choi H, Ferrara K, Barakat A. Modulation of ATP/ADP concentration at the endothelial surface by shear stress: Effect of flow recirculation. *Ann. Biomed. Eng.* 2007; 35:505–16.

[25] Dull RO, Tarbell JM, Davies PF. Mechanisms of flow-mediated signal transduction in endothelial cells: Kinetics of ATP surface concentrations. *J. Vasc. Res.* 1992; 29:410–19.
[26] Davies PF, Reidy MA, Goode TB, Bowyer DE. Scanning electron microscopy in the evaluation of endothelial integrity of the fatty lesion in atherosclerosis. *Atherosclerosis* 1976; 25:125–30.
[27] Dewey CF, Jr., Bussolari SR, Gimbrone MA, Jr., Davies PF. The dynamic response of vascular endothelial cells to fluid shear stress. *J. Biomech. Eng.* 1981; 103:177–88.
[28] Davies PF, Dewey CF, Jr., Bussolari SR, Gordon EJ, Gimbrone MA, Jr. Influence of hemodynamic forces on vascular endothelial function. In vitro studies of shear stress and pinocytosis in bovine aortic cells. *J. Clin. Invest.* 1984; 73:1121–9.
[29] Davies PF, Remuzzi A, Gordon EJ, Dewey CF, Gimbrone MA. Turbulent fluid shear stress induces vascular endothelial cell turnover in vitro. *Proc. Natl. Acad. Sci. USA* 1986; 83:2114–17.
[30] Sprague EA, Steinbach BL, Nerem RM, Schwartz CJ. Influence of a laminar steady-state fluid-imposed wall shear stress on the binding, internalization, and degradation of low-density lipoproteins by cultured arterial endothelium. *Circulation* 1987; 76:648–56.
[31] Ingber D. Integrins as mechanochemical transducers. *Curr. Opin. Cell Biol.* 1991; 3:841–18.
[32] Sims JR, Karp S, Ingber DE. Altering the cellular mechanical force balance results in integrated changes in cell, cytoskeletal and nuclear shape. *J. Cell Sci.* 1992; 103:1215–22.
[33] Davies PF, Robotewskyj A, Griem ML. Endothelial cell adhesion in real time. Measurements in vitro by tandem scanning confocal image analysis. *J. Clin. Invest.* 1993; 91:2640–52.
[34] Davies PF, Robotewskyj A, Griem ML. Quantitative studies of endothelial cell adhesion: Directional remodeling of focal adhesion sites in response to flow forces. *J. Clin. Invest.* 1994; 93:2031–8.
[35] Liu Y, Chen BPC, Lu M, Zhu Y, Stemerman MB, Chien S, et al. Shear stress activation of SREBP1 in endothelial cells is mediated by integrins. *Arterioscler. Thromb. Vasc. Biol.* 2002; 22:76–81.
[36] Fujiwara K, Masuda M, Osawa M, Kano Y, Katoh K. Is PECAM-1 a mechanoresponsive molecule? *Cell Struct. Funct.* 2001; 26:11–17.
[37] Tzima E, Irani-Tehrani M, Kiosses WB, Dejana E, Schultz DA, Engelhardt B, et al. A mechanosensory complex that mediates the endothelial cell response to fluid shear stress. *Nature* 2005; 437:426–31.
[38] Helmke BP, Goldman RD, Davies PF. Rapid displacement of vimentin intermediate filaments in living endothelial cells exposed to flow. *Circ. Res.* 2000; 86:745–52.
[39] Helmke BP, Thakker DB, Goldman RD, Davies PF. Spatiotemporal analysis of flow-induced intermediate filament displacement in living endothelial cells. *Biophys. J.* 2001; 80:184–94.
[40] Mott RE, Helmke BP. Mapping the dynamics of shear stress-induced structural changes in endothelial cells. *Am. J. Physiol.* 2007; 293:C1616–26.
[41] Davies PF, Tripathi SC. Mechanical stress mechanisms and the cell: An endothelial paradigm. *Circ. Res.* 1993; 72:239–45.
[42] Helmke BP, Davies PF. The cytoskeleton under external fluid mechanical forces: Hemodynamic forces acting on the endothelium. *Ann. Biomed. Eng.* 2002; 30:284–96.
[43] Jacobs ER, Cheliakine C, Gebremedhin D, Davies PF, Harder DR. Shear activated channels in cell attached patches of vascular endothelial cells. *Pflügers Arch.* 1995; 431:129–31.

[44] Barbee KA, Davies PF, Lal R. Shear stress-induced reorganization of the surface topography of living endothelial cells imaged by atomic force microscopy. *Circ. Res.* 1994; 74:163–71.

[45] Barbee KA, Davies PF, Lal R. Subcellular distribution of shear stress at the surface of flow aligned and non-aligned endothelial monolayers. *Am. J. Physiol.* 1995; 268:H1765–72.

[46] Weinbaum S, Tarbell JM, Damiano ER. The structure and function of the endothelial glycocalyx layer. *Annu. Rev. Biomed. Eng.* 2007; 9:121–67.

[47] Malek AM, Izumo S. Mechanism of endothelial cell shape change and cytoskeletal remodeling in response to fluid shear stress. *J. Cell Sci.* 1996; 109:713–26.

[48] Helmke BP. Molecular control of cytoskeletal mechanics by hemodynamic forces. *Physiology* 2005; 20:43–53.

[49] Hu S, Chen J, Fabry B, Numaguchi Y, Gouldstone A, Ingber DE, et al. Intracellular stress tomography reveals stress focusing and structural anisotropy in cytoskeleton of living cells. *Am. J. Physiol.* 2003; 285:C1082–90.

[50] Hu S, Wang N. Control of stress propagation in the cytoplasm by prestress and loading frequency. *Mol. Cell. Biomech.* 2006; 3:49–60.

[51] Janmey PA, Euteneuer U, Traub P, Schliwa M. Viscoelastic properties of vimentin compared with other filamentous biopolymer networks. *J. Cell Biol.* 1991; 113:155–60.

[52] Janmey PA, Euteneuer U, Traub P, Schliwa M. Viscoelasticity of intermediate filament networks. *Sub-Cellular Biochem.* 1998; 31:381–97.

[53] Schmid-Schönbein GW, Sung K-LP, Tözeren H, Skalak R, Chien S. Passive mechanical properties of human leukocytes. *Biophys. J.* 1981; 36:243–56.

[54] Theret DP, Levesque MJ, Sato M, Nerem RM, Wheeler LT. The application of a homogeneous half-space model in the analysis of endothelial cell micropipette measurements. *J. Biomech. Eng.* 1988; 110:190–9.

[55] Sato M, Theret DP, Wheeler LT, Ohshima N, Nerem RM. Application of the micropipette technique to the measurement of cultured porcine aortic endothelial cell viscoelastic properties. *J. Biomech. Eng.* 1990; 112:263–8.

[56] Colangelo S, Langille BL, Steiner G, Gotlieb AI. Alterations in endothelial F-actin microfilaments in rabbit aorta in hypercholesterolemia. *Arterioscler. Thromb. Vasc. Biol.* 1998; 18:52–6.

[57] Galbraith CG, Skalak R, Chien S. Shear stress induces spatial reorganization of the endothelial cell cytoskeleton. *Cell Motil. Cytoskel.* 1998; 40:317–30.

[58] Sato M, Ohshima N, Nerem RM. Viscoelastic properties of cultured porcine aortic endothelial cells exposed to shear stress. *J. Biomech.* 1996; 29:461–7.

[59] Satcher RL, Jr., Dewey CF, Jr. Theoretical estimates of mechanical properties of the endothelial cell cytoskeleton. *Biophys. J.* 1996; 71:109–18.

[60] Ferko MC, Bhatnagar A, Garcia MB, Butler PJ. Finite-element stress analysis of a multicomponent model of sheared and focally-adhered endothelial cells. *Ann. Biomed. Eng.* 2007; 35:208–23.

[61] Fuller RB. Tensegrity. *Portfolio Art News Annu.* 1961; 4:112–27.

[62] Wang N, Butler JP, Ingber DE. Mechanotransduction across the cell surface and through the cytoskeleton. *Science* 1993; 260:1124–7.

[63] Fabry B, Maksym GN, Butler JP, Glogauer M, Navajas D, Fredberg JJ. Scaling the microrheology of living cells. *Phys. Rev. Lett.* 2001; 87:148102.

[64] Fabry B, Maksym GN, Butler JP, Glogauer M, Navajas D, Taback NA, et al. Time scale and other invariants of integrative mechanical behavior in living cells. *Phys. Rev. E* 2003; 68(4 pt. 1):041914.

[65] Bursac P, Lenormand G, Fabry B, Oliver M, Weitz DA, Viasnoff V, et al. Cytoskeletal remodeling and slow dynamics in the living cell. *Nat. Mater.* 2005; 4:557–61.

[66] Wootton DM, Ku DN. Fluid mechanics of vascular systems, diseases, and thrombosis. *Annu. Rev. Biomed. Eng.* 1999; 1:299–329.

[67] Friedman MH, Giddens DP. Blood flow in major blood vessels–Modeling and experiments. *Ann. Biomed. Eng.* 2005; 33:1710–13.

[68] Fung YC, Liu SQ. Elementary mechanics of the endothelium of blood vessels. *J. Biomech. Eng.* 1993; 115:1–12.

[69] Haidekker MA, L'Heureux N, Frangos JA. Fluid shear stress increases membrane fluidity in endothelial cells: A study with DCVJ fluorescence. *Am. J. Physiol.* 2000; 278:H1401–6.

[70] Fang Y, Schram G, Romanenko VG, Shi C, Conti L, Vandenberg CA, et al. Functional expression of Kir2.x in human aortic endothelial cells: The dominant role of Kir2.2. *Am. J. Physiol.* 2005; 289:C1134–44.

[71] Sukharev SI, Sigurdson WJ, Kung C, Sachs F. Energetic and spatial parameters for gating of the bacterial large conductance mechanosensitive channel, MscL. *J. Gen. Physiol.* 1999; 113:525–40.

[72] Hamill OP, Martinac B. Molecular basis of mechanotransduction in living cells. *Physiol. Rev.* 2001; 81:685–740.

[73] Martinac B, Hamill OP. Gramicidin A channels switch between stretch activation and stretch inactivation depending on bilayer thickness. *Proc. Natl. Acad. Sci. USA* 2002; 99:4308–12.

[74] Chachisvilis M, Zhang Y-L, Frangos JA. G protein-coupled receptors sense fluid shear stress in endothelial cells. *Proc. Natl. Acad. Sci. USA* 2006; 103:15463–8.

[75] Romanenko VG, Davies PF, Levitan I. Dual effect of fluid shear stress on volume-regulated anion current in bovine aortic endothelial cells. *Am. J. Physiol.* 2002; 282:C708–18.

[76] Balaban NQ, Schwarz US, Riveline D, Goichberg P, Tzur G, Sabanay I, et al. Force and focal adhesion assembly: A close relationship studied using elastic micropatterned substrates. *Nat. Cell Biol.* 2001; 3:466–72.

[77] Reitsma S, Slaaf D, Vink H, van Zandvoort M, oude Egbrink M. The endothelial glycocalyx: Composition, functions, and visualization. *Pflügers Arch.* 2007; 454:345–59.

[78] Smith ML, Long DS, Damiano ER, Ley K. Near-wall μ-PIV reveals a hydrodynamically relevant endothelial surface layer in venules in vivo. *Biophys. J.* 2003; 85:637–45.

[79] Weinbaum S, Zhang X, Han Y, Vink H, Cowin SC. Mechanotransduction and flow across the endothelial glycocalyx. *Proc. Natl. Acad. Sci. USA* 2003; 100:7988–95.

[80] Thi MM, Tarbell JM, Weinbaum S, Spray DC. The role of the glycocalyx in reorganization of the actin cytoskeleton under fluid shear stress: A "bumper-car" model. *Proc. Natl. Acad. Sci. USA* 2004; 101:16483–8.

[81] van den Berg BM, Spaan JAE, Rolf TM, Vink H. Atherogenic region and diet diminish glycocalyx dimension and increase intima-to-media ratios at murine carotid artery bifurcation. *Am. J. Physiol.* 2006; 290:H915–20.

[82] Vink H, Constantinescu AA, Spaan JA. Oxidized lipoproteins degrade the endothelial surface layer: Implications for platelet-endothelial cell adhesion. *Circulation* 2000; 101:1500–2.

[83] Pohl U, Herlan K, Huang A, Bassenge E. EDRF-mediated shear-induced dilation opposes myogenic vasoconstriction in small rabbit arteries. *Am. J. Physiol.* 1991; 261:H2016–23.

[84] Hecker M, Mulsch A, Bassenge E, Busse R. Vasoconstriction and increased flow: Two principal mechanisms of shear stress-dependent endothelial autacoid release. *Am. J. Physiol.* 1993; 265:H828–33.

[85] Pahakis MY, Kosky JR, Dull RO, Tarbell JM. The role of endothelial glycocalyx components in mechanotransduction of fluid shear stress. *Biochem. Biophys. Res. Comm.* 2007; 355:228–33.
[86] Florian JA, Kosky JR, Ainslie K, Pang Z, Dull RO, Tarbell JM. Heparan sulfate proteoglycan is a mechanosensor on endothelial cells. *Circ. Res.* 2003; 93:14.
[87] Gouverneur M, Spaan JAE, Pannekoek H, Fontijn RD, Vink H. Fluid shear stress stimulates incorporation of hyaluronan into endothelial cell glycocalyx. *Am. J. Physiol.* 2006; 290:H458–2.
[88] Yao Y, Rabodzey A, Dewey CF, Jr. Glycocalyx modulates the motility and proliferative response of vascular endothelium to fluid shear stress. *Am. J. Physiol.* 2007; 293:H1023–30.
[89] Kojimahara M. Endothelial cilia in rat mesenteric arteries and intramyocardial capillaries. *Z. Mikrosk. Anat. Forsch.* 1990; 104:412–6.
[90] Iomini C, Tejada K, Mo W, Vaananen H, Piperno G. Primary cilia of human endothelial cells disassemble under laminar shear stress. *J. Cell Biol.* 2004; 164:811–17.
[91] van der Heiden K, Groenendijk BCW, Hierck BP, Hogers B, Koerten HK, Mommaas AM, et al. Monocilia on chicken embryonic endocardium in low shear stress areas. *Developmental Dynamics* 2006; 235:19–28.
[92] van der Heiden K, Hierck BP, Krams R, de Crom R, Cheng C, Baiker M, et al. Endothelial primary cilia in areas of disturbed flow are at the base of atherosclerosis. *Atherosclerosis* 2007; DOI:10.1016/j.atherosclerosis.2007.05.030.
[93] Ando J, Komatsuda T, Kamiya A. Cytoplasmic calcium response to fluid shear stress in cultured vascular endothelial cells. *In Vitro Cell. Dev. Biol.* 1988; 24:871–7.
[94] Isshiki M, Ando J, Korenaga R, Kogo H, Fujimoto T, Fujita T, et al. Endothelial Ca^{2+} waves preferentially originate at specific loci in caveolin-rich cell edges. *Proc. Natl. Acad. Sci. USA* 1998; 95:5009–14.
[95] Isshiki M, Ando J, Yamamoto K, Fujita T, Ying Y, Anderson RGW. Sites of Ca^{2+} wave initiation move with caveolae to the trailing edge of migrating cells. *J. Cell Sci.* 2002; 115:475–84.
[96] Wang Y, Botvinick EL, Zhao Y, Berns MW, Usami S, Tsien RY, et al. Visualizing the mechanical activation of Src. *Nature* 2005; 434:1040–5.
[97] Tzima E, del Pozo MA, Kiosses WB, Mohamed SA, Li S, Chien S, et al. Activation of Rac1 by shear stress in endothelial cells mediates both cytoskeletal reorganization and effects on gene expression. *EMBO J.* 2002; 21:6791–800.
[98] Yamada S, Wirtz D, Kuo SC. Mechanics of living cells measured by laser tracking microrheology. *Biophys. J.* 2000; 78:1736–47.
[99] Crocker JC, Hoffman BD. Multiple-particle tracking and two-point microrheology in cells. *Meth. Cell Biol.* 2007; 83:141–78.
[100] Panorchan P, Lee JSH, Daniels BR, Kole TP, Tseng Y, Wirtz D. Probing cellular mechanical responses to stimuli using ballistic intracellular nanorheology. *Meth. Cell Biol.* 2007; 83:115–40.
[101] Stamenovic D, Mijailovich SM, Tolic-Norrelykke IM, Chen J, Wang N. Cell prestress. II. Contribution of microtubules. *Am. J. Physiol.* 2002; 282:C617–C24.
[102] Wang N, Stamenovic D. Contribution of intermediate filaments to cell stiffness, stiffening, and growth. *Am. J. Physiol.* 2000; 279:C188–94.
[103] Giancotti FG, Ruoslahti E. Integrin signaling. *Science* 1999; 285:1028–33.
[104] Davies PF, Barbee KA, Volin MV, Robotewskyj A, Chen J, Joseph L, et al. Spatial relationships in early signaling events of flow-mediated endothelial mechanotransduction. *Annu. Rev. Physiol.* 1997; 59:527–49.

[105] Li S, Kim M, Hu YL, Jalali S, Schlaepfer DD, Hunter T, et al. Fluid shear stress activation of focal adhesion kinase. Linking to mitogen-activated protein kinases. *J. Biol. Chem.* 1997; 272:30455–62.

[106] Chen K-D, Li Y-S, Kim M, Li S, Yuan S, Chien S, et al. Mechanotransduction in response to shear stress: Roles of receptor tyrosine kinases, integrins, and Shc. *J. Biol. Chem.* 1999; 274:18393–400.

[107] Jalali S, del Pozo MA, Chen K-D, Miao H, Li Y-S, Schwartz MA, et al. Integrin-mediated mechanotransduction requires its dynamic interaction with specific extracellular matrix (ECM) ligands. *Proc. Natl. Acad. Sci. USA* 2001; 98:1042–6.

[108] Hu Y-L, Chien S. Dynamic motion of paxillin on actin filaments in living endothelial cells. *Biochem. Biophys. Res. Comm.* 2007; 357:871–6.

[109] Tai L-K, Okuda M, Abe J-I, Yan C, Berk BC. Fluid shear stress activates proline-rich tyrosine kinase via reactive oxygen species-dependent pathway. *Arterioscler. Thromb. Vasc. Biol.* 2002; 22:1790–6.

[110] Rizzo V, Morton C, DePaola N, Schnitzer JE, Davies PF. Recruitment of endothelial caveolae into mechanotransduction pathways by flow conditioning in vitro. *Am. J. Physiol.* 2003; 285:H1720–H9.

[111] Radel C, Rizzo V. Integrin mechanotransduction stimulates caveolin-1 phosphorylation and recruitment of Csk to mediate actin reorganization. *Am. J. Physiol.* 2005; 288:H936–H45.

[112] Tzima E, del Pozo MA, Shattil SJ, Chien S, Schwartz MA. Activation of integrins in endothelial cells by fluid shear stress mediates Rho-dependent cytoskeletal alignment. *EMBO J.* 2001; 20:4639–47.

[113] Ren XD, Kiosses WB, Sieg DJ, Otey CA, Schlaepfer DD, Schwartz MA. Focal adhesion kinase suppresses Rho activity to promote focal adhesion turnover. *J. Cell Sci.* 2000; 113:3673–8.

[114] Arthur WT, Burridge K. RhoA inactivation by p190RhoGAP regulates cell spreading and migration by promoting membrane protrusion and polarity. *Mol. Biol. Cell* 2001; 12:2711–20.

[115] Maekawa M, Ishizaki T, Boku S, Watanabe N, Fujita A, Iwamatsu A, et al. Signaling from Rho to the actin cytoskeleton through protein kinases ROCK and LIM-kinase. *Science* 1999; 285:895–8.

[116] Orr AW, Sanders JM, Bevard M, Coleman E, Sarembock IJ, Schwartz MA. The subendothelial extracellular matrix modulates NF-κB activation by flow: A potential role in atherosclerosis. *J. Cell Biol.* 2005; 169:191–202.

[117] Orr AW, Ginsberg MH, Shattil SJ, Deckmyn H, Schwartz MA. Matrix-specific suppression of integrin activation in shear stress signaling. *Mol. Biol. Cell* 2006; 17:4686–97.

[118] Orr AW, Stockton R, Simmers MB, Sanders JM, Sarembock IJ, Blackman BR, et al. Matrix-specific p21-activated kinase activation regulates vascular permeability in atherogenesis. *J. Cell Biol.* 2007; 176:719–27.

[119] Kano Y, Katoh K, Masuda M, Fujiwara K. Macromolecular composition of stress fiber-plasma membrane attachment sites in endothelial cells *in situ*. *Circ. Res.* 1996; 79:1000–6.

[120] Osawa M, Masuda M, Kusano K-i, Fujiwara K. Evidence for a role of platelet endothelial cell adhesion molecule-1 in endothelial cell mechanosignal transduction: Is it a mechanoresponsive molecule? *J. Cell Biol.* 2002; 158:773–85.

[121] Tai L-k, Zheng Q, Pan S, Jin Z-G, Berk BC. Flow activates ERK1/2 and endothelial nitric oxide synthase via a pathway involving PECAM1, SHP2, and Tie2. *J. Biol. Chem.* 2005; 280:29620–4.

[122] Shay-Salit A, Shushy M, Wolfovitz E, Yahav H, Breviario F, Dejana E, et al. VEGF receptor 2 and the adherens junction as a mechanical transducer in vascular endothelial cells. *Proc. Natl. Acad. Sci. USA* 2002; 99:9462–7.

[123] Cornhill JF, Roach MR. A quantitative study of the localization of atherosclerotic lesions in the rabbit aorta. *Atherosclerosis* 1976; 23:489–501.
[124] Steinman DA. Image-based computational fluid dynamics modeling in realistic arterial geometries. *Ann. Biomed. Eng.* 2002; 30:483–97.
[125] Caro CG, Fitz-Gerald JM, Schroter RC. Arterial wall shear and distribution of early atheroma in man. *Nature* 1969; 223:1159–61.
[126] Lutz RJ, Cannon JN, Bischoff KB, Dedrick RL, Stiles RK, Fry DL. Wall shear stress distribution in a model canine artery during steady flow. *Circ. Res.* 1977; 41:391–9.
[127] Suo J, Ferrara DE, Sorescu D, Guldberg RE, Taylor WR, Giddens DP. Hemodynamic shear stresses in mouse aortas: Implications for atherogenesis. *Arterioscler. Thromb. Vasc. Biol.* 2007; 27:346–51.
[128] Passerini AG, Polacek DC, Shi C, Francesco NM, Manduchi E, Grant GR, et al. Coexisting proinflammatory and antioxidative endothelial transcription profiles in a disturbed flow region of the adult porcine aorta. *Proc. Natl. Acad. Sci. USA* 2004; 101:2482–7.
[129] Volger OL, Fledderus JO, Kisters N, Fontijn RD, Moerland PD, Kuiper J, et al. Distinctive expression of chemokines and transforming growth factor-β signaling in human arterial endothelium during atherosclerosis. *Am. J. Pathol.* 2007; 171:326–37.
[130] Simmons CA, Grant GR, Manduchi E, Davies PF. Spatial heterogeneity of endothelial phenotypes correlates with side-specific vulnerability to calcification in normal porcine aortic valves. *Circ. Res.* 2005; 96:792–9.
[131] Magid R, Davies PF. Endothelial protein kinase C isoform identity and differential activity of PKCζ in an athero-susceptible region of porcine aorta. *Circ. Res.* 2005; 97:443–9.
[132] Boon RA, Fledderus JO, Volger OL, van Wanrooij , EJA, Pardali E, Weesie F, et al. KLF2 suppresses TGF-β signaling in endothelium through induction of Smad7 and inhibition of AP-1. *Arterioscler. Thromb. Vasc. Biol.* 2007; 27:532–9.
[133] Dai G, Vaughn S, Zhang Y, Wang ET, Garcia-Cardena G, Gimbrone MA, Jr. Biomechanical forces in atherosclerosis-resistant vascular regions regulate endothelial redox balance via phosphoinositol 3-kinase/Akt-dependent activation of Nrf2. *Circ. Res.* 2007; 101:723–33.
[134] Hajra L, Evans AI, Chen M, Hyduk SJ, Collins T, Cybulsky MI. The NF-κB signal transduction pathway in aortic endothelial cells is primed for activation in regions predisposed to atherosclerotic lesion formation. *Proc. Natl. Acad. Sci. USA* 2000; 97:9052–7.
[135] Dekker RJ, van Soest S, Fontijn RD, Salamanca S, de Groot PG, VanBavel E, et al. Prolonged fluid shear stress induces a distinct set of endothelial cell genes, most specifically lung Kruppel-like factor (KLF2). *Blood* 2002; 100:1689–98.
[136] Dekker RJ, Boon RA, Rondaij MG, Kragt A, Volger OL, Elderkamp YW, et al. KLF2 provokes a gene expression pattern that establishes functional quiescent differentiation of the endothelium. *Blood* 2006; 107: 4354–63.
[137] Parmar KM, Larman HB, Dai G, Zhang Y, Wang ET, Moorthy SN, et al. Integration of flow-dependent endothelial phenotypes by Kruppel-like factor 2. *J. Clin. Invest.* 2006; 116:49–58.
[138] Cheng C, van Haperen R, de Waard M, van Damme LCA, Tempel D, Hanemaaijer L, et al. Shear stress affects the intracellular distribution of eNOS: Direct demonstration by a novel in vivo technique. *Blood* 2005; 106:3691–8.

[139] Schoen FJ, Levy RJ. Tissue heart valves: Current challenges and future research perspectives. *J. Biomed. Mater. Res.* 1999; 47:439–65.

[140] Poggianti E, Venneri L, Chubuchny V, Jambrik Z, Baroncini LA, Picano E. Aortic valve sclerosis is associated with systemic endothelial dysfunction. *J. Amer. Coll. Cardiol.* 2003; 41:136–41.

[141] Sacks MS, Yoganathan AP. Heart valve function: A biomechanical perspective. *Philos. Trans. R. Soc. Lond. B Biol. Sci.* 2007; 362:1369–91.

[142] Davies PF, Mundel T, Barbee KA. A mechanism for heterogeneous endothelial responses to flow in vivo and in vitro. *J. Biomech.* 1995; 28:1553–60.

[143] Mott RE, Helmke BP. Control of endothelial cell adhesion by mechanotransmission from cytoskeleton to substrate. In: King MR, ed. *Principles of Cellular Engineering: Understanding the Biomolecular Interface*. Burlington, MA: Elsevier; 2006:25–50.

[144] Karino T, Goldsmith HL. Particle flow behavior in models of branching vessels. II. Effects of branching angle and diameter ratio on flow patterns. *Biorheology* 1985; 22:87–104.

[145] Steinman D, Taylor C. Flow imaging and computing: Large artery hemodynamics. *Ann. Biomed. Eng.* 2005; 33:1704–9.

3

Role of the Plasma Membrane in Endothelial Cell Mechanosensation of Shear Stress

Peter J. Butler and Shu Chien

3.1 Introduction

Mechanotransduction, which is the process by which cells convert mechanical stimuli to biochemical signaling cascades, is involved in the homeostasis of numerous tissues (reviewed in [21] and [56]). The mechanotransduction of hemodynamic shear stress by endothelial cells (ECs) has garnered special attention because of its role in regulating vascular health and disease. In particular, there is intense interest in identifying the primary molecular mechanisms of the EC sensing of shear stress because its (or their) discovery may lead to clinical interventions in atherosclerosis and other diseases related to mechanobiology.

In this chapter, we address the hypothesis that the plasma membrane lipid bilayer is one endothelial cell mechanosensor. Here we define "mechanosensor" as a cellular structure that responds to mechanical stress and initiates mechanotransduction in response to shear stress without involving chemical second messengers. Mechanotransduction, then, is the process by which cells convert this sensory stimulus into changes in biochemical signaling. We define mechanobiology as the study of the entire process of sensation, transduction, and attendant changes in cell phenotype. Because mechanical linkages from the cell surface to lateral, internal, and basal parts of the cell redistribute forces imposed on the cell surface, many structures could serve as mechanosensors. Furthermore, mechanotransduction can involve direct force effects on molecules, diffusion- or convection-mediated transport of molecular second messengers, and the active transport of signaling molecules by molecular motors. Other chapters in this text will address other candidate mechanosensors (e.g., focal adhesions and their integrins). With respect to the membrane, in the context of these definitions, if shear stress induces a perturbation of the membrane constituents, and this perturbation is necessary for mechanotransduction, then the lipid bilayer is considered a mechanosensor. Similarly, if the shear stress acting on the apical portion of the cell leads to the perturbation of the membrane on the basal portion as a result of mechanical linkage, and this membrane perturbation is necessary for subsequent downstream signaling, then we consider the basal membrane also as a mechanosensor.

We first outline the evidence of a role of the plasma membrane in shear stress sensing. We then address the question of whether forces in the membrane are sufficient to elicit changes in lipid dynamics, and we propose novel tools to address whether stresses in the membrane are sufficient to induce lipid-mediated protein signaling. We conclude with a proposal for a unified theory on the role of the plasma membrane in shear sensing. This theory is offered to stimulate discussions of membrane mechanosensing and to foster new research directions.

3.2 Overview of Mechanotransduction

3.2.1 Endothelial Cell Mechanotransduction and Vascular Physiology

The endothelium regulates vascular health by forming a regulated semi-permeable barrier to blood constituents, secreting vasoactive compounds that control vascular caliber, and modulating EC adhesiveness to white blood cells. EC dysfunction can lead to diseases such as atherosclerosis, stroke, and hypertension. The endothelium is subject to hemodynamic forces (e.g., shear stress, τ) that vary temporally, spatially, and in magnitude, depending on the location in the vasculature, the heart rate, and the metabolic demand of tissues (Figure 3.1). Endothelial responses to temporal gradients [5, 11, 12, 33] and spatial gradients [22, 25, 87, 114] in shear stress play a significant role in determining whether ECs exhibit an atherogenic or atheroprotective phenotype [17, 55, 63, 67, 90, 118]. Atherogenic ECs have a higher permeability to low density lipoproteins, greater adhesivity to circulating monocytes, and faster turnover rates. There is a large body of evidence that low and oscillatory shear stress is an atherogenic stimulus, while high and unidirectional shear stress is atheroprotective (e.g., [57]). In addition, EC mechanotransduction plays a role in stent-induced restenosis [71, 113, 120] as well as the development and successful deployment of artificial vascular grafts [89]. The temporal and spatial gradients in shear stress also are important in the coordination of blood flow in the microvasculature [12, 62, 64, 65, 69, 96, 103] and hence in the maintenance of capillary blood pressure and the delivery of oxygen, nutrients, and immunity-related leukocytes to tissues.

3.2.2 Mechanisms of Mechanotransduction

In order to understand the mechanism of the initiation of shear modulation of EC function, there have been many studies searching for primary shear sensors (Figures 3.1 and 3.2). Investigations on the shear modulation of protein products [18], cellular and cytoskeletal orientation [35, 41, 42], production of vasoactive autacoids [31, 47, 66], intracellular calcium concentration [2, 40], gene expression [9, 16, 23, 81, 105], and glycocalyx composition [3] have led to the identification of many molecules involved in mechanotransduction including integrins [57, 60, 76, 86], G-proteins [32, 49, 50, 57, 68], K^+ channels [58, 93, 93, 94], stretch-activated Ca^{2+} channels

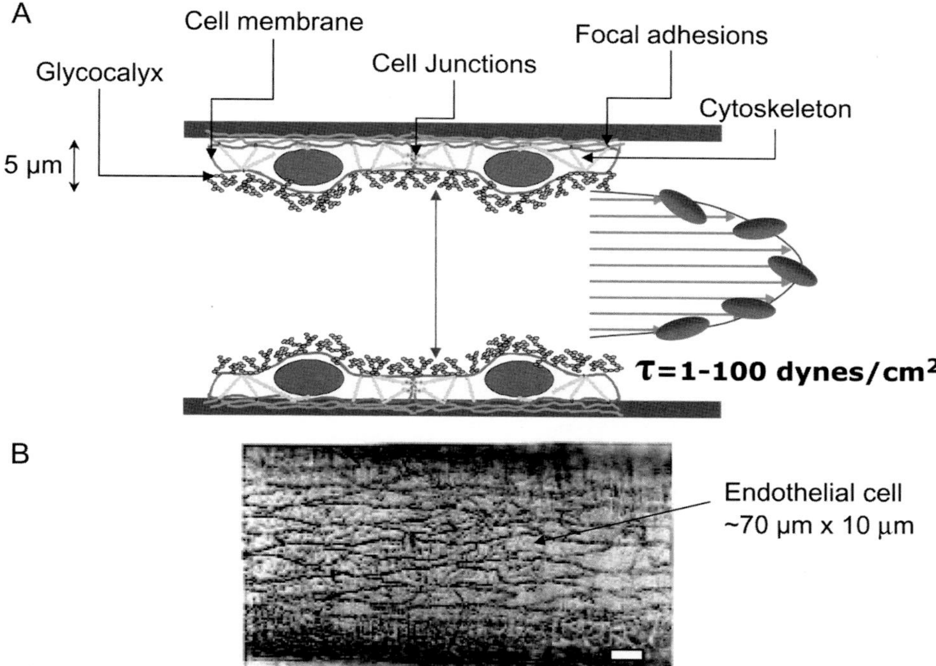

Figure 3.1. Hemodynamics, endothelial cells, and candidate shear sensors. (A) Blood flow induces fluid shear stress on the surfaces of endothelial cells, which distribute the force to multiple structures thought to be involved in mechanotransduction. (B) Silver nitrate staining of endothelial cell borders in isolated arteriole. Bar = 10 μm.

[88, 107], surface proteoglycans [6, 83, 106], and cell–cell junctional proteins [83] (reviewed in [21]). Determining whether these molecules are mechanosensors will require sophisticated experimental methods to detect their direct perturbation by shear stress and their role in the conversion of force to downstream biomechanical signaling.

Experimental methods to elucidate EC mechanosensitivity have included engineering analysis, mechanical testing, molecular biological technology, and fluorescence imaging. For example, approaches to elucidate cell mechanics include characterizing the mechanical properties of the EC membrane by micropipette aspiration [100] and atomic force microscopy (AFM) [15, 99], investigating the effects of shear on EC–membrane lipid lateral diffusion [10] and free volume [52], and analyzing the shear-induced deformation of intermediate filaments [54]. Extensive molecular biological investigations have elucidated the signaling pathways that regulate gene transcription in response to shear stress (reviewed in [17]) and have uncovered the genetic endpoints in EC shear sensitivity [16, 36, 81]. Such studies have provided fundamental insights into the force transduction mechanisms and suggest that ECs use multiple cellular structures to integrate the effects of fluid forces into a coordinated cellular response.

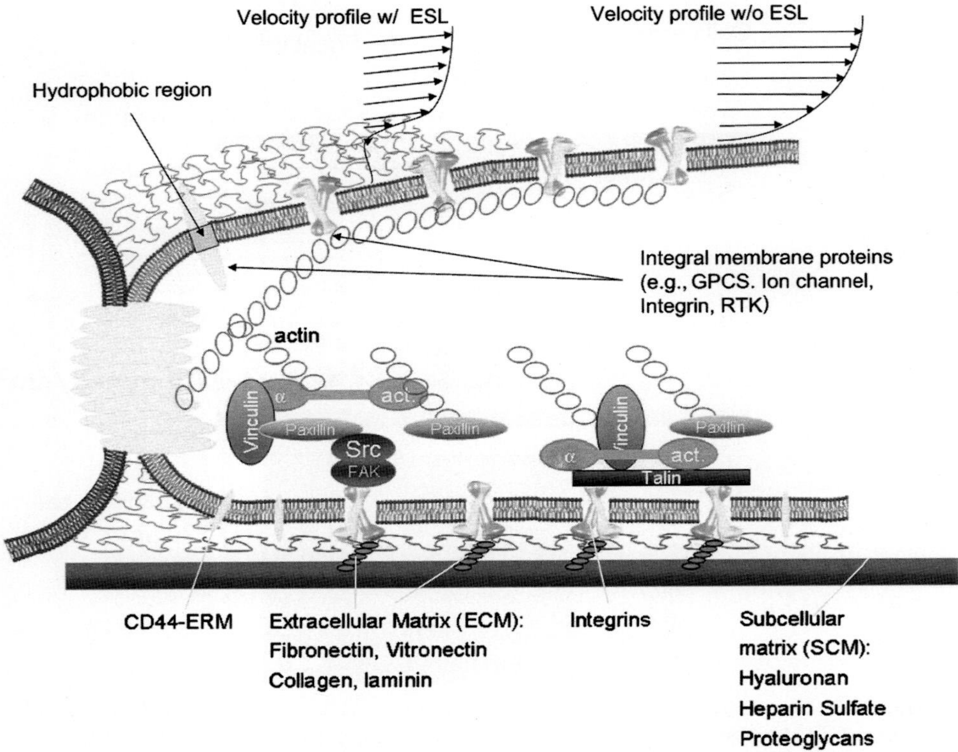

Figure 3.2. Molecular organization in a representative endothelial cell: Membrane, integral membrane proteins, cell junctions, extracellular matrix molecules, and cytoskeleton are represented. Arrows denote velocity near cell membrane. Shear stress (viscosity times velocity gradient) is attenuated in the glycocalyx (endothelial surface layer) compared to areas denuded of the glycocalyx. Stresses may also be transferred to focal adhesions or near focal adhesions where they are amplified.

3.3 The Membrane as a Mechanosensor

3.3.1 Mechanisms of Bilayer Modulation of Protein Function

Circumstantial evidence supports the plausibility that the lipid bilayer is a primary mechanosensor. First, the membrane is exposed to effects of shear stress directly either by its contact with fluid flow, through its interaction with the surface glycocalyx, or by its association with mechanosensitive domains such as focal adhesions (see Figure 3.2). Second, the membrane is a repository for many mechanosensitive proteins [10, 49, 61, 92]. Third, perturbation of EC membrane lipids by shear stress may initiate signaling cascades in ECs leading to altered gene expression [11, 49]. However, the mechanisms of shear perturbation of the membrane or the subsequent molecular mechanisms that link such perturbations with signal transduction have yet to be discovered.

Numerous reviews have been written on the role of the membrane in modulating membrane-associated protein function. For example, Lee [74] discusses the many biophysical and biomolecular mechanisms by which proteins are modulated by their surrounding lipids. These mechanisms involve the interactions of lipid headgroups with proteins; the effects of differences in hydrophobic thicknesses of proteins and bilayers; the effects of lipid structure on protein aggregation and helix–helix interactions; the role of membrane phases and membrane microviscosity on protein–protein association; the effects on proteins of lipid free volume, bilayer curvature, interfacial curvature, and elastic strain; and the effects on protein inclusions of lateral pressure profile and tension. While the exact mechanism of shear-stress–induced modulation of the interaction of lipids with integral membrane proteins remains to be elucidated, in this chapter we evaluate the two hypotheses of shear modulation of EC membranes for which there exist experimental and theoretical support: shear-induced changes in lipid fluidity and hydrophobic mismatch.

3.3.2 Overview of Membrane Fluidity

Singer and Nicholson [109] first proposed a working model of the cell membrane that allowed for in-plane movement of its components. This model hypothesized that the cell membrane is a thin layer of lipid molecules with protein molecules interspersed throughout. Further studies have revealed that the distribution of lipid molecules is heterogeneous and that some integral membrane protein molecules, being bound to an underlying cytoskeleton, are only intermittently free to move [59]. The term "lipid fluidity" has been used to characterize all aspects of lipid mobility, including in-plane and rotational diffusion, and to suggest a mechanism for lipid modulation of protein activity. However, more detailed investigations suggest that lipids modulate protein function through multiple mechanisms [4, 14, 20, 84, 97, 104, 108].

The membrane fluidity of animal cells is determined primarily by the amounts and types of phospholipids, the membrane cholesterol content, and the interaction of lipids with membrane-bound proteins (and the interactions of these proteins with the cytoskeleton). A phospholipid is composed of a hydrophilic head region that is in contact with the extracellular or intracellular space, and two hydrophobic fatty acid chains, which are located in the interior of the membrane and interact with other phospholipid tails. If the tails of the phospholipids are saturated with hydrogen (i.e., with no carbon–carbon double bonds), then they are relatively straight and can pack closely together, leading to a high resistance to lateral movement of membrane constituents and hence a low membrane fluidity. These areas of the membranes tend to be thicker than the more fluid membrane portions. Similarly, increasing the degree of unsaturation, by increasing the number of carbon–carbon double bonds, puts kinks in the otherwise straight chains and makes it more difficult for these types of phospholipids to pack in an orderly fashion. In this case, the membrane is more fluid and thinner. Similarly, cholesterol binds, via a hydrogen bond, to the carboxyl group on the base of the fatty acid and thus intercalates itself into the hydrophilic portion of the membrane [110], thus restricting the movement of lipids and resulting in membrane stiffening.

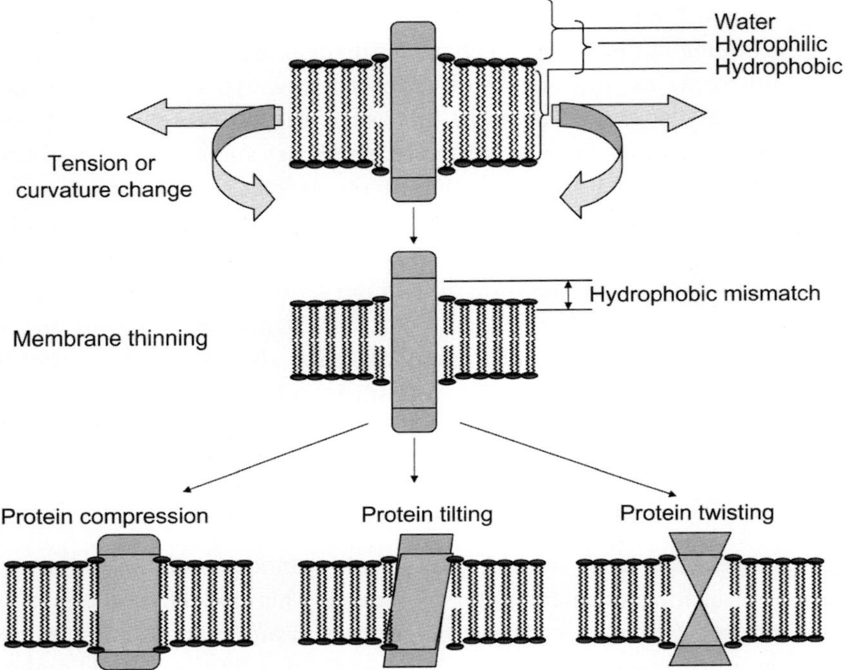

Figure 3.3. Illustration of tension- or curvature-induced changes in hydrophobic thickness of the bilayer: Hydrophobic region of membrane may match hydrophobic thickness of integral membrane proteins. This matching may be perturbed by fluid shear stress leading to membrane tension, which may thin the membrane by direct tension effects or through changes in membrane curvature, leading to alteration of protein conformation by protein compression, tilting, or twisting.

Proteins can also affect membrane fluidity via their interactions with their surrounding lipids [85]. Perturbation of a protein in the lipid medium disturbs the orderly packing of the lipids in an annular region around the protein. Generally, the hydrophilic portion of the lipids (e.g., headgroups) is attracted to the hydrophilic part of the protein. Since the hydrophobic core of the protein may not match the hydrophobic thickness of the lipid bilayer, many proteins protrude out of the membrane, and the lipid environment immediately surrounding the protein is fluidized. Maintaining a separation between fatty acid chains of lipids from opposing leaflets is a high energy process. To minimize the energy required to maintain the membrane integrity around the proteins, often only lipids with longer fatty acid tails will surround the protruding proteins (see Figure 3.3). In this way, proteins determine what types of lipids will surround them. Furthermore, with a high concentration of proteins, these annular regions may overlap, and therefore there is a potential for protein–protein interactions in the annular lipid regions. For proteins that are bound to the cytoskeleton, their presence may lead to an overall stiffening of the membrane.

Each of these properties, that is, fatty acid chain saturation, cholesterol content, and protein content (and protein interaction with the cytoskeleton), are known to be tightly regulated by the cell. When these systems are chronically perturbed, cells can potentially adjust these properties to maintain an "optimal" fluidity. This regulation is known as homeoviscous adaptation [77]. While the notion of a fluidity set-point is still controversial, the homeoviscous adaptation theory has been shown to apply in poikilotherms' adaptation to chronic temperature changes and to cells of hibernating animals. Both poikilotherms and homeotherms adapt to the decreases in cell fluidity induced by cold by synthesizing a greater proportion of unsaturated fatty acids, which, when incorporated into the cells, stabilize membrane fluidity as the temperature falls. If shear stress alters membrane fluidity in a chronic manner, than the cells potentially could adjust their membrane constituents to adapt to this chronic stress. Such adaptation (or lack of) could be a factor, as yet undiscovered, that determines protection from (or propensity for) atherosclerosis.

3.3.3 How Does Lipid Fluidity Affect Proteins?

By restricting (or allowing) movement in the plane of the membrane, changes in lipid fluidity can restrict (or enhance) protein lateral mobility and thus might affect diffusion-dependent protein functions [4]. As an example of diffusion-mediated protein function, G-protein hydrolysis, a process necessary for many agonist-induced changes in the cell, depends on the diffusion of the G-protein from the receptor to a downstream effector. Alterations in this diffusion could potentially affect the equilibrium between receptor/G-protein complexes and effector/G-protein complexes [82]. Since, in equilibrium, fluidity is thought to affect the forward and reverse reaction rates equally [74], it is likely that fluidity changes would affect protein–protein interactions only under nonequilibrium conditions, such as during rapid changes in membrane tension [11].

Finally, lipid fluidity may affect the accessibility of proteins to ligands [104]. According to this concept, tighter packing of lipids causes integral membrane proteins to be pushed out of the membrane and thus make their active sites more available to ligand binding. Increased protrusion of integral proteins out of the membrane could also increase their hydrophobic mismatch and thus alter protein free energy (see next section).

3.3.4 Hydrophobic Mismatch

Because of the energetic cost of exposing hydrophobic amino acid residues to water, it is expected that the hydrophobic region of lipids will match the hydrophobic core of the proteins embedded in the bilayer (Figures 3.2 and 3.3). Equivalently, if the hydrophobic thickness of the membrane changes, then either the lipids will rearrange or the protein will reach a new conformation by tilting, twisting, stretching, or compressing in order to reduce the hydrophobic mismatch (Figure 3.3) [53]. In either case, changes in the membrane thickness will lead to changes in protein conformation. Hydrophobic mismatch can also explain the dimerization of proteins. For example, alterations in lipid fluidity can modulate the protein's local energy environment by either changing the phase of the lipid immediately surrounding the protein to affect the protein conformation or ameliorating the hydrophobic mismatch. The former effect has been hypothesized to cause the anesthetic effects of haloalthane on nerve function; haloalthane fluidizes the entire membrane and negates the phase change from the protein annular region to the remainder of the lipid bilayer, thus preventing the opening of sodium channels in neurons [115]. If protein dimerization leads to a reduction in protein free energy caused by hydrophobic mismatch, then changes in membrane thickness will lead to dimerization. Similarly, changes in lipid fluidity can affect the hydrophobic mismatch between the lipids and the protein and alter its free energy state [84]. Such changes in free energy may be sufficient to cause a conformational change in a receptor protein, for example, to initiate downstream signaling even in the absence of a ligand.

3.3.5 Role of Membrane Microdomains

Recent studies suggest that mechanically induced stresses are transduced to coordinated biochemical pathways in ECs via the perturbation of the plasma membrane in specialized membrane microdomains, for example, lipid rafts, caveoli [97], focal adhesions [76], and cell junctions [34]. For example, shear-sensitive molecules (G-proteins [92], MAPKs [61], eNOS [37]), reside in $\sim$ 100-nm–diameter, cholesterol rich, liquid-ordered–phase membrane microdomains termed caveoli and lipid rafts [26, 97]. G_i-proteins (present in lipid rafts [92]) play a role in shear-induced nitric oxide production [93, 111] and in shear-induced MAPK activity [61] in bovine aortic endothelial cells. Focal adhesions and cell–cell junctions may be mechanosensitive membrane domains that may have unique lipid compositions [44, 48, 116]. Recently, Schnitzer [19] has shown that shear stress activates MAPK in caveoli via lipid-mediated mechanisms involving ceramide and sphingomylinase, which may be activated by G-proteins.

Key unanswered questions, however, remain: Is the local stress induced by fluid shear sufficient to perturb molecules in these domains to a sufficient degree to initiate mechanotransduction events, or are the effects of shear stress the results of activation elsewhere and communicated to the membrane via second messengers? Addressing these questions will require increased use of multiscale mechanical models of cells along with optical tools to detect mechanically induced molecular dynamics changes at the submicrometer length scale.

3.3.6 Evidence for Lipid Domains in Cells

There have been excellent studies using electron microscopy to detect caveoli in ECs [97]. However, lipid rafts have been more elusive because they do not have a characteristic stable shape. Thus, indirect means to detect rafts include the observation that certain parts of the membrane float to the top of a density gradient (hence the term "raft"). Another strategy is to use chain length–sensitive lipoid dyes to take advantage of the fact that lipid rafts and caveoli, because of their increased lipid

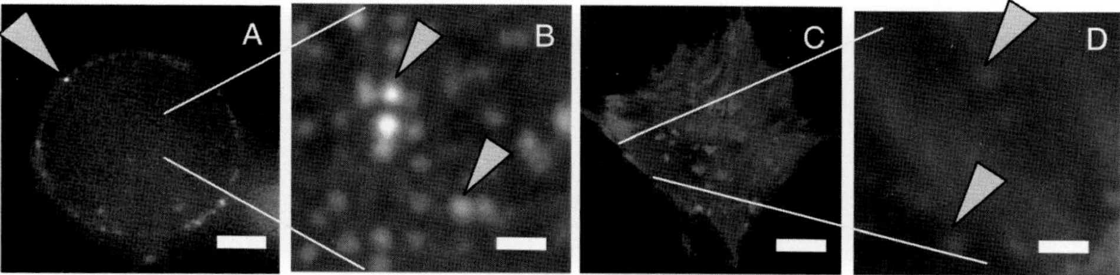

Figure 3.4. Detecting gel-phase lipid microdomains. In order to test whether DiI C_{18} segregates into identifiable gel-phase microdomains, 2-phase vesicles were made from DMPC, DPPC, and DiI C_{18}. At room temperature, dye partitioned into distinct areas presumed to be areas of coalesced gel-phase lipids. Similar structures were identified in BAECs stained with DiI C_{18}. Gel-phase domain formation in GUVs (A, B) and BAECs (C, D). Bars: A = 10 μm, B = 1 μm, C = 10 μm, D = 1 μm.

order, are slightly thicker than bulk plasma membranes. For this chapter, we present the implementation of this strategy in which lipid microdomains are detected using dialkylcarbocyanine, DiI C_{18}, which has the propensity to segregate into the thicker membrane microdomains. First, two-phase vesicles were made from mixtures of dimyristoylphosphatidylcholine (DMPC; liquid at room temperature; phase transition temperature 23.5°C) and dipalmitoylphosphatidylcholine (DPPC; gel at room temperature; phase transition temperature 41°C) [70, 80] and DiI C_{18}. Figure 3.4 illustrates how the dye partitions into distinct areas presumed to be areas of coalesced gel-phase lipids. Similar structures were identified in bovine aortic endothelial cells (BAECs) stained with DiI C_{18}. Similarly, a BAEC culture was stained with both sulfonated (SP-) DiI C_{18} and Alexa-fluor-labeled Cholera-toxin-B (CT-B) (Figure 3.5). CT-B labels GM-1 gangliosides, which are thought to be raft markers [7]. Cells were cultured on borosilicate coverslips and then incubated with a CT-B solution of 25 μg/ml in DPBS on ice for 20 min. The cells were warmed to 37°C and then stained with 10 μM SP-DiI C_{18} using the above procedure. The resulting images were aligned utilizing Autoquant's Autodeblur® software and examined for colocalization of the stains using the NIH ImageJ software. Figure 3.5 (Plate 4) illustrates the colocalization of DiI C_{18} and the raft marker. Thus, it is likely that DiI C_{18}, because of its longer acyl chain length, segregates into lipid domains of comparable thicknesses. Cholesterol is thought to be concentrated in these domains and contributes to its enhanced thickness. We caution that cross-linking of GM-1 with CT-B may induce raft formation. Thus, additional tools are needed to detect rafts in their native state (see section on fluorescence lifetime later in this chapter).

3.3.7 The Glycocalyx May Modulate the Effects of Shear Stress on Membranes

Recent studies support a possible role of the glycocalyx in modulating mechanotransduction events. Weinbaum et al. [119] and Secomb [102] have shown that the glycocalyx attenuates shear stress near the plasma membrane (see Figure 3.2). Such results suggest that the membrane cannot be a primary sensor of shear stress.

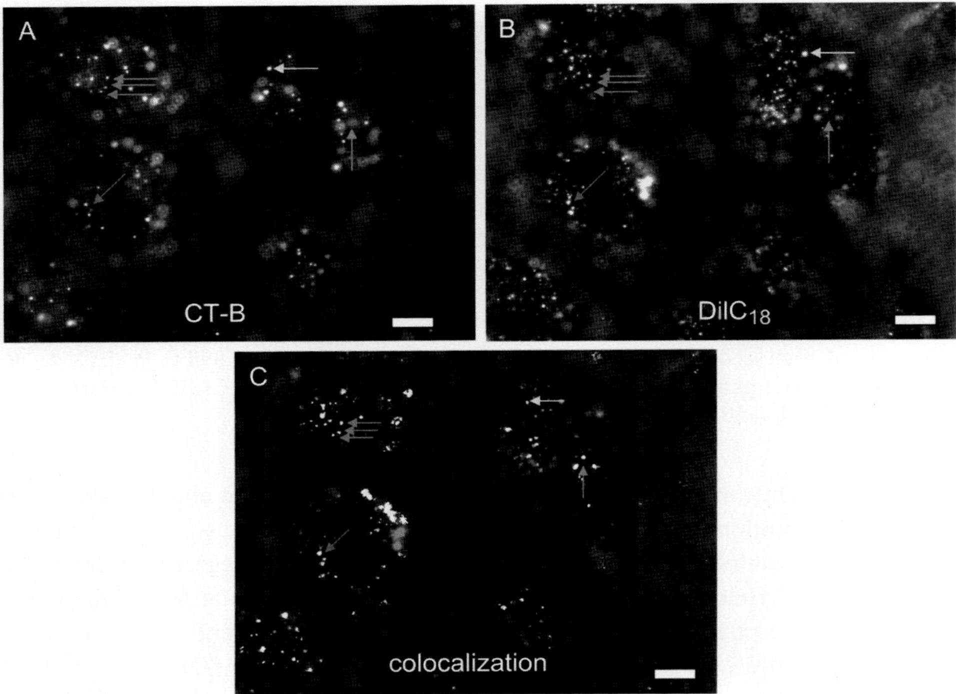

Figure 3.5. Colocalization of long chain lipoid dye and lipid raft markers. BAEC culture was stained with both SP-DiI C_{18} and CT-B. Cells were cultured on borosilicate coverslips, and then incubated with a CT-B solution of 25 μg/ml in DPBS on ice for 20 min. The cells were warmed to 37°C and then stained with 10 μM SP-DiI C_{18}. The resulting images were aligned utilizing Autoquant's Autodeblur® software and examined for colocalization of the stains using NIH's ImageJ software. (A) CT-B staining from channel 1. (B) Sp-DiI C_{18} staining pattern. (C) DiI C_{18} and CT-B colocalization indicated by white pixels.

Consistent with a role of the glycocalyx in mechanotransduction, Mochizuki et al. [83] showed that degradation of the glycocalyx through hyaluronic acid digestion reduces shear-induced nitric oxide production in isolated arterioles. Similarly, Tarbell and colleagues [30] showed that digestion of heparin sulfate proteoglycans abolishes shear-induced nitric oxide production in cultured ECs. Taken together, these studies support the role of the glycocalyx as a primary mechanosensor of shear stress. It is also possible that the glycocalyx may be necessary for the transmission of fluid forces to the plasma membrane. Consistent with this idea, the glycocalyx has been shown to be anchored to the cell membrane via focal attachment points [112] and through glycoclipid linkages [101]. In addition, the integrated drag on the cell is resisted by focal adhesions near which membrane strain may be greatly amplified [28]. Therefore, it is likely that the glycocalyx can play a role in the transmission of force to the lipid bilayer (i) directly, via attachments to membrane lipids; (ii) indirectly, via integral membrane proteins; and (iii) by modulation of the flow near the lipid bilayer surface.

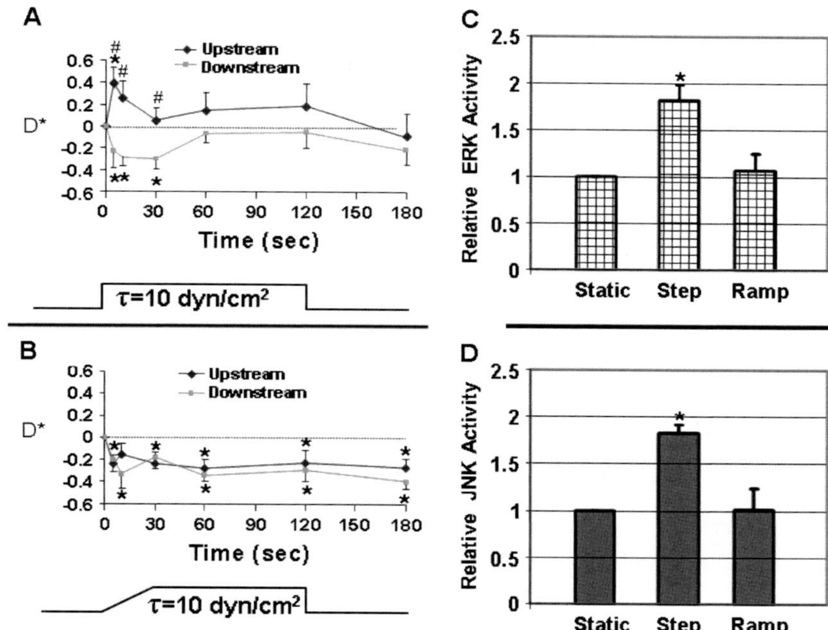

Figure 3.6. Shear stress induces changes in lipid lateral diffusion and activation of MAPK proteins. (A) 10 dynes/cm^2 of shear stress elicit increases and decreases in lipid lateral diffusion on the upstream and downstream sides, respectively, of the EC membrane ($D^* =$ percent change of lipid diffusion from pre-shear values). (B) Shear-induced changes in D^* are absent when shear stress is ramped to maximal values. (C) Comparable shear stresses lead to phosphorylation of extracellular-signal–regulated kinase (ERK). (D) c-jun N-terminal kinase (JNK). (Adapted from [11] with permission.)

3.3.8 Evidence for Shear Stress Perturbation of Membrane Lipids

Shear stress induces a time- and position-dependent change in lipid lateral diffusion [10] (see Figure 3.6), thus supporting the idea that the cell membrane's proximity to the blood flow makes it a candidate shear-sensitive system. By adapting a confocal laser scanning microscope for measurements of fluorescence recovery after photobleaching (FRAP), Butler et al. [10] made quantitative, two-point, subcellular measurements of membrane fluidity (as quantified by the lipid lateral diffusion coefficient, D) on cells while being subjected to shear stress. Their results have shown that (i) shear stress induces a rapid, spatially heterogeneous, and time-dependent increase in D of a fluorescent lipid probe in the BAEC membrane; (ii) the location, magnitude, and persistence of these shear-induced increases in D depend on the shear magnitude; and (iii) shear stress elicits a secondary (7-min) increase in D. They further demonstrated that the lipid lateral diffusion coefficient (D) of the membrane is sensitive to the temporal gradient in shear stress [11]. A step-τ of 10 dynes/cm^2 elicited a rapid (5-sec) increase of D in the portion of the cell upstream of the nucleus, and a concomitant decrease in the downstream portion (Figure 3.6(A)). A ramp-τ with a rate of 10 dynes/cm^2/min elicited a rapid (5-sec) decrease of D in both the upstream and downstream portions (Figure 3.6(B)). Thus, it can be concluded that the lipid bilayer can sense the temporal features of the applied τ with spatial discrimination.

Membrane perturbation may activate signaling cascades responsible for changes in gene expression. Butler et al. [11] have shown that a step-shear stress, which increased *D*, increased the activation of extracellular-signal–regulated kinase (ERK) and c-Jun N-terminal kinase (JNK) (see Figures 3.6(C,D)), which are important signaling molecules for shear stress–related gene expression. Ramp-shear, which failed to increase *D*, did not result in increases in ERK or JNK. Furthermore, the membrane fluidizer, benzyl alcohol, increased ERK and JNK kinase activity. In contrast, cholesterol, which decreased *D*, decreased the activities of these MAP kinases, suggesting that stresses in the membrane, as revealed by changes in *D*, lead to downstream signals that are responsible for altered gene expressions. These results support the hypothesis that the cell membrane plays a role in transducing the magnitude and rate-of-change of shear stress into altered gene expressions, a feature important in areas of complex flow patterns and abrupt changes in shear stress in the microvasculature during exercise and reperfusion after ischemia.

3.3.9 Other Mechanosensitive EC Functions That Are Modulated by Lipid Membrane Fluidity

Recently, Gojova and Barakat [43] showed that vascular endothelial wound closure under shear stress was modulated by membrane fluidity and dependent on flow-sensitive ion channels. They showed that that reducing EC membrane fluidity in cells near an edge of a wounded monolayer significantly slowed down both cell spreading and migration under flow. Interestingly, when they blocked flow-sensitive K^+ and Cl^- channels, cell spreading was reduced but cell migration was unaffected. Although the relationship between membrane fluidity and ion channel activation in response to flow was not assessed in that study, these findings suggest that membrane fluidity modulates shear-sensitive pathways and the consequent cellular functions such as migration and spreading.

Focal adhesions and integrins are known to be involved in mechanotransduction. Recently, Gaus et al. [38] made measurements of membrane order using two-photon microscopy of the fluorescent membrane probe Laurdan to show that focal adhesions are more ordered than caveolae or domains that stain with cholera toxin subunit B (Ct-B) (presumably lipid rafts) [121]. When cells were detached from the substrate, a rapid, caveolin-independent decrease in membrane order occurred. These results show that phospholipids and cholesterol play important roles in focal adhesion assembly and may have a strong influence on signaling at these sites.

In a related study, Gopalakrishna et al. [44] showed that cholesterol modulates $\alpha_5\beta_1$ integrin functions, suggesting that the lateral mobility of integrin molecules in the plasma membrane, which is influenced by cholesterol content (and membrane fluidity), may regulate the clustering of $\alpha_5\beta_1$ integrin molecules in focal adhesions and, subsequently, their adhesion to the extracellular matrix protein fibronectin and

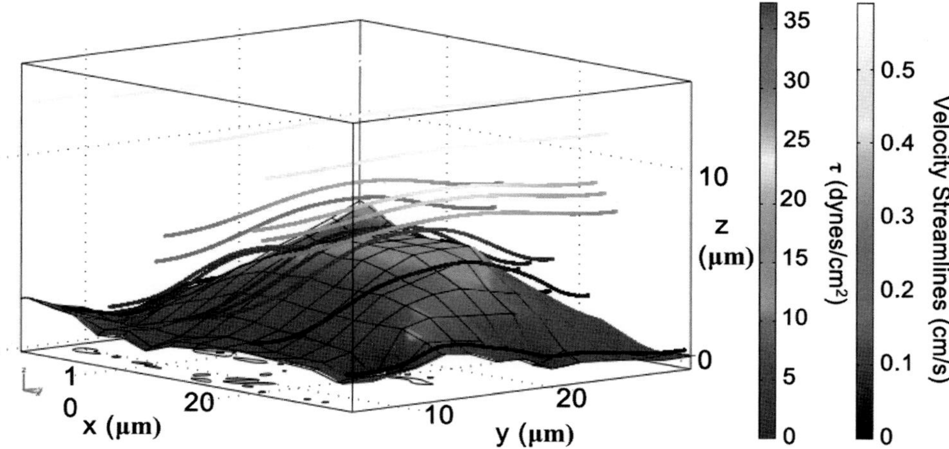

Figure 3.7. Representative subcellular shear stress distribution. A nominal shear stress of 10 dynes/cm^2 was simulated over the solid cell monolayer model in the positive *y*-direction. Model was developed from three-dimensional cellular imaging and quantitative total internal reflection fluorescence imaging of calcein-stained endothelial cells. Finite element analyses of fluid (extracellular) and solid (intracellular) stress distributions were analyzed using computational fluid dynamics and linear elasticity theory. Stress distributions show stress peaks at the apical region over the nucleus while stress is minimum in the valleys between cells. Simulated cell represents a single cell in a monolayer. Simulated velocity field is shown using streamlines. Color plot of shear stress in dynes/cm^2; streamline color corresponds to fluid velocity (cm/s). Axes in μm. (Adapted from [27] with permission.)

intracellular protein talin. In that study, the activation of MAPK pathways by the association of fibronectin with $\alpha_5\beta_1$ integrin was suppressed by cholesterol. The role of membrane fluidity in regulating integrin mobility is supported by models of integrin diffusion laterally in the membrane toward focal adhesions [8]. These results also suggest that focal adhesions, like caveoli and rafts, can be thought of as membrane lipid microdomains.

3.3.10 Quantification of Shear-Induced Membrane Stresses

Ferko et al. [29] recently introduced new integrated methods in fluorescence imaging and image processing for the development of solid models with cell-specific topographies and subcellular organelles (Figure 3.7; Plate 5). The goal of this research was to use these methodologies along with quantitative total internal reflection fluorescence microscopy (qTIRFM) to create a cell-specific, multicomponent, three-dimensional solid elastic continuum model of an EC in a confluent monolayer with experimentally determined FAs [28]. Finite element analysis was used to compute stress transmission throughout the EC due to fluid flow applied at the apical surface. This type of cell-specific modeling based on experimentally determined topographies and boundary conditions may help identify potential sites of force-induced potentiation and directional-biasing of cell signaling.

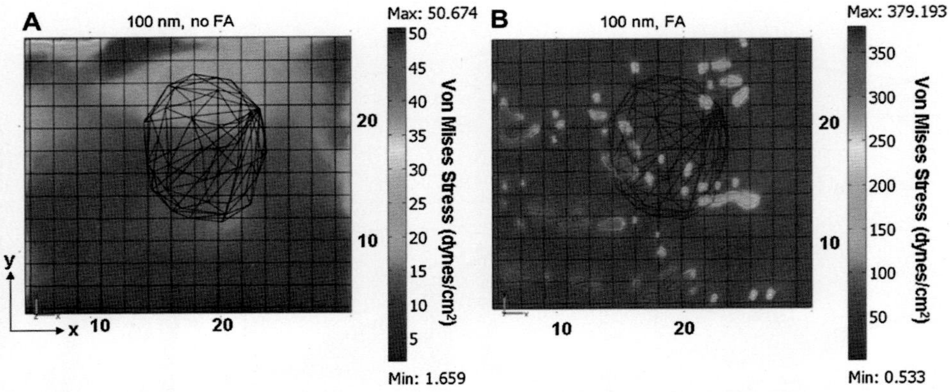

Figure 3.8. Effects of focal adhesions on shear-induced stresses. Von Mises stress distributions were evaluated at $z = 0.1$ μm above the coverslip for (A) model computed without focal adhesions and (B) model solved with focal adhesions. (Adapted from [28] with permission.)

The main contributions of this study were to provide the first quantitative predictions of stress distributions in focally adhered ECs resulting from apically applied fluid flows. These results arose from finite element analysis of a cell-specific model in which surface topography and FA location and area were experimentally determined. The model predicted that shear-induced stresses were generally small but significantly amplified and focused near FAs and the high-modulus nucleus. For example, the inclusion of FAs as attachment locations in a homogeneous linear elastic continuum model resulted in heterogeneous internal stresses, strains, and displacements (see Figure 3.8; Plate 6). Stresses near FAs were nearly 40-fold larger than surface shear stresses, thus supporting the widely held contention that FAs are a means of force amplification of shear stress. The locations and directions of upstream tensile and downstream compressive stresses computed in the vicinity of individual FAs were consistent with the observation of FA growth in the downstream direction of flow and FA retraction in the upstream side, thus providing quantitative information to elucidate the mechanisms of mechanotaxis of ECs [24, 75, 122].

3.3.11 Quantitative Effects of Membrane Tension on Lipid Dynamics and Protein Inclusions

The work of Ferko et al. [28] and Butler et al. [10] support the concept that stresses in the membrane rise to a sufficient degree to induce molecular perturbation. However, the mechanism of shear-induced molecular perturbation of membrane lipids remains unknown. In a theoretical study, Gov [46] provided evidence that changes in the static or dynamic curvatures of the membrane may lead to increased lipid diffusion. To summarize that work, it is postulated that when a flat membrane is curved, the thickness increases, leading to an increased resistance to diffusion. Using molecular diffusion models first proposed by Saffman and Dulbruck [98] in which the diffusion of a cylinder in the plane of the membrane is inversely proportional to the membrane thickness, membrane thickening would lead to decreases in membrane fluidity and diffusion coefficient. Conversely, when a curved membrane is flattened by a surface tension generated by shear flow, the membrane thins and diffusion increases. Gov's [46] analysis of the effects of shear stress on lipid diffusion fit the data from Butler et al. [10] very well. Interestingly, the changes in the membrane thickness from this analysis can be used to estimate the changes in the hydrophobic matching conditions in a candidate shear-sensitive protein embedded in the lipid bilayer. For a shear stress of 20 dynes/cm^2, Equations 8 and 7 from Gov [46] predict that diffusion will increase by about 40% over pre-shear values, consistent with Butler et al. [10]. Using Equation 4 from Gov, and assuming that the membrane goes from curved to flat during shear application, we can predict that shear stress flattens membrane fluctuations from an initial mean squared curvature of 0.0058 nm^{-2}. This change in curvature would decrease the membrane thickness from 5.0 nm to 3.9 nm (see Equation 4 in Gov [46]), corresponding to a 22% strain in the direction of the bilayer normal (strain parallel to the membrane will depend on the Poisson's ratio of the membrane). This is an appreciable strain and is the same order of magnitude needed to sufficiently alter the free energy of ion channel inclusions in bilayers [1, 45, 91]. Membrane strain in the vicinity of focal adhesions may be even more pronounced, and this would lower the shear threshold for protein activation [39].

3.4 Tools for Measurement of Mechanically Induced Lipid Perturbation

3.4.1 Time-Correlated, Single Photon Counting, and Multimodal Microscopy

Our main hypothesis is that the perturbation of cellular structures by force is accompanied by changes in molecular dynamics. In order to address these fundamental issues in mechanosensing and transduction, Gullapalli, Tabouillot, et al. [51] have developed a hybrid multimodal microscopy-time–correlated single photon counting (TCSPC) spectroscopy system to assess time- and position-dependent, mechanically induced changes in the dynamics of molecules in live cells as determined from fluorescence lifetimes and autocorrelation analysis (fluorescence correlation spectroscopy) (see Figures 3.9 and 3.10). Colocalization of cell structures and mechanically induced changes in molecular dynamics can be analyzed in post-processing by comparing TCSPC data with three-dimensional models generated from total internal reflection fluorescence (TIRF), differential interference contrast (DIC), epifluorescence, and deconvolution [28].

3.4.2 Fluorescence Correlation Spectroscopy

Fluorescence correlation spectroscopy is ideally suited for cellular research because diffusion and chemical kinetics can be measured in very small volumes ($\sim$ 1 fl). A pinhole placed in an image plane of the emission optical pathway defines the confocal observation volume into and out of which fluorescently tagged molecules move (Figure 3.9). The resulting fluorescence fluctuations picked up by the photodetector

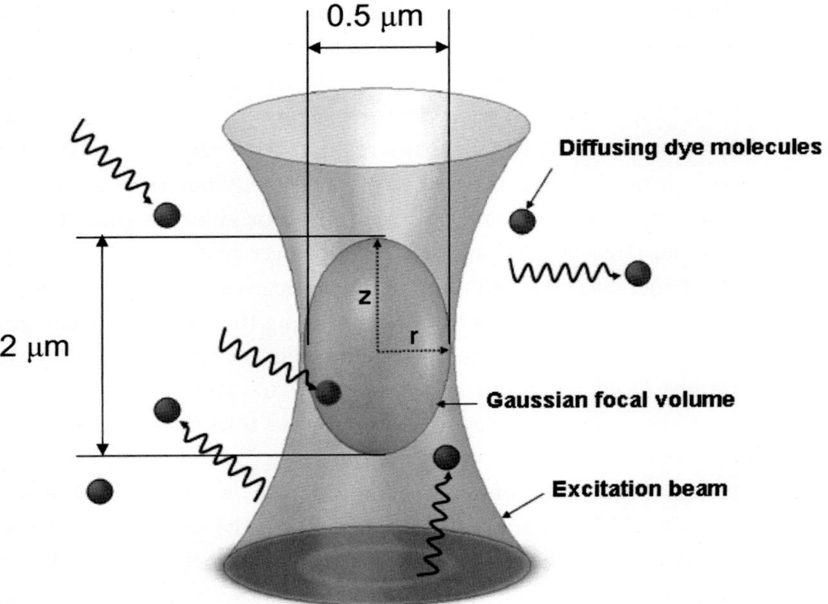

Figure 3.9. Confocal volume for fluorescence correlation spectroscopy. The radial dimension r of the confocal volume is close to the diffraction limit of the objective and is determined by keeping the (r/z) parameter constant (<10) in Equation (3.6) when fitting the autocorrelation curve from R6G molecules in water. The radius thus obtained from experiment is 326 ± 10 nm. Fluorescent molecules are excited by the entire laser beam and fluorescence emission is only collected in the confocal volume. When particles move into and out of the confocal volume in the x, y, and z directions, three-dimensional diffusion is considered. When particles only move in the x-y plane, two-dimensional diffusion is considered. (Adapted from [51] with permission.)

are translated into current intensity fluctuations. The time autocorrelation curve of this signal can be fit to various formulas to extract physical values for transport phenomena (two- or three-dimensional diffusion, anomalous diffusion, translation) or chemical kinetics [27, 78]. For example, the fluorescence autocorrelation funtion $G(\tau_w)$ can be fit for (two-dimensional) diffusion by

$$G(\tau) = \frac{1}{N} \cdot \left(\frac{1}{1 + \left({}^{\tau_w}/_{\tau_D} \right)} \right)$$

where N is the average number of particles, τ_w is the lag time of the autocorrelation function, and τ_D is the characteristic diffusion time, which is related to the diffusion coefficient, D, by $r^2 = 4D\tau_D$.

3.4.3 FCS Measurements of DMPC Giant Unilamellar Vesicles and Endothelial Cell Membranes

In a recent study outlining the use of TCSPC for mechanobiology studies of membranes, Gullapalli et al. [51] performed FCS measurements on DMPC vesicles and EC

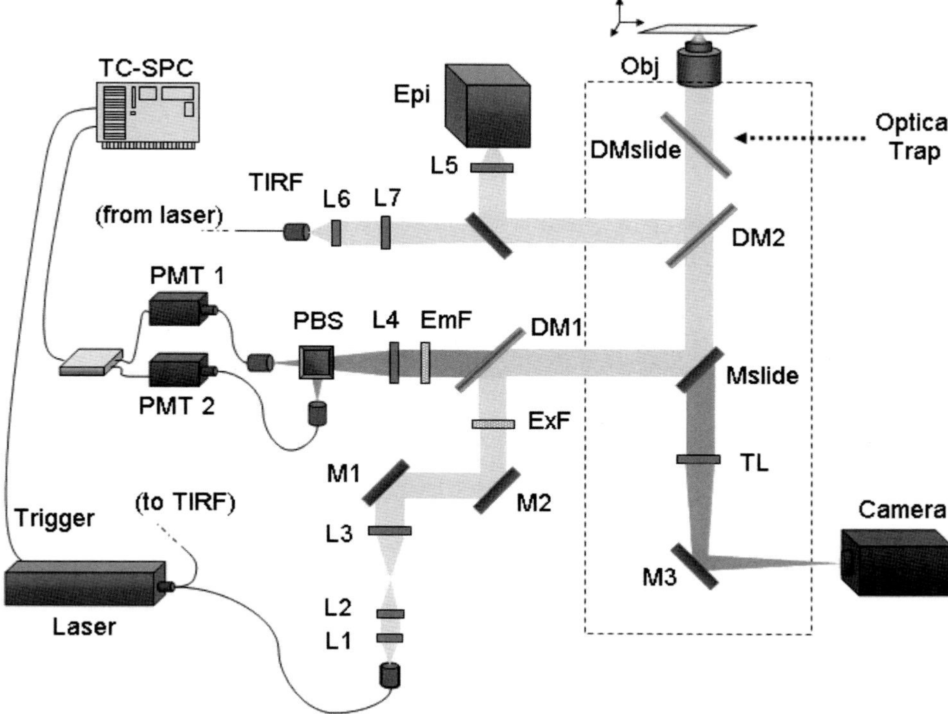

Figure 3.10. Optical setup for TCSPC-multimodal microscopy. The Kr/Ar-ion, diode, or pulsed Nd:YAG laser beam is transmitted via fiber coupling to the TIRF or confocal port. For confocal illumination, upon exiting the fiber, the beam is collimated with lens L1, expanded by L2 and L3, steered by the mirrors M1 and M2, reflected off the dichroic mirror (DM1), and enters the right-side port of the microscope (note that the tube lens for the side port has been removed). After excitation of the sample, the fluorescence emission signal is collimated by the objective and exits the side port, passes through the dichroic mirror, and is focused – using lens L4 – onto the optical fiber, which is connected to the photomultiplier tube (PMT). A polarizing beam splitter (PBS) can be introduced before the fiber to separate light with polarization that is parallel or perpendicular to that of the excitation light. The PMTs convert single photons to electrical pulses, which are routed to the TCSPC board. Laser light from the TIRF system shares the back port of the microscope with the epifluorescence tube (Epi). Lenses L5 and L6 collimate the epifluorescence and TIRF light, respectively. The TIRF illumination is focused at the objective back aperture by lens L7. When the sliding mirror, Mslide1, is removed from the light path, the right-side port is closed and the emission signal can be collected by the camera via the tube lens (TL). In addition, the optical trap can be inserted above the fluorescence cube turret with an infrared dichroic mirror (DMslide) mounted on a custom-built slider (not shown). (Adapted from [51] with permission.)

membranes stained with DiI-C_{18}. Fluorescence of DiI C_{18} was excited with 520-nm light from the CW laser or 532-nm light from the pulsed solid-state laser. Fluorescent light with wavelengths of 545 nm and longer was collected for analysis. The temperature of the media was maintained at 37°C for cells and 26°C for DMPC (phase-transition temperature for DMPC = 24°C). The results were fit using the equation for two-dimensional diffusion. The results suggest that DiI diffusion in homogenous model membranes is uniform, but diffusion of the same dye in cell membranes is more restricted (lower diffusion coefficient) and heterogeneous. These data support the hypothesis that these lipoid dyes segregate into membrane microdomains and that time-resolved fluorescence spectroscopy can be used to detect this segregation.

3.4.4 Fluorescence Lifetime

Time-resolved fluorescence lifetime spectroscopy enables the analysis of subtle changes in the photophysics of fluorescent molecules [72]. When a fluorescent molecule is excited to a higher energy state using a picosecond pulse of laser light, it remains in the excited state for a finite time before it decays to the ground level energy state. Using a high-frequency pulsed laser, histograms of photon emission times relative to excitation times can be generated and fit with a negative exponential (or multiple exponentials) with a characteristic decay time (or lifetime), τ_F (different than the characteristic diffusion time, τ_D, of FCS). The fluorescence lifetime depends on local molecular microenvironmental factors, including ionic strength, hydration, oxygen concentration, binding to macromolecules, and the proximity to other molecules that can deplete the excited state by resonance energy transfer [72]. The fluorescence lifetime and quantum yield are related to the intrinsic photophysical characteristics of a fluorescent molecule such as radiative and nonradiative decay mechanisms. The fluorescence quantum yield, Q, is the ratio of the number of photons emitted to the number of photons absorbed, according to

$$Q = \frac{k_r}{k_r + k_{nr}}$$

where k_r and k_{nr} are the radiative and nonradiative decay rates of the molecule, respectively. Fluorescence lifetime is given by

$$\tau_F = \frac{1}{k_r + k_{nr}}$$

The value for k_{nr} depends on the mode of the nonradiative decay, such as collisional quenching, hydration, and vibrational relaxation. Thus, any alteration of k_{nr} also leads to a detectable change in the value of the fluorescence lifetime.

The value for fluorescence lifetime is obtained by an iterative reconvolution of an instrument response function (IRF), with the fluorescence intensity using an assumed decay law, which can be approximated by a sum of exponentials [72]:

$$I(t) = \textstyle\sum_i \alpha_i \exp(-t/\tau_{Fi})$$

where α_i is the fraction of molecules with lifetime τ_{Fi}, normalized to unity. Fluorescence lifetimes are independent of fluorescence probe concentrations and can provide information not obtainable from intensity variations alone. When polarized light is used to excite a molecule whose excitation dipole is oriented parallel to the polarization of the pulse, it is possible to separate the parallel and perpendicular components of the emitted fluorescence signal and to extract rotational diffusion constants [72].

As an example of how fluorescence lifetime can probe membrane microdomains, BAECs were labeled with either DiI C_{12} or DiI C_{18} and fluorescence lifetime (FL) measurements were taken on the cell membrane (Figure 3.11). Histograms suggest that the FL of DiI C_{18} is twofold greater than the FL of DiI C_{12}, suggesting that nonradiative energy transfer (dictated by k_{nr}) is greater for the short-chain–length dyes. Since the fluorophore of these two probes is identical, these differences in τ_F indicate that DiI C_{18} is probing a more restrictive environment than DiI C_{12}. The absolute value of the DiI C_{18} τ_F is consistent with its probing of gel-phase lipid domains (Figure 3.12) [95]. Further evidence of the association of DiI C_{18} with membrane domains is shown in Figure 3.5 (Plate 5), where DiI C_{18} staining was found to colocalize with FITC-labeled cholera toxin, a raft marker.

3.5 A Proposed Model of Lipid-Mediated Mechanotransduction of Shear Stress

We conclude this chapter by proposing a model for membrane mechanosensing and mechanotransduction. Elements of this model are supported by experimental observations and theoretical predictions. We propose that the membrane is naturally curved or undergoing curvature fluctuations and is flattened when subjected to tension by the integrated effects of fluid shear stress. This effect may be felt on the apical surface of the cell, where the glycocalyx transmits drag to the membrane directly, or where the shear stress impacts on the membrane directly in glycocalyx-deficient cells. Similarly, membrane perturbation by shear stress may occur on the basal membrane near focal adhesions, where membrane stresses are amplified. Flattening of the membrane has three effects that occur simultaneously. First, thinning due to reduction in curvature induces an increased lateral diffusion of lipids and membrane-bound proteins that may influence their interactions with downstream effectors. Second, thinning induces an enhanced hydrophobic mismatch that imposes an energy penalty on transmembrane proteins. These proteins change their conformations to reduce this energy penalty, thus altering their function. The confrontation change may lead to ion channel flux, G-protein–coupled receptor activation [79], or other changes. Third, membrane bending changes the lateral pressure profile in the membrane and increases the pressure in one leaflet while reducing it in the other [13]. This pressure profile can lead directly to changes in protein function and in the distribution of the (more mobile) cholesterol. Cholesterol translocation from one leaflet to another may also function to dissipate the stress in the membranes [73], which might explain the transient nature of shear-induced changes in lipid lateral diffusion [11].

Some elements of this model can be tested experimentally with optical methods based on time-correlated single photon counting. For example, shear-induced changes in lipid diffusion can be assessed with FRAP and FCS, while changes in the interaction with the membrane-domain–selective dyes such as DiI can be elucidated with fluorescence lifetime imaging microscopy (FLIM). More sophisticated experiments in which proteins are engineered with conformation-sensitive fluorescence moieties can be used to detect lipid-mediated conformational changes in

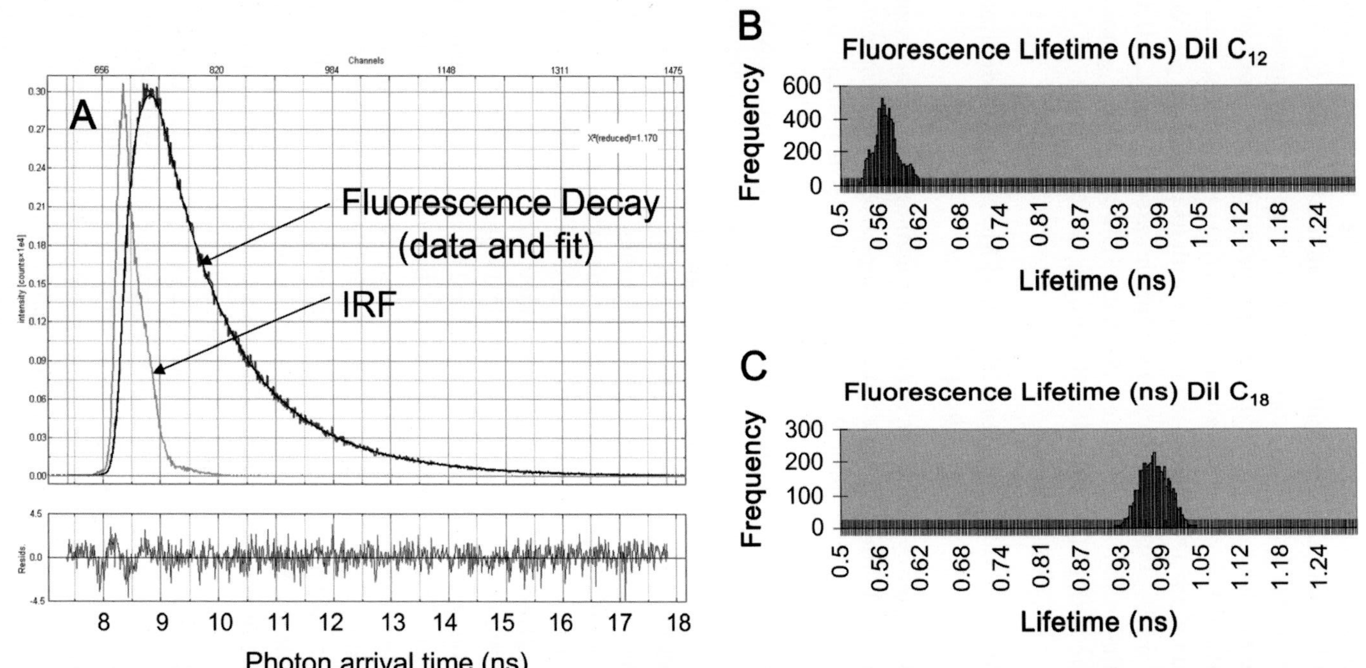

Figure 3.11. Fluorescence lifetime analysis of lipoid dyes. (A) Fluorescence decay and instrument response function for DiD fluorescence in ethanol. The decay histogram is fit with a double exponential (black line) to obtain the fluorescence lifetime. The x^2 value of the fit is 1.17. The full width at half maximum (FWHM) of the instrument response function is ~ 300 ps. (B, C) Fluorescence lifetime of (B) DiI C_{12} and (C) DiI C_{18} in bovine aortic endothelial cells.

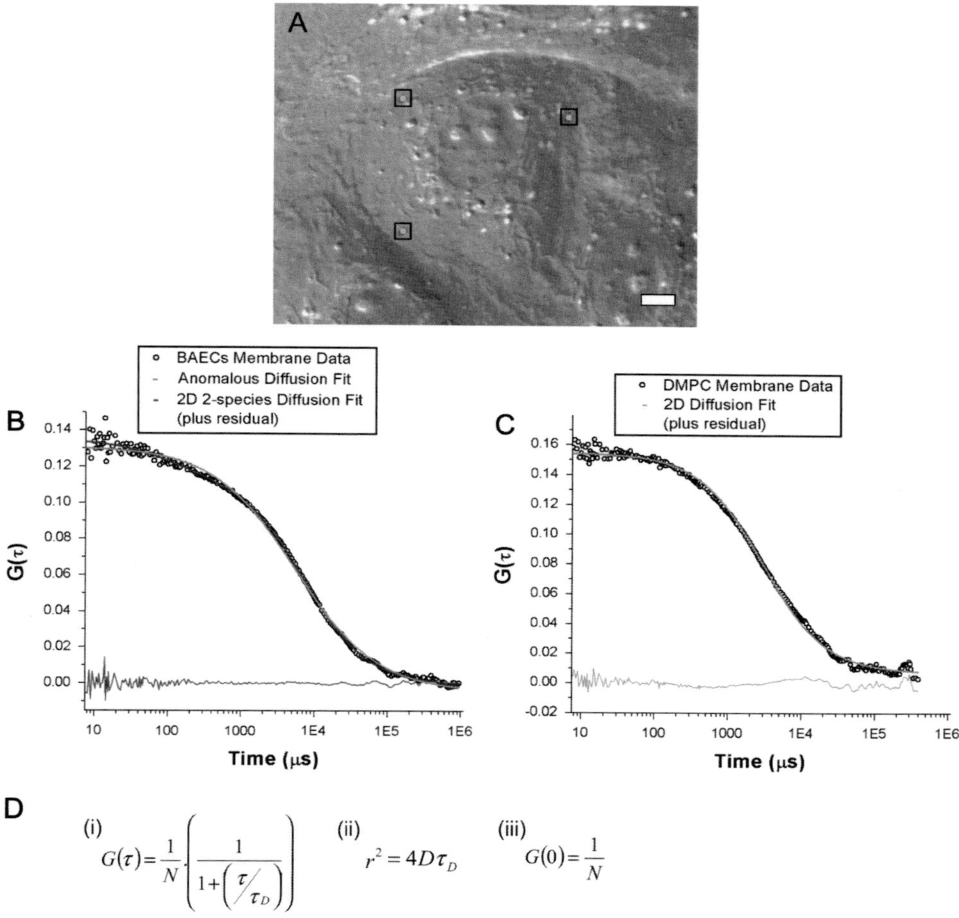

Figure 3.12. Use of fluorescence correlation spectroscopy for membrane mechanobiology. (A) Differential interference contrast images of endothelial cells. (B) Representative autocorrelation curves of fluorescence obtained from model membranes and DiI C_{18}. (C) Representative autocorrelation curves of fluorescence obtained from intact cellular membranes and DiI C_{18}. (D) Equations: (i) equation is fit to autocorrelation curve by adjusting diffusion time, (ii) equation or two-dimensional diffusion to obtain diffusion coefficient from characteristic diffusion time, (iii) *y*-intercept of autocorrelation gives absolute molecular concentration (*N*, number of molecules) in confocal volume.

integral membrane proteins. Other techniques such as fluorescence resonance energy transfer [117] based on spectral, intensity, or fluorescence lifetime changes may reveal time- and membrane-dependent alterations of protein conformation in response to fluid shear stress. While these fluorescence techniques require extensive control experiments in order to interpret molecular-level changes, the accessibility of these techniques makes them the methods of choice for single-molecule mechanobiology experiments [51]. In addition, these experimental observations, coupled with more refined mechanical models based on cellular ultrastructure, will help elucidate the fundamental underpinnings of mechanotransduction.

REFERENCES

[1] Andersen, O. S., et al. Ion channels as tools to monitor lipid bilayer-membrane protein interactions: Gramicidin channels as molecular force transducers. *Methods Enzymol.* **294**, 208–224 (1999).

[2] Ando, J., Ohtsuka, A., Korenaga, R., Kawamura, T., & Kamiya, A. Wall shear stress rather than shear rate regulates cytoplasmic Ca^{++} responses to flow in vascular endothelial cells. *Biochem. Biophys. Res. Commun.* **190**, 716–723 (1993).

[3] Arisaka, T., et al. Effects of shear stress on glycosaminoglycan synthesis in vascular endothelial cells. *Ann. N.Y. Acad. Sci.* **748**, 543–554 (1995).

[4] Axelrod, D. Lateral motion of membrane proteins and biological function. *J. Membr. Biol.* **75**, 1–10 (1983).

[5] Bao, X., Clark, C. B., & Frangos, J. A. Temporal gradient in shear-induced signaling pathway: Involvement of MAP kinase, c-fos, and connexin43. *Am. J. Physiol. Heart Circ. Physiol.* **278**, H1598–H1605 (2000).

[6] Bevan, J. A. & Siegel, G. Blood vessel wall matrix flow sensor: Evidence and speculation. *Blood Vessels* **28**, 552–556 (1991).

[7] Boesze-Battaglia, K. Isolation of membrane rafts and signaling complexes. *Methods Mol. Biol.* **332**, 169–179 (2006).

[8] Broday, D. M. Diffusion of clusters of transmembrane proteins as a model of focal adhesion remodeling. *Bull. Math. Biol.* **62**, 891–924 (2000).

[9] Brooks, A. R., Lelkes, P. I., & Rubanyi, G. M. Gene expression profiling of human aortic endothelial cells exposed to disturbed flow and steady laminar flow. *Physiolog. Genom.* **9**, 27–41 (2002).

[10] Butler, P. J., Norwich, G., Weinbaum, S., & Chien, S. Shear stress induces a time- and position-dependent increase in endothelial cell membrane fluidity. *Am. J. Physiol. Cell Physiol.* **280**, C962–C969 (2001).

[11] Butler, P. J., Tsou, T. C., Li, J. Y., Usami, S., & Chien, S. Rate sensitivity of shear-induced changes in the lateral diffusion of endothelial cell membrane lipids: A role for membrane perturbation in shear-induced MAPK activation 1. *FASEB J.* **16**, 216–218 (2002).

[12] Butler, P. J., Weinbaum, S., Chien, S., & Lemons, D. E. Endothelium-dependent, shear-induced vasodilation is rate-sensitive. *Microcirculation* **7**, 53–65 (2000).

[13] Cantor, R. S. Lipid composition and the lateral pressure profile in bilayers. *Biophys. J.* **76**, 2625–2639 (1999).

[14] Casey, P. J. Protein lipidation in cell signaling. *Science* **268**, 221–225 (1995).

[15] Charras, G. T. & Horton, M. A. Determination of cellular strains by combined atomic force microscopy and finite element modeling. *Biophys. J.* **83**, 858–879 (2002).

[16] Chen, B. P., et al. DNA microarray analysis of gene expression in endothelial cells in response to 24-h shear stress. *Physiol. Genomics* **7**, 55–63 (2001).

[17] Chien, S. Molecular basis of rheological modulation of endothelial functions: importance of stress direction. *Biorheology* **43**, 95–116 (2006).

[18] Chien, S., Li, S., & Shyy, Y. J. Effects of mechanical forces on signal transduction and gene expression in endothelial cells. *Hypertension* **31**, 162–169 (1998).

[19] Czarny, M. & Schnitzer, J. E. Neutral sphingomyelinase inhibitor scyphostatin prevents and ceramide mimics mechanotransduction in vascular endothelium. *Am. J. Physiol. Heart Circ. Physiol.* **287**, H1344–H1352 (2004).

[20] Dan, N. & Safran, S. A. Effect of lipid characteristics on the structure of transmembrane proteins. *Biophys. J.* **75**, 1410–1414 (1998).

[21] Davies, P. F. Flow-mediated endothelial mechanotransduction. *Physiolog. Rev.* **75**, 519–560 (1995).

[22] Davies, P. F., et al. Spatial relationships in early signaling events of flow-mediated endothelial mechanotransduction. [Review] [48 refs]. *Annual Review of Physiology* **59**, 527–549 (1997).

[23] Davies, P. F., Polacek, D. C., Handen, J. S., Helmke, B. P., & DePaola, N. A spatial approach to transcriptional profiling: Mechanotransduction and the focal origin of atherosclerosis. *Trends in Biotechnology* **17**, 347–351 (1999).

[24] Davies, P. F., Robotewskyj, A., & Griem, M. L. Quantitative studies of endothelial cell adhesion. Directional remodeling of focal adhesion sites in response to flow forces. *J. Clin. Invest.* **93**, 2031–2038 (1994).

[25] DePaola, N., Gimbrone, M. A., Jr., Davies, P. F., & Dewey, C. F., Jr. Vascular endothelium responds to fluid shear stress gradients. *Arterioscler. Thromb.* **12**, 1254–1257 (1992).

[26] Dietrich, C., Yang, B., Fujiwara, T., Kusumi, A., & Jacobson, K. Relationship of lipid rafts to transient confinement zones detected by single particle tracking. *Biophys. J.* **82**, 274–284 (2002).

[27] Elson, E. & Magde, D. Fluorescence correlation spectroscopy. I. Conceptual basis and theory. *Biopolymers* **13**, 1–27 (1974).

[28] Ferko, M. C., Bhatnagar, A., Garcia, M. B., & Butler, P. J. Finite-element stress analysis of a multicomponent model of sheared and focally-adhered endothelial cells. *Annals of Biomedical Engineering* **35** (2), 208–223 (2007).

[29] Ferko, M. C., Patterson, B. W., & Butler, P. J. High-resolution solid modeling of biological samples imaged with 3D fluorescence microscopy. *Microsc. Res. Tech.* **69**, 648–655 (2006).

[30] Florian, J. A., et al. Heparan sulfate proteoglycan is a mechanosensor on endothelial cells. *Circ. Res.* **93**, e136–e142 (2003).

[31] Frangos, J. A., Eskin, S. G., McIntire, L. V., & Ives, C. L. Flow effects on prostacyclin production by cultured human endothelial cells. *Science* **227**, 1477–1479 (1985).

[32] Frangos, J. A. & Gudi, S. Shear stress activates reconstituted G-proteins in the absence of protein receptors by modulating lipid bilayer fluidity. *FASEB J.* **11**(3), A521 (1997).

[33] Frangos, J. A., Huang, T. Y., & Clark, C. B. Steady shear and step changes in shear stimulate endothelium via independent mechanisms – superposition of transient and sustained nitric oxide production. *Biochem. Biophys. Res. Commun.* **224**, 660–665 (1996).

[34] Fujiwara, K., Masuda, M., Osawa, M., Kano, Y., & Katoh, K. Is PECAM-1 a mechanoresponsive molecule? *Cell Struct. Funct.* **26**, 11–17 (2001).

[35] Galbraith, C. G., Skalak, R., & Chien, S. Shear stress induces spatial reorganization of the endothelial cell cytoskeleton. *Cell Motil. Cytoskeleton* **40**, 317–330 (1998).

[36] Garcia-Cardena, G., Comander, J. I., Blackman, B. R., Anderson, K. R., & Gimbrone, M. A. Mechanosensitive endothelial gene expression profiles: scripts for the role of hemodynamics in atherogenesis? *Ann. N.Y. Acad. Sci.* **947**, 1–6 (2001).

[37] Garcia-Cardena, G., Oh, P., Liu, J., Schnitzer, J. E., & Sessa, W. C. Targeting of nitric oxide synthase to endothelial cell caveolae via palmitoylation: Implications for nitric oxide signaling. *Proc. Natl. Acad. Sci. U.S.A.* **93**, 6448–6453 (1996).

[38] Gaus, K., Le, L. S., Balasubramanian, N., & Schwartz, M. A. Integrin-mediated adhesion regulates membrane order. *J. Cell Biol.* **174**, 725–734 (2006).

[39] Gautam, M., Shen, Y., Thirkill, T. L., Douglas, G. C., & Barakat, A. I. Flow-activated chloride channels in vascular endothelium: Shear stress sensitivity, desensitization dynamics, and physiological implications. *J. Biol. Chem.* **281** (48), 36492–36500 (2006).

[40] Geiger, R. V., Berk, B. C., Alexander, R. W., & Nerem, R. M. Flow-induced calcium transients in single endothelial cells: Spatial and temporal analysis. *Am. J. Physiol.* **262**, C1411–C1417 (1992).

[41] Girard, P. R. & Nerem, R. M. Endothelial cell signaling and cytoskeletal changes in response to shear stress. *Front. Med. Biol. Eng.* **5**, 31–36 (1993).

[42] Girard, P. R. & Nerem, R. M. Shear stress modulates endothelial cell morphology and F-actin organization through the regulation of focal adhesion-associated proteins. *J. Cell Physiol.* **163**, 179–193 (1995).

[43] Gojova, A. & Barakat, A. I. Vascular endothelial wound closure under shear stress: Role of membrane fluidity and flow-sensitive ion channels. *J. Appl. Physiol.* **98**, 2355–2362 (2005).

[44] Gopalakrishna, P., Chaubey, S. K., Manogaran, P. S., & Pande, G. Modulation of alpha5beta1 integrin functions by the phospholipid and cholesterol contents of cell membranes. *J. Cell Biochem.* **77**, 517–528 (2000).

[45] Goulian, M., et al. Gramicidin channel kinetics under tension. *Biophys. J.* **74**, 328–337 (1998).

[46] Gov, N. S. Diffusion in curved fluid membranes. *Phys. Rev. E. Stat. Nonlin. Soft. Matter Phys.* **73**, 041918 (2006).

[47] Grabowski, E. F., Jaffe, E. A., & Weksler, B. B. Prostacyclin production by cultured endothelial cell monolayers exposed to step increases in shear stress. *Lab. Clin. Med.* **105**, 36–43 (1985).

[48] Green, J. M., et al. Role of cholesterol in formation and function of a signaling complex involving alphavbeta3, integrin-associated protein (CD47), and heterotrimeric G proteins. *J. Cell Biol.* **146**, 673–682 (1999).

[49] Gudi, S., Nolan, J. P., & Frangos, J. A. Modulation of GTPase activity of G proteins by fluid shear stress and phospholipid composition. *Proc. Natl. Acad. Sci. U.S.A.* **95**, 2515–2519 (1998).

[50] Gudi, S. R., Clark, C. B., & Frangos, J. A. Fluid flow rapidly activates G proteins in human endothelial cells. Involvement of G proteins in mechanochemical signal transduction. *Circ. Res.* **79**, 834–839 (1996).

[51] Gullapalli, R. R., Tabouillot, T., Mathura, R., Dangaria, J., & Butler, P. J. Integrated multimodal microscopy, time resolved fluorescence, and optical-trap rheometry: Toward single molecule mechanobiology. *J. Biomed. Opt.* **12**(1), 014012 (2007).

[52] Haidekker, M. A., L'Heureux, N., & Frangos, J. A. Fluid shear stress increases membrane fluidity in endothelial cells: A study with DCVJ fluorescence. *Am. J. Physiol. Heart Circ. Physiol.* **278**, H1401–H1406 (2000).

[53] Hamill, O. P. & Martinac, B. Molecular basis of mechanotransduction in living cells. *Physiol. Rev.* **81**, 685–740 (2001).

[54] Helmke, B. P., Goldman, R. D., & Davies, P. F. Rapid displacement of vimentin intermediate filaments in living endothelial cells exposed to flow. *Circ. Res.* **86**, 745–752 (2000).

[55] Honda, H. M., et al. A complex flow pattern of low shear stress and flow reversal promotes monocyte binding to endothelial cells. *Atherosclerosis* **158**, 385–390 (2001).

[56] Ingber, D. E. Mechanobiology and diseases of mechanotransduction. *Ann. Med.* **35**, 564–577 (2003).

[57] Ishida, T., Takahashi, M., Corson, M. A., & Berk, B. C. Fluid shear stress-mediated signal transduction: How do endothelial cells transduce mechanical force into biological responses? *Ann. N.Y. Acad. Sci.* **811**, 12–23 (1997).

[58] Jacobs, E. R., et al. Shear activated channels in cell-attached patches of cultured bovine aortic endothelial cells. *Pflugers Arch.* **431**, 129–131 (1995).

[59] Jacobson, K., Sheets, E. D., & Simson, R. Revisiting the fluid mosaic model of membranes. *Science* **268**, 1441–1442 (1995).
[60] Jalali, S., et al. Integrin-mediated mechanotransduction requires its dynamic interaction with specific extracellular matrix (ECM) ligands. *Proc. Natl. Acad. Sci. U.S.A.* **98**, 1042–1046 (2001).
[61] Jo, H., et al. Differential effect of shear stress on extracellular signal-regulated kinase and N-terminal Jun kinase in endothelial cells. Gi2- and Gbeta/gamma-dependent signaling pathways. *J. Biol. Chem.* **272**, 1395–1401 (1997).
[62] Kaley, G., Koller, A., Messina, E. J., & Wolin, M. S. Role of endothelium-derived vasoactive factors in the control of the microcirculation, in *Cardiovascular Significance of Endothelium-Derived Vasoactive Factors* (ed. Rubanyi, G. M.), 179–195 (Futura Publishing Co., Mount Kisco, NY, 1991).
[63] Kim, D. W., Langille, B. L., Wong, M. K., & Gotlieb, A. I. Patterns of endothelial microfilament distribution in the rabbit aorta in situ. *Circ. Res.* **64**, 21–31 (1989).
[64] Koller, A. & Bagi, Z. On the role of mechanosensitive mechanisms eliciting reactive hyperemia. *Am. J. Physiol. Heart Circ. Physiol.* **283**, H2250–H2259 (2002).
[65] Koller, A., Sun, D., & Kaley, G. Role of shear stress and endothelial prostaglandins in flow- and viscosity-induced dilation of arterioles in vitro. *Circ. Res.* **72**, 1276–1284 (1993).
[66] Korenaga, R., et al. Laminar flow stimulates ATP- and shear stress-dependent nitric oxide production in cultured bovine endothelial cells. *Biochem. Biophys. Res. Commun.* **198**, 213–219 (1994).
[67] Ku, D. N., Giddens, D. P., Zarins, C. K., & Glagov, S. Pulsatile flow and atherosclerosis in the human carotid bifurcation. Positive correlation between plaque location and low oscillating shear stress. *Arteriosclerosis* **5**, 293–302 (1985).
[68] Kuchan, M. J., Jo, H., & Frangos, J. A. Role of G proteins in shear stress-mediated nitric oxide production by endothelial cells. *Am. J. Physiol.* **267**, C753–C758 (1994).
[69] Kuo, L., Davis, M. J., & Chilian, W. M. Endothelium-dependent, flow-induced dilation of isolated coronary arterioles. *Am. J. Physiol.* **259**, H1063–H1070 (1990).
[70] Ladbrooke, B. D. & Chapman, D. Thermal analysis of lipids, proteins and biological membranes. A review and summary of some recent studies. *Chem. Phys. Lipids* **3**, 304–356 (1969).
[71] LaDisa, J. F., Jr., et al. Three-dimensional computational fluid dynamics modeling of alterations in coronary wall shear stress produced by stent implantation. *Ann. Biomed. Eng.* **31**, 972–980 (2003).
[72] Lakowicz, J. R. *Principles of Fluorescence Spectroscopy* (Springer, New York, 1999).
[73] Lange, Y. The rate of transmembrane movement of cholesterol in the human erythrocyte. *J. Biol. Chem.* **256**, 5321–5323 (1981).
[74] Lee, A. G. How lipids affect the activities of integral membrane proteins. *Biochim. Biophys. Acta* **1666**, 62–87 (2004).
[75] Li, S., et al. The role of the dynamics of focal adhesion kinase in the mechanotaxis of endothelial cells. *Proc. Natl. Acad. Sci. U.S.A.* **99**, 3546–3551 (2002).
[76] Li, S., et al. Fluid shear stress activation of focal adhesion kinase. Linking to mitogen-activated protein kinases. *J. Biol. Chem.* **272**, 30455–30462 (1997).
[77] Macdonald, A. G. The homeoviscous theory of adaptation applied to excitable membranes: A critical evaluation. *Biochim. Biophys. Acta Rev. Biomembr.* **1031**, 291–310 (1990).

[78] Magde, D., Elson, E., & Webb, W. W. Fluorescence correlation spectroscopy. II. An experimental realization. *Biopolymers* **13**, 29–61 (1974).
[79] Makino, A., et al. G protein-coupled receptors serve as mechanosensors for fluid shear stress in neutrophils. *Am. J. Physiol. Cell Physiol.* **290**, C1633–C1639 (2006).
[80] Mateo, C. R. & Douhal, A. A coupled proton-transfer and twisting-motion fluorescence probe for lipid bilayers. *Proc. Natl. Acad. Sci. U.S.A.* **95**, 7245–7250 (1998).
[81] McCormick, S. M., et al. DNA microarray reveals changes in gene expression of shear stressed human umbilical vein endothelial cells. *Proc. Natl. Acad. Sci. U.S.A.* **98**, 8955–8960 (2001).
[82] Mitchell, D. C., Lawrence, J. T., & Litman, B. J. Primary alcohols modulate the activation of the G protein-coupled receptor rhodopsin by a lipid-mediated mechanism. *J. Biol. Chem.* **271**, 19033–19036 (1996).
[83] Mochizuki, S., et al. Role of hyaluronic acid glycosaminoglycans in shear-induced endothelium-derived nitric oxide release. *Am. J. Physiol. Heart Circ. Physiol.* **285**, H722–H726 (2003).
[84] Mouritsen, O. G. & Bloom, M. Mattress model of lipid-protein interactions in membranes. *Biophys.* **46**, 141–153 (1984).
[85] Mouritsen, O. G. & Bloom, M. Models of lipid-protein interactions in membranes. *Ann. Rev. Biophys. & Biomolec. Struct.* **22**, 145–171 (1993).
[86] Muller, J. M., Chilian, W. M., & Davis, M. J. Integrin signaling transduces shear stress–dependent vasodilation of coronary arterioles. *Circ. Res.* **80**, 320–326 (1997).
[87] Nagel, T., Resnick, N., Dewey, C. F., Jr., & Gimbrone, M. A., Jr. Vascular endothelial cells respond to spatial gradients in fluid shear stress by enhanced activation of transcription factors. *Arterioscler. Thromb. Vasc. Biol.* **19**, 1825–1834 (1999).
[88] Naruse, K. & Sokabe, M. Involvement of stretch-activated ion channels in Ca2+ mobilization to mechanical stretch in endothelial cells. *Am. J. Physiol.* **264**, C1037–C1044 (1993).
[89] Nerem, R. M. Role of mechanics in vascular tissue engineering. *Biorheology* **40**, 281–287 (2003).
[90] Nerem, R. M., Levesque, M. J., & Cornhill, J. F. Vascular endothelial morphology as an indicator of the pattern of blood flow. *J. Biomech. Eng.* **103**, 172–176 (1981).
[91] Nielsen, C., Goulian, M., & Andersen, O. S. Energetics of inclusion-induced bilayer deformations. *Biophys. J.* **74**, 1966–1983 (1998).
[92] Oh, P. & Schnitzer, J. E. Segregation of heterotrimeric G proteins in cell surface microdomains. G(q) binds caveolin to concentrate in caveolae, whereas G(i) and G(s) target lipid rafts by default. *Mol. Biol. Cell* **12**, 685–698 (2001).
[93] Ohno, M., Gibbons, G. H., Dzau, V. J., & Cooke, J. P. Shear stress elevates endothelial cGMP. Role of a potassium channel and G protein coupling. *Circulation* **88**, 193–197 (1993).
[94] Olesen, S. P., Clapham, D. E., & Davies, P. F. Haemodynamic shear stress activates a K+ current in vascular endothelial cells. *Nature* **331**, 168–170 (1988).
[95] Packard, B. S. & Wolf, D. E. Fluorescence lifetimes of carbocyanine lipid analogues in phospholipid bilayers. *Biochemistry* **24**, 5176–5181 (1985).
[96] Pries, A. R., Reglin, B., & Secomb, T. W. Structural response of microcirculatory networks to changes in demand: Information transfer by shear stress. *Am. J. Physiol. Heart Circ. Physiol.* **284**, H2204–H2212 (2003).

[97] Rizzo, V., Sung, A., Oh, P., & Schnitzer, J. E. Rapid mechanotransduction in situ at the luminal cell surface of vascular endothelium and its caveolae. *J. Biol. Chem.* **273**, 26323–26329 (1998).

[98] Saffman, P. G. & Delbruck, M. Brownian motion in biological membranes. *Proc. Natl. Acad. Sci. U.S.A.* **72**, 3111–3113 (1975).

[99] Sato, M., Nagayama, K., Kataoka, N., Sasaki, M., & Hane, K. Local mechanical properties measured by atomic force microscopy for cultured bovine endothelial cells exposed to shear stress. *J. Biomech.* **33**, 127–135 (2000).

[100] Sato, M., Theret, D. P., Wheeler, L. T., Ohshima, N., & Nerem, R. M. Application of the micropipette technique to the measurement of cultured porcine aortic endothelial cell viscoelastic properties. *J. Biomechan. Eng.* **112**, 263–268 (1990).

[101] Satoh, A., Toida, T., Yoshida, K., Kojima, K., & Matsumoto, I. New role of glycosaminoglycans on the plasma membrane proposed by their interaction with phosphatidylcholine. *FEBS Lett.* **477**, 249–252 (2000).

[102] Secomb, T. W., Hsu, R., & Pries, A. R. Effect of the endothelial surface layer on transmission of fluid shear stress to endothelial cells. *Biorheology* **38**, 143–150 (2001).

[103] Secomb, T. W. & Pries, A. R. Information transfer in microvascular networks. *Microcirculation* **9**, 377–387 (2002).

[104] Shinitzky, M. The lipid regulation of receptor functions. *Biomembranes and Receptor Mechanisms* **7**, 135–141 (1987).

[105] Shyy, Y. J., Hsieh, H. J., Usami, S., & Chien, S. Fluid shear stress induces a biphasic response of human monocyte chemotactic protein 1 gene expression in vascular endothelium. *Proc. Natl. Acad. Sci. U.S.A.* **91**, 4678–4682 (1994).

[106] Siegel, G., Malmsten, M., & Lindman, B. Flow sensing at the endothelium-blood interface. *Colloids and Surfaces A—Physicochemical and Engineering Aspects* **138**, 345–351 (1998).

[107] Sigurdson, W. J., Sachs, F., & Diamond, S. L. Mechanical perturbation of cultured human endothelial cells causes rapid increases of intracellular calcium. *Am. J. Physiol. Heart Circ. Physiol.* **264**, H1745–H1752 (1993).

[108] Simionescu, M., Simionescu, N., & Palade, G. E. Segmental differentiations of cell junctions in the vascular endothelium: The microvasculature. *J. Cell Biol.* **67**, 863–885 (1975).

[109] Singer, S. J. & Nicolson, G. L. The fluid mosaic model of the structure of cell membranes. *Science* **175**, 720–731 (1972).

[110] Soubias, O., Jolibois, F., Reat, V., & Milon, A. Understanding sterol-membrane interactions, Part II: Complete 1H and 13C assignments by solid-state NMR spectroscopy and determination of the hydrogen-bonding partners of cholesterol in a lipid bilayer. *Chemistry* **10**, 6005–6014 (2004).

[111] Sowa, G., Pypaert, M., & Sessa, W. C. Distinction between signaling mechanisms in lipid rafts vs. caveolae. *Proc. Natl. Acad. Sci. U.S.A.* **98**, 14072–14077 (2001).

[112] Squire, J. M., et al. Quasi-periodic substructure in the microvessel endothelial glycocalyx: A possible explanation for molecular filtering? *J. Struct. Biol.* **136**, 239–255 (2001).

[113] Stone, P. H., et al. Effect of endothelial shear stress on the progression of coronary artery disease, vascular remodeling, and in-stent restenosis in humans: In vivo 6-month follow-up study. *Circulation* **108**, 438–444 (2003).

[114] Tardy, Y., Resnick, N., Nagel, T., Gimbrone, M. A., Jr., & Dewey, C. F., Jr. Shear stress gradients remodel endothelial monolayers in vitro via a cell proliferation-migration-loss cycle. *Arterioscler. Thromb. Vasc. Biol.* **17**, 3102–3106 (1997).

[115] Trudell, J. R. Role of membrane fluidity in anesthetic action in, *Drug and Anesthetic Effects on Membrane Structure and Function* (eds. Aloia, R. C., Curtain, C. C., & Gordon, L. M.), 1–14 (Wiley-Liss, Inc., New York, 1991).

[116] Tzima, E., et al. A mechanosensory complex that mediates the endothelial cell response to fluid shear stress. *Nature* **437**, 426–431 (2005).

[117] Wang, Y., et al. Visualizing the mechanical activation of Src 112. *Nature* **434**, 1040–1045 (2005).

[118] Weinbaum, S. & Chien, S. Lipid transport aspects of atherogenesis. *J. Biomech. Eng.* **115**, 602–610 (1993).

[119] Weinbaum, S., Zhang, X., Han, Y., Vink, H., & Cowin, S. C. Mechanotransduction and flow across the endothelial glycocalyx. *Proc. Natl. Acad. Sci. U.S.A.* **100**, 7988–7995 (2003).

[120] Wentzel, J. J., et al. Shear stress, vascular remodeling and neointimal formation. *J. Biomech.* **36**, 681–688 (2003).

[121] Wolf, A. A., Fujinaga, Y., & Lencer, W. I. Uncoupling of the cholera toxin-G(M1) ganglioside receptor complex from endocytosis, retrograde Golgi trafficking, and downstream signal transduction by depletion of membrane cholesterol. *J. Biol. Chem.* **277**, 16249–16256 (2002).

[122] Zaidel-Bar, R., Kam, Z., & Geiger, B. Polarized downregulation of the paxillin-p130CAS-Rac1 pathway induced by shear flow. *J. Cell Sci.* **118**, 3997–4007 (2005).

4

Mechanotransduction by Membrane-Mediated Activation of G-Protein Coupled Receptors and G-Proteins

Yan-Liang Zhang, John A. Frangos, and Mirianas Chachisvilis

4.1 Introduction

Atherosclerosis, the leading cause of death in the Western world and nearly the leading cause in other developing countries, is associated with systemic risk factors such as hypertension, smoking, diabetes mellitus, and hyperlipidemia (Malek et al., 1999a). Nonetheless, it is clearly a focal disease with atherosclerotic lesions forming in the coronary arteries, major branches of the aortic arch, and the abdominal aorta. Plaque formation preferentially involves the outer walls of vessel bifurcations and points of blood flow recirculation, where hemodynamic shear stress is weaker than in unaffected areas. Detailed analysis of fluid mechanics in these areas revealed a strong spatial correlation between endothelial dysfunction/plaque formation and low mean shear stress and oscillatory flow with recirculation (Nerem, 1992; Malek et al., 1999a). It has been suggested that low mean shear stress and nonlaminar flow stimulate the formation of an atherogenic phenotype (Gibson et al., 1993; Ku et al., 1985) while arterial level shear stress ($\geqslant$15 dynes/cm^2) stimulates atheroprotective gene expression profile and cellular responses that are essential for normal endothelial function (Harrison, 2005; Cunningham and Gotlieb, 2005), implying that the maintenance of physiological, laminar shear stress on endothelial cells is crucial for normal vascular function (Traub and Berk, 1998). The development of atherogenesis can be studied at various levels; however, the origins of the disease have been linked to mechanosensing by a number of studies (Ku et al., 1985; Glagov et al., 1988; Svindland, 1983; Debakey et al., 1985; White et al., 2003), suggesting that investigations at the molecular level are required for further understanding of the atherogenesis.

Mechanical forces in the form of strain, pressure, and fluid shear stress are sensed by cells through so far unidentified mechanoreceptors that are coupled to intracellular signaling pathways (Tzima et al., 2005). Proteomic analysis of bovine aortic endothelial cells (BAEC) in response to shear stress at 15 dynes/cm^2, using isotope-coded affinity and tags and cICAT labeling coupled with liquid chromotography–mass spectrometry/mass spectrometry (LC-MS/MS), found 142, 213, and 186 proteins that were up- or down-regulated by more than twofold after 10 min, 3 h, and 6 h of shear stress, respectively, many of which encompass multiple signaling pathways including integrins, G-protein–coupled receptors, glutamate receptors, PI3K/AKT, apoptosis, Notch, and cAMP-mediated signaling pathways (Wang et al., 2007). In vascular endothelium, shear stress plays an important role in regulating functions such as vasodilation and structure, such as remodeling of the blood vesseles (Frangos et al., 1999; Lehoux et al., 2006). One of the most important aspects of this regulation of the function and structure of the endothelium is the control of nitric oxide (NO) productivity by endothelial nitric oxide synthase (eNOS), whose activity can be closely modulated by shear stress (Fisslthaler et al., 2000).

A major challenge is the identification of the mechanisms by which the mechanical forces are converted into a sequence of intracellular biochemical signals in endothelial cells. Since mechanical perturbation does not involve traditional receptor–ligand interactions, the identification of mechanosensing molecules has been difficult. Studies of mechanosensitivity concentrate either on the membrane bilayer or its associated cytoskeleton (CSK). A central question is, what is the molecular structure of the mechanosensitive center, the exact location, and activation dynamics?

Over the last few decades a number of mechanosensitive biological molecules have been identified; these include mechanically gated channels (Liedtke et al., 2003; Sackin, 1995; Maroto et al., 2005; Patel et al., 2001; Perozo et al., 2002a, 2002b; Martinac and Hamill, 2002; Sukharev et al., 1994), mechanosensitive receptors (Paoletti and Ascher, 1994; Nakamura and Strittmatter, 1996), G-protein coupled receptors (GPCRs) (Chachisvilis et al., 2006; Zou et al., 2004; Makino et al., 2006), G-proteins (Hsieh et al., 1993; Gudi et al., 1996, 1998), neurotransmitters (Chen and Grinnell, 1997), mechanosensitive enzymes (Lehtonen and Kinnunen, 1995; Matsumoto et al., 1995), and CSK (Ingber, 1997). Within seconds following mechanical perturbation a number of processes begin, such as G-protein activation, release of secondary chemical messengers (calcium, cyclic AMP, etc.), protein phosphorylation (Sessa, 2004), and remodeling of cell–cell adhesions. Since each cell may express multiple mechanotransducers, it is difficult to discern the primary mechanochemical step leading to mechanical sensing.

The membranes of mammalian cells are closely integrated with the CSK. Numerous studies have demonstrated that in many cases mechanochemical signal conversion originates at the cell membrane (Sukharev et al., 1994; Gudi et al., 1998; Knudsen and Frangos, 1997). Furthermore, kinetic studies demonstrate that one of the primary events is the activation of G-proteins within 1 s after mechanical perturbation (Gudi et al., 1996). Experiments on purified heterotrimeric G-proteins reconstituted into liposomes have shown that even in the absence of any other potential mechanosensors such as GPCRs, membrane channels, integrins, glycocalyx, or CSK, the G-proteins are activated by fluid shear stress (Gudi et al., 1998). These experiments suggest that the lipid bilayer membrane itself plays a major, possibly primary, role in mediating mechanochemical signal transduction; this would be expected considering the importance of lipid–protein interactions for the function

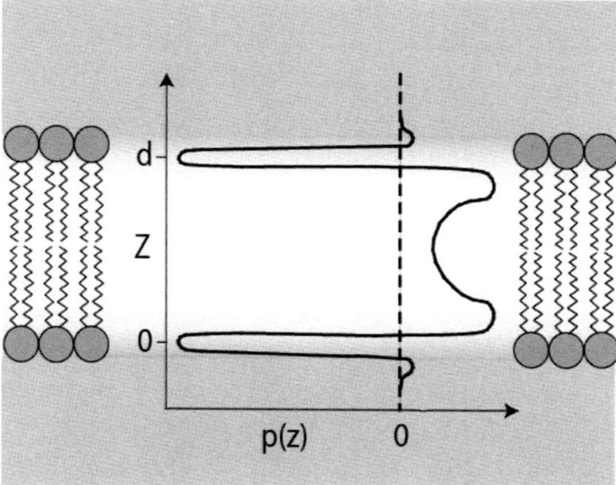

Figure 4.1. Generic transmembrane lateral pressure profile (adapted from Cantor, 1997).

of membrane proteins (Bowie, 2005; McMahon and Gallop, 2005; Lee, 2004). Further support comes from studies with membrane fluidity modulators (e.g., benzyl alcohol) that show receptor-independent activation of G-proteins by an increase in membrane fluidity (Gudi et al., 1998). The association of both GPCRs and G-proteins at the plasma membrane suggests that lipid–protein interactions are crucial to their function, which makes them sensitive to the changes in their lipid environment.

The question we ask is, which membrane property (dynamic or static) is highly sensitive to mechanically induced (possibly small) changes in the bilayer structure and has a clear link to the function of membrane-associated proteins (membrane receptors, enzymes, channels, etc.)? The answer to this question clearly depends on the specific nature of a reactive event that is being modulated by that physical property of the bilayer. In the following we describe a number of expected changes in bilayer properties and discuss their potential role in mechanochemical transduction.

4.2 Lipid Bilayer Membrane as a Mediator of Mechanochemical Response

4.2.1 Structure and Mechanical Properties

The lipid bilayer is a unique composite dynamic structure held together by an exact balance of hydrophobic and lipid–lipid interactions (Ben-Shaul, 1995). Due to the high degree of spatial heterogeneity, the interior of the lipid bilayer is characterized by strong lateral pressure forces that vary greatly with the depth within the bilayer (Cantor, 1997; Gullingsrud and Schulten, 2004). As the bilayer forms, the hydrophilic polar headgroups are pushed together by surface tension to prevent exposure of the hydrophobic lipid tails to water while the entropic repulsion of lipid tails (and headgroups) gives rise to a counteracting force. The transmembrane pressure profile shows (see Figure 4.1) that the strongest negative lateral forces arise in the headgroup area, due to the strong hydrophobic interactions at the interface between the hydrophilic and hydrophobic parts of the bilayer; a quantitative estimation of the operating pressures can be obtained by assuming that surface tension in this region is ~52 dyne/cm, a typical value for a water/oil interface. The resulting pressures inside the larger hydrophobic core of the bilayer reach ~300 atm, while in the thin (~5 Å) headgroup area, where water comes into contact with hydrophobic tails of the lipids, lateral forces of up to ~1000 atm can be expected. Depending on the size of the embedded protein, this translates into forces of up to ~100 pN, which is close in magnitude to the forces involved in the typical ligand–receptor interaction. Molecular dynamics calculations show that when a membrane is stretched by an external force, the steric repulsive interactions are reduced, whereas interfacial tension is increased because of increased exposure of the hydrophobic tails of the lipids to water, leading to an overall change in lateral pressure profile (Gullingsrud and Schulten, 2004). It has been suggested that these changes in intra-bilayer pressure profiles are responsible for the activation of certain membrane proteins such as gated membrane channels (Sukharev et al., 1994; Cantor, 1997; Perozo et al., 2002b; Gullingsrud and Schulten, 2004; Wiggins and Phillips, 2004).

A change in bilayer thickness is another factor that affects both the lateral pressure profile and the so-called hydrophobic matching (Lee, 2004) between the hydrophobic parts of the lipid bilayer and the membrane protein. If the hydrophobic thickness of a membrane protein is smaller in one of the conformational states, then a decrease in the bilayer thickness would shift the equilibrium toward that conformation (Lee, 2004). It has been shown that subnanometer changes in bilayer thickness can reverse the response polarity of a *gramicidin* A ion channel from a stretch-activated to a stretch-inactivated state (Martinac and Hamill, 2002). There are many other examples suggesting that hydrophobic membrane thickness controls the physiological functioning of membrane-bound proteins such as enzymes and receptors. Enzymes like cytochrome c oxidase (Montecucco et al., 1982), Ca^{2+}-ATPase (Johannsson et al., 1981a; Lee, 1998), or (Na+-K+) ATPase (Johannsson et al., 1981b) function optimally when embedded into bilayers of a given thickness while the neural activity of the acetylcholine receptor has also been shown to depend on the membrane thickness (Criado et al., 1984). Experimentally, the bilayer thickness can be changed by either changing the lipid composition or by stretching the membrane mechanically. A recent theoretical study showed that when the lipid bilayer is exposed to flow shear stress, the orientational order of the lipids increases substantialy (Blood et al., 2005); not only is this expected to result in significant changes in the lateral pressure profile, but lateral diffusion rates are also likely to be affected since flow induces noticeable changes in bilayer membrane density profiles (Blood et al., 2005).

Increasing the membrane tension beyond 0.5 dyn/cm leads to a linear regime where an increase in tension results in linear area expansion (increase of area per molecule) (Needham and Nunn, 1990; Evans and Rawicz, 1990). The maximum area expansion of the typical bilayer is limited to 3–5%, beyond which lysis occurs (Evans and Needham, 1987). Accompanying the lateral stretching, the thickness of the bilayer decreases due to weak coupling between the area and the volume. This coupling is weak because lipid molecules can expand laterally and shrink vertically while keeping the bilayer volume constant. This is also observed during the lipid transition from the gel to liquid crystalline phase when the bilayer thickness decreases by ~ 20% and the area per headgroup increases by ~ 25% while the volume increases by only 3 to 4% (Nagle, 1980; Janiak et al., 1976; Jahnig, 1996). Moreover, when the bilayer is stretched the lipids become more loosely packed, while the lipid tails become more disordered (Gullingsrud and Schulten, 2004). The latter disordering effect is similar to that of raising the temperature, which also results in bilayer thining (Nagle, 1980). Thus one may expect the following change in the function of the strained bilayer: increase in fluidity of the membrane due to a moderate increase in the free volume (Cohen and Turnbull, 1959) (despite the small changes in the total volume of the bilayer) or due to a decrease in the thickness and lateral viscosity of the bilayer (Saffman and Delbruck, 1975), depending on the theoretical description used.

4.2.2 Lateral Fluidity

The lateral fluidity (or inverse viscosity) of the lipid bilayer membrane is considered to be one of the key functions of the cell membrane by enabling lateral diffusion, interactions, and signaling by various membrane-associated signalin-proteins. Modulation of the lateral diffusion rates or membrane fluidity by mechanical interaction could lead to a mechanochemical response. The functional significance of modulating the membrane free volume by mechanical perturbation is supported by a number of studies. For example, studies using membrane fluidity modulators such as benzyl alcohol (free-volume maker) or cholesterol (free-volume reducer) showed, respectively, increases and decreases in the activities of G-proteins reconstituted into phospholipid vesicles (Gudi et al., 1998). Other studies have demonstrated that very low added hydrostatic pressure (just a fraction of atmospheric) can stimulate significant responses in vascular epithelial cells and vascular smooth muscle cells (Salwen et al., 1998; Schwartz et al., 1999; Hishikawa et al., 1994). It is at least in principle possible that low hydrostatic pressures can modulate signal transduction by changing the membrane free volume or lateral membrane viscosity. Indeed, the hydrostatic pressure experiments on the lateral diffusion in lipid bilayer membranes suggest that lateral diffusion is slowed down at higher pressures (as determined by the excimer formation technique) (Muller and Galla, 1983), consistent with the free-volume theory. Quantifying the effects of mechanical perturbation on the diffusive dynamics in lipid bilayer membranes in terms of relevant structural changes is one of the important questions that have not been addressed.

Currently there are only a few experimental studies that have attempted to measure the lateral fluidity of the membrane as a function of membrane mechanical perturbation. In some studies enhanced rates of lateral diffusion were reported as deduced from experiments using excimer-forming phospholipids analogs in artificial bilayers (Lehtonen and Kinnunen, 1994; Soderlund et al., 2003) or from fluorescence anisotropy measurements in liposomes and plant protoplasts (Borochov and Borochov, 1979) under osmotic stress conditions; however, these results are inconclusive since osmotic pressure–inducing agents such as high molecular weight poly (ethylene glycol) (PEG) or sucrose can cause dehydration of the lipid bilayer due to exclusion effects (Rand et al., 2000; Parsegian et al., 2000; Arnold et al., 1990; Ito et al., 1989; Luzardo et al., 2000). Fluorescence recovery after photobleaching (FRAP) and fluorescence anisotropy measurements showed that the kosmotropic effect of dehydration by PEG leads to reduced rates of lateral diffusion in lipid bilayers due to increased lateral packing density of the lipids (Yamazaki et al., 1989; Mccown et al., 1981; Lehtonen and Kinnunen, 1994). In another study, a confocal FRAP technique was used to study the diffusion of a lipid probe DiI in the membrane of the cultured endothelial cells exposed to fluid shear stress (Butler et al., 2001); a small, time- and position-dependent increase in probe diffusion coefficients was detected; the results were interpreted as the increase in membrane fluidity in the upstream part of the cell. Despite the advantages of this apporach, it is not clear what effect the deformation of the cell and its membrane under fluid shear stress could have on the observed fluorescence recovery rates obtained using the confocal FRAP technique. A different approach was recently undertaken where the molecular rotors were used to probe the membrane fluidity of the endothelial cells under shear stress (Haidekker et al., 2000, 2001). Molecular rotors exhibit a viscosity-dependent fluorescence quantum yield due to a radiationless deactivation process via intramolecular rotational motion (intramolecular isomerization) (Loutfy, 1986; Chachisvilis et al., 1996). When such molecular rotors are incorporated into the lipid bilayer membrane by covalently attaching hydrocarbon tails, they become membrane fluidity probes that exhibit strong fluorescence dependence on membrane viscosity (Kung and Reed, 1986; Haidekker et al., 2001). Recent experiments indicate enhanced fluidity in the membranes of live cells under shear stress or in the presence of benzyl alcohol (Haidekker et al., 2000, 2001). In Figure 4.2 we show the lateral diffusion of a 1,1′-dioctadecyl-3,3,3′,3′-tetramethylindodicarbocyanine perchlorate (DiD) lipid probe molecule in the membrane of human umbilical vein endothelial cells (HUVECs) under fluid shear stress. The data, obtained using the technique of fluorescence correlation spectroscopy (FCS) (2001), show that the lateral diffusion constant is reduced by a factor of ~2 upon application of shear stress (notice that in the case of lateral diffusion, the diffusion coefficient is inversely proportional to the diffusion time, which characterizes the decay of the fluorescence correlation function $G(t)$ according to the well-known expressions; see Krichevsky and Bonnet (2002). This appears to be in contrast to earlier experimental work that suggested that an increase in fluidity should be expected. However, first, it is not at all obvious that the membrane free volume increases upon mechanical tension. Second, there are other factors involved, such as the thinning of the bilayer, the disordering and intertwining of lipid tails, and changes in the lateral pressure profile. We have estimated the importance of the membrane free volume in our control experiments by measuring the effects of temperature and benzyl alcohol on the lateral diffusion of the lipid probe DiD; note that both higher temperature and benzyl alcohol increase the membrane free volume and fluidity. As expected, our data in Figure 4.3 clearly show that a temperature increase of 17°C leads to ~2.3 times faster diffusion (which is equivalent to ~2.17 times

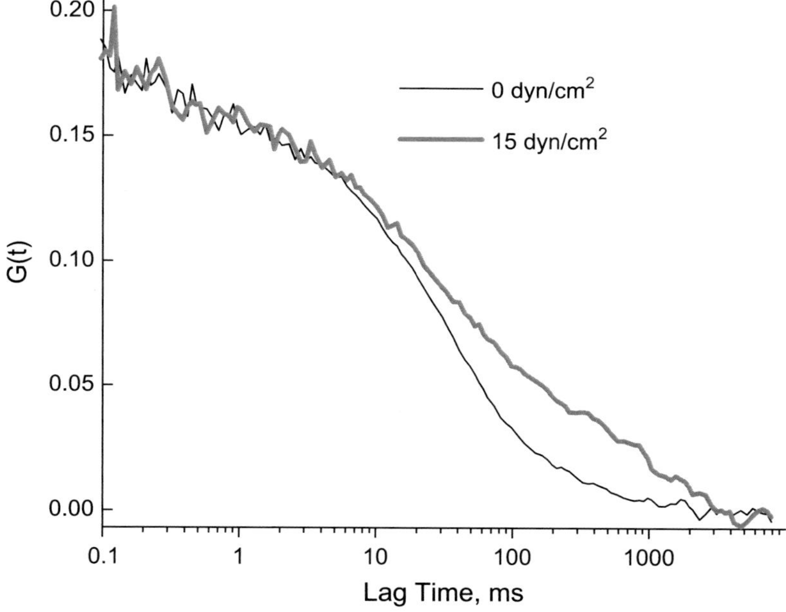

Figure 4.2. Fluorescence correlation traces from a membrane of endothelial cell (HUVEC) subjected to fluid shear stress.

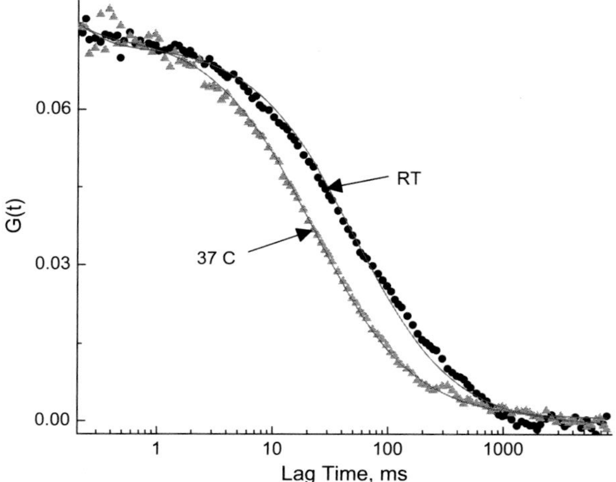

Figure 4.3. Effect of temperature on diffusion of DiD lipid probe in the membrane of HUVEC cell as measured using FCS technique.

higher fluidity), while Figure 4.4 shows the expected decrease in DiD diffusion times in supported bilayers at higher concentrations of benzyl alcohol, implying higher membrane fluidity, too. Thus it seems that membrane tension effect on lateral diffusion might include more factors in addition to the membrane-free volume change. It must be pointed out that the FCS technique directly measures the diffusion of single lipid probe molecules on a time-scale ranging from microseconds to seconds, while molecular rotor techniques rely on processes happening on very different time-scales – some molecular

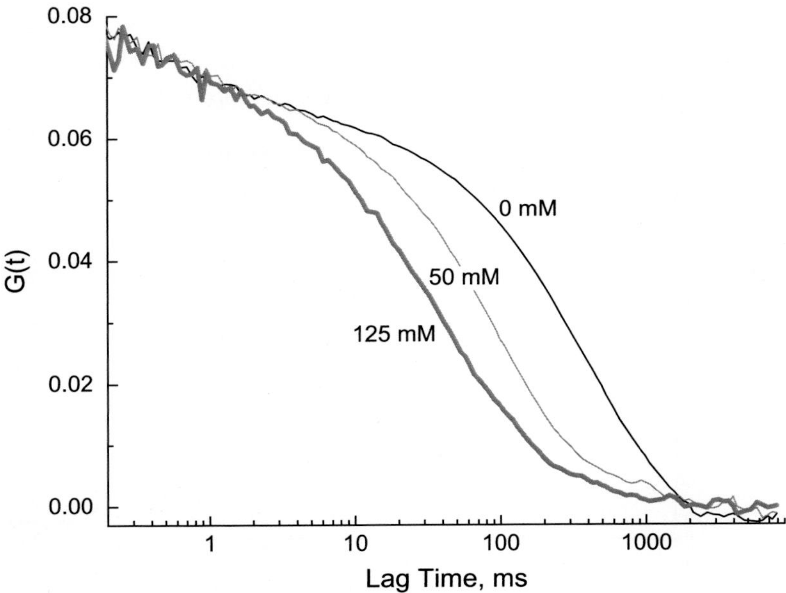

Figure 4.4. Effect of membrane fluidity enhancer – benzyl alcohol on diffusion of DiD lipid probe in DOPC supported bilayer as measured using FCS technique.

rotors can deactivate on the picosecond (1 ps $= 10^{-12}$ s) time-scale, excimer formation happens on a 100-ps time-scale (Novikov and Visser, 2001) while lateral diffusion measured by FCS and FRAP operates on a microsecond to second time-scale. Thus another question that needs to be addressed is the significance of the time-scale on which the change in membrane property is measured and the time-scale of the primary reactive step in mechanochemical transduction. This question is important because some of the phenomenological parameters used to describe reactive or diffusive dynamics in the lipid bilayer environment (or any condensed phase environment), such as viscosity, microscopic friction, and free volume, can be dependent on the time-scale and the nature (e.g., spatial extent) of the reactive event (Grote and Hynes, 1980, 1981; Baskin et al., 1998). As an example, the microscopic friction experienced during the protein isomerization reaction leading to the activation of a GPCR or an ion channel could depend on the time-scale of this reaction and the type of reactive motion involved [in the case of rhodopsin, a GPCR, the activation time-scale spans from 200 femtoseconds (1 fs $= 10^{-15}$ s) to milliseconds (Okada et al., 2001)]. In this case, a description of the activation process using Kramers reaction rate theory (Hanggi et al., 1990) may not be straightforward; that is, the activation rate may not correlate with the microscopic friction or lateral viscosity measured on a different time-scale, for example, using FRAP, FCS, or other techniques. Thus, in order to establish a link between the change in a given lipid bilayer property and mechanochemical activation, a number of spectroscopic techniques have to be used to measure mechanically induced changes in the membrane properties at different time-scales.

4.2.3 Free Volume

The microviscosity of the membrane can be quantified using a concept of the membrane free volume v_f (Cohen and Turnbull, 1959). It characterizes the differences in molecular volumes of the membrane at a given temperature with respect to molecular volume at absolute zero $v_f = v - v_0$. The membrane microviscosity η can be directly related to the membrane free volume according to the expression (Doolittle, 1951) $\eta = A\, e^{B\frac{v_0}{v_f}}$, where $A$and B are constants. In the case of orientational motion the membrane free volume defines the relative volume available for the reorientational motion of the probe molecule. Molecular rotors exhibit viscosity-dependent fluorescence quantum yield due to a nonradiative deactivation channel. The nonradiative deactivation of molecular rotors is mediated by isomerization (intramolecular rotation of a part of the moelcule), therefore higher viscosity tends to inhibit this rotation, leading to a higher fluorescence quantum yield. The changes in the quantum yield in this case are mostly determined by the nonradiative deactivation rate k_{nonrad}, which can be related to the free volume v_f as

$$k_{nonrad} = k_o e^{-\beta\frac{v_0}{v_f}},$$

where k_0 and β are constants (Kung and Reed, 1986). By measuring the quantum yield dependence of a molecular rotor incorporated into a membrane, one can in principle determine the membrane free volume changes. A similar concept is also used in fluorescence anisotropy measurements that detect the orientational motion of the whole molecule. For example, the rotational motion of a DPH membrane probe has been used to assess membrane free volume in studying the effects of lipid composition (polyunsaturation) on membrane protein function (Litman and Mitchell, 1996). However, it has been shown that when anisotropy experiments are done on endothelial cells under fluid shear stress, the cell membrane deformation can produce artifacts that can be misinterpreted as changes in membrane fluidity (Abdul-Rahim and Bouchy, 1998). The molecular rotor approach is free from this problem.

We have recently performed fluid shear stress experiments with a lipid molecular rotor 9-(2-carboxy-farnesylester-2-cyanovinyl-julolidine) (FCVJ). The fluorescence intensity of FCVJ incorporated into membranes of confluent layers of HUVEC cells exposed to variable fluid shear stress has been measured. Data indicate that, upon application of a shear stress ramp, FCVJ fluorescence intensity proportionally decreases, suggesting that membrane tension is leading to a larger membrane-free volume (higher fluidity) (White and Frangos, 2007).

In order to test the lipid molecular rotor approach we have used a new molecular rotor SC1-20A (see Figure 4.5), which shows excellent partitioning into the lipid bilayer membrane; this molecule is similar to molecular rotors like DCVJ, CCVJ, and FCVJ, which were shown to be sensitive to fluid flow–induced shear stress (Haidekker et al., 2000, 2002). Our unpublished experiments on SC1-20A incorporated into a membrane of a giant unilamellar vesicle (GUV) showed that osmotically induced membrane stretch has an effect on fluorescence decay kinetics, suggesting

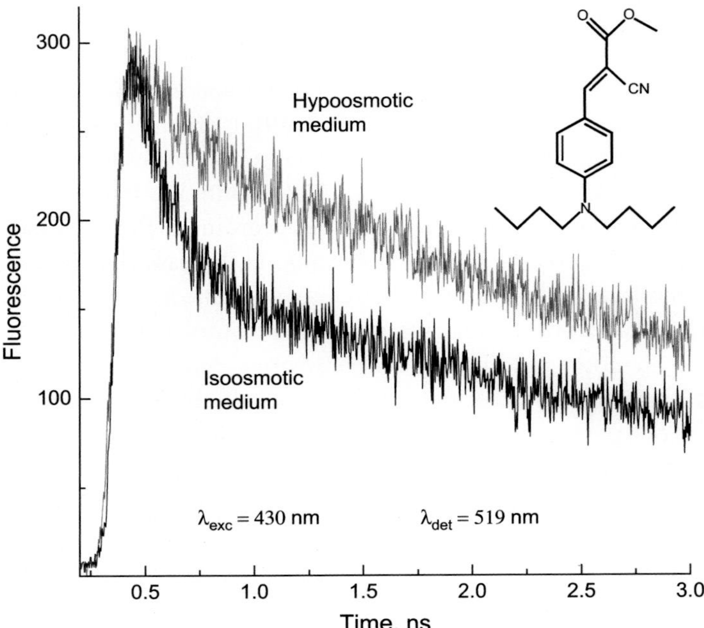

Figure 4.5. Osmotic stress dependence of fluorescence decay kinetics of molecular rotor SC1-20A in the membrane of a single DOPC GUV.

that membrane tension effects are detectable at the molecular level (Figure 4.5). Note that the molecular rotors of the above-mentioned type react to a change in viscosity with a change in intensity (quantum yield), and to a change in the environment's polarity with a change in emission wavelength (Loutfy and Law, 1980; Law, 1980). Therefore, more experiments are needed to attribute these data to a specific bilayer property change.

4.2.4 Polarity

A small variation in the composition of the cell membrane can strongly influence the activities of the membrane proteins (Parola, 1993; Mouritsen and Bloom, 1993; Lee, 2004). In many cases it is not known whether membrane components (various lipids, cholesterol, gangliosides, or small solutes such as general anesthetics) act directly by binding to the membrane protein or indirectly through changes in the physical properties of the bilayer such as viscosity, membrane free volume, transmembrane lateral force, or polarity. In particular, the hydration of the hydrophilic headgroups plays an important role in the structure and function of the lipid bilayers (Jendrasiak, 1996; Simon and Mcintosh, 1986). Considering the high sensitivity of proteins and especially integral membrane proteins to the environment, it is quite reasonable to expect functionally significant conformational changes in membrane proteins in response to a change in the hydration of the lipid bilayer.

Recently we have used the membrane environment–sensitive probe Laurdan to assess hydration changes in the lipid bilayers of small and large unilamellar vesicles

(SUVs and LUVs) under mechanical tension induced by osmotic gradient (Zhang et al., 2006). Earlier studies by other groups showed that Laurdan emission is extremely sensitive to the composition, phase state (gel or liquid-crystalline), and hydration of the lipid bilayer membrane (Parasassi et al., 1990, 1991, 1998; Bagatolli et al., 1998). In our study, the generalized fluorescence polarization (GP) of the Laurdan probe was measured as a function of osmolality difference across the bilayer membrane for a suspension of liposomes (Zhang et al., 2006). The emission GP value characterizes the relative distribution of intensities in the bimodal fluorescence emission spectrum from the locally excited and charge transfer states of the Laurdan (for a definition see Parasassi et al., 1990, 1993; Zhang et al., 2006). The distinct asymmetry and difference in slope around the isotonic point (when hydrostatic pressure difference across the membrane is zero) clearly indicates that this probe is sensitive to membrane tension and not to some specific interaction with the solute. Based on earlier Laurdan photophysical studies (Parasassi et al., 1998; Bagatolli et al., 1998), we interpret these results as an indication of the change in hydration due to increased water permeability into the bilayer under mechanical stress (Zhang et al., 2006). Note that the overall decrease of GP values with liposome size suggests that the strain due to higher curvature in smaller liposomes leads to higher polarity (due to deeper water penetration) of the bilayer. Thus these data suggest that mechanical membrane tension leads to significant changes in polarity (i.e., level of hydration) of the lipid bilayer membrane. To quantify the bilayer response to fluid shear stress on a picosecond-nanosecond time-scale, time-resolved fluorescence techniques could be used to study the solvation dynamics of small lipid probes, such as Laurdan (Viard et al., 1997; Parasassi et al., 1991; Bagatolli et al., 1998) and Prodan (Parasassi et al., 1998; Krasnowska et al., 1998; Hutterer et al., 1998).

It is known that lipid composition, organization, and physical state can regulate the peripheral binding of proteins to the lipid surface (Kinnunen et al., 1994). Therefore, changes in the bilayer structure under mechanical stress could potentially have an effect on the binding affinity of various peripheral membrane proteins to the membrane. For example, it has been shown that properly designed surfactant molecules, so-called "molecular harpoons," can recognize membranes under osmotic stress (Naka et al., 1992). Such changes in membrane-binding properties could directly affect the functioning of the membrane-associated G-proteins.

4.2.5 Membrane Tension and Cell Membrane Fluctuations (CMF)

In an elastic membrane, an isotropic lateral tension will result in a change in membrane area. The fractional area (i.e., strain) dependence on tension (i.e., stress) is a nonlinear function of membrane tension. The molecularly thin bilayer membranes are extremely flexible and therefore are constantly undulating due to thermal and metabolically driven excitations (CMFs) (Tuvia et al., 1997, 1998). For tension values below ~ 0.5 dyn/cm (depending on bilayer composition) the tension is primarily entropy-driven, that is, applied stresses go into smoothing out cell membrane undulations (Evans and Rawicz, 1990). CMFs, which are submicron, out-of-plane fluctuations of the cell membrane, were first discovered in red blood cells (Brochard and Lennon, 1975) and then observed in various types of cells (Krol et al., 1990; Mittelman et al., 1994; Levin and Korenstein, 1991). It has been shown that CMFs are partly metabolically driven (i.e., their amplitude is reduced in ATP depleted cells) (Tuvia et al., 1997, 1998). The presence of coupling between CMFs and biochemical processes in a cell raises a question about whether there may be an opposite effect, when mechanically induced changes in CMFs would result in biochemical responses. The fact that CMFs are smoothed out at very low membrane tension values could potentially enable very high mechanosensitivity. One potential mechanism that could couple CMFs to biochemical responses is discussed next.

Diffusive dynamics in membranes is typically described by either the free-volume theory (Cohen and Turnbull, 1959) or the Saffman hydrodynamic theory (Saffman and Delbruck, 1975). The former is believed to be appropriate for describing the lateral diffusion of small, lipid-like molecules (Vaz et al., 1985) while the latter is considered to be more appropriate for larger and more slowly diffusing integral membrane proteins (Peters and Cherry, 1982; Vaz et al., 1984). In the Saffman theory the membrane thickness is a major factor that determines the lateral diffusion coefficients of membrane proteins. Therefore any membrane perturbation that affects membrane thickness can potentially affect signaling dynamics, as discussed previously. It has been shown that deformation of the bilayer, or, more precisely, changes in membrane curvature, can lead to a change in membrane thickness (Safran, 2003). More specifically, the presence of curvature leads to an overall thickening of the bilayer and therefore slower lateral diffusion. Recently the membrane curvature induced by CMFs has been suggested as a potential mechanism for mechanosensing of fluid shear stress and, generally, of any membrane mechanical perturbation that increases membrane tension (Gov, 2006); this study has shown theoretically that the membrane tension induced by fluid shear stress reduces the amplitudes of CMFs and thereby increases the lateral diffusion coefficients, which can potentially account for the experimentally observed changes in membrane fluidity under shear stress.

4.2.6 Membrane Heterogeneity

Multiple experiments suggest that lipids exist in several coexisting phases in model lipid bilayers (Brown and London, 1998; Korlach et al., 1999; Bagatolli and Gratton, 2000; Munro, 2003). These phases include gel, liquid-ordered, and liquid-disordered bilayer states in order of increasing lateral fluidity. The lateral diffusion coefficient increases by a few orders of magnitude between the gel and liquid crystalline phases. The existence of raft domains in living cells is more elusive; however, many indirect experiments support their existence (Simons and Vaz, 2004; Munro, 2003). The current view of the membrane is that it is "patchy": There are functional protein complexes, lipid compositional areas, and regions with functional specializations (Engelman, 2005). The importance of such lateral membrane inhomogeneities for mechanochemical transduction comes from the experimental evidence that rafts can selectively sequester certain membrane proteins into membrane microdomains and affect signal transduction (Simons and Toomre, 2000). More specifically, proteins with high raft affinity include doubly acylated proteins such as α-subunits of G-proteins or Src kinases, GPI-anchored proteins, palmitoylated transmembrane proteins, cholesterol-linked and palmitoylated proteins, etc. (Brown and London, 1998; Rietveld et al., 1999). It has also been shown that some membrane proteins promote the formation of secondary, nonlamellar lipid structures such as inverted hexagonal phases (H_{II}) or cubic phases, which are attributed to the presence of hexagonal phase–prone phospolipid phosphatidylethanolamine (PE) in natural membranes (Epand, 1998). Nonlamellar prone lipids modulate the physical properties of lipid bilayer membranes by introducing membrane defects. Particularly significant is the finding that the inactive heterotrimeric $G_{\alpha\beta\gamma}$-proteins and $G_{\beta\gamma}$ dimers have been shown to have higher binding affinities to nonlamellar phases while upon activation the G-protein α-subunit changes its binding preference to the bilayer lamellar phase (Vogler et al., 2004). This implies that upon activation by a GPCR, the G_{α}-subunit leaves the nonlamellar area leaving the GPCR and $G_{\beta\gamma}$ dimmers behind – thus an additional lateral diffusion step is involved. Moreover, in yet another study it was shown that the membrane lipid structure not only determines the G-protein localization but also regulates the function of G-proteins (Yang et al., 2005). Recently it has been reported that the lateral diffusion of the small G-protein Ras is sensitive to the viscosity of the plasma membrane; cholesterol was used to change the membrane fluidity (Goodwin et al., 2005). These observations support the hypothesis that mechanically induced changes in the lateral fluidity of the membrane could also have a significant effect on signaling efficiency.

Caveolae, which are plasma membrane microdomains rich in cholesterol, sphingolipids, and coat proteins (such as caveolin-1), have also been implicated in sensing steady hemodynamic forces (Rizzo et al., 1998, 2003; Yu et al., 2006). In these studies caveolin-1 was shown to be involved in the flow-mediated dilation in carotid arteries, indicating that caveloae are required for mechanostransduction in blood vessels on both the short (vasodilaton) and long time-scales (remodeling of the cytoskeleton).

4.3 The Role of G-Protein–Coupled Receptors in Mechanosensing

4.3.1 Structure and Dynamics of Conformational Changes

Rhodopsin is a particularly important example of a GPCR that is known to be very sensitive to its lipid environment (Botelho et al., 2002; Feller and Gawrisch, 2005; Mitchell et al., 1992; Polozova and Litman, 2000). It has been shown that rhodopsin can be made functional in an artificial membrane without any of the lipids native to its natural environment by making a membrane with the correct balance of pressures in the lipid headgroup and tail regions (Botelho et al., 2002; Wang et al., 2002). GPCRs constitute the largest family of membrane receptors and exhibit a common structure containing seven transmembrane α-helices (Morris and Malbon, 1999).

The fact that certain G-proteins are activated within seconds following mechanical perturbation by shear stress (Gudi et al., 1996; Kuchan et al., 1994) leads us to suggest that corresponding GPCRs may be involved in the mediating mechanochemical signal transduction in endothelial cells. It has been well established that mechanical signals in the form of shear stress and cyclic strain cause the activation of multiple signal transduction pathways such as Raf-MEK-ERK (Werry et al., 2005) and PI3K-PDK-PKB (Dimmeler et al., 1999). Very recently it has been suggested that mechanical stretching can activate angiotensin II type 1 GPCR (Zou et al., 2004) without direct stimulation by the ligand; although promising, this study did not completely exclude the contribution by the ligand from an intracrine pathway (Hunyady and Turu, 2004). Typically the activity of the GPCR is initiated when an extracellular ligand induces or binds to an active conformation (Gether, 2000) (except in the case of rhodopsin [Okada et al., 2001]), which in turn can activate hundreds of G-proteins causing a strong signal amplification (Roberts and Waelbroeck, 2004; Heck and Hofmann, 2001; Gierschik et al., 1991). Since the interaction of GPCRs and G-proteins involves lateral diffusion steps of G-protein, this may further increase the sensitivity of the overall signaling process to changes in lateral membrane fluidity. It is important to note that receptor-mediated signaling pathways operate as signal amplification cascades; therefore, even a small change in a rate of one of the primary amplification steps (factor of ~ 2) will have a significant physiological response.

Much of the information regarding conformational changes involved in GPCR activation is derived from biophysical studies on rhodopsin (Farrens et al., 1996), the β_2-adrenergic receptor (Ghanouni et al., 2001a, 2001b; Peleg et al., 2001; Vilardaga et al., 2005), the PTH receptor (Vilardaga et al., 2003), and the adenosine receptor (Hoffmann et al., 2005). It has been shown that certain key amino acid residues in transmembrane helices, interacting with each other, keep the receptor preferentially locked in an inactive state. Ligand binding or isomerization of the preloaded ligand retinal (in the case of rhodopsin) changes the helix–helix interactions in the GPCR and causes/allows the movement of helix-5, -6, and -7, translating into conformational change in the third cytoplasmic loop that subsequently interacts with and activates G-proteins (Gether, 2000). It has been shown that the GPCR structure is rather "plastic," that is, the GPCR can assume a number of different conformations depending on the specific ligand (Ghanouni et al., 2001a; Perez and Karnik, 2005; Gether, 2000). More importantly, many GPCRs also exhibit agonist-independent constitutive activity, suggesting that GPCRs can adopt active conformation spontaneously (Leurs et al., 1998; Lefkowitz et al., 1993; Gether, 2000). The inherent plasticity of the GPCR (Perez and Karnik, 2005) makes it likely that changes in transmembrane lateral force profile, level of hydration, polarity, and membrane fluidity due to external mechanical force could also lead to the GPCR conformational change required for activation.

The nonspecific nature of membrane mechanical perturbations such as fluid shear stress or direct stretching raises a question about interpretations of experimental studies that try to find a primary sensor responsible for mechanosensation. Here

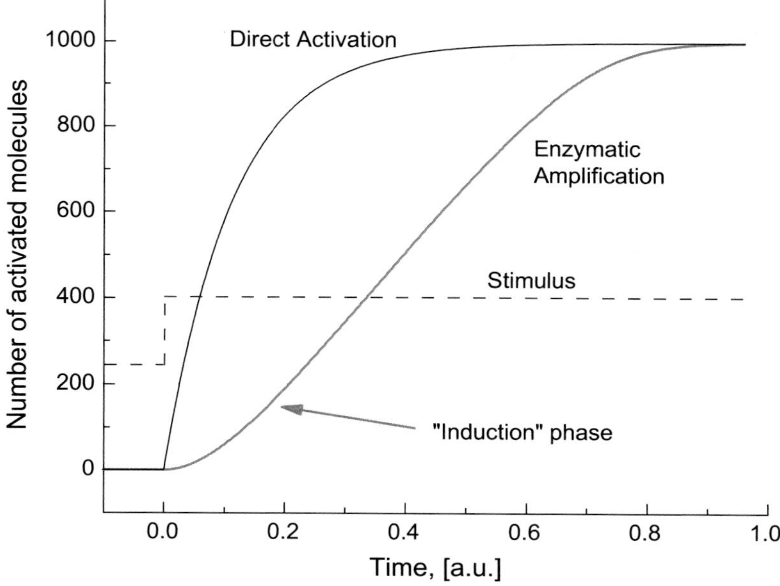

Figure 4.6. The dependence of the number of activated molecules on the mechanism of activation.

we would like to point out that the temporal dynamics of a specific activation process, when experimentally available, could be used to draw conclusions about the sequence of events during mechanochemical responses.

The differences in temporal dynamics of direct (primary) activation versus downstream (secondary) enzymatic amplification are illustrated in Figure 4.6. In the case of a primary response, a step stimulus (for example, application of shear stress at time = 0) leads to an exponential rise of the activated receptor concentration, as described by

$$\frac{dR(t)^*}{dt} = S(t)\,(R_{Total} - R(t)^*). \tag{4.1}$$

In the case of detection of a secondary response, such as enzymatic amplification by the activated receptor, like the one encountered in the GPCR–G-protein activation cycle, the concentration of activated proteins is described by the following simplified equation (for a more general equation see Detwiler et al. [2000]):

$$\frac{dG(t)^*}{dt} = \frac{k_{cat}R(t)^*(G_{Total} - G(t)^*)}{K_M + G_{Total} - G(t)^*} \tag{4.2}$$

As Figure 4.6 shows, the enzymatic amplification leads to the appearance of an induction period in the temporal dynamics of the detected concentration of activated proteins that is due to the fact that a sufficient increase of the activated receptor concentration is needed before enzymatic amplification can result in a noticeable rise in the concentration of the activated proteins.

4.3.2 Angiotensin II, Bradykinin B_2, and FPR Receptors

Endothelial dysfunction has been strongly implicated in atherogenesis and is characterized by diminished flow-dependent dilation (Davies, 1995; Lehoux et al., 2006). Thus, elucidating the mechanisms of mechanosensing by endothelial cells is fundamental for understanding the origins of atherogenesis. In order to test the hypothesis that membrane mechanical perturbation can induce conformational transitions in a GPCR, we have recently performed studies of human B_2 bradykinin GPCR (Chachisvilis et al., 2006), which is known to be coupled to similar activation pathways as those induced by blood shear stress in endothelial cells (e.g., vasodilation) (Leeb-Lundberg et al., 2005). The involvement of bradykinin receptors in coronary vasomotor control is well established (Groves et al., 1995; Leeb-Lundberg et al., 2005); an impaired flow-dependent response was observed in arteries from knockout mice lacking bradykinin B_2 receptors (Bergaya et al., 2001). Moreover, it has been reported that the B_2 receptor directly participates in reversible inhibitory interactions with eNOS (Ju et al., 1998; Marrero et al., 1999; Golser et al., 2000). Stimulation of the B_2 receptor with a ligand leads to eNOS activation via phosphorylation (Sessa, 2004); eNOS is a known source of nitric oxide (NO), a potent vasodilator. eNOS has also been implicated in playing an important role in inflammatory responses (Cirino et al., 2003). Our more recent data suggest that mechanically induced changes in membrane properties may in fact lead to the activation of a membrane protein, such as the bradykinin B_2 GPCR (Chachisvilis et al., 2006). These experiments used FRET to directly monitor the dynamics of membrane protein conformational change upon mechanical stimulation of the membrane. Since the flow-dependent dilation is a fundamental mechanism by which large arteries ensure adequate supplies of blood to the tissues (Bevan et al., 1995; Davies, 1995), our observation that the B_2 receptor can respond to mechanical perturbation of the cell membrane without involvement of ligand suggests a new molecular mechanism by which fluid shear stress can be sensed, which could have significant implications for understanding the molecular origins of atherogenesis.

Vasoactive peptides such as angiotensin II (Ang II) and endothelin, and cytokines and growth factors as well, are known factors involved in the development of cardiomyocyte hypertrophy, which is an adaptive response of the heart to hemodynamic overload such as hypertension. Hemodynamic overload has also been identified as a major cause of cardiac hypertrophy. There has been evidence suggesting that ion channels, integrins, and angiotensin II type 1 (AT1) receptors may be the mechanoreceptors, the activation of which lead to cardiac hypertrophy (Komuro, 2000; Yamazaki et al., 1999).

In a recent study by Zou et al. (2004), the AT1 receptor was found to play a crucial role in mechanical stress–induced cardiac hypertrophy. In a set of *in vitro* experiments using HEK293 and COS7 cells that do not express detectable angiotensinogen by RT-PCR, only when these cells are transfected with AT1 GPCR and express AT1 GPCR, can the ERKs be activated by mechanical stretch; when HEK293 and COS7 cells are transfected with a mutant AT1 GPCR that does not bind Ang II, mechanical stretch still strongly activates the ERKs while Ang II does not; this AT1 GPCR activation by mechanical force in HEK293 and COS7 is also illustrated by stretch-induced AT1 GPCR associations with Janus kinase 2 (JAK2), the cytosolic translocation of G-proteins, and phosphorylation cascades of protein kinases such as PKC and ERKs; furthermore, all those biochemical responses can be inhibited by the AT1 GPCR blocker candesartan or uncoupled by mutations that disrupt the AT1 GPCR/G-protein and AT1 GPCR/JAK2 interactions. Moreover, *in vivo* data from angiotensin knockout mice (ATG−/−) (to rule out the possibility that the effect is due to autocrine or paracrine AngII) showed that pressure overloading can induce cardiac hypertrophy through AT1 receptors in the animals, in contrast to the animals treated with the AT1 GPCR inhibitior candesartan that had much attenuated cardiac hypertrophy development. These findings strongly indicate that AT1 GPCR can be activated by mechanical stress and can function as a mechanosensor.

In blood vessels, fluid shear stress not only plays an important role in regulating the structure and function of endothelium, but it also controls pseudopod formation and the migration of circulating neutrophiles. Neutrophiles respond to shear stress with the retraction of the pseudopod and the reduction of the projected area (Makino et al., 2005). The receptors for the mechanosignal transduction of this process were studied by detecting FRET signal changes of heterotrimeric G-protein complexes that were genetically labeled with fluorescent proteins (Janetopoulos and Devreotes, 2002; Bunemann et al., 2003; Makino et al., 2006). It was found that fluid shear stress lowered the G-protein activity in neutrophiles, which was attributed to the inhibition by shear stress of the constitutive activity of the formyl peptide receptor (FPR), and possibly a couple of other GPCRs in neutrophiles (Makino et al., 2006). Apparently, mechanical stress can cause the constitutive activity of GPCRs to be increased, as in the cases of B_2 bradykinin receptors in endothelial cells and AT1 receptors in myocardiocytes, and decreased in the case of FPR in neutrophiles.

4.4 G-Protein Activation Pathways

4.4.1 GPCR-Triggered G-Protein Signal Transduction Pathway

The commonly accepted mechanism of activation of heterotrimeric G-proteins is by the GEF (guanine nucleotide exchange factor) activity of GPCRs. There are no major conformational changes in the G-protein α-subunit structure when the high-resolution crystal structures of the G_α-subunit in the inactive and active states are compared (Oldham and Hamm, 2006). There are only three locations in the G_α-subunit where local conformational changes are identified, which are thought of as switches for triggering G-protein activation (Oldham and Hamm, 2006; Bourne, 1997). Although our knowledge of how GPCR interacts with and consequently activates G-proteins to initiate the intracellular biological responses is limited, the consensus model of G-protein and GPCR interaction indicates that the GDP bound in a G-protein is about 30 Å from the membrane interface where the

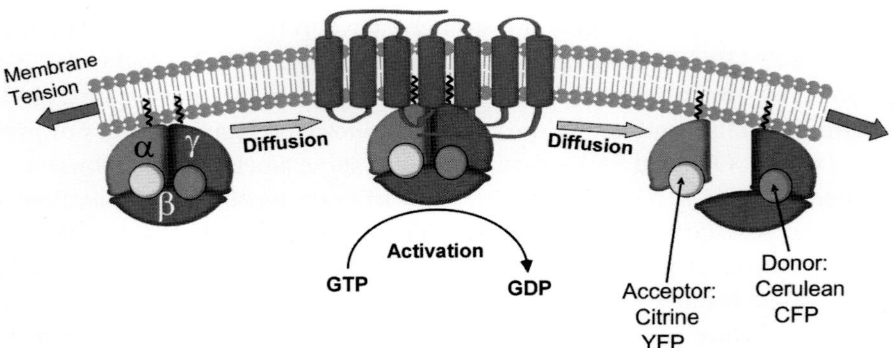

Figure 4.7. Catalytic activation process of membrane-associated heterotrimeric G-proteins involves interaction with GPCRs, dissociation/rearrangement of α- and βγ-subunits, and multiple lateral diffusion steps in the membrane. Modulation of lateral membrane viscosity by membrane tension (e.g., shear stress) can have a direct effect on the overall activation efficiency.

GPCR and G-protein interact (Hamm et al., 1988; Bourne, 1997), and predicts that the release of GDP from the G-protein has to involve conformational changes in the G_α-subunit that are induced by activated GPCR (Bourne, 1997). The molecular mechanism of GDP release in G-proteins caused by GPCR has been revealed in a recent study that showed that the C-terminal tail of a G_α-subunit in contact with a GPCR undergoes a rigid-body rotational-translational conformational change in response to activation of the GPCR and the rigid-body movement (of the α5 helix of GTPase domain) facilitates the release of GDP located 30 Å away from the interaction site (Oldham et al., 2006).

4.4.2 G-Protein Activation Mechanisms That Do Not Involve GPCRs

Since G-proteins are peripheral membrane proteins, the next question we would like to address is, what is the effect of membrane tension on the activation process of G-proteins themselves, irrespective of how the corresponding GPCRs are activated. The Activation of G-proteins is a catalytic process (Roberts and Waelbroeck, 2004; Heck and Hofmann, 2001; Gierschik et al., 1991) involving binding to the GPCR and the dissociation of a large number of trimeric G-protein complexes (Figure 4.7; Plate 7). It has been shown that GPCRs and heterotrimeric G-proteins interact freely in the membrane, that is, they do not preform complexes (Hein et al., 2005; Azpiazu and Gautam, 2004) (it is known that G-proteins have low affinity to GPCR receptors). Within this model activation of G-proteins involves lateral diffusion and binding/dissociation events of heterotrimeric G-proteins and a GPCR (see Figure 4.7; Plate 7) and therefore must be sensitive to mechanical perturbations of membrane physical properties that affect lateral diffusivity, such as membrane fluidity, thickness, and polarity. Such a mechanosensitivity mechanism would not depend on the nature of GPCR activation, but would still be able to modulate overall downstream signaling by affecting the lateral diffusion of a heterotrimeric G-protein and its subunits via changes in membrane fluidity.

It has been shown that G-proteins can also be activated in the absence of GPCRs in the liposomes exposed to fluid shear stress (Gudi et al., 1998). In this study, the activation of purified G-proteins reconstituted into phospholipid vesicles loaded with [γ-^{32}P]GTP were subjected to physiological levels of fluid shear stress in a cone-and-plate viscometer. Steady state GTP hydrolysis was measured as an indication of G-protein activation. The study has shown that shear stress (0–30 dynes/cm^2) activated G-proteins in a dose-dependent manner (0.48–4.6 pmol/min per microg of protein). The molecular mechanisms of the G-protein activation by the shear stress are thought to be linked or coupled to physical property changes in the bilayer caused by the shear stress. This concept was supported by the study on liposomes containing lysophosphatidylcholine (30 mol%) or treated with membrane fluidizer benzyl alcohol (40 mM); the results showed that basal GTPase activity was enhanced three- to five-fold. Conversely, the incorporation of cholesterol (24 mol%) into liposomes reduced the activation of G-proteins by shear. These results demonstrate the ability of the phospholipid bilayer to mediate the shear stress–induced activation of membrane-bound G-proteins in the absence of protein receptors and that bilayer physical properties modulate this response (Gudi et al., 1998; Haidekker et al., 2000, 2001). Thus, there is sufficient evidence to indicate that G-proteins can be activated directly by mechanical stimulation.

It seems that while the molecular mechanisms of G-protein activation by GPCR is clearly a matter of highly specific and finely coordinated protein–protein interactions, that of G-protein activation by direct mechanical forces is a matter of protein–lipid interactions (Gudi et al., 1998). The conformational changes required for the activation of G-proteins caused by mechanical forces could be the same as that by GPCR, in that they both should induce a conformation of the G-protein that enhances the release of G-protein bound GDP. However, detailed information is lacking at the moment regarding how mechanical forces achieve G-protein activation without the involvement of GPCRs. The question about how changes in lipid–protein interactions can lead to the relatively large motion of the $\alpha 5$ helix of the GTPase domain that is required to release GDP (located ~ 30 Å away from the interaction site and from the lipid bilayer interface) (Oldham et al., 2006) still needs to be addressed.

It has been speculated that membrane proteins could be activated by mechanical waves propagating in the lipid bilayer membrane (Haines, 1982). In this hypothesis a sudden conformational transition in the membrane protein receptor (such as rhodopsin) launches a mechanical wave in the lipid bilayer membranes, which modulates membrane thickness as it propagates away from the receptor. When such a wave passes by other membrane-associated proteins (such as transducin) it alters their conformation, leading to activation. Although such membrane dynamics hypothesis is consistent with high activation rates observed for rhodopsin-transducin cycles and observation that the G-protein can be activated by perturbation of the lipid bilayer membrane (Gudi et al., 1998), it still awaits experimental support.

In endothelial cells, shear stress–induced NO production has two phases: the initial phase and the sustained phase. The initial phase of burst NO production is more responsive to the change in shear stress rather than the shear stress itself and is dependent on G-protein activation since GDPβS, a nonhydrolyzable GDP analog used as a general G-protein inhibitor, blocked the flow-mediated burst in cGMP and NO_x ($NO_2 + NO_3$) production (Gudi et al., 2003; Malek et al., 1999b; Kuchan et al., 1994), whereas the sustained phase of NO production is independent of G-protein activation (Kuchan et al., 1994). This temporal order of responses from endothelial cells to shear stress is also reflected in calcium transients and activation of ERK1/2, which are also G-protein dependent (Ishida et al., 1997; Tseng et al., 1995; Jo et al., 1997), as the immediate response and activation of focal adhesion kinase (FAK) by phosphorylation during sustained response (Takahashi et al., 1997). The G-protein dependence of shear stress–induced NO production and ERK1/2 activation indicates that the G-protein is an important transducer in mediating the biological effects of mechanical forces in endothelial cells. The G-proteins mediating shear stress response are both pertussis toxin-sensitive (Ohno et al., 1993) and refractory (Kuchan et al., 1994), indicating that Gi/Go and Gαq subunits are involved. Shear stress–induced mobilization of calcium (Schwarz et al., 1992); activation of MAPKs such as ERK1/2 (White et al., 2005; Butler et al., 2002; Bao et al., 2000; Sumpio et al., 2005, p. 38; Kadohama et al., 2006; Surapisitchat et al., 2001), JNK (Berk et al., 2001; Butler et al., 2002), and ERK5 (Pi et al., 2004); and activation of protein kinases such as PKA (Boo et al., 2002a, 2002b; Dixit et al., 2005; Boo and Jo, 2003), PKB (Akt) (Fleming et al., 2005; Jin et al., 2005; Boo and Jo, 2003; Sumpio et al., 2005), and PKC (Kuchan and Frangos, 1993; Wedgwood et al., 2001; Traub et al., 1997; Hsieh et al., 1992) that lead to the activation of eNOS (Sessa, 2004; Dimmeler et al., 1999) by phosphorylation are certainly consistent with the involvement of G-protein activation by shear stress (Hsieh et al., 1992, 1993). G-protein activation by shear stress is also demonstrated in other cell types such as osteoblasts (Reich et al., 1997) and osteoclasts (McAllister and Frangos, 1999), chondrocytes (Das et al., 1997), and even such marine micro-organisms as dinoflagellates (Chen et al., 2003, 2007).

Thus, although we now know that the mechanical force is transduced via multiple pathways involving G-proteins, the molecular mechanisms of these processes are not completely understood (Gudi et al., 1998; Jo et al., 1997; Sessa, 2004; Kuchan et al., 1994), particularly when the primary mechanosensor is concerned.

4.5 Conclusion

Although many biochemical mechanical force transduction pathways have been discovered, the primary mechanosensor(s) remain elusive. Since the lipid bilayer membrane constitutes the first barrier separating the cellular contents from their environment, it is reasonable to expect that the receptors of external mechanical stimuli may be located in the membrane. This hypothesis is further strengthened by the fact that the plasma membrane contains the majority of the cell's receptors for sensing very diverse extracellular signals such as light, odorants, tastant, and other ligands. Multiple experimental studies indicate that mechanical perturbation of the lipid bilyer membrane can lead to conformational transitions in membranes or membrane-associated proteins. However, only in very few studies was the causal relationship demonstrated between the increased membrane tension and conformational transitions in membrane proteins such as in studies on bacterial mechanosensitive channels (Perozo et al., 2002b). The majority of studies of the effects of fluid shear stress are performed on cells, which makes it difficult to relate the activation of specific biochemical pathways to a particular receptor due to crosstalk between various pathways. Future experiments on reconstituted systems based on liposomes or supported bilayers may provide more definite answers. Even more challenging will be the determination of effects of mechanically induced changes in the physical properties of lipid bilayer membranes on the dynamics of membrane proteins such as GPCRs. The membrane physical properties such as fluidity, free volume, polarity, lateral pressure profile, lipid ordering, and thickness are all expected to change to various degrees upon mechanical stretching or fluid shear stress. Deconvolution of all these changes and determination of the relevant physical membrane properties that control membrane protein conformational dynamics in mechanosensing will require carefully designed series of experiments in both model systems and cells.

4.6 Acknowledgments

This work was supported by National Institutes of Health grants HL040696 (JAF) and HL086943 (MC).

REFERENCES

Abdul-Rahim H and Bouchy M (1998) Analysis of the Fluorescence Anisotropy of Labelled Membranes Submitted to a Shear Stress. *J Photochem Photobiol B* **47**: 95–108.

Arnold K, Zschoernig O, Barthel D and Herold W (1990) Exclusion of Poly (Ethylene Glycol) From Liposome Surfaces. *Biochimica et Biophysica Acta* **1022**: 303–310.

Azpiazu I and Gautam N (2004) A Fluorescence Resonance Energy Transfer-Based Sensor Indicates That Receptor Access to a G-protein Is Unrestricted in a Living Mammalian Cell. *J Biol Chem* **279**: 27709–27718.

Bagatolli L A and Gratton E (2000) Two Photon Fluorescence Microscopy of Coexisting Lipid Domains in Giant Unilamellar Vesicles of Binary Phospholipid Mixtures. *Biophys J* **78**: 290–305.

Bagatolli L A, Gratton E and Fidelio G D (1998) Water Dynamics in Glycosphingolipid Aggregates Studied by LAURDAN Fluorescence. *Biophys J* **75**: 331–341.

Bao X, Clark C B and Frangos J A (2000) Temporal Gradient in Shear-Induced Signaling Pathway: Involvement of MAP Kinase, C-Fos, and Connexin4319. *Am J Physiol Heart Circ Physiol* **278**: H1598–H1605.

Baskin J S, Chachisvilis M, Gupta M and Zewail A H (1998) Femtosecond Dynamics of Solvation: Microscopic Friction and Coherent Motion in Dense Fluids. *J Phys Chem A* **102**: 4158–4171.

Ben-Shaul A (1995) in *Structure and Dynamics of Membranes* (Lipowsky R and Sackman E, eds.), pp. 359–401, Elsevier, Amsterdam.

Bergaya S, Meneton P, Bloch-Faure M, Mathieu E, Alhenc-Gelas F, Levy B I and Boulanger C M (2001) Decreased Flow-Dependent Dilation in Carotid Arteries of Tissue Kallikrein-Knockout Mice. *Circ Res* **88**: 593–599.

Berk B C, Abe J I, Min W, Surapisitchat J and Yan C (2001) Endothelial Atheroprotective and Anti-Inflammatory Mechanisms. *Ann NY Acad Sci* **947**: 93–109.

Bevan J A, Kaley G and Rubanyi G M (1995) *Flow-Dependent Regulation of Vascular Function*. Oxford University Press, New York.

Blood P D, Ayton G S and Voth G A (2005) Probing the Molecular-Scale Lipid Bilayer Response to Shear Flow Using Nonequilibrium Molecular Dynamics. *J Phys Chem B* **109**: 18673–18679.

Boo Y C and Jo H (2003) Flow-Dependent Regulation of Endothelial Nitric Oxide Synthase: Role of Protein Kinases. *Am J Physiol Cell Physiol* **285**: C499–C508.

Boo Y C, Hwang J, Sykes M, Michell B J, Kemp B E, Lum H and Jo H (2002a) Shear Stress Stimulates Phosphorylation of ENOS at Ser (635) by a Protein Kinase A-Dependent Mechanism. *Am J Physiol Heart Circ Physiol* **283**: H1819–H1828.

Boo Y C, Sorescu G, Boyd N, Shiojima I, Walsh K, Du J and Jo H (2002b) Shear Stress Stimulates Phosphorylation of Endothelial Nitric-Oxide Synthase at Ser1179 by Akt-Independent Mechanisms: Role of Protein Kinase A. *J Biol Chem* **277**: 3388–3396.

Borochov A and Borochov H (1979) Increase in Membrane Fluidity in Liposomes and Plant Protoplasts Upon Osmotic Swelling. *Biochimica et Biophysica Acta* **550**: 546–549.

Botelho A V, Gibson N J, Thurmond R L, Wang Y and Brown M F (2002) Conformational Energetics of Rhodopsin Modulated by Nonlamellar-Forming Lipids. *Biochemistry* **41**: 6354–6368.

Bourne H R (1997) How Receptors Talk to Trimeric G-proteins. *Curr Opin Cell Biol* **9**: 134–142.

Bowie J U (2005) Solving the Membrane Protein Folding Problem. *Nature* **438**: 581–589.

Brochard F and Lennon J F (1975) Frequency Spectrum of Flicker Phenomenon in Erythrocytes. *J Physique* **36**: 1035–1047.

Brown D A and London E (1998) Functions of Lipid Rafts in Biological Membranes. *Ann Rev Cell and Developmental Bio* **14**: 111–136.

Bunemann M, Frank M and Lohse M J (2003) Gi Protein Activation in Intact Cells Involves Subunit Rearrangement Rather Than Dissociation. *Proceedings of the National Academy of Sciences of the United States of America* **100**: 16077–16082.

Butler P J, Norwich G, Weinbaum S and Chien S (2001) Shear Stress Induces a Time- and Position-Dependent Increase in Endothelial Cell Membrane Fluidity. *Amer J Physiol – Cell Physiol* **280**: C962–C969.

Butler P J, Tsou T C, Li J Y, Usami S and Chien S (2002) Rate Sensitivity of Shear-Induced Changes in the Lateral Diffusion of Endothelial Cell Membrane Lipids: A Role for Membrane Perturbation in Shear-Induced MAPK Activation. *FASEB J* **16**: 216–218.

Cantor R S (1997) Lateral Pressures in Cell Membranes: A Mechanism for Modulation of Protein Function. *J Phys Chem B* **101**: 1723–1725.

Chachisvilis M, Chirvony V S, Shulga A M, Kallebring B, Larsson S and Sundstrom V (1996) Spectral and Photophysical Properties of Ethylene-Bridged Side-to-Side Porphyrin Dimers. 2. Femtosecond Transient Absorption and Picosecond Fluorescence Study of Trans-1,2-Bis (Meso-Octaethylporphyrinyl)Ethene. *J Phys Chem* **100**: 13867–13873.

Chachisvilis M, Zhang Y L and Frangos J A (2006) G-protein-Coupled Receptors Sense Fluid Shear Stress in Endothelial Cells. *Proceedings of the National Academy of Sciences of the United States of America* **103**: 15463–15468.

Chen A K, Latz M I and Frangos J A (2003) The Use of Dinoflagellate Bioluminescence to Characterize Cell Stimulation in Bioreactors. *Biotechnol Bioeng* **83**: 93–103.

Chen A K, Latz M I, Sobolewski P and Frangos J A (2007) Evidence for the Role of G-Proteins in Flow Stimulation of Dinoflagellate Bioluminescence. *Am J Physiol Regul Integr Comp Physiol.*

Chen B M and Grinnell A D (1997) Kinetics, Ca2+ Dependence, and Biophysical Properties of Integrin-Mediated Mechanical Modulation of Transmitter Release From Frog Motor Nerve Terminals. *J Neurosci* **17**: 904–916.

Cirino G, Fiorucci S and Sessa W C (2003) Endothelial Nitric Oxide Synthase: the Cinderella of Inflammation? *Trends Pharmaco Sci* **24**: 91–95.

Cohen M H and Turnbull D (1959) Molecular Transport in Liquids and Glasses. *J Chem Phys* **31**: 1164–1169.

Criado M, Eibl H and Barrantes F J (1984) Functional Properties of the Acetylcholine Receptor Incorporated in Model Lipid-Membranes – Differential Effects of Chain-Length and Head Group of Phospholipids on Receptor Affinity States and Receptor-Mediated Ion Translocation. *J Bio Chem* **259**: 9188–9198.

Cunningham K S and Gotlieb A I (2005) The Role of Shear Stress in the Pathogenesis of Atherosclerosis. *Lab Invest* **85**: 9–23.

Das P, Schurman D J and Smith R L (1997) Nitric Oxide and G-proteins Mediate the Response of Bovine Articular Chondrocytes to Fluid-Induced Shear. *J Orthop Res* **15**: 87–93.

Davies P F (1995) Flow-Mediated Endothelial Mechanotransduction. *Physiol Rev* **75**: 519–560.

Debakey M E, Lawrie G M and Glaeser D H (1985) Patterns of Atherosclerosis and Their Surgical Significance. *Annals Surgery* **201**: 115–131.

Detwiler P B, Ramanathan S, Sengupta A and Shraiman B I (2000) Engineering Aspects of Enzymatic Signal Transduction: Photoreceptors in the Retina. *Biophys J* **79**: 2801–2817.

Dimmeler S, Fleming I, Fisslthaler B, Hermann C, Busse R and Zeiher A M (1999) Activation of Nitric Oxide Synthase in Endothelial Cells by Akt-Dependent Phosphorylation. *Nature* **399**: 601–605.

Dixit M, Loot A E, Mohamed A, Fisslthaler B, Boulanger C M, Ceacareanu B, Hassid A, Busse R and Fleming I (2005) Gab1, SHP2, and Protein Kinase A Are Crucial for the Activation of the Endothelial NO Synthase by Fluid Shear Stress. *Circ Res* **97**: 1236–1244.

Doolittle A K (1951) Studies in Newtonian Flow. II. The Dependence of the Viscosity of Liquids on Free-Space. *J Appl Phys* 1471–1475.

Engelman D M (2005) Membranes Are More Mosaic Than Fluid. *Nature* **438**: 578–580.

Epand R M (1998) Lipid Polymorphism and Protein-Lipid Interactions. *Biochim Biophysi Acta – Revi Biomembranes* **1376**: 353–368.

Evans E and Needham D (1987) Physical Properties of Surfactant Bilayer Membranes – Thermal Transitions, Elasticity, Rigidity, Cohesion, and Colloidal Interactions. *J Phys Chem* **91**: 4219–4228.

Evans E and Rawicz W (1990) Entropy-Driven Tension and Bending Elasticity in Condensed-Fluid Membranes. *Physical Review Letters* **64**: 2094–2097.

Farrens D L, Altenbach C, Yang K, Hubbell W L and Khorana H G (1996) Requirement of Rigid-Body Motion of Transmembrane Helices for Light Activation of Rhodopsin. *Science* **274**: 768–770.

Feller S E and Gawrisch K (2005) Properties of Docosahexaenoic-Acid-Containing Lipids and Their Influence on the Function of Rhodopsin. *Current Opinion Struct Biol* **15**: 416–422.

Fisslthaler B, Dimmeler S, Hermann C, Busse R and Fleming I (2000) Phosphorylation and Activation of the Endothelial Nitric Oxide Synthase by Fluid Shear Stress. *Acta Physiol Scand* **168**: 81–88.

Fleming I, Fisslthaler B, Dixit M and Busse R (2005) Role of PECAM-1 in the Shear-Stress-Induced Activation of Akt and the Endothelial Nitric Oxide Synthase (ENOS) in Endothelial Cells. *J Cell Sci* **118**: 4103–4111.

Frangos S G, Gahtan V and Sumpio B (1999) Localization of Atherosclerosis: Role of Hemodynamics. *Arch Surg* **134**: 1142–1149.

Gether U (2000) Uncovering Molecular Mechanisms Involved in Activation of G-protein-Coupled Receptors. *Endocrine Reviews* **21**: 90–113.

Ghanouni P, Gryczynski Z, Steenhuis J J, Lee T W, Farrens D L, Lakowicz J R and Kobilka B K (2001a) Functionally Different Agonists Induce Distinct Conformations in the G-protein Coupling Domain of the Beta (2) Adrenergic Receptor. *J Biol Chem* **276**: 24433–24436.

Ghanouni P, Steenhuis J J, Farrens D L and Kobilka B K (2001b) Agonist-Induced Conformational Changes in the G-Protein-Coupling Domain of the Beta (2) Adrenergic

Receptor. *Proceedings of the National Academy of Sciences of the United States of America* **98**: 5997–6002.

Gibson C M, Diaz L, Kandarpa K, Sacks F M, Pasternak R C, Sandor T, Feldman C and Stone P H (1993) Relation of Vessel Wall Shear Stress to Atherosclerosis Progression in Human Coronary Arteries. *Arterioscler Thromb* **13**: 310–315.

Gierschik P, Moghtader R, Straub C, Dieterich K and Jakobs K H (1991) Signal Amplification in HL-60 Granulocytes. Evidence That the Chemotactic Peptide Receptor Catalytically Activates Guanine-Nucleotide-Binding Regulatory Proteins in Native Plasma Membranes. *Eur J Biochem* **197**: 725–732.

Glagov S, Zarins C, Giddens D P and Ku D N (1988) Hemodynamics and Atherosclerosis – Insights and Perspectives Gained from Studies of Human Arteries. *Archives of Pathology & Laboratory Medicine* **112**: 1018–1031.

Golser R, Gorren A C F, Leber A, Andrew P, Habisch H J, Werner E R, Schmidt K, Venema R C and Mayer B (2000) Interaction of Endothelial and Neuronal Nitric-Oxide Syntheses With the Bradykinin B2 Receptor – Binding of an Inhibitory Peptide to the Oxygenase Domain Blocks Uncoupled NADPH Oxidation. *J Bio Chem* **275**: 5291–5296.

Goodwin J S, Drake K R, Remmert C L and Kenworthy A K (2005) Ras Diffusion Is Sensitive to Plasma Membrane Viscosity. *Biophys J* **89**: 1398–1410.

Gov N S (2006) Diffusion in Curved Fluid Membranes. *Physical Review* **73**.

Grote R F and Hynes J T (1980) The Stable States Picture of Chemical Reactions. 2. Rate Constants for Condensed and Gas-Phase Reaction Models. *J Chem Phys* **73**: 2715–2732.

Grote R F and Hynes J T (1981) Saddle-Point Model for Atom Transfer-Reactions in Solution. *J Chem Phys* **75**: 2191–2198.

Groves P, Kurz S, Just H and Drexler H (1995) Role of Endogenous Bradykinin in Human Coronary Vasomotor Control. *Circulation* **92**: 3424–3430.

Gudi S, Huvar I, White C R, McKnight N L, Dusserre N, Boss G R and Frangos J A (2003) Rapid Activation of Ras by Fluid Flow Is Mediated by Galpha (q) and Gbetagamma Subunits of Heterotrimeric G-proteins in Human Endothelial Cells. *Arterioscler Thromb Vasc Biol* **23**: 994–1000.

Gudi S, Nolan J P and Frangos J A (1998) Modulation of GTPase Activity of G-proteins by Fluid Shear Stress and Phospholipid Composition. *Proceedings of the National Academy of Sciences of the United States of America* **95**: 2515–2519.

Gudi S R P, Clark C B and Frangos J A (1996) Fluid Flow Rapidly Activates G-proteins in Human Endothelial Cells – Involvement of G-proteins in Mechanochemical Signal Transduction. *Circ Res* **79**: 834–839.

Gullingsrud J and Schulten K (2004) Lipid Bilayer Pressure Profiles and Mechanosensitive Channel Gating. *Biophys J* **86**: 3496–3509.

Haidekker M A, Brady T, Wen K, Okada C, Stevens H Y, Snell J M, Frangos J A and Theodorakis E A (2002) Phospholipid-Bound Molecular Rotors: Synthesis and Characterization. *Bioorganic & Medicinal Chemistry* **10**: 3627–3636.

Haidekker M A, L'Heureux N and Frangos J A (2000) Fluid Shear Stress Increases Membrane Fluidity in Endothelial Cells: A Study with DCVJ Fluorescence. *Amer J Physiol – Heart and Circulatory Physiol* **278**: H1401–H1406.

Haidekker M A, Ling T T, Anglo M, Stevens H Y, Frangos J A and Theodorakis E A (2001) New Fluorescent Probes for the Measurement of Cell Membrane Viscosity. *Chemistry & Biology* **8**: 123–131.

Haines T H (1982) A Model for Transition-State Dynamics in Bilayers – Implications for the Role of Lipids in Biomembrane Transport. *Biophys J* **37**: 147–148.

Hamm H E, Deretic D, Arendt A, Hargrave P A, Koenig B and Hofmann K P (1988) Site of G-protein Binding to Rhodopsin Mapped with Synthetic Peptides from the Alpha Subunit. *Science* **241**: 832–835.

Hanggi P, Talkner P and Borkovec M (1990) Reaction-Rate Theory – 50 Years After Kramers. *Rev Modern Phys* **62**: 251–341.

Harrison D G (2005) The Shear Stress of Keeping Arteries Clear. *Nat Med* **11**: 375–376.

Heck M and Hofmann K P (2001) Maximal Rate and Nucleotide Dependence of Rhodopsin-Catalyzed Transducin Activation: Initial Rate Analysis Based on a Double Displacement Mechanism. *J Biol Chem* **276**: 10000–10009.

Hein P, Frank M, Hoffmann C, Lohse M J and Bunemann M (2005) Dynamics of Receptor/G-protein Coupling in Living Cells. *EMBO J* **24**: 4106–4114.

Hishikawa K, Nakaki T, Marumo T, Hayashi M, Suzuki H, Kato R and Saruta T (1994) Pressure Promotes DNA-Synthesis in Rat Cultured Vascular Smooth-Muscle Cells. *J Clinical Investigation* **93**: 1975–1980.

Hoffmann C, Gaietta G, Bunemann M, Adams S R, Oberdorff-Maass S, Behr B, Vilardaga J P, Tsien R Y, Eisman M H and Lohse M J (2005) A FlAsH-Based FRET Approach to Determine G-protein – Coupled Receptor Activation in Living Cells. *Nature Methods* **2**: 171–176.

Hsieh H J, Li N Q and Frangos J A (1993) Pulsatile and Steady Flow Induces C-Fos Expression in Human Endothelial-Cells. *J Physiol* **154**: 143–151.

Hsieh H J, Li N Q and Frangos J A (1992) Shear-Induced Platelet-Derived Growth Factor Gene Expression in Human Endothelial Cells Is Mediated by Protein Kinase C. *J Cell Physiol* **150**: 552–558.

Hunyady L and Turu G (2004) The Role of the AT1 Angiotensin Receptor in Cardiac Hypertrophy: Angiotensin II Receptor or Stretch Sensor? *Trends Endocrinol Metab* **15**: 405–408.

Hutterer R, Parusel A B J and Hof M (1998) Solvent Relaxation of Prodan and Patman: A Useful Tool for the Determination of Polarity and Rigidity Changes in Membranes. *J Fluorescence* **8**: 389–393.

Ingber D E (1997) Tensegrity: The Architectural Basis of Cellular Mechanotransduction. *Ann Rev Physiol* **59**: 575–599.

Ishida T, Takahashi M, Corson M A and Berk B C (1997) Fluid Shear Stress-Mediated Signal Transduction: How Do Endothelial Cells Transduce Mechanical Force into Biological Responses? *Ann NY Acad Sci* **811**: 12–23.

Ito T, Yamazaki M and Ohnishi S (1989) Osmoelastic Coupling in Biological Structures – A Comprehensive Thermodynamic Analysis of the Osmotic Response of Phospholipid-Vesicles and a Reevaluation of the Dehydration Force Theory. *Biochemistry* **28**: 5626–5630.

Jahnig F (1996) What Is the Surface Tension of a Lipid Bilayer Membrane? *Biophys J* **71**: 1348–1349.

Janetopoulos C and Devreotes P (2002) Monitoring Receptor-Mediated Activation of Heterotrimeric G-Proteins by Fluorescence Resonance Energy Transfer. *Methods* **27**: 366–373.

Janiak M J, Small D M and Shipley G G (1976) Nature of Thermal Pre-Transition of Synthetic Phospholipids – Dimyristoyllecithin and Dipalmitoyllecithin. *Biochemistry* **15**: 4575–4580.

Jendrasiak G L (1996) The Hydration of Phospholipids and Its Biological Significance. *J Nutritional Biochem* **7**: 599–609.

Jin Z G, Wong C, Wu J and Berk B C (2005) Flow Shear Stress Stimulates Gab1 Tyrosine Phosphorylation to Mediate Protein Kinase B and Endothelial Nitric-Oxide Synthase Activation in Endothelial Cells. *J Biol Chem* **280**: 12305–12309.

Jo H, Sipos K, Go Y M, Law R, Rong J and McDonald J M (1997) Differential Effect of Shear Stress on Extracellular Signal-Regulated Kinase and N-Terminal Jun Kinase in Endothelial Cells. Gi2- and Gbeta/Gamma-Dependent Signaling Pathways. *J Biol Chem* **272**: 1395–1401.

Johannsson A, Keightley C A, Smith G A, Richards C D, Hesketh T R and Metcalfe J C (1981a) The Effect of Bilayer Thickness and Normal-Alkanes on the Activity of the (Ca2++Mg2+)-Dependent Atpase of Sarcoplasmic-Reticulum. *J Biolog Chem* **256**: 1643–1650.

Johannsson A, Smith G A and Metcalfe J C (1981b) The Effect of Bilayer Thickness on the Activity of (Na+ + K+)-Atpase. *Biochimica et Biophysica Acta* **641**: 416–421.

Ju H, Venema V J, Marrero M B and Venema R C (1998) Inhibitory Interactions of the Bradykinin B2 Receptor with Endothelial Nitric-Oxide Synthase. *J Biol Chem* **273**: 24025–24029.

Kadohama T, Akasaka N, Nishimura K, Hoshino Y, Sasajima T and Sumpio B E (2006) P38 Mitogen-Activated Protein Kinase Activation in Endothelial Cell Is Implicated in Cell Alignment and Elongation Induced by Fluid Shear Stress. *Endothelium* **13**: 43–50.

Kinnunen P K J, Koiv A, Lehtonen J Y A, Rytomaa M and Mustonen P (1994) Lipid Dynamics and Peripheral Interactions of Proteins with Membrane Surfaces. *Chem Phys Lipids* **73**: 181–207.

Knudsen H L and Frangos J A (1997) Role of Cytoskeleton in Shear Stress-Induced Endothelial Nitric Oxide Production. *Amer J Physiol – Heart and Circul Phys* **42**: H347–H355.

Komuro I (2000) Molecular Mechanism of Mechanical Stress-Induced Cardiac Hypertrophy. *Jpn Heart J* **41**: 117–129.

Korlach J, Schwille P, Webb W W and Feigenson G W (1999) Characterization of Lipid Bilayer Phases by Confocal Microscopy and Fluorescence Correlation Spectroscopy. *Proceedings of the National Academy of Sciences of the United States of America* **96**: 8461–8466.

Krasnowska E K, Gratton E and Parasassi T (1998) Prodan as a Membrane Surface Fluorescence Probe: Partitioning Between Water and Phospholipid Phases. *Biophys J* **74**: 1984–1993.

Krichevsky O and Bonnet G (2002) Fluorescence Correlation Spectroscopy: The Technique and Its Applications. *Reports on Progress in Physics* **65**: 251–297.

Krol A Y, Grinfeldt M G, Levin S V and Smilgavichus A D (1990) Local Mechanical Oscillations of the Cell-Surface Within the Range 0.2–30 Hz. *European Biophys J* **19**: 93–99.

Ku D N, Giddens D P, Zarins C K and Glagov S (1985) Pulsatile Flow and Atherosclerosis in the Human Carotid Bifurcation. Positive Correlation Between Plaque Location and Low Oscillating Shear Stress. *Arteriosclerosis* **5**: 293–302.

Kuchan M J and Frangos J A (1993) Shear Stress Regulates Endothelin-1 Release via Protein Kinase C and CGMP in Cultured Endothelial Cells. *Am J Physiol* **264**: H150–H156.

Kuchan M J, Jo H and Frangos J A (1994) Role of G-Proteins in Shear Stress-Mediated Nitric-Oxide Production by Endothelial-Cells. *Am J Physiol* **267**: C753–C758.

Kung C E and Reed J K (1986) Microviscosity Measurements of Phospholipid-Bilayers Using Fluorescent Dyes That Undergo Torsional Relaxation. *Biochemistry* **25**: 6114–6121.

Law K Y (1980) Fluorescence Probe for Micro-Environments – Anomalous Viscosity Dependence of the Fluorescence Quantum Yield of Para-N,N-Dialkylaminobenzylidenemalononitrile in 1-Alkanols. *Chem Phys Lett* **75**: 545–549.

Lee A G (1998) How Lipids Interact with an Intrinsic Membrane Protein: The Case of the Calcium Pump. *Biochim Biophys Acta* **1376**: 381–390.

Lee A G (2004) How Lipids Affect the Activities of Integral Membrane Proteins. *Biochim Biophys Acta* **1666**: 62–87.

Leeb-Lundberg L M F, Marceau F, Muller-Esterl W, Pettibone D J and Zuraw B L (2005) International Union of Pharmacology. XLV. Classification of the Kinin Receptor Family: From Molecular Mechanisms to Pathophysiological Consequences. *Pharmacol Rev* **57**: 27–77.

Lefkowitz R J, Cotecchia S, Samama P and Costa T (1993) Constitutive Activity of Receptors Coupled to Guanine-Nucleotide Regulatory Proteins. *Trends Pharmacol Sci* **14**: 303–307.

Lehoux S, Castier Y and Tedgui A (2006) Molecular Mechanisms of the Vascular Responses to Haemodynamic Forces. *J Internal Med* **259**: 381–392.

Lehtonen J Y A and Kinnunen P K J (1994) Changes in the Lipid Dynamics of Liposomal Membranes Induced by Poly (Ethylene Glycol) – Free-Volume Alterations Revealed by Intermolecular and Intramolecular Excimer-Forming Phospholipid Analogs. *Biophys J* **66**: 1981–1990.

Lehtonen J Y A and Kinnunen P K J (1995) Phospholipase A (2) as a Mechanosensor. *Biophys J* **68**: 1888–1894.

Leurs R, Smit M J, Alewijnse A E and Timmerman H (1998) Agonist-Independent Regulation of Constitutively Active G-Protein-Coupled Receptors. *Trends Biochem Sci* **23**: 418–422.

Levin S and Korenstein R (1991) Membrane Fluctuations in Erythrocytes Are Linked to Mgatp-Dependent Dynamic Assembly of the Membrane Skeleton. *Biophys J* **60**: 733–737.

Liedtke W, Tobin D M, Bargmann C I and Friedman J M (2003) Mammalian TRPV4 (VR-OAC) Directs Behavioral Responses to Osmotic and Mechanical Stimuli in Caenorhabditis Elegans. *Proceedings of the National Academy of Sciences of the United States of America* **100**: 14531–14536.

Litman B J and Mitchell D C (1996) A Role for Phospholipid Polyunsaturation in Modulating Membrane Protein Function. *Lipids* **31**: S193–S197.

Loutfy R O (1986) Fluorescence Probes for Polymer Free-Volume. *Pure Appl Chem* **58**: 1239–1248.

Loutfy R O and Law K Y (1980) Electrochemistry and Spectroscopy of Intramolecular Charge-Transfer Complexes – Para-N,N-Dialkylaminobenzylidenemalononitriles. *J Phys Chem* **84**: 2803–2808.

Luzardo M D, Amalfa F, Nunez A M, Diaz S, de Lopez A C B and Disalvo E A (2000) Effect of Trehalose and Sucrose on the Hydration and Dipole Potential of Lipid Bilayers. *Biophys J* **78**: 2452–2458.

Makino A, Glogauer M, Bokoch G M, Chien S and Schmid-Schonbein G W (2005) Control of Neutrophil Pseudopods by Fluid Shear: Role of Rho Family GTPases. *Am J Phys – Cell Physiol* **288**: C863–C871.

Makino A, Prossnitz E R, Bunemann M, Wang J M, Yao W and Schmid-Schonbein G W (2006) G-protein-Coupled Receptors Serve as Mechanosensors for Fluid Shear Stress in Neutrophils. *Am J Physiol – Cell Physiol* **290**: C1633–C1639.

Malek A M, Alper S L and Izumo S (1999a) Hemodynamic Shear Stress and Its Role in Atherosclerosis. *JAMA* **282**: 2035–2042.

Malek A M, Jiang L, Lee I, Sessa W C, Izumo S and Alper S L (1999b) Induction of Nitric Oxide Synthase MRNA by Shear Stress Requires Intracellular Calcium and G-Protein Signals and Is Modulated by PI 3 Kinase. *Biochem Biophys Res Commun* **254**: 231–242.

Maroto R, Raso A, Wood T G, Kurosky A, Martinac B and Hamill O P (2005) TRPC1 Forms the Stretch-Activated Cation Channel in Vertebrate Cells. *Nature Cell Biol* **7**: 179–U99.

Marrero M B, Venema V J, Ju H, He H, Liang H Y, Caldwell R B and Venema R C (1999) Endothelial Nitric Oxide Synthase Interactions With G-Protein-Coupled Receptors. *Biochem J* **343**: 335–340.

Martinac B and Hamill O P (2002) Gramicidin A Channels Switch Between Stretch Activation and Stretch Inactivation Depending on Bilayer Thickness. *Proceedings of the National Academy of Sciences of the United States of America* **99**: 4308–4312.

Matsumoto H, Baron C B and Coburn R F (1995) Smooth-Muscle Stretch-Activated Phospholipase-C Activity. *Am J Physiol – Cell Physiol* **37**: C458–C465.

McAllister T N and Frangos J A (1999) Steady and Transient Fluid Shear Stress Stimulate NO Release in Osteoblasts Through Distinct Biochemical Pathways. *J Bone Miner Res* **14**: 930–936.

Mccown J T, Evans E, Diehl S and Wiles H C (1981) Degree of Hydration and Lateral Diffusion in Phospholipid Multibilayers. *Biochemistry* **20**: 3134–3138.

McMahon H T and Gallop J L (2005) Membrane Curvature and Mechanisms of Dynamic Cell Membrane Remodelling. *Nature* **438**: 590–596.

Mitchell D C, Straume M and Litman B J (1992) Role of Sn-1-Saturated,Sn-2-Polyunsaturated Phospholipids in Control of Membrane-Receptor Conformational Equilibrium – Effects of Cholesterol and Acyl Chain Unsaturation on the Metarhodopsin-I-Metarhodopsin-Ii Equilibrium. *Biochemistry* **31**: 662–670.

Mittelman L, Levin S, Verschueren H, Debaetselier P and Korenstein R (1994) Direct Correlation Between Cell-Membrane Fluctuations, Cell Filterability and the Metastatic Potential of Lymphoid-Cell Lines. *Biochem Biophys Res Commun* **203**: 899–906.

Montecucco C, Smith G A, Dabbenisala F, Johannsson A, Galante Y M and Bisson R (1982) Bilayer Thickness and Enzymatic-Activity in the Mitochondrial Cytochrome C-Oxidase and Atpase Complex. *FEBS Letters* **144**: 145–148.
Morris A J and Malbon C C (1999) Physiological Regulation of G-protein-Linked Signaling. *Physiolog Rev* **79**: 1373–1430.
Mouritsen O G and Bloom M (1993) Models of Lipid-Protein Interactions in Membranes. *Ann Rev Biophys Biomolec Structure* **22**: 145–171.
Muller H J and Galla H J (1983) Pressure Variation of the Lateral Diffusion in Lipid Bilayer-Membranes. *Biochimica et Biophysica Acta* **733**: 291–294.
Munro S (2003) Lipid Rafts: Elusive or Illusive? *Cell* **115**: 377–388.
Nagle J F (1980) Theory of the Main Lipid Bilayer Phase-Transition. *Ann Rev Phys Chem* **31**: 157–195.
Naka K, Sadownik A and Regen S L (1992) Molecular Harpoons – Membrane-Disrupting Surfactants That Recognize Osmotic-Stress. *J Am Chem Soc* **114**: 4011–4013.
Nakamura F and Strittmatter S M (1996) P2Y (1) Purinergic Receptors in Sensory Neurons: Contribution to Touch-Induced Impulse Generation. *Proceedings of the National Academy of Sciences of the United States of America* **93**: 10465–10470.
Needham D and Nunn R S (1990) Elastic-Deformation and Failure of Lipid Bilayer-Membranes Containing Cholesterol. *Biophys J* **58**: 997–1009.
Nerem R M (1992) Vascular Fluid Mechanics, the Arterial Wall, and Atherosclerosis. *J Biomech Eng* **114**: 274–282.
Novikov E G and Visser A J W G (2001) Inter- and Intramolecular Dynamics of Pyrenyl Lipids in Bilayer Membranes from Time-Resolved Fluorescence Spectroscopy. *J Fluorescence* **11**: 297–305.
Ohno M, Gibbons G H, Dzau V J and Cooke J P (1993) Shear Stress Elevates Endothelial CGMP. Role of a Potassium Channel and G-protein Coupling. *Circulation* **88**: 193–197.
Okada T, Ernst O P, Palczewski K and Hofmann K P (2001) Activation of Rhodopsin: New Insights from Structural and Biochemical Studies. *Trends in Biochem Sci* **26**: 318–324.
Oldham W M and Hamm H E (2006) Structural Basis of Function in Heterotrimeric G-proteins 28. *Q Rev Biophys* **39**: 117–166.
Oldham W M, Van E N, Preininger A M, Hubbell W L and Hamm H E (2006) Mechanism of the Receptor-Catalyzed Activation of Heterotrimeric G-proteins 30. *Nat Struct Mol Biol* **13**: 772–777.
Paoletti P and Ascher P (1994) Mechanosensitivity of Nmda Receptors in Cultured Mouse Central Neurons. *Neuron* **13**: 645–655.
Parasassi T, Destasio G, Dubaldo A, Rusch R and Gratton E (1990) Phase Fluctuation in Phospholipids Revealed by Laurdan Fluorescence. *Biophys J* **57**: A272.
Parasassi T, Destasio G, Ravagnan G, Rusch R M and Gratton E (1991) Quantitation of Lipid Phases in Phospholipid-Vesicles by the Generalized Polarization of Laurdan Fluorescence. *Biophys J* **60**: 179–189.
Parasassi T, Krasnowska E K, Bagatolli L and Gratton E (1998) LAURDAN and PRODAN as Polarity-Sensitive Fluorescent Membrane Probes. *J Fluorescence* **8**: 365–373.
Parasassi T, Loiero M, Raimondi M, Ravagnan G and Gratton E (1993) Absence of Lipid Gel-Phase Domains in Seven Mammalian Cell Lines and in Four Primary Cell Types. *Biochim Biophys Acta* **1153**: 143–154.
Parola A H (1993) in *Biomembranes Physical Aspects* (Shinitsky M, ed.), pp. 159–277, VCH, New York.
Parsegian V A, Rand R P and Rau D C (2000) Osmotic Stress, Crowding, Preferential Hydration, and Binding: A Comparison of Perspectives. *Proceedings of the National Academy of Sciences of the United States of America* **97**: 3987–3992.
Patel A J, Lazdunski M and Honore E (2001) Lipid and Mechano-Gated 2P Domain K+ Channels. *Current Opinion in Cell Biology* **13**: 422–427.

Peleg G, Ghanouni P, Kobilka B K and Zare R N (2001) Single-Molecule Spectroscopy of the Beta (2) Adrenergic Receptor: Observation of Conformational Substates in a Membrane Protein. *Proceedings of the National Academy of Sciences of the United States of America* **98**: 8469–8474.

Perez D M and Karnik S S (2005) Multiple Signaling States of G-Protein-Coupled Receptors. *Pharmacol Rev* **57**: 147–161.

Perozo E, Cortes D M, Sompornpisut P, Kloda A and Martinac B (2002a) Open Channel Structure of MscL and the Gating Mechanism of Mechanosensitive Channels. *Nature* **418**: 942–948.

Perozo E, Kloda A, Cortes D M and Martinac B (2002b) Physical Principles Underlying the Transduction of Bilayer Deformation Forces During Mechanosensitive Channel Gating. *Nature Structural Biology* **9**: 696–703.

Peters R and Cherry R J (1982) Lateral and Rotational Diffusion of Bacteriorhodopsin in Lipid Bilayers – Experimental Test of the Saffman-Delbruck Equations. *Proceedings of the National Academy of Sciences of the United States of America – Biological Sciences* **79**: 4317–4321.

Pi X, Yan C and Berk B C (2004) Big Mitogen-Activated Protein Kinase (BMK1)/ERK5 Protects Endothelial Cells from Apoptosis. *Circ Res* **94**: 362–369.

Polozova A and Litman B J (2000) Cholesterol Dependent Recruitment of Di22: 6-PC by a G-protein-Coupled Receptor into Lateral Domains. *Biophys J* **79**: 2632–2643.

Rand R P, Parsegian V A and Rau D C (2000) Intracellular Osmotic Action. *Cellular and Molecular Life Sciences* **57**: 1018–1032.

Reich K M, Mcallister T N, Gudi S and Frangos J A (1997) Activation of G-proteins Mediates Flow-Induced Prostaglandin E (2) Production in Osteoblasts. *Endocrinology* **138**: 1014–1018.

Rietveld A, Neutz S, Simons K and Eaton S (1999) Association of Sterol- and Glycosylphosphatidylinositol-Linked Proteins with Drosophila Raft Lipid Microdomains. *J Biol Chem* **274**: 12049–12054.

Rigler R and Elson ES (2001) *Fluorescence Correlation Spectroscopy*. Springer, Berlin.

Rizzo V, McIntosh D P, Oh P and Schnitzer J E (1998) In Situ Flow Activates Endothelial Nitric Oxide Synthase in Luminal Caveolae of Endothelium with Rapid Caveolin Dissociation and Calmodulin Association. *J Biol Chem* **273**: 34724–34729.

Rizzo V, Morton C, DePaola N, Schnitzer J E and Davies P F (2003) Recruitment of Endothelial Caveolae into Mechanotransduction Pathways by Flow Conditioning in Vitro. *Am J Physiol – Heart and Circulatory Physiol* **285**: H1720–H1729.

Roberts D J and Waelbroeck M (2004) G-protein Activation by G-protein Coupled Receptors: Ternary Complex Formation or Catalyzed Reaction? *Biochem Pharmacol* **68**: 799–806.

Sackin H (1995) Mechanosensitive Channels. *Ann Rev Physiol* **57**: 333–353.

Saffman P G and Delbruck M (1975) Brownian-Motion in Biological Membranes. *Proceedings of the National Academy of Sciences of the United States of America* **72**: 3111–3113.

Safran A S (2003) *Statistical Thermodynamics of Surfaces, Interfaces and Membranes*. Westview Press, Boulder, Co.

Salwen S A, Szarowski D H, Turner J N and Bizios R (1998) Three-Dimensional Changes of the Cytoskeleton of Vascular Endothelial Cells Exposed to Sustained Hydrostatic Pressure. *Medical & Biological Engineering & Computing* **36**: 520–527.

Schwartz E A, Bizios R, Medow M S and Gerritsen M E (1999) Exposure of Human Vascular Endothelial Cells to Sustained Hydrostatic Pressure Stimulates Proliferation – Involvement of the Alpha (v) Integrins. *Circ Res* **84**: 315–322.

Schwarz G, Callewaert G, Droogmans G and Nilius B (1992) Shear Stress-Induced Calcium Transients in Endothelial Cells from Human Umbilical Cord Veins. *J Physiol* **458**: 527–538.

Sessa W C (2004) ENOS at a Glance. *J Cell Sci* **117**: 2427–2429.

Simon S A and Mcintosh T J (1986) Depth of Water Penetration into Lipid Bilayers. *Methods in Enzymology* **127**: 511–521.

Simons K and Toomre D (2000) Lipid Rafts and Signal Transduction. *Nature Rev Mol Cell Biol* **1**: 31–39.

Simons K and Vaz W L C (2004) Model Systems, Lipid Rafts, and Cell Membranes. *Ann Rev Biophys Biomolec Structure* **33**: 269–295.

Soderlund T, Alakoskela J M I, Pakkanen A L and Kinnunen P K J (2003) Comparison of the Effects of Surface Tension and Osmotic Pressure on the Interfacial Hydration of a Fluid Phospholipid Bilayer. *Biophys J* **85**: 2333–2341.

Sukharev S I, Blount P, Martinac B, Blattner F R and Kung C (1994) A Large-Conductance Mechanosensitive Channel in E. Coli Encoded by Mscl Alone. *Nature* **368**: 265–268.

Sumpio B E, Yun S, Cordova A C, Haga M, Zhang J, Koh Y and Madri J A (2005) MAPKs (ERK1/2, P38) and AKT Can Be Phosphorylated by Shear Stress Independently of Platelet Endothelial Cell Adhesion Molecule-1 (CD31) in Vascular Endothelial Cells 45. *J Biol Chem* **280**: 11185–11191.

Surapisitchat J, Hoefen R J, Pi X, Yoshizumi M, Yan C and Berk B C (2001) Fluid Shear Stress Inhibits TNF-Alpha Activation of JNK but Not ERK1/2 or P38 in Human Umbilical Vein Endothelial Cells: Inhibitory Crosstalk Among MAPK Family Members. *Proceedings of the National Academy of Sciences of the United States America* **98**: 6476–6481.

Svindland A (1983) The Localization of Sudanophilic and Fibrous Plaques in the Main Left Coronary Bifurcation. *Atherosclerosis* **48**: 139–145.

Takahashi M, Ishida T, Traub O, Corson M A and Berk B C (1997) Mechanotransduction in Endothelial Cells: Temporal Signaling Events in Response to Shear Stress. *J Vasc Res* **34**: 212–219.

Traub O and Berk B C (1998) Laminar Shear Stress: Mechanisms by Which Endothelial Cells Transduce an Atheroprotective Force. *Arterioscler Thromb Vasc Biol* **18**: 677–685.

Traub O, Monia B P, Dean N M and Berk B C (1997) PKC-Epsilon Is Required for Mechano-Sensitive Activation of ERK1/2 in Endothelial Cells. *J Biol Chem* **272**: 31251–31257.

Tseng H, Peterson T E and Berk B C (1995) Fluid Shear Stress Stimulates Mitogen-Activated Protein Kinase in Endothelial Cells. *Circ Res* **77**: 869–878.

Tuvia S, Almagor A, Bitler A, Levin S, Korenstein R and Yedgar S (1997) Cell Membrane Fluctuations Are Regulated by Medium Macroviscosity: Evidence for a Metabolic Driving Force. *Proceedings of the National Academy of Sciences of the United States of America* **94**: 5045–5049.

Tuvia S, Levin S, Bitler A and Korenstein R (1998) Mechanical Fluctuations of the Membrane-Skeleton Are Dependent on F-Actin ATPase in Human Erythrocytes. *J Cell Biol* **141**: 1551–1561.

Tzima E, Irani-Tehrani M, Kiosses W B, Dejana E, Schultz D A, Engelhardt B, Cao G, DeLisser H and Schwartz M A (2005) A Mechanosensory Complex That Mediates the Endothelial Cell Response to Fluid Shear Stress. *Nature* **437**: 426–431.

Vaz W L C, Clegg R M and Hallmann D (1985) Translational Diffusion of Lipids in Liquid-Crystalline Phase Phosphatidylcholine Multibilayers – A Comparison of Experiment with Theory. *Biochemistry* **24**: 781–786.

Vaz W L C, Goodsaidzalduondo F and Jacobson K (1984) Lateral Diffusion of Lipids and Proteins in Bilayer-Membranes. *FEBS Letters* **174**: 199–207.

Viard M, Gallay J, Vincent M, Meyer O, Robert B and Paternostre M (1997) Laurdan Solvatochromism: Solvent Dielectric Relaxation and Intramolecular Excited-State Reaction. *Biophys J* **73**: 2221–2234.

Vilardaga J P, Bunemann M, Krasel C, Castro M and Lohse M J (2003) Measurement of the Millisecond Activation Switch of G-protein-Coupled Receptors in Living Cells. *Nature Biotechnol* **21**: 807–812.

Vilardaga J P, Steinmeyer R, Harms G S and Lohse M J (2005) Molecular Basis of Inverse Agonism in a G-protein-Coupled Receptor. *Nature Chemical Biol* **1**: 25–28.

Vogler O, Casas J, Capo D, Nagy T, Borchert G, Martorell G and Escriba P V (2004) The G Beta Gamma Dimer Drives the Interaction of Heterotrimeric G (i) Proteins with Nonlamellar Membrane Structures. *J Biol Chem* **279**: 36540–36545.

Wang X L, Fu A, Raghavakaimal S and Lee H C (2007) Proteomic Analysis of Vascular Endothelial Cells in Response to Laminar Shear Stress. *Proteomics* **7**: 588–596.

Wang Y, Botelho A V, Martinez G V and Brown M F (2002) Electrostatic Properties of Membrane Lipids Coupled to Metarhodopsin II Formation in Visual Transduction. *J Am Chem Soci* **124**: 7690–7701.

Wedgwood S, Bekker J M and Black S M (2001) Shear Stress Regulation of Endothelial NOS in Fetal Pulmonary Arterial Endothelial Cells Involves PKC. *Am J Physiol Lung Cell Mol Physiol* **281**: L490–L498.

Werry T D, Sexton P M and Christopoulos A (2005) "Ins and Outs" of Seven-Transmembrane Receptor Signalling to ERK. *Trends in Endocrinology and Metabolism* **16**: 26–33.

White C R, Dusserre N and Frangos J A (2003) Steady and Unsteady Fluid Shear Control of Inflammation, in *Molecular Basis of Microcirculatory Disorders* (Schmid-Schoenbein G W and Granger N D, eds.), pp. 141–160, Springer-Verlag, New York.

White C R and Frangos J A (2007) The Shear Stress of It All: The Cell Membrane and Mechanochemical Transduction. *Philos Trans R Soc Lond B Biol Sci.*

White C R, Stevens H Y, Haidekker M and Frangos J A (2005) Temporal Gradients in Shear, but Not Spatial Gradients, Stimulate ERK1/2 Activation in Human Endothelial Cells. *Am J Physiol – Heart Circ Physiol* **289**: H2350–H2355.

Wiggins P and Phillips R (2004) Analytic Models for Mechanotransduction: Gating a Mechanosensitive Channel. *Proceedings of the National Academy of Sciences of the United States of America* **101**: 4071–4076.

Yamazaki M, Ohnishi S and Ito T (1989) Osmoelastic Coupling in Biological Structures – Decrease in Membrane Fluidity and Osmophobic Association of Phospholipid-Vesicles in Response to Osmotic-Stress. *Biochemistry* **28**: 3710–3715.

Yamazaki T, Komuro I, Shiojima I and Yazaki Y (1999) The Molecular Mechanism of Cardiac Hypertrophy and Failure. *Ann NY Acad Sci* **874**: 38–48.

Yang Q, Alemany R, Casas J, Kitajka K, Lanier S M and Escriba P V (2005) Influence of the Membrane Lipid Structure on Signal Processing via G-protein-Coupled Receptors. *Molecular Pharmacology* **68**: 210–217.

Yu J, Bergaya S, Murata T, Alp I F, Bauer M P, Lin M I, Drab M, Kurzchalia T V, Stan R V and Sessa W C (2006) Direct Evidence for the Role of Caveolin-1 and Caveolae in Mechanotransduction and Remodeling of Blood Vessels. *J Clin Investig* **116**: 1284–1291.

Zhang Y L, Frangos J A and Chachisvilis M (2006) Laurdan Fluorescence Senses Mechanical Strain in the Lipid Bilayer Membrane. *Biochem Biophys Res Communic* **347**: 838–841.

Zou Y Z, Akazawa H, Qin Y J, Sano M, Takano H, Minamino T, Makita N, Iwanaga K, Zhu W D, Kudoh S, Toko H, Tamura K, Kihara M, Nagai T, Fukamizu A, Umemura S, Iiri T, Fujita T and Komuro I (2004) Mechanical Stress Activates Angiotensin II Type 1 Receptor Without the Involvement of Angiotensin II. *Nature Cell Biology* **6**: 499–506.

5

Cellular Mechanotransduction: Interactions with the Extracellular Matrix

Andrew D. Doyle and Kenneth M. Yamada

5.1 Introduction

Much like whole organisms, single cells have the ability to "sense" and respond to their surroundings. This "sensing" not only includes monitoring and responding to changes in extracellular chemical messages, but also the physical nature of the cell's microenvironment, particularly the components of the extracellular matrix (ECM). Anchorage to the surrounding ECM is important for many cellular functions and is mediated primarily by the integrin family, a group of heterodimeric transmembrane proteins that provide physical links of the cell to the external environment. Although integrins were once viewed as structural membrane proteins providing anchor points involved in cell adhesion and movement, they are now known to be centrally important for sensing the external environment and regulating the precise intracellular responses necessary for proper mechanotransduction.

Recent evidence suggests that besides its biochemical composition, the dimensional and rheological properties of the ECM are involved in signaling processes that not only affect cell motility, but also a multitude of intracellular second messenger pathways and gene regulation. In this chapter, we review how cells and their surrounding ECM interact, particularly focusing on integrins and fibronectin, and examine how their points of contact are involved in inside-out and outside-in signaling for setting the stage for mechanotransduction. In addition, a second major focus will be on the most recent findings regarding cellular mechanosensing and its relationship to the ECM. Furthermore, we describe how alterations to matrix components can lead to altered cellular motility, phenotype, and cellular responses.

The associations of cells with the extracellular environment are not only important for regulating cell homeostasis and longevity, but they also act as the "glue" that interconnects single cells into tissues and tissues into organs that sustain multicellular organisms. This "glue" or extracellular matrix is comprised mainly of glycoproteins and proteoglycans produced by cells for both cell–matrix associations and connective tissue or skeletal structures. While cytoskeletal proteins provide cells with an internal structure, ECM components offer cells external support, as well as providing physical and chemical cues to which cells react and respond. While many matrix components contribute to the unique environment of each ECM, relatively few of these components are found in all tissues. The specific components of the ECM surrounding cells are dependent on the tissue type and the individual cells. While endothelial and epithelial cells grow on a basal lamina consisting of laminin, collagen IV, and other molecules, tissues such as tendons that require an ECM with high tensile strength are composed mainly of collagens (Alberts et al., 2002; Hay, 1981).

Most cell types, with few exceptions, generate their own matrices to provide the optimal constituents necessary for cell/tissue survival. A key function of the ECM related to mechanotransduction is to provide anchorage points and a physical scaffolding for cell proliferation and differentiation. Although there are many different ECM proteins, in keeping with the theme of this book, we will concentrate on integrin ligands involved in mechanotransduction, mainly fibronectin and collagens.

5.2 Integrins

The integration of cells into their physical surroundings requires distinct structural connections to the ECM. It is through these connections that cells process information about their environment and respond with an appropriate reaction. Abercrombie and Dunn (1975) used electron microscopy and interference reflection microscopy to show that cells attach at specific sites to the underlying substratum. The molecular compositions and functions of these cell adhesions have been elucidated through the work of many laboratories over the past two decades (Danen and Yamada, 2001; DeMali et al., 2003; Geiger and Bershadsky, 2001; Giancotti and Ruoslahti, 1999; Kaverina et al., 2002; Wozniak et al., 2004).

The integrin family of adhesion receptors consists of single α- and β-subunits combined into specific heterodimers. Different combinations of these subunits result in integrins with different binding affinities to ECM ligands such as fibronectin, collagen, and laminin (Hynes, 1987). To date, 18 α-subunits and 8 β-subunits have been identified, which together can form 24 different heterodimer combinations. Although other adhesion receptors can function in a similar manner, most vertebrate cells utilize integrins for cell–substrate adhesion formation. On the cytoplasmic side of the cell membrane, integrins bind directly to α-actinin (Otey et al., 1993; Sampath et al., 1998), talin (Buck et al., 1986), tensin (Davis et al., 1991; Miyamoto et al., 1995a), focal adhesion kinase (FAK) (Lipfert et al., 1992; Schaller et al., 1993), and other accessory proteins. These components of integrin-based adhesion complexes provide anchorage of cytoskeletal molecules such as F-actin, regulate integrin function, and activate or modulate many signal transduction pathways relevant to mechanotransduction (Arnaout et al., 2005; Bershadsky et al., 2003; French-Constant and Colognato, 2004; Giancotti and Ruoslahti, 1999; Ginsberg et al., 2005; Larsen et al., 2006; Miranti and Brugge, 2002; Schoenwaelder and Burridge, 1999; Schwartz and Ginsberg, 2002; Yamada and Geiger, 1997).

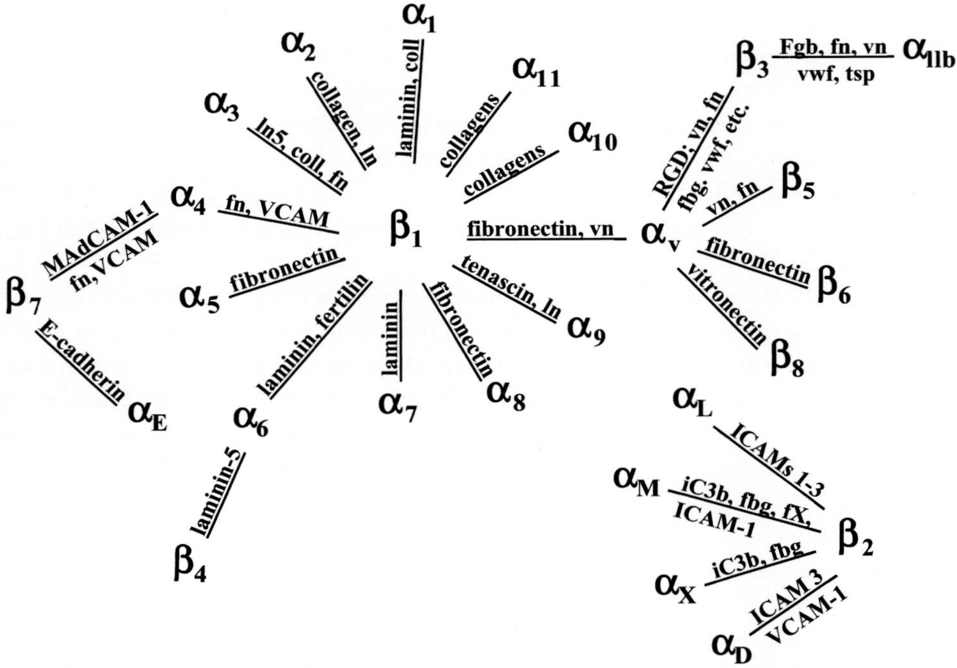

Figure 5.1. Integrins and their ligands. This schematic shows all 18 α- and 8 β-subunits that comprise the 24 known integrin pairs. Representative molecules bound by each pair of integrin α- and β-subunits are shown above the line drawn between specific subunits. Abbreviations are as follows: fn, fibronectin; vn, vitronectin; ln, laminin; coll, collagen; fbg, fibrinogen; vwf, von Wildebrand factor; RGD, arginine-glycine-aspartate tripeptide; iC3b, activated C3 complement fragment; fX, factor X.

5.3 Integrin Binding to Ligands

Integrins bind to specific peptide sequences or protein conformations in ligands. The best-characterized peptide recognition sequence is the tripeptide alanine-glycine-aspartate (RGD peptide) found in fibronectin, vitronectin, and other adhesion molecules, which is required for ligation to the extracellular N-terminus of a number of β_1- and β_3-integrins (Hynes, 1987; Yamada, 1991; Ruoslahti, 1996). Although the majority of integrins bind to other sequences, the RGD integrin binding site is important for a number of ECM–integrin interactions. A single amino acid substitution of glutamate for the aspartate (D) in the RGD sequence abrogates integrin binding and is commonly used as a negative control for experiments in which RGD peptides are used to probe the role of RGD-binding integrins in cell–matrix interactions.

While the β-integrins bind directly to the RGD sequence of several ECM proteins, the α-subunit conveys ECM specificity as well as being involved in regulation of the β-subunit (Arnaout et al., 2005; Hynes, 2002). For example, Figure 5.1 portrays the 8 known β and the 18 α chains and major ligands. Although the β_1-subunit is well known for its binding to fibronectin in combination with α_5, it can also bind via α_v and α_8. In contrast, when paired with the α_2-subunit, β_1 binds various collagens.

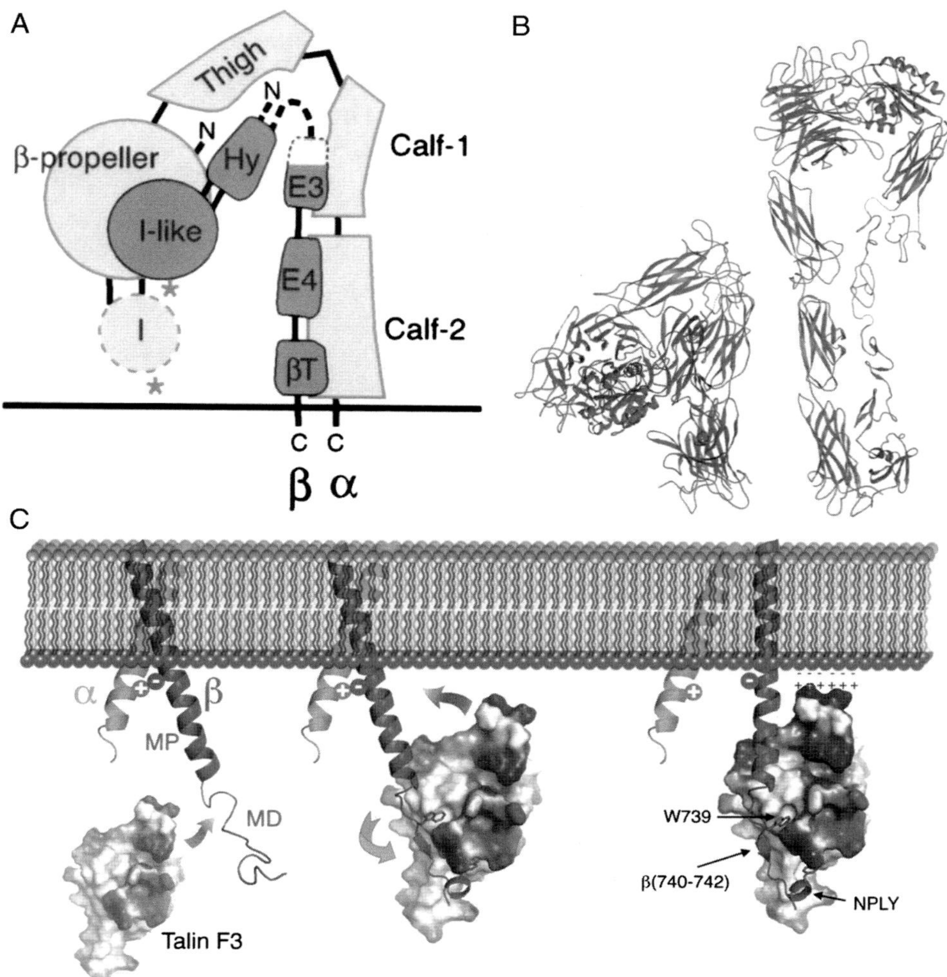

Figure 5.2. Conformational states of α- and β-integrin subunits. (A) Cartoon representing the bent/V-shaped conformation of the complexed $\alpha_v\beta_3$-integrin. The large α-subunit physically contours along the β-chain extracellular domain, blocking extension and activation. Note the proximity of the α-subunit's β-propeller to the I-like region of the β-subunit, which places the synergy site close to the RGD peptide binding site. (B) X-ray crystallography structures of the α- (blue) and the β- (red) subunits in the unbound/inactive (left) and ligand-bound/active (right) conformation. (C) Representation of the changes in the intracellular domains of α- and β-chains when talin binds to the NPLY motif of the β-chain, leading to an extension of the β-chain into the active conformational state. From Takagi et al. (2002) (A), Zhong et al. (2001) (B), and Wegener et al. (2007) (C).

Within the extracellular domain of α-subunits lies a seven-bladed β-propeller region that complexes with the I/A domain of the β-subunit to form the binding head of the heterodimeric integrin (Figure 5.2; Plate 8) (Arnaout et al., 2005; Hynes, 2002; Springer, 1997; Xiao et al., 2004). In the case of the $\alpha_5\beta_1$-integrin, this interaction juxtaposes the RGD binding site within the β-I/A domain with the fibronectin synergy site in the propeller domain of the α-subunit (Mould et al., 1997). The synergy site, which contains the pentapeptide sequence Pro-His-Ser-Arg-Asn (PHSRN), is found in the central cell-binding domain of fibronectin along with the RGD binding site (Aota et al., 1994). Interaction of α_5 with the synergy site not only increases β_1 specificity for fibronectin but also increases affinity and hence a stronger interaction with the ECM (Garcia et al., 2002); both sites are required for cell spreading, migration, and fibronectin matrix assembly (Nagai et al., 1991). Danen et al. (1995) showed that while the synergy site is important for binding to fibronectin, its requirement in ECM–integrin interactions depends on the integrins involved and, more importantly, on the activation state of the integrins. The addition of manganese (Mn^{2+}) or stimulatory antibodies results in integrin activation and strong binding to mutated polypeptides lacking the synergy site.

Boettiger and coworkers (Friedland et al., 2009) established recently that the $\alpha_5\beta_1$ integrin can serve as a mechanosensitive switch activating adhesion and signaling via the synergy site of fibronectin. Extracellular matrix stiffness and cytoskeletal force generated by myosin II can activate this integrin switch, which increases cell–matrix bond strength via the synergy site and activates signal transduction through tyrosine phosphorylation of focal adhesion kinase.

Interestingly, the α-subunit is thought to hinder cytoskeletal attachment to an unligated β_1-subunit. In an early study, Laflamme et al. (1992) showed that chimeras of the β_1-subunit, where the extracellular and transmembrane domains were replaced with those of the human interleukin-2 receptor, localized to regions of the cell containing ligand-occupied integrins, suggesting the intracellular region of the β_1-subunit is sufficient for targeting to adhesion complexes in the absence of the intracellular domain of an α-subunit. In fact, deletion of, or mutations to, an α-subunit also mimic the cytoskeletal interactions of the isolated β intracellular domain (Huttenlocher et al., 1996; Hynes, 2002; Shimaoka et al., 2002; Ylanne et al., 1993). From these and many other studies, it appears that the α-subunit not only regulates the β-subunit's extracellular specificity and avidity but also the interactions of the cytoplasmic β tail with cytoskeletal and signaling molecules.

5.4 Integrin Activation: Opening the Door to Ligand Binding

The capacity of a particular integrin to bind to a ligand can be regulated by both intracellular and extracellular mechanisms. For example, interactions with the ECM and constituents of extracellular fluid can modulate integrin binding. The extracellular N-terminus of the β_1 and β_3 (as well as other β-subunits) chains undergoes a conformational change before or during ligation with the ECM. This alteration can be detected using an antibody to a region of the integrin extracellular domain that is normally hidden until integrin activation exposes the activation epitope. For example, antibodies such as PAC1, TS2/16 or 9EG7, and SNAKA51 detect, and in some cases can directly induce, the activated state of β_3-, β_1-, or α_5-integrins, respectively (Bazzoni et al., 1995; Clark et al., 2005).

The molecular structural basis of integrin activation has been revealed by analyses of the crystal structures of $\alpha_v\beta_3$- and $\alpha_5\beta_1$-integrins. These integrins have either a V-shaped bent/closed conformation or an extended/open conformation that is associated with activation. As shown in Figure 5.2 (Plate 8), in the bent/closed conformation the α-subunit covers or wraps around the smaller β-subunit with its β-propeller structure. Binding of the ligand results in a "switch-blade"–like action, extending the N-terminus more than 80° from the closed conformation into the high-affinity state (Luo et al., 2004; Takagi et al., 2002; Xiao et al., 2004).

The activation of integrins by intracellular mechanisms has been described as "inside-out signaling," where cytoskeletal elements interact with the short β cytoplasmic domain and induce conformational changes in the extracellular domain that lead to integrin activation. Work of the Ginsberg laboratory revealed that binding of the actin-binding protein talin to the phosphotyrosine binding (PTB) domain of the cytoplasmic β-tail leads to integrin activation (Calderwood, 2004b; Calderwood et al., 2002). The FERM domain within the head region of talin binds directly to the NPLY motif of the PTB domain and leads to the conformational changes in the extracellular N-terminus of all β-subunits (Tadokoro et al., 2003). The FERM domain of talin often lies hidden and can be exposed via talin head interaction with phosphatidylinositol-4,5-bisphosphate or proteolysis by calpain, which cleaves the head from the rod-like domain (Calderwood et al., 1999; Yan et al., 2001). Many other kinases such as FAK and Src are known to interact with the PTB domain, yet they do not activate integrins and can in fact lead to inhibition when overexpressed (Calderwood et al., 2003; Garcia-Alvarez et al., 2003).

After the F3 portion of the talin FERM domain initially binds to the PTB domain of the distal region of the cytoplasmic β_3-tail, a conformational change occurring between the two proteins results in the separation of the putative salt bridge between the α- and β-subunits (Figure 5.2(C); Plate 8) (Wegener et al., 2007). This separation of the salt bridge occurs concomitantly with the association of the F3 domain of talin with the membrane-proximal region of the β_3-tail while maintaining an interaction with the distal region. Electrostatic forces between membrane lipids (negatively charged) and the F3 domain (positively charged) favor the conformational change that results in the activated state.

Besides talin, cytoplasmic proteins termed kindlins have been found recently to play essential roles in regulating integrin activation (Harburger et al., 2009; Larjava et al., 2008; Montanez et al., 2008; Moser et al., 2008, 2009; Ussar et al., 2008). Even in the presence of functional talin, genetic ablation of kindlin-1, kindlin-2, or kindlin-3 causes major disruptions of integrin activation, cell adhesion, and the function of a wide variety of cell and tissue types. Kindlins can bind directly to the carboxyl-terminal region of integrin β-subunit cytoplasmic domains to modulate integrin activation. Although the precise molecular mechanisms of kindlin function and modes of cooperation with talin remain to be determined, it is clear that integrin activation is a complex process involving multiple proteins.

Another important mechanism for regulating integrin activation involves the concentrations of three divalent cations found in the extracellular milieu: manganese (Mn^{2+}), magnesium (Mg^{2+}), and calcium (Ca^{2+}). On both the α- and β-subunits are several divalent cation-binding sites that regulate the binding affinity of integrins for the ECM (e.g., fibronectin and vitronectin) (Mould et al., 1995a, 1995b). Both Mn^{2+} and Mg^{2+} support ligand binding, whereas Ca^{2+} modulates the other cations in a noncompetitive and competitive manner, respectively. All three cations bind to a specific metal ion–dependent adhesion site (MIDAS) found in all integrins (Shimaoka et al., 2002). The presence of high [Ca^{2+}] decreases Mg^{2+} binding to its putative site, yet at lower concentrations it can promote binding. The ratio of Mg^{2+} to Ca^{2+} determines the binding affinity of β_1-integrins for fibronectin (Grzesiak et al., 1992; Onley et al., 2000). While examining conformational rearrangements associated with outside-in and inside-out signaling through $\alpha_v\beta_3$-integrins, Takagi et al. (2002) found that purification of the heterodimer in 5 mM Ca^{2+} resulted in a V-shaped bent or closed conformation as well as a conformational change in the head piece that is associated with ligand binding. Purification in the absence of Ca^{2+} and the presence of Mn^{2+}, which mimics inside-out activation similar to talin binding (Calderwood, 2004a, 2004b), causes extension into an open conformation. Interestingly, when the $\alpha_v\beta_3$-integrin was purified with Ca^{2+} and RGD peptide, the configuration shifted to an extended/open conformation, indicating that ligand binding induces conformational changes in the extracellular domain. Further evidence for Mn^{2+} involvement in β_1-integrin activation was shown by its ability to promote fibronectin fibril formation in a recombinant form of fibronectin lacking the synergy site (Sechler et al., 2001). These changes associated with the Mn^{2+} activation of integrins have also been visualized in live cells with fluorescence time-lapse imaging. The Wehrle-Haller laboratory discovered that the addition of Mn^{2+} to melanoma cells expressing a GFP-tagged β_3-integrin induced clustering at the cell periphery as well as aberrant clustering beneath the cell body (Cluzel et al., 2005). Similar results were obtained when mutations were generated in the cytoplasmic salt bridge and an extracellular glycosylation site of the β_3-subunit.

Together, multiple entities, including extracellular cations, the presence of ECM–ligand binding and synergy sites, the specific subunits involved, and several key cytoskeletal proteins all are involved in regulating integrin function and activation states. However, once the ligand is physically bound and the talin associates with the β-subunit cytoplasmic domain to promote integrin activation, the formation of the structural attachment between the ECM and the cell has only just begun. For mechanotransduction-based signaling to be transmitted (i.e., through contractile forces) back and forth across the cell membrane, numerous actin-binding proteins, adaptor proteins, kinases, and phosphatases are required. In the next section, we will briefly describe events that occur in direct response to integrin engagement, known as outside-in signaling.

5.5 Complex Formation: How Outside-In Signaling Is Initiated through Ligand Binding

Although integrin activation and attachment to ECM ligands is obviously crucial for the interaction of cells with the surrounding ECM, this phase is simply the "tip of the iceberg" in the process of adhesion formation. Increased clustering of integrins bound to ligands leads to rapid incorporation of other adhesion-complex proteins such as vinculin and a number of other anchoring and signaling molecules (Geiger et al., 2001; Miyamoto et al., 1995a). Other chapters focus directly on many of the constituents that are involved and their specified functions, so we will only briefly review proteins that initially accumulate at contact sites. Irrespective of integrin ligand occupancy, clustering of integrins is an important first step toward adhesion complex formation; this process alone can induce FAK activation (Akiyama, 1996; Akiyama et al., 1989; Miyamoto et al., 1995a). Ligand binding and tyrosine phosphorylation stimulate the accumulation of other accessory molecules. For example, in the case of ligand binding (e.g., fibronectin), the actin-binding proteins α-actinin, talin, and vinculin accumulate, while tyrosine phosphorylation promotes the co-clustering of paxillin, Src kinases, extracellular-signal–regulated kinase (ERK), and others. Once the ligand is bound to the integrin and tyrosine phosphorylation occurs, the cytoskeletal link to actin is formed.

It is now widely accepted that adhesions "mature," since the incorporation of many adhesion molecules into an adhesion is dependent upon the presence of others in a hierarchical fashion (Miyamoto et al., 1996). As adhesions mature, more and more proteins are incorporated into their macromolecular complexes. As a result, adhesions become more stable, yet they remain dynamic in the sense that proteins are continually incorporating and leaving the adhesion site. Because adhesions can proceed through numerous stages of maturity, specific names have been given to the major types of adhesions (Figure 5.3). First are the newly formed cell adhesions termed focal complexes (FC) that are transient in nature and contain $\alpha_v\beta_3$-integrin, α-actinin, paxillin, vinculin, and FAK, while displaying high levels of tyrosine phosphorylation (Zaidel-Bar et al., 2003). The conversion of FC to focal adhesions (FAs) is highlighted by the addition of zyxin, the departure of paxillin, and the linking of stress fibers to adhesion sites (Zaidel-Bar et al., 2003; Zamir et al., 1999). Paxillin molecules within FCs were shown to leave adhesion sites and to be reincorporated into nascent adhesions at the leading edge (Webb et al., 2004). Not all cells use β_3-integrin, such as fish keratocytes, which are dependent on β_1-integrin for adhesion formation (Lee and Jacobson, 1997).

This conversion of FCs to FAs depends on RhoA activation (Ballestrem et al., 2001) or the application of force (Galbraith et al., 2002; Riveline et al., 2001). Blocking RhoA activity leads to stress fiber and adhesion disassembly, indicating that the regulation of this process is vital for cytoskeletal linkage and the maturation of adhesions (Chrzanowska-Wodnicka and Burridge, 1996). Fibrillar adhesions (FX) found beneath the cell body with an elongated appearance are marked by tensin (Pankov et al., 2000; Zaidel-Bar et al., 2003; Zamir et al., 1999). FXs also show a transition from the $\alpha_v\beta_3$- to the $\alpha_5\beta_1$-integrin specific for fibronectin (Pankov et al., 2000). FXs have little to no tyrosine phosphorylation (Cukierman et al., 2001, 2002; Pankov et al., 2000; Zaidel-Bar et al., 2003) and assemble along actin stress fibers while regulating or mediating fibronectin fibrillogenesis (Pankov et al., 2000). These three

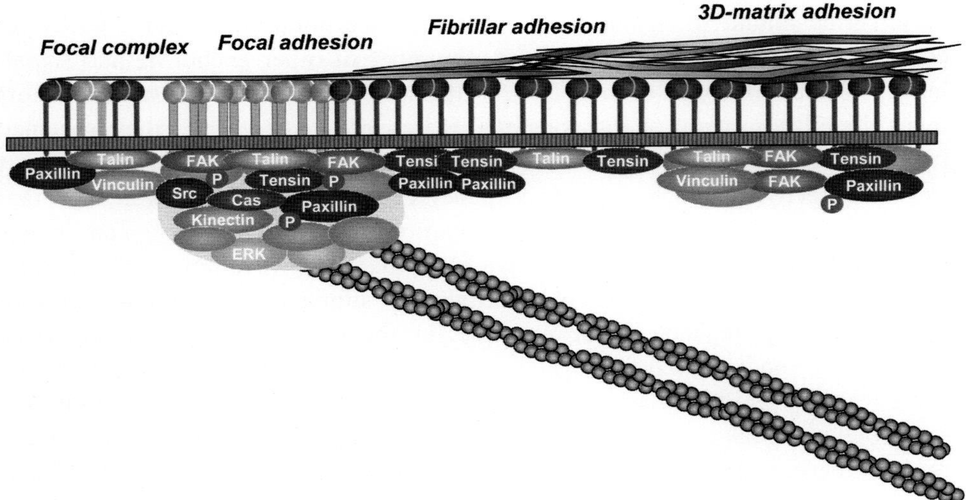

Figure 5.3. From focal complexes to three-dimensional-matrix adhesions. Starting from the left, focal complexes form with the clustering of integrin receptors, ligand occupancy, and talin binding to the NPxY of the β-subunit. Vinculin bound to talin and actin, as well as the adaptor protein paxillin, are also incorporated early. As complexes continue to assemble and enlarge, other focal adhesion proteins aggregate (FAK, tensin, and kinectin) and anchor F-actin in stress fibers. In addition, signaling proteins such as ERK, Src, and Cas ($p130^{Cas}$) accumulate at focal adhesion sites, which have prominent tyrosine phosphorylated proteins including FAK, Src, and paxillin. Tensin and talin remain as prominent constituents as focal adhesions elongate to form fibrillar adhesions containing the $\alpha_5\beta_1$-integrin. Cells in three-dimensional fibronectin-containing matrices for three-dimensional matrix adhesions containing paxillin, tensin, FAK, talin, and vinculin colocalized with $\alpha_5\beta_1$-integrins. FAK becomes nonphosphorylated in these adhesions, whereas paxillin remains phosphorylated.

(FC, FA, FX) adhesion types are involved in cell movement and matrix assembly in two-dimensional cell culture conditions.

In contrast, long and slender three-dimensional matrix adhesions (Cukierman et al., 2001, 2002) form when cells are surrounded by a three-dimensional ECMs *in vitro* or *in vivo* (Figure 5.3). This different type of adhesion in three-dimensional settings adds another layer of complexity to adhesion formation. Because three-dimensional ECMs are naturally elastic compared to the rigid surfaces used for routine cell cultures, the rigidity and dimensionality of the matrix may affect the signaling within adhesion complexes and produce different distributions of adhesion components. It seems likely that all of these adhesion types could be intricately involved in cell–matrix interactions depending on the timing and nature of the local ECM microenvironment. Besides the stable attachments that cell adhesions provide, they also provide the mechanical and signaling links for force sensing, generation, and propagation that are important in integrin–ECM contacts.

5.6 Traction Forces: Transmitting Contractility to the ECM

Cellular mechanotransduction depends on four elements: the cytoskeleton, the extracellular matrix, the transmembrane proteins that link these entities, and the numerous signaling molecules that associate within adhesion complexes. The dynamic nature of the cytoskeleton provides a mechanism for altering cell shape, while still providing cytoskeletal stabilization. Through the interconnection of the numerous parts of the cytoskeleton with transmembrane adhesion receptors (particularly integrins), cells are linked to their external environment. It is at these points of contact that contractile forces, and hence mechanotransduction, generated by the actin cytoskeleton is transmitted to the ECM. Contractile forces and mechanotransduction are required for cell movement and contribute to cell signaling and gene regulation. There are three general types of forces that are produced by cells: 1) traction forces transmitted through integrin connections to the ECM, 2) protrusive forces produced by actin polymerization against the cell membrane, and 3) cell-to-cell forces transmitted through cell-to-cell contacts such as cadherins and intercellular structures such as desmosomes and tight junctions. We will focus solely on traction forces because of their direct relevance to the ECM.

Regardless of the surrounding environment, the force-generating capability of the cell depends on the interactions of motor proteins such as myosins, dyneins, and kinesins with cytoskeletal filaments. Of these motor proteins, the myosin superfamily is the largest, consisting of 17 different classes that together contain 139 members (Hodge and Cope, 2000). Among the variants in this superfamily, class II (i.e., myosin II) is the predominant type that binds F-actin and is involved in cytoskeletal contractility. Myosin II is expressed in most cells, ranging from muscle (cardiac, skeletal, and smooth) to nonmuscle cells. Myosin II exists as a dimer consisting of two heavy chains, each associated with a pair of light chains (Alberts et al., 2002). Three myosin II (MII) isoforms have been identified (A, B, and C), which are defined by their associated heavy chain. Only MIIA and MIIB are ubiquitously expressed in most tissues, and they demonstrate a distinct localization on two-dimensional substrates; MIIA is located more toward the leading edge, whereas MIIB is at the cell rear (Cai et al., 2006; Vicente-Manzanares et al., 2008). The head region that contains light chains has ATPase activity. The rod-like tail is formed from a coiled-coil interaction of the two α-helices of the heavy chains. Between these two regions is a hinge/neck that changes conformation during contraction. One crucial element in the ability of myosin II to move along actin filaments is the interaction of the two myosin tails. It is this coupling that gives myosin II its cross-linking capability and the ability to "walk" toward the + ends of adjacent anti-parallel actin filaments. It was established by Finer et al. (1994) that a single step by a single myosin II molecule generates on average 3.4×10^{-12} N while moving approximately 11 nm. It is this walking mechanism that leads to the shortening of the actomyosin filaments and ultimately to cytoskeletal contraction.

The regulation of actomyosin contractility in smooth and nonmuscle cells is dependent on the phosphorylation of the essential light-chains (ELCs), which involves ATP hydrolysis (Alberts et al., 2002; Citi and Kendrick-Jones, 1988). While many signaling molecule pathways feed into and regulate cell contractility, there are only a few central kinases necessary for force transduction. Through a chain of events starting with increased $[Ca^{2+}]_i$, myosin light-chain kinase (MLCK) phosphorylates a serine of the ELC. Rho kinase, a downstream effector of the RhoA signaling pathway that is involved in stress fiber formation (Chrzanowska-Wodnicka and Burridge, 1996; Katoh et al., 1998, 2001a), can phosphorylate the ELC at serine 19 in a Ca^{2+}-independent manner. Rho kinase is also important for the indirect regulation of contractility through its inhibition of myosin phosphatase, which dephosphorylates the ELC (Kimura et al., 1996). PAK, which is often stimulated downstream of Rac, regulates force production in the cell in two ways: first, through the inhibition of MLCK binding to serine 18 of the ELC, and, second, by increasing the dimerization of the myosin II heavy chains (Parrini et al., 2005; Zhao and Manser, 2005). Interestingly, Rac is known to locally down-regulate RhoA-dependent activities (Ridley, 2001; Ridley and Hall, 1992a, 1992b; Ridley et al., 2003). MLCK and Rho kinase pathways have been implicated in traction force regulation (Doyle et al., 2004; Doyle and Lee, 2002) and may be cell-type–dependent.

Interactions of the cytoskeleton with integral membrane proteins such as integrins allow transmission of the contractile force generated by the coupling of myosin II and F-actin to the surrounding extracellular matrix (ECM) and are termed traction forces. In early studies of cell movement, investigators visualized a microfibrillar network (actin cytoskeleton) in contact with adhesion plaques attached to the underlying ECM (Abercrombie and Dunn, 1975; Abercrombie et al., 1971; Dunn, 1980; Mould et al., 2003). Although research suggested that cells exerted rearward-directed forces emanating from the lamella for movement, culturing cells on non-deformable glass cannot reveal the magnitude and localization of these forces.

Albert Harris circumvented this problem using silicone oil that, upon rapid heating with a glow-discharge apparatus, formed a thin, cross-linked elastic film. After replating chick heart fibroblasts onto the cross-linked silicone substrata, Harris et al. (1980) observed a wrinkle formation beneath the motile cells, demonstrating that cells generate "traction" forces exerted behind the leading edge of the lamellipodia. Because the elasticity of the thin film could be quantified, it was possible to estimate that these cells produce a shear force of 1 dyne/μm ($\sim 1.0 \times 10^{-7}$ N/μm).

Further investigations of force production have used improved techniques (Oliver et al., 1994, 1995), while others have used different technical approaches for differing purposes: calibrated microneedles for studying growth cones (Lamoureux et al., 1990) and forces during fibroblast cytokinesis; micromachined quartz cantilevers for studying both fibroblast and fish keratocyte traction forces (Galbraith and Sheetz, 1997, 1999); and optical laser-traps to study the forces generated and the interaction of the cytoskeleton at single focal adhesions (Choquet et al., 1997; Galbraith and Sheetz, 1999; Galbraith et al., 2002). Furthermore, advances have been made in calculating and processing data. Dembo and Wang (1999) created pattern-recognition software for analyzing bead-labeled or patterned substrata, which has been applied to thin polyacrylamide gels (Discher et al., 2005) and gelatin-based assays (Doyle et al., 2004; Doyle and Lee, 2002). This software utilizes an optical flow algorithm for rapidly calculating hundreds to thousands of data points to estimate the locations, directionality, and magnitudes of traction stresses generated by moving cells in mere minutes. Other investigators have also developed procedures that have greatly increased the rate at which temporal data from time-lapse imaging can be calculated (Balaban et al., 2001; Numaguchi et al., 2003; Tan et al., 2003; Wang et al., 2001).

The magnitudes of forces produced by different cell types vary greatly, as do the patterns and locations from which these forces originate. The initial research by Harris showed that motile fibroblasts deform silicone substrata perpendicular to the axis of migration (Harris et al., 1980, 1981). When a "pinching" force is applied to a "wrinkling" substratum, similar wrinkles form. These findings indicate that fibroblasts generate their major cell–substratum force parallel to the direction of movement (Figure 5.4; Plate 9). Most traction forces generated by fibroblasts are directed inward toward the cell body (Beningo et al., 2001; Dembo and Wang, 1999; Munevar et al., 2001a, 2001b; Pelham and Wang, 1999; Wang et al., 2001). The traction forces generated by a cell consist of propulsive and frictional forces. Propulsive forces are found at the leading edge and act to "propel" cells forward. Frictional forces originate at the rear of cells from mature focal adhesions and are directed toward the cell body along the axis of travel and are hypothesized to be passive anchoring points crucial for maintaining a spread cell morphology (Munevar et al., 2001a). Because the propulsive forces are larger than the frictional forces, the result is a net forward movement, with asymmetry between the leading and trailing edges.

Although sites of cell interaction with the substratum are generally thought to be the sites of force transmission, the locations of high-magnitude tractions and their role in cell movement are still uncertain. There are several models of localized force generation at the front and rear of cells. Beningo et al. (2001) described small nascent FAs at the leading edge that generate very large traction forces, while larger, more mature adhesions behind the leading edge exert weaker forces. FAs in the trailing edge or tail of migrating cells were found to generate extremely large traction stresses, but were thought to act as passive stabilizing anchors. However, Chen (1981) previously demonstrated that retraction of the trailing edge is dependent on ATP hydrolysis, most likely due to myosin ATPase activity. In the Beningo model, the largest traction forces were found at the leading edge, and calculations of shear stress implied that this region was mechanically different from the rest of the cell. Their model postulates that nascent adhesions at the leading edge provide propulsive forces, while adhesions in the trailing edge act passively to anchor the cell.

Contrary to these findings, several laboratories have argued against the idea that focal adhesion size is inversely proportional to localized traction forces (Balaban et al., 2001; Chrzanowska-Wodnicka and Burridge, 1996; Galbraith and Sheetz, 1997; Sawada and Sheetz, 2002). Balaban et al. (2001) showed high correlation between fluorescence intensity of GFP-vinculin and force, FA area and

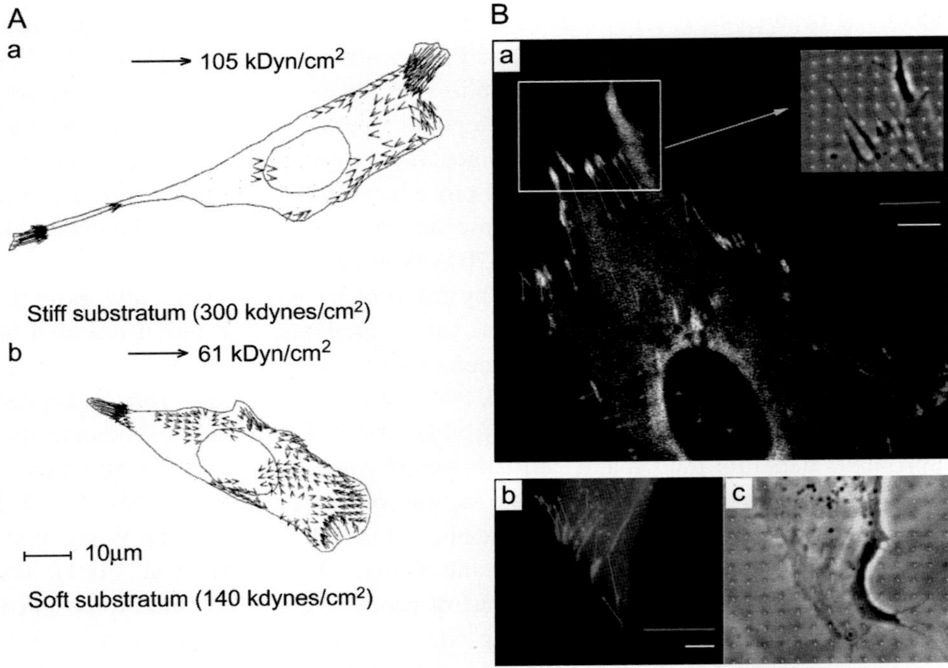

Figure 5.4. Traction force development in cells. (Aa, Ab) Traction force mapping in migrating 3T3 fibroblasts plated on two-dimensional pliable polyacrylamide gels coated with collagen. (Aa and Ab) represent cells on stiff (300 kdynes/cm^2) and soft (140 kdynes/cm^2) gels, respectively. (Ba–c) vector mapping of forces generated at focal adhesion sites. (Ba) GFP-vinculin expressed in human foreskin fibroblasts. (Bb, Bc) Immunofluorescence staining for paxillin (Bb) and the corresponding phase-contrast image of a cell on polydimethylsiloxane (PDMS). Red arrows indicate the direction and magnitude of local forces generated against the underlying substratum. From Lo et al. (2000) (A), and Balaban et al. (2001) (B).

force, and FA orientation and force orientation. Furthermore, after treatment with butadiene monoxime (BDM) to inhibit myosin II activity, the force decreased over time, and the correlation between the FA area and the force declined, indicating an interdependence of traction stress on actin-myosin contractility. Galbraith and Sheetz (1997) observed forces within the tails of migrating fibroblasts that were nearly four-fold higher than forces generated within the lamellipodium. Others have also suggested that the large forces produced in the trailing edges of cells represent contractility within this region to promote cell-substratum detachment (Galbraith and Sheetz, 1997, 1999). This interpretation is consistent with computational evidence that, for forward migration, the sum of all cellular forces is in the direction of movement (DiMilla et al., 1991, 1993; Galbraith and Sheetz, 1997).

Nevertheless, several lines of evidence partially support the Beningo model. Although migration was not directly investigated, Reinhart-King et al. (2005) found that endothelial cells generate significant traction forces in the initial phase of cell spreading prior to the formation of stress fibers and the aggregation of vinculin in focal complexes. Interestingly, during cell spreading, a direct relationship exists between the total traction forces generated by a cell and spreading. Furthermore, Tan et al. (2003) showed that FAs >1 μm^2 show a high degree of correlation with tractions, in agreement with Balaban. However, a subset of very small adhesions below 1 μm^2 displayed abnormally high traction forces. Tan and coworkers suggest that these nonlinear relationships in small adhesions may represent another subset of force-generating adhesions.

MIIA and MIIB isoforms contribute approximately 60% and 30% of the total traction forces generated by a cell, respectively (Cai et al., 2006). Genetic ablation of MIIA results in a loss of stress fibers and large focal adhesions (Even-Ram et al., 2007), which are replaced by small, phosphotyrosine-rich adhesions at the leading edge (Vicente-Manzanares et al., 2007). Ablation of MIIA also stimulates cell migration, protrusion, and cell spreading (Cai et al., 2006; Even-Ram et al., 2007; Sandquist et al., 2006). Conversely, even though the loss of MIIB has little effect on the cytoskeleton, it results in reduced cell migration rates and a more retractile cellular phenotype (Even-Ram et al., 2007; Sandquist et al., 2006). Together, these results suggest that the force regulation through the different MII isoforms strongly affects processes such as cell migration that depend on mechanotransduction with the ECM.

It is important to note that traction forces can be altered directly by the elastic component of the underlying ECM. Elasticity is one of the parameters involved in force calculation on a pliable substratum (Dembo and Wang, 1999). The relative elasticity of a substratum directly correlates with traction force production, with more rigid surfaces leading to higher force transmission (Discher et al., 2005; Engler et al., 2004, 2006). Due to a positive-feedback mechanism where high tension, whether dynamic or static, will increase clustering of adhesion components, less deformable ECMs will promote physical signaling between the ECM and cells more than highly pliable substrates. ECM pliability also plays a role in a number of diseases, including cardiovascular disease (Polte et al., 2004) and breast cancer (Nelson and Bissell, 2006), in which increased ECM deposition and remodeling are altered. We will elaborate on these themes in later sections.

5.7 Ligand Density and ECM Elasticity: How ECM Density and Rigidity Regulate Cellular Mechanics

From a physiological or *in vivo* standpoint, the ECMs undergo constant flux. Cells such as fibroblasts continually remodel their local environment, generating new ECMs. These activities can be observed directly during embryonic development and remodeling of normal adult tissues. For example, during wound healing after injury, ECM proteins such as fibrinogen, fibronectin, and vitronectin that are abundant in serum and normally low in connective tissues appear in the provisional matrix and then in healing tissues. These new components of the ECM together with chemo-attractant growth factors released locally can alter cellular functions by both haptotactic and chemotactic mechanisms, respectively.

Fibroblasts, in turn, respond to these alterations and further change the surrounding matrix through the generation of fibronectin and collagenous fibrils while degrading fibrin clots at the healing wound site (Clark, 1996). Pathological states such as fibrotic disorders (Radisky et al., 2007), breast cancer (Nelson and Bissell, 2006; Wang et al., 1998; Wozniak et al., 2003), and cardiovascular disease involve altered local ECM protein deposition and increased ECM rigidity. They also result in altered integrin expression and increased cellular phosphorylation, further exemplifying how the ECM can control outside-in signaling events and gene expression. In this section, we concentrate on two important biophysical parameters associated with the ECM: ligand density and ECM rigidity.

5.7.1 Ligand Density Regulation of Cellular Mechanotransduction and Cell Migration

Carter (1967) first demonstrated that cells react to differences in ligand density. When presented with a gradient of an adhesive substrate, cells migrate from areas of low to higher concentrations of adhesive molecules (Figure 5.5). This concept, known as haptotaxis, has been the basis for a wide array of assays and areas of study.

The ability of cells to attach to an underlying substratum often relies on the interactions of specific amino acid residues in the α and β chains of integrins with specific ligand sites such as the RGD and synergy sites found on fibronectin or analogous functional sites on other matrix glycoproteins. The binding affinity or the dissociation constant of receptors for a particular ligand can regulate not only short-term local interactions of a single adhesion site, but also cell migration as a whole. During haptotaxis, cells migrate up an increasing ECM gradient, where they ultimately become paralyzed due to high cell adhesion. Rates of migration can be tuned by altering the avidity of integrins for specific matrix proteins. This concept has been illustrated using anti-β_1–integrin antibodies with different affinities. Antibodies attached to glass or tissue culture plastic that bind the extracellular domain of

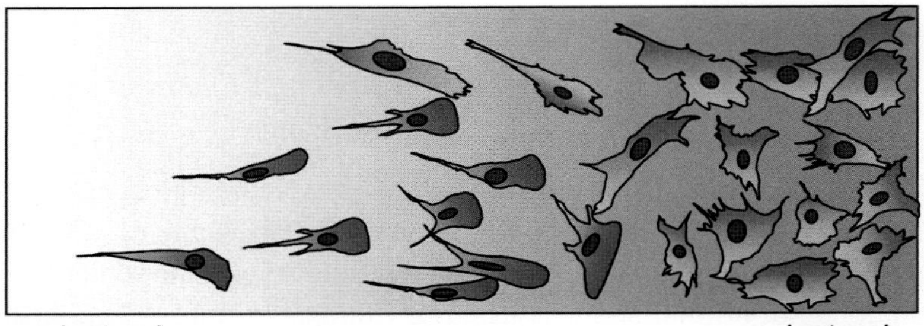

Figure 5.5. Cellular haptotaxis. The cartoon depicts the response of cells when plated on a two-dimensional surface coated with a gradient of ECM ligand. At low ligand density (left), fibroblasts show little spreading and are more directional and elongated (anisotropic cell shape). Cells migrate toward the higher ligand densities (far right), where they become more spread and isotropic in appearance and migrate very little.

integrins can promote cell spreading and motility (Duband et al., 1986). Using this methodology, antibodies with high affinity (greater than that of fibronectin to the fibronectin receptor integrin, i.e., $K_d > 10^{-6}$ M [Akiyama and Yamada, 1985]) limited neural crest cell migration from explants, which was delayed when compared with fibronectin (Duband et al., 1991). Antibodies with moderate affinity permitted migration, and vinculin localized in focal contacts similarly to cells on fibronectin. Interestingly, the extent of migration was also associated with antibody concentration (μg/ml), with biphasic behavior in which an intermediate concentration promoted the largest migration away from the explants. These initially surprising results indicated that both the affinity and density of the substratum adhesion molecule, whether natural or artificial, play an important role in regulating the speed of cell migration.

Computational modeling has provided insight into the question of how physical extracellular cues (outside-in signaling) can dictate aspects of cell migration. DiMilla et al. (1991) presented a computational model of cell migration rate that was biphasic and dependent on cell adhesiveness. Cell adhesivity, or the relative strength of bonding between cells and the underlying substratum, cannot be directly measured but can be inferred from traction forces or detachment forces. Both of these forces are measures of contractile strength exerted against the substratum, which depend on the dissociation constant of the receptors for ligands, the relative amount of integrins remaining attached to the ligand after cell retraction, and cellular dynamics.

The model concentrated solely on one-dimensional movement (x) and disregarded lateral (y) movement, as well as the organization of integrin receptors on the cell surface and any intracellular signaling mechanisms. It proposed that, for optimal migration, an asymmetry in cell adhesiveness must exist between the front and rear of the cell. Under these stringent conditions, cell adhesiveness depends on two factors: 1) the ligand/receptor equilibrium constant or ligand-receptor binding affinity and 2) the ligand density as measured by the amount of ECM ligand deposited on a two-dimensional surface. It was suggested that free receptors at the leading edge have a greater affinity to bind ligands, which was recently demonstrated (Galbraith et al., 2007; Jiang et al., 2006; Zaidel-Bar et al., 2003), and leads in part to greater adhesivity at the lamella. The biphasic curve of motility is directly related to several cellular factors: 1) cell contractility, 2) cell adhesiveness, and 3) cell rheology or plasticity. Interestingly, it has now been shown that these three cell biophysical factors are all interrelated; increasing cell contractility also increases adhesiveness and cell rigidity (or decreases plasticity) (Engler et al., 2006; Peyton and Putnam, 2005).

Ligand density directly affects the likelihood of integrin binding to ligands. This binding is analogous to an enzymatic reaction, where equilibrium is reached between the substrate and enzyme for any given amount of substrate. For example, at low ligand density there is an overabundance of receptors for binding a limited quantity of ligands. At a given dissociation constant, receptors cannot bind more ligands than is readily available (Duband et al., 1988). This situation leads to limited adhesiveness that in turn limits cellular contractility and keeps the cell in a flaccid state with low cortical rigidity.

At intermediate or optimal ligand densities, there is a balance between ligand density and the ligand/receptor binding affinity: Contractility, adhesiveness, and cell rigidity all increase to a state where contractile strength is sufficiently above the threshold needed for ECM/integrin bond breakage at the rear to permit its release, yet below the threshold for detaching high-affinity bonds at the leading edge. Above this optimal ligand density, however, contractility is thought to be high enough to break all ECM/integrin bonds at the front and back of the cell and to disrupt the asymmetry between the front and back of the cell. With such high contractility, the cells are rigid and deformation of the membrane is limited, possibly reducing further protrusions due to a higher required protrusive force (Palecek et al., 1998). Figure 5.6 (Plate 10) summarizes the effects of cell adhesiveness, ligand density, and receptor/ligand binding affinity on cell migration rates.

The lateral diffusion coefficient of integrins within the cell membrane can differ markedly depending on cell–substratum binding. High lateral diffusion of integrins is associated with rapid migration, in contrast to the low diffusion observed for integrins in the adhesions of stationary cells (Duband et al., 1988). Diffusion also depends on ligand–receptor binding affinities and changes dynamically with perturbations of cell adhesion.

Under conditions of high ligand density, cells become firmly attached and the migration rate is low. Detachment from the substratum only occurs when cell adhesivity is either lowered by some biochemical action or through increases in localized force production until contractility can exceed adhesiveness. As discussed earlier, shortly after ligand occupancy, integrins begin to cluster and in turn promote the aggregation of numerous other cytoskeletal and signaling proteins (Miyamoto et al., 1995a, 1995b; Plopper et al., 1995). As these components of cell adhesions are added, the connections between integrins and filamentous actin grow stronger. This latter process has been shown directly by Balaban et al. (2001) by a strong correlation between focal adhesion area and local force production and by Riveline et al. (2001), who observed increased focal adhesion growth following the local application of force via a micropipette, as well as with other techniques used by other investigators (Galbraith et al., 2002). This suggests that a positive feedback loop exists: Following the initial formation of a focal adhesion, local force generation potentiates the growth of the adhesion and hence enhanced local cell adhesiveness.

It was suggested in the original DiMilla computational model that changes in cell adhesivity and associated parameters (ligand density and ligand/receptor binding affinity) regulate the cell migration rate. In other words, changing the binding affinity of integrins to their specific ligand should alter the rate of cell migration at a given ligand density. Huttenlocher et al. (1996) demonstrated this property using nonperturbing activating antibodies and introducing mutations into the integrin β-chain cytoplasmic domain. Activation of $\alpha_{IIb}\beta_3$, an integrin specific for binding to fibrinogen, with the activating antibody LIBS6 decreased migration on 10 μg/ml fibrinogen in a manner similar to the deletion of the membrane-proximal sequences of the α or β chains.

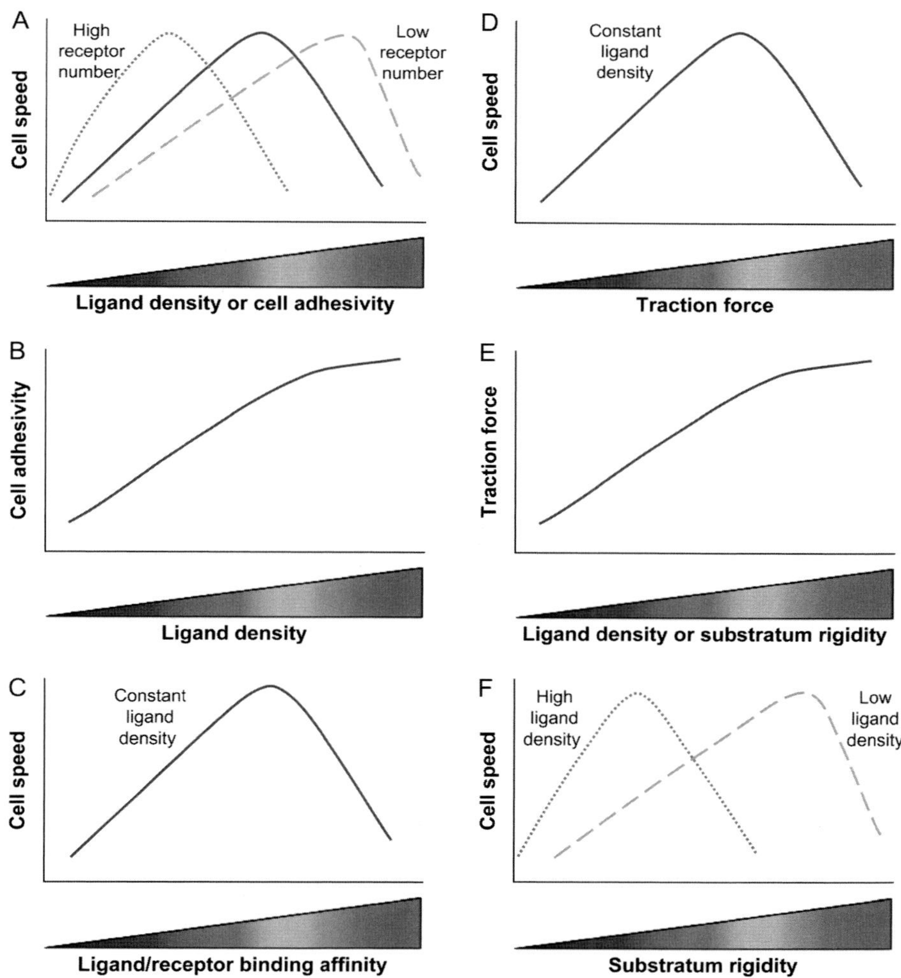

Figure 5.6. Control of cellular responses through regulation of ligand density, ligand/receptor binding affinity, and substratum rigidity. The graphs depict known effects of ligand density or cell adhesivity (A), ligand/receptor binding affinity (C), traction force (D), and substratum rigidity (F) on cell migration rates. (B) The quasi-linear effect of ligand density on cell adhesivity. (E) The effect of ligand density or substratum rigidity on the magnitude of global cell traction forces. In (A), the effects of relative overexpression (red dotted line) and under/low expression (green dashed line) of integrins are shown, with the maximal migration rate shifting to the left (high integrin expression) or to the right (low expression). In (F), the two curves illustrate the effects of high (red dotted line) versus low (green dashed line) ligand density on cell migration rate at different substrate compliances (rigidity).

α chain mutants displayed a complete loss of motility, while only haptotactic or ligand-dependent migration was affected in the β chain mutants. Furthermore, alterations to the ability of the β-subunit to bind cytoskeletal adaptor proteins also decreased migration rates. Taken together, these findings illustrate that the binding affinity of integrins for their ligands directly controls cellular migratory events. Although not directly documented, the relative dissociation constants (on/off rates) between antibody-activated and mutant $\alpha_{IIb}\beta_3$-integrins likely increased, and hence a lower ligand density was necessary to promote optimal migratory rates on fibrinogen.

While ligand density directly controls motility, the same can be said for the number of available receptors. Increasing the relative expression level of α_5-integrin affects migration rates on a fibronectin substratum. There is a direct inverse relationship between α_5 expression and the ligand density required to achieve maximal rates of migration (i.e., low α_5-expressing cells required approximately one order-of-magnitude higher ligand density to attain a similar maximal migration rate when compared to high expressors) (Palecek et al., 1997). In addition, cells expressing higher levels of α_5-integrin required higher shear forces for detachment from the underlying substratum independent of ligand concentration. From a mechanotransduction perspective, cells with an abundance of receptors can cluster integrins to a greater extent at a given ligand density, which in turn generates stronger linkages to the cytoskeletal contractile apparatus. Increased clustering of cytoskeletal components in ECM–cell contacts increases the local force-generating capacity of contacts (Galbraith et al., 2002), leading to higher cell adhesivity.

It has been well established that proper cell migration requires 1) protrusion of the lamellipodium, 2) adhesion of new protrusions to the substratum, 3) cell body translocation, and 4) retraction of the trailing edge. In *Dictyostelium discoideum,* the protrusion rate of the lamellipodium is known to limit the rate of motility in three dimensions (Wessels et al., 1994), yet when the cells are plated on high concentrations of poly-L-lysine (100 μg/ml), retraction becomes the rate-limiting step in migration (Jay et al., 1995).

Fibroblasts plated on different concentrations of fibronectin show distinct differences in spread morphology, protrusion rates, and retraction rates (Huttenlocher et al., 1996; Palecek et al., 1997, 1998, Palecek et al., 1999; Schmidt et al., 1993). Low ligand densities limit cell protrusion and adhesion, whereas at high densities, retraction becomes the rate-limiting step for cell movement, and calpain, a neutral cellular protease, is needed for tail detachment (Huttenlocher et al., 1997; Palecek et al. 1998). Cells expressing the calpain ligand talin (Wells et al., 2005), mutated to lack its calpain cleavage site, display reduced ability to migrate and retract tail regions similar to control cells treated with calpain II inhibitors; the adhesion turnover rates decrease, suggesting that at high cell adhesivity, force production alone cannot effectively initiate retraction and requires additional biochemical mechanisms to disassemble cell–ECM adhesions (Huttenlocher et al., 1997; Palecek et al., 1998).

Under most conditions, cells leave behind some cellular debris, mostly comprised of portions of the cell membrane still attached to the ECM surface after the cell has retracted its tail regions (Schmidt et al., 1993). Interestingly, the amount of residual attached integrins correlates directly with ligand density and mean detachment forces, as well as biphasically with cell speed as shown by time-lapse imaging of fluorescently labeled anti-$\alpha_{IIb}\beta_3$ antibodies (Palecek et al., 1996). It was recently found that myosin IIB and its activation through myosin light-chain phosphorylation is associated with tail formation in migrating cells, where it assembles into robust actin bundles and mediates the formation of stable focal adhesions that reduce local protrusive activity (Vicente-Manzanares et al., 2008). These findings suggest that as the ligand density increases, so does the force generation required for the physical dissociation of receptors from the ligand. Furthermore, they suggest that ligand/receptor binding affinities can be greater than local force transduction through the cytoskeleton, resulting in membrane disruption.

Although we have focused primarily on integrin/ECM interactions in this section, including the effects of ligand density and outside-in signaling mechanisms, other intriguing research has shown that cytoskeletal and integrin protein dynamics can change in direct response to ECM alterations. By combining high-speed fluorescence imaging with computational modeling of F-actin flow, plus integrin and focal adhesion protein dynamics, the Waterman-Storer laboratory (2006) discovered that in analogy to migration events studied at the whole-cell level, ECM density can dictate 1) focal adhesion size and localization, 2) cell spreading, 3) actin flow rates, 4) myosin light-chain phosphorylation, and 5) focal adhesion turnover rates and lifetimes. Modification of actomyosin contractility using blebbistatin (a myosin II ATPase inhibitor) and calyculin A (a myosin phosphatase inhibitor) to decrease and increase contractility, respectively, showed that the effects of ligand density can be tuned. At an optimal or intermediate ligand density (or optimal cell adhesion) either increasing or decreasing contractility decreases motility rates. Increasing contractility at high ligand densities increased adhesion turnover together with actin flow and migration rate, while lowering contractility on low ligand densities increased cell migration.

The overwhelming evidence for the importance of outside-in integrin signaling suggests that dynamic physical interactions between cells and the extracellular environment play a major role in cell decision-making and in determining how intracellular signals are transduced and force production is regulated.

5.7.2 Compliance of the Extracellular Matrix and Mechanosensing

As suggested throughout this chapter, the components of the ECM and their physical properties directly affect cellular mechanosensing. ECM ligand density can regulate cell attachment and speed through effects on specific ligand/receptor binding avidities and cell adhesiveness, and density can also alter cytoskeletal dynamics. This regulation by the quantity of ligand per unit area can be explained by the ability of higher ligand densities to bind integrins more rapidly due to greater local availability, resulting in higher effective avidity.

A second way in which the ECM can influence cellular interactions is through its rheological properties or compliance. Every substance has several physical parameters that together comprise its Young's or elastic modulus (E). The E of a material can be calculated from stress–strain curves, where the stress = force/cross-sectional area and strain = change in length/original length. Other factors can be taken into account, including material density, Poisson's ratio (the amount x decreases as y increases through physical stretching) and shear modulus, a measure of rigidity that describes a material's response to force parallel to its surface, such as the shear stress encountered by endothelial cells in blood vessels. Hooke's law of elasticity states that the amount of deformation a material undergoes (stress) is directly related to the amount of force (strain) causing the deformation. Hence, E is given in values of Newtons per meter squared or, in general, force per unit area. A Hookean relationship refers to stress–strain curves and presumes a linear-elastic material. Most tissue matrices are generally considered to be Hookean in nature as opposed to viscoelastic; for the sake of simplicity, we will use a Hookean model.

Excluding osteocytes and certain other cells associated with bone, the majority of cells and tissues are associated with flexible, compliant matrices. As mentioned previously, elastic polymer gels and thin cross-linked silicone films originally intended for studying traction forces during cell motility have increasingly proven useful for understanding the effects of ECM rigidity on cellular mechanics. The elastic properties of hydrogels such as polyacrylamide gels are easily altered by simply changing the ratio of the monomer to cross-linker (the acrylamide/*bis*-acrylamide ratio) (Pelham and Wang, 1997, 1998, 1999; Wang and Pelham, 1998). At constant acrylamide concentrations, increasing the percentage of *bis*-acrylamide increases the E of the hydrogel and hence the rigidity of the gel. For biological assays, ECM molecules such as collagen are covalently attached in equal amounts to gels of varying rigidity. As the rigidity of a substratum increases, cells display increased spreading and their migration becomes less directional, often with multiple lamellipodia (Lo et al., 2000; Polte et al., 2004; von Wichert et al., 2003). Interestingly, similar phenotypic effects are observed when cells are plated onto increasingly dense ECMs, suggesting a possible interplay between ligand density and substratum rigidity. Lamellar ruffling shows an inverse relationship with substratum rigidity. In addition, fibroblasts display smaller focal adhesions that are limited to the cell edge with increasing elasticity, implying that these structures are also influenced by substratum compliance.

Cells can sense the rheological attributes of surfaces they encounter. Choquet et al. (1997) established that the link between integrins and cytoskeletal components is strengthened as ECM rigidity increases. Using a laser trap to physically "hold" polystyrene beads coated with fibronectin fragments bound to the dorsal side of fibroblasts, cells were shown to generate larger forces to resist the localized forces, often pulling beads from the trap. Interestingly, the rate at which cells respond to increased localized stiffness is on the time-scale of seconds.

The ability of a cell to sense differences in substrate stiffness has been dubbed durotaxis (Lo et al., 2000). The Wang laboratory has characterized the behavior of cells migrating on polyacrylamide hydrogels of varying elasticity and their ability to generate traction forces for migration. Plating fibroblasts onto polyacrylamide gels of soft versus stiff compliance with constant ligand density revealed that cells prefer a rigid substratum. When cells migrating on soft regions of gels interacted with less compliant areas, migration increased rapidly onto the rigid surface. Conversely, when cells advanced onto soft polyacrylamides from a stiff region, protrusion at the leading edge quickly diminished, and the cells began migrating along the boundary in a direction perpendicular to the original axis of travel. As expected, cells on stiff polyacrylamide gels generated higher traction forces and displayed increased spreading with migration rates reduced by nearly half. Introducing strain into soft gels near the cell edge caused cells to reorient and move toward the strain field, similar to Dictyostelium chemotaxis toward a source of cAMP (Xiao et al., 1997). The theory of durotaxis, much like haptotaxis, suggests that cells can sense their physical environment and adapt to the stresses that it imposes by directional migration.

As discussed previously, the regulation of cell migration speed displays a biphasic dependence on ligand density, in which optimal migration is attained at an intermediate density (DiMilla et al., 1991, 1993). A similar regulatory response occurs with substratum rigidity: Cells attached to exceedingly soft gels (i.e., <1 kPa or 10 kdynes/cm^2) or rigid gels (>300 kPa or 3000 kdynes/cm^2) display slower migration rates compared to cells on gels of intermediate stiffness (~14 kPa or 140 kdynes/cm^2) with the same ligand density (Peyton and Putnam, 2005). Remarkably, the cellular response to substrate stiffness can be tuned for optimal migration rates by varying ligand density. Pairing high ligand densities with soft gels or low ligand densities with stiff gels enhances migration rates. Thus, cellular responses can be honed solely by the external physical parameters of the surrounding environment. It is important to note that, as with ligand density, the cellular response to a defined level of ECM rigidity is cell-type–specific. As a result, relative (soft or rigid) values are commonly used in describing and modeling these mechanisms.

Cellular responses to differences in elasticity of a substratum, which can range from nearly viscoelastic films and gels (<1 kPa) to rigid polystyrene or glass surfaces (~3,000,000 and 72,000,000 kPa, respectively) can vary substantially. Mechanistically, a cell's ability to sense rigidity may stem from the dynamic nature of soft gels. For instance, elastic surfaces react to physical strain (L_o) like a compressed spring (extracellular spring). With the application of stress via a pull, the tension within the elastic substratum increases proportionally (assuming a Hookean model). Hence, the greater the force application by a cell to the ECM (extracellular spring), the greater the ECM reaction with an equal and opposite level of tension, leading to dynamic changes in the overall tension and the cell–ECM interface. This dynamic interplay probably alters adhesion protein dynamics; although such specific protein dynamics remain to be documented, published findings suggest they will be found; for example, force is known to affect focal adhesion size, and there are similar effects of ligand density on both cell adhesions and migration rates.

Cells are unable to sense differences in stiffness above 300 kPa, which may reflect an upper limit of their force-generating capacity to physically deform a substrate. Such rigid surfaces appear to be perceived by the cell as static and immovable, as if cellular mechanosensing is turned off on extremely rigid substrates. How and why do cells react to compliant substrata and seek out stiff matrices? Schwarz et al. (2006) propose a two-spring model based on the linear elasticity of hydrogels. In this model, the ECM and cytoskeletal components are each represented by springs with different spring constants acting in series (Figure 5.7) with the key feature being force–velocity relationships. If the extracellular "spring" is softer (i.e., stretches with less applied strain), a cell must invest more energy to attain the threshold of force

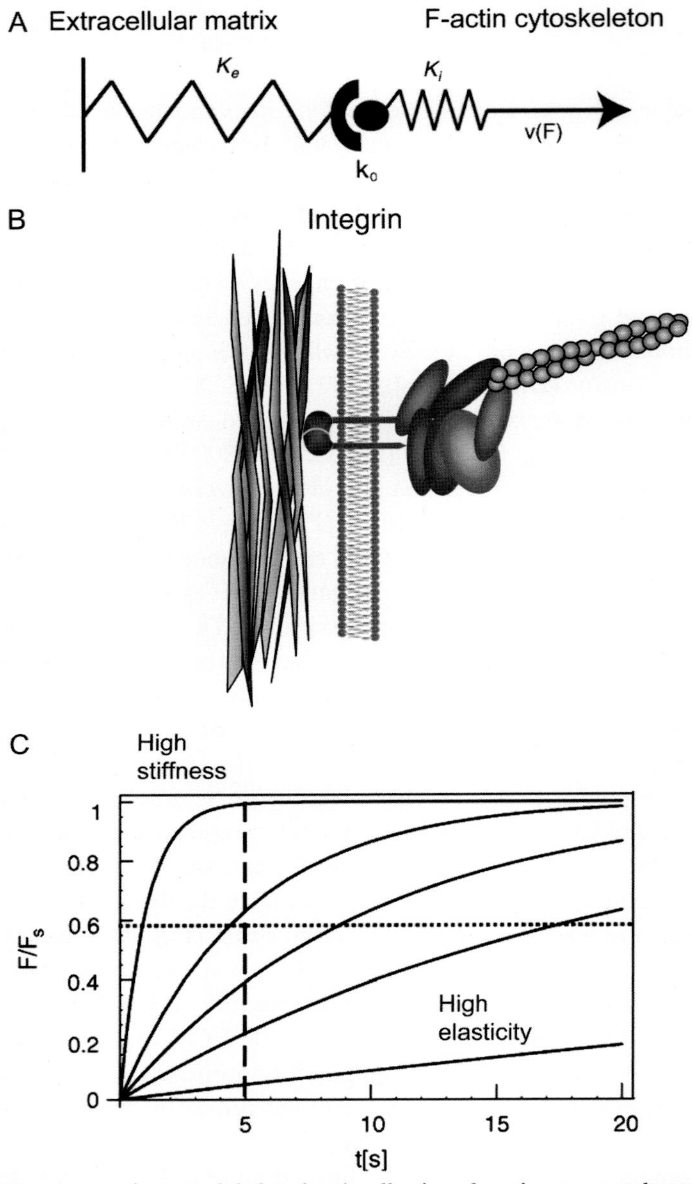

Figure 5.7. The two-spring model for focal adhesion function as mechanosensors. (A) Schwarz and Bischofs' (2006) representation of cytoskeletal-integrin-matrix interactions. (B) A visual molecular depiction of the two-spring model. (C) Theoretical force/velocity curves that demonstrate the dynamic effect that varying the elasticity of an extracellular matrix has on force development. They hypothesize that the extracellular matrix (K_e) can be considered a spring of variable tension that is more dynamic on elastic surfaces and more static on rigid ones. The F-actin–based cytoskeleton (K_i) acts as a second spring in series with integrins, which link the two dynamic structures. Under rigid (static) conditions, the ECM is tensed and hence the cytoskeleton needs to perform little relative work to reach the threshold of force necessary to maintain stable adhesion complexes (dotted line in C). However, under elastic (dynamic) conditions, as the cytoskeleton pulls against the integrin–ECM complex, the ECM yields and begins to stretch. To maintain a stable adhesion, the cytoskeleton must pull at a faster rate. As tension within the ECM increases in response to cytoskeletal tension, a balance of outside and inside forces is reached to stabilize adhesions. From Schwarz and Bischofs (2006) (A and C).

needed to maintain the stability of an adhesion. When the ECM is rigid, however, the amount of force necessary for stabilization is reduced because all of the force generated by actin-myosin contraction can be used for stabilization of focal adhesions. Furthermore, the force–velocity relationship suggests that the time during which force is generated is substantially shorter on rigid, immovable substrates. The net result would be the observed preference of cells for stiff substrates. It has also been suggested that cells generate traction forces to match their surroundings, which is based on the relative amount of tension necessary to stabilize adhesion components (Bischofs et al., 2004; Bischofs and Schwarz, 2003; Evans and Calderwood, 2007; Lo et al., 2000; Schwarz and Bischofs, 2005; Schwarz et al., 2006).

Although many studies to date have concentrated on understanding the biophysical properties of traction forces (directionality, force amplitude, etc.) and on the general effects of elastic surfaces on cell migration and their phenotypic responses, investigations into the molecular nature of rigidity sensing have opened an intriguing new field of research exploring how substrate compliance regulates intracellular signaling pathways. The phosphorylation state of numerous signaling molecules that localize to adhesion sites (e.g., FAK and paxillin) or are part of the cells' contractile apparatus (the myosin essential light-chain, MLC) can regulate adhesion and motility events. As rigidity decreases, FAK, paxillin, and MLC all display a concomitant reduction in phosphorylation (Gupton and Waterman-Storer, 2006; Pelham and Wang, 1997; Polte et al., 2004), all contributing to the regulation of focal adhesion turnover. FAK-null cells fail to upregulate tractions on more rigid polyacrylamide gels, while overexpression of constitutively active F397 FAK enhances contractility, suggesting a possible mechanosensory role for FAK (Wang et al., 2001). Several phosphatases, including receptor-like tyrosine phosphatase α (RPTP-α), members of the Src family kinases (Src, Yes, and Fyn), and p130^{CAS}, have all been implicated in mechanosensing and the cellular rigidity response (see Chapter 7 for more details) (Jiang et al., 2006; Kostic and Sheetz, 2006; von Wichert et al., 2003). RPTP-α null cells display an aberrant spreading response on soft substrata and remain relatively spread, suggesting an inability to sense differences in ECM rigidity (Jiang et al., 2006; Kostic and Sheetz, 2006; von Wichert et al., 2003). Clearly, rigidity sensing by cells can lead to the differential activation of numerous signaling cascades, depending on substrate compliance.

In vivo, the matrix surrounding each cell and tissue type varies not only by its physical composition of proteoglycans and organization (e.g., differences in planar epithelial basement membrane ECM versus the three-dimensional ECM of connective tissue), but also in their relative rigidity or compliance. Stem cells are the primordial cell types from which all other cells (neuronal, muscle, bone, etc.) are derived. As they differentiate, stem cells acquire a distinct gene expression pattern, giving each cell type unique functional capabilities. Matrix elasticity alone can influence gene expression patterns in mesenchymal stem cells and can determine cell fate. Engler et al. (2006) demonstrated that neurogenic (β3 tubulin), myogenic (MyoD), and osteogenic (CBFα1) transcription markers are up-regulated differentially by differences in substrate rigidity. Soft (<1 kPa), intermediate (~10 kPa), and stiff (~30 kPa) polyacrylamide gels induced cell phenotypes similar to neurons, myoblasts, and osteocytes, respectively. Inhibition of myosin II activity with blebbistatin during cell growth on these phenotype-inducing elastic substrata suppressed the expression of these specific markers. Consequently, contractility generated by nonmuscle myosin II and substratum elasticity regulates the transcription of tissue-specific genes and differentiated cell fates. Interestingly, expression levels of both A and B myosin II isoforms were increased in cells cultured on more rigid substrates (>~10 KPa), further underscoring the role of contractility in cell-type specification. Both cell prestress, the amount of intracellular tension generated by a cell to remain adherent, and cortical stiffness were linear and correlated well with substrate elasticity.

Collectively, recent research into the effects of substrate elasticity illustrates roles in the outside-in signaling that regulates cellular signaling cascades, migration, adhesion formation and turnover, gene expression, and cell fate.

5.8 Bringing It Together: Three-Dimensionality of Cell–Matrix Interactions

Although most current research on cell migration and mechanotransduction is based on two-dimensional cell culture systems, cells *in vivo* generally interact with ECM lattices in three dimensions. Studying cells in two-dimensional environments imposes an artificial dorsal-ventral polarity that limits their matrix interactions to a single plane. In three-dimensional culture systems, this problem is resolved by the presence of ECM on all sides. Other differences between two- and three-dimensional systems include: 1) ligand distribution and density, 2) ECM rigidity, 3) physical structure, and 4) matrix content. All of these factors can contribute to how cells physically sense their environment and therefore regulate cellular outside-in signaling and mechanotransduction.

One of the most remarkable observable characteristics of cells in three-dimensional matrices is their characteristic cell shape. In contrast to the well-spread isotropic morphologies that fibroblasts and epithelial cells display on two-dimensional surfaces, cells within three-dimensional matrices are generally anisotropic, polarized, and either elongated or organized into structures such as acini (Figure 5.8; Plate 11). Other physical characteristics in three dimensions include reduced lamellae, both in number and size (Beningo et al., 2004), and few stress fibers (Beningo et al., 2004; Grinnell, 2003; Jiang and Grinnell, 2005; Ruoslahti et al., 1985; Walpita and Hay, 2002). In fibroblasts, a decrease in overall Rac activity can account for the reduction in lamella number in three dimensions, while the reduced rigidity of three-dimensional ECMs may account for the absence of a highly organized contractile apparatus because of the cellular requirement to match their physical environment (Bischofs and Schwarz, 2003). Interestingly, siRNA knockdown of Rac1 in a variety of cell types on two-dimensional surfaces can mimic the anisotropic morphology and persistent movement of fibroblasts in three dimensions, consistent with ECM regulation of small GTPase activity (Pankov et al., 2005).

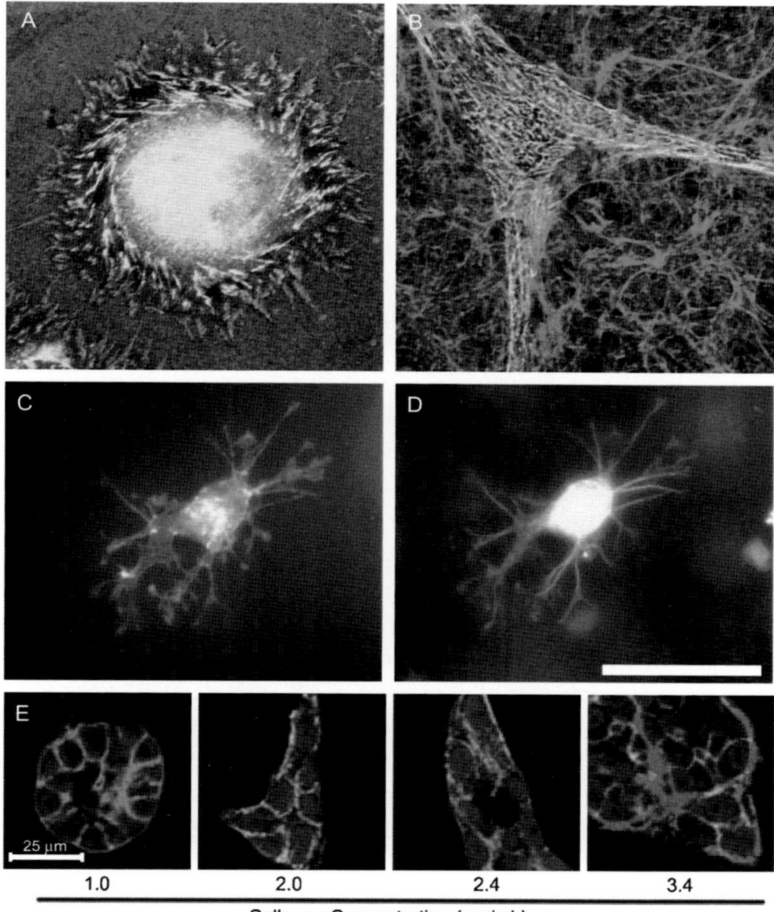

Figure 5.8. Effects of three-dimensional matrices on cell morphology and protein localization. (A) Human foreskin fibroblast (HFF) plated on a rigid two-dimensional surface pre-coated with 10 μg/ml fibronectin showing fibronectin (green) and α_5 integrin (red) immunostaining. (B) HFFs in a three-dimensional cell-derived matrix immunostained as in (A). Note the complete colocalization of fibronectin and α_5 integrin in (B) compared to (A). (C and D) Immunostaining for actin (C) and β_1-integrin in a three-dimensional collagen gel. (E) Differences in collagen concentration greatly affect cell phenotype and localization of $\alpha_6\beta_4$-integrin (red) and β-catenin (green) in mammary epithelial cells. At 1.0 mg/ml collagen, mammary cells polarize and form acini resembling *in vivo* acini. As collagen density and subsequently ECM rigidity increase, the cellular phenotype changes, and cells begin to move out of the acinar structure. From Pankov et al. (2000) (A), Cukierman et al. (2001) (B), Jiang et al. (2005) (C and D), and Paszek et al. (2006) (E).

It could be argued that the anisotropic, elongated shape of fibroblasts is due to physical restraints and constriction imposed by a three-dimensional fibrillar matrix. However, Beningo et al. (2004) demonstrated that simply positioning a second elastic, collagen-coated hydrogel coverslip on the dorsal surfaces of fibroblasts induces the characteristic three-dimensional fibroblast morphology, with long cell extensions instead of broad lamellae, together with a rearrangement of the cytoskeleton and localization of the adhesion proteins. In these studies, there was a noteworthy absence of the cycles of protrusion/retraction at the leading edge that normally accompany cell migration on two-dimensional planar surfaces. Although not yet proven, mimicking the structure and elasticity of a three-dimensional matrix on both dorsal and ventral surfaces of cells may alter the intracellular signaling mechanisms that regulate cell morphology and migration.

Alterations in integrin expression can modify cell migration. Most cell types express multiple α- and β-subunits that can form numerous heterodimeric binding partners to interact with various ECM ligands. Palecek et al. (1997) demonstrated that overexpression of the α_5-integrin increases cell migration rates at low fibronectin concentrations. The $\alpha_v\beta_3$-integrin, a binding partner for vitronectin, fibrinogen, and fibronectin, is prominently clustered in focal adhesions; several recent reports have suggested that it is a key transducer in mechanotransduction (Kostic and Sheetz, 2006). Remarkably, however, $\alpha_v\beta_3$ is often absent from three-dimensional matrix adhesions, only being found at the cell periphery in small puncta (Cukierman et al., 2001). Several cancers, including human multiple myeloma (Ria et al., 2002) and highly differentiated ovarian carcinomas (Carreiras et al., 1996), display high levels of both integrins, which may contribute to malignant behavior.

In cells grown on soft three-dimensional matrices, levels of expression of focal adhesion components such as talin, paxillin, FAK, and p130^{CAS} are reduced, probably due to the reduced need for a cytoskeletal response to the more-pliable matrix (Wang et al., 2003). In fibroblasts migrating through three-dimensional cell-derived matrices, the α_5-integrin displays a unique co-localization with focal adhesion proteins. Cukierman et al. (2001) discovered that α_5 co-localizes with FAK, vinculin, and paxillin in elongated adhesion structures termed three-dimensional matrix adhesions. This co-localization was unexpected, since α_5 is normally restricted to fibrillar adhesions (Pankov et al., 2000) on two-dimensional substrates, whereas FAK, vinculin, and paxillin are clustered in focal complexes and focal adhesion structures. Co-localization of α_5 with paxillin and FAK is only seen transiently in early focal adhesions (Zaidel-Bar et al., 2003; Zamir et al., 1999) or in the large aberrant focal adhesions that form on artificially immobilized two-dimensional fibronectin substrates (Katz et al., 2000).

Both FAK and paxillin show high levels of phosphorylation within focal adhesions, but only paxillin phosphorylation at Tyr-31 remained elevated in three dimensions. In contrast, FAK displayed little phosphorylation at Tyr-397. This finding in three dimensions contrasts with the high phosphorylation at Tyr-397 on two-dimensional substrates, including fibronectin. This high phosphorylation of FAK in two dimensions may be due to the rigidity of routine two-dimensional cell culture substrates. Another unusual finding was a switch of integrin utilization to only $\alpha_5\beta_1$ in three-dimensional matrices: function-blocking antibodies against α_5 selectively disrupted cellular attachment to three-dimensional cell-derived matrices, as well as spreading, migration, and enhanced proliferation, whereas this antibody alone did not inhibit function on any other two- or three-dimensional substrate tested. These findings suggest differences in integrin utilization and in mechanosensing during cell interactions with three-dimensional matrices. Such studies again demonstrate that when cells probe their environment, external cues feed back to elicit unique cellular responses. It should be mentioned that while we have focused here on $\alpha_5\beta_1$-integrin interactions with fibronectin three-dimensional matrices, fibroblasts migrating through collagen gels use $\alpha_2\beta_1$-integrins for engagement. While it has been suggested that fibronectin is a candidate ECM molecule for mechanotransduction (Kostic and Sheetz, 2006), adhesions containing other integrin pairs in three-dimensional matrices require further investigation.

Ligand density can dictate the cellular signaling response, as discussed earlier. As opposed to the uniform ligand presented to cells by a two-dimensional surface, the fibrous architecture of three-dimensional matrices submits cells to a varied, non-homogeneous array of matrix contact sites that often contain multiple ECM molecules. While collagens, laminin, and fibrinogen can undergo self-polymerization at certain concentrations and temperatures, fibronectin requires a physical cell-derived force to form fibers in a process known as fibrillogenesis (Mao and Schwarzbauer, 2005; Pankov et al., 2000; Sechler et al., 2001). The mechanical tension provided by cells unfolds the molecule, exposing matrix self-assembly sites (Zhong et al., 1998). In addition, tension has been reported to expose several cryptic cysteine residues within the fibronectin C-terminus that can then form disulfide bonds with other fibronectin molecules due to the disulfide isomerase activity of these sites (Baneyx et al., 2002; Langenbach and Sottile, 1999). Because physical force is needed to generate insoluble fibronectin fibers from their soluble form, fibronectin matrix assembly has been of interest to the mechanotransduction community as a key ECM protein involved in sensing (Vogel, 2006; Vogel and Sheetz, 2006). Vitronectin does not form fibers, but it can attach to other ECM molecules such as collagen through its collagen-binding domain (Gebb et al., 1986), for example, for promoting rapid motility.

The fibers that comprise three-dimensional matrices can be viewed as dense and compact islands of ligand, which are in some ways similar to two-dimensional linear micropatterns, which can physically direct adhesion formation and cell migration.

In the 1950s, the Weiss laboratory described the concept of contact guidance, in which the physical structure of the surrounding ECM can regulate cell migration and morphology (Weiss and Garber, 1952). Their findings suggest that cells can physically "sense" the topography of fibers. More recently, it was discovered that fibroblast migration along aligned three-dimensional cell-derived matrices can be mimicked by a simple micropatterned 1.5-micron-wide one-dimensional line (Doyle et al., 2009). Interestingly, even though the one-dimensional lines are rigid – unlike the fibers in three-dimensional cell-derived matrices – treatment of fibroblasts with blebbistatin to inhibit actomyosin-based contractility inhibits cell migration velocity in both one- and three-dimensional systems. In contrast, reducing contractility in rigid two-dimensional systems normally results in increased migration (Even-Ram et al., 2007), suggesting that the topography of the ECM alone can regulate or evoke a mechanosensitive response. Similar studies using highly porous collagen-glycoaminoglycan scaffolds show that the intersections of large fiber-like struts within the scaffolds strongly affect fibroblast decision-making and migration rate (Harley et al., 2008).

In vivo systems also demonstrate how the physical structure of the ECM can regulate cellular responses. Self-assembled nanofibers reduce glial cell infiltration and promote motor and sensory neuron outgrowth after spinal cord injuries in mice, suggesting that physical guidance cues influence cell behavior (Tysseling-Mattiace et al., 2008). In addition, two-photon imaging in mouse mammary fat pads has revealed that both carcinoma cells and macrophages migrate along bundles of collagen fibers, which promote rapid migration (Sidani et al., 2006), while others have shown that metastatic cancers actually remodel the surrounding ECM into parallel aligned fibers for migration (Provenzano et al., 2008).

Although it is still not entirely clear how cells sense physical structures, the transmembrane proteoglycan syndecan-4 may play a key role (Bass et al., 2007). Synedan-4 along with $\alpha_5\beta_1$-integrin can bind fibronectin and help assemble intracellular cell adhesion structures. Syndecan-4 may facilitate cellular topography sensing by locally activating Rac downstream of protein kinase C (PKC), restricting Rac activation and cell protrusion to discrete regions of the ECM, while suppressing Rac activity elsewhere in the cell. Knockdown of syndecan-4 results in increased Rac activity and an inability of cells to discern fibers in a three-dimensional cell-derived matrix (Bass et al., 2007). Although more studies focusing on cellular responses to ECM structure are required, these findings together suggest that the structure in which ECM molecules are presented to cells (i.e., as fibers or two-dimensional surfaces) can markedly influence intracellular responses.

Although cells can migrate or invade efficiently through a non-crosslinked collagenous matrix without the use of proteases (Wolf et al., 2003), they require MMP-mediated proteolysis to migrate in native collagen gels that retain physiological crosslinking (Packard et al., 2009; Sabeh et al., 2009).

Although two-dimensional surfaces with high ligand density can produce slow rates of migration, the inherent elastic nature of three-dimensional matrices may decrease this tendency; for example, Peyton and Putnam (2005) confirmed that decreasing two-dimensional substratum rigidity can increase migration rates at high ligand densities. Conversely, rigidifying three-dimensional matrices through chemical cross-linking can counteract this effect and prevent the formation of three-dimensional matrix adhesions (Cukierman et al., 2001).

DiMilla et al. (1991) originally presented a simplified computational model for the dependence of cell migration on several factors including ligand density, receptor–ligand binding affinity, and cell contractility (detailed in a previous section). Since then, a new computational model has emerged to update the contributions of several parameters controlling cell migration, while adding other components such as substratum elasticity to generate a model for movement in three dimensions. Zaman, Lauffenburger, and coworkers (2005) designed their model to focus on both the matrix and the cell as opposed to being cell-centric or

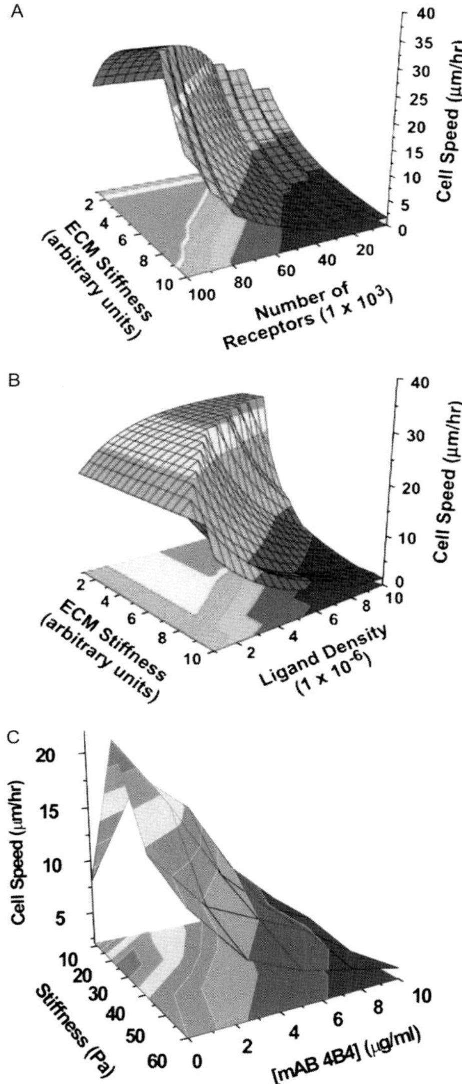

Figure 5.9. Model of the regulation of the rate of cell migration in three dimensions. (A) Three-dimensional graph of computation-derived cell speed as a function of ECM stiffness and receptor expression level. (B) Three-dimensional graph of computational modeled cell speed as a function of ECM stiffness and ligand density. (C) Experimental data, which are found to mimic the computational model shown in (A). mAB 4B4 is a function-blocking antibody against β_1-integrin.

matrix-centered. The model (Figure 5.9) suggests a similar biphasic effect as in the two-dimensionality of cell adhesivity on cell migration; the major factors influencing cell adhesiveness in three dimensions were ECM elasticity, ligand density, and receptor (integrin) expression level. These predicted effects mirror those computed and demonstrated for two-dimensional interactions. Consequently, cellular mechanosensing may be regulated by the same outside-in signals from the ECM in two- and three-dimensional microenvironments, but with possible differences in the relative importance of specific parameters (e.g., elasticity). One important aspect that may deserve more attention is the relationship between three-dimensional matrix density and elasticity. Zaman et al. (2005, 2006), as well as other investigators, have used gel matrices, such as collagen, fibrinogen, and Matrigel (a tumor basement membrane extract produced by Engelbrecht-Holm-Swarm sarcoma cells consisting mainly of laminin and several collagens), to establish that increasing ligand density leads to a nonlinear increase in matrix rigidity and a decrease in matrix pore size. These effects can be explained by the self-cross-linking capabilities of these matrix proteins.

Local proteolysis of matrices by matrix metalloproteinases (MMPs) may also be necessary for three-dimensional cell migration through densely cross-linked collagens. Inhibition of several MMPs, particularly MT-1 MMP and other transmembrane-type MMPs, inhibits the migration of invasive tumor cells (Hotary et al., 2006; Sabeh et al., 2004). Although matrix degradation might initially appear to be unrelated to mechanotransduction with the ECM, this type of local proteolysis is thought to be a key mechanism by which cells physically alter their surroundings, for example, by increasing pore size, decreasing crosslinking, and reducing local ligand density, all of which will directly affect mechanical interactions.

Other potential modulators of mechanotransduction may be the numerous growth factors and cytokines that directly or indirectly associate with ECM proteins. For example, fibronectin can bind TNF-α and VEGF, while heparan sulfate proteoglycans bind FGFs in the matrix (Alon et al., 1994; Wijelath et al., 2006). Growth factors and integrins display molecular cross-talk and mutually enhance signaling. FGF, PDGF, and EGF can cooperate with integrins to upregulate phosphorylation pathways, including the MAP kinase ERK1/2 (Miyamoto et al., 1996). Activation of these phosphorylation cascades requires both integrin aggregation and ligand occupancy. Although these experiments were performed on two-dimensional rigid cell culture dishes, ERK1/2 phosphorylation can be up-regulated in three-dimensional cell-derived matrices. The Bissell laboratory (2003) demonstrated that in three-dimensional cultures, reciprocal interactions occur between the β_1-integrin and the EGF receptor (EGFR) in HMT-3522 cells. Inhibitory antibodies against either protein or inhibition of MAP kinase decreased β_1 and/or EGFR expression levels. Likewise, overexpression of EGFR elevated β_1 expression. These and many other studies have identified a complex web of synergistic interactions between growth factor/cytokine and integrin receptors involved in cell migration and signaling. These relationships between extracellular chemical and physical signals are likely to be important to mechanotransduction.

While much of this chapter has focused on single cells and their interactions with their surrounding environment, the composition and elasticity of the matrix in tissues may regulate not only local signal cascades, but also the overall cellular differentiation and development of tissues and organs by controlling gene expression patterns and cell fate (Engler et al., 2006). ECM rigidity, density, and molecular composition vary greatly between tissues; ligaments and tendons consist of highly dense and compact collagens whereas mammary tissue structure is quite different.

Cells grown in extremely elastic collagen matrices (< 200 Pa) similar to the ECM of mammary tissue (~ 167 Pa) display little activation of Rho, an activator of stress fiber assembly (Chrzanowska-Wodnicka and Burridge, 1996), low ERK activation, low FAK phosphorylation at Y397, few stress fibers, and little vinculin association with adhesions among other adhesion associated factors (Paszek et al., 2005). In striking contrast, tumors within mammary tissue display a Young's modulus of ~ 4000 Pa, high levels of FAK Tyr-397 phosphorylation, and ERK1/2 activation. In terms of morphology, cells under more normal conditions of elasticity aggregate to form mammary acinar or ductal glandular structures similar to *in vivo*, with a basement membrane on the exterior of a well-organized multicellular structure and a definitive lumen at the center. Increasing the stiffness of collagen gels disrupted the glandular structure and resulted in the absence of a basement membrane and an increase in apparent cellular invasiveness into the surrounding matrix resembling tumor invasion. These findings suggest that ECM structure and elasticity may contribute to malignancy through mechanotransduction.

Acknowledgment

The research in this chapter was supported by the Intramural Program of the National Institute of Dental and Craniofacial Research, National Institutes of Health.

REFERENCES

Abercrombie, M., and G. A. Dunn. 1975. Adhesions of fibroblasts to substratum during contact inhibition observed by interference reflection microscopy. *Exp Cell Res.* **92**: 57–62.

Abercrombie, M., J. E. Heaysman, and S. M. Pegrum. 1971. The locomotion of fibroblasts in culture. IV. Electron microscopy of the leading lamella. *Exp Cell Res.* **67**: 359–67.

Akiyama, S. K. 1996. Integrins in cell adhesion and signaling. *Hum Cell.* **9**: 181–6.

Akiyama, S. K., and K. M. Yamada. 1985. The interaction of plasma fibronectin with fibroblastic cells in suspension. *J Biol Chem.* **260**: 4492–500.

Akiyama, S. K., S. S. Yamada, W. T. Chen, and K. M. Yamada. 1989. Analysis of fibronectin receptor function with monoclonal antibodies: Roles in cell adhesion, migration, matrix assembly, and cytoskeletal organization. *J Cell Biol.* **109**: 863–75.

Alberts, B., A. Johnson, J. Lewis, M. Raff, K. Roberts, P. Walter. 2002. *Molecular Biology of the Cell.* Garland Science, New York.

Alon, R., L. Cahalon, R. Hershkoviz, D. Elbaz, B. Reizis, D. Wallach, S. K. Akiyama, K. M. Yamada, and O. Lider. 1994. TNF-alpha binds to the N-terminal domain of fibronectin and augments the beta 1-integrin-mediated adhesion of CD4+ T lymphocytes to the glycoprotein. *J Immunol.* **152**: 1304–13.

Aota, S., M. Nomizu, and K. M. Yamada. 1994. The short amino acid sequence Pro-His-Ser-Arg-Asn in human fibronectin enhances cell-adhesive function. *J Biol Chem.* **269**: 24756–61.

Arnaout, M. A., B. Mahalingam, and J. P. Xiong. 2005. Integrin structure, allostery, and bidirectional signaling. *Annu Rev Cell Dev Biol.* **21**: 381–410.

Balaban, N. Q., U. S. Schartz, D. Riveline, P. Goichberg, G. Tzur, I. Sabanay, D. Mahalu, S. Safran, A. Bershadsky, L. Addad, and B. Geiger. 2001. Force and focal adhesion assembly: A close relationship studied using elastic micropatterned substrates. *Nature Cell Biol.* **3**: 466–472.

Ballestrem, C., B. Hinz, B. A. Imhof, and B. Wehrle-Haller. 2001. Marching at the front and dragging behind: Differential alphaVbeta3-integrin turnover regulates focal adhesion behavior. *J Cell Biol.* **155**: 1319–32.

Baneyx, G., L. Baugh, and V. Vogel. 2002. Fibronectin extension and unfolding within cell matrix fibrils controlled by cytoskeletal tension. *Proc Natl Acad Sci USA.* **99**: 5139–43.

Bass, M. D., K. A. Roach, M. R. Morgan, Z. Mostafavi-Pour, T. Schoen, T. Muramatsu, U. Mayer, C. Ballestrem, J. P. Spatz, and M. J. Humphries. 2007. Syndecan-4-dependent Rac1 regulation determines directional migration in response to the extracellular matrix. *J Cell Biol.* **177**: 527–38.

Bazzoni, G., D. T. Shih, C. A. Buck, and M. E. Hemler. 1995. Monoclonal antibody 9EG7 defines a novel beta 1 integrin epitope induced by soluble ligand and manganese, but inhibited by calcium. *J Biol Chem.* **270**: 25570–7.

Beningo, K. A., M. Dembo, I. Kaverina, J. V. Small, and Y. L. Wang. 2001. Nascent focal adhesions are responsible for the generation of strong propulsive forces on migrating fibroblasts. *J Cell Biol.* **153**: 881–7.

Beningo, K. A., M. Dembo, and Y. L. Wang. 2004. Responses of fibroblasts to anchorage of dorsal extracellular matrix receptors. *Proc Natl Acad Sci USA.* **101**: 18024–9.

Bershadsky, A. D., N. Q. Balaban, and B. Geiger. 2003. Adhesion-dependent cell mechanosensitivity. *Annu Rev Cell Dev Biol.* **19**: 677–95.

Bischofs, I. B., S. A. Safran, and U. S. Schwarz. 2004. Elastic interactions of active cells with soft materials. *Phys Rev E Stat Nonlin Soft Matter Phys.* **69**: 021911.

Bischofs, I. B., and U. S. Schwarz. 2003. Cell organization in soft media due to active mechanosensing. *Proc Natl Acad Sci USA.* **100**: 9274–9.

Buck, C. A., E. Shea, K. Duggan, and A. F. Horwitz. 1986. Integrin (the CSAT antigen): Functionality requires oligomeric integrity. *J Cell Biol.* **103**: 2421–8.

Cai, Y., N. Biais, G. Giannone, M. Tanase, G. Jiang, J. M. Hofman, C. H. Wiggins, P. Silberzan, A. Buguin, B. Ladoux, and M. P. Sheetz. 2006. Nonmuscle myosin IIA-dependent force inhibits cell spreading and drives F-actin flow. *Biophys J.* **91**: 3907–20.

Calderwood, D. A. 2004a. Integrin activation. *J Cell Sci.* **117**: 657–66.

Calderwood, D. A. 2004b. Talin controls integrin activation. *Biochem Soc Trans.* **32**: 434–7.

Calderwood, D. A., Y. Fujioka, J. M. de Pereda, B. Garcia-Alvarez, T. Nakamoto, B. Margolis, C. J. McGlade, R. C. Liddington, and M. H. Ginsberg. 2003. Integrin beta cytoplasmic domain interactions with phosphotyrosine-binding domains: A structural prototype for diversity in integrin signaling. *Proc Natl Acad Sci USA.* **100**: 2272–7.

Calderwood, D. A., B. Yan, J. M. de Pereda, B. G. Alvarez, Y. Fujioka, R. C. Liddington, and M. H. Ginsberg. 2002. The phosphotyrosine binding-like domain of talin activates integrins. *J Biol Chem.* **277**: 21749–58.

Calderwood, D. A., R. Zent, R. Grant, D. J. Rees, R. O. Hynes, and M. H. Ginsberg. 1999. The Talin head domain binds to integrin beta subunit cytoplasmic tails and regulates integrin activation. *J Biol Chem.* **274**: 28071–4.

Carreiras, F., Y. Denoux, C. Staedel, M. Lehmann, F. Sichel, and P. Gauduchon. 1996. Expression and localization of alpha v integrins and their ligand vitronectin in normal ovarian epithelium and in ovarian carcinoma. *Gynecol Oncol.* **62**: 260–7.

Carter, S. B. 1967. Haptotaxis and the mechanism of cell motility. *Nature.* **213**: 256–60.

Chen, W. 1981. Mechanism of retraction of the trailing edge during fibroblast movement. *J Cell Biol.* **90**: 187–200.

Choquet, D., D. P. Felsenfeld, and M. P. Sheetz. 1997. Extracellular matrix rigidity causes strengthening of integrin-cytoskeleton linkages. *Cell.* **18**: 39–48.

Chrzanowska-Wodnicka, M., and K. Burridge. 1996. Rho-stimulated contractility drives the formation of stress fibers and focal adhesions. *J Cell Biol.* **133**: 1403–15.

Citi, S., and J. Kendrick-Jones. 1988. Regulation of non-muscle myosin structure and function. *BioEssays*. **7**: 155–159.

Clark, K., R. Pankov, M. A. Travis, J. A. Askari, A. P. Mould, S. E. Craig, P. Newham, K. M. Yamada, and M. J. Humphries. 2005. A specific alpha5beta1-integrin conformation promotes directional integrin translocation and fibronectin matrix formation. *J Cell Sci.* **118**: 291–300.

Clark, R. A. F. 1996. *The Molecular and Cellular Biology of Wound Repair*. Plenum Press, New York. 611 pp.

Cluzel, C., F. Saltel, J. Lussi, F. Paulhe, B. A. Imhof, and B. Wehrle-Haller. 2005. The mechanisms and dynamics of (alpha)v(beta)3 integrin clustering in living cells. *J Cell Biol.* **171**: 383–92.

Cukierman, E., R. Pankov, D. R. Stevens, and K. M. Yamada. 2001. Taking cell-matrix adhesions to the third dimension. *Science*. **294**: 1708–12.

Cukierman, E., R. Pankov, and K. M. Yamada. 2002. Cell interactions with three-dimensional matrices. *Curr Opin Cell Biol.* **14**: 633–9.

Danen, E. H., S. Aota, A. A. van Kraats, K. M. Yamada, D. J. Ruiter, and G. N. van Muijen. 1995. Requirement for the synergy site for cell adhesion to fibronectin depends on the activation state of integrin alpha 5 beta 1. *J Biol Chem.* **270**: 21612–8.

Danen, E. H., and K. M. Yamada. 2001. Fibronectin, integrins, and growth control. *J Cell Physiol.* **189**: 1–13.

Davis, S., M. L. Lu, S. H. Lo, S. Lin, J. A. Butler, B. J. Druker, T. M. Roberts, Q. An, and L. B. Chen. 1991. Presence of an SH2 domain in the actin-binding protein tensin. *Science*. **252**: 712–5.

DeMali, K. A., K. Wennerberg, and K. Burridge. 2003. Integrin signaling to the actin cytoskeleton. *Curr Opin Cell Biol.* **15**: 572–82.

Dembo, M., and Y. L. Wang. 1999. Stresses at the cell-to-substrate interface during locomotion of fibroblasts. *Biophys. J.* **76**: 2307–16.

DiMilla, P. A., K. Barbee, and D. A. Lauffenburger. 1991. Mathematical model for the effects of adhesion and mechanics on cell migration speed. *Biophys J.* **60**: 15–37.

DiMilla, P. A., J. A. Stone, J. A. Quinn, S. M. Albelda, and D. A. Lauffenburger. 1993. Maximal migration of human smooth muscle cells on fibronectin and type IV collagen occurs at an intermediate attachment strength. *J Cell Biol.* **122**: 729–37.

Discher, D. E., P. Janmey, and Y. L. Wang. 2005. Tissue cells feel and respond to the stiffness of their substrate. *Science*. **310**: 1139–43.

Doyle, A., W. Marganski, and J. Lee. 2004. Calcium transients induce spatially coordinated increases in traction force during the movement of fish keratocytes. *J Cell Sci.* **117**: 2203–14.

Doyle, A. D., and J. Lee. 2002. Simultaneous, real-time imaging of intracellular calcium and cellular traction force production. *Biotechniques*. **33**: 358–64.

Doyle, A. D., F. W. Wang, K. Matsumoto, and K. M. Yamada. 2009. One-dimensional topography underlies three-dimensional fibrillar cell migration. *J Cell Biol.* **184**: 481–90.

Duband, J. L., S. Dufour, S. S. Yamada, K. M. Yamada, and J. P. Thiery. 1991. Neural crest cell locomotion induced by antibodies to beta 1 integrins. A tool for studying the roles of substratum molecular avidity and density in migration. *J Cell Sci.* **98**(Pt 4):517–32.

Duband, J. L., G. H. Nuckolls, A. Ishihara, T. Hasegawa, K. M. Yamada, J. P. Thiery, and K. Jacobson. 1988. Fibronectin receptor exhibits high lateral mobility in embryonic locomoting cells but is immobile in focal contacts and fibrillar streaks in stationary cells. *J Cell Biol.* **107**: 1385–96.

Duband, J. L., S. Rocher, W. T. Chen, K. M. Yamada, and J. P. Thiery. 1986. Cell adhesion and migration in the early vertebrate embryo: Location and possible role of the putative fibronectin receptor complex. *J Cell Biol.* **102**: 160–78.

Dunn, G. A. 1980. The locomotory machinery of fibroblasts. *Eur J Cancer*. **16**: 6–8.

Engler, A., L. Bacakova, C. Newman, A. Hategan, M. Griffin, and D. Discher. 2004. Substrate compliance versus ligand density in cell on gel responses. *Biophys J.* **86**: 617–28.

Engler, A. J., S. Sen, H. L. Sweeney, and D. E. Discher. 2006. Matrix elasticity directs stem cell lineage specification. *Cell.* **126**: 677–89.

Evans, E. A., and D. A. Calderwood. 2007. Forces and bond dynamics in cell adhesion. *Science.* **316**: 1148–53.

Even-Ram, S., V. Artym, and K. M. Yamada. 2006. Matrix control of stem cell fate. *Cell.* **126**: 645–7.

Even-Ram, S., A. D. Doyle, M. A. Conti, K. Matsumoto, R. S. Adelstein, and K. M. Yamada. 2007. Myosin IIA regulates cell motility and actomyosin-microtubule crosstalk. *Nat Cell Biol.* **9**: 299–309.

Finer, J. T., R. M. Simmons, and J. A. Spudich. 1994. Single myosin molecule mechanics: piconewton forces and nanometre steps. *Nature.* **368**: 113–9.

French-Constant, C., and H. Colognato. 2004. Integrins: Versatile integrators of extracellular signals. *Trends Cell Biol.* **14**: 678–86.

Friedland, J. C., M. H. Lee, and D. Boettiger. 2009. Mechanically activated integrin switch controls alpha5beta1 function. *Science.* **323**: 642–4.

Galbraith, C., and M. Sheetz. 1997. A micromachined device provides a new bend on fibroblast traction forces. *Proc Nat Assoc Sci* **94**: 9114–9118.

Galbraith, C., and M. Sheetz. 1999. Keratocytes pull with similar forces on their dorsal and ventral surfaces. *J. Cell Biol.* **147**: 1313–23.

Galbraith, C. G., K. M. Yamada, and J. A. Galbraith. 2007. Polymerizing actin fibers position integrins primed to probe for adhesion sites. *Science.* **315**: 992–5.

Galbraith, C. G., K. M. Yamada, and M. P. Sheetz. 2002. The relationship between force and focal complex development. *J Cell Biol.* **159**: 695–705.

Garcia-Alvarez, B., J. M. de Pereda, D. A. Calderwood, T. S. Ulmer, D. Critchley, I. D. Campbell, M. H. Ginsberg, and R. C. Liddington. 2003. Structural determinants of integrin recognition by talin. *Mol Cell.* **11**: 49–58.

Garcia, A. J., J. E. Schwarzbauer, and D. Boettiger. 2002. Distinct activation states of alpha5-beta1 integrin show differential binding to RGD and synergy domains of fibronectin. *Biochemistry.* **41**: 9063–9.

Gebb, C., E. G. Hayman, E. Engvall, and E. Ruoslahti. 1986. Interaction of vitronectin with collagen. *J Biol Chem.* **261**: 16698–703.

Geiger, B., and A. Bershadsky. 2001. Assembly and mechanosensory function of focal contacts. *Curr Opin Cell Biol.* **13**: 584–92.

Giancotti, F. G., and E. Ruoslahti. 1999. Integrin signaling. *Science.* **285**: 1028–32.

Ginsberg, M. H., A. Partridge, and S. J. Shattil. 2005. Integrin regulation. *Curr Opin Cell Biol.* **17**: 509–16.

Grinnell, F. 2003. Fibroblast biology in three-dimensional collagen matrices. *Trends Cell Biol.* **13**: 264–9.

Grzesiak, J. J., G. E. Davis, D. Kirchhofer, and M. D. Pierschbacher. 1992. Regulation of alpha 2 beta 1-mediated fibroblast migration on type I collagen by shifts in the concentrations of extracellular Mg2+ and Ca2+. *J Cell Biol.* **117**: 1109–17.

Gupton, S. L., and C. M. Waterman-Storer. 2006. Spatiotemporal feedback between actomyosin and focal-adhesion systems optimizes rapid cell migration. *Cell.* **125**: 1361–74.

Harburger, D. S., M. Bouaouina, and D. A. Calderwood. 2009. Kindlin-1 and -2 directly bind the C-terminal region of beta integrin cytoplasmic tails and exert intergrin-specific activation effects. *J Biol Chem.* **284**: 11485–97.

Harley, B. A., H. D. Kim, M. H. Zaman, I. V. Yannas, D. A. Lauffenburger, and L. J. Gibson. 2008. Microarchitecture of three-dimensional scaffolds influences cell migration behavior via junction interactions. *Biophys J.* **95**: 4013–24.

Harris, A. K., D. Stopak, and P. Wild. 1981. Fibroblast traction as a mechanism for collagen morphogenesis. *Nature.* **290**: 249–51.

Harris, A. K., P. Wild, and D. Stopak. 1980. Silicone rubber substrata: a new wrinkle in the study of cell locomotion. *Science.* **208**: 177–9.

Hay, E. D. 1981. Extracellular matrix. *J Cell Biol.* **91**: 205s–223s.

Hodge, T., and M. J. Cope. 2000. A myosin family tree. *J Cell Sci.* **113**(Pt 19):3353–4.

Hotary, K., X. Y. Li, E. Allen, S. L. Stevens, and S. J. Weiss. 2006. A cancer cell metalloprotease triad regulates the basement membrane transmigration program. *Genes Dev.* **20**: 2673–86.

Huttenlocher, A., M. H. Ginsberg, and A. F. Horwitz. 1996. Modulation of cell migration by integrin-mediated cytoskeletal linkages and ligand-binding affinity. *J Cell Biol.* **134**: 1551–62.

Huttenlocher, A., S. P. Palecek, Q. Lu, W. Zhang, R. L. Mellgren, D. A. Lauffenburger, M. H. Ginsberg, and A. F. Horwitz. 1997. Regulation of cell migration by the calcium-dependent protease calpain. *J Biol Chem.* **272**: 32719–22.

Hynes, R. O. 1987. Integrins: A family of cell surface receptors. *Cell.* **48**: 549–54.

Hynes, R. O. 2002. Integrins: Bidirectional, allosteric signaling machines. *Cell.* **110**: 673–87.

Jay, P. Y., P. A. Pham, K. Wong, and E. Elson. 1995. A mechanical function of myosin II in cell motility. *J Cell Sci.* **108**: 387–93.

Jiang, G., A. H. Huang, Y. Cai, M. Tanase, and M. P. Sheetz. 2006. Rigidity sensing at the leading edge through alphavbeta3 integrins and RPTPalpha. *Biophys J.* **90**: 1804–9.

Jiang, H., and F. Grinnell. 2005. Cell-matrix entanglement and mechanical anchorage of fibroblasts in three-dimensional collagen matrices. *Mol Biol Cell.* **16**: 5070–6.

Katoh, K., Y. Kano, M. Amano, K. Kaibuchi, and K. Fujuiwara. 2001a. Stress fiber organization regulated by MLCK and Rho-kinase in culture human fibroblasts. *Am J Physiol Cell Physiol.* **280**: C1669–C1679.

Katoh, K., Y. Kano, M. Masuda, H. Onishi, and K. Fujiwara. 1998. Isolation and contraction of the stress fiber. *Mol Biol Cell.* **9**: 1919–38.

Katz, B. Z., E. Zamir, A. Bershadsky, Z. Kam, K. M. Yamada, and B. Geiger. 2000. Physical state of the extracellular matrix regulates the structure and molecular composition of cell-matrix adhesions. *Mol Biol Cell.* **11**: 1047–60.

Kaverina, I., O. Krylyshkina, and J. V. Small. 2002. Regulation of substrate adhesion dynamics during cell motility. *Int J Biochem Cell Biol.* **34**: 746–61.

Kimura, K., M. Ito, M. Amano, K. Chihara, Y. Fukata, M. Nakafuku, B. Yamamori, J. Feng, T. Nakano, K. Okawa, A. Iwamatsu, and K. Kaibuchi. 1996. Regulation of myosin phosphates by rho and rho-associated kinase (rho-kinase). *Science.* **273**: 245–8.

Kostic, A., and M. P. Sheetz. 2006. Fibronectin rigidity response through Fyn and p130Cas recruitment to the leading edge. *Mol Biol Cell.* **17**: 2684–95.

LaFlamme, S. E., S. K. Akiyama, and K. M. Yamada. 1992. Regulation of fibronectin receptor distribution. *J Cell Biol.* **117**: 437–47.

Lamoureux, P., V. L. Steel, C. Regal, L. Adgate, R. E. Buxbaum, and S. R. Heidemann. 1990. Extracellular matrix allows PC12 neurite elongation in the absence of microtubules. *J Cell Biol.* **110**: 71–9.

Langenbach, K. J., and J. Sottile. 1999. Identification of protein-disulfide isomerase activity in fibronectin. *J Biol Chem.* **274**: 7032–8.

Larjava, H., E. F. Plow, and C. Wu. 2008. Kindlins: Essential regulators of integrin signalling and cell-matrix adhesion. *EMBO Rep.* **9**: 1203–8.

Larsen, M., V. V. Artym, J. A. Green, and K. M. Yamada. 2006. The matrix reorganized: extracellular matrix remodeling and integrin signaling. *Curr Opin Cell Biol.* **18**: 463–71.

Lee, J., and K. Jacobson. 1997. The composition and dynamics of cell-substratum adhesions in locomoting fish keratocytes. *J Cell Sci.* **110**(Pt 22):2833–44.

Lipfert, L., B. Haimovich, M. D. Schaller, B. S. Cobb, J. T. Parsons, and J. S. Brugge. 1992. Integrin-dependent phosphorylation and activation of the protein tyrosine kinase pp125FAK in platelets. *J Cell Biol.* **119**: 905–12.

Lo, C., H. B. Wang, M. Dembo, and Y. L. Wang. 2000. Cell movement is guided by the rigidity of the substrate. *Biophys J.* **79**: 144–52.

Luo, B. H., K. Strokovich, T. Walz, T. A. Springer, and J. Takagi. 2004. Allosteric beta1 integrin antibodies that stabilize the low affinity state by preventing the swing-out of the hybrid domain. *J Biol Chem.* **279**: 27466–71.

Mao, Y., and J. E. Schwarzbauer. 2005. Fibronectin fibrillogenesis, a cell-mediated matrix assembly process. *Matrix Biol.* **24**: 389–99.

Miranti, C. K., and J. S. Brugge. 2002. Sensing the environment: A historical perspective on integrin signal transduction. *Nat Cell Biol.* **4**: E83–90.

Miyamoto, S., S. K. Akiyama, and K. M. Yamada. 1995a. Synergistic roles for receptor occupancy and aggregation in integrin transmembrane function. *Science.* **267**: 883–5.

Miyamoto, S., H. Teramoto, O. A. Coso, J. S. Gutkind, P. D. Burbelo, S. K. Akiyama, and K. Yamada. 1995b. Integrin function: Molecular hierarchies of cytoskeletal and signaling molecules. *J. Cell Biol.* **131**: 791–805.

Miyamoto, S., H. Teramoto, J. S. Gutkind, and K. M. Yamada. 1996. Integrins can collaborate with growth factors for phosphorylation of receptor tyrosine kinases and MAP kinase activation: Roles of integrin aggregation and occupancy of receptors. *J Cell Biol.* **135**: 1633–42.

Montanez, E., S. Ussar, M. Schifferer, M. Bosl, R. Zent, M. Moser, and R. Fassler. 2008. Kindlin-2 controls bidirectional signaling of integrins. *Genes Dev.* **22**: 1325–30.

Moser, M., M. Bauer, S. Schmid, R. Ruppert, S. Schmidt, M. Sixt, H. V. Wang, M. Sperandio, and R. Fassler. 2009. Kindlin-3 is required for beta2 integrin-mediated leukocyte adhesion to endothelial cells. *Nat Med.* **15**: 300–5.

Moser, M., B. Nieswandt, S. Ussar, M. Pozgajova, and R. Fassler. 2008. Kindlin-3 is essential for integrin activation and platelet aggregation. *Nat Med.* **14**: 325–30.

Mould, A. P., S. K. Akiyama, and M. J. Humphries. 1995a. Regulation of integrin alpha 5 beta 1-fibronectin interactions by divalent cations. Evidence for distinct classes of binding sites for Mn2+, Mg2+, and Ca2+. *J Biol Chem.* **270**: 26270–7.

Mould, A. P., J. A. Askari, S. Aota, K. M. Yamada, A. Irie, Y. Takada, H. J. Mardon, and M. J. Humphries. 1997. Defining the topology of integrin alpha5beta1-fibronectin interactions using inhibitory anti-alpha5 and anti-beta1 monoclonal antibodies. Evidence that the synergy sequence of fibronectin is recognized by the amino-terminal repeats of the alpha5 subunit. *J Biol Chem.* **272**: 17283–92.

Mould, A. P., A. N. Garratt, J. A. Askari, S. K. Akiyama, and M. J. Humphries. 1995b. Regulation of integrin alpha 5 beta 1 function by anti-integrin antibodies and divalent cations. *Biochem Soc Trans.* **23**: 395S.

Mould, A. P., E. J. Symonds, P. A. Buckley, J. G. Grossmann, P. A. McEwan, S. J. Barton, J. A. Askari, S. E. Craig, J. Bella, and M. J. Humphries. 2003. Structure of an integrin-ligand complex deduced from solution x-ray scattering and site-directed mutagenesis. *J Biol Chem.* **278**: 39993–9.

Munevar, S., Y. L. Wang, and M. Dembo. 2001a. Distinct roles of frontal and rear cell-substrate adhesions in fibroblast migration. *Mol Biol Cell.* **12**: 3947–54.

Munevar, S., Y. L. Wang, and M. Dembo. 2001b. Traction force microscopy of migrating normal and H-ras transformed 3T3 fibroblasts. *Biophys J.* **80**: 1744–57.

Nagai, T., N. Yamakawa, S. Aota, S. S. Yamada, S. K. Akiyama, K. Olden, and K. M. Yamada. 1991. Monoclonal antibody characterization of two distant sites required for function of the central cell-binding domain of fibronectin in cell adhesion, cell migration, and matrix assembly. *J Cell Biol.* **114**: 1295–305.

Nelson, C. M., and M. J. Bissell. 2006. Of extracellular matrix, scaffolds, and signaling: Tissue architecture regulates development, homeostasis, and cancer. *Annu Rev Cell Dev Biol.* **22**: 287–309.

Numaguchi, Y., S. Huang, T. R. Polte, G. S. Eichler, N. Wang, and D. E. Ingber. 2003. Caldesmon-dependent switching between capillary endothelial cell growth and apoptosis through modulation of cell shape and contractility. *Angiogenesis.* **6**: 55–64.

Oliver, T., M. Dembo, and K. Jacobson. 1995. Traction forces in locomoting cells. *Cell Mot Cyotskelet.* **31**: 225–40.

Oliver, T., J. Lee, and K. Jacobson. 1994. Forces exerted by locomoting cells. *Sem Cell Biol.* **5**: 139–194.

Onley, D. J., C. G. Knight, D. S. Tuckwell, M. J. Barnes, and R. W. Farndale. 2000. Micromolar Ca2+ concentrations are essential for Mg2+-dependent binding of collagen by the integrin alpha 2beta 1 in human platelets. *J Biol Chem.* **275**: 24560–4.

Otey, C. A., G. B. Vasquez, K. Burridge, and B. W. Erickson. 1993. Mapping of the alpha-actinin binding site within the beta 1 integrin cytoplasmic domain. *J Biol Chem.* **268**: 21193–7.

Packard, B. Z., V. V. Artym, A. Komoriya, and K. M. Yamada. 2009. Direct visualization of protease activity on cells migrating in three-dimensions. *Matrix Biol.* **28**: 3–10.

Palecek, S. P., A. F. Horwitz, and D. A. Lauffenburger. 1999. Kinetic model for integrin-mediated adhesion release during cell migration. *Ann Biomed Eng.* **27**: 219–35.

Palecek, S. P., A. Huttenlocher, A. F. Horwitz, and D. A. Lauffenburger. 1998. Physical and biochemical regulation of integrin release during rear detachment of migrating cells. *J Cell Sci.* **111**(Pt 7):929–40.

Palecek, S. P., J. C. Loftus, M. H. Ginsberg, D. A. Lauffenburger, and A. F. Horwitz. 1997. Integrin-ligand binding properties govern cell migration speed through cell-substratum adhesiveness. *Nature.* **385**: 537–40.

Palecek, S. P., C. E. Schmidt, D. A. Lauffenburger, and A. F. Horwitz. 1996. Integrin dynamics on the tail region of migrating fibroblasts. *J Cell Sci.* **109** (Pt 5):941–52.

Pankov, R., E. Cukierman, B. Z. Katz, K. Matsumoto, D. C. Lin, S. Lin, C. Hahn, and K. M. Yamada. 2000. Integrin dynamics and matrix assembly: Tensin-dependent translocation of alpha(5)beta(1) integrins promotes early fibronectin fibrillogenesis. *J Cell Biol.* **148**: 1075–90.

Pankov, R., Y. Endo, S. Even-Ram, M. Araki, K. Clark, E. Cukierman, K. Matsumoto, and K. M. Yamada. 2005. A Rac switch regulates random versus directionally persistent cell migration. *J Cell Biol.* **170**: 793–802.

Parrini, M. C., M. Matsuda, and J. de Gunzburg. 2005. Spatiotemporal regulation of the Pak1 kinase. *Biochem Soc Trans.* **33**: 646–8.

Paszek, M. J., N. Zahir, K. R. Johnson, J. N. Lakins, G. I. Rozenberg, A. Gefen, C. A. Reinhart-King, S. S. Margulies, M. Dembo, D. Boettiger, D. A. Hammer, and V. M. Weaver. 2005. Tensional homeostasis and the malignant phenotype. *Cancer Cell.* **8**: 241–54.

Pelham, R. J., Jr., and Y. L. Wang. 1998. Cell locomotion and focal adhesions are regulated by the mechanical properties of the substrate. *Biol Bull.* **194**: 348–9; discussion 349–50.

Pelham, R. J., and Y. L. Wang. 1997. Cell locomotion and focal adhesions are regulated by substrate flexibility. *PNAS.* **94**: 13661–5.

Pelham, R. J., and Y. L. Wang. 1999. High-resolution detection of mechanical forces exerted by locomoting fibroblasts on the substrate. *Mol Biol Cell.* **10**: 935–45.

Peyton, S. R., and A. J. Putnam. 2005. Extracellular matrix rigidity governs smooth muscle cell motility in a biphasic fashion. *J Cell Physiol.* **204**: 198–209.

Plopper, G. E., H. P. McNamee, L. E. Dike, K. Bojanowski, and D. E. Ingber. 1995. Convergence of integrin and growth factor receptor signaling pathways within the focal adhesion complex. *Mol Biol Cell.* **6**: 1349–65.

Polte, T. R., G. S. Eichler, N. Wang, and D. E. Ingber. 2004. Extracellular matrix controls myosin light chain phosphorylation and cell contractility through modulation of cell shape and cytoskeletal prestress. *Am J Physiol Cell Physiol.* **286**: C518–28.

Provenzano, P. P., D. R. Inman, K. W. Eliceiri, S. M. Trier, and P. J. Keely. 2008. Contact guidance mediated three-dimensional cell migration is regulated by Rho/ROCK-dependent matrix reorganization. *Biophys J.* **95**: 5374–84.

Radisky, D. C., P. A. Kenny, and M. J. Bissell. 2007. Fibrosis and cancer: Do myofibroblasts come also from epithelial cells via EMT? *J Cell Biochem.* **101**: 830–9.

Reinhart-King, C. A., M. Dembo, and D. A. Hammer. 2005. The dynamics and mechanics of endothelial cell spreading. *Biophys J.* **89**: 676–89.

Ria, R., A. Vacca, D. Ribatti, F. Di Raimondo, F. Merchionne, and F. Dammacco. 2002. Alpha(v)beta(3) integrin engagement enhances cell invasiveness in human multiple myeloma. *Haematologica*. **87**: 836–45.

Ridley, A. J. 2001. Rho proteins: Linking signaling with membrane trafficking. *Traffic*. **2**: 303–10.

Ridley, A. J., and A. Hall. 1992a. Distinct patterns of actin organization regulated by the small GTP-binding proteins Rac and Rho. *Cold Spring Harb Symp Quant Biol*. **57**: 661–71.

Ridley, A. J., and A. Hall. 1992b. The small GTP-binding protein rho regulates the assembly of focal adhesions and actin stress fibers in response to growth factors. *Cell*. **70**: 389–99.

Ridley, A. J., M. A. Schwartz, K. Burridge, R. A. Firtel, M. H. Ginsberg, G. Borisy, J. T. Parsons, and A. R. Horwitz. 2003. Cell migration: Integrating signals from front to back. *Science*. **302**: 1704–9.

Riveline, D., E. Zamir, N. Q. Balaban, U. S. Schwarz, T. Ishizaki, S. Narumiya, Z. Kam, B. Geiger, and A. D. Bershadsky. 2001. Focal contacts as mechanosensors: Externally applied local mechanical force induces growth of focal contacts by an mDia1-dependent and ROCK-independent mechanism. *J Cell Biol*. **153**: 1175–86.

Ruoslahti, E. 1996. RGD and other recognition sequences for integrins. *Annu Rev Cell Dev Biol*. **12**: 697–715.

Ruoslahti, E., E. G. Hayman, and M. D. Pierschbacher. 1985. Extracellular matrices and cell adhesion. *Arteriosclerosis*. **5**: 581–94.

Sabeh, F., I. Ota, K. Holmbeck, H. Birkedal-Hansen, P. Soloway, M. Balbin, C. Lopez-Otin, S. Shapiro, M. Inada, S. Krane, E. Allen, D. Chung, and S. J. Weiss. 2004. Tumor cell traffic through the extracellular matrix is controlled by the membrane-anchored collagenase MT1-MMP. *J Cell Biol*. **167**: 769–81.

Sabeh, F., R. Shimizu-Hirota, and S. J. Weiss. 2009. Protease-dependent versus -independent cancer cell invasion programs: Three-dimensional amoeboid movement revisited. *J Cell Biol*. **185**: 11–19.

Sampath, R., P. J. Gallagher, and F. M. Pavalko. 1998. Cytoskeletal interactions with the leukocyte integrin beta2 cytoplasmic tail. Activation-dependent regulation of associations with talin and alpha-actinin. *J Biol Chem*. **273**: 33588–94.

Sandquist, J. C., K. I. Swenson, K. A. Demali, K. Burridge, and A. R. Means. 2006. Rho kinase differentially regulates phosphorylation of nonmuscle myosin II isoforms A and B during cell rounding and migration. *J Biol Chem*. **281**: 35873–83.

Sawada, Y., and M. P. Sheetz. 2002. Force transduction by triton cytoskeletons. *J Cell Biol*. **156**: 609–15.

Schaller, M. D., C. A. Borgman, and J. T. Parsons. 1993. Autonomous expression of a non-catalytic domain of the focal adhesion-associated protein tyrosine kinase pp125FAK. *Mol Cell Biol*. **13**: 785–91.

Schmidt, C. E., A. F. Horwitz, D. A. Lauffenburger, and M. P. Sheetz. 1993. Integrin-cytoskeletal interactions in migrating fibroblasts are dynamic, asymmetric, and regulated. *J Cell Biol*. **123**: 977–91.

Schoenwaelder, S. M., and K. Burridge. 1999. Bidirectional signaling between the cytoskeleton and integrins. *Curr Opin Cell Biol*. **11**: 274–86.

Schwartz, M. A., and M. H. Ginsberg. 2002. Networks and crosstalk: Integrin signalling spreads. *Nat Cell Biol*. **4**: E65–8.

Schwarz, U. S., and I. B. Bischofs. 2005. Physical determinants of cell organization in soft media. *Med Eng Phys*. **27**: 763–72.

Schwarz, U. S., T. Erdmann, and I. B. Bischofs. 2006. Focal adhesions as mechanosensors: The two-spring model. *Biosystems*. **83**: 225–32.

Sechler, J. L., H. Rao, A. M. Cumiskey, I. Vega-Colon, M. S. Smith, T. Murata, and J. E. Schwarzbauer. 2001. A novel fibronectin binding site required for fibronectin fibril growth during matrix assembly. *J Cell Biol*. **154**: 1081–8.

Shimaoka, M., J. Takagi, and T. A. Springer. 2002. Conformational regulation of integrin structure and function. *Annu Rev Biophys Biomol Struct*. **31**: 485–516.

Sidani, M., J. Wyckoff, C. Xue, J. E. Segall, and J. Condeelis. 2006. Probing the microenvironment of mammary tumors using multiphoton microscopy. *J Mammary Gland Biol Neoplasia*. **11**: 151–63.

Springer, T. A. 1997. Folding of the N-terminal, ligand-binding region of integrin alpha-subunits into a beta-propeller domain. *Proc Natl Acad Sci USA*. **94**: 65–72.

Tadokoro, S., S. J. Shattil, K. Eto, V. Tai, R. C. Liddington, J. M. de Pereda, M. H. Ginsberg, and D. A. Calderwood. 2003. Talin binding to integrin beta tails: A final common step in integrin activation. *Science*. **302**: 103–6.

Takagi, J., B. M. Petre, T. Walz, and T. A. Springer. 2002. Global conformational rearrangements in integrin extracellular domains in outside-in and inside-out signaling. *Cell*. **110**: 599–11.

Tan, J. L., J. Tien, D. M. Pirone, D. S. Gray, K. Bhadriraju, and C. S. Chen. 2003. Cells lying on a bed of microneedles: An approach to isolate mechanical force. *Proc Natl Acad Sci USA*. **100**: 1484–9.

Tysseling-Mattiace, V. M., V. Sahni, K. L. Niece, D. Birch, C. Czeisler, M. G. Fehlings, S. I. Stupp, and J. A. Kessler. 2008. Self-assembling nanofibers inhibit glial scar formation and promote axon elongation after spinal cord injury. *J Neurosci*. **28**: 3814–23.

Ussar, S., M. Moser, M. Widmaier, E. Rognoni, C. Harrer, O. Genzel-Boroviczeny, and R. Fassler. 2008. Loss of Kindlin-1 causes skin atrophy and lethal neonatal intestinal epithelial dysfunction. *PLoS Genet*. **4**:e1000289.

Vicente-Manzanares, M., M. A. Koach, L. Whitmore, M. L. Lamers, and A. F. Horwitz. 2008. Segregation and activation of myosin IIB creates a rear in migrating cells. *J Cell Biol*. **183**: 543–54.

Vicente-Manzanares, M., J. Zareno, L. Whitmore, C. K. Choi, and A. F. Horwitz. 2007. Regulation of protrusion, adhesion dynamics, and polarity by myosins IIA and IIB in migrating cells. *J Cell Biol*. **176**: 573–80.

Vogel, V. 2006. Mechanotransduction involving multimodular proteins: Converting force into biochemical signals. *Annu Rev Biophys Biomol Struct*. **35**: 459–88.

Vogel, V., and M. Sheetz. 2006. Local force and geometry sensing regulate cell functions. *Nat Rev Mol Cell Biol*. **7**: 265–75.

von Wichert, G., G. Jiang, A. Kostic, K. De Vos, J. Sap, and M. P. Sheetz. 2003. RPTP-alpha acts as a transducer of mechanical force on alphav/beta3-integrin-cytoskeleton linkages. *J Cell Biol*. **161**: 143–53.

Walpita, D., and E. Hay. 2002. Studying actin-dependent processes in tissue culture. *Nat Rev Mol Cell Biol*. **3**: 137–41.

Wang, F., V. M. Weaver, O. W. Petersen, C. A. Larabell, S. Dedhar, P. Briand, R. Lupu, and M. J. Bissell. 1998. Reciprocal interactions between beta1-integrin and epidermal growth factor receptor in three-dimensional basement membrane breast cultures: A different perspective in epithelial biology. *Proc Natl Acad Sci USA*. **95**: 14821–6.

Wang, H. B., M. Dembo, S. K. Hanks, and Y. Wang. 2001. Focal adhesion kinase is involved in mechanosensing during fibroblast migration. *Proc Natl Acad Sci USA*. **98**: 11295–300.

Wang, Y. K., Y. H. Wang, C. Z. Wang, J. M. Sung, W. T. Chiu, S. H. Lin, Y. H. Chang, and M. J. Tang. 2003. Rigidity of collagen fibrils controls collagen gel-induced down-regulation of focal adhesion complex proteins mediated by alpha2beta1 integrin. *J Biol Chem*. **278**: 21886–92.

Wang, Y. L., and R. J. Pelham, Jr. 1998. Preparation of a flexible, porous polyacrylamide substrate for mechanical studies of cultured cells. *Methods Enzymol*. **298**: 489–96.

Webb, D. J., K. Donais, L. A. Whitmore, S. M. Thomas, C. E. Turner, J. T. Parsons, and A. F. Horwitz. 2004. FAK-Src signalling through paxillin, ERK and MLCK regulates adhesion disassembly. *Nat Cell Biol*. **6**: 154–61.

Wegener, K. L., A. W. Partridge, J. Han, A. R. Pickford, R. C. Liddington, M. H. Ginsberg, and I. D. Campbell. 2007. Structural basis of integrin activation by talin. *Cell*. **128**: 171–82.

Weiss, P., and B. Garber. 1952. Shape and movement of mesenchyme cells as functions of the physical structure of the medium: Contributions to a quantitative morphology. *Proc Natl Acad Sci USA*. **38**: 264–80.

Wells, A., A. Huttenlocher, and D. A. Lauffenburger. 2005. Calpain proteases in cell adhesion and motility. *Int Rev Cytol*. **245**: 1–16.

Wessels, D., H. Vawter-Hugart, J. Murray, and D. R. Soll. 1994. Three-dimensional dynamics of pseudopod formation and the regulation of turning during the motility cycle of Dictyostelium. *Cell Motil Cytoskeleton*. **27**: 1–12.

Wijelath, E. S., S. Rahman, M. Namekata, J. Murray, T. Nishimura, Z. Mostafavi-Pour, Y. Patel, Y. Suda, M. J. Humphries, and M. Sobel. 2006. Heparin-II domain of fibronectin is a vascular endothelial growth factor-binding domain: Enhancement of VEGF biological activity by a singular growth factor/matrix protein synergism. *Circ Res*. **99**: 853–60.

Wolf, K., I. Mazo, H. Leung, K. Engelke, U. H. von Andrian, E. I. Deryugina, A. Y. Strongin, E. B. Brocker, and P. Friedl. 2003. Compensation mechanism in tumor cell migration: Mesenchymal-amoeboid transition after blocking of pericellular proteolysis. *J Cell Biol*. **160**: 267–77.

Wozniak, M. A., R. Desai, P. A. Solski, C. J. Der, and P. J. Keely. 2003. ROCK-generated contractility regulates breast epithelial cell differentiation in response to the physical properties of a three-dimensional collagen matrix. *J Cell Biol*. **163**: 583–95.

Wozniak, M. A., K. Modzelewska, L. Kwong, and P. J. Keely. 2004. Focal adhesion regulation of cell behavior. *Biochim Biophys Acta*. **1692**: 103–19.

Xiao, T., J. Takagi, B. S. Coller, J. H. Wang, and T. A. Springer. 2004. Structural basis for allostery in integrins and binding to fibrinogen-mimetic therapeutics. *Nature*. **432**: 59–67.

Xiao, Z., N. Zhang, D. B. Murphy, and P. N. Devreotes. 1997. Dynamic distribution of chemoattractant receptors in living cells during chemotaxis and persistent stimulation. *J Cell Biol*. **139**: 365–74.

Yamada, K., and B. Geiger. 1997. Molecular interactions in cell adhesion complexes. *Curr Opin Cell Biol*. **9**: 76–85.

Yamada, K. M. 1991. Adhesive recognition sequences. *J Biol Chem*. **266**: 12809–12.

Yan, B., D. A. Calderwood, B. Yaspan, and M. H. Ginsberg. 2001. Calpain cleavage promotes talin binding to the beta 3 integrin cytoplasmic domain. *J Biol Chem*. **276**: 28164–70.

Ylanne, J., Y. Chen, T. E. O'Toole, J. C. Loftus, Y. Takada, and M. H. Ginsberg. 1993. Distinct functions of integrin alpha and beta subunit cytoplasmic domains in cell spreading and formation of focal adhesions. *J Cell Biol*. **122**: 223–33.

Zaidel-Bar, R., C. Ballestrem, Z. Kam, and B. Geiger. 2003. Early molecular events in the assembly of matrix adhesions at the leading edge of migrating cells. *J Cell Sci*. **116**: 4605–13.

Zaman, M. H., R. D. Kamm, P. Matsudaira, and D. A. Lauffenburger. 2005. Computational model for cell migration in three-dimensional matrices. *Biophys J*. **89**: 1389–97.

Zaman, M. H., L. M. Trapani, A. L. Sieminski, D. Mackellar, H. Gong, R. D. Kamm, A. Wells, D. A. Lauffenburger, and P. Matsudaira. 2006. Migration of tumor cells in 3D matrices is governed by matrix stiffness along with cell-matrix adhesion and proteolysis. *Proc Natl Acad Sci USA*. **103**: 10889–94.

Zamir, E., B. Z. Katz, S. Aota, K. M. Yamada, B. Geiger, and Z. Kam. 1999. Molecular diversity of cell-matrix adhesions. *J Cell Sci*. **112**(Pt 11):1655–69.

Zhao, Z. S., and E. Manser. 2005. PAK and other Rho-associated kinases–effectors with surprisingly diverse mechanisms of regulation. *Biochem J*. **386**: 201–14.

Zhong, C., M. Chrzanowska-Wodnicka, J. Brown, A. Shaub, A. M. Belkin, and K. Burridge. 1998. Rho-mediated contractility exposes a cryptic site in fibronectin and induces fibronectin matrix assembly. *J Cell Biol*. **141**: 539–51.

6

Role of Ion Channels in Cellular Mechanotransduction – Lessons from the Vascular Endothelium

Abdul I. Barakat and Andrea Gojova

6.1 Introduction

Two essential functions of arterial endothelium are flow-mediated vasoregulation in response to acute changes in blood flow and vascular wall remodeling in response to chronic hemodynamic alterations [1, 2]. Both of these functions require arterial endothelial cells (ECs) to be capable of sensing the mechanical forces associated with blood flow and of transducing these forces into biochemical signals that mediate structural and functional responses. Mechanosensing and -transduction in arterial endothelium also play a critical role in the development and localization of atherosclerosis. The topography of early atherosclerotic lesions is highly focal and correlates with arterial regions that are exposed to low and/or oscillatory shear stress [3, 4]. There is mounting evidence that low and oscillatory shear stress elicit a pro-inflammatory and adhesive EC phenotype, whereas relatively high and non-reversing pulsatile shear stress induce a phenotype that is largely anti-inflammatory [5–9]. In light of the central role of EC inflammation in atherogenesis [9–14], the key to understanding the involvement of flow in the development of atherosclerosis may lie in determining the mechanisms governing the differential responsiveness of ECs to different types of flows.

The current concept of EC mechanotransduction postulates that it involves a sequential progression of events involving sensing of the mechanical stimulus, transduction of the stimulus to a biochemical signal, and cellular reaction and subsequent possible adaptation to the new mechanical environment [15–19]. Consistent with this construct, a number of candidate mechanosensors have been proposed. These include stretch- and flow-sensitive ion channels [20–27], cell-surface integrins at both the luminal and basal cell surfaces [19, 28], the cellular cytoskeletal network [15], subregions of the cell membrane or the entire membrane [29, 30], membrane-associated GTP-binding proteins (or G-proteins) [31, 32] and G-protein–coupled receptors [33], cell–cell junction constituents including platelet–EC adhesion molecule-1 (PECAM-1) [34], and the glycocalyx at the cell luminal surface [35–37]. The rationale for classifying these various structures as candidate mechanosensors is threefold: 1) They are associated with the cell membrane, where the effects of an externally applied force would likely be most immediately felt; 2) they generally respond very rapidly following the onset of the mechanical stimulus; and 3) interfering with the activation of these structures abrogates, or at least significantly diminishes, some of the downstream responses induced by the applied mechanical force. It remains unclear, however, how these various structures interact with one another to potentially form an integrated mechanosensory system. The mechanisms by which these candidate mechanosensors are activated by a mechanical force also remain to be established. Furthermore, it is unknown which, if any, of these candidate mechanosensors is truly involved in *sensing* the mechanical force versus in the early stages of force signaling. Finally, how mechanosensors transduce the mechanical force to a biochemical signal remains to be elucidated.

Despite the uncertainties associated with how ECs *sense* a mechanical force, there is ample evidence that mechanical loads on the cellular surface are propagated within the intracellular space, resulting in the activation of an array of signaling cascades whose induction profoundly impacts EC structure and function. Two early mechanosignaling events are the mobilization of intracellular calcium [38, 39] and the induction of mitogen-activated protein (MAP) kinase [40, 41]. These responses are rapidly followed by the activation of the transcription factors NFκB and AP-1 [42], leading ultimately to the altered expression of a large number of mechanosensitive genes and proteins spanning the range of vasoactive and growth factors, structural molecules, and mediators of cell growth, survival, and inflammation [7, 15, 19, 43, 44]. Concurrent with the activation of these signaling cascades, extensive cytoskeletal remodeling occurs progressively and leads ultimately to morphological changes and cellular alignment in the direction of net flow [45–47].

In the presence of sustained stimulation, ECs adapt to the new mechanical environment. Examples of this adaptation include the desensitization of mechanosensitive ion channels [48] and of intracellular calcium oscillations [49] as well as the return of mRNA and protein levels to their pre-load baseline values [15]. It has been suggested that the extensive cytoskeletal remodeling that occurs in response to flow is a form of structural adaptation that allows ECs to adopt a more "streamlined" conformation, thereby reducing the magnitude of peak shear stresses and spatial gradients of shear stress on the cell surface [15, 50]. The precise triggers for adaptive responses to sustained mechanical loading remain to be defined.

The present chapter focuses specifically on the role of ion channels in mechanosensing and mechanosignaling in ECs. Other chapters address the role of other candidate sensors as well as the transduction pathways activated by mechanical stimulation. We will describe the different types of mechanosensitive ion channels and will discuss the role that these channels play in the overall endothelial responsiveness to flow. We will also address candidate pathways by which these channels are activated by mechanical loading as well as the potential role that these channels play in endowing ECs with the ability to discriminate among different types of mechanical loadings.

6.2 Mechanosensitive Ion Channels in ECs

There are broadly two categories of mechanosensitive ion channels in ECs: stretch-sensitive channels and flow-sensitive channels. The commonalities and differences between the two categories of channels have not been studied systematically. Because several excellent reviews on stretch-sensitive ion channels are available [22, 23, 26, 27], these channels will only be described briefly, and the primary focus of the present chapter will be on flow-sensitive channels.

Stretch-sensitive ion channels: Membrane stretch activates a channel that is permeable to both monovalent cations (with a single-channel conductance of ~50 pS) and Ca^{2+} (~20 pS) [51]. In endocardial and microvascular ECs, stretch activates a nonspecific channel with a unitary conductance of 20–30 pS for monovalent cations and 10–20 pS for Ca^{2+} and Ba^{2+} [52, 53]. Activation of these channels leads to a sufficiently large increase in intracellular Ca^{2+} concentration to induce Ca^{2+}-activated K^+ (BK_{Ca}) currents and to hyperpolarize the cell membrane. Stretch forces have also been reported to activate a K^+ channel in both intact rat aortic endothelium and isolated rat aortic ECs [54]. Interestingly, the channel appears to be equally permeable to Na^+ as it is to K^+ (22 pS) and also exhibits limited permeability to Ca^{2+} (4 pS). Virtually all stretch-sensitive ion channels are effectively blocked by the small lanthanide gadolinium (Gd^{3+}). In neurons, stretch has been reported to inactivate a class of low-conductance spontaneously active channels [55]. It remains to be determined if stretch-inactivated channels are also present in ECs.

How stretch-activated ion channels are gated by the mechanical force remains incompletely understood and is under intense investigation. There is evidence that this gating is mediated by stretch-induced alterations in the interactions between the channels and either membrane lipids or components of the cortical cytoskeleton, most notably actin [22, 23, 26, 27]. Stretch-induced alterations in channel–lipid interactions that contribute to channel gating may be in the form of changes in cell membrane tension and/or changes in membrane curvature [22, 26].

Flow-sensitive ion channels: Flow-activated ion channels are present in various cell types. For instance, epithelial Na^+ channels (ENaC) in the renal collecting tubule [56, 57] and transient receptor potential (TRP) cation channels in smooth muscle cells are activated by flow [58]. In ECs, several types of flow-activated channels have been reported. In both human aortic ECs (HAECs) and human umbilical vein ECs (HUVECs), shear stress has been reported to activate a cation channel that is more permeable to Ca^{2+} than to other cations [59]. Interestingly, these channels have been reported to be absent in human capillary ECs [60]. There is conflicting evidence for the presence of flow-activated Na^+ channels in ECs. Such channels have been reported in rat cardiac microvascular ECs [61] and in bovine aortic ECs (BAECs) [62]. Furthermore, Na^+ has been shown to enter HUVECs through flow-sensitive cation channels that, though more permeable to Ca^{2+}, are also permeable to Na^+ [59]. On the other hand, other studies have reported that flow-sensitive Na^+ channels are absent in both BAECs [20] and HAECs [63].

The two types of channels that are most prominently activated by flow are K^+ and Cl^- channels. In aortic ECs, flow-activated K^+ and Cl^- currents entirely account for flow-induced changes in cell membrane potential. Because these two currents appear to be the dominant flow-induced currents in large-vessel ECs, we will discuss them in more detail.

6.2.1 K^+ Channels

Flow activates several types of K^+ channels in ECs. Inward-rectifying K^+ channels were the first type of flow-activated ion channel reported in ECs [25]. Single-channel recordings subsequently established that these channels are present on the luminal EC surface [64]. Typical of inward-rectifying K^+ channels, the flow-activated current is completely blocked by either Ba^{2+} or Cs^+. In BAECs, the K^+ current is activated by a steady flow at a very low shear stress level, as low as 0.1 dyne/cm^2. The current attains half-maximal activation at ~ 0.7 dyne/cm^2 and saturates above ~ 10–15 dyne/cm^2. Because oscillatory flow induces EC dysfunction and correlates with the development of early atherosclerotic lesions [3, 4], it is important to establish ion channel responsiveness to oscillatory flow. K^+ channels have been shown to be equally sensitive to oscillatory flow with a frequency of 0.2 or 1 Hz as they are to steady flow but are not responsive to a 5-Hz oscillation [65]. An inward-rectifying K^+ channel belonging to the Kir2.1 family was subsequently cloned from BAECs [66]. Expressing the Kir2.1 channel in either *Xenopus* oocytes or mammalian HEK293 cells resulted in a large flow-activated K^+ current in these cells that was absent in nontransfected cells [21].

In addition to Kir2.1 channels, a steady shear stress of 10 dyne/cm^2 for 24 h significantly increases the expression of ATP-activated K^+ channels (Kir6.2 or K_{ATP}) in rat pulmonary microvascular ECs and bovine pulmonary artery ECs [67]. A steady shear stress of 5 dyne/cm^2 for 24 h causes a threefold increase in the expression of Ca^{2+}-activated K^+ channels (IK_{Ca}) in HUVECs [68]. In response to a higher shear stress (15 dyne/cm^2), up-regulation of IK_{Ca} expression is eightfold following 4 h of flow but decreases to fourfold at the 24 h time point. In the cases of both Kir6.2 and IK_{Ca}, the flow-induced increase in channel expression leads directly to increased whole-cell currents in response to ATP.

6.2.2 Cl^- Channels

Flow also activates outward-rectifying Cl^- channels in BAECs, as demonstrated by whole-cell patch clamp recordings and measurements using fluorescent potentiometric dyes [20, 63]. These channels are effectively blocked by the Cl^- channel antagonists 4,4′-diisothiocyanatostilbene-2,2′-disulfonic acid (DIDS) and nitrophenylpropylbenzoic acid (NPPB). In response to steady flow, the Cl^- current initiates at a low shear stress of ~ 0.3 dyne/cm^2, a value similar to that observed for inward-rectifying K^+ channels; however, unlike K^+ channels, the Cl^- current saturates above 3.5 dyne/cm^2 [48]. Outward-rectifying Cl^- channels are equally sensitive to nonreversing pulsatile flow as they are to steady flow; they are insensitive to oscillatory flow with frequencies of 1 or 5 Hz, but partially and progressively regain their sensitivity as the frequency of oscillation is decreased [48, 65].

ECs *in vivo* are exposed to shear stress continuously. In the presence of sustained steady flow, outward-rectifying Cl^- channels desensitize slowly – desensitization initiates within ~ 3–4 min of flow onset and occurs progressively over a period of several minutes [48]. There is significant cell-to-cell variability in the extent of desensitization: ~ 50% of cells exhibit pronounced desensitization (defined as the return of whole-cell conductance to near-baseline pre-flow levels), ~ 25% exhibit moderate desensitization, and the remaining ~ 25% of the cells exhibit very modest desensitization. It remains unclear if this variability in desensitization characteristics reflects a fundamental difference among cells or simply slower desensitization dynamics in some cells. In other systems, there is evidence for the existence of cellular subpopulations exhibiting different desensitization characteristics. For example, nicotinic acetylcholine receptors in rat sympathetic ganglion neurons exhibit rapid desensitization in 90% of the cells while the remaining 10% show slower desensitization [69]. The two distinct desensitization patterns are due to the differences in subunit compositions of the channels. It remains to be determined if analogous differences are present in the case of flow-activated Cl^- channels in ECs.

The fact that Cl^- channels desensitize in the presence of a sustained flow stimulus suggests that, *in vivo*, these channels would be "closed" under baseline conditions but would be activated by sudden changes in shear stress as occurs following changes in stress and/or activity level. When BAECs are subjected in a stepwise manner to consecutive episodes of progressively increasing shear stress, the magnitude of the Cl^- current progressively increases if the shear stress levels fall below the shear stress required for maximal current induction (i.e., <3.5 dyne/cm^2) [48]. On the other hand, if the shear stresses are above the level that induces maximal current, the Cl^- channels exhibit desensitization following the first flow stimulus and consequent insensitivity to subsequent shear stress steps [48]. These results suggest that flow-activated Cl^- channels in ECs respond to sudden changes in shear stress only if the shear stress levels are below those eliciting maximal current.

There have been other reports of flow-induced Cl^- currents in ECs. Fluid flow has been reported to induce a Cl^- current in HAECs that is blocked by niflumic acid, is insensitive to changes in extracellular Na^+, and modulates Ca^{2+} influx into the cells [63]. Flow has also been demonstrated to modulate the volume-regulated anion current (VRAC), which is responsible for the regulation of cellular volume in response to an osmotic challenge. Although it appears that flow per se does not induce VRAC in ECs, flow potentiates the current in the presence of an osmotic challenge [70]. A step increase in shear stress from 0.1 to 1 dyne/cm^2 enhances VRAC activation in response to an osmotic stress, whereas a step to 3 dyne/cm^2 leads to a transient increase followed by current inhibition. At even higher shear stresses (5–10 dyne/cm^2), the current is rapidly suppressed. The response of VRAC to shear stress is attenuated following repetitive application of shear, indicating a level of desensitization. The impact of flow on VRAC currents suggests that shear stress is an important factor in regulating the ability of vascular ECs to maintain volume homeostasis.

How the various flow-regulated Cl^- currents in ECs are related to one another remains to be determined. It is certainly possible that the current reported in HAECs is the same as the outward-rectifying current in BAECs. It is likely, however, that these currents are different from VRAC for several reasons. First, as already described, VRAC is regulated by flow only in the presence of an osmotic challenge, whereas the other Cl^- currents are activated by flow even under iso-osmolar conditions. Second, VRAC responsiveness to the magnitude of applied shear stress differs greatly from that of the outward-rectifying current as described previously. Finally, while VRAC activation requires the presence of ATP [71], the outward-rectifying Cl^- current has been measured in the absence of ATP in the patch pipette intracellular solution.

6.3 Consequences of Ion Channel Activation by Flow

The most immediate consequence of ion channel activation by flow in ECs is a change in cell membrane potential. Flow-induced changes in membrane potential occur immediately upon the onset of flow (within the resolution of measurement systems). For steady flow at shear stress levels sufficiently large to activate both inward-rectifying K^+ channels and outward-rectifying Cl^- channels, the membrane potential change takes the form of initial membrane hyperpolarization that is reversed to depolarization within 35–160 s of flow initiation [20, 65] (Figure 6.1(A)). The hyperpolarization is due to the activation of the K^+ channels, whereas the depolarization is due to the activation of the Cl^- channels. This assertion is supported by data demonstrating that the depolarization is absent in the presence of specific inhibitors of flow-sensitive Cl^- channels and the hyperpolarization is absent when flow-sensitive K^+ channels are blocked (Figures 6.1(B) and (C)). Further support is provided by the observation that 1-Hz oscillatory flow, which activates K^+ channels but not Cl^- channels, as already described, elicits EC hyperpolarization with no depolarization [65]. The lack of depolarization is attributable not to the inability of the cells to depolarize but rather to the nature of the imposed flow stimulus. Stimulating the same cells with steady flow leads to the expected hyperpolarization followed by robust depolarization [65].

The channel inhibition data in Figure 6.1 lead additionally to three important conclusions. First, activation of K^+ and Cl^- channels is sufficient to account for the membrane potential changes induced by a steady flow in ECs; other channels appear to play a minimal or no role. Second, both K^+ and Cl^- channels are activated very rapidly upon flow initiation, and the two channels are activated independently of one another; thus, the activation of either channel is not dependent on the activation of the other. Third, the observation that membrane hyperpolarization precedes depolarization in spite of the fact that both channels are activated simultaneously and that the electrochemical driving force is larger for Cl^- than for K^+ suggests that K^+

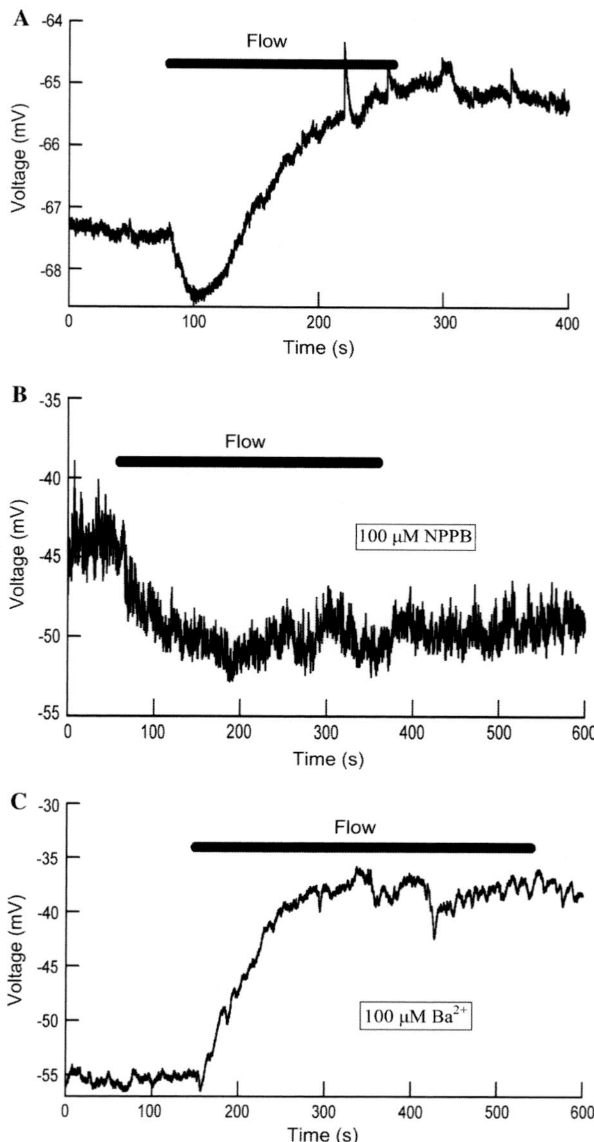

Figure 6.1. EC membrane potential changes in response to steady flow are mediated by flow-sensitive K^+ and Cl^- channels. (A) Flow induces initial membrane hyperpolarization that is eventually reversed to depolarization. (B) Membrane depolarization is absent in the presence of the Cl^- channel blocker NPPB (100 μM). (C) Membrane hyperpolarization is absent in the presence of the K^+ channel blocker Ba^{2+} (100 μM). Modified from [65].

channels are activated more rapidly than Cl^- channels; therefore, both channels are activated immediately upon flow onset, but the K^+ current attains its peak faster than the Cl^- current. Peak hyperpolarization occurs when the ratio of K^+ to Cl^- currents is at its maximum. As the Cl^- current develops further, the extent of hyperpolarization progressively decreases and is eventually reversed to depolarization.

If we are to postulate that flow-sensitive ion channels play an important role in regulating flow-mediated signaling in ECs, then an important question to address is how flow-induced changes in cell membrane potential affect EC structure and function. The most direct link is through the impact of changes in membrane voltage on the electrochemical driving force for Ca^{2+} across the cell membrane and thus on the intracellular concentration of Ca^{2+}, an important second messenger whose mobilization influences an array of cellular signaling events. Indeed, as already described, flow-activated Cl^- currents in HAECs do modulate Ca^{2+} influx into the cells [63]. In rat megacaryocytes, membrane depolarization has been demonstrated to directly induce Ca^{2+} release from internal stores, leading to oscillations in intracellular Ca^{2+} concentrations [72]. An additional interesting coupling in this regard is provided by the observation that Ca^{2+} mobilization may feed back onto certain types of ion channels. For instance, shear stress–induced oscillations in intracellular Ca^{2+} in BAECs activate a K^+ spontaneous transient outward current (STOC) [73]. It remains to be established if the flow-induced activation of ion channels directly modulates the changes in intracellular Ca^{2+} concentration that have been reported in ECs [38, 39].

In addition to their direct impact on intracellular Ca^{2+} levels, ion channels modulate a number of other flow-induced responses in ECs; however, the mechanisms governing this modulation remain unknown. For instance, flow-sensitive K^+ channels regulate the flow-induced release of both cGMP [74, 75] and NO [74], the up-regulation of transforming growth factor beta-1 (TGF-β1) [76] and endothelial nitric oxide synthase (eNOS) [77] mRNA, and the down-regulation of endothelin-1 protein expression [44]. Flow-sensitive K^+ and Cl^- channels modulate flow-induced increases in endothelial Na-K-Cl cotransport proteins [78] as well as EC spreading during wound closure following mechanical injury [79]. Recent data demonstrate that flow-sensitive Cl^- channels but not K^+ channels regulate phosphorylation by flow of the serine/threonine kinase Akt [48]; Akt phosphorylation plays an integral role in limiting endothelial apoptosis and in inducing the expression of the anti-inflammatory and atheroprotective enzyme eNOS. These results suggest that activation of flow-sensitive ion channels plays an important role in regulating normal EC physiology and may also be involved in the endothelial dysfunction associated with the onset of atherosclerosis.

6.4 Mechanisms of Ion Channel Activation by Flow

Understanding the role that ion channels play in flow sensing and transduction in ECs requires elucidating how these channels are activated by flow. Furthermore, determining the role that ion channels play in atherogenesis requires establishing the involvement of these channels in the endothelium's ability to respond differently to different types of flows. As already described, oscillatory flow induces a pro-inflammatory and atherogenic EC phenotype, whereas steady flow induces an anti-inflammatory and atheroprotective profile. Therefore, it is essential to determine if ion channels contribute to the endothelium's differential sensitivity to steady and oscillatory flows.

6.4.1 How Are Ion Channels Activated by Flow?

To date, virtually nothing is known about how flow activates ion channels. A critical question in this regard is whether flow activates the channels in a "direct" or an "indirect" manner. The two scenarios are depicted schematically in Figure 6.2. Direct activation is defined here as ion channel transition from an inactive to an active conformation as a result of the impact of the fluid mechanical force on the channel protein itself or on other structures to which the channel is *physically* coupled, such as the intracellular cytoskeleton, the cell-surface glycocalyx, or cell membrane proteins and/or lipids. Indirect activation is defined as ion channel activation as a result of the effect of flow on structures that have no physical connection to the channel. For instance, if flow-activated channels are gated by the binding of particular ligands to receptors on the EC surface, then flow might activate these channels by altering the kinetics of ligand delivery to and/or removal from the cell surface. Another type of indirect activation is channel gating by signaling molecules that are stimulated by flow.

There is some experimental evidence for both direct and indirect ion channel activation by flow in ECs. The only report of indirect activation of flow-sensitive ion channels to date is the regulation of flow-sensitive Kir2.1 currents in BAECs by tyrosine kinase phosphorylation; the kinase inhibitor genistein led to >90% inhibition of the current [21]. Another potential pathway for indirect channel activation is provided by evidence that flow elicits ATP release in ECs [80]. This release may provide an explanation for the sensitivity of K_{ATP} channels to flow [67], although this remains to be demonstrated definitively. Data in support of direct activation are provided by recent experiments demonstrating through changes in the viscosity of the extracellular solution that the extent of activation of flow-sensitive Cl^- channels in BAECs is dependent on the magnitude of applied shear stress rather than the applied shear rate [48]. Although this finding suggests that the Cl^- current correlates with the amplitude of the applied fluid mechanical force rather than the rate of agonist transport to or from the EC surface, it remains unclear whether the applied force acts on the channel or on other channel-associated structures. Moreover, one cannot exclude the possibility that the force first induces a shear stress–dependent signaling pathway that subsequently activates the channels, in which case the activation would be indirect.

How might direct ion channel activation by flow occur? The simplest construct would consist of the fluid mechanical stimulus imparting sufficient energy to the channel protein to bias it from a "closed" to an "open" conformation. This would be analogous to other mechanosensitive structures such as the stereocilia in the hair cells of the inner ear [81–83]. However, this model is unlikely to apply to flow-sensitive ion channels in ECs for at least two reasons. First, as detailed elsewhere [84, 85], given typical ion channel dimensions ($\sim$ 10 nm) and typical shear stresses required for ion channel activation (0.1–1 dyne/cm^2), the energy generated by the flow is expected to be $\sim$ 0.01–0.1 kT, or one to two orders of magnitude smaller than the channel's thermal energy; thus, the effect of the flow will not be "felt" by the channel. The second argument against ion channel gating via the direct impact of flow on the

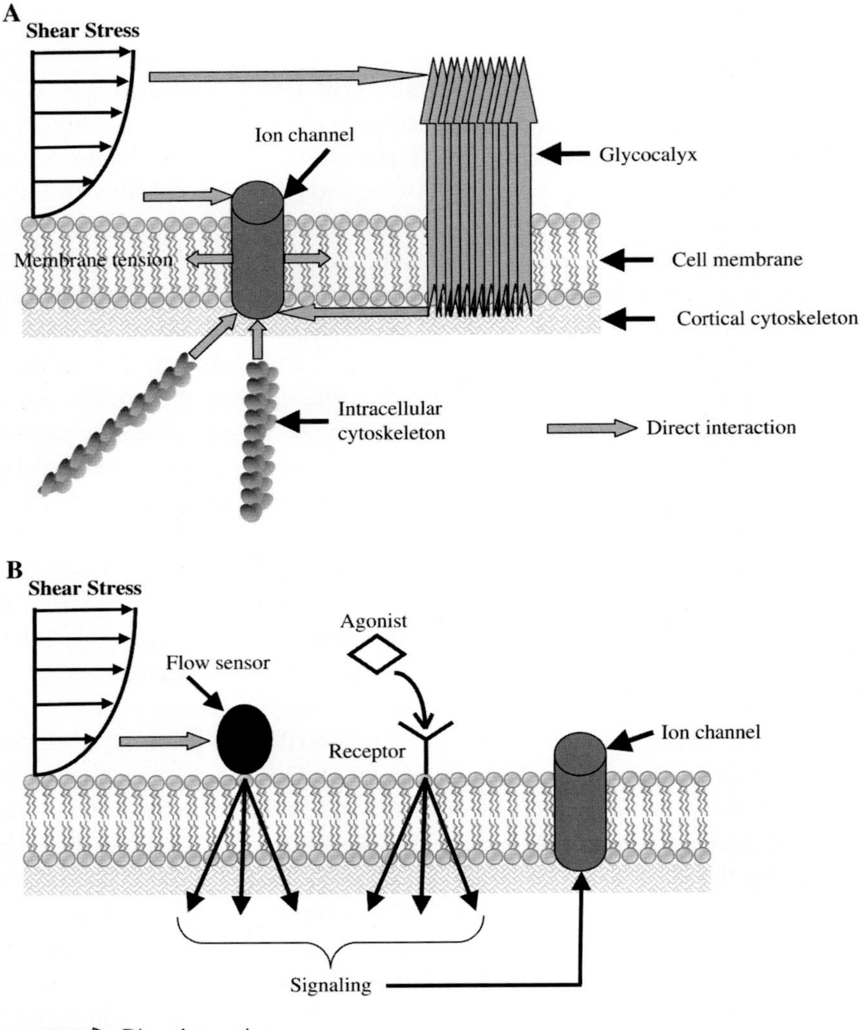

Figure 6.2. Possible mechanisms of ion channel activation by flow. (A) "Direct" activation involving interactions between the flow and the channel protein itself or other structures to which the channel is physically coupled, including the intracellular cytoskeleton, the glycocalyx (through its coupling to the cortical cytoskeleton), or the cell membrane. (B) "Indirect" activation whereby channel activation occurs as a result of the effect of flow on intracellular signaling by structures that have no physical connection to the channel.

channel is that in single-channel recordings of K^+ channels [64], the channel from which the recording is acquired is physically shielded from the flow by the patch pipette; hence, an alternative pathway for force transmission to the channel is needed.

If ion channel activation as a result of the impact of flow on the channel itself is unlikely, then how might direct channel activation occur? It is possible that the fluid mechanical force acts on discrete structures that are physically coupled to the channels and is subsequently transmitted to the channel to affect gating. The most obvious candidate structures would be the intracellular cytoskeleton and the cell-surface glycocalyx. We will next consider each of these two possibilities.

In eukaryotic cells, mechanical force transmission often occurs through the cytoskeleton [86]. Consistent with this notion, ion channel association with the cytoskeleton, either directly or through linker proteins such as ankyrin, has been demonstrated in several cell types [27]; however, no such coupling has yet been reported for flow-sensitive ion channels. Pharmacological disruption of the cytoskeleton has been shown to affect ion channel activity in *Xenopus* oocytes, leukocytes, Jurkat cells, and rat colon sensory neurons [87]. There is some suggestion that filamentous actin (F-actin) disruption in BAECs diminishes flow-activated K^+ currents; however, the effect is small and does not attain statistical significance [21]. The role that the cytoskeleton potentially plays in regulating other flow-sensitive ion channels in ECs remains to be determined. Furthermore, in light of the nonspecific effects that cytoskeletal disruption agents have, it would be particularly useful to devise less invasive methods for testing the involvement of the cytoskeleton in the activation of flow-sensitive ion channels. Finally, understanding the possible modulation of flow-activated ion channels by the cytoskeleton requires elucidating how the flow signal is transmitted to the cytoskeleton.

There is mounting evidence that the glycocalyx plays an important role in EC responsiveness to flow [35–37]. The glycocalyx appears to extend a distance of more than ~200 nm above the cell surface [88]. A structure that large provides sufficient surface area for flow-derived drag forces to be directly "felt." To our knowledge, there is no evidence to date of direct coupling between the glycocalyx and any type of ion channel; therefore, any potential role of the glycocalyx in regulating flow-activated ion currents remains to be determined. A relevant observation, however, is that many of the patch clamp recordings of flow-sensitive ion channels have been performed using serum-free solutions [20, 48, 65]. In light of data suggesting that in the absence of serum proteins, the glycocalyx exists in a highly collapsed or degraded state, the presence of flow-activated ion currents even in serum-free solutions may indicate that activation of flow-sensitive channels does not require the presence of the glycocalyx (or at least a completely intact glycocalyx).

Flow-sensitive ion channel regulation may also occur via direct interactions of the channel with the cell membrane [22, 26]. For instance, in accordance with the bilayer model of mechanosensitive ion channel regulation and as occurs in mechanosensitive channels of large conductance (MscL channels) in prokaryotic cells, flow might gate the channels by inducing an increase in EC membrane tension [89–91]. The amount of membrane tension that would be required to activate flow-sensitive ion channels is unknown; however, some data exist on tension levels needed to gate other types of mechanosensitive channels. Activation of MscL channels has been reported to require a membrane tension of ~10 dyne/cm [89, 92, 93] while nonselective mechanosensitive cation channels in osteoblasts do not appear to be activated below ~0.4 dyne/cm [94]. Application of a shear stress τ to a cell surface increases membrane tension by $\sim \tau L$, where L is the cell length [95]. In light of the fact that flow-sensitive ion channels are activated by shear stresses of ~0.1–10 dyne/cm^2 and that the length of an EC is ~10 μm, the flow-induced increase in EC membrane tension is expected to be ~10^{-4}–10^{-2} dyne/cm. Therefore, in order for flow-sensitive ion channels in ECs to be activated by increased cell membrane tension, these channels need to be sensitive to significantly lower levels of membrane tension than other types of mechanosensitive ion channels. In line with this reasoning, a steady shear stress of 8.5 dyne/cm^2, which activates robust K^+ and Cl^- currents, has been shown to lead to negligible changes in cell membrane surface area [21]. Furthermore, the fact that flow-sensitive K^+ channels are not affected by inhibitors of stretch-activated mechanosensitive ion channels including gadolinium and amiloride [21] suggests that flow-activated ion channels are distinct from their stretch-activated counterparts and that activation of stretch-activated ion channels is not involved in the activation of flow-sensitive channels.

Aside from the flow-induced increase in cell membrane tension, subjecting ECs to flow also increases cell membrane fluidity (i.e., decreases viscosity) [29, 30], which may impact interactions between the membrane and ion channels and hence modulate channel gating. Changes in membrane viscosity induced by cholesterol enrichment or depletion have indeed been reported to modulate Kir2.1 channels in ECs [96]. Because altering membrane cholesterol content affects various aspects of cellular signaling [19, 97–99], one cannot exclude the possibility that the effects of cholesterol on ion currents may be attributable to effects other than changes in cell membrane viscosity. Therefore, it remains to be established if flow-induced changes in membrane fluidity directly affect flow-activated ion currents in ECs and, if so, how this effect occurs.

6.4.2 What Role Do Ion Channels Play in EC Differential Responsiveness to Different Types of Flows?

As already described, steady flow induces a markedly different EC phenotype from oscillatory flow with a physiological frequency of 1 Hz. Nothing is known about how ECs discriminate among and respond differently to different types of flows. A flow sensing model for this differential responsiveness based on the integration of signals from flow-sensitive K^+ and Cl^- channels has recently been proposed [84] and is schematically depicted in Figure 6.3. This model is based on data described previously indicating that steady flow activates both K^+ and Cl^- channels, whereas 1-Hz oscillatory flow activates K^+ channels but not Cl^- channels [48, 65]. Cl^- channel sensitivity is progressively recovered as the frequency of oscillation is decreased to subphysiological levels. On the other hand, neither channel is capable of sensing oscillations at a superphysiological frequency of 5 Hz. Thus, the model postulates that over a spectrum of frequencies spanning the physiological range, the relative magnitudes of K^+ and Cl^- currents endow an EC with the ability to distinguish between steady and oscillatory flows and to sense differences among different frequencies of oscillation. The predictions of this model await experimental verification. In any case, it is noteworthy that flow-sensitive ion channels are the only

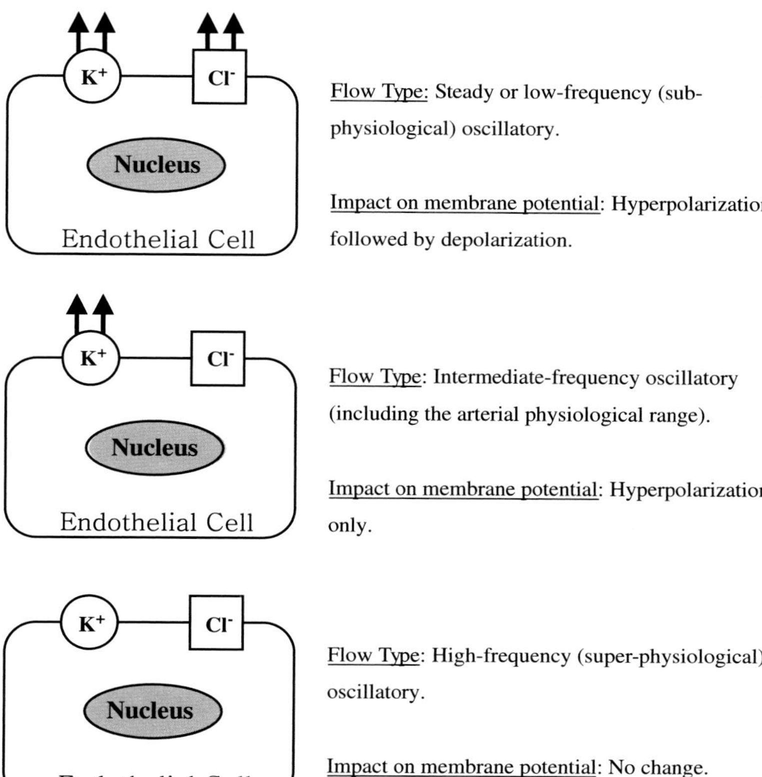

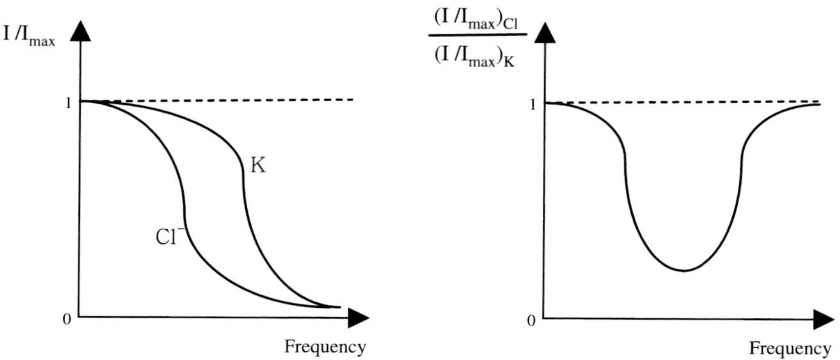

Figure 6.3. Hypothesized model for how ECs distinguish between steady flow and oscillatory flow. As a consequence of the different sensitivities of flow-activated K^+ and Cl^- channels to the frequency of oscillation (see text), the K^+ to Cl^- current ratio may provide an indication of this frequency. Steady flow can be thought of as oscillatory flow in the limit of zero frequency of oscillation. Modified from [84].

candidate flow sensors that may provide a basis for explaining EC differential responsiveness to different types of flows.

Why a 1-Hz oscillatory flow activates K^+ channels but fails to activate Cl^- channels in ECs remains to be determined. One possibility is that the two types of channels are regulated differently, and this difference in regulation translates into a difference in channel gating. For instance, in view of the fact that mechanical stress relaxation occurs significantly more slowly in the cellular cytoskeleton than in the cell membrane (relaxation time constant of seconds for cytoskeleton and micro- to milliseconds in the membrane) [100–104], it is possible that flow-sensitive K^+ channels are regulated primarily by the effect of flow on the interactions of the channels with the membrane while Cl^- channels are regulated by flow-induced changes in channel–cytoskeleton interactions. This construct needs to be tested experimentally.

6.5 Conclusion

Our understanding of the role that flow-sensitive ion channels play in regulating EC structure and function under both normal and pathological conditions remains in its infancy, and much remains to be established. There is a critical need for determining the molecular identity of most flow-sensitive ion channels. As described previously, a flow-activated Kir2.1 channel has been cloned from BAECs, but no other flow-sensitive channels have been cloned to date. Cloning additional channels will open the doors for developing antibodies targeting these channels and for conducting a range of fundamental studies aimed at establishing the mechanism(s) of channel activation. Such studies would include x-ray crystallography investigations as well as experiments in which the impact of expressing channels in other cell types or inserting them in lipid bilayers on channel activity can be probed. There is also a need for more systematic investigations of which types of flow-sensitive ion channels are present in which types of ECs.

As described previously, steady flow activates flow-sensitive K^+ and Cl^- channels, whereas oscillatory flow at physiological frequencies activates K^+ but not Cl^- channels. This observation highlights the need for a more complete understanding of the impact of physiologically relevant flow signals on flow-sensitive ion channels. *In vivo*, arterial flow fields are often complex and multidirectional, particularly in regions that are particularly prone to the development of atherosclerosis. The potential impact of this added complexity on flow-sensitive ion channels merits future study. Another *in vivo* complication is that arterial ECs are in constant communication with smooth muscle cells (SMCs). In light of data showing that EC responses in cells co-cultured with SMCs may differ from those of ECs cultured alone, it would be important to determine the potential impact of EC–SMC communication on flow-sensitive ion channels in ECs. These investigations will necessitate the development of novel EC–SMC co-culture systems that permit exposure of ECs to controlled flow environments while simultaneously allowing electrophysiological recordings.

Ion channels represent only one of a number of candidate flow sensors in ECs. An interesting question is how ion channels interact with other candidate flow sensors in ECs. Elucidating these interactions promises to provide a more integrated understanding of how ECs respond to flow under both normal and pathological conditions.

REFERENCES

[1] Langille B. L. and O'Donnell F. Reductions in arterial diameter produced by chronic decreases in blood flow are endothelium-dependent. *Science*. **231**: 405–407, 1986.

[2] Pohl U., Holtz J., Busse R., and Bassenge E. Crucial role of endothelium in the vasodilator response to increased flow in vivo. *Hypertension*. **8**: 37–44, 1986.

[3] Ku D. N., Giddens D. P., Zarins C. K., and Glagov S. Pulsatile flow and atherosclerosis in the human carotid bifurcation: Positive correlation between plaque location and low oscillating shear stress. *Arteriosclerosis*. **5**: 293–302, 1985.

[4] Nerem R. M. Vascular fluid mechanics, the arterial wall, and atherosclerosis. *J Biomech Eng*. **114**: 274–282, 1992.

[5] Barakat A. I. and Lieu D. K. Differential responsiveness of vascular endothelial cells to different types of fluid mechanical shear stress. *Cell Biochem Biophys*. **38**: 323–343, 2003.

[6] Cunningham K. S. and Gotlieb A. I. The role of shear stress in the pathogenesis of atherosclerosis. *Lab Invest*. **85**: 9–23, 2005.

[7] Dai G., Kaazempur-Mofrad M. R., Natarajan S., Zhang Y., Vaughn S., Blackman B. R., Kamm R. D., Garcia-Cardena G., and Gimbrone M. A., Jr. Distinct endothelial phenotypes evoked by arterial waveforms derived from atherosclerosis-susceptible and -resistant regions of human vasculature. *Proc Natl Acad Sci USA*. **101**: 14871–14876, 2004.

[8] Passerini A. G., Polacek D. C., Shi C., Francesco N. M., Manduchi E., Grant G. R., Pritchard W. F., Powell S., Chang G. Y., Stoeckert C. J., Jr., and Davies P. F. Coexisting proinflammatory and antioxidative endothelial transcription profiles in a disturbed flow region of the adult porcine aorta. *Proc Natl Acad Sci USA*. **101**: 2482–2487, 2004.

[9] Tedgui A. and Mallat Z. Anti-inflammatory mechanisms in the vascular wall. *Circ Res*. **88**: 877–887, 2001.

[10] Libby P. Inflammation in atherosclerosis. *Nature*. **420**: 868–874, 2002.

[11] Libby P. Atherosclerosis: The new view. *Sci Am*. **286**: 46–55, 2002.

[12] Libby P., Ridker P. M., and Maseri A. Inflammation and atherosclerosis. *Circulation*. **105**: 1135–1143, 2002.

[13] Ross R. Atherosclerosis is an inflammatory disease. *Am Heart J*. **138**: S419–420, 1999.

[14] Ross R. Atherosclerosis – An inflammatory disease. *N Engl J Med*. **340**: 115–126, 1999.

[15] Davies P. F. Flow-mediated endothelial mechanotransduction. *Physiol Rev*. **75**: 519–560, 1995.

[16] Huang H., Kamm R. D., and Lee R. T. Cell mechanics and mechanotransduction: Pathways, probes, and physiology. *Am J Physiol*. **287**: C1–11, 2004.

[17] Kaazempur Mofrad M. R., Abdul-Rahim N. A., Karcher H., Mack P. J., Yap B., and Kamm R. D. Exploring the molecular basis for mechanosensation, signal transduction, and cytoskeletal remodeling. *Acta Biomater*. **1**: 281–293, 2005.

[18] Kamm R. D. and Kaazempur-Mofrad M. R. On the molecular basis for mechanotransduction. *Mech Chem Biosyst*. **1**: 201–209, 2004.

[19] Li Y. S., Haga J. H., and Chien S. Molecular basis of the effects of shear stress on vascular endothelial cells. *J Biomech*. **38**: 1949–1971, 2005.

[20] Barakat A. I., Leaver E. V., Pappone P. A., and Davies P. F. A flow-activated chloride-selective membrane current in vascular endothelial cells. *Circ Res*. **85**: 820–828, 1999.

[21] Hoger J. H., Ilyin V. I., Forsyth S., and Hoger A. Shear stress regulates the endothelial Kir2.1 ion channel. *Proc Natl Acad Sci USA*. **99**: 7780–7785, 2002.

[22] Kung C. A possible unifying principle for mechanosensation. *Nature.* **436**: 647–654, 2005.
[23] Martinac B. Mechanosensitive ion channels: Molecules of mechanotransduction. *J Cell Sci.* **117**: 2449–2460, 2004.
[24] Nilius B. and Droogmans G. Ion channels and their functional role in vascular endothelium. *Physiol Rev.* **81**: 1415–1459, 2001.
[25] Olesen S. P., Clapham D. E., and Davies P. F. Hemodynamic shear stress activates a K^+ current in vascular endothelial cells. *Nature.* **331**: 168–170, 1988.
[26] Perozo E. Gating prokaryotic mechanosensitive channels. *Nat Rev Mol Cell Biol.* **7**: 109–119, 2006.
[27] Sachs F. and Morris C. E. Mechanosensitive ion channels in nonspecialized cells. *Rev Physiol Biochem Pharmacol.* **132**: 1–77, 1998.
[28] Shyy J. Y. and Chien S. Role of integrins in cellular responses to mechanical stress and adhesion. *Curr Opin Cell Biol.* **9**: 707–713, 1997.
[29] Butler P. J., Norwich G., Weinbaum S., and Chien S. Shear stress induces a time- and position-dependent increase in endothelial cell membrane fluidity. *Am J Physiol.* **280**: C962–969, 2001.
[30] Haidekker M. A., L'Heureux N., and Frangos J. A. Fluid shear stress increases membrane fluidity in endothelial cells: A study with DCVJ fluorescence. *Am J Physiol.* **278**: H1401–1406, 2000.
[31] Gudi S., Nolan J. P., and Frangos J. A. Modulation of GTPase activity of G proteins by fluid shear stress and phospholipid composition. *Proc Natl Acad Sci USA.* **95**: 2515–2519, 1998.
[32] Gudi S. R., Clark C. B., and Frangos J. A. Fluid flow rapidly activates G proteins in human endothelial cells. Involvement of G proteins in mechanochemical signal transduction. *Circ Res.* **79**: 834–839, 1996.
[33] Chachisvilis M., Zhang Y. L., and Frangos J. A. G protein-coupled receptors sense fluid shear stress in endothelial cells. *Proc Natl Acad Sci USA.* **103**: 15463–15468, 2006.
[34] Tzima E., Irani-Tehrani M., Kiosses W. B., Dejana E., Schultz D. A., Engelhardt B., Cao G., DeLisser H., and Schwartz M. A. A mechanosensory complex that mediates the endothelial cell response to fluid shear stress. *Nature.* **437**: 426–431, 2005.
[35] Pahakis M. Y., Kosky J. R., Dull R. O., and Tarbell J. M. The role of endothelial glycocalyx components in mechanotransduction of fluid shear stress. *Biochem Biophys Res Commun.* **355**: 228–233, 2007.
[36] Thi M. M., Tarbell J. M., Weinbaum S., and Spray D. C. The role of the glycocalyx in reorganization of the actin cytoskeleton under fluid shear stress: a "bumper-car" model. *Proc Natl Acad Sci USA.* **101**: 16483–16488, 2004.
[37] Weinbaum S., Zhang X., Han Y., Vink H., and Cowin S. C. Mechanotransduction and flow across the endothelial glycocalyx. *Proc Natl Acad Sci USA.* **100**: 7988–7995, 2003.
[38] Geiger R. V., Berk B. C., Alexander R. W., and Nerem R. M. Flow-induced calcium transients in single endothelial cells: spatial and temporal analysis. *Am J Physiol.* **262**: C1411–1417, 1992.
[39] Shen J., Luscinskas F. W., Connolly A., Dewey C. F., Jr., and Gimbrone M. A., Jr. Fluid shear stress modulates cytosolic free calcium in vascular endothelial cells. *Am J Physiol.* **262**: C384–390, 1992.
[40] Tseng H., Peterson T. E., and Berk B. C. Fluid shear stress stimulates mitogen-activated protein kinase in endothelial cells. *Circ Res.* **77**: 869–878, 1995.
[41] Yan C., Takahashi M., Okuda M., Lee J. D., and Berk B. C. Fluid shear stress stimulates big mitogen-activated protein kinase 1 (BMK1) activity in

endothelial cells: Dependence on tyrosine kinases and intracellular calcium. *J Biol Chem.* **274**: 143–150, 1999.

[42] Lan Q., Mercurius K. O., and Davies P. F. Stimulation of transcription factors NF kappa B and AP1 in endothelial cells subjected to shear stress. *Biochem Biophys Res Commun.* **201**: 950–956, 1994.

[43] Garcia-Cardena G., Comander J., Anderson K. R., Blackman B. R., and Gimbrone M. A., Jr. Biomechanical activation of vascular endothelium as a determinant of its functional phenotype. *Proc Natl Acad Sci USA.* **98**: 4478–4485, 2001.

[44] Malek A. M. and Izumo S. Molecular aspects of signal transduction of shear stress in the endothelial cell. *J Hypertens.* **12**: 989–999, 1994.

[45] Dewey C. F., Jr., Bussolari S. R., Gimbrone M. A., Jr., and Davies P. F. The dynamic response of vascular endothelial cells to fluid shear stress. *J Biomech Eng.* **103**: 177–185, 1981.

[46] Eskin S. G., Ives C. L., McIntire L. V., and Navarro L. T. Response of cultured endothelial cells to steady flow. *Microvasc Res.* 28: 87–94, 1984.

[47] Nerem R. M., Levesque M. J., and Cornhill J. F. Vascular endothelial morphology as an indicator of the pattern of blood flow. *J Biomech Eng.* **103**: 172–176, 1981.

[48] Gautam M., Shen Y., Thirkill T. L., Douglas G. C., and Barakat A. I. Flow-activated chloride channels in vascular endothelium – Shear stress sensitivity, desensitization dynamics, and physiological implications. *J Biol Chem.* **281**: 36492–36500, 2006.

[49] Dull R. O. and Davies P. F. Flow modulation of agonist (ATP)-response (Ca^{2+}) coupling in vascular endothelial cells. *Am J Physiol.* **261**: H149–154, 1991.

[50] Barbee K. A., Mundel T., Lal R., and Davies P. F. Subcellular distribution of shear stress at the surface of flow-aligned and nonaligned endothelial monolayers. *Am J Physiol.* **268**: H1765–1772, 1995.

[51] Lansman J. B., Hallam T. J., and Rink T. J. Single stretch-activated ion channels in vascular endothelial cells as mechanotransducers? *Nature.* **325**: 811–813, 1987.

[52] Hoyer J., Distler A., Haase W., and Gogelein H. Ca^{2+} influx through stretch-activated cation channels activates maxi K^+ channels in porcine endocardial endothelium. *Proc Natl Acad Sci USA.* **91**: 2367–2371, 1994.

[53] Popp R., Hoyer J., Meyer J., Galla H. J., and Gogelein H. Stretch-activated non-selective cation channels in the antiluminal membrane of porcine cerebral capillaries. *J Physiol.* **454**: 435–449, 1992.

[54] Hoyer J., Kohler R., Haase W., and Distler A. Up-regulation of pressure-activated Ca^{2+}-permeable cation channel in intact vascular endothelium of hypertensive rats. *Proc Natl Acad Sci USA.* **93**: 11253–11258, 1996.

[55] Morris C. E. and Sigurdson W. J. Stretch-inactivated ion channels coexist with stretch-activated ion channels. *Science.* **243**: 807–809, 1989.

[56] Carattino M. D., Sheng S., and Kleyman T. R. Epithelial Na^+ channels are activated by laminar shear stress. *J Biol Chem.* **279**: 4120–4126, 2004.

[57] Satlin L. M., Sheng S., Woda C. B., and Kleyman T. R. Epithelial Na^+ channels are regulated by flow. *Am J Physiol.* **280**: F1010–1018, 2001.

[58] Kohler R., Heyken W. T., Heinau P., Schubert R., Si H., Kacik M., Busch C., Grgic I., Maier T., and Hoyer J. Evidence for a functional role of endothelial transient receptor potential V4 in shear stress-induced vasodilatation. *Arterioscler Thromb Vasc Biol.* **26**: 1495–1502, 2006.

[59] Schwarz G., Droogmans G., and Nilius B. Shear stress induced membrane currents and calcium transients in human vascular endothelial cells. *Pflugers Archiv.* **421**: 394–396, 1992.

[60] Jow F. and Numann R. Fluid flow modulates calcium entry and activates membrane currents in cultured human aortic endothelial cells. *J Memb Biol.* **171**: 127–139, 1999.
[61] Moccia F., Villa A., and Tanzi F. Flow-activated Na^+ and K^+ current in cardiac microvascular endothelial cells. *J. Mol Cell Cardiol.* **32**: 1589–1593, 2000.
[62] Traub O., Ishida T., Ishida M., Tupper J. C., and Berk B. C. Shear stress-mediated extracellular signal-regulated kinase activation is regulated by sodium in endothelial cells. *J Biol Chem.* **274**: 20144–20150, 1999.
[63] Nakao M., Ono K., Fujisawa S., and Iijima T. Mechanical stress-induced Ca^{2+} entry and Cl^- current in cultured human aortic endothelial cells. *Am J Physiol.* **276**: C238–C249, 1999.
[64] Jacobs E. R., Cheliakine C., Gebremedhin D., Birks E. K., Davies P. F., and Harder D. R. Shear activated channels in cell-attached patches of cultured bovine aortic endothelial cells. *Pflugers Arch.* **431**: 129–131, 1995.
[65] Lieu D. K., Pappone P. A., and Barakat A. I. Differential membrane potential and ion current responses to different types of shear stress in vascular endothelial cells. *Am J Physiol.* **286**: C1367–C1375, 2004.
[66] Forsyth S. E., Hoger A., and Hoger J. H. Molecular cloning and expression of a bovine endothelial inward rectifier potassium channel. *FEBS Lett.* **409**: 277–282, 1997.
[67] Chatterjee S., Al-Mehdi A., Levitan I., Stevens T., and Fisher A. B. Shear stress increases expression of a K_{ATP} Channel in rat and bovine pulmonary vascular endothelial cells. *Am J Physiol Cell Physiol.* **285**: C959–C967, 2003.
[68] Brakemeier S., Kersten A., Eichler I., Grgic I., Zakrzewicka A., Hopp H., Kohler R., and Hoyer J. Shear stress-induced up-regulation of the intermediate-conductance Ca^{2+}-activated K^+ channel in human endothelium. *Cardiovasc Res.* **60**: 488–496, 2003.
[69] Britt J. C. and Brenner H. R. Rapid drug application resolves two types of nicotinic receptors on rat sympathetic ganglion cells. *Pflugers Arch.* **434**: 38–48, 1997.
[70] Romanenko V. G., Davies P. F., and Levitan I. Dual effect of fluid shear stress on volume-regulated anion current in bovine aortic endothelial cells. *Am J Physiol.* **282**: C708–C718, 2002.
[71] Nilius B., Oike M., Zahradnik I., and Droogmans G. Activation of a Cl^- current by hypotonic volume increase in human endothelial cells. *J Gen Physiol.* **103**: 787–805, 1994.
[72] Mason M. J., Hussain J. F., and Mahaut-Smith M. P. A novel role for membrane potential in the modulation of intracellular Ca^{2+} oscillations in rat megakaryocytes. *J Physiol.* **524**(Pt 2): 437–446, 2000.
[73] Hoyer J., Kohler R., and Distler A. Mechanosensitive Ca^{2+} oscillations and STOC activation in endothelial cells. *FASEB J.* **12**: 359–366, 1998.
[74] Cooke J. P., Rossitch E., Jr., Andon N. A., Loscalzo J., and Dzau V. J. Flow activates an endothelial potassium channel to release an endogenous nitrovasodilator. *J Clin Invest.* **88**: 1663–1671, 1991.
[75] Ohno M., Gibbons G. H., Dzau V. J., and Cooke J. P. Shear stress elevates endothelial cGMP – Role of a potassium channel and G-protein coupling. *Circulation.* **88**: 193–197, 1993.
[76] Ohno M., Cooke J. P., Dzau V. J., and Gibbons G. H. Fluid shear stress induces endothelial transforming growth factor beta-1 transcription and production. Modulation by potassium channel blockade. *J Clin Invest.* **95**: 1363–1369, 1995.
[77] Uematsu M., Ohara Y., Navas J. P., Nishida K., Murphy T. J., Alexander R. W., Nerem R. M., and Harrison D. G. Regulation of endothelial cell nitric oxide synthase mRNA expression by shear stress. *Am J Physiol.* **269**: C1371–1378, 1995.

[78] Suvatne J., Barakat A. I., and O'Donnell M. E. Flow-induced expression of endothelial Na-K-Cl cotransport: Dependence on K^+ and Cl^- channels. *Am J Physiol.* **280**: C216–C227, 2001.

[79] Gojova A. and Barakat A. I. Vascular endothelial wound closure under shear stress: role of membrane fluidity and flow-sensitive ion channels. *J Appl Physiol.* **98**: 2355–2362, 2005.

[80] Milner P., Kirkpatrick K. A., Ralevic V., Toothill V., Pearson J., and Burnstock G. Endothelial cells cultured from human umbilical vein release ATP, substance P and acetylcholine in response to increased flow. *Proc Biol Sci.* **241**: 245–248, 1990.

[81] Denk W., Holt J. R., Shepherd G. M., and Corey D. P. Calcium imaging of single stereocilia in hair cells: Localization of transduction channels at both ends of tip links. *Neuron.* **15**: 1311–1321, 1995.

[82] Fettiplace R. and Hackney C. M. The sensory and motor roles of auditory hair cells. *Nat Rev Neurosci.* **7**: 19–29, 2006.

[83] Hudspeth A. J. How the ear's works work. *Nature.* **341**: 397–404, 1989.

[84] Barakat A. I., Lieu D. K., and Gojova A. Secrets of the code: Do vascular endothelial cells use ion channels to decipher complex flow signals? *Biomaterials.* **27**: 671–678, 2006.

[85] Gautam M., Gojova A., and Barakat A. I. Flow-activated ion channels in vascular endothelium. *Cell Biochem Biophys.* **46**: 277–284, 2006.

[86] Hamill O. P. and Martinac B. Molecular basis of mechanotransduction in living cells. *Physiol Rev.* **81**: 685–740, 2001.

[87] Hamill O. P. and McBride D. W., Jr. Mechanogated channels in Xenopus oocytes: Different gating modes enable a channel to switch from a phasic to a tonic mechanotransducer. *Biol Bull.* **192**: 121–122, 1997.

[88] Squire J. M., Chew M., Nneji G., Neal C., Barry J., and Michel C. Quasi-periodic substructure in the microvessel endothelial glycocalyx: A possible explanation for molecular filtering? *J Struct Biol.* **136**: 239–255, 2001.

[89] Bilston L. E. and Mylvaganam K. Molecular simulations of the large conductance mechanosensitive (MscL) channel under mechanical loading. *FEBS Lett.* **512**: 185–190, 2002.

[90] Chang G., Spencer R. H., Lee A. T., Barclay M. T., and Rees D. C. Structure of the MscL homolog from Mycobacterium tuberculosis: A gated mechanosensitive ion channel. *Science.* **282**: 2220–2226, 1998.

[91] Gullingsrud J., Kosztin D., and Schulten K. Structural determinants of MscL gating studied by molecular dynamics simulations. *Biophys J.* **80**: 2074–2081, 2001.

[92] Moe P. and Blount P. Assessment of potential stimuli for mechano-dependent gating of MscL: Effects of pressure, tension, and lipid headgroups. *Biochemistry.* **44**: 12239–12244, 2005.

[93] Sukharev S. Mechanosensitive channels in bacteria as membrane tension reporters. *FASEB J.* **13**(Suppl): S55–S61, 1999.

[94] Charras G. T., Williams B. A., Sims S. M., and Horton M. A. Estimating the sensitivity of mechanosensitive ion channels to membrane strain and tension. *Biophys J.* **87**: 2870–2884, 2004.

[95] Fung Y. C. and Liu S. Q. Elementary mechanics of the endothelium of blood vessels. *J Biomech Eng.* **115**: 1–12, 1993.

[96] Romaneneko V. G., Rothblat G. H., and Levitan I. Modulation of endothelial inward-rectifier K^+ current by optical isomers of cholesterol. *Biophys J.* **83**: 3211–3222, 2002.

[97] Simmons C. A., Grant G. R., Manduchi E., and Davies P. F. Spatial heterogeneity of endothelial phenotypes correlates with side-specific vulnerability to calcification in normal porcine aortic valves. *Circ Res.* **96**: 792–799, 2005.

[98] Simmons C. A., Zilberberg J., and Davies P. F. A rapid, reliable method to isolate high quality endothelial RNA from small spatially-defined locations. *Ann Biomed Eng*. **32**: 1453–1459, 2004.

[99] Yang M. B., Vacanti J. P., and Ingber D. E. Hollow fibers for hepatocyte encapsulation and transplantation: Studies of survival and function in rats. *Cell Transplant*. **3**: 373–385, 1994.

[100] Bausch A. R., Moller W., and Sackmann E. Measurement of local viscoelasticity and forces in living cells by magnetic tweezers. *Biophys J*. **76**: 573–579, 1999.

[101] Feneberg W., Aepfelbacher M., and Sackmann E. Microviscoelasticity of the apical cell surface of human umbilical vein endothelial cells (HUVEC) within confluent monolayers. *Biophys J*. **87**: 1338–1350, 2004.

[102] Lo C. M. and Ferrier J. Electrically measuring viscoelastic parameters of adherent cell layers under controlled magnetic forces. *Eur Biophys J*. **28**: 112–118, 1999.

[103] Shkulipa S. A., den Otter W. K., and Briels W. J. Simulations of the dynamics of thermal undulations in lipid bilayers in the tensionless state and under stress. *J Chem Phys*. **125**: 234905, 2006.

[104] Wohlert J. and Edholm O. Dynamics in atomistic simulations of phospholipid membranes: Nuclear magnetic resonance relaxation rates and lateral diffusion. *J Chem Phys*. **125**: 204703, 2006.

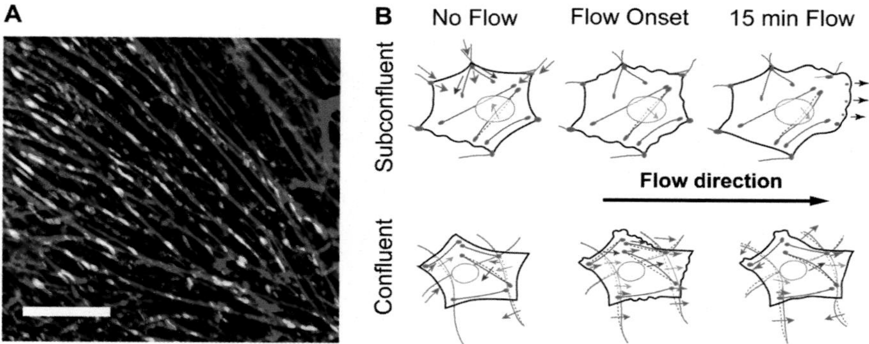

Plate 1. See Figure 2.3.

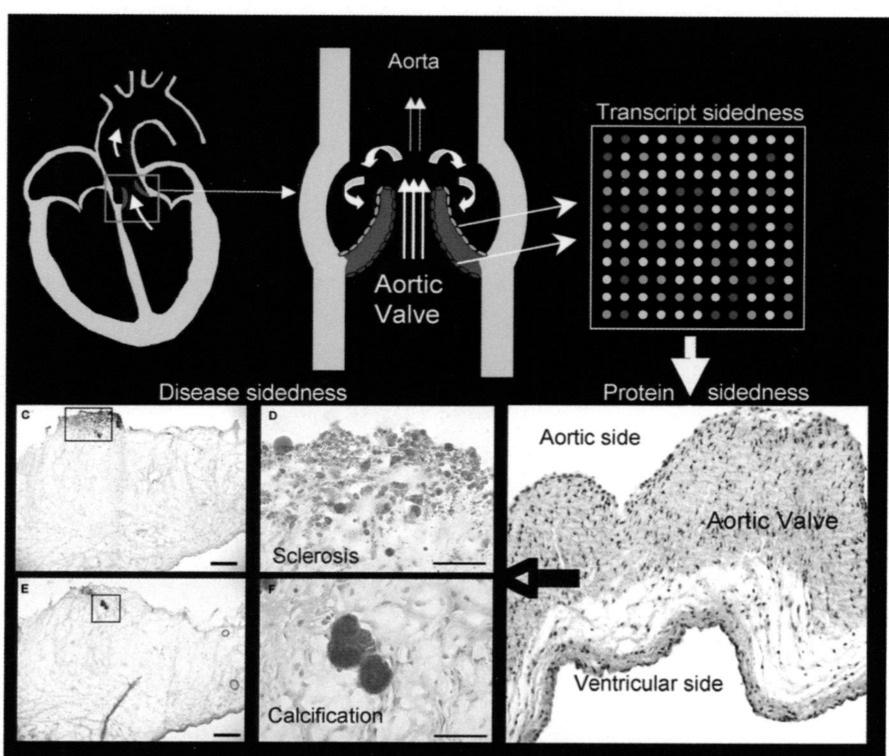

Plate 2. See Figure 2.6.

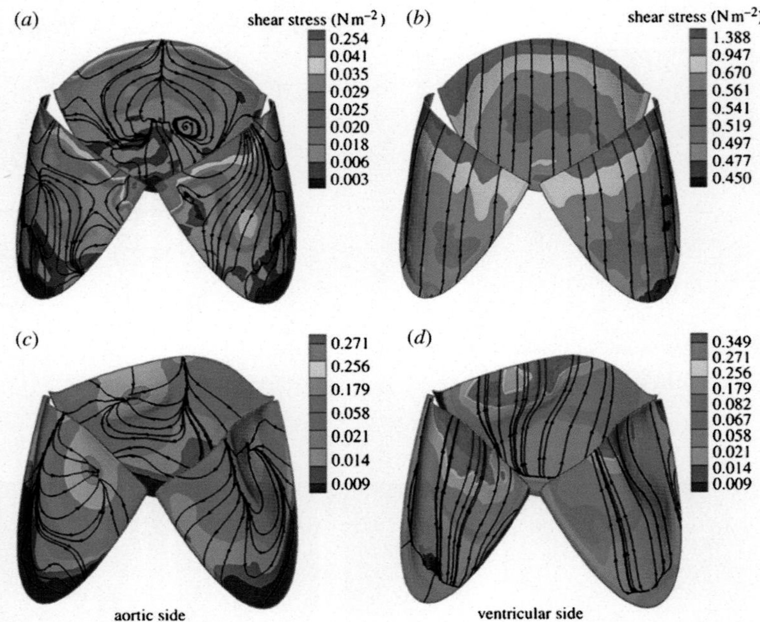

Plate 3. See Figure 2.7.

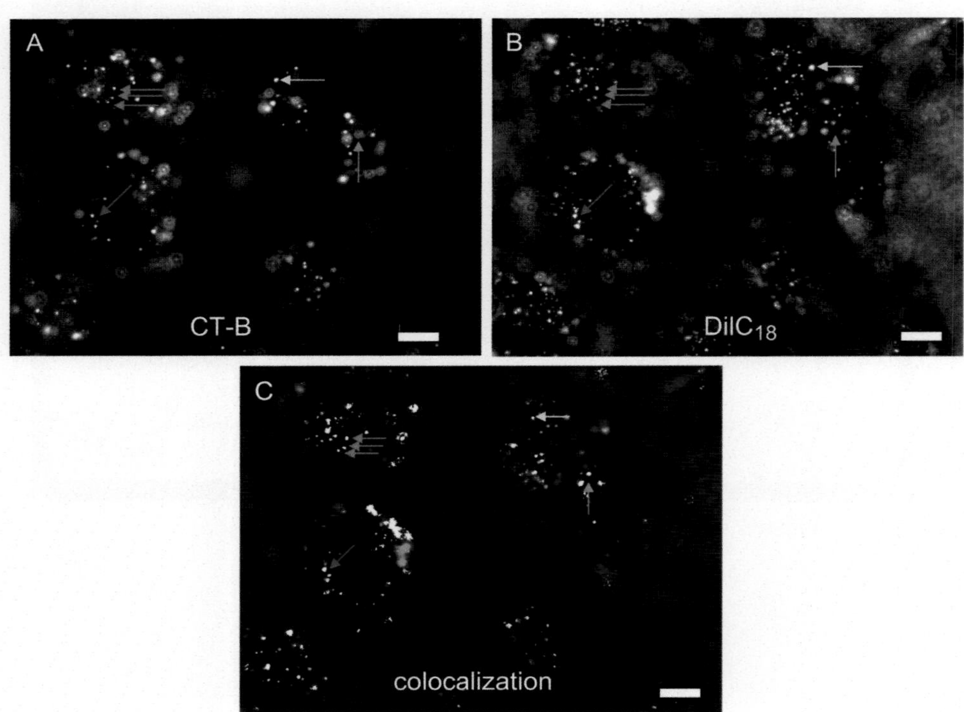

Plate 4. See Figure 3.5.

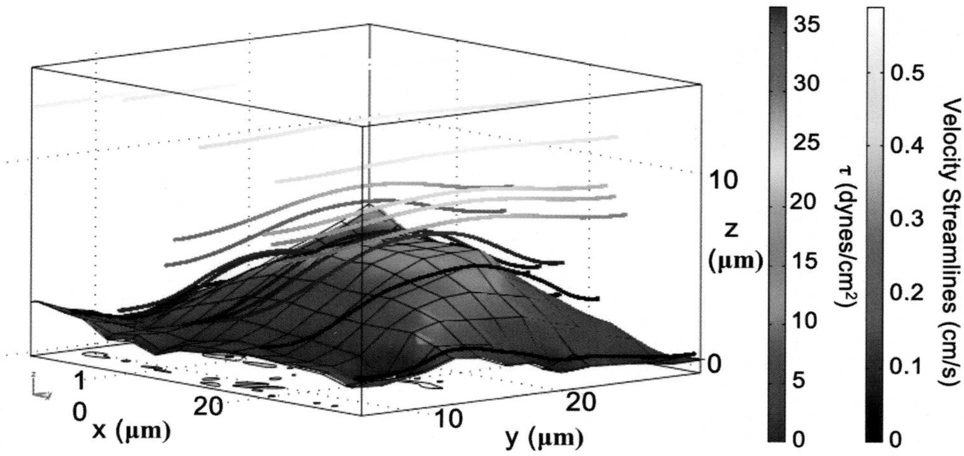

Plate 5. See Figure 3.7.

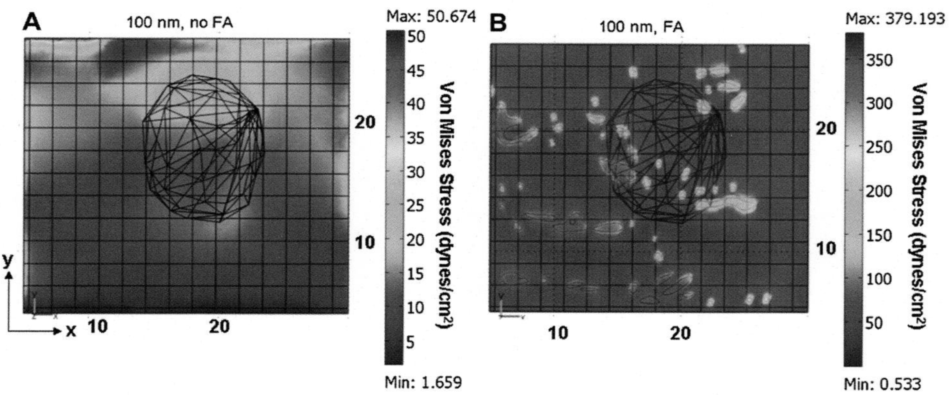

Plate 6. See Figure 3.8.

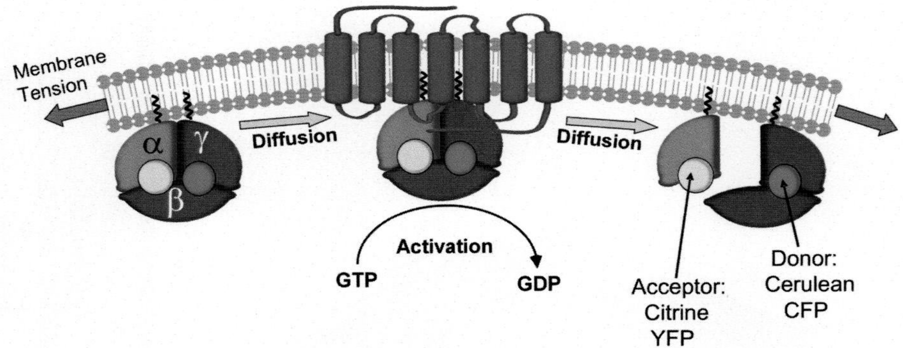

Plate 7. See Figure 4.7.

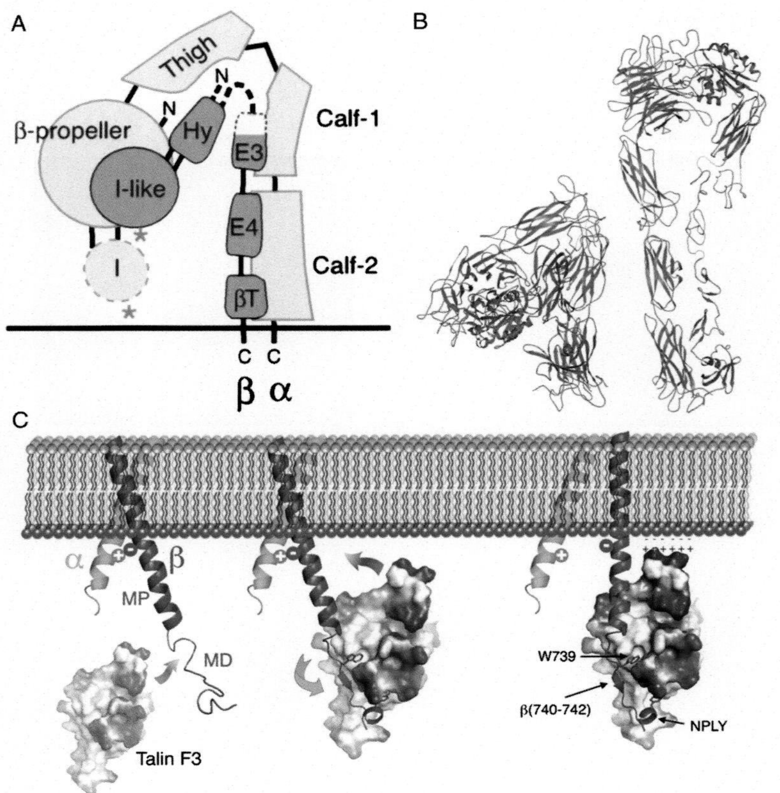

Plate 8. See Figure 5.2.

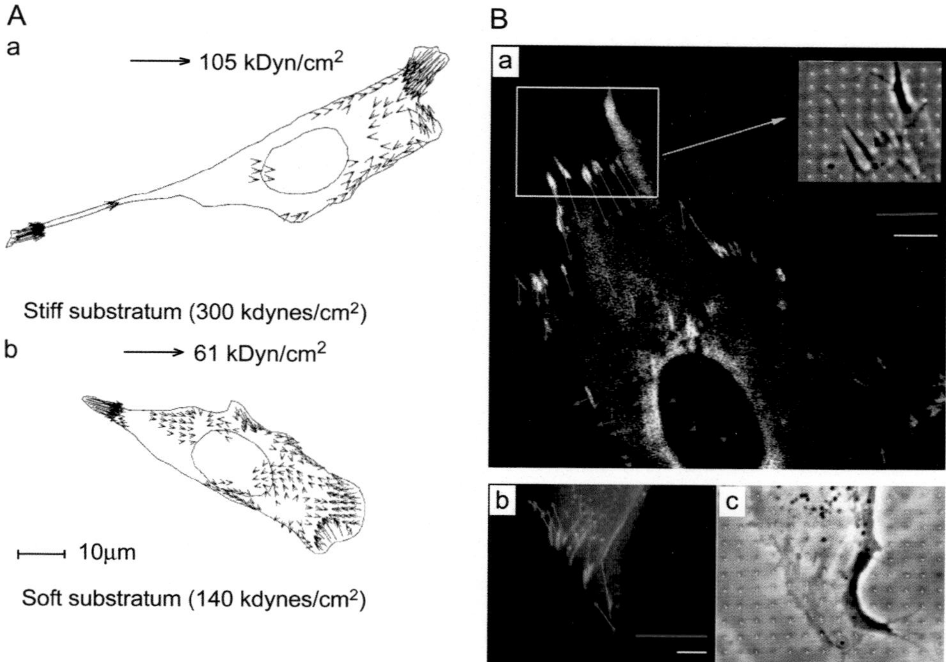

Plate 9. See Figure 5.4.

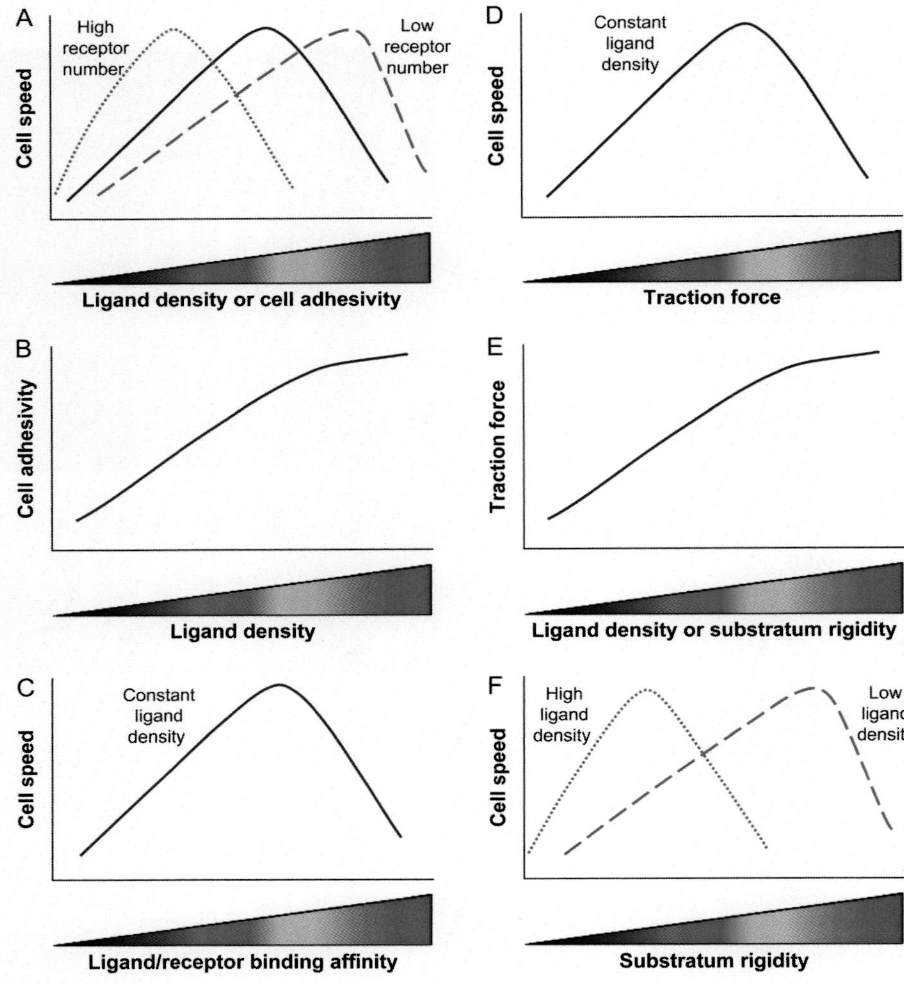

Plate 10. See Figure 5.6.

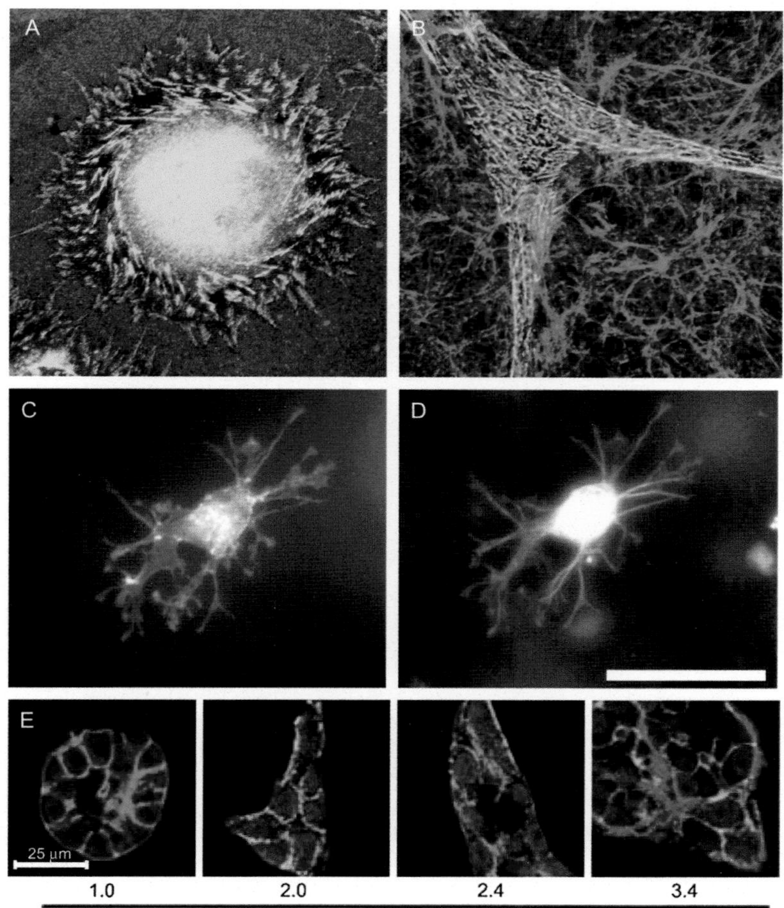

Plate 11. See Figure 5.8.

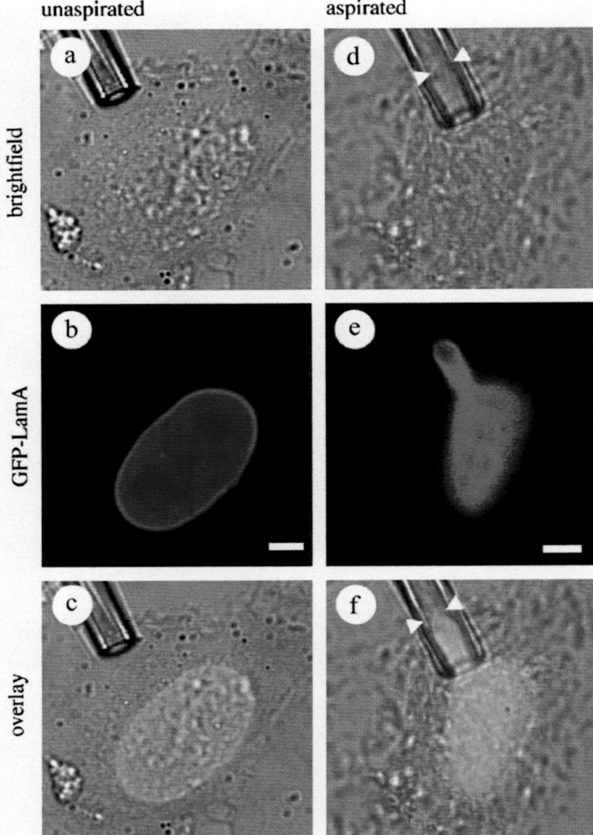

Plate 12. See Figure 9.3.

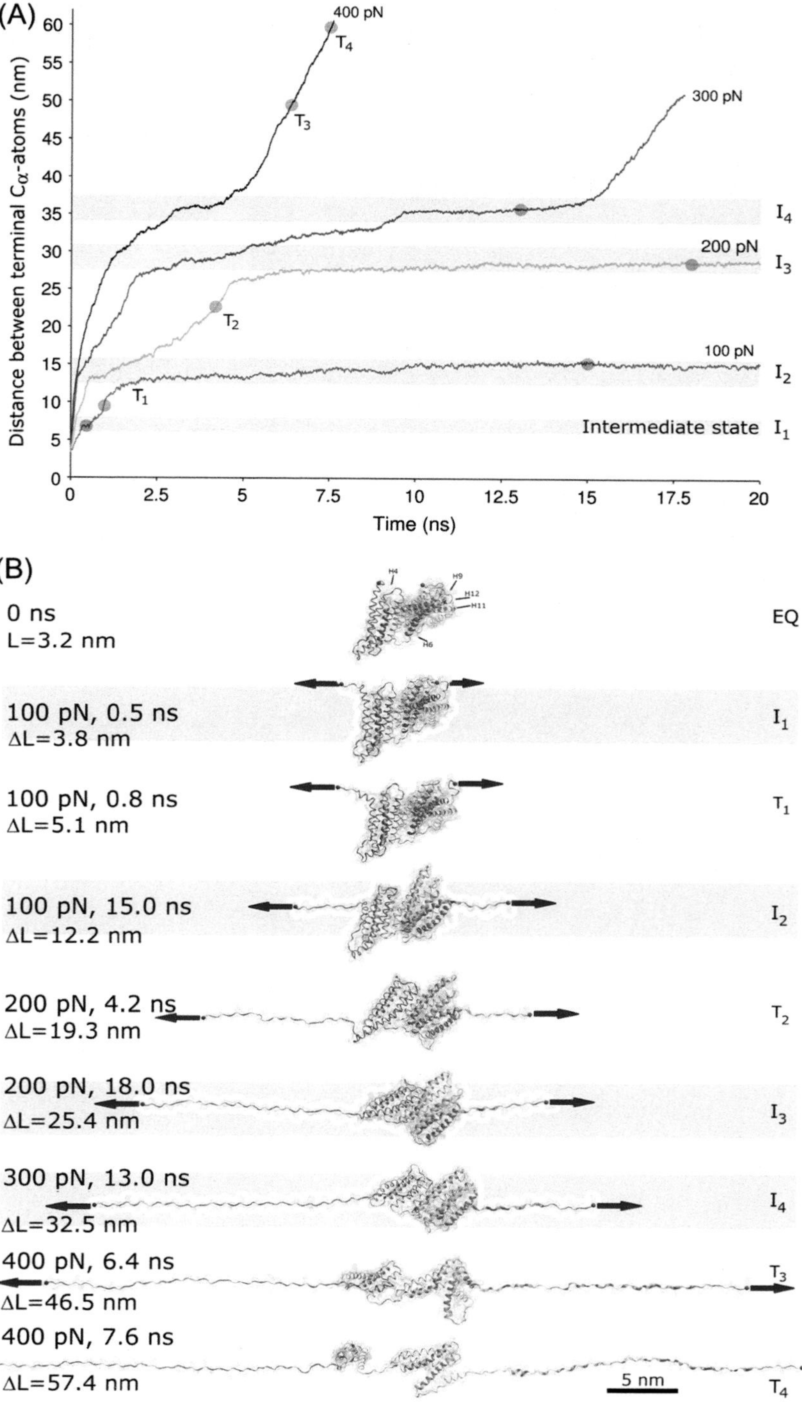

Plate 13. See Figure 11.4.

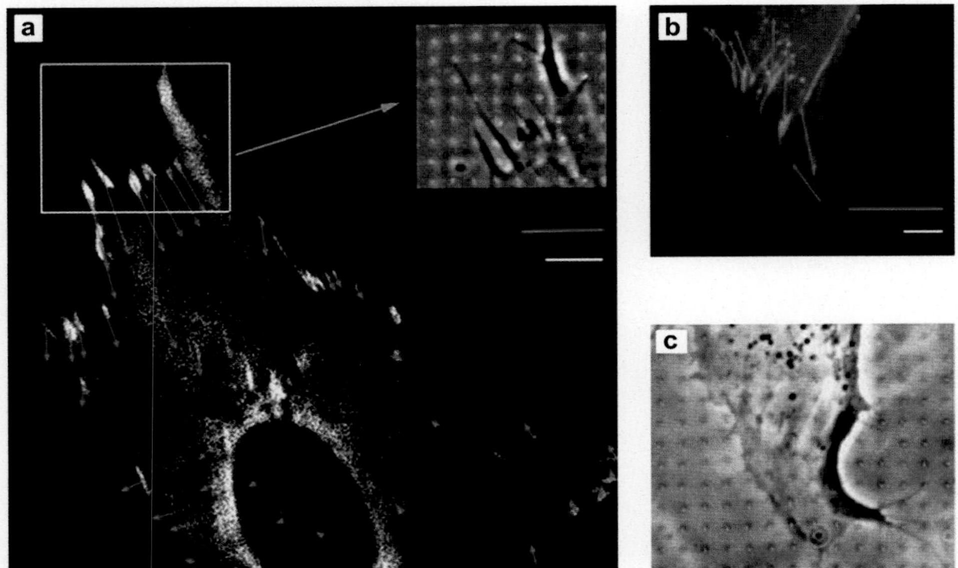

Plate A. See Figure 13.1.

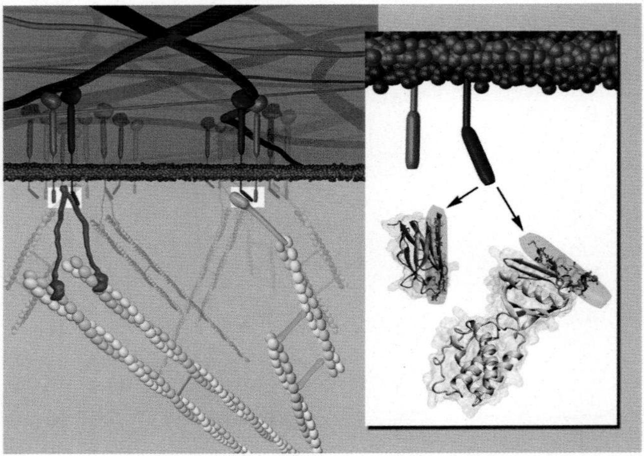

Plate B. See Figure 13.15.

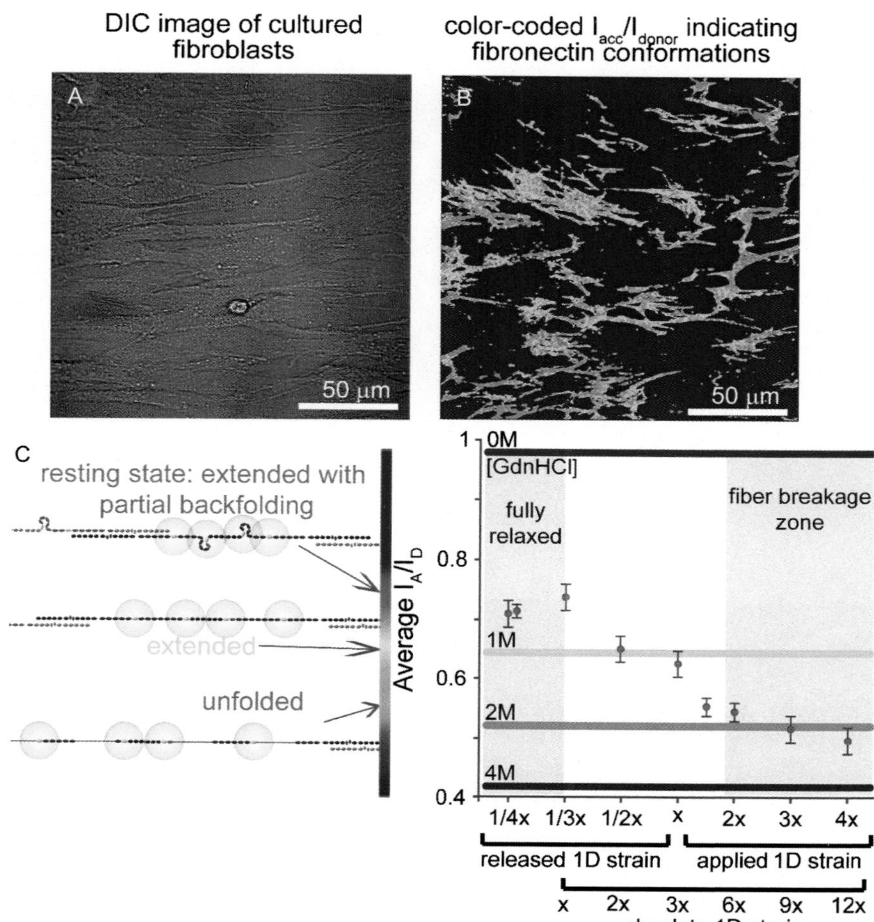

Plate 14. See Figure 13.3.

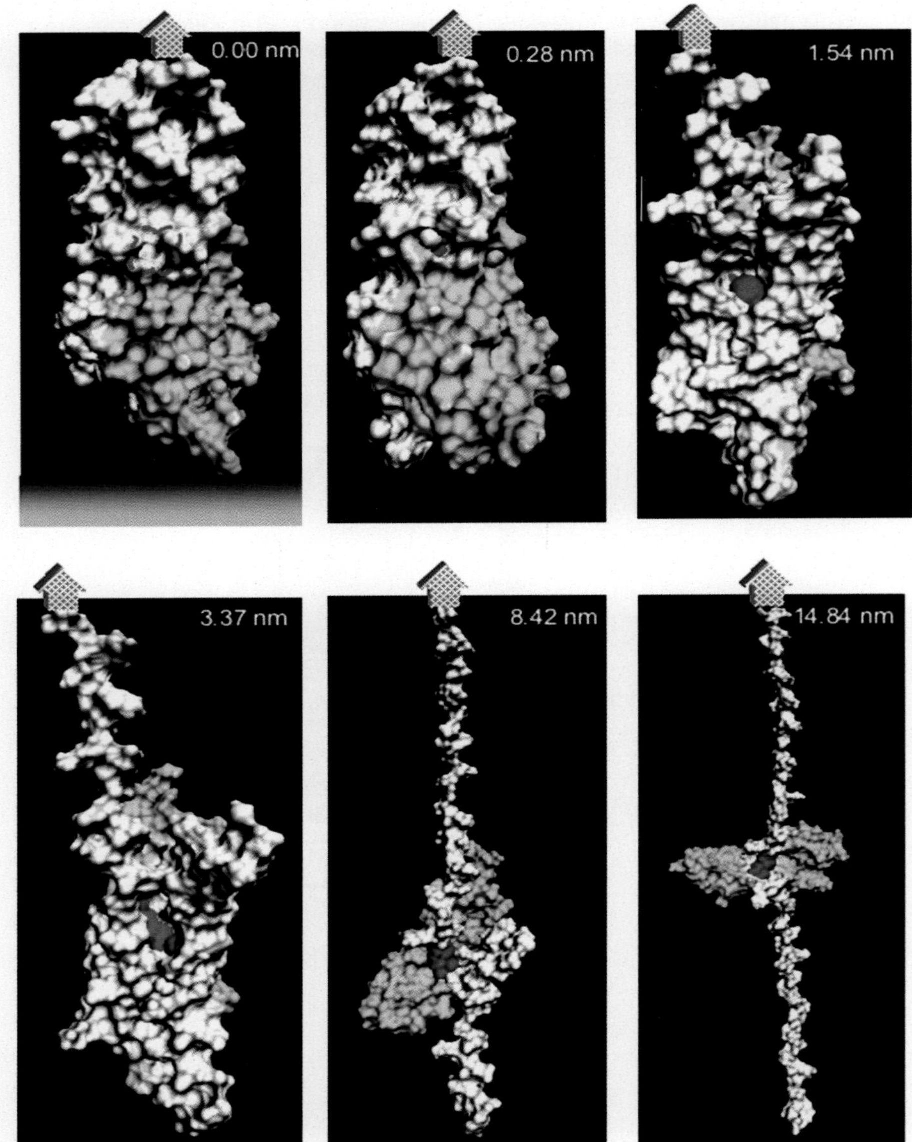

Plate 15. See Figure 13.5.

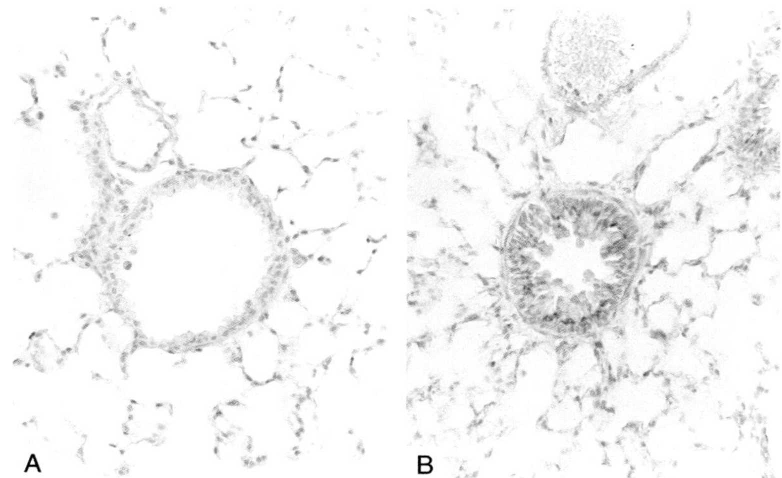

Plate 16. See Figure 14.1.

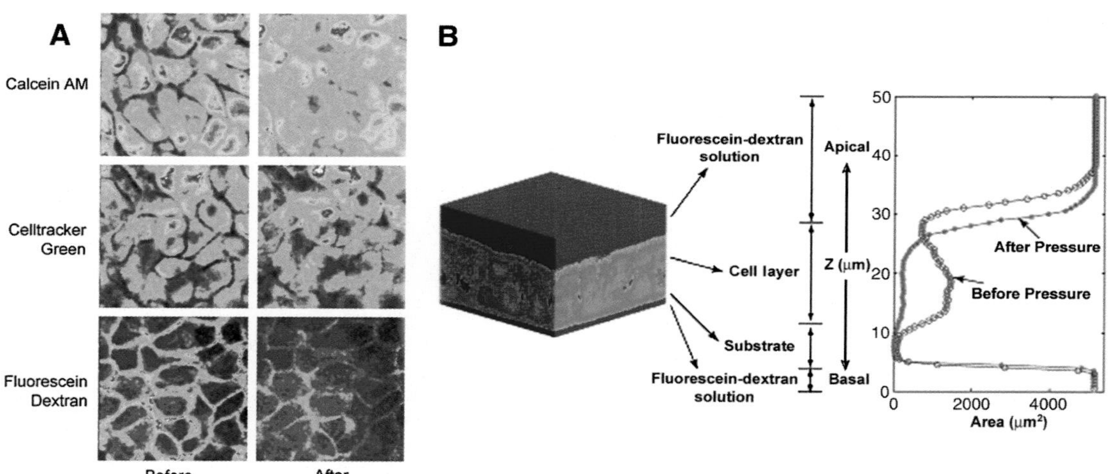

Plate 17. See Figure 14.3.

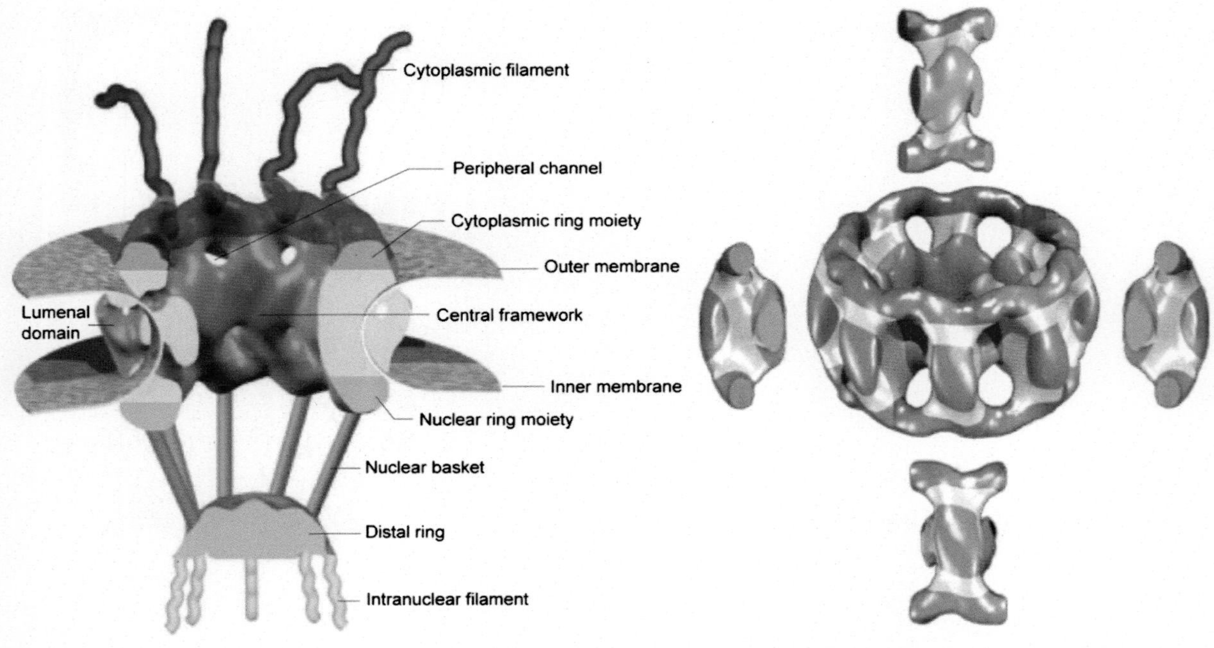

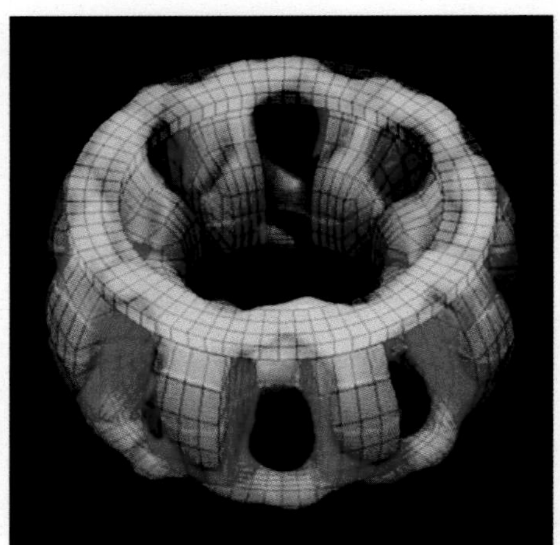

Plate 18. See Figure 18.1.

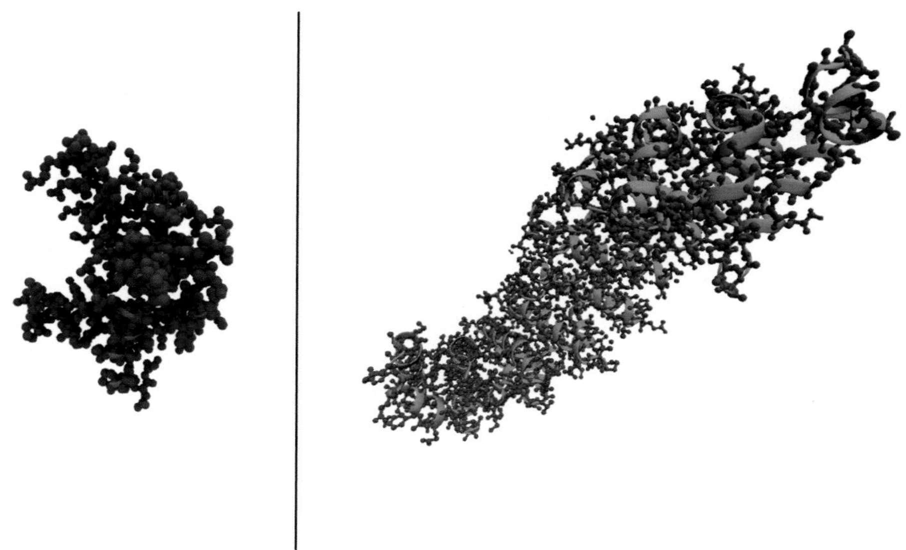

Plate 19. See Figure 18.8.

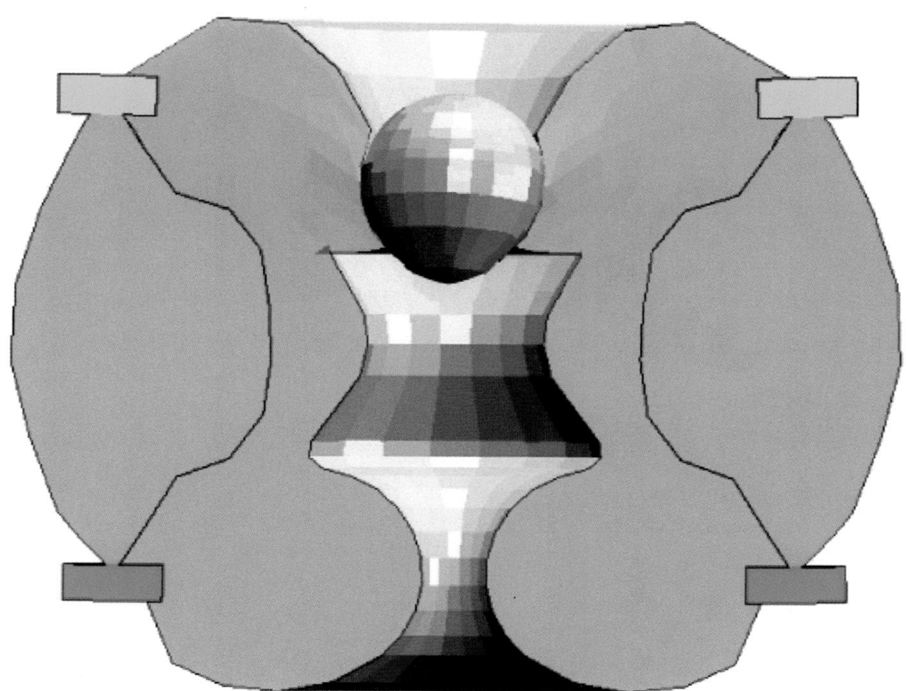

Plate 20. See Figure 18.9.

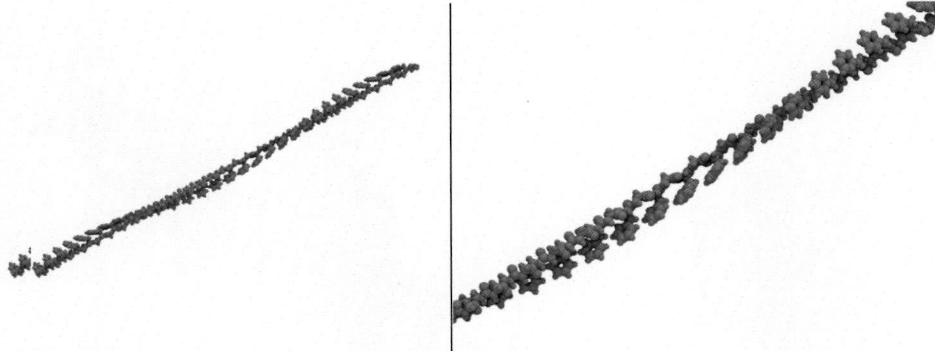

Plate 21. See Figure 18.10.

Plate 22. See Figure 18.13.

7

Toward a Modular Analysis of Cell Mechanosensing and Mechanotransduction: A Manual for Cell Mechanics

Benjamin J. Dubin-Thaler and Michael P. Sheetz

7.1 Introduction

Cellular mechanosensing and -transduction are critical functions in the shaping of cells and tissues. Although an increasing literature details the proteins and complexes involved in mechanotransduction, how these mechanisms generate the mechanochemical functions of cell motility is often poorly understood. This is a result of the fact that cells can exhibit a number of different types of motility depending on factors such as cell type and local chemical and mechanical perturbations. Due to these factors, even a genetically homogeneous cell population presents a confusing array of different motility phenotypes to the experimentalist. Therefore, we suggest a new approach to understanding cell mechanical functions through reverse systems engineering. Through quantitative analysis, we have observed that, though motility over a population of cells is heterogeneous, at a particular time and location at the cell edge, a cell exhibits only one of a limited number of modular, morphodynamic states of the acto-myosin cytoskeleton. Furthermore, a single motility module can exhibit a heterogeneous cycle of individual steps, with chemical and mechanical interactions changing over the course of this cycle. Thus, much in the way an engineer would describe the functions of components in a car engine, we should be able to approach many problems in cell motility by first describing the molecular steps involved in the basic motility modules and then showing how signaling pathways regulate those modules in order to perform cell-wide functions. In the case of cell motility, we believe there are less than thirty distinct motility modules. With a detailed, quantitative understanding of normal cell motility functions, it will be possible to understand how their malfunction can result in disease processes and to develop therapies that target specific motility modules.

7.2 Is Cytoskeletal Organization a Continuous or Discrete Function?

It has recently become clear that force is commonly transduced into biochemical signals via the cytoskeleton [1–4]. A major mechanism of transduction appears to be the unfolding of cytoskeleton-linked proteins [5, 6]. This can lead to the direct binding of other proteins, as in the case of talin unfolding causing vinculin binding [7]. Alternatively, the phosphorylation of substrates can be stimulated by *substrate priming*, the unfolding of a substrate in the presence of an active kinase, as in the case of p130Cas substrate domain phosphorylation by Src family kinases [5]. In addition, bulk mechanical signals, such as changes in membrane tension, can alter the rate of actin polymerization in the cell [8, 9]. While we are still at the early stages of understanding how mechanical signals give rise to molecular signals and how these molecular signals in turn alter cytoskeletal motility, basic design principles of motile systems as a whole are beginning to emerge. For example, relatively few basic types of motility are observed in cells (e.g., filopodial extension, lamellipodial extension, blebbing), in spite of depending on a large variety of continuously varying input signals, including cytoplasmic- and membrane-bound biochemical signals as well as mechanical signals. Through the description and perturbation of motile cells, we suggest that this continuously varying mechanochemical input space is mapped to discrete modules of cytoskeletal organization. Each module is defined by a stereotypical state of actin regulation, adhesion formation, and myosin activity that is stable over a certain domain in the mechanochemical input space, with each state corresponding to an observed *motility module*. Sharp transitions between two motility modules occur when crossing the boundary between two domains of stability, giving rise to a discrete description of cellular motility. We propose that motility modules represent conserved, stable states of actin-cytoskeletal organization.

Many recent experiments have led to this picture of a discrete, modular organization of acto-myosin motility. The systematic overexpression of dozens of members of the Rho GTPase family could be categorized, based on their steady-state edge morphology, into only 10 distinct states [10]. In *Dictyostelium*, two mutually exclusive and mechanically distinct forms of motility are observed, either blebbing or filopodial-lamellipodial–dominated [11]. We have observed two discrete modes of cell spreading, filopodial-dominated anisotropic spreading and lamellipodial isotropic spreading, and have described the bimodal distribution of these phenotypes in a genetically homogenous fibroblast background [12]. Isotropic spreading itself consists of three distinct types of fibroblast motility, with sharp temporal transitions between different types occurring on the second time-scale [13]. Switching between discrete modes of migration has been observed in neurons [14, 15], the immune synapse [16], and in tumor cells [17, 18]. These results suggest that, in spite of a complex regulatory machinery consisting of hundreds and perhaps thousands of proteins and molecules, the cytoskeleton itself possesses a limited number of stable configurations, where each configuration corresponds to a microscopically measurable motility module.

Each motility module represents a stable, organization of the acto-myosin cytoskeleton and adhesion proteins, along with the proteins and processes that directly modify their dynamics. Examples include the actin polymerization factors VASP [19, 20] and mDia2 [21], actin depolymerization factors such as cofilin [22], actin-binding proteins such as alpha-actinin [12] and tropomyosin [23], the mechanical functionality of adhesion proteins, and the mechanical effects of the membrane [9] – all molecules and processes that directly participate in generating motility. A particular combination of these molecules gives rise to a particular motility module, resulting in a stereotypical kinematic signature of the cell edge [12, 13, 24, 25], as a result of a particular balance of actin polymerization, adhesion formation, myosin activity, filament cross-linking, and any number of other processes that alter the local cytoskeletal environment [19, 26]. In theory, these proteins could be reconstituted *in vitro* and would be able to generate different motility modules, depending on their relative concentrations and activities. While some of these motility modules possess intrinsic mechanisms for the propagation over long distances along the cell edge [19, 27], we believe that motility modules are, in general, local cytoskeletal phenomenon.

While we have enumerated some of the many molecules that directly regulate motility modules on a local level, global coordination of motility is required for proper motile function [28]. We propose that rapid spatial and temporal control of motility modules is achieved by a regulatory network that modulates the localization and activity of the molecules involved in the direct control of motility modules. Molecules such as MAP kinases and the Ras family GTPases may be at the top of this signaling hierarchy, while PKC [28], the Rho family GTPases [29], and calcium signals [30] may play intermediate roles. We propose that such global signals influence the probability that a given motility module is activated by affecting the molecules and processes that directly control the local motility module. This hypothesis of modular, hierarchical control of the cytoskeleton provides a framework by which diverse signals and highly variable signals can be integrated into reproducible motile functions via the conserved motility modules.

This hierarchy of locally defined motility modules combined with global regulation could allow an evolutionary independence between motility modules and their regulation. Thus robust, conserved, low-level cytoskeletal functions would be subordinate to a more diverse set of regulatory mechanisms dictating cellular functions. For instance, when a cell migrates, global signals generate a polarized phenotype, where the front of the cell exhibits one of several protrusive motility modules while the back of the cell retracts. Cell spreading, however, involves the global coordination of protrusive motility modules. Thus, the same protrusive motility modules can result in different cell functions through altered global regulation. In this way, the cell can achieve a wide variety of high-level functions using the same low-level cytoskeletal machinery.

Modularity and nonlinearity are properties of a variety of evolved systems, and these properties are readily achieved through hierarchical organization [31]. For example, the modular nature of genes in which *cis*-regulatory sequences and protein coding sequences are independently altered through evolutionary processes results in the potential for the development of complex spatial and temporal expression patterning while using similar fundamental protein building blocks [32]. In this case, evolutionarily conserved motility modules play a role analogous to the expressed proteins while global motility regulatory signals play a role analogous to the *cis*-regulatory elements, giving the cell the ability to use the same low-level motility machinery to carry out diverse functions depending on the specific cellular context.

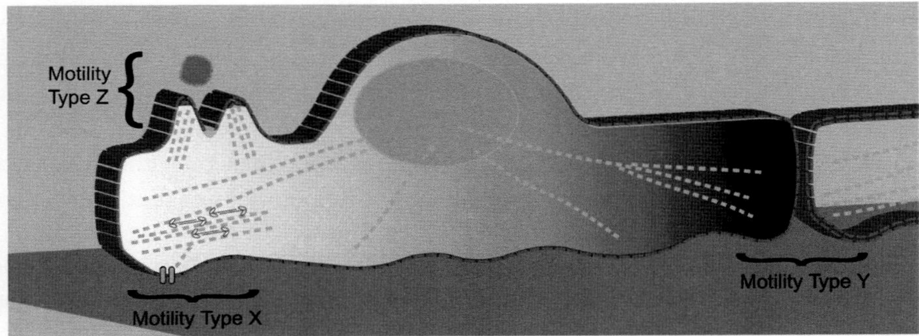

Figure 7.1. Schematic of a cell exhibiting several different motility types.

Our experimental approach has been to address both the large-scale biochemical and mechanical organizations of motility modules while searching for novel molecular-level mediators of mechanical transduction, with the intention of eventually joining the two into a coherent, biomechemical biological + mechanical + chemical pathway for motility regulation. Recognizing that the system itself is modular allows one to reduce experimental complexity by focusing just on the subset of molecules involved in a particular motility module.

7.3 Toward a "Manual" of Cell Function

The mechanical properties of cells define the shape of the organism and are critical for many different functions. Although many studies have focused on the component processes of motility such as actin filament assembly or myosin contraction, relatively few have considered the general types of motility and mechanical behaviors that are exhibited by cells and how best to describe them (Figure 7.1). After a search of the literature and examination of thousands of fibroblastic cells, we have classified about twenty different motility modules (see Box Summary of Motility types). This is a lower limit because more detailed descriptions will undoubtedly identify additional types of motility. Here we present a first version of a manual of cell function for cell motility. The term *manual* was chosen because motile processes involve very complex mechanisms not easily described by either classical enzymatic pathways or classical considerations of filament assembly–disassembly processes.

Some of the most quantitative descriptions of motility have come from the analysis of the spreading of cells on matrix-coated glass surfaces. Those surfaces are not physiological, and understanding how motility modules in two dimensions relate to three-dimensional motility is crucial in making a more general description of cellular motility. However, we suggest that many of the motility processes seen on those surfaces involve physiological steps and components that are employed by cells *in vivo*. Thus, it is useful to describe the processes in detail to understand the complexities of the types of motility such as long-range mechanical actions, physical force, matrix rigidity, and a variety of other factors that affect mechanical functions of cells. From the few cases where the details of the mechanical functions have been

Summary of motility types

1. Bleb extension and retraction
2. Cytokinesis
3. Amoeboid extension and retraction
4. Lamellipodial extension with transient retraction (ruffling)
5. Lamellipodial extension between filopodia
6. Lamellipodial extension without adhesions
7. Lamellipodial retraction with loss of adhesions
8. Smooth edge retraction
9. Filopodial extensions from stationary edges
10. Filopodial extensions from lamellipodia
11. Filopodial retractions
12. Stress fiber assembly and contraction
13. Invadipodia
14. Traction force generation
15. Fiber binding and movement
16. Cell–cell movement
17. Immune synapse formation
18. Crown (corona) actin filament assembly
19. Endothelial cell tube formation
20. Phagocytosis

mapped, there is evidence of spatial differentiation on molecular dimensions that is mechanically controlled. In addition, the interactions of components are very transient and the dynamics control the overall behavior. Cells do not seem to rely on the assembly of static structures. Thus, they have to continually develop force on a moving actin cytoskeleton. Despite the dynamic nature of the motile processes and the stochastic nature of many of the steps in each type of motility, the types of motility are reproducibly observed and easily distinguished. In the list of different types of motility, some types are poorly defined, such as cell–cell movement, because the relative movements of cells in a monolayer or other aggregates is slow and has not been clearly described. This list should be further qualified in that the criteria for each different type of motility has not been rigidly defined and will evolve over time so that the list will most likely grow. As specific protein modules are linked to specific types of motility, relationships between different motility modules will become more clear, leading to merging two modules into one or dividing a single module into multiple modules.

In this first version of the manual, we will focus on the motility modules observed in cells spreading and crawling in two dimensions. These motility modules are defined by their dependence on mechanical and/or chemical signals from the substrate. The modules include blebbing, lamellar protrusion, lamellipodial extension, and filopodial extension.

7.3.1 Filopodial Motility

Filopodia are long, thin, actin-based protrusions, and are probably the easiest motility module of the four spreading motility modules to identify visually. Filopodia have been observed in a broad range of cells, including migrating fibroblasts and *Dictyostelium* [21, 33], and the microvilli of epithelial cells may be very similar to filopodia, although their length is constant and they are much less dynamic [34, 35]. While we have studied filopodial extension in the context of anisotropic cell spreading, the molecular players involved in filopodial extension have been studied in great detail elsewhere. In order for a filopodium to form, it must 1) polymerize actin more rapidly than the surrounding filaments and 2) bundle many filaments together in order to maintain structural integrity, and a variety of molecular players are involved in these processes.

Members of the Ena/VASP protein family are often localized at the tip of a filopodium [33] and are proposed to anchor filaments to the membrane and promote polymerization [36]. Formins are also believed to catalyze actin polymerization at the tip of the growing filopodia [21, 37, 38], and it is unclear what interactions exist between formins and the Ena/VASP proteins, although a theory for their cooperative activity has been proposed [39].

In addition, actin filament bundling in filopodia is required for the maintenance of structural integrity. The persistence length of a single F-actin is relatively short at 17 μm, compared to several millimeters for microtubules [40], and without bundling F-actin could not form extensions that can reach tens of micrometers in length. The bundled filopodial structure was most recently observed *in vivo* using cryo electron tomography [41], and the actin cross-linker fascin seemed to be the primary molecule responsible for this bundling [34, 42]. Without fascin, filopodials were broader and less straight [42]. VASP may also have played a role in bundling at the tip [36].

Myosin X has been implicated in filopodial extension independent of VASP activity [43]. In addition to the forces from polymerization inducing an outward bend in the membrane, a convex membrane curvature may be stabilized or induced by a pair of membrane-binding proteins [44]. However, the major initiator of filopodial protrusion is most likely local, rapid polymerization. Capping protein has a negative effect on filopodial formation both *in vivo* [45] and *in vitro* [46, 47], supporting the idea that prolonged polymerization of individual filaments is important for filopodial growth.

While the adhesion molecule paxillin is found beneath some filopodial extensions [48], presumably playing a stabilizing role, the long, thin filopodial protrusions are not generally well adhered to the substrate and present a major challenge in designing a general kinematic image analysis algorithm. First, because they are so thin, they are more difficult to reliably detect in an image segmentation scheme. Second, because they are not attached to the substrate, the motion of a filopodium does not necessarily mean protrusion – it could mean that the entire structure is pivoting at its base. Algorithms that quantify protrusion based on an arc-length parameterization are particularly unsuited for analyzing filopodia. First, if a filopodium is detected in one frame and is not detected in the next frame, the length of the contour will fluctuate greatly. Parameterizing contours based on arc-length means that discontinuities in the total arc-length will result in discontinuities on the resulting velocity map. Second, even if the contour is drawn with perfect fidelity with respect to the filopodia, when a filopodium moves laterally, as they often do, the velocity map will show a protrusion corresponding to the leading edge of the swinging filopodium and a retraction corresponding to the trailing edge of the filopodium. While this is mathematically accurate, it is not the most relevant measurement. The rate of growth or shrinking of the tip of the filopodium, thought to be a more relevant measurement in terms of chemotactic guidance [49], would be difficult to extract in the midst of a swath-like representation of filopodial motion on a velocity map.

To avoid these problems, a polar coordinate parameterization is used to analyze filopodia-dominated anisotropic spreading. However, the result is that such measurements focus on the lamella that stretches between adjacent filopodia. Interestingly, the dynamics of lamellipodial protrusion in the filopodial-rich neuronal growth cone [50] were found to be similar with those of filopodial-dominated, anisotropic-spreading fibroblasts [12]. These results underscore the similarity in motility modules across cell types.

Many of the same players present in lamellipodia (LP) motility are also present in growth cone filopodia (FP) motility [51]. The thick, bundled FPs may limit the ability of the LP actin to organize a coherent structure and thus generate coordinated periodicity. This loss of coherence may relate to mechanical edge effects owing to interactions between LP molecules and FP molecules such as fascin. We propose that protrusion of the lamella during filopodial-dominated motility modules in fibroblasts is very similar to the protrusion in growth cones.

7.3.2 Blebbing Motility

Apoptotic blebbing occurs when the connection between the actin cytoskeleton and the membrane is disrupted and the positive osmotic pressure of the cell generates a force in which a hemispherical-shaped piece of membrane extends outward. In healing blebs, an acto-myosin cortex quickly reforms on the membrane surface, exerting forces that retract the bleb [52].

While blebbing can be a sign of apoptosis, it has also been reported in the early phase of cell spreading [53], mitosis [54], and tumor cell migration [17, 55]. In general, bleb formation is a Myosin II (MII) dependent process, although a variety of pathways leading to the requisite myosin light-chain phosphorylation [56] have been identified. In some cases, Rho kinase is targeted by caspases and cleaved into an active form [57, 58], a pathway that can be blocked with Rho kinase inhibitors (A3) [59]. Caspase-independent blebbing pathways also exist [60], and these pathways may act through either myosin light-chain kinase [56, 61, 62] or through the family of DAP kinases that are also thought to phosphorylate myosin regulatory light-chains (MRLC) [63]. In some blebbing cells, the activation of p38MAPK is important, possibly acting through the actin-polymerization factor HSP27 [64–66]. There is a second type of blebbing, necrotic blebbing, that occurs during acute cell death leading to lysis, and this form of blebbing is not thought to be sensitive to any of the aforementioned kinases [62].

In *Dictyostelium*, when blebbing is blocked by Rho kinase inhibitors, its blebbing motility transitions to a filopodial-type behavior [11] and we have observed a similar transition in P0 spreading cells upon the inhibition of blebbing by Rho kinase inhibitors. While some studies have observed that the apoptotic mode of blebbing is blocked by the inhibition of myosin light-chain kinase [62], we found that MLCK inhibition did not inhibit blebbing in P0. The spatially segregated actions of these two kinases, both of which can phosphorylate MRLC and activate myosin II activity [67], have been well established in polarized fibroblasts [68], and these results suggest that this segregation exists even in rounded cells. Interestingly, even when blebbing is inhibited in P0, the P0–P1 transition was undisturbed [25]. We believe this indicates that the elements regulating the P0–P1 transition are "upstream" from the myosin-dependent blebbing or filopodial motility. In general, cells will continue to bleb in P0 until the P0–P1 initiation occurs, a result that has been independently confirmed in cells with the gene for integrin-linked kinase affixin deleted [69].

While we found that FN accelerates the initiation of the P0–P1 transition through integrins, what molecules downstream of integrins are involved, and what mechanism could generate such an explosive increase in protrusion? In carcinoma cells acutely exposed to EGF, a similar explosive increase in protrusion occurs [70]. In this case, it was found that phospholipase C (PLC), an enzyme that can cleave PIP_2 into the diffusible second messenger IP_3 and the membrane-bound DAG, could be responsible for the increase in protrusion [22, 71]. In addition, there was a time lag between the beginning of activation of PLC and the initiation of protrusion, suggesting an integration of signal was required before a cytoskeleton effect was achieved. Indeed, in preliminary experiments with the PLC inhibitor U46619 and PLC-γ knockout fibroblasts [72], we have observed a delay in the P0–P1 transition. This delay could be overcome by increasing the concentration of fibronectin, suggesting that PLC may contribute a signal to a thresholded response function generated by other downstream molecules.

7.3.3 Continuous Protrusion

Broad regions of continuous protrusion lacking retraction or ruffles give rise to fast, isotropic spreading in P1. Edge velocity in this motility module is on average 5 μm/min (A1), occurs in uniform regions of the cell edge (A1, A3), and is associated with VASP localization at the tip of the protruding edge (A3). It is likely that continuous protrusion is a self-catalyzing process, since it is not acutely affected by blocking integrin binding (A1). Focal adhesions are not formed during continuous protrusion, and the rearward flow of the actin cytoskeleton, as indicated by α-actinin speckle tracking, is also slow. Although treatment with CD decreases the probability of continuous protrusion in a given region of the cell, regions exhibiting continuous protrusions are relatively unaffected by CD up to 100 nM.

Motility very similar to P1 continuous spreading has been observed in post-mitotic cell spreading (Gauthier et al., in submission) and keratocyte migration [73], as well as in tumor-derived epithelial cell lines acutely exposed to an epidermal growth factor, which undergo a 2-min-long period of rapid actin polymerization [70]. In addition, we have observed continuous protrusion without ruffling in limited spatial regions during P2 with the same VASP distribution as well as in migrating cells, and this is likely to be the same type of Rac-dependent motility that has been described in a variety of different circumstances [74].

The molecular players in continuous protrusion have yet to be fully investigated, although a large variety of molecules have been shown to block P1 from spreading, including WAVE1 [75] and affixin [69].

7.3.4 Periodic Lamellipodial Contractions

Our elucidation of the basic underlying mechanism of periodic lamellipodial contractions represents a prime example of the power of high-resolution analysis of cell motility in two dimensions, as well as the power of cell spreading as a model system that allows the study of its basic motility modules in isolation.

Periodic contractions are exhibited by many of cells within 20 min after plating onto fibronectin. Each step in the process is repeated from one edge retraction to the next, making the process periodic. From the slope of the edge movement in the kymograph, the velocity of extension is about 60 nm/s, and extensions last for about 20 s. Retractions are shorter (about 5 s) and of much lower velocity. Thus, the edge moves outward about 1000 nm per cycle of 24 s. This movement is restricted to the lamellipodium, which we define primarily by the presence of an α-actinin–rich band at the edge of the lamella, where the lamella in turn is defined as the vesicle-free region at the cell periphery and includes the lamellipodium. Crucially, the α-actinin–rich band does not define a steady width of the lamellipodium – indeed, it is regenerated during each cycle, growing from the tip of the lamella back roughly 2000 nm, until it collapses back at the end of the retraction phase. This collapsing lamellipodial actin is also visible as a dense "wave" in the contrast microscope, moving back rapidly after the retraction ends and then slowing its rearward movement as the *contractile wave* condensed at a line 2000 nm back from where the lamellipodium was when the retraction ended. MII aggregates are at this line as well, increasing as the wave condenses. In addition, we found myosin light-chain kinase co-localized with the contractile waves, while the adhesion proteins VASP and paxillin formed early adhesions just under the tip during cell edge retraction.

In modeling the underlying mechanism of the periodic contraction cycle, we sought to explain three major steps: 1) the leading edge extension, 2) the transient retraction of the leading edge, and 3) the collapse and transport contractile wave from the edge to the back of the lamellipodium. Each of these major steps could be further broken down into eight putative molecular and mechanical steps: 1) lamellar actin assembly; 2) lamellipodial actin assembly; 3) myosin filament assembly; 4) edge retraction; 5) adhesion site assembly; 6) upward bending of the lamellipodia during edge retraction; 7) lamellipodial actin release and condensation, concomitant with resumption of edge protrusion; and 8) the occasional generation of a ruffle instead of stable adhesions.

We propose the simplest model that explains each of these steps. 1) Lamella extension occurs via the normal polymerization mechanism of the lamella, perhaps including a formin or VASP anti-capping polymerization mechanism. 2) As the lamella grows forward, lamellipodial (LP) actin grows back from the tip, probably via an Arp2/3 mediated polymerization mechanism. 3) LP actin reaches a myosin-rich region 1500 nm behind the edge of the lamellipodium, and myosin heads begin pulling on the LP actin. LP actin may be more sensitive to myosin because it lacks tropomyosin [23]. 4) Myosin pulls on the LP actin, which in turn pulls on the tip of the lamella, causing a retraction of the edge. 5) As the LP actin pulls back on the front of the cell, nascent, force-induced adhesions begin to form. These adhesions are labile in that they are not firmly attached to the actin filaments, allowing the front edge of the cell to slide back while remaining attached to the substrate. 6) The lamellipodium is then pulled back from the front and held in place in the back by the rigidity of the lamellar actin network, and thus buckles upwards. 7) Eventually, myosin generates such high forces on the LP actin that the bonds holding it to the front edge rupture, and the edge resumes protrusion with step 1. 8) If the nascent adhesion sites do not form fast enough because the substrate is too soft for the required force-dependent reinforcement or because the chemical composition does not allow enough integrins to bind, then step 6 will result in ruffling instead of buckling.

We have discovered the mechanism that coordinates MII activity with adhesion formation and edge retraction. Previous models based on biochemical signaling [75–77] could not explain the complex pattern of edge retraction. Likewise, it was not understood how the timing and location of MII activity related to adhesion-site formation, particularly in light of evidence showing that MII clusters were found well behind newly forming adhesion sites [26, 78]. We propose a model in which MII activity and adhesion-site initiation are coordinated by a physical signal. LP actin serves as this physical signal by transmitting forces between the MII and the cell edge. Thus, our work shows that a mechanical mechanism coordinates spatially isolated biochemical processes to shape cell motility.

7.3.5 Process: Elements of a Quantitative Description of a Motility Module

Here we provide the elements required for a thorough and general description of the constituent biochemical and biomechanical steps of a given motility module using fibroblast motility as an example.

1. Form a quantitative description of the motility module.
2. Define the mechanical steps that generate this motility module.
3. Define of the molecules involved in each step.
4. Develop a testable, quantitative model for the mechanism underlying each step.

5. Define how transitions from one step to another are controlled.
6. Develop a quantitative model for the overall function, including steps and their transitions.

7.4 Conclusion

The motility modules described in fibroblasts appear in other cell types and are quite standard. We propose that motility modules represent universal modes of cellular motility arising from conserved core protein complexes and basic motility functions. Thus, if the mechanism of motility is understood in detail in one cell system, the mechanism in other cells should be readily understood. Although we expect our initial enumeration of motility modules to change over time, we believe that there are relatively few unique modules. The detailed understanding of the mechanisms of each module will provide useful tools for identifying motility in a wide variety of contexts, including the role of mechanical factors in cancer progression as well as in the process of cell differentiation.

REFERENCES

[1] Kostic, A., and M. P. Sheetz. 2006. Fibronectin rigidity response through Fyn and p130Cas recruitment to the leading edge. *Mol Biol Cell* **17**: 2684–2695.

[2] Giannone, G., and M. P. Sheetz. 2006. Substrate rigidity and force define form through tyrosine phosphatase and kinase pathways. *Trends Cell Biol* **16**: 213–223.

[3] Vogel, V., and M. Sheetz. 2006. Local force and geometry sensing regulate cell functions. *Nature Revs* **7**: 265–275.

[4] Glogauer, M., P. Arora, G. Yao, I. Sokholov, J. Ferrier, and C. A. McCulloch. 1997. Calcium ions and tyrosine phosphorylation interact coordinately with actin to regulate cytoprotective responses to stretching. *J Cell Sci* **110**(Pt 1): 11–21.

[5] Sawada, Y., M. Tamada, B. J. Dubin-Thaler, O. Cherniavskaya, R. Sakai, S. Tanaka, and M. P. Sheetz. 2006. Force sensing by mechanical extension of the Src family kinase substrate p130Cas. *Cell* **127**: 1015–1026.

[6] Tamada, M., M. P. Sheetz, and Y. Sawada. 2004. Activation of a signaling cascade by cytoskeleton stretch. *Dev Cell* **7**: 709–718.

[7] Fillingham, I., A. R. Gingras, E. Papagrigoriou, B. Patel, J. Emsley, D. R. Critchley, G. C. Roberts, and I. L. Barsukov. 2005. A vinculin binding domain from the talin rod unfolds to form a complex with the vinculin head. *Structure* **13**: 65–74.

[8] Raucher, D., and M. P. Sheetz. 1999. Characteristics of a membrane reservoir buffering membrane tension. *Biophys J* **77**: 1992–2002.

[9] Raucher, D., and M. P. Sheetz. 2000. Cell spreading and lamellipodial extension rate is regulated by membrane tension. *J Cell Bio* **148**: 127–136.

[10] Heo, W. D., and T. Meyer. 2003. Switch-of-function mutants based on morphology classification of Ras superfamily small GTPases. *Cell* **113**: 315–328.

[11] Yoshida, K., and T. Soldati. 2006. Dissection of amoeboid movement into two mechanically distinct modes. *J Cell Sci* **119**: 3833–3844.

[12] Dubin-Thaler, B. J., G. Giannone, H. G. Dobereiner, and M. P. Sheetz. 2004. Nanometer analysis of cell spreading on matrix-coated surfaces reveals two distinct cell states and STEPs. *Biophys J* **86**: 1794–1806.

[13] Döbereiner, H. G., B. Dubin-Thaler, G. Giannone, H. S. Xenias, and M. P. Sheetz. 2004. Dynamic phase transitions in cell spreading. *Phys Rev Lett* **93**: 108105.
[14] Ayala, R., T. Shu, and L. H. Tsai. 2007. Trekking across the brain: The journey of neuronal migration. *Cell* **128**: 29–43.
[15] Marin, O., M. Valdeolmillos, and F. Moya. 2006. Neurons in motion: Same principles for different shapes? *Trends Neurosci* **29**: 655–661.
[16] Friedl, P., A. T. den Boer, and M. Gunzer. 2005. Tuning immune responses: Diversity and adaptation of the immunological synapse. *Nat Rev Immunol* **5**: 532–545.
[17] Sahai, E., and C. J. Marshall. 2003. Differing modes of tumour cell invasion have distinct requirements for Rho/ROCK signalling and extracellular proteolysis. *Nat Cell Biol* **5**: 711–719.
[18] Wang, W., J. B. Wyckoff, S. Goswami, Y. Wang, M. Sidani, J. E. Segall, and J. S. Condeelis. 2007. Coordinated regulation of pathways for enhanced cell motility and chemotaxis is conserved in rat and mouse mammary tumors. *Cancer Res* **67**: 3505–3511.
[19] Giannone, G., B. J. Dubin-Thaler, O. Rossier, Y. Cai, O. Chaga, G. Jiang, W. Beaver, H. G. Dobereiner, Y. Freund, G. Borisy, and M. P. Sheetz. 2007. Lamellipodial actin mechanically links myosin activity with adhesion-site formation. *Cell* **128**: 561–575.
[20] Dubin-Thaler, B., J. M. Hofman, H. Xenias, I. Spielman, A. V. Shneidman, L. A. David, H. G. Dobereiner, C. H. Wiggins, and M. P. Sheetz. 2008. Quantification of cell edge velocities and fraction forces reveals distinct motility modules during cell spreading. *PLOS ONE* **3**(11): e 3735.
[21] Schirenbeck, A., T. Bretschneider, R. Arasada, M. Schleicher, and J. Faix. 2005. The Diaphanous-related formin dDia2 is required for the formation and maintenance of filopodia. *Nat Cell Biol* **7**: 619–625.
[22] DesMarais, V., M. Ghosh, R. Eddy, and J. Condeelis. 2005. Cofilin takes the lead. *J Cell Sci* **118**: 19–26.
[23] Gupton, S. L., K. L. Anderson, T. P. Kole, R. S. Fischer, A. Ponti, S. E. Hitchcock-DeGregori, G. Danuser, V. M. Fowler, D. Wirtz, D. Hanein, and C. M. Waterman-Storer. 2005. Cell migration without a lamellipodium: Translation of actin dynamics into cell movement mediated by tropomyosin. *J Cell Biol* **168**: 619–631.
[24] Machacek, M., and G. Danuser. 2006. Morphodynamic profiling of protrusion phenotypes. *Biophys J* **90**: 1439–1452.
[25] Dubin-Thaler, B. J., J. M. Hofman, Y. Cai, H. Xenias, I. Spielman, A. V. Shneidman, L. A. David, H. G. Dobereiner, C. H. Wiggins, and M. P. Sheetz. 2008. Quantification of cell edge velocities and traction forces reveals distinct motility modules during cell spreading. *PloS One* **3**: e 3735.
[26] Gupton, S. L., and C. M. Waterman-Storer. 2006. Spatiotemporal feedback between actomyosin and focal-adhesion systems optimizes rapid cell migration. *Cell* **125**: 1361–1374.
[27] Johnson, K. E. 1976. Circus movements and blebbing locomotion in dissociated embryonic cells of an amphibian, *Xenopus laevis*. *J Cell Sci* **22**: 575–583.
[28] Sims, T. N., T. J. Soos, H. Xenias, B. Dubin-Thaler, J. Hofman, J. Waite, T. O. Cameron, V. K. Thomas, R. Varma, C. Wiggins, M. P. Sheetz, D. R. Littman, and M. L. Dustin. 2007. Opposing effects of PKCtheta and WASp on symmetry breaking and relocation of the immunological synapse. *Cell* **129**: 773–785.
[29] Vial, E., E. Sahai, and C. J. Marshall. 2003. ERK-MAPK signaling coordinately regulates activity of Rac1 and RhoA for tumor cell motility. *Cancer Cell* **4**: 67–79.

[30] Munevar, S., Y. L. Wang, and M. Dembo. 2004. Regulation of mechanical interactions between fibroblasts and the substratum by stretch-activated Ca2+ entry. *J Cell Sci* **117**: 85–92.
[31] Kirschner, M., and J. Gerhart. 2005. *The Plausibility of Life: Resolving Darwin's Dilemma.* Yale University Press, New Haven, CT.
[32] Wittkopp, P. J. 2006. Evolution of cis-regulatory sequence and function in Diptera. *Heredity* **97**: 139–147.
[33] Han, Y. H., C. Y. Chung, D. Wessels, S. Stephens, M. A. Titus, D. R. Soll, and R. A. Firtel. 2002. Requirement of a vasodilator-stimulated phosphoprotein family member for cell adhesion, the formation of filopodia, and chemotaxis in dictyostelium. *J Biol Chem* **277**: 49877–49887.
[34] Edwards, R. A., and J. Bryan. 1995. Fascins, a family of actin bundling proteins. *Cell Motility and the Cytoskeleton* **32**: 1–9.
[35] Kureishy, N., V. Sapountzi, S. Prag, N. Anilkumar, and J. C. Adams. 2002. Fascins, and their roles in cell structure and function. *Bioessays* **24**: 350–361.
[36] Applewhite, D. A., M. Barzik, S. I. Kojima, T. M. Svitkina, F. B. Gertler, and G. G. Borisy. 2007. Ena/VASP proteins have an anti-capping independent function in filopodia formation. *Mol Biol Cell.* **18(7)**: 2579–2591.
[37] Pruyne, D., M. Evangelista, C. Yang, E. Bi, S. Zigmond, A. Bretscher, and C. Boone. 2002. Role of formins in actin assembly: Nucleation and barbed-end association. *Science* **297**: 612–615.
[38] Kovar, D. R., E. S. Harris, R. Mahaffy, H. N. Higgs, and T. D. Pollard. 2006. Control of the assembly of ATP- and ADP-actin by formins and profilin. *Cell* **124**: 423–435.
[39] Schirenbeck, A., R. Arasada, T. Bretschneider, M. Schleicher, and J. Faix. 2005. Formins and VASPs may co-operate in the formation of filopodia. *Biochem Soc Trans* **33**: 1256–1259.
[40] Brangwynne, C. P., G. H. Koenderink, E. Barry, Z. Dogic, F. C. Mackintosh, and D. A. Weitz. 2007. Bending dynamics of fluctuating biopolymers probed by automated high-resolution filament tracking. *Biophys J* **93(1)**: 346–59.
[41] Medalia, O., M. Beck, M. Ecke, I. Weber, R. Neujahr, W. Baumeister, and G. Gerisch. 2007. Organization of actin networks in intact filopodia. *Curr Biol* **17**: 79–84.
[42] Vignjevic, D., S. Kojima, Y. Aratyn, O. Danciu, T. Svitkina, and G. G. Borisy. 2006. Role of fascin in filopodial protrusion. *J Cell Biol* **174**: 863–875.
[43] Bohil, A. B., B. W. Robertson, and R. E. Cheney. 2006. Myosin-X is a molecular motor that functions in filopodia formation. *Proc Nat Acad Sci USA* **103**: 12411–12416.
[44] Mattila, P. K., A. Pykalainen, J. Saarikangas, V. O. Paavilainen, H. Vihinen, E. Jokitalo, and P. Lappalainen. 2007. Missing-in-metastasis and IRSp53 deform PI(4,5)P2-rich membranes by an inverse BAR domain-like mechanism. *J Cell Biol* **176**: 953–964.
[45] Mejillano, M. R., S. Kojima, D. A. Applewhite, F. B. Gertler, T. M. Svitkina, and G. G. Borisy. 2004. Lamellipodial versus filopodial mode of the actin nanomachinery: Pivotal role of the filament barbed end. *Cell* **118**: 363–373.
[46] Haviv, L., Y. Brill-Karniely, R. Mahaffy, F. Backouche, A. Ben-Shaul, T. D. Pollard, and A. Bernheim-Groswasser. 2006. Reconstitution of the transition from lamellipodium to filopodium in a membrane-free system. *Proc Nat Acad Sci USA* **103**: 4906–4911.
[47] Vignjevic, D., D. Yarar, M. D. Welch, J. Peloquin, T. Svitkina, and G. G. Borisy. 2003. Formation of filopodia-like bundles in vitro from a dendritic network. *J Cell Biol* **160**: 951–962.

[48] Bukharova, T., G. Weijer, L. Bosgraaf, D. Dormann, P. J. van Haastert, and C. J. Weijer. 2005. Paxillin is required for cell-substrate adhesion, cell sorting and slug migration during Dictyostelium development. *J Cell Sci* **118**: 4295–4310.

[49] Mallavarapu, A., and T. Mitchison. 1999. Regulated actin cytoskeleton assembly at filopodium tips controls their extension and retraction. *J Cell Biol* **146**: 1097–1106.

[50] Betz, T., D. Lim, and J. A. Kas. 2006. Neuronal growth: A bistable stochastic process. *Phys Rev Lett* **96**: 098103–098104.

[51] Medeiros, N. A., D. T. Burnette, and P. Forscher. 2006. Myosin II functions in actin-bundle turnover in neuronal growth cones. *Nat Cell Biol* **8**: 215–226.

[52] Charras, G. T., C. K. Hu, M. Coughlin, and T. J. Mitchison. 2006. Reassembly of contractile actin cortex in cell blebs. *J Cell Biol* **175**: 477–490.

[53] Bereiter-Hahn, J., M. Luck, T. Miebach, H. K. Stelzer, and M. Voth. 1990. Spreading of trypsinized cells: Cytoskeletal dynamics and energy requirements. *J Cell Sci* **96**(Pt 1):171–188.

[54] Boss, J. 1955. Mitosis in cultures of newt tissues. IV. The cell surface in late anaphase and the movements of ribonucleoprotein. *Exp Cell Res* **8**: 181–187.

[55] Keller, H., and P. Eggli. 1998. Protrusive activity, cytoplasmic compartmentalization, and restriction rings in locomoting blebbing Walker carcinosarcoma cells are related to detachment of cortical actin from the plasma membrane. *Cell Motil Cytoskeleton* **41**: 181–193.

[56] Mills, J. C., N. L. Stone, J. Erhardt, and R. N. Pittman. 1998. Apoptotic membrane blebbing is regulated by myosin light chain phosphorylation. *J Cell Biol* **140**: 627–636.

[57] Sebbagh, M., C. Renvoize, J. Hamelin, N. Riche, J. Bertoglio, and J. Breard. 2001. Caspase-3-mediated cleavage of ROCK I induces MLC phosphorylation and apoptotic membrane blebbing. *Nat Cell Biol* **3**: 346–352.

[58] Coleman, M. L., E. A. Sahai, M. Yeo, M. Bosch, A. Dewar, and M. F. Olson. 2001. Membrane blebbing during apoptosis results from caspase-mediated activation of ROCK I. *Nat Cell Biol* **3**: 339–345.

[59] Charras, G. T., J. C. Yarrow, M. A. Horton, L. Mahadevan, and T. J. Mitchison. 2005. Non-equilibration of hydrostatic pressure in blebbing cells. *Nature* **435**: 365–369.

[60] McCarthy, N. J., M. K. Whyte, C. S. Gilbert, and G. I. Evan. 1997. Inhibition of Ced-3/ICE-related proteases does not prevent cell death induced by oncogenes, DNA damage, or the Bcl-2 homologue Bak. *J Cell Biol* **136**: 215–227.

[61] Mills, J. C., N. L. Stone, and R. N. Pittman. 1999. Extranuclear apoptosis. The role of the cytoplasm in the execution phase. *J Cell Biol* **146**: 703–708.

[62] Barros, L. F., T. Kanaseki, R. Sabirov, S. Morishima, J. Castro, C. X. Bittner, E. Maeno, Y. Ando-Akatsuka, and Y. Okada. 2003. Apoptotic and necrotic blebs in epithelial cells display similar neck diameters but different kinase dependency. *Cell Death and Differentiation* **10**: 687–697.

[63] Kogel, D., J. H. Prehn, and K. H. Scheidtmann. 2001. The DAP kinase family of pro-apoptotic proteins: novel players in the apoptotic game. *Bioessays* **23**: 352–358.

[64] Deschesnes, R. G., J. Huot, K. Valerie, and J. Landry. 2001. Involvement of p38 in apoptosis-associated membrane blebbing and nuclear condensation. *Mol Biol Cell* **12**: 1569–1582.

[65] Huot, J., F. Houle, F. Marceau, and J. Landry. 1997. Oxidative stress-induced actin reorganization mediated by the p38 mitogen-activated protein kinase/heat shock protein 27 pathway in vascular endothelial cells. *Circ Res* **80**: 383–392.

[66] Huot, J., F. Houle, S. Rousseau, R. G. Deschesnes, G. M. Shah, and J. Landry. 1998. SAPK2/p38-dependent F-actin reorganization regulates early membrane blebbing during stress-induced apoptosis. *J Cell Biol* **143**: 1361–1373.

[67] Totsukawa, G., Y. Yamakita, S. Yamashiro, D. J. Hartshorne, Y. Sasaki, and F. Matsumura. 2000. Distinct roles of ROCK (Rho-kinase) and MLCK in spatial regulation of MLC phosphorylation for assembly of stress fibers and focal adhesions in 3T3 fibroblasts. *J Cell Biol* **150**: 797–806.
[68] Totsukawa, G., Y. Wu, Y. Sasaki, D. J. Hartshorne, Y. Yamakita, S. Yamashiro, and F. Matsumura. 2004. Distinct roles of MLCK and ROCK in the regulation of membrane protrusions and focal adhesion dynamics during cell migration of fibroblasts. *J Cell Biol* **164**: 427–439.
[69] Yamaji, S., A. Suzuki, H. Kanamori, W. Mishima, R. Yoshimi, H. Takasaki, M. Takabayashi, K. Fujimaki, S. Fujisawa, S. Ohno, and Y. Ishigatsubo. 2004. Affixin interacts with alpha-actinin and mediates integrin signaling for reorganization of F-actin induced by initial cell-substrate interaction. *J Cell Biol* **165**: 539–551.
[70] Bailly, M., J. S. Condeelis, and J. E. Segall. 1998. Chemoattractant-induced lamellipod extension. *Microsc Res Tech* **43**: 433–443.
[71] Ghosh, M., X. Song, G. Mouneimne, M. Sidani, D. S. Lawrence, and J. S. Condeelis. 2004. Cofilin promotes actin polymerization and defines the direction of cell motility. *Science* **304**: 743–746.
[72] Horstman, D. A., K. DeStefano, and G. Carpenter. 1996. Enhanced phospholipase C-gamma1 activity produced by association of independently expressed X and Y domain polypeptides. *Proc Nat Acad Sci USA* **93**: 7518–7521.
[73] Theriot, J. A., and T. J. Mitchison. 1991. Actin microfilament dynamics in locomoting cells. *Nature* **352**: 126–131.
[74] Azuma, T., W. Witke, T. P. Stossel, J. H. Hartwig, and D. J. Kwiatkowski. 1998. Gelsolin is a downstream effector of rac for fibroblast motility. *EMBO J* **17**: 1362–1370.
[75] Yamazaki, D., T. Fujiwara, S. Suetsugu, and T. Takenawa. 2005. A novel function of WAVE in lamellipodia: WAVE1 is required for stabilization of lamellipodial protrusions during cell spreading. *Genes Cells* **10**: 381–392.
[76] Gov, N. S., and A. Gopinathan. 2006. Dynamics of membranes driven by actin polymerization. *Biophys J* **90**: 454–469.
[77] Wolgemuth, C. W. 2005. Lamellipodial contractions during crawling and spreading. *Biophys J* **89**: 1643–1649.
[78] Ponti, A., M. Machacek, S. L. Gupton, C. M. Waterman-Storer, and G. Danuser. 2004. Two distinct actin networks drive the protrusion of migrating cells. *Science* **305**: 1782–1786.

8

Tensegrity as a Mechanism for Integrating Molecular and Cellular Mechanotransduction Mechanisms

Donald E. Ingber

8.1 Introduction

Large-scale mechanical forces due to gravity, movement, air flow, and hemodynamic forces have been recognized to be important regulators of tissue form and function for more than a century. The effects of compression on bone, tension on muscle, respiratory motion on lung, and shear on blood vessels are a few prime examples. Although interest in mechanics waned when biology shifted its focus to chemicals and genes in the middle of the last century, there has been a recent renaissance in the field of mechanical biology. Physical forces are now known to be key regulators of virtually all facets of molecular and cellular behavior, as well as developmental control and wound repair [1]. Impaired mechanical signaling also underlies many diseases, and numerous clinical therapies utilize mechanical stimulation to produce their healing effects [2]. However, we still do not fully understand "mechanotransduction" – the process by which individual cells sense and respond to mechanical forces by altering biochemistry and gene expression.

To understand how individual cells respond to mechanical forces, we need to define how stresses are borne and distributed within cells, as well as how they are focused on critical molecular elements that mediate mechanochemical conversion. Early studies assumed that mechanotransduction might be mediated through generalized deformation of the cell's surface membrane that produced changes in membrane-associated signal transduction. Although this may occur, it is now clear that multiple molecules and structures distributed throughout the membrane, cytoplasm, cytoskeleton, and nucleus contribute to the mechanotransduction response that governs cell behavioral control [1, 3, 4]. Thus, it is critical that we understand how cells are structured so that mechanical stresses are channeled and focused simultaneously on the various key conversion molecules and structures that mediate the transduction response.

Another critical challenge is to understand mechanoregulation in the physiological context of the whole organism. Most research in this area involves applying exogenous mechanical loads (e.g., fluid shear, tension, compression, and strain) to isolated tissues or cultured cells. But cells experience forces as part of larger organ structures with highly defined architectures and tissue-specific material properties in our bodies. Multicellular organisms are constructed as tiers of systems within a system within a system: The various organs (bone, muscle, blood vessels, and the nervous system and the interconnecting tendons, ligaments, joints, and fascia) are constructed from tissues (e.g., muscle fibers, vascular endothelium, nerves, and connective tissues), which are composed of groups of living cells and their associated extracellular matrices (ECMs). Each cell contains intracellular organelles, a nucleus, membranes, and a filamentous cytoskeleton permeated by a viscous cytosol. Each of these subcellular components is, in turn, composed of different molecules organized in distinct three-dimensional arrangements. These structural hierarchies are unique in that mechanical deformation results in coordinated structural rearrangements on many different size scales – from the whole organ to the molecular level. Thus, the question of how the body senses and responds to mechanical stresses is not simply an issue of the material properties of its components, or of the chemical activities of specific molecular transducers; rather, it is a problem of architecture.

Finally, most investigators assume that any change in cell metabolism or behavior that follows the application of a load is a direct result of the mechanical stimulus. However, all cells actively generate their own internal mechanical (tensional) forces within actomyosin filaments in their cytoskeletons, and hence the reality is that any applied stress is imposed on a preexisting force balance. These cytoskeletal forces are distributed to virtually all of the organelles in the cytoplasm, as well as to the central nucleus and overlying surface membrane, where they concentrate on specialized anchoring sites that join the cytoskeleton to ECM scaffolds and junctional complexes that hold cells together within tissues. These external adhesion sites resist cell traction forces and thereby place the whole cell, tissue, and organ, as well as key load-bearing molecules, in a state of isometric tension or "prestress" before any external load is applied. It is therefore impossible to understand mechanotransduction without taking into account these internal forces, just as it is not possible to determine the tone generated by a violin string when it is strummed without knowing the level of tension in the wire.

In this chapter, I briefly summarize work from my laboratory and others carried out over the past 25 years that has focused on the possibility that our bodies, organs, tissues, cells, and molecules use a tension-dependent form of architecture known as "tensegrity" to stabilize themselves. This fundamental design principle facilitates force focusing and channeling across multiple size scales, and thereby orchestrates and tunes the mechanotransduction response. Because of the structural complexity of living materials, and the fact that bodies are neither homogeneous nor isotropic, cells and tissues cannot easily be described using existing, conventional continuum-based engineering models. Importantly, computational models based on tensegrity are beginning to offer new and more effective methods to simulate cell mechanical behavior, while providing the novel ability to relate mechanics to specific structural elements at the molecular scale. Pursuing mechanotransduction from this architectural perspective has drastically changed our view of cell structure. It also has led to novel insights into how cells and tissues integrate structural and information processing networks, and thereby control cell behaviors that are critical for tissue formation, maintenance, and repair.

8.2 The Mechanotransduction Problem

Most scientific publications that focus on mechanotransduction include verbiage in their introduction along the lines of "little is known about how cells convert mechanical signals into a chemical response," or "the molecular basis of mechanochemical conversion is poorly understood." Although this was true 20 years ago, we now know of numerous molecules that change their activities when cells are mechanically loaded, as described in recent reviews [1–5] and in the other chapters of this book. For example, stress-activated (SA) ion channels are ubiquitous mediators of mechanoelectrical conversion that alter ion transport when cells are mechanically stressed, and much is known about how this mechanical gating occurs at the molecular level [5]. Integrin receptors that mediate cell adhesion to the ECM also have been shown to function as mechanoreceptors by providing a preferential path for the transfer of mechanical forces across the cell surface and to the cytoskeleton [6, 7]. Forces transmitted across integrins become focused on specialized cytoskeletal anchoring complexes, known as "focal adhesions," that link clustered integrin receptors to the actin-microtubule-intermediate filament lattice of the deep cytoskeleton [4]. Importantly, many signal transduction molecules that mediate mechanochemical conversion (e.g., Src, FAK, growth factor receptors, and small and large G-proteins) are oriented on the cytoskeletal backbone of the focal adhesion [8, 9], and some change their chemical activities when they experience stresses that are transmitted through these structural linkages [4, 10–14]. For this reason, the focal adhesion is now thought of as a "mechanosensory organelle" [3, 4, 15].

But the mechanotransduction process is more complex that this. Mechanically sensitive cellular components other than SA channels, integrins, and focal adhesion molecules include caveolae, cadherins, growth factor receptors, myosin motors, cytoskeletal filaments, nuclear ion channels, chromatin, and nucleoli, among others [1]. Conventional engineering models of the cell assume that the cytoskeleton behaves as if it were a mechanical continuum, such as a viscoelastic gel, which would distribute forces homogeneously and exhibit a rapid drop-off of force with distance in the cell. But the reality is that forces transferred over integrins to the focal adhesion are channeled along the discrete load-bearing elements of the filamentous cytoskeleton and are thereby distributed simultaneously to many sites in the cell. For example, when mechanical stresses are applied locally to integrins on one side of the cell membrane, they are transmitted over filamentous cytoskeletal connections to the nucleus and to the opposite pole of the cell, and these stresses can sometimes be concentrated locally at these distant sites [16–18]. This behavior can explain why many different molecules and structures alter their form and function when mechanical forces are applied to cells.

The other key aspect of the mechanotransduction problem that is largely ignored in most experimental studies is that cells within the depths of our tissues and organs constantly sense and respond to physical forces that are exerted at the level of our whole bodies (e.g., due to gravity or movement). Thus, we need to develop a unified model of cellular mechanotransduction that can explain how single cells can sense and respond similarly to mechanical stresses, whether isolated in a dish or living as part of a complex hierarchical organ structure.

8.3 Tensegrity

In the early 1980s, we first proposed that cells and tissues might use "Tensegrity" architecture to structure themselves so that they can optimally sense and respond to physical forces, and thereby integrate structure and function during development and throughout adult life [19–21]. This building system depends on tensional integrity, rather than a continuous transmission of compressive forces, for its mechanical stability [22]. Since that time, we and others have experimentally confirmed that cells are structured according to the rules of tensegrity, and that computational tensegrity models predict many mechanical behaviors of cells [1, 23]. In the following I briefly summarize the principles that underly this form of architecture and provide evidence indicating its use by cells, as well as by smaller and larger structures in the hierarchy of life. In the following section, I describe how the use of this building system orchestrates cellular mechanotransduction and provides a mechanism to integrate mechanical signal processing across all multiple size scales.

8.3.1 Tensegrity Architecture

Tensegrity is a form of architecture that describes structures that self-stabilize and resist shape distortion because they are composed of opposing tension and compression elements that balance each other and thereby create an internal prestress or resting state of isometric tension. These structures are distinct from most elastic materials (e.g., rubber, polymers, and metals) that do not require prestress for their mechanical stability, and they differ from compressionally prestressed structures (e.g., prestressed concrete). Artists and architects have long recognized that tensegrity is used at the macro scale, for example, in the way pulling forces that are generated in muscles and resisted by bones generate isometric tension (muscle tone) that stabilizes the shape of our bodies.

The simplest embodiment of a tensegrity force balance can be seen in the sculptures of Kenneth Snelson, which are composed of a tensed network of high-tension cables that are interconnected by a series of isolated compression struts in the form of free-standing metal columns (Figure 8.1). The tension in the cables pulls in on the ends of each column, thereby compressing the struts, whereas the struts push out and tense the interconnected web of cables. Prestressed tensegrity structures of this type develop a restoring stress when exposed to external mechanical loads primarily through geometric rearrangements of their pre-tensed members. The greater the

Figure 8.1. A tensegrity sculpture ("needle tower") by Kenneth Snelson composed of aluminum struts suspended by tensed stainless steel cables. This structure is composed of multiple tensegrity modules that are interconnected by similar rules.

prestress carried by these members, the less rearrangements they undergo, and, hence, the more rigid they become. This feature leads to one of the hallmark properties of tensegrity structures: Structural rigidity (stiffness) of the structure is proportional to the level of prestress that supports it [24, 25].

Part or all of the pre-tension in tensed members of a tensegrity is balanced by a subset of other structural elements within the structure that resist being compressed. However, it is critical to define precisely which members are "internal" or "external" to a tensegrity structure. Tensegrity structures must exhibit the ability to stabilize themselves, and thus all members required for self-stability are "internal" elements of the tensegrity, regardless of their relative positions. For example, a spider web is a pure tensile cable net in the absence of adhesion to stiffer anchoring structures (e.g., blades of grass and tree branches); however, it only exhibits self-stability when the tensed silk cables are balanced and prestressed by these ostensibly external anchoring sites. In other words, a detached silk web that exhibits no shape stability is not a tensegrity structure, whereas a tensionally stabilized spider web combined with its stiffer anchoring components comprise a tensegrity. In more general terms, a portion of the prestress in a tensegrity may be borne by its adhesion sites at points where the structure attaches to external scaffolds that themselves resist shape distortion. Importantly, these external resistance sites may be rigid or elastic

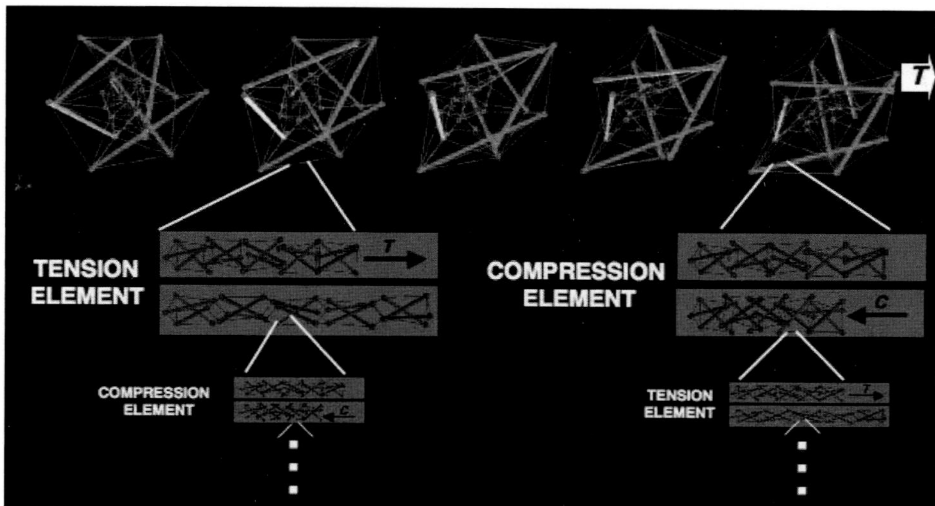

Figure 8.2. Computer models depicting multiscale structural rearrangements with a prestressed tensegrity hierarchy. Top: A two-tier hierarchical tensegrity composed of concentric larger and smaller spherical (polyhedral) structures composed of struts and cables connected by tension elements. Note that the structure exhibits coordinated structural rearrangements of its internal elements as it extends to the right in response to tension (*T*) application (movies showing dynamic movements in tensegrities can be seen at: www.childrenshospital.org/research/Site2029/mainpageS2029P23sublevel24.html). Lower panels show how individual struts and cables of the structure may themselves be organized as compressed (*C*) and tensed (*T*) tensegrity mast structures at smaller and smaller size scales ad infinitum. A stress applied at the macroscale will result in global rearrangements at multiple size scales, rather than local bending or breakage, as long as tensional integrity and stabilizing prestress are maintained throughout the hierarchical network. Reprinted with permission from [1].

materials, or they may be other prestressed networks at smaller and larger size scales within tensegrity-based structural hierarchies (Figure 8.2) [1, 21–23].

In both single and hierarchical tensegrities containing multiple linked structural modules stabilized by the same rules, a local stress is borne by many elements that rearrange at multiple size scales, rather than deform or yield locally (Figure 8.2). These geometric rearrangements occur without compromising the physical integrity of the structure, and instead, the whole network strengthens when it is mechanically stressed [6, 23]. Specifically, as the applied stress is raised, the stiffness of these structures increases in direct proportion (i.e., they exhibit "linear stiffening behavior") as more and more of the stiff elements rearrange along the main axis of applied tension [6, 24, 25]. Because loads are borne by networks composed of multiple support elements in tensegrities, mechanical forces also are transmitted along specific paths that extend across multiple size scales inside these structures, and these stresses can be focused or concentrated on distant sites based on the geometric forms of these discrete structures (e.g., where multiple elements intersect) [3, 23]. Moreover, the efficiency of long-distance force transfer will be determined by the level of the prestress in a tensegrity structure.

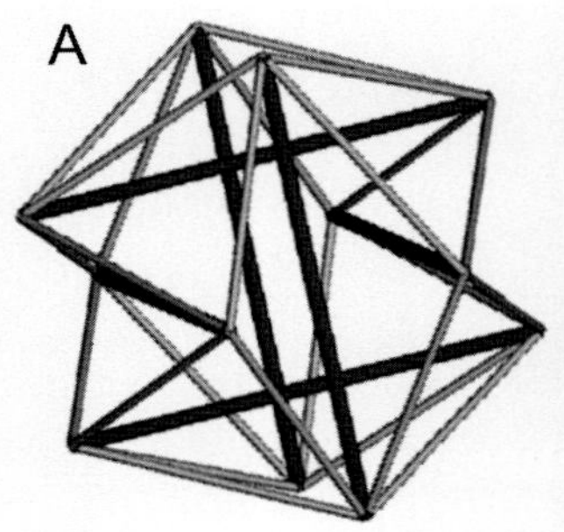

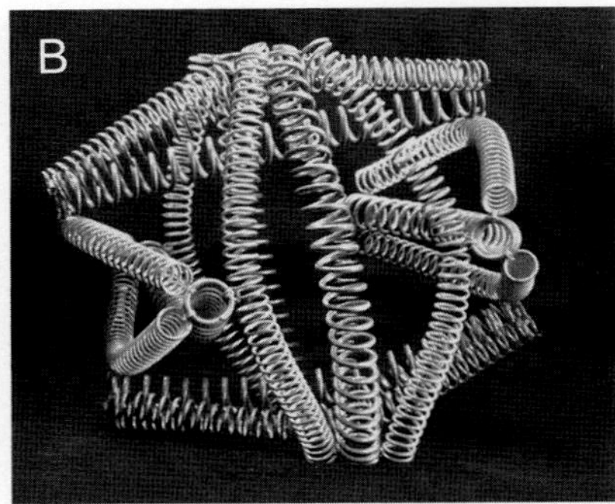

Figure 8.3. A spherical tensegrity configuration composed of 6 struts and 24 cables may be constructed with struts and cables (A), or using springs with different elasticities (B).

Tensegrities do *not* have to be built from strings and struts; for example, we have constructed tensegrities composed entirely of metal springs (Figure 8.3). The only requirement is that a subset of components are stiff enough so that they can resist the level of compressive forces exerted on them by the surrounding tensed elements; the lower the compressive force, the more flexible these "compression-resistant" members may be. Thus, the most important concept to fall out of the pursuit of the tensegrity model is that *prestress is the key unifying principle that governs the shape stability* of this class of structures.

8.3.2 Cells as Tensegrity Structures

We first proposed that cells are tensegrity structures based on the observation that living cells actively generate tensile forces in their contractile actomyosin cytoskeleton, and that the overall shape of the cell is governed by the ability of the ECM substrate to resist being compressed [19, 20]. In early studies, we demonstrated that spherical tensegrities built with sticks and elastic strings mimicked cell spreading on an ECM substrate, as well as the cell retraction and rounding observed when the cell–ECM adhesions are dislodged (e.g., during trypsinization) [20, 21]. A similar three-dimensional cell model containing a tensegrity nucleus composed of sticks and elastic strings connected to the model's surface by additional tensile connectors also predicted the cell and nuclear shape will change in a coordinated manner, and the nucleus will polarize (move closer to the basal cell membrane), when cells spread (Figure 8.4). Living cells were shown experimentally to exhibit these same behaviors when they adhere to and spread on an ECM substrate [26–28]. We also showed that membrane-permeabilized cells exhibit similar dynamic structural couplings between cell, cytoskeletal, and nuclear shape when levels of tension are changed in their cytoskeletons by directly altering actomyosin–filament interactions [29]. Thus, the cytoskeleton of living cells that govern cell shape behaves as if it were a tensegrity structure.

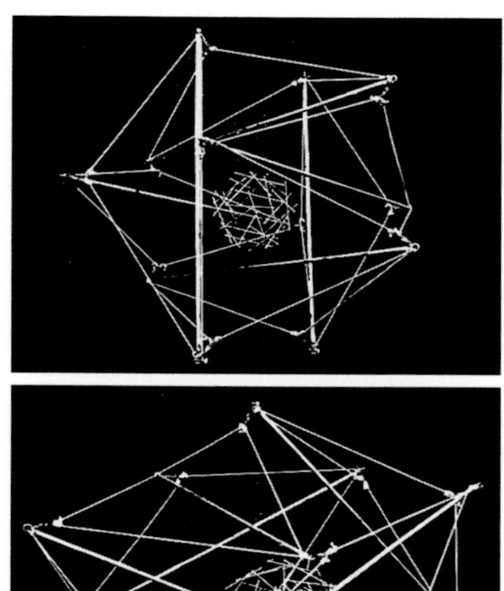

Figure 8.4. A tensegrity model of a nucleated cell in two shape configurations. The cell and nucleus are round when the cell is unanchored and round (top), whereas they both spread in a coordinated manner and the nucleus polarizes toward the base when the cell model is anchored to a rigid substrate (bottom). The cell model is composed of metal struts and elastic cord; the nucleus contains sticks and elastic strings; and the nucleus is coupled to the cell surface by black tensile cables (not visible). In this cell model, the large struts conceptually represent microtubules; the elastic cords correspond to microfilaments and intermediate filaments that carry tensional forces in the cytoskeleton; and the substrate represents the ECM. Reprinted with permission from [21].

To explore the tensegrity hypothesis further, we carried out mechanical engineering analysis at the single cell level. If cells are tensegrity structures, then mechanical forces will not be transmitted equally at all points across their surface membrane. Instead, these forces should be preferentially transmitted to the internal tensed support network (cytoskeleton) at points where it is physically coupled to the anchoring substrate, much like how forces are preferentially transmitted across the anchoring pegs of a tent relative to other points on its canvas membrane. To test this prediction, we allowed cultured cells to bind to ferromagnetic microbeads (1- to 10-μm diameter) coated with ligands for different transmembrane receptors, and then we applied magnetic fields to magnetically twist the beads in place or to pull on them while still bound on the cell surface. By measuring the rotation or displacement of the receptor-bound beads, respectively, in response to varying levels of applied stress, we were able to measure stress–strain curves in single living cells [6, 12, 30–33]. Other groups have executed similar measurements using optical tweezers to pull on receptor-bound beads [7].

Using this approach, we and others have unequivocally demonstrated that transmembrane ECM receptors, known as integrins, function as mechanoreceptors in that they are among the first molecules to sense an applied mechanical stress, and they transmit this mechanical signal across the cell surface and to the cytoskeleton in a specific manner [6, 7, 30]. Integrins mechanically couple to the cytoskeleton by forming specialized cytoskeletal complexes, called "focal adhesions," that contain proteins (e.g., talin, vinculin, paxillin, and α-actinin, among others) that bind to each other as well as to both the cytoplasmic domain of integrins and to actin filaments [4, 8, 9]. Importantly, when similar stresses were applied to other receptors (e.g., metabolic receptors, growth factor receptors, and histocompatibility antigens) that also spanned the plasma membrane, but failed to form focal adhesions and tightly couple to the deep tensed cytoskeleton, little stiffening behavior was observed [6, 34, 35]. Moreover, cells that lacked the focal adhesion protein vinculin failed to mount an effective stiffening response [31, 36], as did cells in which similar stresses were applied to integrins using nonactivating ligands (specific antibodies) that ligate the receptors but fail to induce focal adhesion assembly [12]. The normal mechanical response to stress applied to activated integrins also could be inhibited by disrupting any one of the three major cytoskeletal filament systems of the cell–actin microfilaments, microtubules, intermediate filaments – and all three needed to be disrupted simultaneously to fully suppress this stiffening response [6]. Thus, these three filament systems interact mechanically with each other and with integrins (and ECM) in living cells.

Importantly, these studies confirmed that when the level of applied stress is raised, cell stiffness increases in direct proportion, a result that was predicted by stick and string tensegrity models [6]. Later analysis of mathematical tensegrity models developed from first mechanical principles revealed that this property is a fundamental feature of tensegrity structures [24, 25, 37, 38]. Cytoskeletal stiffness also was shown to rise immediately when internal prestress was raised by increasing actomyosin-based contraction or rapidly applying mechanical strain to cells, whereas it decreased when internal tension was dissipated [17, 39–42]. Permeabilized cells that lost the integrity of their surface membranes also exhibited similar mechanical behavior [29, 30], hence clearly demonstrating that these novel mechanical properties of cells are due to the physical behavior of their tensed internal cytoskeleton, and not to surface membrane tension.

In addition, application of mechanical stresses directly to cell surface integrin receptors using magnetic beads or glass microneedles coated with integrin ligands was shown to produce long-range force transfer inside the cell. Pulling on surface integrin receptors induced immediate structural rearrangements in the cytoskeleton deep within the cytoplasm, as well as inside the nucleus and nucleolus, in living cells [17, 43]. More recent studies using a three-dimensional mapping technique have confirmed that stresses applied to integrins are not only transmitted throughout the cell, but they are concentrated at distant sites [18, 44–46]. Force transfer in the cytoplasm can be inhibited by either dissipating tensile prestress in the cell or by disrupting the discrete cytoskeletal networks that carry these loads (Figure 8.5) [17, 18, 43–47].

One of the most fundamental properties of living cells is their ability to switch between solid- and fluid-like behaviors. When cells need to crawl, divide, or invade ECMs, they must modify their cytoskeleton to become highly deformable and almost fluid-like, whereas to maintain stable tissue form when mechanically stressed, the cytoskeleton must behave like an elastic solid. These responses have been generally assumed to be due to controlled cycles of polymerization and depolymerization of the cytoskeleton that lead to "sol–gel" transitions. However, cells can change shape from round to spread without altering the total amount of microfilament or microtubule polymers in the cell [48, 49]. Individual cytoskeletal filament bundles

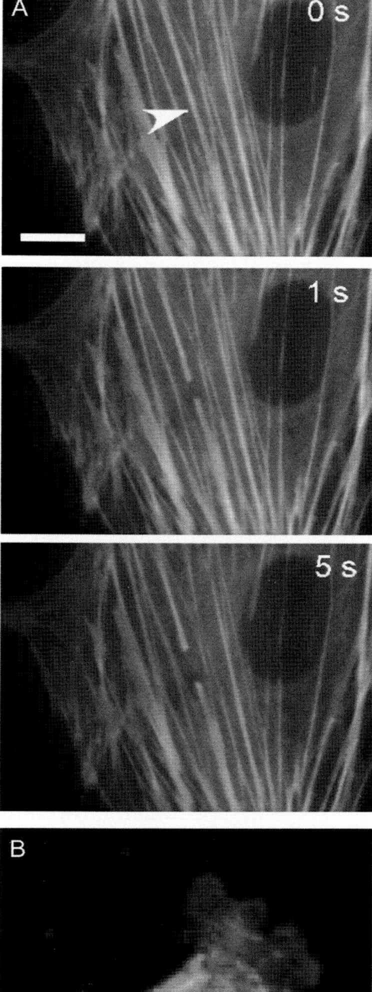

Figure 8.5. Prestress in the cytoskeleton revealed by incising single stress fibers using a laser nanoscissor in living cells expressing EYFP-actin. (A) When the center of a single stress fiber is cut using a femtosecond laser, both ends spontaneously retract within seconds, thus revealing pretension in these cytoskeletal elements (bar, 10 mm). (B) Overlay of fluorescence microscopic images showing the effects of incision of a single stress fiber on cytoskeletal organization and global shape in a cell anchored on a flexible polyacrylamide ECM substrate (stiffness = 3.75 kPa). Note that stress fiber incision resulted in global cytoskeletal rearrangements, including outward translation of the whole cell and cytoskeleton along the main axis of the cut fiber. The two vertical white lines indicate a vertically oriented stress fiber located many micrometers away from the site of incision in the right portion of the cytoskeleton that undergoes large-scale lateral displacement in response to stress fiber ablation (bar, 10 μm). Reprinted with permission from [50].

also can remain structurally intact for periods of minutes to hours, even though individual molecule subcomponents bind and unbind over faster time frames [50]. In fact, rapid waves of new actin filament polymerization and depolymerization are largely restricted to small regions of the cell that undergo rapid extension and retraction, such as within ruffles at the leading edge of actively migrating cells.

It is now clear that mechanical state transitions in cells are also governed by the passive material properties of a prestressed cytoskeleton. For example, rheological measurements of living cells using various methods have revealed that the mechanical behavior of living cells conforms to a weak power law that depends on both the frequency and time of loading [51]; moreover, this power-law dependence is inversely related to cytoskeletal prestress [25, 52–54]. This is important because the power-law behavior is directly related to cell deformability: The cell behaves as a solid when the power-law exponent approaches zero, and as a fluid when it approaches one. Thus, cells apparently use mechanical prestress – the fundamental principle of tensegrity architecture – to regulate their transition between elastic and fluid behaviors [52]. Internal prestress is also now recognized to be absolutely critical for the mechanical stability of all cell types, including erythrocytes, which lack nuclei or a well-developed internal cytoskeleton [55, 56]. Prestress is even required for artificial cytoskeletal filament gels to mimic the mechanical behavior of living cells [57, 58]. Taken together, these findings clearly show that cells do not behave like simple viscoelastic or elastic gels, and they do behave as tensegrity structures.

8.3.3 Visualizing Tensegrity at the Molecular Level in Living Cells

The experimental observations described in the previous section confirm that prestress is fundamental to cell and cytoskeletal shape stability, and hence they provide strong general support for the cellular tensegrity theory. However, the greatest value of the tensegrity model lies in its potential ability to define the molecular basis of cell mechanical behavior. The tensegrity model predicts that the cytoskeletal prestress that stabilizes global cell shape is generated both actively by the cell's contractile apparatus through the action of motor proteins, and passively by physical distension of the cell due to its adhesions to a distended ECM. The cell, cytoskeleton, and ECM are effectively one prestressed, interconnected, structural network.

More specifically, the tensegrity model assumes that basal actomyosin-containing "stress fibers" contribute to cell shape control by generating tensional forces, transmitting them to the remainder of the entire cytoskeleton and underlying ECM, and bringing these forces into balance. A major limitation in evaluating this prediction is that it has not been possible to analyze the load-bearing properties of individual stress fibers in living cells. We recently confirmed that actin stress fibers are prestressed in living cells by using a femtosecond laser to vaporize small regions of these cytoskeletal bundles in cells expressing GFP-actin that make these filaments fluorescent [50].

First, using the fluorescent recovery after the photobleaching (FRAP) technique, we found that these stress fibers are extremely stable over extended periods of time (hours), even though individual fluorescent actin monomers continuously bind and unbind to these filaments. So cytoskeletal structures can carry significant mechanical loads over extended periods of time and function like "hard-wiring" in living cells, as tension cables and struts do in tesegrity structures.

Second, when we made a 300-nm punch hole with the laser in a single stress fiber, the round hole rapidly deformed and took on an elliptical form, revealing the internal prestress that preexisted in the intact structure, and when the whole stress fiber was cut through completely, both ends retracted spontaneously (Figure 8.5(A)), much like when a muscle is surgically incised and its cut ends retract toward its insertion sites. Most importantly, when similar studies were repeated in cells cultured on compliant ECM gels that exhibited flexibilities closer to those of living materials, cutting a single stress fiber resulted in the global rearrangements of the entire actin cytoskeleton as well as a significant change in the overall shape of the cell (Figure 8.5(B)). Thus, living cells clearly do use the ECM as an external compression element to resist actomyosin-based tensile forces and thereby generate the prestress that stabilizes cell and cytoskeletal form. The tensile cell and its adhesions to a compression-resistant ECM that are required for shape stability therefore comprise a self-stabilizing tensegrity system, similar to a spider web and its anchors to blades of grass.

But do cells also possess internal compression struts? Many critics of cellular tensegrity feel that cells must contain linear struts inside the cell for this model to have validity [59]. However, rigid compression struts are *not* a prerequisite for a tensegrity structure. For example, a balloon is a tensegrity because it uses isolated compression-resistant elements to bring membrane tension into balance and prestress the entire structure; however, thousands of independent gas molecules that cannot pass through its porous elastic membrane serve this role in this tensegrity [22]. Nevertheless, we and others have found that cells sometimes do use internal molecular filaments as discrete compression struts.

In early formulations of the tensegrity model, we proposed that microtubules might act as compression elements in cells [20, 21] because, as hollow polymers with a larger diameter (25 nm vs 7–9 nm for actin filaments and intermediate filaments), they should be more rigid than the other cytoskeletal biopolymers. Microtubules have a bending rigidity approximately 100 times that of actin filaments and a persistence length on the order of millimeters [60]. Yet, cytoplasmic microtubules are often highly curved, so that they appear to be buckled due to compressive loads in living cell. But *in vitro* studies have revealed that isolated microtubules exhibit buckling at much larger length scales, and that they can only withstand exceedingly small compressive forces. Thus, this discrepancy brought the tensegrity model into question.

Studies with various cells expressing GFP-tubulin have demonstrated that individual growing microtubules buckle when they impinge end-on on other cellular structures or the cortical membrane [17, 61]. Thus, at least a subset of microtubules are loaded compressively within living cells. But, more recently, we were able to demonstrate that individual intracellular microtubules can bear much higher compressive loads then ever thought possible, and that most microtubules bear compression in cells; however, their buckling wavelength is reduced significantly because of mechanical coupling to the surrounding elastic network [62]. Quantitative analysis and physical modeling studies revealed that individual microtubules can bear levels of compressive force that are 100 times greater in whole cells than *in vitro* because of these lateral connections. Microtubules can therefore make a more significant structural contribution to the mechanical behavior of whole cells than previously thought possible because they are positioned within a larger tensegrity-based cytoskeletal network. So cells appear to stabilize themselves much like a tensegrity-based camping tent that uses both tent pegs in the ground (ECM adhesions) and tent poles (microtubules) to prestress and thereby stabilize its surface membrane, as well as its entire internal framework.

The tensegrity force balance that endows these structures with self-stabilizing properties suggests that internal microtubule struts and the cell's anchors to external ECM resistance sites should function in a complementary manner. For example, if an anchored cell is mechanically equilibrated, then detachment of its ECM anchors should transfer the unbalanced forces onto the cell's internal compression struts (i.e., microtubules), whereas disruption of these supporting struts should transfer increased loads onto remaining external adhesions. In fact, this exact type of force balance has been demonstrated experimentally in various cell types [63–66]. This can be seen when microtubules are disrupted in cells attached to flexible ECM substrates, as this results in increased traction on the cell's adhesions [17]. Normally curved microtubules also straighten when tension is dissipated or the membrane is pulled outward, whereas microtubule buckling increases when contractility is increased or the membrane is compressed (e.g., using a micropipette) [17, 62]. The relative mechanical contribution of microtubules to cytoskeletal prestress also increases as the number of ECM adhesions is lowered and these internal struts bear more of the compressive load [67]. Furthermore, if cells use this tensegrity balance, then when the cell forms a new ECM adhesion near the end of a microtubule that is polymerizing outward and pushing on the surface membrane, some of the compressive load should be transferred from the end of the microtubule to the external adhesive substrate. Because microtubules polymerize when they are decompressed, tension application to ECM adhesions should therefore facilitate microtubule polymerization and promote cell outgrowth, and this is exactly what is observed in cultured nerve cells [63, 64].

In summary, these data, combined with results from many other studies that have been reviewed elsewhere [1, 21, 23], provide convincing evidence that shows that living cells use the tensegrity principle to mechanically stabilize their cytoskeletons and control their overall three-dimensional form. In all cells studied, shape stability is governed by prestress, which in turn is determined through a balance between internal tensile and compressive forces.

8.3.4 Tensegrity-Based Structural Hierarchies

Tensegrity systems may be constructed as structural hierarchies in which the tension or compression elements that comprise the structure at one level are tensegrity systems composed of multiple components on a smaller scale (Figure 8.2) [22]. The "stick-and-string" tensegrity model of a nucleated cell, in which the entire nuclear tensegrity lattice is itself a tension element in the larger structure, provides a tangible example that shows how deformation at the macro scale translates into geometric rearrangements of structural members at multiple size scales (Figure 8.4). All organs are similarly organized as structural hierarchies composed of tissues, cells, multimolecular structures, and individual molecules on smaller size scales [1, 23, 68].

Importantly, whole organs and tissues exhibit immediate mechanical responsiveness and increase their stiffness in direct proportion to the applied mechanical stress [69], as do both living cells and tensegrity structures [6, 24, 25, 37, 38]. However, this prestress that governs the mechanical behavior of living materials arises differently in different organs.

In the musculoskeletal system, for example, prestress is generated as a result of a balance between muscle contractility and the ability of a rigid bone matrix to resist these forces. Cartilage imparts pretension by using osmotic swelling forces to push out against its nonextensible collagen lattice and to oppose the contractile forces of its constituent cells. In the lung, residual air pressure left at the end of expiration tenses and stiffens the basement membranes, collagen fibers, and elastin bundles that surround each alveolus, and resist surface tension forces acting on the epithelium, thereby stabilizing the alveoli in an open form [70].

At a smaller size scale, the shape stability of individual muscles, epithelial tissues, blood vessels, and nerves requires that a similar force balance must be established between tractional forces exerted by the cells and resisting forces that are exerted by stiffened ECMs, neighboring contractile parenchymal cells, surrounding connective tissue cells, and other microenvironmental forces (e.g., gravity, movement, and hemodynamic stresses). Tissues adapt to stresses applied at the macro scale and protect themselves against injury by rearranging on many size scales, including altering the positions of the molecular components that comprise these tensed ECMs and interconnected cytoskeletal elements inside individual cells [71, 72]. For example, compression in tissues such as cartilage results in the reorientation of proteoglycans throughout the ECM network [73], in addition to changes in cell shape, cytoskeletal organization, and nuclear form within individual cells [74]. Forces due to breathing similarly produce complex micromechanical responses in lung parenchyma, including lengthening and shortening (tension and compression) of alveolar walls by extension and linearization of some collagen fibers on inspiration, as well as buckling of the same fibers on expiration [75]. It also causes the lateral intercellular spaces between the epithelial cells to reversibly shrink and expand without compromising the structural integrity of the tissue [76].

The point here is that because they are prestressed, the ECMs that hold cells together within tissues react immediately as a tensionally integrated structural network that is both hierarchical and multimodular when the whole organ is stressed. Integrins connect these tensed ECMs to the prestressed cytoskeleton, which in turn responds mechanically to forces transferred over these connections by rearranging its interlinked actin microfilaments, microtubules, and intermediate filaments, as well as associated organelles and nuclei, on even smaller scales. These multiscale rearrangements stiffen and strengthen the whole cell so as to minimize mechanical injury [6, 12, 17, 23, 32, 43, 71].

The cortical cytoskeleton that underlies the cell's surface membrane is yet another prestressed network that is composed of actin filaments linked by large spectrin molecules arranged in a triangulated geodesic array. Computational models of the red blood cell membrane that incorporate the tensegrity principle have revealed that this structure mechanically stabilizes itself because the rigid actin protofilaments are held in place by multiple spectrin molecules that are suspended from the overlying, noncompressible lipid bilayer, and that act like tensed cables, thereby creating a tensegrity force balance [55].

On a smaller size scale, nuclear lamins and linked nuclear pore complexes form a spherical lattice at the cell's center that is pretensed due to the pull of surrounding cytoskeletal filaments in spread cells, and by the swelling forces of folded chromatin in round cells [23, 43, 77]. Although the mitotic spindle lacks a nuclear membrane, it is also stabilized through the establishment of a mechanical force balance between compressed microtubules that push out against a tensed network of chromosomes interlinked by nuclear scaffolds [16, 78].

These different subcellular tensegrity structures (e.g., microfilament-microtubule-intermediate filament lattice, submembranous cytoskeleton, and nucleus) may act independently if structural connections are weak or not present (e.g., in the absence of focal adhesion formation). But when they are mechanically coupled, they function as one integrated, hierarchical tensegrity system. For example, application of mechanical stresses to cell surface ECM receptors produces immediate changes in nuclear shape as well as nucleolar organization on progressively smaller size scales [43]. This hard-wiring between integrins and the nucleus is largely mediated by intermediate filaments that connect these structures, and to a lesser degree by the actin cytoskeleton [43, 47]. Importantly, the efficiency of this multiscale mechanical response is once again governed by cytoskeletal prestress [44], as predicted by the tensegrity model.

Cells also use tensegrity to stabilize specialized subcellular structures. For instance, stick and string tensegrity models can predict how tensed actin nets can transform between linear bundles and actin geodesic domes in living cells, and the predictions of these models exhibit strut-for-strut and vertex-for-vertex identities with native actin structures at the nanometer scale [21, 82]. In filopodia of migrating cells, stiffened bundles of cross-linked actin filaments push out against the tensed surface membrane [79] and thereby create another tensegrity force balance. Cross-linked microtubule bundles similarly stabilize the shape of membrane cilia, as well as long cell processes in nerve, epithelial, and endothelial cells [63, 80, 81]. Note, however, that the same molecular element (e.g., actin filament or microtubule) can act as either a compression strut (when cross-linked into stiffened bundles) or a tension cable at different times or at different sites in the same cell (e.g., actin fibers within contractile microfilaments and spindle microtubules that pull on chromosomes during their segregation) [82].

Prestress is also used for shape stability at even smaller size scales. Viruses [83], microfilament polymers [84], and lipid micelles [85, 86], as well as individual proteins and RNA and DNA molecules, all have been described as prestressed tensegrity structures [68, 86–88]. In proteins, for example, stiffened peptide domains (e.g., α-helices and β-strands) act locally to resist inward-directed compressive forces generated by attractive (tensile) intramolecular bonding interactions. Thus, a protein is analogous to a tensegrity composed entirely of springs that have different elasticities (Figure 8.3), except that intramolecular binding forces obviate the need for physical tensile connections in molecules. Tensegrity-based molecular models are now being used to understand and predict the unfolding behavior of proteins subjected to external forces [88]. Interestingly, molecular force spectroscopy studies show that individual DNA molecules exhibit linear stiffening behavior [89] that is similarly displayed by tensegrities as well as living cells and tissues. Thus, these observations suggest that whole organisms composed of organs, tissues, cells, organelles, and individual molecules may exhibit integrated mechanical behavior because they are organized as prestressed, hierarchical, tensegrity structures [1, 20, 21, 23, 68, 82, 90, 91].

8.4 Implications for Mechanotransduction

Most work on mechanotransduction has focused on the identification of molecules, genes, or intracellular signaling pathways that alter their activities when cells are mechanically stressed. Once a key molecule is identified, biophysicists seek to understand how mechanical distortion can alter its biochemical activity. Great advances have been made in this area. For example, we and others have shown that integrins function as mechanoreceptors and focus these stresses on signal transducing molecules in focal adhesions, including SA channels, heterotrimeric G-proteins, small GTPases, and protein kinases (src, p190Cas), that alter their activity when mechanically stressed [4, 10–14]. The mechanical gating properties of some SA ion channels are also now well characterized [5], and various molecules both inside and outside of the cell (e.g., p130cas and fibronectin) have been shown to change their activities by exposing new binding sites when physically deformed by applied mechanical stresses [13, 92]. However, the bigger challenge we now face is to explain how these molecules sense and respond to mechanical forces in the context of whole living cells.

It would be extremely difficult to understand how cells integrate chemical and mechanical signals if cellular biochemistry were carried out in solution. However, the reality is that most of the enzymes and substrates that mediate cellular metabolism and signal transduction act when physically immobilized on insoluble cytoskeletal scaffolds within the cytoplasm and nucleus [93]. Tensegrity is only an architectural principle that describes how structures, including living cells and tissues, control shape stability autonomously. However, use of this type of building system has important implications for how cells sense physical signals and convert them into a chemical response because it provides a novel mechanism to distribute mechanical stresses to discrete elements located throughout these hierarchical structures, and to channel forces from the macro scale to the micro scale. More specifically, because living organisms are organized as hierarchical tensegrity networks, mechanical forces applied at the macro scale are distributed and channeled over discrete molecular paths through the ECM and across cell membrane adhesion receptors to many sites inside the cytoplasm and nucleus, where these stresses may be focused on by different mechanotransducer molecules simultaneously.

Use of this novel mechanism for structural integration and force transmission may explain why so many different molecules and subcellular structures contribute to the mechanotransduction response [1]. It also explains why the level of prestress (isometric tension) at the different system levels in this network governs how the system (tissue, cell, or molecule) will respond to external mechanical loads. For example, sound sensation is influenced by the level of tension in the tympanic membrane, the basilar membrane, and the hair cell cytoskeleton, as well as in the tip links that insert on mechanotransducing SA channels [1]. The level of prestress in cartilage similarly governs fluid flow in its ECM lattice, which can shear cell surfaces and alter electrochemical potentials when the tissue is compressed [94, 95]. Reversible expansion and contraction of intercellular spaces in the prestressed epithelium of the lung also can alter the local concentration of growth factors in these spaces [76]. All of these structural changes can alter cellular signal transduction; however, the response will differ depending on the baseline level of prestress in the system. These changes in cellular biochemistry also can influence connective tissue remodeling and tissue mechanics, which can then feed back to further modify the cell mechanotransduction response.

8.4.1 The Whole Cell as the Mechanotransducer

We and others have shown that cell shape distortion and cytoskeletal prestress govern how whole living cells behave, as well as how specialized tissues form in the embryo [96–100]. For example, many anchorage-dependent cells, such as capillary endothelial cells, hepatocytes, and smooth muscle cells, proliferate when spread, whereas they shut off growth and switch on differentiation (e.g., tube formation, production of serum proteins, and enhanced contractility, respectively) when spreading is prevented [96, 97, 101]. Furthermore, when capillary cells are induced to retract and round by inhibiting ECM adhesion formation, these cells turn on apoptosis (cellular suicide program) [97]. Localized cell distortion also governs the direction in which cells will move by influencing the spatial distribution of actin polymerization and motile process (filopodia, lamellipodia, and microspike) formation within individual cells [99]. Interestingly, recent work suggests that cell lineage switching also can be controlled in human mesenchymal stem cells solely by varying the cell shape or ECM mechanics [102, 103]. Thus, mechanical cues are indeed central to cell and developmental control.

We and others have shown that signal transduction and gene transcription can be activated by applying mechanical stresses directly to cell surface integrin receptors [11, 104, 105]. Thus, cell or ECM distortion may influence cell fate switching by altering stresses transferred across cell surface integrins and their focal adhesion connections. However, we have found that local application of stress to integrins produces similar early signaling responses in round and spread cells [11], yet the former die whereas the latter proliferate. These are important observations because they suggest that the cell is able to sense the overall degree of distortion of its whole cytoskeleton, to integrate this information with other chemical and mechanical cues, and then to decide which behavior to pursue. Even cells that have been shown to sense mechanical forces (e.g., apical fluid shear) through changes in the activities of specific SA channels (polycystins) as a result of stress-induced deformation of mechanosensory primary cilia on their apical surface respond differently to the same mechanical stimulus depending on the integrity of the deep cytoskeleton and their basal cell–ECM adhesions [106]. Thus, the physiological response of the cell to any mechanical stress is governed by the physical state of the whole cell and cytoskeleton, and not by changes in any single signaling molecule.

Finally, while maintenance of the cytoskeletal architecture and network connections is important for integrating the cell's various mechanotransduction pathways, many cell behavioral responses can be altered solely by varying the level of prestress in these tensed networks. Examples of prestress-sensitive biological responses include ECM fibril formation and remodeling, cell proliferation, cell motility, contractility, and differentiation, to name a few. Altering the balance of forces between the cytoskeleton and the ECM also can modulate epithelial tissue development and angiogenesis during embryogenesis [100]. And, of course, we know from human physiology and sports medicine that lack of appropriate muscle tone or cartilage hydration can lead to significant negative impact on the cells and tissues of our bodies. Thus, the prestress that is so central to tensegrity is also a unifying principle of mechanical biology.

8.5 Conclusion

Mechanotransduction is critical for tissue formation, maintenance, and repair, and thus it is important to understand its molecular and biophysical basis. Great advances have been made over the past few decades in terms of identifying the molecules and signaling pathways that mediate mechanosensitivity and the cellular response to mechanical stress. However, to fully understand mechanoregulation, we must explain how cells simultaneously activate many different signal transducers, as well as how they orchestrate these cues with chemical and genetic signals that also convey regulatory information to the cell [107]. We also must explain how mechanotransduction is carried out in cells in their normal physiological context, as part of more complex tissues and organs within whole living organisms.

Tensegrity is a useful paradigm because it provides a mechanism to explain how molecules, cells, and tissues stabilize their shapes and respond mechanically when exposed to physical loads. More importantly, use of this architectural system provides a physical mechanism to distribute mechanical stresses through tissues and cells and across size scales so that they focus on multiple relevant mechanotransducer molecules simultaneously.

The importance of tensile prestress for mechanical stability is the central principle of tensegrity. This same principle now appears to be key for the shape stability of many different types of living materials, as well as for the control of mechanosensitivity and mechanochemical conversion. Importantly, before the introduction of the tensegrity model, prestress had never even entered the conversation in the field of mechanotransduction. Theoretical models based on tensegrity also have been developed from first principles, and computer simulations using these models predict many static and dynamic mechanical behaviors of various types of living cells and viruses, as well as subcellular components (e.g., cortical cytoskeleton, cell membrane, and lipid vacuoles). The novelty of these models is that they potentially provide a mechanism to both predict global mechanical behavior and explain its structural basis at the molecular level. Understanding the fundamental relation between mechanical structure and biochemical function in living materials through the pursuit of tensegrity theory, and translation of this knowledge into computational models, may open entirely new avenues for drug development and potentially lead to the creation of artificial biologically inspired materials and devices that mimic the complex behaviors of living cells and tissues.

8.6 Acknowledgments

This work was supported by grants from NIH, NASA, and DARPA. The computer simulations of tensegrity structures in Figure 8.2 were kindly provided by E. Xuan (Biomedical Communications, University of Toronto).

REFERENCES

[1] Ingber, D. E. Cellular mechanotransduction: Putting all the pieces together again, *FASEB J* **20**, 2006, 811–827.

[2] Ingber, D. E. Mechanobiology and diseases of mechanotransduction, *Ann Med* **35**, 2003, 564–577.

[3] Ingber, D. E. Tensegrity: The architectural basis of cellular mechanotransduction, *Annu Rev Physiol* **59**, 1997, 575–599.

[4] Geiger, B., Bershadsky, A., Pankov, R. and Yamada, K. M. Transmembrane crosstalk between the extracellular matrix – Cytoskeleton crosstalk, *Nat Rev Mol Cell Biol* **2**, 2001, 793–805.

[5] Sukharev, S. and Corey, D. P. Mechanosensitive channels: Multiplicity of families and gating paradigms, *Sci STKE* **2004**, 2004, re4.

[6] Wang, N., Butler, J. P. and Ingber, D. E. Mechanotransduction across the cell surface and through the cytoskeleton, *Science* **260**, 1993, 1124–1127.

[7] Choquet, D., Felsenfeld, D. P. and Sheetz, M. P. Extracellular matrix rigidity causes strengthening of integrin-cytoskeleton linkages, *Cell* **88**, 1997, 39–48.

[8] Plopper, G. E., McNamee, H. P., Dike, L. E., Bojanowski, K. and Ingber, D. E. Convergence of integrin and growth factor receptor signaling pathways within the focal adhesion complex, *Mol Biol Cell* **6**, 1995, 1349–1365.

[9] Miyamoto, S., Teramoto, H., Coso, O. A., Gutkind, J. S., Burbelo, P. D., Akiyama, S. K. and Yamada, K. M. Integrin function: Molecular hierarchies of cytoskeletal and signaling molecules, *J Cell Biol* **131**, 1995, 791–805.

[10] Chicurel, M. E., Singer, R. H., Meyer, C. J. and Ingber, D. E. Integrin binding and mechanical tension induce movement of mRNA and ribosomes to focal adhesions, *Nature* **392**, 1998, 730–733.

[11] Meyer, C. J., Alenghat, F. J., Rim, P., Fong, J. H., Fabry, B. and Ingber, D. E. Mechanical control of cyclic AMP signalling and gene transcription through integrins, *Nat Cell Biol* **2**, 2000, 666–668.

[12] Matthews, B. D., Overby, D. R., Mannix, R. and Ingber, D. E. Cellular adaptation to mechanical stress: Role of integrins, Rho, cytoskeletal tension, and mechanosensitive ion channels, *J Cell Sci* **119**, 2006, 508–518.

[13] Sawada, Y., Tamada, M., Dubin-Thaler, B. J., Cherniavskaya, O., Sakai, R., Tanaka, S. and Sheetz, M. P. Force sensing by mechanical extension of the Src family kinase substrate p130Cas, *Cell* **127**, 2006, 1015–1026.

[14] Wang, Y., Botvinick, E. L., Zhao, Y., Berns, M. W., Usami, S., Tsien, R. Y. and Chien, S. Visualizing the mechanical activation of Src, *Nature* **434**, 2005, 1040–1045.

[15] Ingber, D. Integrins as mechanochemical transducers, *Curr Opin Cell Biol* **3**, 1991, 841–848.

[16] Maniotis, A. J., Bojanowski, K. and Ingber, D. E. Mechanical continuity and reversible chromosome disassembly within intact genomes removed from living cells, *J Cell Biochem* **65**, 1997, 114–130.

[17] Wang, N. et al. Mechanical behavior in living cells consistent with the tensegrity model, *Proc Natl Acad Sci USA* **98**, 2001, 7765–7770.

[18] Hu, S. et al. Intracellular stress tomography reveals stress focusing and structural anisotropy in cytoskeleton of living cells, *Am J Physiol Cell Physiol* **285**, 2003, C1082–C1090.

[19] Ingber, D. E., Madri, J. A. and Jamieson, J. D. Role of basal lamina in neoplastic disorganization of tissue architecture, *Proc Natl Acad Sci USA* **78**, 1981, 3901–3905.

[20] Ingber, D. and Jamieson, J. D. (1985) in: *Gene Expression During Normal and Malignant Differentiation*, pp. 13–32 (Andersson, L. C., Gahmberg, C. G. and Ekblom, P., Eds.), Academic Press, Orlando.

[21] Ingber, D. E. Cellular tensegrity: Defining new rules of biological design that govern the cytoskeleton, *J Cell Sci* **104(Pt 3)**, 1993, 613–627.

[22] Fuller, B. Tensegrity, *Portfolio Artnews Annual* **4**, 1961, 112–127.

[23] Ingber, D. E. Tensegrity I. Cell structure and hierarchical systems biology, *J Cell Sci* **116**, 2003, 1157–1173.

[24] Stamenovic, D., Fredberg, J. J., Wang, N., Butler, J. P. and Ingber, D. E. A microstructural approach to cytoskeletal mechanics based on tensegrity, *J Theor Biol* **181**, 1996, 125–136.

[25] Stamenovic, D. and Ingber, D. E. Models of cytoskeletal mechanics of adherent cells, *Biomech Model Mechanobiol* **1**, 2002, 95–108.

[26] Ingber, D. E., Madri, J. A. and Jamieson, J. D. Basement membrane as a spatial organizer of polarized epithelia. Exogenous basement membrane reorients pancreatic epithelial tumor cells in vitro, *Am J Pathol* **122**, 1986, 129–139.

[27] Ingber, D. E., Madri, J. A. and Folkman, J. Endothelial growth factors and extracellular matrix regulate DNA synthesis through modulation of cell and nuclear expansion, *In Vitro Cell Dev Biol* **23**, 1987, 387–394.

[28] Ingber, D. E. Fibronectin controls capillary endothelial cell growth by modulating cell shape, *Proc Natl Acad Sci USA* **87**, 1990, 3579–3583.

[29] Sims, J. R., Karp, S. and Ingber, D. E. Altering the cellular mechanical force balance results in integrated changes in cell, cytoskeletal and nuclear shape, *J Cell Sci* **103(Pt 4)**, 1992, 1215–1222.

[30] Wang, N. and Ingber, D. E. Control of cytoskeletal mechanics by extracellular matrix, cell shape, and mechanical tension, *Biophys J* **66**, 1994, 2181–2189.

[31] Alenghat, F. J., Fabry, B., Tsai, K. Y., Goldmann, W. H. and Ingber, D. E. Analysis of cell mechanics in single vinculin-deficient cells using a magnetic tweezer, *Biochem Biophys Res Commun* **277**, 2000, 93–99.

[32] Matthews, B. D., Overby, D. R., Alenghat, F. J., Karavitis, J., Numaguchi, Y., Allen, P. G. and Ingber, D. E. Mechanical properties of individual focal adhesions probed with a magnetic microneedle, *Biochem Biophys Res Commun* **313**, 2004, 758–764.

[33] Overby, D. R., Matthews, B. D., Alsberg, E. and Ingber, D. E. Novel dynamic rheological behavior of focal adhesions measured within single cells using electromagnetic pulling cytometry, *Acta Biomateriala* **3**, 2005, 295–303.

[34] Yoshida, M., Westlin, W. F., Wang, N., Ingber, D. E., Rosenzweig, A., Resnick, N. and Gimbrone, M. A., Jr. Leukocyte adhesion to vascular endothelium induces E-selectin linkage to the actin cytoskeleton, *J Cell Biol* **133**, 1996, 445–455.

[35] Overby, D. R., Alenghat, F. J., Montoya-Zavala, M., Bei, H. C., Oh, P., Karavitis, J. and Ingber, D. E. Magnetic cellular switches, *IEEE Trans Magnetics* **40**, 2004, 2958–2960.

[36] Ezzell, R. M., Goldmann, W. H., Wang, N., Parasharama, N. and Ingber, D. E. Vinculin promotes cell spreading by mechanically coupling integrins to the cytoskeleton, *Exp Cell Res* **231**, 1997, 14–26.

[37] Stamenovic, D. and Coughlin, M. F. The role of prestress and architecture of the cytoskeleton and deformability of cytoskeletal filaments in mechanics of adherent cells: A quantitative analysis, *J Theor Biol* **201**, 1999, 63–74.

[38] Stamenović, D., Wang, N. and Ingber, D. E. Tensegrity architecture and the mammalian cell cytoskeleton. In: *Multiscaling in Molecular and Continuum Mechanics: Integration of Time and Size from Macro to Nano* (Sih, C. G., Ed.), Springer, New York, 2006.

[39] Pourati, J. et al. Is cytoskeletal tension a major determinant of cell deformability in adherent endothelial cells?, *Am J Physiol* **274**, 1998, C1283–1289.

[40] Wang, N., Tolic-Norrelykke, I. M., Chen, J., Mijailovich, S. M., Butler, J. P., Fredberg, J. J. and Stamenovic, D. Cell prestress. I. Stiffness and prestress are closely associated in adherent contractile cells, *Am J Physiol Cell Physiol* **282**, 2002, C606–C616.

[41] Stamenovic, D., Mijailovich, S. M., Tolic-Norrelykke, I. M. and Wang, N. Experimental tests of the cellular tensegrity hypothesis, *Biorheology* **40**, 2003, 221–225.

[42] Rosenblatt, N., Hu, S., Suki, B., Wang, N. and Stamenovic, D. Contributions of the active and passive components of the cytoskeletal prestress to stiffening of airway smooth muscle cells, *Ann Biomed Eng* **35**, 2007, 224–234.

[43] Maniotis, A. J., Chen, C. S. and Ingber, D. E. Demonstration of mechanical connections between integrins, cytoskeletal filaments, and nucleoplasm that stabilize nuclear structure, *Proc Natl Acad Sci USA* **94**, 1997, 849–854.

[44] Hu, S., Chen, J., Butler, J. P. and Wang, N. Prestress mediates force propagation into the nucleus, *Biochem Biophys Res Commun* **329**, 2005, 423–428.

[45] Hu, S. and Wang, N. Control of stress propagation in the cytoplasm by prestress and loading frequency, *Mol Cell Biomech* **3**, 2006, 49–60.

[46] Wang, N., Hu, S. and Butler, J. P. Imaging stress propagation in the cytoplasm of a living cell, *Methods Cell Biol* **83**, 2007, 179–198.

[47] Eckes, B. et al. Impaired mechanical stability, migration and contractile capacity in vimentin-deficient fibroblasts, *J Cell Sci* **111(Pt 13)**, 1998, 1897–1907.

[48] Mooney, D. J., Langer, R. and Ingber, D. E. Cytoskeletal filament assembly and the control of cell spreading and function by extracellular matrix, *J Cell Sci* **108(Pt 6)**, 1995, 2311–2320.

[49] Bereiter-Hahn, J., Luck, M., Miebach, T., Stelzer, H. K. and Voth, M. Spreading of trypsinized cells: Cytoskeletal dynamics and energy requirements, *J Cell Sci* **96(Pt 1)**, 1990, 171–188.

[50] Kumar, S., Maxwell, I. Z., Heisterkamp, A., Polte, T. R., Lele, T. P., Salanga, M., Mazur, E. and Ingber, D. E. Viscoelastic retraction of single living stress fibers and its impact on cell shape, cytoskeletal organization, and extracellular matrix mechanics, *Biophys J* **90**, 2006, 3762–3773.

[51] Fabry, B., Maksym, G. N., Shore, S. A., Moore, P. E., Panettieri, R. A., Jr., Butler, J. P. and Fredberg, J. J. Selected contribution: Time course and heterogeneity of contractile responses in cultured human airway smooth muscle cells, *J Appl Physiol* **91**, 2001, 986–994.

[52] Stamenovic, D., Suki, B., Fabry, B., Wang, N. and Fredberg, J. J. Rheology of airway smooth muscle cells is associated with cytoskeletal contractile stress, *J Appl Physiol* **96**, 2004, 1600–1605.

[53] Rosenblatt, N., Hu, S., Chen, J., Wang, N. and Stamenovic, D. Distending stress of the cytoskeleton is a key determinant of cell rheological behavior, *Biochem Biophys Res Commun* **321**, 2004, 617–622.

[54] Stamenovic, D. Effects of cytoskeletal prestress on cell rheological behavior, *Acta Biomater* **1**, 2005, 255–262.

[55] Vera, C., Skelton, R., Bossens, F. and Sung, L. A. 3-d nanomechanics of an erythrocyte junctional complex in equibiaxial and anisotropic deformations, *Ann Biomed Eng* **33**, 2005, 1387–1404.

[56] Zhu, Q., Vera, C., Asaro, R. J., Sche, P. and Sung, L. A. A hybrid model for erythrocyte membrane: A single unit of protein network coupled with lipid bilayer, *Biophys J* **93**, 2007, 386–400.

[57] Gardel, M. L., Nakamura, F., Hartwig, J. H., Crocker, J. C., Stossel, T. P. and Weitz, D. A. Prestressed F-actin networks cross-linked by hinged filamins replicate mechanical properties of cells, *Proc Natl Acad Sci USA* **103**, 2006, 1762–1767.

[58] Gardel, M. L., Nakamura, F., Hartwig, J., Crocker, J. C., Stossel, T. P. and Weitz, D. A. Stress-dependent elasticity of composite actin networks as a model for cell behavior, *Phys Rev Lett* **96**, 2006, 88–102.

[59] Heidemann, S. R., Lamoureaux, P. and Buxbaum, R. E. Opposing views on tensegrity as a structural framework for understanding cell mechanics, *J Appl Physiol* **89**, 2000, 1670–1678.

[60] Gittes, F., Mickey, B., Nettleton, J. and Howard, J. Flexural rigidity of microtubules and actin filaments measured from thermal fluctuations in shape, *J Cell Biol* **120**, 1993, 923–934.

[61] Kaech, S., Ludin, B. and Matus, A. Cytoskeletal plasticity in cells expressing neuronal microtubule-associated proteins, *Neuron* **17**, 1996, 1189–1199.

[62] Brangwynne C. M. F., Kumar, S., Geisse, N. A., Mahadevan, L., Parker, K. K., Ingber, D. E. and Weitz, D. Microtubules can bear enhanced compressive loads in living cells due to lateral reinforcement, *J Cell Biol* **175**(5), 2006, 733–741.

[63] Dennerll, T. J., Joshi, H. C., Steel, V. L., Buxbaum, R. E. and Heidemann, S. R. Tension and compression in the cytoskeleton of PC-12 neurites. II: Quantitative measurements, *J Cell Biol* **107**, 1988, 665–674.

[64] Buxbaum, R. E. and Heidemann, S. R. A thermodynamic model for force integration and microtubule assembly during axonal elongation, *J Theor Biol* **134**, 1988, 379–390.

[65] Mooney, D. J., Hansen, L. K., Langer, R., Vacanti, J. P. and Ingber, D. E. Extracellular matrix controls tubulin monomer levels in hepatocytes by regulating protein turnover, *Mol Biol Cell* **5**, 1994, 1281–1288.

[66] Putnam, A. J., Schultz, K. and Mooney, D. J. Control of microtubule assembly by extracellular matrix and externally applied strain, *Am J Physiol Cell Physiol* **280**, 2001, C556–C564.

[67] Hu, S., Chen, J. and Wang, N. Cell spreading controls balance of prestress by microtubules and extracellular matrix, *Front Biosci* **9**, 2004, 2177–2182.

[68] Ingber, D. E. The architecture of life, *Sci Am* **278**, 1998, 48–57.
[69] McMahon, T. A. in: *Muscles, Reflexes, and Locomotion*, Princeton University Press, Princeton, NJ, 1984.
[70] Stamenovic, D. Micromechanical foundations of pulmonary elasticity, *Physiol Rev* **70**, 1990, 1117–1134.
[71] Ralphs, J. R., Waggett, A. D. and Benjamin, M. Actin stress fibres and cell-cell adhesion molecules in tendons: Organisation in vivo and response to mechanical loading of tendon cells in vitro, *Matrix Biol* **21**, 2002, 67–74.
[72] Komulainen, J., Takala, T. E., Kuipers, H. and Hesselink, M. K. The disruption of myofibre structures in rat skeletal muscle after forced lengthening contractions, *Pflugers Arch* **436**, 1998, 735–741.
[73] Quinn, T. M., Dierickx, P. and Grodzinsky, A. J. Glycosaminoglycan network geometry may contribute to anisotropic hydraulic permeability in cartilage under compression, *J Biomech* **34**, 2001, 1483–1490.
[74] Guilak, F. Compression-induced changes in the shape and volume of the chondrocyte nucleus, *J Biomech* **28**, 1995, 1529–1541.
[75] Toshima, M., Ohtani, Y. and Ohtani, O. Three-dimensional architecture of elastin and collagen fiber networks in the human and rat lung, *Arch Histol Cytol* **67**, 2004, 31–40.
[76] Tschumperlin, D. J. et al. Mechanotransduction through growth-factor shedding into the extracellular space, *Nature* **429**, 2004, 83–86.
[77] Hutchison, C. J. Lamins: Building blocks or regulators of gene expression?, *Nat Rev Mol Cell Biol* **3**, 2002, 848–858.
[78] Pickett-Heaps, J. D., Forer, A. and Spurck, T. Traction fibre: Toward a "tensegral" model of the spindle, *Cell Motil Cytoskeleton* **37**, 1997, 1–6.
[79] Sheetz, M. P., Wayne, D. B. and Pearlman, A. L. Extension of filopodia by motor-dependent actin assembly, *Cell Motil Cytoskeleton* **22**, 1992, 160–169.
[80] Domnina, L. V., Rovensky, J. A., Vasiliev, J. M. and Gelfand, I. M. Effect of microtubule-destroying drugs on the spreading and shape of cultured epithelial cells, *J Cell Sci* **74**, 1985, 267–282.
[81] Ingber, D. E., Prusty, D., Sun, Z., Betensky, H. and Wang, N. Cell shape, cytoskeletal mechanics, and cell cycle control in angiogenesis, *J Biomech* **28**, 1995, 1471–1484.
[82] Ingber, D. E. et al. Cellular tensegrity: Exploring how mechanical changes in the cytoskeleton regulate cell growth, migration, and tissue pattern during morphogenesis, *Int Rev Cytol* **150**, 1994, 173–224.
[83] Caspar, D. L. Movement and self-control in protein assemblies. Quasi-equivalence revisited, *Biophys J* **32**, 1980, 103–138.
[84] Schutt, C. E., Kreatsoulas, C., Page, R. and Lindberg, U. Plugging into actin's architectonic socket, *Nat Struct Biol* **4**, 1997, 169–172.
[85] Butcher, J. A., Jr. and Lamb, G. W. The relationship between domes and foams: Application of geodesic mathematics to micelles, *J Am Chem Soc* **106**, 1984, 1217–1220.
[86] Farrell, H. M., Jr., Qi, P. X., Brown, E. M., Cooke, P. H., Tunick, M. H., Wickham, E. D. and Unruh, J. J. Molten globule structures in milk proteins: Implications for potential new structure-function relationships, *J Dairy Sci* **85**, 2002, 459–471.
[87] Ingber, D. E. The origin of cellular life, *Bioessays* **22**, 2000, 1160–1170.
[88] Zanotti, G. and Guerra, C. Is tensegrity a unifying concept of protein folds?, *FEBS Lett* **534**, 2003, 7–10.
[89] Smith, S. B., Finzi, L. and Bustamante, C. Direct mechanical measurements of the elasticity of single DNA molecules by using magnetic beads, *Science* **258**, 1992, 1122–1126.

[90] Pienta, K. J. and Coffey, D. S. Cell motility as a chemotherapeutic target, *Cancer Surv* **11**, 1991, 255–263.

[91] Pienta, K. J., Murphy, B. C., Getzenberg, R. H. and Coffey, D. S. The effect of extracellular matrix interactions on morphologic transformation in vitro, *Biochem Biophys Res Commun* **179**, 1991, 333–339.

[92] Baneyx, G., Baugh, L. and Vogel, V. Fibronectin extension and unfolding within cell matrix fibrils controlled by cytoskeletal tension, *Proc Natl Acad Sci USA* **99**, 2002, 5139–5143.

[93] Ingber, D. E. The riddle of morphogenesis: A question of solution chemistry or molecular cell engineering?, *Cell* **75**, 1993, 1249–1252.

[94] Lai, W. M., Hou, J. S. and Mow, V. C. A triphasic theory for the swelling and deformation behaviors of articular cartilage, *J Biomech Eng* **113**, 1991, 245–258.

[95] Grodzinsky, A. J. Electromechanical and physicochemical properties of connective tissue, *Crit Rev Biomed Eng* **9**, 1983, 133–199.

[96] Singhvi, R., Kumar, A., Lopez, G. P., Stephanopoulos, G. N., Wang, D. I., Whitesides, G. M. and Ingber, D. E. Engineering cell shape and function, *Science* **264**, 1994, 696–698.

[97] Chen, B. M. and Grinnell, A. D. Kinetics, Ca2+ dependence, and biophysical properties of integrin-mediated mechanical modulation of transmitter release from frog motor nerve terminals, *J Neurosci* **17**, 1997, 904–916.

[98] Huang, S., Chen, C. S. and Ingber, D. E. Control of cyclin D1, p27(Kip1), and cell cycle progression in human capillary endothelial cells by cell shape and cytoskeletal tension, *Mol Biol Cell* **9**, 1998, 3179–3193.

[99] Parker, K. K. et al. Directional control of lamellipodia extension by constraining cell shape and orienting cell tractional forces, *FASEB J* **16**, 2002, 1195–1204.

[100] Moore, K. A., Polte, T., Huang, S., Shi, B., Alsberg, E., Sunday, M. E. and Ingber, D. E. Control of basement membrane remodeling and epithelial branching morphogenesis in embryonic lung by Rho and cytoskeletal tension, *Developmental Dynamics* **232**, 2005, 268–281.

[101] Polte, T. R., Eichler, G. S., Wang, N. and Ingber, D. E. Extracellular matrix controls myosin light chain phosphorylation and cell contractility through modulation of cell shape and cytoskeletal prestress, *Am J Physiol Cell Physiol* **286**, 2004, C518–C528.

[102] McBeath, R., Pirone, D. M., Nelson, C. M., Bhadriraju, K. and Chen, C. S. Cell shape, cytoskeletal tension, and RhoA regulate stem cell lineage commitment, *Dev Cell* **6**, 2004, 483–495.

[103] Engler, A. J., Griffin, M. A., Sen, S., Bonnemann, C. G., Sweeney, H. L. and Discher, D. E. Myotubes differentiate optimally on substrates with tissue-like stiffness: Pathological implications for soft or stiff microenvironments, *J Cell Biol* **166**, 2004, 877–887.

[104] Resnick, N., Collins, T., Atkinson, W., Bonthron, D. T., Dewey, C. F., Jr. and Gimbrone, M. A., Jr. Platelet-derived growth factor B chain promoter contains a cis-acting fluid shear-stress-responsive element, *Proc Natl Acad Sci USA* **90**, 1993, 4591–4595.

[105] Chen, J., Fabry, B., Schiffrin, E. L. and Wang, N. Twisting integrin receptors increases endothelin-1 gene expression in endothelial cells, *Am J Physiol Cell Physiol* **280**, 2001, C1475–C1484.

[106] Alenghat, F. J., Nauli, S. M., Kolb, R., Zhou, J. and Ingber, D. E. Global cytoskeletal control of mechanotransduction in kidney epithelial cells, *Exp Cell Res* **301**, 2004, 23–30.

9

Nuclear Mechanics and Mechanotransduction

Shinji Deguchi and Masaaki Sato

9.1 Introduction

Many studies have shown that mechanical forces inherent in the living body, such as fluid shear stress due to blood flow, play critical roles in regulating cellular physiology and pathogenesis (Davies 1995; Li et al. 2005; Haga et al. 2007). These extracellular forces are sensed at the cellular level, and inside the cells the forces are somehow transduced into changes in gene expression responsible for the cellular responses (Ingber 1997). The mechanism of this force-sensing process or mechanotransduction still remains unclear. Signaling pathways following mechanical force loadings have been identified by means of biochemistry (Li et al. 2005; Haga et al. 2007). In addition to the involvement of these signaling molecules, direct intracellular force transmission from the cell membrane or extracellular matrix to the nucleus via, probably, cytoskeletal filaments may be another possible pathway through which cells respond to extracelullar forces (Wang et al. 1993; Davies 1995; Ingber 1997; Maniotis et al. 1997; Janmey 1998): Loaded forces may cause a deformation of the nucleus that is the principal site of DNA and RNA synthesis, alter the spatial positioning or dynamics of the chromatin that is a complex of DNA and proteins (such as histone) making up chromosomes, and then affect gene expression because such changes in chromatin organization could expose new sites for transcriptional regulation (Dahl et al. 2004; Lammerding et al. 2004).

There is actually considerable evidence that the cell nucleus is deformed or remodeled in response to extracellular forces when the cell adapts to the local mechanical environment by the reorganization of cytoskeletons. Flaherty et al. (1972) demonstrated by *in vivo* experiments that endothelial nuclei elongate and orient in the direction of blood flow, as do the whole cells (Figure 9.1). Lee et al. (2005) found, based on *in vitro* observations, that movement of the nucleus in the cytoplasm is enhanced by fluid flow and regulated by mediators of cytoskeleton reorganization. Deguchi et al. (2005a) suggested that not only the endothelial cell cytoskeleton but also its nucleus remodels structure under shear stress applied to the cell. This study suggested that the shear stress applied to the cell might induce structural rearrangement in the nucleus structure, which leads to a permanent

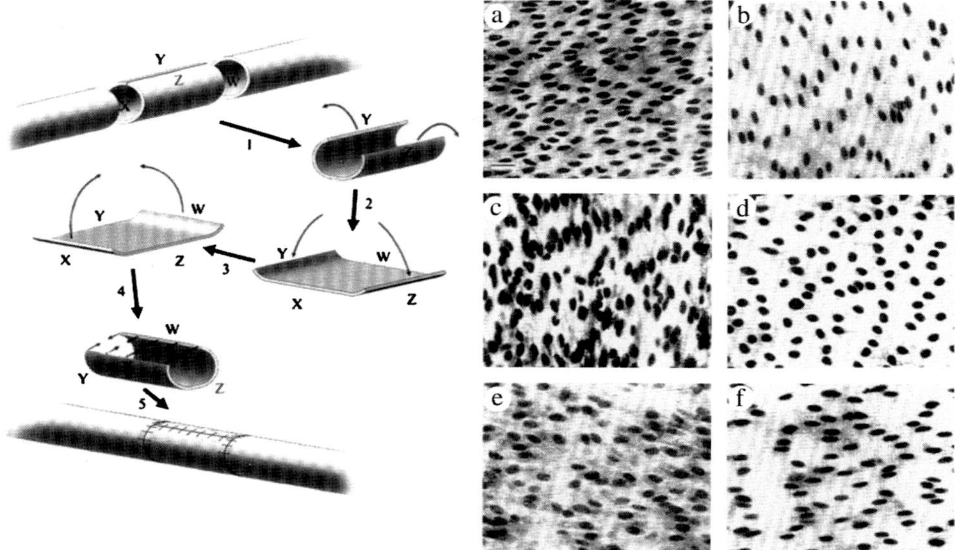

Figure 9.1. Orientation and elongation of endothelial nuclei in the direction of blood flow. The left figure shows the operative procedure. One portion of the blood vessel is excised, placed at a right angle to the original direction, and then sutured. Right: (a) Control. Nuclei are stained. The flow direction is from the left to right. (b) Nuclei immediately after the operation. (c and d) Nuclei on the third day observed at different places. (e) The tenth day. (f) The seventieth day. (Figures are modified from Flaherty et al. (1972). Copyright ©1972, American Heart Association.)

alteration in its overall shape and stiffness. Thus, the nucleus deformation and remodeling appear during the force-loading and resultant cellular responses.

Understanding at the molecular (i.e., DNA) level of the mechanism underlying the possible force-induced change in gene expression will be the subject of future investigation. To reveal how mechanosensing sites in the cell nucleus work, we must know the mechanical properties (such as viscoelasticity) of the nucleus itself and its subnuclear structural components. Knowledge of the nuclear and subnuclear mechanical properties will be vital to quantitatively evaluate how much stress the nucleus is actually bearing to keep the cell architecture under loading and how the stress is distributed within the nucleus, which may contain unidentified mechanosensing sites. The force–deformation relationship of biological tissues including cells (Sato et al. 1987; Desprat et al. 2005; Fernandez et al. 2006; Kasza et al. 2006) and subcellular components (Deguchi et al. 2006) is in general highly nonlinear, indicating that the nucleus may also have a strain-dependent resistance to loading. Hence determination of the mechanical strain state of the nucleus is required in order to obtain its physiologically meaningful mechanical properties (such as elastic modulus) at the various states.

Moreover, the mechanical connection with surrounding cytoplasmic constituents (i.e., how extracellular forces are transmitted to a surface of the nucleus through which subcellular structural components in living cells) also remains elusive; yet, it is central to revealing the origin of nuclear deformations inside cells. Understanding of these three factors, namely, nuclear mechanical properties, strain, and connection with surrounding constituents (or boundary condition for the nucleus), is still insufficient and developing. In the present review, particular attention is paid to advances in knowledge of the mechanical properties of the nucleus as well as the limitations we face in both experiments and interpretation of the results. Experimental data on the mechanical properties of the nucleus must be actually considered together with its strain magnitude and boundary connection with surrounding cytoplasmic constituents, although they are often ambiguous.

9.2 Macroscopic Mechanical Properties of the Nucleus

Approaches to the mechanical properties of the nucleus can be classified into investigations on nuclei inside of living cells and those on individual nuclei isolated from cells (Dong et al. 1991; Guilak 1995; Caille et al. 1998, 2002; Guilak et al. 2000; Dahl et al. 2004, 2005, 2006; Tseng et al. 2004; Deguchi et al. 2005a, 2007; Rowat et al. 2005, 2006; Vaziri et al. 2006). These investigations were conducted, for both nuclei in living cells and for isolated nuclei, by means of micropipette aspiration tests (Dahl et al. 2004; Deguchi et al. 2005a; Rowat et al. 2005), scanning probe microscopy (Vaziri et al. 2006), compression tests (Caille et al. 2002; Knight et al. 2002), and stretching tests (Deguchi et al. 2007) to obtain the loading–deformation relationship. Nanoparticle tracking measurements were also done to examine detailed motion behaviors of the nuclei in cells under mechanical loading (Lee et al. 2005).

In measurements on nuclei in living cells, we can observe intact nucleus behaviors subjected to mechanical loadings. As pointed out previously, however, these measurements are sometimes complicated because estimated loading onto the nucleus within cells is likely to be skewed due to force redistribution by the cytoskeletons and force dissipation by other cytoplasmic constituents. Thus accurate quantification of the nuclear mechanical properties, which are determined from the relationship between loading and deformation, is difficult. In contrast, in measurements on nuclei isolated from surrounding cytoplasmic constituents, characteristics of the individual nuclei can be identified.

The isolation of nuclei from cells is realized by either mechanical (Guilak et al. 2000) or chemical extraction (Muramatsu et al. 1963). In a mechanical extraction, a thin pipette is used to pierce and aspirate parts of the cytoplasm and then remove the outside structure of the nucleus. Such manipulations, however, need technical skills. The chemical extraction of the nucleus is carried out after cell lysis and chemical digestion of the surrounding cytoskeleton. The isolated nuclei are collected after centrifugation to remove the detergent that may damage the nucleus. At least, the mechanical properties of isolated nuclei are extremely sensitive to external buffer conditions (Dahl et al. 2004). The possible damage has not been fully understood so far, and future investigation should check its influence on the nuclear mechanical properties in more detail. In particular, it is conceivable that such a chemical digestion affects the nuclear envelope. The nuclear envelope that forms a boundary

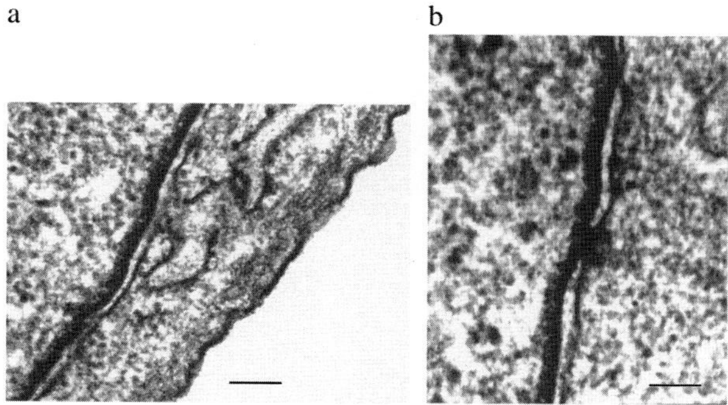

Figure 9.2. Transmission electron micrographs of the nuclear envelope. Heterochromatin aggregates in the nuclear periphery and forms thick layers of approximately 30 nm. (a) The endoplasmic reticulum continuous with the outer nuclear envelope. (b) The nuclear pore complex. Scale bars: 0.1 μm.

between the nucleoplasm and the cytoplasm consists of two concentric membranes. The double-membrane envelope is continuous with the endoplasmic reticulum (Figure 9.2(a)) and is penetrated by nuclear pore complexes (Figure 9.2(b)) that span both the inner and outer membranes and help to regulate nuclear composition. Fluorescence microscopy gives macroscopic views of the distinct outer edges of isolated nuclei including nuclear pore complexes (Rowat et al. 2005) (Figure 9.3; Plate 12), whereas microscopic observation by electron microscopy showed partly unrecognizable outer nuclear membranes with apparently intact nuclear pore complexes (Deguchi et al. 2005a). Furthermore, the process of isolation itself might induce fiber-like artifacts in the nuclear matrix. To assess the contribution of each component to the mechanical properties of the whole nucleus (Vaziri et al. 2007a), the intactness at the molecular level of the isolated nuclei should be a critical factor.

Despite the limitations described previously, many measurements have so far provided qualitatively consistent macroscopic mechanical properties of the nucleus: The nucleus is many times stiffer and more viscous than the cells, and thus nuclear properties are distinct from those of the surrounding cytoplasm (Dong et al., 1991; Guilak et al., 2000; Caille et al., 2002). For instance, micropipette aspirations of isolated chondrocyte nuclei gave an elastic modulus of ~1 kPa, which is 3–4 times stiffer than the elasticity of the cytoplasmic region (Guilak et al. 2000). It was estimated that microdissected or osmotically isolated neutrophil nuclei were approximately 10 times stiffer than the cytoplasm (Dong et al., 1991). Compression tests, using parallel plates, of nuclei isolated from endothelial cells provided an elastic modulus of 5–8 kPa, which was on average 10 times more rigid than the cytoplasm (Caille et al. 2002). Similar differences in elasticity between the nucleus and cytoplasm were obtained by scanning probe microscopy: Miyazaki and Hayashi (1999) found a greater stiffness at locations above the nucleus of endothelial cells. Maniotis et al. (1997) estimated, by pulling a local part of the nucleus through transmembrane proteins and actin filaments, that the nucleus was 9 times stiffer than the cytoplasm. Deguchi et al. (2007) stretched single whole cells using a pair of cantilevers and

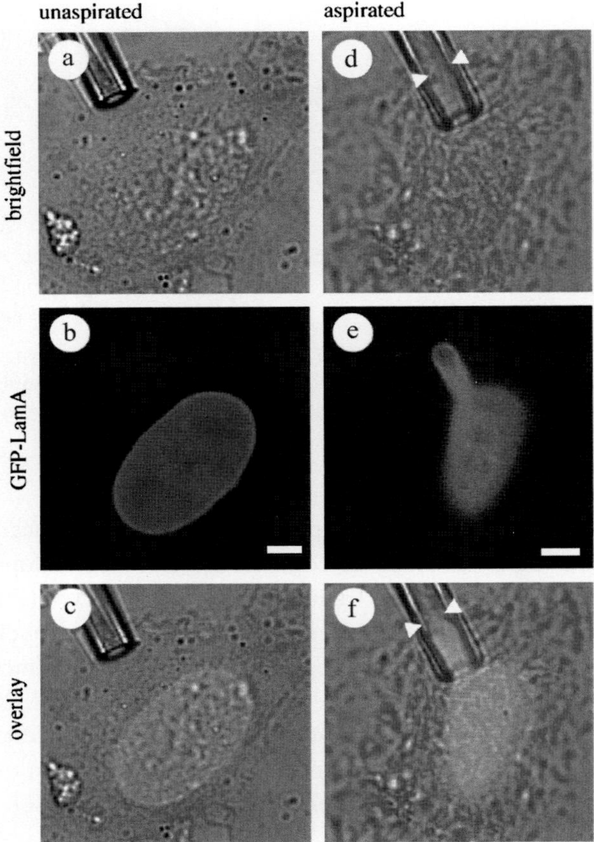

Figure 9.3. Large deformability of the nuclear envelope. The nucleus of a HeLa cell transiently expressing GFP-LamA in cell (a–c) is partly aspirated (d–f). Arrowheads show that the edge of the nuclear envelope only slightly precedes the GFP-LamA fluorescence in the micropipette (d and f), indicating that the lamina remains associated with the nuclear envelope. Scale bars: 5 mm. (Figures are modified from Rowat et al. (2005). Copyright ©2005, The Royal Society.)

tracked the responses of the nuclei inside of the cells. They predicted the intracellular stress distribution during the cell stretching based on systematic finite element analysis, yielding almost 10 times stiffer nuclear elasticity compared with the cytoplasm.

Thus, the macroscopic mechanical properties of the nucleus are qualitatively similar between different studies; yet, they vary in quantitative terms by approximately an order of magnitude, as shown above. The variation in values is probably due to the differences in the cell types as well as the measurement methods (including experimental parameters such as strain range and loaded region). Recent studies are further exploring the contribution of subnuclear components to the macroscopic nuclear mechanical properties. These studies are linked to several diseases caused by deficiencies in the subnuclear components, which will be presented in Section 9.5.

9.3 Mechanical State of the Nucleus Inside of Adhering Cells

As for the intrinsic (macroscopic) strain of the nuclei in living cells, it is known that the nuclei that originally have spheroidal shapes spontaneously alter their shape to rounder ones when the adherent cells detach from the substrate and have spherical cellular shapes (Sims et al. 1992). Since this morphological change of the nucleus at the macroscopic scale occurs soon after it is detached, it is not likely that the change is caused only by molecular reorganization or remodeling. Mechanical deformation is surely associated with the morphological change. This rounding of the nucleus proposes some mechanical models of the nucleus inside the adhering cells. For instance, the nuclei inside adhering cells may have residual or preexisting strains even without external loadings, and the pre-strain may be released during the cell rounding (Ingber 1997). This might be why the nucleus becomes, after cell detachment, round (at which less stress is borne in the nucleus compared with the adhered condition). Alternatively, the nucleus, in adhering cells that have a spheroidal shape without preexisting stress may be compressed laterally by surrounding cytoplasmic constituents, including microtubules, that may be able to bear the compressive force (Brangwynne et al. 2006) when cells are dislodged and rounded. Then, the nucleus can become rounder while bearing the compressive stress.

Interpretation of the strain state of the nucleus within cells is thus elusive. However, this intrinsic strain of the nucleus is important in that it may affect its mechanical resistance to loading, possibly leading to changes in gene expression as described in the Introduction. Selective regulation or disruption of the activity of cytoskeletal components that would keep cell structural integrity induces alterations in the nuclear morphology (unpublished data, S.D.). Experiments enabled us to partly remove the stress that the nucleus may bear, thereby leading to the accurate interpretation of the strain state of the nucleus within adhering cells.

The origin of the preexisting strain of the nucleus within cells may come from a passive or active force. Ingber (1997) proposed the cell tensegrity model, in which an interconnected network composed of cytoskeletons, the cell membrane, and the nucleus plays a structural role in stabilizing the cell architecture. In the interconnected network, each structural element bears forces even without external loading onto the adherent cells. For example, actomyosin contraction produces tension, whereas microtubules bear compressive stresses. The possible preexisting stress in the nucleus may be produced under the intracellular force balance between the nucleus and the surrounding cytoplasm, including cytoskeletons. Meanwhile, it is unclear whether the nucleus actively generates tension, although nuclear actin and myosin have been identified (Shumaker et al. 2003; Bettinger et al. 2004).

9.4 Contribution of Subnuclear Components to the Macroscopic Mechanical Properties of the Nucleus

Understanding the mechanical properties of subnuclear structural componets and their contributions to the macroscopic nuclear mechanical properties will elucidate the intranuclear force transmission in mechanotransduction as well as the biomechanical mechanism underlying nuclear physiological activities such as mitosis. Moreover, those studies at the subnuclear scale can be linked to nucleus-related diseases and viral infections. Recent studies have attempted to quantitatively characterize the nuclear envelope (Dahl et al. 2004, 2005, 2006; Rowat et al. 2005, 2006; Vaziri et al. 2006, 2007) and single chromosomes (Houchmandzadeh et al. 1997), which are touched upon in this section. Before that, an overview of the subnuclear components is presented in the following paragraph.

The nucleus is separated from the cytoplasm by the nuclear envelope (Figure 9.2), which comprises inner and outer nuclear membranes, nuclear pore complexes, and nuclear lamina. Each of the inner and outer nuclear membranes is a lipid bilayer separated by an electron transparent region of approximately 10–40 nm (Figure 9.2(a)). Nuclear lamina are stable filamentous networks composed of intermediate filaments made of lamin (Aebi et al. 1986) that line the inner nuclear membrane and are thought to provide the nucleus with mechanical stability. Mammalian cells produce two lamin protein types, referred to as A-type and B-type. A-type lamins, which are further categorized as lamin A and C, are the product of a single gene LMNA and are expressed mainly in differentiated cells. B-type lamins, lamin B1 and B2, are expressed from the LMNB1 and LMNB2 genes that are present in all cell types and are vital for cell viability and function (Burke and Stewart 2002). The nuclear envelope has physical connections to chromatin via lamins, which bind directly to chromatin (Newport and Forbes 1987; Haque et al. 2006) and to chromatin-binding nuclear membrane proteins (Burke and Stewart 2002). Individual chromatin fibers contain wound coils of DNA. The nuclear matrix is thought to serve as the scaffolding responsible for the higher order chromosome packing and nuclear organization (Replogle-Schwab et al. 1996). Although the exact function of this matrix is still disputed, there is evidence that many nuclear-specific proteins are associated with the nuclear matrix (Nakayasu and Berezney 1991) and that the matrix is involved in the regulation of gene expression (Tetko et al. 2006).

Experiments were done to separately characterize the contributions of the subnuclear components to the whole nucleus. Dahl et al. (2004, 2005) investigated, with micropipette aspiration tests and scanning probe microscope indentation tests, the elastic characteristics of the nuclear envelope that is treated to become independent of the nucleoplasm by osmotic swelling with high molecular mass dextrans. This experiment showed that chromatin is a primary force-bearing element at low strain, whereas the lamina network composed of B-type lamin sustains much of the load at high strain, yielding quantitative measures of nuclear envelope mechanics. Moreover, they found that the chromatin has a very viscous nature over large strains. Rowat et al. (2005, 2006) combined micropipette aspiration tests with confocal laser scanning microscopy to capture the mechanical characteristics of fluorescently visualized A-type lamin and nuclear pore complexes (Figure 9.3; Plate 12). They demonstrated that the nuclear envelope undergoes large deformations, in which the embedded nuclear pore complexes are also displaced, and maintains its structural stability when exposed to mechanical loading.

Vaziri et al. (2006, 2007a) performed continuum mechanics-based computer simulations of mechanical tests on isolated nuclei to examine the respective roles of the envelope components. This study showed that the response of isolated nuclei in both scanning probe microscopy indentation and micropipette aspiration is highly sensitive to the nuclear lamina stiffness, suggesting that testing nuclear mechanical properties with mechanical tests can be effective in evaluating the influence of the various kinds of envelopathies described in the following section.

To understand, at the chromosome level, the mechanics of mitosis, Houchmandzadeh et al. (1997) measured by micromanipulation the elastic modulus of single chromosomes isolated from newt lung cells. The elastic modulus varied from approximately 1 to 5 kPa, depending on the stage of mitosis: Chromosomes are highly extensible in the metaphase, in which condensed chromosomes, carrying genetic information, align in the middle of the cell before cell division into each of the two daughter cells. Knowing the mechanical properties thus allows us to estimate the force exerted by the chromosome and accordingly to approach the relationship between chromatin organization and chromatin dynamics (Hay and De Boni 1991; Panorchan et al. 2004). Then, how do the chromatins forming the chromosomes affect the intranuclear force transmission in interphase cells? In cells, two types of chromatins are distinguished: euchromatins and heterochromatins. Euchromatins consist of active DNA and have a relatively loose structure. In contrast, heterochromatins consisting of inactive DNA line the inner nuclear membrane and exist as condensed patches of ~30-nm thickness, which is thicker than the layer of the nuclear lamina (Figure 9.2). Since chromatin has an elastic modulus comparable in magnitude to the whole nucleus, heterochromatin in particular may be significant in bearing the force loaded onto the nucleus.

9.5 Linkage between Nuclear Mechanics and Nuclear Envelope-Related Diseases

Although the exact mechanisms governing force transduction are unknown, nuclear envelope mechanical properties are a hypothesized regulator of the effects of forces on chromatin or gene expressions. This motivated the study of the linkage between nuclear mechanics and nuclear envelopathies. At least 10 diseases have been linked to mutations in lamins or lamin-binding proteins (Dahl et al. 2005; Lammerding et al. 2004, 2005). These diseases include Emery-Dreifuss muscular dystrophy (Burke and Stewart 2002; Ostlund and Worman 2003) and Hutchinson-Gilford progeria syndrome (Dahl et al. 2006). Emery-Dreifuss muscular dystrophy is a neuromuscular degenerative condition with an associated dilated cardiomyopathy and cardiac conduction defect. Hutchinson-Gilford progeria syndrome is a rare genetic disease that causes segmental premature aging in children. The mechanisms of these diseases have been related to defects in gene expression coupled to mechanical weakness of the nuclear envelope (Wilson 2005). Recently, Dahl et al. (2006) demonstrated by live-cell imaging and biochemical analysis that lamins A and C are trapped at the nuclear periphery in Hutchinson-Gilford progeria syndrome patient cells. Using micropipette aspiration, they showed that the lamina in the cells has a significantly reduced ability to rearrange under mechanical stress. Thus, the changes in mechanical properties associated with diseases were actually found. Understanding the detailed mechanism underlying the emergence of these diseases in terms of mechanics will be the subject of future investigation. For more details, readers can refer to more complete discussions from the mechanical viewpoints of Lammerding (2005), Dahl (2006), and Vaziri (2007b).

9.6 Future Directions

In the future, to emphasize the significance of nuclear mechanics in mechanotransduction, a more thorough understanding of the two different force transmission pathways in living cells, namely, from the cell membrane to the nuclear envelope (intracellular pathway) and from the nuclear envelope to the subnuclear structural components including DNA (intranuclear pathway), is inevitable. To identify the intracellular force transmission pathway, RGD-coated beads are bound to integrins in the cell membrane. Through the beads, mechanical stimuli are loaded onto the cells, and the resultant intracellular strain is detected (Wang et al. 1993; Maniotis et al. 1997). The use of beads may, however, enhance the local development of the actin cytoskeletal structure through the recruitment of integrins to the bound area (Kawakami et al. 2001). The force transmission obtained in that case might therefore be more or less different in magnitude as well as in pattern from the actual force transmission to the nucleus subjected to physiological forces such as fluid shear stress.

The nucleus in adhering endothelial cells occupies almost the entire height of the cell as viewed with transmission electron microscopy (Figures 9.4(a) and 9.4(b)) or confocal microscopy. The profile of the spread cells within their vertical cross section seems to be determined by the size of the nucleus. In contrast, the nucleus in isolated and floating cells is located around the cell center apart from the peripheral cell membrane (Figure 9.3; Plate 12). The influence of external loading on the nucleus would therefore be different between these two states of the nucleus. Studies employing isolated and therefore rounded cells or nuclei are surely valuable in examining the intranuclear force transmission and the involvement of subnuclear components in all of the nuclear mechanical properties; however, in terms of mechanotransduction, estimation of the intracellular force transmission in adhering cells is also necessary.

In living adherent cells, how are physiological forces transmitted to the nucleus? Tension via actin filaments and compressive stress via microtubules are conventional candidates for the physical pathway (Ingber 1997; Janmey 1998). In addition to the involvement of the cytoskeletal network, the vertical morphology of the adherent cells and their nuclei pointed out above (Figure 9.4(b)) suggests that the force loaded onto the plasma membrane may directly displace the upper part of the nucleus because the cell membrane and nucleus nearly touch each other. In that case, a fluid shear stress–induced decrease in cell height (Yamaguchi and Hanai 1988; Barbee et al. 1994) might directly compress the nucleus.

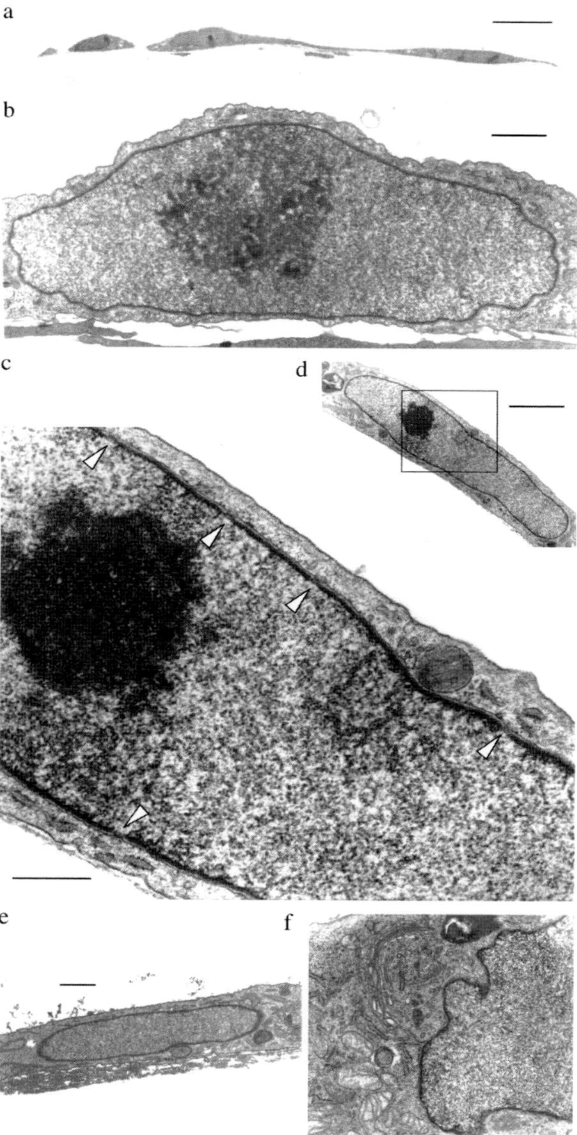

Figure 9.4. Transmission electron micrographs of endothelial cells. (a) Cells cut in around their middle cross section. The profile of spread endothelial cells seems to be determined by the size of their nuclei. (b) One whole nucleus cut in the middle cross section. (c–f) Images of a part of the nuclei cut in the middle cross section. The rectangular area in (d) is enlarged and shown in (c). Note that the nucleus seems to be locally compressed by a membrane-bound organelle located between the nuclear membrane and the luminal (c) or abluminal membrane (e). In addition to conventional intracellular force-bearing components, cytoskeletons, these organelles may affect local mechanical conditions of the nucleus, e.g., tension in adjacent nuclear pores (arrowheads). Scale bars: 10 μm (a); 1 μmin (b, c, e, and f); and 5 μm (d).

We note another possible intracellular loading that may affect the nucleus in a mechanical way. The cytoplasm is packed with numerous organelles, including membrane-bound structures, and each organelle would have its own stiffness dependent on the shape. From transmission electron microscopy observing the cross section of the boundary between the cell membrane and the nuclear membrane (Figures 9.4(c–e)), small membrane-bound organelles seem to push "mechanically" a local part of the nucleus, although the nucleus is stiffer than the cytoplasmic region according to macroscopic studies, as already described in this review. This may indicate that the stiffness of some small membrane-bound organelles is actually larger than that of the nucleus, probably because of the size difference between them, as a larger spring has a smaller stiffness. The structural components of the nucleus are in many cases unique to this organelle, and therefore the scaling effect cannot be simply applied. However, at the moment, we cannot explain without the mechanical viewpoint why such locally curved portions appear in the nucleus where the cell and nuclear membranes run parallel and straight except at the place when the small organelle locates. The local incurve of the nucleus is different in morphology from the endocytosis-like incurve (Figure 9.4(f)) that appears around the lateral side of the nucleus even in interphase cells, depending on physiological activities. The mechanism about the possible changes in the local mechanical properties of such live nuclei may become an interesting biomechanics-research theme. Again, a hypothesis could be that membrane-bound organelles in the cytoplasm, which are a soft biomaterial at the macroscopic scale, might deform local parts of the nucleus and hence chromatin by compression, although it is commonly considered that what can bear intracellular stresses is only cytoskeletal filaments based on a view that the cytoskeleton is the only cellular structure directly linking the cell surface to the nucleus (Ingber 1997; Janmey 1998). The evidence for this hypothesis is still indirect; however, it indicates that a continuum mechanical approach for the estimation of intracellular force transmission to the nucleus at the micron scale without considering the cell structural heterogeneity may be misleading (Deguchi et al. 2005b). Surely such microscopic studies are required, in addition to macroscopic studies by fluorescence microscopy, for a thorough understanding of the involvement of nuclear mechanics in mechanotransduction.

REFERENCES

Aebi U, Cohn J, Buhle L, and Gerace, L. 1986. The nuclear lamina is a meshwork of intermediate-type filaments. *Nature* **323**, 560–564.

Barbee K A, Davies P F, and Lal R. 1994. Shear stress-induced reorganization of the surface topography of living endothelial cells imaged by atomic force microscopy. *Circul Res* **74**, 163–171.

Bettinger B T, Gilbert D M, and Amberg D C. 2004. Actin up in the nucleus. *Nature* **5**, 410–415.

Brangwynne C P, MacKintosh F C, Kumar S, Geisse N A, Talbot J, Mahadevan L, Parker K K, Ingber D E, and Weitz D A. 2006. Microtubules can bear enhanced compressive loads in living cells because of lateral reinforcement. *J Cell Biol* **173**, 733–741.

Burke B and Stewart C L. 2002. Life at the edge: The nuclear envelope and human disease. *Nat Rev Mol Cell Bio* **3**, 575–585.

Caille N, Tardy Y, and Meister J J. 1998. Assessment of strain field in endothelial cells subjected to uniaxial deformation of their substrate. *Ann Biomed Eng* **26**, 409–416.
Caille N, Thoumine O, Tardy Y, and Meister J J. 2002. Contribution of the nucleus to the mechanical properties of endothelial cells. *J Biomech* **35**, 177–187.
Dahl K N, Kahn S M, Wilson K L, and Discher D E. 2004. The nuclear envelope lamina network has elasticity and a compressibility limit suggestive of a molecular shock absorber. *J Cell Sci* **117**, 4779–4786.
Dahl K N, Engler A J, Pajerowski J D, and Discher D E. 2005. Power-law rheology of isolated nuclei with deformation mapping of nuclear substructures. *Biophys J* **89**, 2855–2864.
Dahl K N, Scaffidi P, Islam M F, Yodh A G, Wilson K L, and Misteli T. 2006. Distinct structural and mechanical properties of the nuclear lamina in Hutchinson-Gilford progeria syndrome. *Proc Natl Acad Sci USA* **103**, 10271–10276.
Davies P F. 1995. Flow-mediated endothelial mechanotransduction. *Physiol Rev* **75**, 519–560.
Deguchi S, Maeda K, Ohashi T, and Sato M. 2005a. Flow-induced hardening of endothelial nuclei as an intracellular stress-bearing organelle. *J Biomech* **38**, 1751–1759.
Deguchi S, Ohashi T, and Sato M. 2005b. Intracellular stress transmission through actin stress fiber network in adherent vascular cells. *Mol Cell Biomech* **2**, 205–216.
Deguchi S, Ohashi T, and Sato M. 2006. Tensile properties of single stress fibers isolated from cultured vascular smooth muscle cells. *J Biomech* **39**, 2603–2610.
Deguchi S, Yano M, Hashimoto K, Fukamachi H, Washio S, and Tsujioka K. 2007. Assessment of the mechanical properties of the nucleus inside a spherical endothelial cell based on microtensile testing. *J Mech Mater Struct* **2**, 1087–1102.
Desprat N, Richert A, Simeon J, and Asnacios A. 2005. Creep function of a single living cell. *Biophys J* **88**, 2224–2233.
Dong C, Skalak R, and Sung K L P. 1991. Cytoplasmic rheology of passive neutrophils. *Biorheology* **28**, 557–567.
Fernandez P, Pullarkat P A, and Ott A. 2006. A master relation defines the nonlinear viscoelasticity of single fibroblasts. *Biophys J* **90**, 3796–3805.
Flaherty J T, Pierce J E, Ferrans V J, Patel D J, Tucker W K, and Fry D L. 1972. Endothelial nuclear patterns in the canine arterial tree with particular reference to hemodynamic events. *Circ Res* **30**, 23–33.
Guilak F. 1995. Compression-induced changes in the shape and volume of the chondrocyte nucleus. *J Biomech* **28**, 1529–1541.
Guilak F, Tedrow J R, and Burgkart R. 2000. Viscoelastic properties of the cell nucleus. *Biochem Biophys Res Commun* **269**, 781–786.
Haga J H, Li Y J, and Chien S. 2007. Molecular basis of the effects of mechanical stretch on vascular smooth muscle cells. *J Biomech* **40**, 947–960.
Haque F, Lloyd D J, Smallwood D T, Dent C L, Shanahan C M, Fry A M, Trembath R C, Shackleton S. 2006. SUN1 interacts with nuclear lamin A and cytoplasmic nesprins to provide a physical connection between the nuclear lamina and the cytoskeleton. *Mol Cell Biol* **26**, 3738–3751.
Hay M and De Boni U. 1991. Chromatin motion in neuronal interphase nuclei: Changes induced by disruption of intermediate filaments. *Cell Motil Cytoskel* **18**, 63–75.
Houchmandzadeh B, Marko J F, Chatenay D, and Libchaber A. 1997. Elasticity and structure of eukaryote chromosomes studied by micromanipulation and micropipette aspiration. *J Cell Biol* **139**, 1–12.
Ingber D E. 1997. Tensegrity: The architectural basis of cellular mechanotransduction. *Annu Rev Physiol* **59**, 575–599.
Janmey P A. 1998. The cytoskeleton and cell signaling: Component localization and mechanical coupling. *Physiol Rev* **78**, 763–781.
Kan H C, Shyy W, Udaykumar H S, Vigneron P, and Tran-Son-Tay R. 1999. Effects of nucleus on leukocyte recovery. *Ann Biomed Eng* **27**, 648–655.

Kasza K E, Rowat A C, Liu J, Angelini T E, Brangwynne C P, Koenderink G H, and Weitz D A. 2006. The cell as a material. *Curr Opin Cell Biol* **19**, 1–7.

Kawakami K, Tatsumil H, and Sokabe M. 2001. Dynamics of integrin clustering at focal contacts of endothelial cells studied by multimode imaging microscopy. *J Cell Sci* **114**, 3125–3135.

Knight M M, van de Breevaart Bravenboer J, Lee D A, van Osch G J V M, Weinans H, and Bader D L. 2002. Cell and nucleus deformation in compressed chondrocyte-alginate constructs: Temporal changes and calculation of cell modulus. *BBA* **1570**, 1–8.

Lammerding J, Schulze P C, Takahashi T, Kozlov S, Sullivan T, Kamm R D, Stewart C L, and Lee R T. 2004. Lamin A/C deficiency causes defective nuclear mechanics and mechanotransduction. *J Clin Invest* **113**, 370–378.

Lammerding J, Hsiao J, Schulze PCh., Kozlov S, Stewart C L, Lee R T. 2005. Abnormal nuclear shape and impaired mechanotransduction in emerin-deficient cells. *J Cell Biol*, **170**, 781–791.

Lee JS, Chang M I, Tseng Y, and Wirtz D. 2005. Cdc42 mediates nucleus movement and MTOC polzarization in Swiss 3T3 fibroblasts uder mechanical shear stress. *Mol Biol Cell* **16**, 871–880.

Li Y J, Haga J H, and Chien S. 2005. Molecular basis of the effects of shear stress on vascular endothelial cells. *J Biomech* **38**, 1949–1971.

Maniotis A J, Chen C S, and Ingber D E. 1997. Demonstration of mechanical connections between integrins, cytoskeletal filaments, and nucleoplasm that stabilize nuclear structure. *Proc Natl Acad Sci USA* **94**, 849–854.

Miyazaki H and Hayashi K. 1999. Atomic force microscopic measurement of the mechanical properties of intact endothelial cells in fresh arteries. *Med Biol Eng Comput* **37**, 530–536.

Muramatsu M, Smetana K, and Busch H. 1963. Quantitative aspects of isolation of nucleoli of the walker carcinosarcoma and liver of the rat. *Cancer Res* **25**, 693–697.

Nakayasu H and Berezney R. 1991. Nuclear matrins: Identification of the major nuclear matrix proteins. *Proc Natl Acad Sci USA* **88**, 10312–10316

Newport J W and Forbes D J. 1987. The nucleus: Structure, function, and dynamics. *Ann Rev Biochem* **56**, 535–65.

Ostlund C and Worman H J. 2003. Nuclear envelope proteins and neuromuscular diseases. *Muscle Nerve* **27**, 393–406.

Panorchan P, Schafer B W, Wirtz D, and Tseng Y. 2004, Nuclear envelope breakdown requires overcoming the mechanical integrity of the nuclear lamina. *J Biol Chem* **279**, 43462–43467.

Replogle-Schwab T S, Getzenberg R H, Donat T L, and Pienta K J. 1996. Effect of organ site on nuclear matrix protein composition. *J Cell Biochem* **62**, 132–141.

Rowat A C, Foster L J, Nielsen M M, Weiss M, and Ipsen J H. 2005. Characterization of the elastic properties of the nuclear envelop. *J Royal Soc Interface* **63**, 63–69.

Rowat A C, Lammerding J, and Ipsen J H. 2006. Mechanical properties of the cell nucleus and the effect of emerin deficiency. *Biophys J* **91**, 4649–4664.

Sato, M, Levesque, M J, and Nerem, R M. 1987. Micropipette aspiration of cultured bovine aortic endothelial cells exposed to shear stress. *Arteriosclerosis* **7**, 276–286.

Shumaker D K, Kuczmarski E R, and Goldman R D. 2003. The nucleoskeleton: Lamins and actin are major players in essential nuclear functions. *Curr Opin Cell Biol* **16**, 358–366.

Sims J R, Karp S, and Ingber D E. 1992. Altering the cellular mechanical force balance results in integrated changes in cell, cytoskeletal and nuclear shape. *J Cell Sci* **103**, 1215–1222.

Tetko I V, Haberer G, Rudd S, Meyers B, Mewes H W, and Mayer KFX. 2006. Spatiotemporal expression control correlates with intragenic scaffold matrix attachment regions (S/MARs) in Arabidopsis thaliana. *PLoS Comput Biol* **2**, 131–145.

Tseng Y, Lee J S, Kole T P, Jiang I, and Wirtz D. 2004. Micro-organization and viscoelasticity of the interphase nucleus revealed by particle nanotracking. *J Cell Sci* **117**, 2159–2167.

Vaziri A, Lee H, and Kaazempur-Mofrad M R. 2006. Deformation of the cell nucleus under indentation: Mechanics and mechanisms. *J Mater Res* **21**, 2126–2135.

Vaziri A and Kaazempur-Mofrad M R. 2007. Mechanics and deformation of the nucleus in micropipette aspiration experiment. *J Biomech* **40**, 2053–2062.

Vaziri A, Gopinath A, and Deshpande V S. 2007. Continuum-based computational models for cell and nuclear mechanics. *J Mech Mater Struct* **2**, 1169–1191.

Wang N, Butler J P, and Ingber D E. 1993. Mechanotransduction across the cell surface and through the cytoskeleton. *Science* **260**, 1124–1127.

Wilson K. 2005. Integrity matters: Linking nuclear architecture to lifespan. *Proc Natl Acad Sci USA* **102**, 18767–18768.

Yamaguchi T and Hanai S. 1988. To what extent does a minimal atherosclerotic plaque alter the arterial wall shear stress distribution? – A model study by an electrochemical method. *Biorheology* **25**, 31–36.

10

Microtubule Bending and Breaking in Cellular Mechanotransduction

Andrew D. Bicek, Dominique Seetapun, and David J. Odde

10.1 Introduction

Cellular mechanotransduction is the mechanism by which living cells respond to mechanical signals from their environment. As early as 1892, Julius Wolff described the ability of bone to be deposited and resorbed in accordance with the mechanical stresses placed upon it, implying that the bone must have some internal mechanical stress or strain sensors (Huiskes and Verdonschot, 1997; Roesler, 1987; Wolff, 1892). More recently, investigating the precise biochemical mechanisms by which a direct mechanical stimulus is converted into a cellular response has become an area of interest, and the macro-scale effects of mechanotransduction, such as the alignment of load-bearing components, are now widely recognized. For example, the extracellular matrix protein, collagen, is organized into a hierarchy of fibrillar structures by tenocytes to form a tendon that functionally transmits mechanical tension (Kastelic et al., 1978). Additionally, vascular endothelial cells have been observed to align and alter their morphology in response to an applied fluid shear stress (Levesque and Nerem, 1985). In another example of cells sensing a mechanical stimulus, neuronal cells are capable of responding directly to a tensile force through neurite initiation and extension, a phenomenon termed "towed growth" (Bray, 1984; Fass and Odde, 2003; Fischer et al., 2005; Heidemann and Buxbaum, 1990; Pfister et al., 2004). Since individual cells are capable of responding directly to an applied force via secreting, organizing, and remodeling the extracellular matrix, or through morphological and gene expression changes, mechanotransduction is presumably controlled and integrated into a response at the cellular level.

Perhaps the best documentation of cellular mechanotransduction is the role of mechanically gated ion channels in hearing (Hudspeth, 1989). The stereocilia of the auditory hair cells vibrate and bend with incoming sound waves. As the stereocilia bend, a linker protein filament is tensed between two adjacent cilia and the tension generated opens a mechanically gated ion channel. Opening of the ion channel causes an influx of positive charges that depolarize the hair cell and lead to an electrical signal that the brain interprets as sound. While this is a clear example of a mechanotransduction event, it is also clear that mechanically gated ion channels are not the sole mechanism for mechanotransduction in every cell. Other structures within the cell therefore need to be identified and investigated for their mechanosensory features, with the cytoskeleton being a leading candidate.

A potential cellular-based mechanosensory apparatus is the microtubule cytoskeleton. The microtubule cytoskeleton, which is comprised of individual microtubules typically nucleated from a microtubule organizing center near the nucleus, usually forms a radial network that radiates outward to the cell periphery. The microtubules within this radial array are composed of individual αβ-tubulin heterodimers, which self-assemble head-to-tail and side-to-side to form a rigid tubular structure 25 nm in outer diameter, and typically micrometers long (Desai and Mitchison, 1997). These microtubules are inherently polarized due to their head-to-tail assembly, and this allows for the processive movement of cargo via motor proteins that move specifically toward either the plus or the minus end. In addition, microtubules have an unusual self-assembly behavior where assembling (polymerizing) microtubules switch stochastically to a disassembling state (depolymerizing), and back again, a process known as dynamic instability. These characteristics allow the microtubule cytoskeleton to serve diverse roles within the cell, which include structural resistance to mechanical deformation, transportation of intracellular cargo via coupling to molecular motors, and the ability to probe the intracellular environment to act as a signaling structure via plus-end–tracking proteins. For these reasons, the microtubule cytoskeleton is a reasonable candidate for integrating a mechanical stimulus into a cellular response, and providing a mechanism for cellular mechanotransduction.

In this chapter, we discuss how the diverse roles of the microtubule, including structural, transport, and signaling, could contribute to cellular mechanotransduction. We also explore the potential origins of mechanical forces that deform and break microtubules and identify events that could act as a mechanical "trigger." Finally, we consider the cellular consequences of mechanical deformation and breaking of microtubules.

10.2 Roles of the Microtubule

10.2.1 Structural

The microtubule has long been thought to be important for structural integrity in the cell because its rigidity allows it to resist deformation under compressive loads. The rigidity of a microtubule arises from its structure. Due to its hollow tubular structure, the mass of the microtubule is distributed in its wall, thereby increasing the second moment of the cross-sectional area I. When compared to a solid rod of the same mass, this structural feature greatly increases the microtubule's resistance to bending. The flexural rigidity EI is given by the product of the elastic modulus E and the second moment of the cross-sectional area I and is a measure of the microtubules' bending resistance. The flexural rigidity has been estimated for microtubules *in vitro*, and the reported values range over two orders of magnitude from 1 to 200 pN-μm^2, depending on the experimental conditions and the measurement technique

(Cassimeris et al., 2001; Felgner et al., 1996, 1997; Fygenson et al., 1997; Gittes et al., 1993; Janson and Dogterom, 2004; Kis et al., 2002; Kurachi et al., 1995; Kurz and Williams, 1995; Mickey and Howard, 1995; Pampaloni et al., 2006; Takasone et al., 2002; Venier et al., 1994). As an example, the flexural rigidity of a microtubule was measured *in vitro* to be 26 pN-μm^2 (Mickey and Howard, 1995), which corresponds to an elastic modulus of 1.9 Gpa (Howard, 2001) and is comparable to Plexiglas®. From this comparison, it is apparent that microtubules are potentially able to serve as rigid support structures.

In specialized eukaryotic cells that contain cilia or flagella, microtubules serve a structural role, and the cell utilizes their stiff nature to its advantage. Cilia and flagella are made up of specialized microtubule structures that are used to propel the cell through its environment or to move extracellular fluid relative to the cell. Through coupling with dynein molecular motors, the stiff microtubule structure is "whipped" for cell locomotion in fluid environments. In this case, the stiffness of the microtubule backbone is roughly matched to the amount of force produced when the cilia or flagella are whipped. If the microtubules were either much more or much less flexible, then the cilia or flagella would be less efficient in propelling the fluid relative to the cell.

However, in other cell types that lack cilia or flagella, the structural role of microtubules remains controversial. In one view, microtubules are thought of as rigid struts that resist compressive loading in the cell and balance the tensional forces produced by actomyosin contractility (Buxbaum and Heidemann, 1988; Ingber, 1993; Wang et al., 1993). This view is supported by the slow retrograde flow of the actin cytoskeleton, the contraction of the cell, and the externally applied force all compressively loading the microtubules and causing observed buckling (Brangwynne et al., 2006; Schaefer et al., 2002; Wang et al., 2001). However, in another view, microtubules are thought to be too weak to resist large compressive forces, and therefore they are observed to be mechanically buckled and fluid in nature (Heidemann et al., 1999). Disruption of the microtubule cytoskeleton only transfers about 13% more stress to the extracellular adhesion sites, indicating that the structural role is limited (Stamenovic et al., 2001). Indeed, microtubules have been observed to break after buckling in localized regions of the cell, promoting spatial regulation of microtubule turnover (Gupton et al., 2002). Similarly, in another study, microtubules are more likely to break when highly curved, suggesting they are susceptible to being overpowered by compressive forces acting on them within the cell (Odde et al., 1999). Since the lengths of microtubules are often relatively long in living cells (10–30 μm), the force estimated to buckle them in endwise compression (Euler buckling) is very small (in the range of a single picoNewton, pN). The relative magnitude of this force is small when compared to molecular motors, which can generate 1–10 pN of force per motor (Howard, 2001). In addition, the sheer number of microtubules per cell (~ 1000) does not seem to be sufficient to significantly resist the forces generated through the actomyosin machinery (~ 30,000 myosin II motors assuming 1 μM myosin II expression in 10 μm diameter cell), which could generate on the order of ~ 100 nN of force given an estimate of 4 pN per motor force generation capability.

To complicate the matter, a study reports that individual microtubules can potentially bear enhanced compressive loads due to the lateral reinforcement of elastic elements in the cell (Brangwynne et al., 2006) and thereby increase by ~ 100-fold the critical buckling force of a microtubule. This finding supports the role of microtubules as structural elements; however, without a complete understanding of the origin and magnitude of all the forces acting in the cellular system, it is difficult to make accurate predictions with the Euler buckling model. Also, Brangwynne et al. reported that in cardiac myocytes some microtubules buckle upon cell contraction while other nearby microtubules remain unbuckled (Brangwynne et al., 2006). This suggests that only a fraction of all microtubules are coupled to elastic elements in the cell, which may limit the ability of the microtubule array to bear enhanced loads. It will be interesting to determine quantitatively the loads that microtubules actually bear *in vivo*.

Regardless of the degree to which microtubules serve a structural role, it is significant that microtubules clearly respond to forces by deforming and breaking. Bending implies that the microtubule is storing potential energy, which could potentially be harnessed to do work. This raises an important question in the field: What is the significance of microtubule deformation and breaking in the cell? More specifically, is microtubule deformation and breaking capable of being a sensor that triggers a mechanotransduction event?

10.2.2 Transport

Microtubules also serve as tracks to transport cargo around the cell via molecular motors. The molecular structure of the microtubule is important for its transport properties. The α-subunit is similar to the β-subunit, and the two are joined together in a head-to-tail fashion. This inherently causes polarity in the microtubule, with the plus end found on the β-tubulin and the minus end on the α-tubulin (Mitchison, 1993). Usually, microtubule plus ends point outward toward the cell periphery, and minus ends point inward and attach to the centrosome, the microtubule-organizing center within the cell; however, the opposite has been found in limited cases (Baas et al., 1988).

The polarity of microtubules in the cell is important for the transport of cargo. First, the polarity serves as a guide to the direction of microtubule motor movement. Kinesin, generally a plus-end–directed motor (Vale et al., 1985) moves cargo such as membrane-bound vesicles and organelles away from the cell center, while dynein, a minus-end–directed motor (Vallee et al., 1988), moves cargo toward the cell center. Also, the assembly rates for microtubules are higher at the plus ends when compared with the minus ends. This promotes plus-end growth and allows the microtubule tip to probe the local environment in the cell (McNally, 2001).

The ability to move cargo around the cell could be part of a mechanotransduction process. If the microtubule cytoskeleton is capable of structurally reorganizing in response to a mechanical stimulus, the microtubule's cargo may act as a signaling mechanism to integrate the mechanical signal into a cellular response. For example, during nerve growth cone advance, a key step in the motility is the engorgement of the growth cone with vesicles from the axon via a microtubule-based transport mechanism (Goldberg and Burmeister, 1986; Suter and Forscher, 1998). In this model, the microtubule cytoskeleton senses the mechanical stimulus, and the vesicle-based cargo serves as the signaling mechanism to change the cell's phenotype.

A potential benefit of the microtubule cytoskeleton reorganizing in response to mechanical force is that it can maintain its radial array, which is important for the efficient transport of cargo within the cell. If microtubules are allowed to polymerize freely and deform by applied forces, the radial array of microtubules could soon look similar to a worm-like chain. This would make intracellular transport inefficient and could potentially cause transportation problems if the microtubule cytoskeleton were allowed to become so tortuous that the polarity of the microtubule network in the cell was compromised.

10.2.3 Dynamic Instability and Signaling

In living cells, microtubules are constantly changing back and forth between growing and shrinking states. This phenomenon is known as dynamic instability, and it describes the stochastic switching between slow growth phases and rapid shortening phases (Mitchison and Kirschner, 1984). Dynamic instability can be described by four parameters: the growth and shortening velocities, and the frequency of switching between those two states, also known as the catastrophe and rescue frequencies. *In vivo*, these dynamics occur most prominently at the plus ends, which extend outward to the cell periphery. Dynamic instability has been proposed as the mechanism for the capture of microtubules by kinetochores during mitosis. If a microtubule plus end makes contact with the kinetochore during a random growth phase, then the microtubule is captured and the connection is kept. However, if a kinetochore connection is not made, the microtubule eventually undergoes catastrophe, depolymerizes, and a new randomly directed microtubule grows from the spindle pole. In this way, the microtubule array is able to explore the region around the chromosome to form persistent attachments to kinetochores. This model has also been proposed as a general means for the cell to reorganize and regulate its cytoskeleton. In this model, microtubules grow radially outward in random directions from the centrosome (the microtubule-organizing center where microtubule minus ends are embedded) into the cytoplasmic periphery, where the plus ends can be locally stabilized (Kirschner and Mitchison, 1986).

Dynamic instability allows for the microtubule plus end to probe its local environment, which is potentially important in cell signaling. In particular, the establishment and maintenance of cell polarity have been attributed to the presence of a dynamic, polarized microtubule array. In epithelial and fibroblast cells cultured *in vitro*, cell migration is achieved through the protrusion of a large lamella at the leading edge and a polarized stable microtubule array pointing toward the direction of migration (Gundersen and Bulinski, 1988; Nagasaki et al., 1992). Leading edge (lamellipodial) protrusion decreases dramatically when drugs that inhibit microtubule dynamic instability are applied (Liao et al., 1995). One possible explanation is that microtubules participate in a feedback loop that maintains the leading edge, thereby sustaining forward migration, through the activation of a signaling cascade that stimulates cell migration (Fukata et al., 2002, 2003; Siegrist and Doe, 2007; Waterman-Storer et al., 1999).

The microtubule's role in the signaling cascade depends on proteins associated with their plus ends. These proteins are capable of altering microtubule assembly dynamics and can provide a signal to a local part of the cell, such as the cell cortex. In mammalian cells, CLIP-170 is an example of a plus-end–localizing protein that incorporates into the growing tip of the microtubule, and then dissociates from the microtubule lattice as the tip continues to grow, a phenomenon known as "tip-tracking." In this way, CLIP-170 is able to treadmill on the growing microtubule and stays concentrated at the leading edge (Perez et al., 1999).

The maintenance of the leading edge in migrating cells involves microtubule plus-end–binding proteins. Microtubule plus ends are captured by IQGAP1, a GTPase binding protein, at cortical actin sites along the leading edge of the cell. IQGAP1 can bind and cross-link actin filaments and is known to increase Rac1 activity (Briggs and Sacks, 2003; Fukata et al., 2003). Rac1 can stimulate leading edge protrusion, which is essential for cell migration. The linkage between microtubules and IQGAP1 is mediated by CLIP-170. Through CLIP-170, it has been posited that microtubules can activate or stabilize IQGAP1, stimulate Rac1 activity, and therefore affect lamellipodial protrusion and cell migration (Fukata et al., 2002; Siegrist and Doe, 2007). Additionally, microtubule capture by IQGAP1 via CLIP-170 has been shown to increase the interaction time between microtubules and cortical actin, temporarily stabilizing the microtubules and maintaining the polarized microtubule array (Fukata et al., 2002; Watanabe et al., 2004). In this way microtubules are able to promote migration and maintain polarization through their plus ends.

Microtubule plus ends are also capable of delivering signaling proteins that polarize the cell. For example, cultured neurons undergo a transition from an unpolarized state, having multiple immature neurites, to a polarized state by specifying one of the neurites as the axon (Dotti et al., 1988). An evolutionary conserved complex of proteins, referred to as the polarity complex, first identified in the *C. elegans* embryo, has been shown to be important for neuron polarization (Etienne-Manneville and Hall, 2003; Schneider and Bowerman, 2003). The highly conserved polarity complex, comprising of one protein kinase, an atypical protein kinase C (aPKC), and two scaffold proteins, Par3 and Par6, accumulates at the tip of the nascent axon to specify the axon (Nishimura et al., 2004; Shi et al., 2003). Accumulation of the polarity complex at the tip of the nascent axon is mediated through a direct interaction between the microtubules, adenomatous polyposis coli (APC), another microtubule plus-end–binding protein that promotes microtubule growth, and KIF3, a kinesin motor with plus-end–directed movement (Nishimura et al., 2004; Shi et al., 2004). By selectively delivering the polarity complex and APC to the microtubule plus end, microtubules are able to help specify which immature neurite is to become the future axon. Since microtubules clearly play a role in cell signaling, therefore, any mechanical stimulus that causes a change in the microtubule array could lead to changes in cell signaling and ultimately cellular phenotype.

10.3 Origin of Mechanical Forces That Deform MTs

10.3.1 Thermal Fluctuations

Despite being relatively stiff filaments, individual microtubules in living cells are often highly bent, the origin of which is not entirely clear. One possibility is that bending originates solely from random thermal forces, which are necessarily present. Thermal forces are clearly capable of deforming microtubules, as observed from *in vitro* assays (Cassimeris et al., 2001; Felgner et al., 1996, 1997; Fygenson et al., 1997; Gittes et al., 1993; Janson and Dogterom, 2004; Kurz and Williams, 1995; Mickey and Howard, 1995); however, this explanation as the sole deformation mechanism seems unlikely because microtubules *in vitro* are far less curved than microtubules observed in living cells at the same temperature. For thermal fluctuations to account for all the deformation observed in living cells, the *in vivo* flexural rigidity EI would have to be orders of magnitude lower than it is *in vitro*. However, given that EI has not been estimated *in vivo*, thermally driven bending remains a formal possibility. It is more likely, however, that numerous force mechanisms act on microtubules and deform them, only one of which is a thermal fluctuation.

10.3.2 Polymerization Forces

Microtubules are dynamic polymers that stochastically switch between persistent growth and shortening states. The addition of tubulin subunits at the plus ends can generate a force that is large enough to deform microtubules *in vitro* so that they bend and buckle when they polymerize end-on into an impenetrable barrier (Dogterom and Yurke, 1997). In a living cell, if a polymerizing microtubule encountered a similar barrier, the polymerization force would presumably cause microtubule buckling. A potential cellular-based barrier could be formed by the actin network, provided the density of actin filaments is high enough to inhibit microtubule penetration (pore size ~ 25 nm or less). Experimental evidence for such a barrier has been described in newt lung epithelial cells as a zone 5–10 μm back from the leading edge, where microtubules only occasionally penetrate beyond and more commonly bend and start growing parallel to the leading edge (Waterman-Storer and Salmon, 1997). In addition, treatment of nerve growth cones with cytochalasin B (an actin filament disrupting drug) resulted in microtubule penetration into the peripheral domain (P-domain), implying that the actin network was behaving as a barrier to microtubule polymerization (Forscher and Smith, 1988; Zhou et al., 2002).

10.3.3 Molecular Motors

The main force generators in living cells are cytoskeleton-based molecular motors such as myosin, kinesin, and dynein. These motor proteins bind to either the actin or the microtubule cytoskeleton and are capable of exerting force as well as transporting cargo around the cell.

In the case of actin filaments and myosin II motors, force is generated via sliding of the actin filaments with respect to each other, resulting in a net contractile stress on the actin network. In addition, continuous actin polymerization at the cell periphery, in many cases coupled to myosin motor activity, results in a retrograde flow of the actin network rearward toward the cell center. Microtubules have been observed to be synchronously coupled to this network and swept back at similar rates (Salmon et al., 2002). The divergence of the actin network retrograde flow rate results in large-scale microtubule buckling and subsequent breaking in some cases (Gupton et al., 2002; Schaefer et al., 2002). Finally, cardiac myocyte microtubules have been observed to cyclically buckle and unbuckle in synchrony with each contraction and relaxation of the beating cell (Brangwynne et al., 2006).

Just as actomyosin-based contractility can exert compressive forces on microtubules resulting in bending and breaking, microtubule-based motors may also directly exert forces that could potentially deform microtubules. In the axon of neurons, dynein, a minus-end–directed motor, is used to transport short segments (a few microns long) of microtubules toward the growth cone by coupling with the actin network via dynactin (Dillman et al., 1996; Pfister, 1999; Wang and Brown, 2002). Similarly in epithelial cells, anterograde (outward) microtubule transport, most likely mediated by molecular motors, results in microtubule bending and buckling (Bicek et al., 2009). Further, a complex force balance between plus-end (N-terminal kinesins)– and minus-end (both dynein and C-terminal kinesins)–directed motor proteins, actin filaments, and microtubules has been proposed as a model for the net transport of microtubules down the axon (Baas et al., 2006). If this model is applied to other cell types such as interphase epithelial cells that lack the geometric constraints of the long and slender axon, then microtubule motors such as dyneins and kinesins could conceivably bind to the actin network via their cargo domain and also cross-link to a nearby microtubule via their motor domain, or potentially cross-link between two microtubules. These motors would then be capable of exerting forces on the cytoskeleton and promoting bending. In this example, the actin network would presumably act as an immobile substrate for the motor to exert force upon, resulting in microtubule deformation (Bicek et al., 2009).

10.3.4 Externally Applied Force

In addition to the internal mechanisms of force generation utilizing cellular machinery, externally applied force is capable of altering microtubule dynamics and causing microtubule deformation. Putnam et al. cultured rat aortic smooth muscle cells on silicone rubber substrates and then exposed the rubber substrates to step changes in strain (Putnam et al., 1998, 2001, 2003). By stretching or relaxing the elastic membrane to which the cells were adhered, they were able to exert an externally applied load on the cells in both tension and compression. While they did not directly observe microtubule deformation, they showed that by applying a 10% tensional strain to the rubber substrate, microtubules responded by assembling, and, conversely, by applying a 10% compressive strain to the rubber substrate, microtubules responded by disassembling. They concluded that assembly and disassembly of microtubules can be regulated by externally applied mechanical forces, whereby relieving compressive forces on the microtubule induces assembly, and applying compressive forces to the microtubule induces disassembly. Consistent with these observations is the report that tensile stress stimulates microtubule outgrowth (Kaverina et al., 2002).

However, in a contradictory report, Heidemann et al. did not find evidence of compression-induced microtubule disassembly in rat embryo fibroblasts when the cells were directly observed rounding up from treatment with EDTA; rather, they found large-scale buckling of microtubules, suggesting microtubules in fibroblasts are limited in supporting compressive loads (Heidemann et al., 1999). In addition, application of a force to the cell through a glass microneedle is capable of directly deforming the underlying microtubules (Brangwynne et al., 2006; Heidemann et al., 1999). While the contradictory nature of these results highlights the controversy in the structural role of microtubules, it provides evidence that microtubules are capable of responding to externally applied mechanical forces through both buckling and altered assembly dynamics.

10.4 Consequences of Mechanical Deformation of the MT

10.4.1 Microtubule Bending

Even though microtubules are usually characterized as stiff, rigid polymers, they are more commonly observed in living cells in a bent or buckled conformation, rather than a straight conformation. It is clear that applied force on the microtubule lattice results in microtubule bending and deformation (as outlined previously); however, the important question remains: What are the consequences of microtubule deformation?

One consequence of microtubule deformation may be its effect on dynamic instability. For example, compressive forces applied at growing microtubule tips *in vitro* indicate that the growth velocity of microtubules that are polymerizing against a barrier is reduced from 1.2 μm/min with zero load to 0.2 μm/min for loads of 3–4 pN (Dogterom and Yurke, 1997). The reduction in the polymerization rate is presumably a result of limiting tubulin access to the microtubule tip and thereby limiting the on-rate of the tubulin addition (Janson et al., 2003). Inhibition of the on-rate of tubulin is similar to, but not identical with, lowering the tubulin concentration, which will both slow assembly and promote catastrophe (Janson et al., 2003; Walker et al., 1988). This simple mechanism could select the mechanically deformed (i.e., bent) microtubules for depolymerization and lead to reorganization of the cell to a more mechanically relaxed configuration (i.e., straight). Experimental *in vivo* evidence has shown that the microtubule catastrophe rate is higher at the cell periphery than in the cell body (Brunner and Nurse, 2000; Drummond and Cross, 2000; Komarova et al., 2002; Tran et al., 2001), and these findings support a force-induced catastrophe mechanism in living cells (Janson et al., 2003). One feature of this model is that it argues against a major structural role for microtubules, as it predicts that microtubules under compression would be short-lived and would soon catastrophe.

10.4.2 Microtubule Breaking

In addition to bending, mechanical forces promote microtubule breaking (Gupton et al., 2002; Odde et al., 1999; Waterman-Storer and Salmon, 1997). It has been shown that increased curvature of the microtubule lattice correlates with increases in the rate of microtubule breaking of up to $\sim$40-fold (Odde et al., 1999). Bending presumably destabilizes the microtubule lattice by increasing the elastic strain energy and could lead to failure of the microtubule if the tubulin–tubulin bonds are broken. In addition, an active process for breaking via the microtubule severing enzyme katanin (Vale, 1991; McNally and Vale, 1993) or spastin (Errico et al., 2002; Evans et al., 2005) could be associated with increased microtubule curvature. Katanin is an ATP-dependent severing enzyme that has been shown to play a role in cell division (McNally and Thomas, 1998), axonal outgrowth (Ahmad et al., 1999), flagellar resorption (Lohret et al., 1999), and spindle length in dividing cells (McNally et al., 2002, 2006; Yang et al., 2003). However, the mechanism for katanin-mediated severing and the effect of microtubule curvature on katanin's activity still remains unclear. One hypothesis would be that local microtubule curvature could change the off-rate for tubulin- and microtubule-associated proteins (MAPs), resulting in a net dissociation of both MAPs and tubulin from the microtubule lattice. This action could clear a binding site for katanin and lead to severing (Odde et al., 1999). This hypothesis is supported by the findings that MAPs can inhibit katanin activity (McNally et al., 2002), and that microtubule lattice defects provide a preferred site for katanin severing (Davis et al., 2002). In addition, the sizes of these defects could be increased through the deformation of the microtubule, which might alter the on- and off-rate constants of tubulin enough to promote defect growth and thus provide more access for katanin to sever the microtubule. In addition, lattice defects have also been proposed to affect the rigidity of microtubules and produce more flexible microtubules (Janson and Dogterom, 2004), thereby potentially increasing their sensitivity to bending and severing via katanin. This model could provide a mechanism for the cell to replace mechanically stressed microtubules by marking them as "defective," with subsequent degradation via katanin or spastin.

When microtubules break, one consequence is that they depolymerize from the newly exposed ends (Keating et al., 1997; McNally and Vale, 1993; Odde et al., 1999; Waterman-Storer and Salmon, 1997). Microtubule depolymerization increases the free tubulin concentration in the cytoplasm so that the released tubulin subunits can be incorporated into another growing microtubule. Through curvature-induced breaking, the microtubule network could be remodeled over time to accommodate changes in cellular shape (Gupton et al., 2002; Odde et al., 1999). Additionally, microtubule depolymerization can trigger NF-κB gene expression, which results in increased synthesis of platelet-derived growth factor (PDGF) (Khachigian et al., 1995; Rosette and Karin, 1995). PDGF is important in cell proliferation and wound healing. Microtubule-breaking events may therefore initiate a cellular signaling cascade in response to mechanical stress.

In addition, when a microtubule breaks, a new plus end is formed that is free to probe the cell through its growth and shortening dynamics. This new plus end may deliver signals that affect the cell, perhaps through a microtubule-associated guanine-nucleotide-exchange factor (GEF) regulating Rho protein activity (Wittmann and Waterman-Storer, 2001). Interestingly, polymerization of microtubules has been shown to activate Rac1, while depolymerization of microtubules results in an increase of RhoA (Ren et al., 1999; Waterman-Storer et al., 1999). A mechanical breaking event could therefore alter the growth and shortening dynamics and provide G-protein–mediated signals to the cell's actin cytoskeleton.

Finally, in neurons, microtubule breaking is important in the axonal transport of microtubules, and axonal branching (Ahmad et al., 1999; Baas et al., 2006). Only short microtubules (a few microns) are transported down the axon (Wang and Brown, 2002) and at axonal branching sites large numbers of short microtubules are present (Yu et al., 1994). These data support a model for microtubule transport down the axon where microtubules are cut into short sections and then transported by motor forces to the growth cone (Baas et al., 2005).

10.5 Conclusions

While the structural role of microtubules remains controversial, it seems clear that microtubules are subject to mechanical forces. This is important because for the cell to recognize a mechanical stimulus, its mechanosensory apparatus must be sensitive enough to respond to the ranges of applied force. Not only are microtubules physically located in a radial array that may help to sense a mechanical force, but they commonly are observed to bend, buckle, and break in response to both internally and externally applied forces. Many pathways have been identified that are capable of exerting forces large enough to deform microtubules; however, the specific details of force transmission and the contribution of each pathway remain unclear.

There are two leading mechanisms by which the microtubule cytoskeleton responds to a mechanical stimulus. The first mechanism is by altering the dynamic instability parameters, and the second is via microtubule breaking. In the first case, polymerization dynamics of the microtubule provide a self-assembly mechanism that is inherently sensitive to mechanical forces. Through self-assembly into physical barriers or the application of a compressive force in an endwise manner, the microtubule could alter its growth dynamics, leading to selective depolymerization. This simple mechanism could lead to the reorganization of the microtubule cytoskeleton in response to an applied force in a short time. In addition, there is mounting evidence that there are physical barriers in the cell that inhibit microtubule polymerization, in particular a dense actin network at the cell cortex.

In the second case, microtubule breaking could also be a response to a mechanical event. Breaking has many consequences, including polymerization and depolymerization of the newly exposed ends. Both of these pathways are capable of signaling to the cell that it is under stress. In addition, the new plus end is free to explore the cell via random growth and shortening, and could provide signaling to the cell through the small Rho–GTPase pathways. Alternatively, plus-end–binding proteins could play a signaling role, and reorganization of the plus ends changes the destination of motor-based cargo, perhaps affecting the cell. Therefore, through force-induced bending and breaking leading to altered assembly dynamics, the microtubule network could remodel over time, provide mechanical force integration and subsequent signaling, and thereby ultimately serve a mechanotransduction function in the cell.

REFERENCES

Ahmad, F. J., Yu, W., McNally, F. J. and Baas, P. W. (1999). An essential role for katanin in severing microtubules in the neuron. *J Cell Biol* **145**, 305–15.

Baas, P. W., Deitch, J. S., Black, M. M. and Banker, G. A. (1988). Polarity orientation of microtubules in hippocampal neurons: Uniformity in the axon and nonuniformity in the dendrite. *Proc Natl Acad Sci USA* **85**, 8335–9.

Baas, P. W., Karabay, A. and Qiang, L. (2005). Microtubules cut and run. *Trends Cell Biol* **15**, 518–24.

Baas, P. W., Vidya Nadar, C. and Myers, K. A. (2006). Axonal transport of microtubules: The long and short of it. *Traffic* **7**, 490–8.

Bicek, A. D., Tuzel, E., Demtchouk, A., Uppalapati, M., Hancock, W. O., Kroll, D. M., and Odde, D. J. (2009). Anterograde microtubule transport drives microtubule bending in LLC-PK1 epithelial cells. *Mol Biol Cell* **20**, 2943–53.

Brangwynne, C. P., MacKintosh, F. C., Kumar, S., Geisse, N. A., Talbot, J., Mahadevan, L., Parker, K. K., Ingber, D. E. and Weitz, D. A. (2006). Microtubules can bear enhanced compressive loads in living cells because of lateral reinforcement. *J Cell Biol* **173**, 733–41.

Bray, D. (1984). Axonal growth in response to experimentally applied mechanical tension. *Dev Biol* **102**, 379–89.

Briggs, M. W. and Sacks, D. B. (2003). IQGAP proteins are integral components of cytoskeletal regulation. *EMBO Rep* **4**, 571–4.

Brunner, D. and Nurse, P. (2000). CLIP170-like tip1p spatially organizes microtubular dynamics in fission yeast. *Cell* **102**, 695–704.

Buxbaum, R. E. and Heidemann, S. R. (1988). A thermodynamic model for force integration and microtubule assembly during axonal elongation. *J Theor Biol* **134**, 379–90.

Cassimeris, L., Gard, D., Tran, P. T. and Erickson, H. P. (2001). XMAP215 is a long thin molecule that does not increase microtubule stiffness. *J Cell Sci* **114**, 3025–33.

Davis, L. J., Odde, D. J., Block, S. M. and Gross, S. P. (2002). The importance of lattice defects in katanin-mediated microtubule severing *in vitro*. *Biophys J* **82**, 2916–27.

Desai, A. and Mitchison, T. J. (1997). Microtubule polymerization dynamics. *Annu Rev Cell Dev Biol* **13**, 83–117.

Diamantopoulos, G. S., Perez, F., Goodson, H. V., Batelier, G., Melki, R., Kreis, T. E. and Rickard, J. E. (1999). Dynamic localization of CLIP-170 to microtubule plus ends is coupled to microtubule assembly. *J Cell Biol* **144**, 99–112.

Dillman, J. F., III, Dabney, L. P., Karki, S., Paschal, B. M., Holzbaur, E. L. and Pfister, K. K. (1996). Functional analysis of dynactin and cytoplasmic dynein in slow axonal transport. *J Neurosci* **16**, 6742–52.

Dogterom, M. and Yurke, B. (1997). Measurement of the force-velocity relation for growing microtubules. *Science* **278**, 856–60.

Dotti, C. G., Sullivan, C. A. and Banker, G. A. (1988). The establishment of polarity by hippocampal neurons in culture. *J Neurosci* **8**, 1454–68.

Drummond, D. R. and Cross, R. A. (2000). Dynamics of interphase microtubules in Schizosaccharomyces pombe. *Curr Biol* **10**, 766–75.

Errico, A., Ballabio, A. and Rugarli, E. I. (2002). Spastin, the protein mutated in autosomal dominant hereditary spastic paraplegia, is involved in microtubule dynamics. *Hum Mol Genet* **11**, 153–63.

Etienne-Manneville, S. and Hall, A. (2003). Cell polarity: Par6, aPKC and cytoskeletal cross-talk. *Curr Opin Cell Biol* **15**, 67–72.

Evans, K. J., Gomes, E. R., Reisenweber, S. M., Gundersen, G. G. and Lauring, B. P. (2005). Linking axonal degeneration to microtubule remodeling by Spastin-mediated microtubule severing. *J Cell Biol* **168**, 599–606.

Fass, J. N. and Odde, D. J. (2003). Tensile force-dependent neurite elicitation via anti-beta1 integrin antibody-coated magnetic beads. *Biophys J* **85**, 623–36.

Felgner, H., Frank, R., Biernat, J., Mandelkow, E. M., Mandelkow, E., Ludin, B., Matus, A. and Schliwa, M. (1997). Domains of neuronal microtubule-associated proteins and flexural rigidity of microtubules. *J Cell Biol* **138**, 1067–75.

Felgner, H., Frank, R. and Schliwa, M. (1996). Flexural rigidity of microtubules measured with the use of optical tweezers. *J Cell Sci* **109(Pt 2),** 509–16.

Fischer, T. M., Steinmetz, P. N. and Odde, D. J. (2005). Robust micromechanical neurite elicitation in synapse-competent neurons via magnetic bead force application. *Ann Biomed Eng* **33**, 1229–37.

Forscher, P. and Smith, S. J. (1988). Actions of cytochalasins on the organization of actin filaments and microtubules in a neuronal growth cone. *J Cell Biol* **107**, 1505–16.

Fukata, M., Nakagawa, M. and Kaibuchi, K. (2003). Roles of Rho-family GTPases in cell polarisation and directional migration. *Curr Opin Cell Biol* **15**, 590–7.

Fukata, M., Watanabe, T., Noritake, J., Nakagawa, M., Yamaga, M., Kuroda, S., Matsuura, Y., Iwamatsu, A., Perez, F. and Kaibuchi, K. (2002). Rac1 and Cdc42 capture microtubules through IQGAP1 and CLIP-170. *Cell* **109**, 873–85.

Fygenson, D. K., Elbaum, M., Shraiman, B. and Libchaber, A. (1997). Microtubules and vesicles under controlled tension. *Phys Rev E* **55**, 850–59.

Gittes, F., Mickey, B., Nettleton, J. and Howard, J. (1993). Flexural rigidity of microtubules and actin filaments measured from thermal fluctuations in shape. *J Cell Biol* **120**, 923–34.

Goldberg, D. J. and Burmeister, D. W. (1986). Stages in axon formation: Observations of growth of Aplysia axons in culture using video-enhanced contrast-differential interference contrast microscopy. *J Cell Biol* **103**, 1921–31.

Gundersen, G. G. and Bulinski, J. C. (1988). Selective stabilization of microtubules oriented toward the direction of cell migration. *Proc Natl Acad Sci USA* **85**, 5946–50.

Gupton, S. L., Salmon, W. C. and Waterman-Storer, C. M. (2002). Converging populations of f-actin promote breakage of associated microtubules to spatially regulate microtubule turn-over in migrating cells. *Curr Biol* **12**, 1891–9.

Heidemann, S. R. and Buxbaum, R. E. (1990). Tension as a regulator and integrator of axonal growth. *Cell Motil Cytoskeleton* **17**, 6–10.

Heidemann, S. R., Kaech, S., Buxbaum, R. E. and Matus, A. (1999). Direct observations of the mechanical behaviors of the cytoskeleton in living fibroblasts. *J Cell Biol* **145**, 109–22.

Howard, J. (2001). *Mechanics of Motor Proteins and the Cytoskeleton*. Sunderland, MA: Sinauer Associates.

Hudspeth, A. J. (1989). How the ear's works work. *Nature* **341**, 397–404.

Huiskes, R. and Verdonschot, N. (1997). Biomechanics of Artificial Joints: The Hip. In *Basic Orthopaedic Biomechanics,* 2nd ed. (ed. V. C. Mow and W. C. Hayes). Philadelphia: Lippincott-Raven Publishers.

Ingber, D. E. (1993). Cellular tensegrity: Defining new rules of biological design that govern the cytoskeleton. *J Cell Sci* **104(Pt 3),** 613–27.

Janson, M. E., de Dood, M. E. and Dogterom, M. (2003). Dynamic instability of microtubules is regulated by force. *J Cell Biol* **161**, 1029–34.

Janson, M. E. and Dogterom, M. (2004). A bending mode analysis for growing microtubules: Evidence for a velocity-dependent rigidity. *Biophys J* **87**, 2723–36.

Kastelic, J., Galeski, A. and Baer, E. (1978). The multicomposite structure of tendon. *Connect Tissue Res* **6**, 11–23.

Kaverina, I., Krylyshkina, O., Beningo, K., Anderson, K., Wang, Y. L. and Small, J. V. (2002). Tensile stress stimulates microtubule outgrowth in living cells. *J Cell Sci* **115**, 2283–91.

Keating, T. J., Peloquin, J. G., Rodionov, V. I., Momcilovic, D. and Borisy, G. G. (1997). Microtubule release from the centrosome. *Proc Natl Acad Sci USA* **94**, 5078–83.

Khachigian, L. M., Resnick, N., Gimbrone, M. A., Jr. and Collins, T. (1995). Nuclear factor-kappa B interacts functionally with the platelet-derived growth factor B-chain shear-stress response element in vascular endothelial cells exposed to fluid shear stress. *J Clin Invest* **96**, 1169–75.

Kirschner, M. W. and Mitchison, T. (1986). Microtubule dynamics. *Nature* **324**, 621.

Kis, A., Kasas, S., Babic, B., Kulik, A. J., Benoit, W., Briggs, G. A. D., Schonenberger, C., Catsicas, S. and Forro, L. (2002). Nanomechanics of microtubules. *Phys Rev Lett* **89**, 248101.

Komarova, Y. A., Akhmanova, A. S., Kojima, S., Galjart, N. and Borisy, G. G. (2002). Cytoplasmic linker proteins promote microtubule rescue in vivo. *J Cell Biol* **159**, 589–99.

Kurachi, M., Hoshi, M. and Tashiro, H. (1995). Buckling of a single microtubule by optical trapping forces: Direct measurement of microtubule rigidity. *Cell Motil Cytoskeleton* **30**, 221–8.

Kurz, J. C. and Williams, R. C., Jr. (1995). Microtubule-associated proteins and the flexibility of microtubules. *Biochemistry* **34**, 13374–80.

Levesque, M. J. and Nerem, R. M. (1985). The elongation and orientation of cultured endothelial cells in response to shear stress. *J Biomech Eng* **107**, 341–7.

Liao, G., Nagasaki, T. and Gundersen, G. G. (1995). Low concentrations of nocodazole interfere with fibroblast locomotion without significantly affecting microtubule level: Implications for the role of dynamic microtubules in cell locomotion. *J Cell Sci* **108**, 3473–83.

Lohret, T. A., Zhao, L. and Quarmby, L. M. (1999). Cloning of Chlamydomonas p60 katanin and localization to the site of outer doublet severing during deflagellation. *Cell Motil Cytoskeleton* **43**, 221–31.

McNally, F. J. (2001). Cytoskeleton: CLASPing the end to the edge. *Curr Biol* **11**, R477–80.

McNally, F. J. and Thomas, S. (1998). Katanin is responsible for the M-phase microtubule-severing activity in Xenopus eggs. *Mol Biol Cell* **9**, 1847–61.

McNally, F. J. and Vale, R. D. (1993). Identification of katanin, an ATPase that severs and disassembles stable microtubules. *Cell* **75**, 419–29.

McNally, K., Audhya, A., Oegema, K. and McNally, F. J. (2006). Katanin controls mitotic and meiotic spindle length. *J Cell Biol* **175**, 881–91.

McNally, K. P., Buster, D. and McNally, F. J. (2002). Katanin-mediated microtubule severing can be regulated by multiple mechanisms. *Cell Motil Cytoskeleton* **53**, 337–49.

Mickey, B. and Howard, J. (1995). Rigidity of microtubules is increased by stabilizing agents. *J Cell Biol* **130**, 909–17.

Mitchison, T. and Kirschner, M. (1984). Dynamic instability of microtubule growth. *Nature* **312**, 237–42.

Mitchison, T. J. (1993). Localization of an exchangeable GTP binding site at the plus end of microtubules. *Science* **261**, 1044–7.

Nagasaki, T., Chapin, C. J. and Gundersen, G. G. (1992). Distribution of detyrosinated microtubules in motile NRK fibroblasts is rapidly altered upon cell-cell contact: Implications for contact inhibition of locomotion. *Cell Motil Cytoskeleton* **23**, 45–60.

Nishimura, T., Kato, K., Yamaguchi, T., Fukata, Y., Ohno, S. and Kaibuchi, K. (2004). Role of the PAR-3-KIF3 complex in the establishment of neuronal polarity. *Nat Cell Biol* **6**, 328–34.

Odde, D. J., Ma, L., Briggs, A. H., DeMarco, A. and Kirschner, M. W. (1999). Microtubule bending and breaking in living fibroblast cells. *J Cell Sci* **112(Pt 19)**, 3283–8.

Pampaloni, F., Lattanzi, G., Jonas, A., Surrey, T., Frey, E. and Florin, E. L. (2006). Thermal fluctuations of grafted microtubules provide evidence of a length-dependent persistence length. *Proc Natl Acad Sci USA* **103**, 10248–53.

Perez, F., Diamantopoulos, G. S., Stalder, R. and Kreis, T. E. (1999). CLIP-170 highlights growing microtubule ends in vivo. *Cell* **96**, 517–27.

Pfister, B. J., Iwata, A., Meaney, D. F. and Smith, D. H. (2004). Extreme stretch growth of integrated axons. *J Neurosci* **24**, 7978–83.

Pfister, K. K. (1999). Cytoplasmic dynein and microtubule transport in the axon: The action connection. *Mol Neurobiol* **20**, 81–91.

Putnam, A. J., Cunningham, J. J., Dennis, R. G., Linderman, J. J. and Mooney, D. J. (1998). Microtubule assembly is regulated by externally applied strain in cultured smooth muscle cells. *J Cell Sci* **111(Pt 22)**, 3379–87.

Putnam, A. J., Cunningham, J. J., Pillemer, B. B. and Mooney, D. J. (2003). External mechanical strain regulates membrane targeting of Rho GTPases by controlling microtubule assembly. *Am J Physiol Cell Physiol* **284**, C627–39.

Putnam, A. J., Schultz, K. and Mooney, D. J. (2001). Control of microtubule assembly by extracellular matrix and externally applied strain. *Am J Physiol Cell Physiol* **280**, C556–64.

Ren, X. D., Kiosses, W. B. and Schwartz, M. A. (1999). Regulation of the small GTP-binding protein Rho by cell adhesion and the cytoskeleton. *Embo J* **18**, 578–85.

Roesler, H. (1987). The history of some fundamental concepts in bone biomechanics. *J Biomech* **20**, 1025–34.

Rosette, C. and Karin, M. (1995). Cytoskeletal control of gene expression: Depolymerization of microtubules activates NF-kappa B. *J Cell Biol* **128**, 1111–9.

Salmon, W. C., Adams, M. C. and Waterman-Storer, C. M. (2002). Dual-wavelength fluorescent speckle microscopy reveals coupling of microtubule and actin movements in migrating cells. *J Cell Biol* **158**, 31–7.

Schaefer, A. W., Kabir, N. and Forscher, P. (2002). Filopodia and actin arcs guide the assembly and transport of two populations of microtubules with unique dynamic parameters in neuronal growth cones. *J Cell Biol* **158**, 139–52.

Schneider, S. Q. and Bowerman, B. (2003). Cell polarity and the cytoskeleton in the Caenorhabditis elegans zygote. *Annu Rev Genet* **37**, 221–49.

Shi, S. H., Cheng, T., Jan, L. Y. and Jan, Y. N. (2004). APC and GSK-3beta are involved in mPar3 targeting to the nascent axon and establishment of neuronal polarity. *Curr Biol* **14**, 2025–32.

Shi, S. H., Jan, L. Y. and Jan, Y. N. (2003). Hippocampal neuronal polarity specified by spatially localized mPar3/mPar6 and PI 3-kinase activity. *Cell* **112**, 63–75.

Siegrist, S. E. and Doe, C. Q. (2007). Microtubule-induced cortical cell polarity. *Genes Dev* **21**, 483–96.

Stamenovic, D., Mijailovich, S. M., Tolic-Norrelykke, I. M., Chen, J. and Wang, N. (2001). Cell prestress. II. Contribution of microtubules. *Am J Physiol Cell Physiol* **282**, C617–C624.

Suter, D. M. and Forscher, P. (1998). An emerging link between cytoskeletal dynamics and cell adhesion molecules in growth cone guidance. *Curr Opin Neurobiol* **8**, 106–16.

Takasone, T., Juodkazis, S., Kawagishi, Y., Yamaguchi, A., Matsuo, S., Sakakibara, H., Nakayama, H. and Misawa., H. (2002). Flexural rigidity of a single microtubule. *Jpn J Appl Phys* **41**, 3015–19.

Tran, P. T., Marsh, L., Doye, V., Inoue, S. and Chang, F. (2001). A mechanism for nuclear positioning in fission yeast based on microtubule pushing. *J Cell Biol* **153**, 397–411.

Vale, R. D. (1991). Severing of stable microtubules by a mitotically activated protein in Xenopus egg extracts. *Cell* **64**, 827–39.

Vale, R. D., Reese, T. S. and Sheetz, M. P. (1985). Identification of a novel force-generating protein, kinesin, involved in microtubule-based motility. *Cell* **42**, 39–50.

Vallee, R. B., Wall, J. S., Paschal, B. M. and Shpetner, H. S. (1988). Microtubule-associated protein 1C from brain is a two-headed cytosolic dynein. *Nature* **332**, 561–3.

Venier, P., Maggs, A. C., Carlier, M. F. and Pantaloni, D. (1994). Analysis of microtubule rigidity using hydrodynamic flow and thermal fluctuations. *J Biol Chem* **269**, 13353–60.

Vorobjev, I. A., Rodionov, V. I., Maly, I. V. and Borisy, G. G. (1999). Contribution of plus and minus end pathways to microtubule turnover. *J Cell Sci* **112(Pt 14)**, 2277–89.

Walker, R. A., O'Brien, E. T., Pryer, N. K., Soboeiro, M. F., Voter, W. A., Erickson, H. P. and Salmon, E. D. (1988). Dynamic instability of individual microtubules analyzed by video light microscopy: Rate constants and transition frequencies. *J Cell Biol* **107**, 1437–48.

Wang, L. and Brown, A. (2002). Rapid movement of microtubules in axons. *Curr Biol* **12**, 1496–501.

Wang, N., Butler, J. P. and Ingber, D. E. (1993). Mechanotransduction across the cell surface and through the cytoskeleton. *Science* **260**, 1124–7.

Wang, N., Naruse, K., Stamenovic, D., Fredberg, J. J., Mijailovich, S. M., Tolic-Norrelykke, I. M., Polte, T., Mannix, R. and Ingber, D. E. (2001). Mechanical behavior in living cells consistent with the tensegrity model. *Proc Natl Acad Sci USA* **98**, 7765–70.

Watanabe, T., Wang, S., Noritake, J., Sato, K., Fukata, M., Takefuji, M., Nakagawa, M., Izumi, N., Akiyama, T. and Kaibuchi, K. (2004). Interaction with IQGAP1 links APC to Rac1, Cdc42, and actin filaments during cell polarization and migration. *Dev Cell* **7**, 871–83.

Waterman-Storer, C. M. and Salmon, E. D. (1997). Actomyosin-based retrograde flow of microtubules in the lamella of migrating epithelial cells influences microtubule dynamic instability and turnover and is associated with microtubule breakage and treadmilling. *J Cell Biol* **139**, 417–34.

Waterman-Storer, C. M., Worthylake, R. A., Liu, B. P., Burridge, K. and Salmon, E. D. (1999). Microtubule growth activates Rac1 to promote lamellipodial protrusion in fibroblasts. *Nat Cell Biol* **1**, 45–50.

Wittmann, T. and Waterman-Storer, C. M. (2001). Cell motility: Can Rho GTPases and microtubules point the way? *J Cell Sci* **114**, 3795–803.

Wolff, J. (1892). *Das Gesetz der Transformation der knochen* [The Law of Bone Remodeling]. Berlin: A. Hirschwald.

Yang, H. Y., McNally, K. and McNally, F. J. (2003). MEI-1/katanin is required for translocation of the meiosis I spindle to the oocyte cortex in C elegans. *Dev Biol* **260**, 245–59.

Yu, W., Ahmad, F. J. and Baas, P. W. (1994). Microtubule fragmentation and partitioning in the axon during collateral branch formation. *J Neurosci* **14**, 5872–84.

Zhou, F. Q., Waterman-Storer, C. M. and Cohan, C. S. (2002). Focal loss of actin bundles causes microtubule redistribution and growth cone turning. *J Cell Biol* **157**, 839–49.

11

A Molecular Perspective on Mechanotransduction in Focal Adhesions

Seung E. Lee, Roger D. Kamm, and Mohammad R. K. Mofrad

11.1 Introduction

11.1.1 Mechanotransduction and Focal Adhesions

Living cells respond to mechanical stimulation in a variety of ways that shape their phenotype in health and disease. Such a mechanosensing character is essential for cells to probe their environment and respond accordingly to their fate of cell growth, differentiation, or death. Although the biochemical signaling pathways activated by mechanical stimuli have been extensively studied, little is known of the basic underlying mechanisms. As discussed throughout this book, several mechanisms of mechanotransduction have been proposed. It is conceivable that the mechanical signal may be transduced into a chemical signal through protein activation, leading to the upregulation of intracellular signaling proteins. Alternatively, the forces may be transmitted via individual proteins either at the site of cell adhesion to its surroundings or within the stress-bearing members of the cytoskeleton and can cause conformational changes that alter their binding affinity to other intracellular molecules. This altered equilibrium state can subsequently initiate a biochemical signaling cascade or produce more immediate and local structural changes; see reviews [1–4].

One extensively studied example of mechanotransduction is force regulation of the focal adhesion assembly. Focal adhesions have important cellular functions, and their study can provide useful insight into understanding mechanotransduction pathways. Many studies have looked into the force response of focal adhesions with the aim of understanding their mechanisms and implications. The focus of this chapter will be on the direct mechanical response of focal adhesions at the cellular and molecular levels. In the remainder of this section, the role of focal adhesions and their maturation stages are discussed. In the following section, experimental studies on force regulation of focal adhesions are described. Then, the numerical works on the effect of force on adhesion proteins and their molecular mechanisms are examined. Finally, future outlook and perspectives are presented in the last section.

11.1.2 Role of Focal Adhesions

A focal adhesion is a protein complex forming a physical linkage between the cytoskeleton and the extracellular matrix (ECM). A cell can use focal adhesions to gain traction on the ECM during the process of spreading and migration. The assembly of focal adhesions is a dynamic process that is closely regulated by the mechanical and chemical cues the cell experiences. Both the intracellular and extracellular mechanical stresses transmitted through focal adhesions are shown to be important in the formation of a focal adhesion complex [5–7], whereas the release of stress results in the turnover of a focal adhesion [5]. The myosin-mediated contractile force transmitted to the ECM and the tension applied to the adhesion complex are necessary for promoting focal adhesion development [5] whereas disruption of myosin activity effectively inhibits the formation of focal adhesions. In the absence of myosin contractile force, externally applied mechanical forces can also promote the formation of focal adhesions [6, 8]. This force-regulated focal adhesion assembly allows a cell to probe the mechanical stiffness of its surroundings and to respond accordingly, for example, by migrating in the direction of increasing substrate stiffness [7, 9].

A GTPase protein Rho is involved in the formation and regulation of focal adhesions in response to growth factors [10, 11]. Two downstream targets of Rho are Rho-associated kinase and Dia1, which are responsible for myosin II activation [12] and nucleation of actin polymerization [13], respectively. By promoting cell spreading through focal adhesion regulation, Rho may be a key regulator of cell proliferation [14]. In the absence of either cell adhesion or growth factors, the cell may be directed down the path of apoptosis [15]. Tension applied to focal adhesions can activate a number of intracellular signaling proteins, including focal adhesion kinase (FAK) [16], which can in turn phosphorylate other focal adhesion proteins. FAK may be involved in promoting cell migration as it activates Rac, another Rho family GTPase, which is involved in lamellipodia formation [10]. With the evidence that FAK suppresses Rho activity [17], it is likely that a cell uses this collection of signaling proteins and focal adhesions in deciding on spreading, migration, or programmed death.

11.1.3 Focal Adhesion Maturation Stages

Focal adhesion complexes contain a rich mixture of structural as well as signaling proteins and are often used by a cell to probe the stiffness of its environment [14, 18]. A sequence of maturation stages leads to the formation of focal adhesion complexes. An *initial adhesion* consists of the minimum essential set of proteins needed to link the ECM with the cytoskeleton and is only able to withstand tensile forces on the order of 2 pN [19]. When the initial contact is activated or sustains a force, it grows into a *focal complex*, which is a short-lived dot-like ($\sim$1 μm) adhesion structure containing, for example, vinculin, paxillin, α-actinin, and Arp2/3 [20–22]. The focal complex can then mature into a *focal adhesion* under sustained force, which is characterized by its larger size (1–10 μm), elongated shape, and association with stress fibers [14]. Arp2/3, which is related to actin nucleation and branching, is absent in focal adhesions, making it relatively more static in nature compared to the focal complex [20].

11.2 Experimental Studies on Focal Adhesion Mechanotransduction

Various experiments were performed to examine the regulation of focal adhesions by mechanical forces using flexible substrates [5, 7, 9] and extracellular force manipulations [6, 23]. Although these studies provide strong support for the force-dependent assembly of focal adhesions, the molecular basis of the mechanotransduction mechanism remains elusive. The forces exerted by cells through focal adhesions have been measured through various methods, usually through flexible substrates [5]. By dynamically regulating focal adhesion assembly, cells can use focal adhesions to gain traction to migrate and spread, as well as probe for mechanical stiffness of their surroundings. In this section, a number of these experimental studies are reviewed and their implications are discussed.

11.2.1 Force-Induced Focal Adhesion Assembly

Maturation of a focal complex into a focal adhesion is dependent on the application of mechanical force to the adhesion site [5, 6, 23]. Inhibition of myosin-driven contractility results in an inability of cells to form stable focal adhesions [5, 23] and can also lead to rapid degeneration of already-formed focal adhesions into focal complexes [5]. Blocking Rho or Rho-associated kinase activities, which in turn regulates myosin contractility, also inhibits the formation of focal adhesions [23]. Another requirement for the formation of a focal adhesion is the integrity of the cytoskeleton [23]. These observations suggest the requirement of myosin-driven tensile stress transmitted to adhesion complexes through the cytoskeletal network for the formation of stable focal adhesions, but they fail to identify the precise molecular mechanism responsible for these phenomena.

Interestingly, focal adhesion assembly can be restored by application of external mechanical force [6, 23] even in cells in which Rho-associated kinase or myosin activity has been inhibited [23]. Galbraith et al. [6] employed fibronectin-coated microbeads of various diameters to study the force response of adhesion strengthening. When small-diameter beads (1 μm) are attached to lamellipodia of fibroblasts, no focal complexes are formed in the submembrane beneath the bead. However, focal complexes are formed beneath the bead when external tensile force is applied using an optical laser trap. Riveline et al. [23] treated 3T3 fibroblasts with an inhibitor of actomyosin contraction, 2,3-butanedione monoxime (BDM), and observed no focal adhesion formation. The BDM-treated cells, however, recovered elongated focal adhesions after an adhesive ligand-coated micropipette was used to apply a localized external force.

Force-dependent focal adhesion assembly is supported further by investigating the role of substrate stiffness dependence on focal adhesion formation [7, 9]. Pelham et al. [7] developed a polyacrylamide-based, collagen-coated flexible substrate to investigate the response of rat kidney epithelial cells and 3T3 fibroblasts on varying substrate stiffnesses. They determined that the cells on a flexible substrate displayed irregular focal adhesions, reduced spreading, and increased motility compared to the ones on a stiffer substrate, which may be due to destabilized adhesion complexes. Cells migrate from a soft to a stiff substrate when plated on a substrate with a gradient in stiffness [7, 9, 24]. Motile 3T3 fibroblasts can be triggered to change the direction of migration by a locally induced alteration of substrate stiffness [9]. The general finding is that cells form stable focal adhesions on stiffer substrates and migrate in the direction of higher stiffness. It appears that once an adhesion is formed, myosin-driven contractility probes the mechanical properties of the substrate, and only when the substrate can withstand a certain level of tensile stress can a stable focal adhesion be formed.

These observations provide strong support for the need of tensile stress on the adhesion complexes for the formation of stable focal adhesions. The soft substrate cannot support a sufficient level of tensile stress on the adhesion complex even with active myosin contractility. The presence of a force-sensitive molecular switch within an adhesion complex that promotes the formation of a focal adhesion is one of the more plausible explanations for these observations. However, the molecular mechanisms of the force-sensing adhesion proteins have not been identified by these studies.

11.2.2 Force Exerted by Focal Adhesion

Force exerted via focal adhesions allows cells to generate intracellular tension during spreading and to gain traction during migration, and the magnitude of the exerted force is dependent on the size of the adhesion complex [5, 24]. Balaban et al. [5] developed a micropatterned elastomer substrate and, in combination with fluorescence imaging, measured the force exerted by human foreskin fibroblasts on the deformable substrate. Using these measurements of adhesion area and force magnitude on adhesions larger than 1 μm^2 in area, the stress was observed to be remarkably constant at a value of 5.5 ± 2 nN/μm^2. When a myosin contractility inhibitor, BDM, was applied to the adherent cell, the adhesion area and force magnitude both decreased simultaneously so as to maintain the ratio of 5.5 ± 2 nN/μm^2. A similar stress level (4.8 nN/μm^2) has been measured for smooth muscle cells by using substrates with elastic microposts [24]. It is interesting to find that the measured stress levels are consistent despite significant morphological differences seen in these cells.

The growing size of an adhesion complex occurs by means of aggregation of integrin-mediated ECM–cytoskeleton linkages and the recruitment of other adhesion molecules to the adhesion complex. If a single integrin-mediated adhesion complex is isolated from a focal adhesion, the observed withstanding force arises from noncovalent interactions between the adhesion molecules comprising an integrin-mediated linkage. Assuming close packing of integrins in the focal adhesion, force exerted through a single integrin molecule is estimated to be ~1 pN [5]. For comparison, forces generated by a single myosin are ~4 pN [25] and forces needed to rupture a talin–F-actin bond are ~2 pN [19]. The force generated by even a single myosin might therefore be sufficient to actively regulate adhesion dynamics and sense mechanical stimuli if a substrate of sufficient rigidity is present. In the case of soft substrates, stress transmitted to the adhesion by a single myosin may not be sufficient to signal turnover of the focal adhesion. On the extracellular side, the force required to break a single integrin–fibronectin bond is in the range of ~20–100 pN [26], which suggests that the force sensing of an adhesion complex occurs in the adhesion plaque connected to the cytoplasmic domain of the integrin.

11.3 Numerical Studies on Focal Adhesion Mechanotransduction

Many experimental studies have explored the role of mechanical forces on cellular behavior. However, studies to identify a molecular basis for force regulation of focal adhesions are needed to fully understand the focal adhesion mechanotransduction pathways. This can be achieved by either experimental or numerical single molecule studies with proper forcing schemes. Molecular dynamics (MD) has been used to study the force response on various molecular structures [27–29]. Mostly, unfolding pulling forces were applied between two atoms (e.g., C- and N-termini), and the results were compared with the corresponding atomic force microscopy (AFM) experiments [30, 31]. In this section, MD studies are presented investigating the force response and binding mechanisms of talin and vinculin, two key focal adhesion proteins that may be involved in the force regulation of focal adhesion assembies. In the first part, a number of binding and structural studies of talin and vinculin are reviewed, which provides the basis for the subsequent MD simulations. Then, recent studies identifying force-induced activation of talin and vinculin–talin binding mechanisms are discussed.

11.3.1 Force Sensing Proteins in Focal Adhesions

Applied tension on the initial adhesion between the ECM and the cytoskeleton allows the recruitment of vinculin to reinforce the linkage to form a focal complex [6]. Jiang et al. [19] identified that the initial adhesion consists of an ECM–integrin–talin–F-actin linkage. Separate from the force-regulated signaling pathways, the local immediate force response of adhesion reassembly is thought to be through conformational changes in the linking proteins that enhance their binding to other reinforcing proteins [3, 4]. Indeed, talin1 is critical in force-dependent vinculin recruitment to adhesion sites independent of Src family kinase and focal adhesion kinase activities [32]. Talin is present in both the initial adhesion and the focal complex, whereas vinculin is only found in the focal complex [6]. Vinculin has binding sites for both talin [33] and F-actin [34], making it a candidate for strengthening an

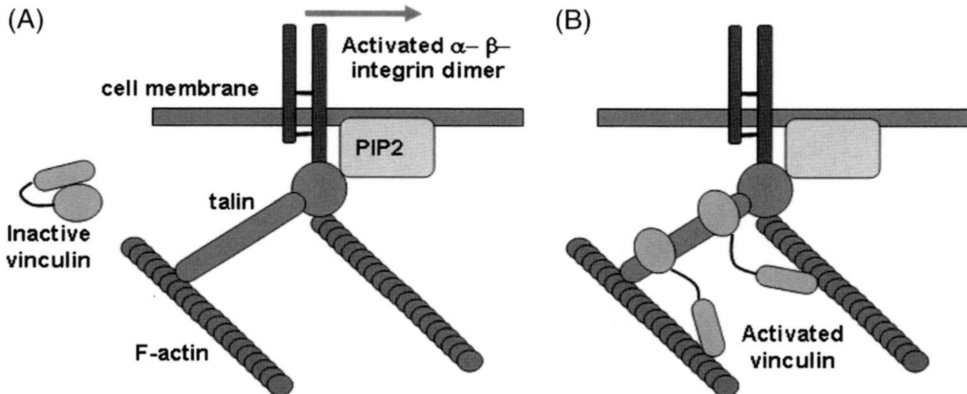

Figure 11.1. (A) Model of the initial adhesion consisting of integrin–talin–F-actin linkage. Vinculin is present in the cytosol in an inactive, autoinhibitory conformation, and tensile force is applied to the integrin dimer from the outside of the cell membrane. (B) Transmitted force through the linkage alters the talin configuration and recruits vinculin to reinforce the initial adhesion linkage.

initial adhesion (see Figure 11.1). Interestingly, talin contains 11 potential vinculin binding sites with a number of them known to be cryptic [35–37].

Vinculin is a highly conserved cytoplasmic protein that functions as a structural reinforcing link for cell–cell and cell–matrix junctions. It consists of a globular head domain (Vh), a proline-rich neck region, and a rod-like tail domain (Vt) and contains binding sites for numerous other cytoplasmic proteins [38, 39]. In its inactive state, Vh binds to the vinculin tail, masking its cryptic binding sites for many of its ligands [40, 41]. A cell with vinculin disruption can still form focal adhesions, yet it displays reduced ability to spread and increased cell motility [42]. Therefore, along with its structural function, vinculin has also been suggested to be a regulatory protein in cell adhesion [43].

It has been shown that PIP2 can disrupt the autoinhibitive Vt–Vh interaction [40]. However, recent studies suggest that talin might also play a role in vinculin activation [44, 45]. Bakolitsa et al. [46] suggested a combinatorial pathway in activating vinculin, wherein PIP2 partially releases Vt from Vh, exposing the talin binding site, and the head and tail interaction is completely severed as the talin binds to the Vh (see Figure 11.1). The crystal structure of full-length vinculin in the inactive autoinhibitory conformation provides many insights into vinculin function [46]. Vinculin tail domain (Vt; residue 897–1066) is bound most strongly to the D1 domain (residue 1–258) and relatively weakly to other domains [46]. Izard et al. [44] reported the crystal structure of the Vh subdomain (residues 1–258; same as D1 domain in ref [46]) in complex with the Vt domain.

Talin is another cytoplasmic protein with a globular head domain and an elongated rod domain that provides a structural link between the integrin and actin cytoskeleton [43] (see Figure 11.1). Initially, the talin rod domain was characterized as having three high-affinity vinculin binding sites where each binding sequence forms an amphipathic helix [47]. However, a recent study showed that talin may have a total of 11 potential vinculin binding sites (VBSs) [35]. Bass et al. [48] showed that three initially identified vinculin binding sites are mutually exclusive, hence suggesting that, given that they are all amphipathic helices, they bind to the same site on vinculin through an identical binding mechanism. Since the vinculin binding sites share only partial sequence homology [48] and α-actinin, another cytoplasmic protein, is shown to bind at the same binding site on vinculin [44], this suggests that the binding site on vinculin for the vinculin binding sites of talin or α-actinin is not highly specific. Unlike vinculin disrupted cells, which still form focal adhesions [42], mouse embryonic stem cells with disrupted talin genes did not, and also exhibited spreading defects [49].

The crystal structure of a full-length talin is not available, but subdomains of the talin rod (talin residues 482–789) containing the VBS1 [36] and talin rods (talin residues 755–889) containing the VBS2 [50] have been reported (see Figure 11.2). Although VBS1 can strongly bind to vinculin [51], the five-helix bundle containing VBS1 is unable to bind to vinculin as the binding surface is cryptic [37].

VBSs can bind to a subdomain of vinculin head Vh1 [37, 44] at high binding affinities [51], but longer full-length talin only weakly binds to Vh1 [37], indicating that many of the VBSs are cryptic. In a thorough mutational study on VBS1 binding to Vh1, most of the hydrophobic residues of VBS1 that are embedded within Vh1 were shown to be important in stable binding to Vh1 [36]. Interestingly, these same hydrophobic residues are also embedded within the hydrophobic core of *N*-terminal five-helix bundle of talin rod (TAL5) [36]. Experiments have shown that isolated TAL5 has a low binding affinity for Vh1, whereas a four-helix bundle with helix-5 (H5) removed from TAL5 [37], a mutated TAL5 with an unstable hydrophobic core [36], or the wild-type TAL5 molecule in elevated temperature solvent [37] can each disrupt TAL5 stability and strongly bind to Vh1. One working hypothesis is that the tensile force transmitted through TAL5 is the destabilizing cue that exposes the cryptic binding surface of VBS1, but direct evidence has been lacking.

11.3.2 A Potential Mechanism for Force-Induced Talin Activation

Structural data and binding experiments indicate that talin is a likely candidate as the force-sensing protein within the adhesion complex, yet these studies alone do not provide the molecular mechanism of its functionality. Lee et al. [52] employed molecular dynamics simulations that attempted to realistically mimic the transmission of force acting on the focal adhesion protein talin (Figures 11.2 and 11.3). They found that this leads to a conformational change that exposes the cryptic vinculin-binding residues of VBS1, thereby providing molecular insight into a potential force-sensing mechanism. Atoms of polar residues on one end of TAL5 were harmonically constrained in space, and then atoms of polar residues on the other end of TAL5 were subjected to both constant velocity and constant force pulling.

In the TAL5 simulations displaying VBS1 activation, hydrophobic residues of VBS1 form a tight hydrophobic core within the TAL5 structure before extension.

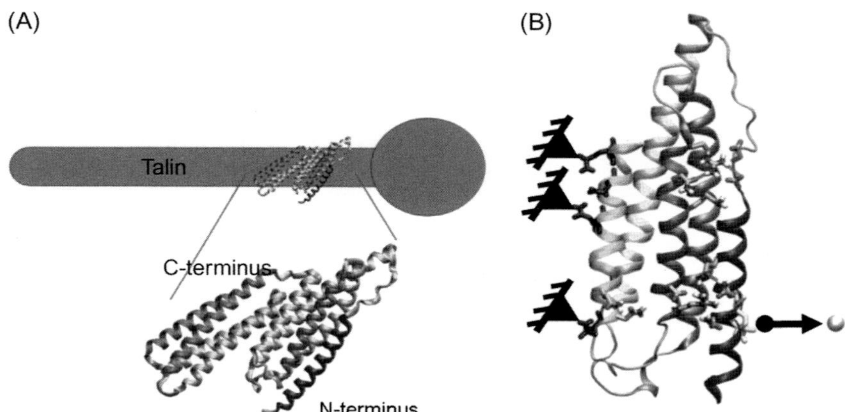

Figure 11.2. (A) Crystal structure of TAL9 (PDB ID: 1SJ8) in ribbon representation is shown superimposed on a hypothetical talin model. The talin rod has tandem repeats of helical bundles, where TAL9 is aligned such that the centers of mass of the two helical bundles lie on the talin rod axis. (B) Detailed view of the N-terminal five-helix bundle (TAL5) used in the TAL5 simulations. These five helices are denoted H1, H2, H3, H4 (or VBS1), and H5. Some important polar residues are shown in stick representations. H5 polar side chains are harmonically constrained in space (constraints shown as triangles). H1 polar side chains are pulled toward the dummy atom (effective pulling direction shown as an arrow).

Upon application of force, hydrogen-bonding and salt bridge interactions between VBS1 and an adjacent α-helix, denoted "RQK and ER handles," after the residues involved, transmit the tensile force and induce a rigid-body rotation of VBS1 to expose its hydrophobic residues to the solvent. Interestingly, the exposed hydrophobic residues were previously identified as the key residues involved in VBS1 binding to vinculin [36]. Simulations with alanine mutations of the "RQK and ER handles" were demonstrated to impair the VBS1 activation mechanism as anticipated. In addition, similar sequences of these force-transmitting handles are found at other locations in the talin molecule, suggesting that this may be a general force-transmitting, or force-sensing, mechanism within talin [47, 52].

In order to check for consistency, multiple pulling simulations with different initial conditions were performed, and 71.4% of the TAL5 simulations ($n = 20$ out of 28) exhibited VBS1 activation. Analyzing only the simulations with VBS1 activation, VBS1 rotated by $62.0 \pm 9.5°$ and a mean force of 13.2 ± 8.0 pN was required for activation. The effective pulling rate of 0.125 Å/ns is still orders of magnitude faster than the pulling rates we might expect *in vivo* or with AFM experiments (~1 nm/ms ~10^{-5} Å/ns). Such rapid pulling results in significantly larger force levels in bond rupture [53] or protein unfolding [54] compared to the corresponding AFM measurements. In both cases, the forces measured by AFM were ~30% of the force computed using MD [53, 54]. Using this value as a very rough approximation, the force needed to activate VBS1 at more realistic, slower rates of pulling would lie in the range of ~4 pN. This estimated lower force at

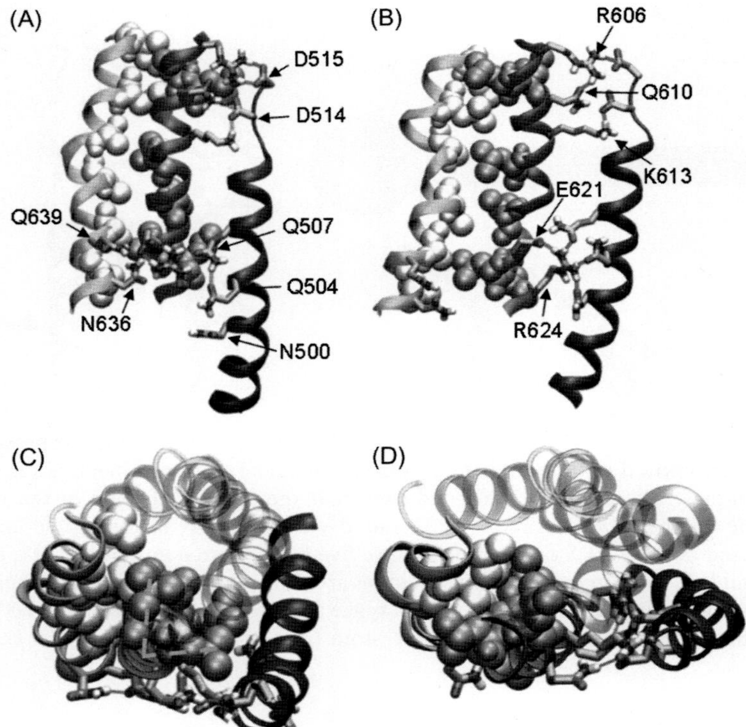

Figure 11.3. Drawings depicting VBS1 activation from the TAL5 simulation: H1, H2, H3, VBS1, H5, hydrophobic residues of VBS1 (darker VDW sphere representation; also the vinculin-binding residues), hydrophobic residues of H5 (lighter VDW), and some important polar residues (stick representation with shading denoting the atom type). Polar residues are labeled on the figures. (A) Conformation at $t = 2.08$ ns. The hydrophobic residues of VBS1 (darker VDW) are hidden in the hydrophobic core. (B) Conformation at $t = 7.40$ ns showing the hydrophobic residues of VBS1 being exposed to solvent. Hydrogen bonds between H5 and VBS1 are broken. The hydrophobic residues, or the vinculin-binding residues, point into the page in (A) and point to the left in (B). (C) Conformation at $t = 0.86$ ns viewed from the top. The V-shaped VBS1 hydrophobic residues are packed within the hydrophobic core of TAL5 (dotted lines). (D) Conformation at $t = 9.24$ ns showing VBS1 rotation. Hydrophobic residues H5 (lighter VDW) fit into the "V" of the VBS1 hydrophobic residues (darker VDW).

a slower pulling rate is on the order of (i) forces generated by a single myosin, ~4 pN [25]; (ii) forces needed to rupture a talin–F-actin bond, ~2 pN [19]; and (iii) the estimated force experienced by a single integrin linkage, based on close packing in a focal contact, ~1 pN [5]. On the extracellular side, the force required to break a single integrin–fibronectin bond has been reported to be ~20 pN [26], or up to ~100 pN [55].

In this study an empirically based method was used to represent solvent effects, namely, the implicit EEF1 model was employed, which is characterized by high efficiency [56]. Other implicit solvent methods [57, 58] are theory-based and tend to be 5–10 times slower than EEF1. EEF1 has been demonstrated to produce reasonable MD trajectories [56, 59, 60]. To verify the validity of the EEF1 results, a corresponding constant velocity simulation with an explicit solvent was performed, which confirmed that similar critical interactions are present in the explicit simulations as found in EEF1 simulations.

This study identified a potential mechanism for VBS1 activation, involving a force-induced conformational change causing the hydrophobic vinculin-binding residues on VBS1 within TAL5 to become accessible for vinculin binding. This would then constitute the initiating event leading to force-induced focal adhesion strengthening by vinculin recruitment. Sequence homology of VBS1 with other VBSs suggests that the proposed mechanism may be a general force-induced activation mechanism of cryptic VBSs, and perhaps even be one of the general mechanotransduction mechanisms of helical bundles.

An alternative mechanism was recently proposed by [64], which involves the unfolding of the rod domain in order to expose the vinculin binding site VBS1 (see Chapter 13). Using steered molecular dynamics [64] demonstrates that the talin rod can be fragmented into three helix subbundles, which is followed by the sequential exposure of vinculin-binding helices to water (see Figure 11.4; Plate 13). The unfolding of a vinculin-binding helix into a completely stretched polypeptide might then inhibit further binding of vinculin (see Chapter 13).

11.3.3 Talin and Vinculin Binding Mechanism

Abundant structural data are available for vinculin and talin subdomains; however, these data do not provide the molecular basis of their binding mechanism. A recent molecular dynamics study has been conducted to shed light on the binding mechanism of vinculin to talin, and proposed a link between force-induced activation of talin [52] and force regulation of focal adhesion strengthening [6].

Crystal structures of vinculin bound [36] and unbound [44] to VBS1 indicate that vinculin undergoes large conformational changes upon binding to VBS1 (Figures 11.4–11.6; Plate 13). In the simulation, the vinculin subdomain, Vh1, and VBS1 are initially separated by 12 Å [61]. As the simulation progressed, Vh1 and VBS1 formed an encounter complex with hydrophobic residues of VBS1 being inserted into helices 1 and 2 of Vh1. With time, VBS1 displaced helices 1 and 2 of Vh1, and eventually embedded itself into the Vh1 hydrophobic core. The final bound conformation is very similar to the existing Vh1–VBS1 crystal structure. From analyzing the solution, it appears that the initial hydrophobic insertion of VBS1 into the Vh1 groove is critical for Vh1–VBS1 binding. Binding was observed in one simulation with no constraints, but in half of the simulations when limited constraints to guide the initial hydrophobic insertion were imposed during the first 800 ps of simulation. This points up one of the major limitations of binding simulations: If the actual binding time-scale is in the order of microseconds, it should be unlikely to observe Vh1–VBS1 binding in ~ 50 ns MD simulations in the absence of any guiding constraints.

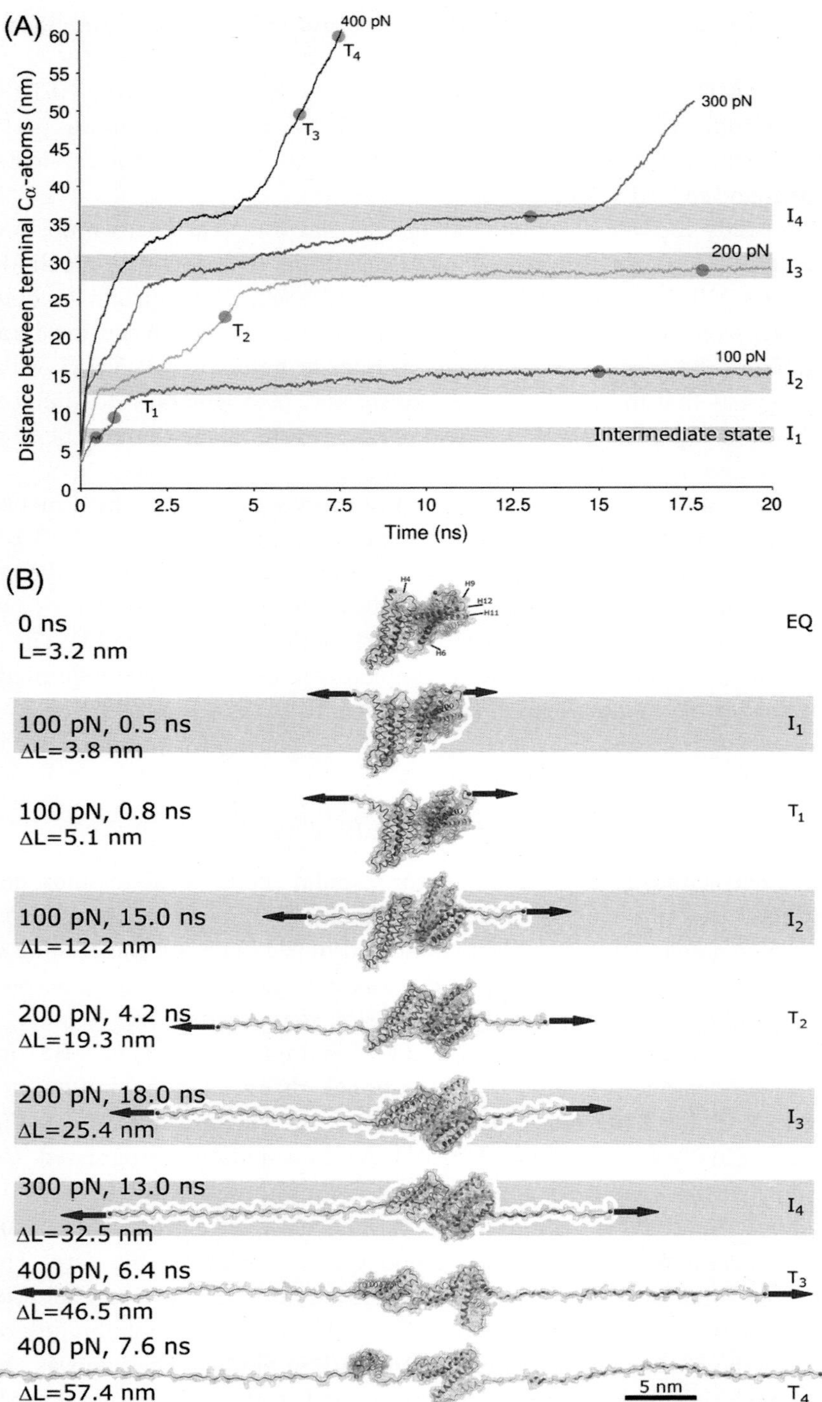

Figure 11.4. Sequential unfolding trajectories and associated structures of the talin rod H1–H12 extended under constant force applied to terminal C_α-atoms (reproduced from Hytonen and Vogel, 2008). (A) Extension-time plots for different constant force pulls, from 100 to 400 pN. Plateaus in the extension-time curves indicate the existence of multiple intermediate states (I_1–I_4). (B) Intermediate states (I_1–I_4, gray bars) and transitional snapshots (T_1–T_4) seen at indicated time points in the unfolding trajectories in (A).

The hydrophobic residues involved in the binding mechanism are the same ones identified to be exposed in the force-induced activation of TAL5 [52]. Molecular dynamics simulations with alanine mutations on these critical residues also impaired the successful binding of the two molecules and are found to be consistent with the mutational experiments [36]. Simulations between Vh1 with VBS2, VBS3, and αVBS all underwent complete binding similar to that observed with Vh1–VBS1. Interestingly, the residue sequence in αVBS is reversed to that of talin VBSs, but the critical hydrophobic residues are still found in the corresponding positions needed to undergo the proposed binding mechanism to Vh1. Similar to talin, α--actinin contains a cryptic VBS and is possibly subjected to mechanical force within cell–cell junctions or in actin cross-linking [62]. Therefore, vinculin and α-actinin binding may occur by an analogous mechanism within both cell–cell junctions and cell–matrix junctions. Also, this generality provides a critical insight into how talin, containing 11 potential VBSs [35], might modify its conformation when subjected to tensile force to recruit multiple vinculins with a concomitant increase in adhesion strength, as has been observed experimentally [63].

This work proposes the Vh1–VBS binding mechanism, which involves hydrophobic insertion of VBS1 into Vh1, separation of H1 and H2 of Vh1, and VBS1 rotation to snap in exposed hydrophobic residues into the hydrophobic core (Figure 11.5). Results from mutational simulations and binding simulations with other VBSs suggest that the proposed mechanism may be more generally valid (Figure 11.7). This work constitutes the potential early stages of force-induced focal adhesion strengthening by vinculin recruitment immediately following the force-induced talin activation mechanism [52].

11.4 Conclusion

Experimental studies on focal adhesions have provided strong support for the force regulation of focal adhesion assembly. Tensile stress transmitted through adhesion complexes is essential for the formation of stable focal adhesions. The magnitude of the force exerted by a focal adhesion of a cell is linearly dependent on the size of the focal adhesion, typically greater than 1 μm^2 in area. A cell can continuously regulate the dynamics of adhesion complexes to gain traction to migrate or spread. Focal adhesions are also used by cells as a way of probing the mechanical state of their surrounding, and the preference of the cell to stiff substrates causes it to migrate toward the direction of the stiffer substrate. However, these studies mostly look at cellular-level characteristics and fail to provide the molecular basis for the observed mechanotransduction mechanisms.

Molecular dynamics provides a tool to investigate protein structural dynamics at the molecular scale not possible experimentally. However, few major advancements had been made in protein mechanics other than in the forced unfolding of proteins [3, 4] due to limitations in available computational power and a lack of structural information (i.e., crystal structures). Force regulation of focal adhesion has been in the spotlight recently, and a large number of experiments have been conducted on

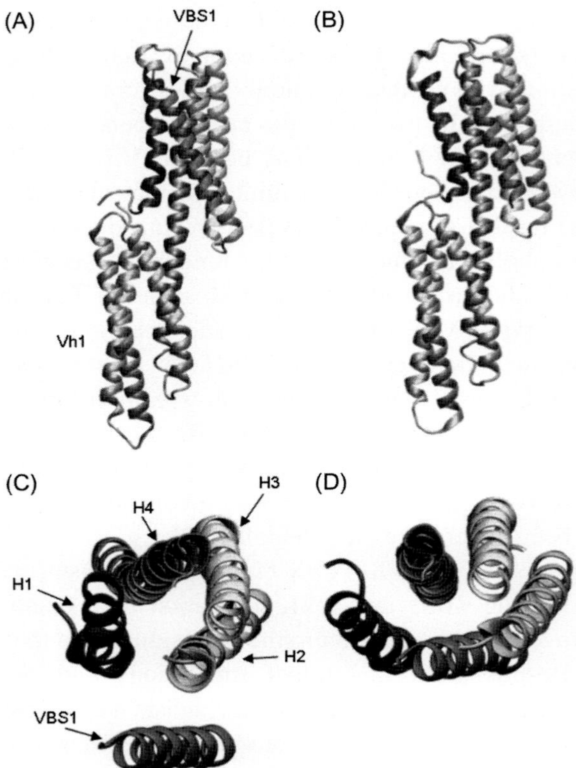

Figure 11.5. (A) Vh1 (obtained from PDB ID: 1RKE) and VBS1 (obtained from PDB ID: 1T01) unbound structures viewed from the front. VBS1 is translated by 12 Å from its corresponding position bound within the Vh1–VBS1 complex. (B) Vh1 and VBS1 bound complex (PDB ID: 1T01) viewed from the front. (C) Vh1 and VBS1 unbound structures viewed from the top. Only the first four helices of Vh1 (a seven-helix bundle) are shown for clarity. (D) Vh1 and VBS1 bound complex viewed from the top.

vinculin and talin interactions [35, 37, 44, 46, 48, 50, 51]. Using these experimental findings as the basis, recent MD studies shed light on the early adhesion assembly mechanism. These studies identified a potential mechanism for force-induced talin activation and the subsequent vinculin binding mechanism. This not only provides a big step toward understanding the force response mechanism of focal adhesions, but it also demonstrates the possibility to use MD for studies on intracellular force-sensitive proteins with subtle conformational changes.

Together, these results shed light on two of the key steps in the force-induced reinforcement of an initial contact. This is the critical first step in gaining a full molecular understanding of the force-sensitive processes within the adhesion plaque, and it suggests a possible pathway leading to the next set of numerical and experimental studies to further elucidate focal adhesion maturation. To test and shed light on different mechanisms proposed for the activation of talin and vinculin, and their subsequent binding trajectories and recruitment of focal adhesion binding proteins it is important to plan further computational and experimental studies involving

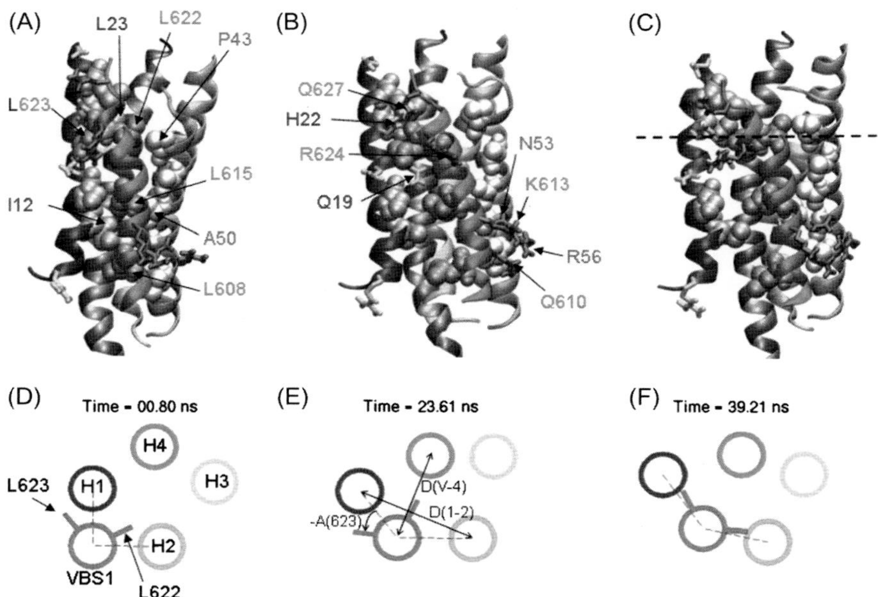

Figure 11.6. Snapshots from one Vh1–VBS1 binding simulation with ribbon representations for helical backbone, stick representations for polar and charged residues, and spherical representations for hydrophobic residues at (A) VBS1 hydrophobic insert ($t = 0.8$ ns) between the hydrophobic patch of H1 and H2, (B) Vh1's H1 and H2 separation ($t = 23.6$ ns), and (C) VBS1 rotation ($t = 39.2$ ns) to snap in the exposed hydrophobic residues, i.e., L623, into the hydrophobic core. Some of the critical residues in Vh1–VBS1 binding are labeled: residues on VBS1, residues on H1, and residues on H2. (D–F) Corresponding cross-sectional views to (A)–(C) at the plane represented by the dashed line in (C).

in vivo and *in vitro* systems with devised mutations and perturbations. Another potentially fruitful direction would be to use MD to investigate the activation mechanism of vinculin. It is still not clear whether vinculin binds to talin only when activated, or if talin binding to vinculin itself activates vinculin. Evidence shows that talin may be involved in the activation of vinculin [44, 46]. Crystal structures of the autoinhibitory vinculin head–tail complex are available for the study of potential vinculin activation through talin binding [44, 46].

More experimental studies are needed for the validation of the proposed molecular mechanisms. Mutational studies in living cells would be useful as a means of verifying the current findings. For example, the torque applied to the polar and charged residues of VBS1 in talin activation is essential for the force response of talin in vinculin recruitment. Cells transfected with vectors to produce talin with mutations in these critical residues, mutated to nonpolar residues, would therefore be expected to show decreased force sensitivity compared to cells with wild-type talin. Care should be taken in isolating the role of this talin domain in the overall force sensitivity of talin, however, since talin has redundancies such as the 11 total potential VBSs [35]. Single molecule force spectroscopy can also be envisioned to validate

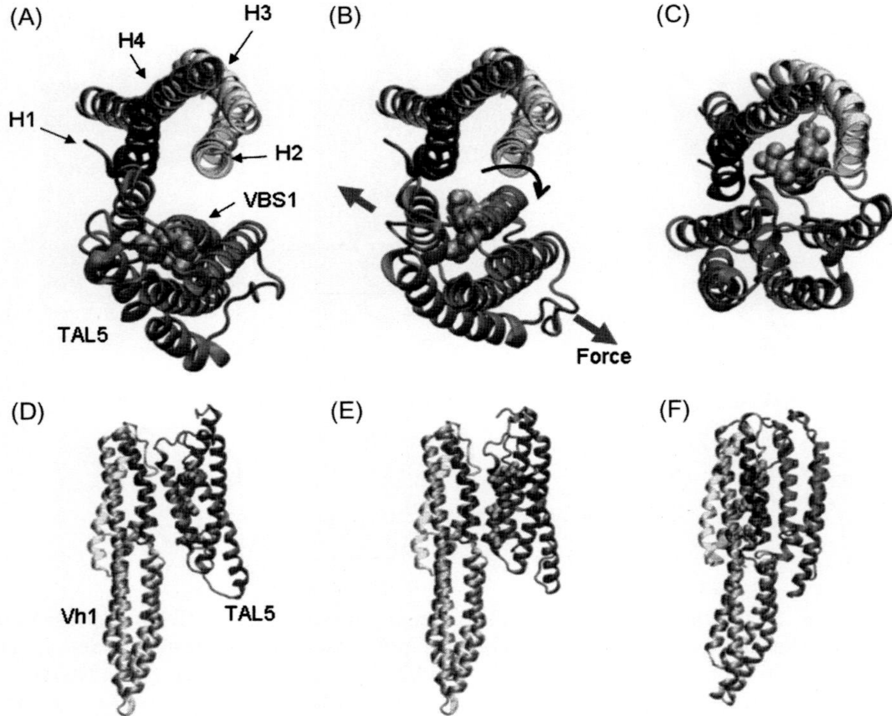

Figure 11.7. Proposed binding model for Vh1 and TAL5. (A) The N-terminal four-helix bundle of Vh1. Inactivated TAL5 containing VBS1 as viewed from the top. The hydrophobic residues necessary for Vh1 binding are shown as gray spheres. (B) With application of mechanical force on TAL5, VBS1 undergoes a rigid body rotation to expose the hydrophobic residues [52]. The VBS1 rotation in response to the applied force is indicated by the curved arrow. (C) Final configuration of Vh1 with TAL5 in a binding simulation. (D–E) The corresponding configurations to (A–C) viewed from the side.

the predicted mechanism of activation. The ability to apply force to a single talin molecule or just the TAL5 domain (or tandem repeats of it) would enable the mentioned mutational studies in a more controlled and isolated environment. Application of force with simultaneous visualization of vinculin (e.g., by tethering a fluorescent antibody) would allow direct detection of talin activation. With this revealing MD study as the stepping stone, numerical methods, experiments, and coarse-grained modeling should be employed concurrently for the fundamental understanding of focal adhesion dynamics, which can eventually lead to a variety of clinical applications. The potential now exists to use a combined experimental/ computational approach to gain new insights into these essential phenomena in which force can regulate or activate intracellular signaling. Recognition of these "mechanical" signaling pathways holds the further potential for the use of "mechanical therapeutics" or ways of controlling cell behavior through mechanical, as opposed to biochemical, approaches.

REFERENCES

[1] Janmey, P. A. and D. A. Weitz. 2004. Dealing with mechanics: Mechanisms of force transduction in cells. *Trends Biochem Sci* **29**:364–370.
[2] Kamm, R. D. and M. R. Kaazempur-Mofrad. 2004. On the molecular basis for mechanotransduction. *Mech Chem Biosyst* **1**:201–209.
[3] Vogel, V. 2006. Mechanotransduction involving multimodular proteins: Converting force into biochemical signals. *Annu Rev Biophys Biomol Struct* **35**:459–488.
[4] Vogel, V. and M. Sheetz. 2006. Local force and geometry sensing regulate cell functions. *Nat Rev Mol Cell Biol* **7**:265–275.
[5] Balaban, N. Q., U. S. Schwarz, D. Riveline, P. Goichberg, G. Tzur, I. Sabanay, D. Mahalu, S. Safran, A. Bershadsky, L. Addadi, and B. Geiger. 2001. Force and focal adhesion assembly: A close relationship studied using elastic micropatterned substrates. *Nat Cell Biol* **3**:466–472.
[6] Galbraith, C. G., K. M. Yamada, and M. P. Sheetz. 2002. The relationship between force and focal complex development. *J Cell Biol* **159**:695–705.
[7] Pelham, R. J., Jr. and Y. Wang. 1997. Cell locomotion and focal adhesions are regulated by substrate flexibility. *Proc Natl Acad Sci USA* **94**:13661–13665.
[8] Cramer, L. P. and T. J. Mitchison. 1995. Myosin is involved in postmitotic cell spreading. *J Cell Biol* **131**:179–189.
[9] Lo, C. M., H. B. Wang, M. Dembo, and Y. L. Wang. 2000. Cell movement is guided by the rigidity of the substrate. *Biophys J* **79**:144–152.
[10] Nobes, C. D. and A. Hall. 1995. Rho, rac, and cdc42 GTPases regulate the assembly of multimolecular focal complexes associated with actin stress fibers, lamellipodia, and filopodia. *Cell* **81**:53–62.
[11] Ridley, A. J. and A. Hall. 1992. The small GTP-binding protein rho regulates the assembly of focal adhesions and actin stress fibers in response to growth factors. *Cell* **70**:389–399.
[12] Fukata, Y., M. Amano, and K. Kaibuchi. 2001. Rho-Rho-kinase pathway in smooth muscle contraction and cytoskeletal reorganization of non-muscle cells. *Trends Pharmacol Sci* **22**:32–39.
[13] Pruyne, D., M. Evangelista, C. Yang, E. Bi, S. Zigmond, A. Bretscher, and C. Boone. 2002. Role of formins in actin assembly: Nucleation and barbed-end association. *Science* **297**:612–615.
[14] Geiger, B. and A. Bershadsky. 2001. Assembly and mechanosensory function of focal contacts. *Curr Opin Cell Biol* **13**:584–592.
[15] Frisch, S. M. and H. Francis. 1994. Disruption of epithelial cell-matrix interactions induces apoptosis. *J Cell Biol* **124**:619–626.
[16] Domingos, P. P., P. M. Fonseca, W. Nadruz, Jr., and K. G. Franchini. 2002. Load-induced focal adhesion kinase activation in the myocardium: Role of stretch and contractile activity. *Am J Physiol Heart Circ Physiol* **282**: H556–564.
[17] Ren, X. D., W. B. Kiosses, D. J. Sieg, C. A. Otey, D. D. Schlaepfer, and M. A. Schwartz. 2000. Focal adhesion kinase suppresses Rho activity to promote focal adhesion turnover. *J Cell Sci* **113**(Pt 20):3673–3678.
[18] Zamir, E. and B. Geiger. 2001. Molecular complexity and dynamics of cell-matrix adhesions. *J Cell Sci* **114**:3583–3590.
[19] Jiang, G. Y., G. Giannone, D. R. Critchley, E. Fukumoto, and M. P. Sheetz. 2003. Two-piconewton slip bond between fibronectin and the cytoskeleton depends on talin. *Nature* **424**:334–337.
[20] DeMali, K. A., C. A. Barlow, and K. Burridge. 2002. Recruitment of the Arp2/3 complex to vinculin: Coupling membrane protrusion to matrix adhesion. *J Cell Biol* **159**:881–891.

[21] Laukaitis, C. M., D. J. Webb, K. Donais, and A. F. Horwitz. 2001. Differential dynamics of alpha 5 integrin, paxillin, and alpha-actinin during formation and disassembly of adhesions in migrating cells. *J Cell Biol* **153**:1427–1440.

[22] Rottner, K., A. Hall, and J. V. Small. 1999. Interplay between Rac and Rho in the control of substrate contact dynamics. *Curr Biol* **9**:640–648.

[23] Riveline, D., E. Zamir, N. Q. Balaban, U. S. Schwarz, T. Ishizaki, S. Narumiya, Z. Kam, B. Geiger, and A. D. Bershadsky. 2001. Focal contacts as mechanosensors: Externally applied local mechanical force induces growth of focal contacts by an mDia1-dependent and ROCK-independent mechanism. *J Cell Biol* **153**:1175–1186.

[24] Tan, J. L., J. Tien, D. M. Pirone, D. S. Gray, K. Bhadriraju, and C. S. Chen. 2003. Cells lying on a bed of microneedles: An approach to isolate mechanical force. *Proc Natl Acad Sci USA* **100**:1484–1489.

[25] Finer, J. T., A. D. Mehta, and J. A. Spudich. 1995. Characterization of single actin-myosin interactions. *Biophys J* **68**:291S–296S; discussion 296S–297S.

[26] Thoumine, O. and J. J. Meister. 2000. Dynamics of adhesive rupture between fibroblasts and fibronectin: Microplate manipulations and deterministic model. *Eur Biophys J* **29**:409–419.

[27] Carrion-Vazquez, M., H. Li, H. Lu, P. E. Marszalek, A. F. Oberhauser, and J. M. Fernandez. 2003. The mechanical stability of ubiquitin is linkage dependent. *Nat Struct Biol* **10**:738–743.

[28] Gao, M., D. Craig, V. Vogel, and K. Schulten. 2002. Identifying unfolding intermediates of FN-III(10) by steered molecular dynamics. *J Mol Biol* **323**:939–950.

[29] Lu, H., B. Isralewitz, A. Krammer, V. Vogel, and K. Schulten. 1998. Unfolding of titin immunoglobulin domains by steered molecular dynamics simulation. *Biophys J* **75**:662–671.

[30] Marszalek, P. E., H. Lu, H. Li, M. Carrion-Vazquez, A. F. Oberhauser, K. Schulten, and J. M. Fernandez. 1999. Mechanical unfolding intermediates in titin modules. *Nature* **402**:100–103.

[31] Oberhauser, A. F., C. Badilla-Fernandez, M. Carrion-Vazquez, and J. M. Fernandez. 2002. The mechanical hierarchies of fibronectin observed with single-molecule AFM. *J Mol Biol* **319**:433–447.

[32] Giannone, G., G. Jiang, D. H. Sutton, D. R. Critchley, and M. P. Sheetz. 2003. Talin1 is critical for force-dependent reinforcement of initial integrin-cytoskeleton bonds but not tyrosine kinase activation. *J Cell Biol* **163**:409–419.

[33] Gilmore, A. P., P. Jackson, G. T. Waites, and D. R. Critchley. 1992. Further characterization of the talin-binding site in the cytoskeletal protein vinculin. *J Cell Sci* **103**:719–731.

[34] Menkel, A. R., M. Kroemker, P. Bubeck, M. Ronsiek, G. Nikolai, and B. M. Jockusch. 1994. Characterization of an F-actin-binding domain in the cytoskeletal protein vinculin. *J Cell Biol* **126**:1231–1240.

[35] Gingras, A. R., W. H. Ziegler, R. Frank, I. L. Barsukov, G. C. Roberts, D. R. Critchley, and J. Emsley. 2005. Mapping and consensus sequence identification for multiple vinculin binding sites within the talin rod. *J Biol Chem* **280**:37217–37224.

[36] Papagrigoriou, E., A. R. Gingras, I. L. Barsukov, N. Bate, I. J. Fillingham, B. Patel, R. Frank, W. H. Ziegler, G. C. K. Roberts, D. R. Critchley, and J. Emsley. 2004. Activation of a vinculin-binding site in the talin rod involves rearrangement of a five-helix bundle. *Embo J.* **23**:2942–2951.

[37] Patel, B. C., A. R. Gingras, A. A. Bobkov, L. M. Fujimoto, M. Zhang, R. C. Liddington, D. Mazzeo, J. Emsley, G. C. Roberts, I. L. Barsukov, and D. R. Critchley. 2006. The activity of the vinculin binding sites in talin is influenced by

the stability of the helical bundles that make up the talin rod. *J Biol Chem* **281**(11):7458–67.

[38] Bakolitsa, C., J. M. de Pereda, C. R. Bagshaw, D. R. Critchley, and R. C. Liddington. 1999. Crystal structure of the vinculin tail suggests a pathway for activation. *Cell* **99**:603–613.

[39] Winkler, J., H. Lunsdorf, and B. M. Jockusch. 1996. The ultrastructure of chicken gizzard vinculin as visualized by high-resolution electron microscopy. *J Struct Biol* **116**:270–277.

[40] Gilmore, A. P. and K. Burridge. 1996. Regulation of vinculin binding to talin and actin by phosphatidyl-inositol-4-5-bisphosphate. *Nature* **381**:531–535.

[41] McGregor, A., A. D. Blanchard, A. J. Rowe, and D. R. Critchley. 1994. Identification of the vinculin-binding site in the cytoskeletal protein alpha-actinin. *Biochem J* **301**:225–233.

[42] Xu, W. M., J. L. Coll, and E. D. Adamson. 1998. Rescue of the mutant phenotype by reexpression of full-length vinculin in null F9 cells; effects on cell locomotion by domain deleted vinculin. *J Cell Sci* **111**:1535–1544.

[43] Critchley, D. R. 2000. Focal adhesions – the cytoskeletal connection. *Curr Opin Cell Biol* **12**:133–139.

[44] Izard, T., G. Evans, R. A. Borgon, C. L. Rush, G. Bricogne, and P. R. J. Bois. 2004. Vinculin activation by talin through helical bundle conversion. *Nature* **427**:171–175.

[45] Ling, K., R. L. Doughman, A. J. Firestone, M. W. Bunce, and R. A. Anderson. 2002. Type I gamma phosphatidylinositol phosphate kinase targets and regulates focal adhesions. *Nature* **420**:89–93.

[46] Bakolitsa, C., D. M. Cohen, L. A. Bankston, A. A. Bobkov, G. W. Cadwell, L. Jennings, D. R. Critchley, S. W. Craig, and R. C. Liddington. 2004. Structural basis for vinculin activation at sites of cell adhesion. *Nature* **430**: 583–586.

[47] Bass, M. D., B. J. Smith, S. A. Prigent, and D. R. Critchley. 1999. Talin contains three similar vinculin-binding sites predicted to form an amphipathic helix. *Biochem J* **341**:257–263.

[48] Bass, M. D., B. Patel, I. G. Barsukov, I. J. Fillingham, R. Mason, B. J. Smith, C. R. Bagshaw, and D. R. Critchley. 2002. Further characterization of the interaction between the cytoskeletal proteins talin and vinculin. *Biochem J* **362**:761–768.

[49] Priddle, H., L. Hemmings, S. Monkley, A. Woods, B. Patel, D. Sutton, G. A. Dunn, D. Zicha, and D. R. Critchley. 1998. Disruption of the talin gene compromises focal adhesion assembly in undifferentiated but not differentiated embryonic stem cells. *J Cell Biol* **142**:1121–1133.

[50] Fillingham, I., A. R. Gingras, E. Papagrigoriou, B. Patel, J. Emsley, D. R. Critchley, G. C. K. Roberts, and I. L. Barsukov. 2005. A vinculin binding domain from the talin rod unfolds to form a complex with the vinculin head. *Structure* **13**:65–74.

[51] Izard, T. and C. Vonrhein. 2004. Structural basis for amplifying vinculin activation by talin. *J Biol Chem* **279**:27667–27678.

[52] Lee, S. E., R. D. Kamm, and M. R. Mofrad. 2007. Force-induced activation of Talin and its possible role in focal adhesion mechanotransduction. *J Biomech* **40**:2096–2106.

[53] Evans, E. and K. Ritchie. 1997. Dynamic strength of molecular adhesion bonds. *Biophys J* **72**:1541–1555.

[54] Hummer, G. and A. Szabo. 2003. Kinetics from nonequilibrium single-molecule pulling experiments. *Biophys J* **85**:5–15.

[55] Litvinov, R. I., H. Shuman, J. S. Bennett, and J. W. Weisel. 2002. Binding strength and activation state of single fibrinogen-integrin pairs on living cells. *Proc Natl Acad Sci USA* **99**:7426–7431.

[56] Lazaridis, T. and M. Karplus. 1999. Effective energy function for proteins in solution. *Proteins* **35**:133–152.

[57] Im, W., M. S. Lee, and C. L. Brooks, 3rd. 2003. Generalized born model with a simple smoothing function. *J Comput Chem* **24**:1691–1702.

[58] Schaefer, M., C. Bartels, F. Leclerc, and M. Karplus. 2001. Effective atom volumes for implicit solvent models: Comparison between Voronoi volumes and minimum fluctuation volumes. *J Comput Chem* **22**:1857–1879.

[59] Brockwell, D. J., E. Paci, R. C. Zinober, G. S. Beddard, P. D. Olmsted, D. A. Smith, R. N. Perham, and S. E. Radford. 2003. Pulling geometry defines the mechanical resistance of a beta-sheet protein. *Nat Struct Biol* **10**:731–737.

[60] Paci, E. and M. Karplus. 2000. Unfolding proteins by external forces and temperature: The importance of topology and energetics. *Proc Natl Acad Sci USA* **97**:6521–6526.

[61] Lee, S. E., S. Chunsrivirot, R. D. Kamm, and M. R. Mofrad. 2008. Molecular dynamics study of talin-vinculin binding. *Biophys J* **95**:2027–2036.

[62] Bois, P. R., R. A. Borgon, C. Vonrhein, and T. Izard. 2005. Structural dynamics of alpha-actinin-vinculin interactions. *Mol Cell Biol* **25**:6112–6122.

[63] Ziegler, W. H., R. C. Liddington, and D. R. Critchley. 2006. The structure and regulation of vinculin. *Trends Cell Biol* **16**:453–460.

[64] Hytonen, V. P. and V. Vogel. 2008. How force might activate talin's vinculin binding sites: SMD reveals a structural mechanism. *PLoS Comput Biol* **4**:e24.

12

Protein Conformational Change: A Molecular Basis of Mechanotransduction

Gang Bao

12.1 Cells Can Sense and Respond to Mechanical Forces

The biological cell constitutes the basic unit of life and performs a large variety of functions through synthesis, sorting, storage, and transport of biomolecules; expression of genetic information; recognition, transmission, and transduction of signals; and converting different forms of energy [1]. Many of the cellular processes involve mechanical force, or deformation, at the cellular, subcellular, and molecular levels [2, 3]. For example, biomolecular motors and machines convert chemical energy into mechanical motion in performing their diverse range of functions [4, 5]. During cell migration, contractile forces must be generated within the cell in order for the cell body to move forward [6]. Adhesion of cells to an extracellular matrix (ECM) through focal adhesion complexes is sensitive to the stiffness of the substrate [7, 8]. All living cells on Earth are constantly under physical force (gravitational and other forms of force), and many normal and diseased conditions of cells are dependent on or regulated by their mechanical environment. Some cells, such as bone and endothelial cells, are subjected to specific forces as part of their "native" physiological environment. Some other cells, such as muscle and cochlear outer hair cells [9], perform a mechanical function by converting an electrical or chemical stimulus into a mechanical motion.

Of particular importance is the ability of cells to sense mechanical force or deformation and transduce this mechanical signal into a biological response [10–14]. It is well established that vascular smooth muscle cells in the arterial wall remodels when subject to hemodynamic forces [15]. Bone keeps adjusting its structure to adapt to its mechanical environment [16]. Stem cells, on the other hand, can sense the elasticity of the substrate and differentiate into different phenotypes accordingly [17–19]. Endothelial cells can recognize the magnitude, mode (steady or pulsatile), type (laminar or turbulent), and duration of applied shear flow and respond accordingly, maintaining healthy endothelium or leading to vascular diseases including thrombosis and atherosclerosis [20]. As an example, Figure 12.1 shows shear stress–induced changes in the expression level of the Krüppel-like factor 2 (KLF2) mRNA in human umbilical vein endothelial cells (HUVEC) cells under

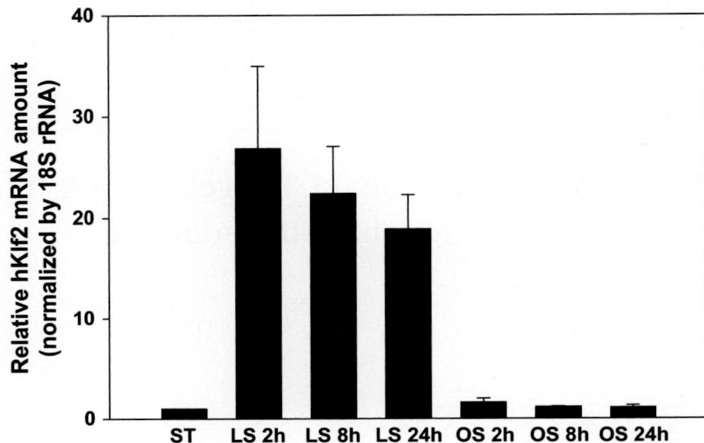

Figure 12.1. Shear stress–induced changes in the expression level of Krüppel-like factor 2 (KLF2) mRNA in HUVEC under static, laminar flow, and oscillatory flow conditions for different shear durations. It reveals a significant increase of KLF2 mRNA level under laminar shear (LS); however, the KLF2 expression remains at the control (static) level under oscillatory shear (OS).

static, laminar shear, and oscillatory shear conditions, with applied shear stress of 15 dyn/cm^2. These and other examples clearly demonstrate that cells can sense and respond to their local mechanical environment, including applied mechanical force and deformation, the stiffness of the substrate and surrounding media, vibration, and even sound waves.

But how do cells sense mechanical force (or deformation)? How do cells transduce mechanical signals into biochemical or biological responses? To date, the molecular mechanisms of force sensing and mechanotransduction remain elusive. One may speculate that there are specialized "force sensors" on the cell membrane that sense the mechanical signal; however, little is know about how these sensors work. One may think that there are stretch-activated ion channels that sense and transduce the mechanical signal [21], but it is unlikely that the force-induced activation of ion channels is the dominant mechanism for mechanotransduction. One may propose that it is the interaction between the cell and the extracellular matrix (ECM) through integrin molecules that serves as force sensors and transducers [22]; however, the underlying molecular mechanism is not well understood. There is a need to identify definitively the molecular mechanisms of mechanotransduction in living cells that are applicable to most, if not all, cell types [18].

12.2 Protein Deformation and Its Biological Consequences

Proteins perform a broad range of tasks in living cells, including metabolic and catalytic functions (enzymes), signal transduction (signaling proteins), transport of biomolecules (transporters), transmission of genetic information (nucleoprotein machines), and structural support (filaments and their networks) [1]. With a specific polypeptide sequence of up to 20 different amino acids, most proteins (globular

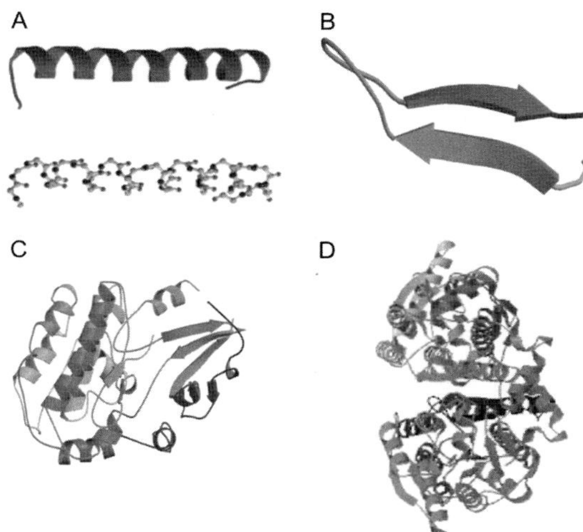

Figure 12.2. Basic structural features of proteins. Proteins are polypeptides constructed from 20 different amino acids. The polypeptide folds into secondary structures such as a-helices (A) and β-sheets (B), which, together with polypeptide loops, form globular domains (C). Multi-domain globular proteins typically have a tertiary structure (D).

proteins) form by folding the polypeptide into a tertiary structure (Figure 12.2(D)) of single or multiple globular domains (Figure 12.2(C)) consisting of α-helices (Figure 12.2(A)), β-sheets (Figure 12.2(B)), and polypeptide loops. Proteins may also have rod-or wire-like tertiary structures (fibrous proteins). Polypeptide chains of more than ~200 amino acids usually fold into two or more globular domains. These structurally independent domains have an average size of ~ 2.5 nm and possess unique three-dimensional geometries. Most protein domains have specific functions such as the binding of small molecules (ligands) and other protein domains. The ligand-binding sites in multidomain proteins are often located in the clefts between domains, facilitating the interactions with different ligands [1, 23].

It has been well established that the three-dimensional conformation of a protein largely determines its function [1]. However, the conformation of a protein can be altered by applied mechanical force, resulting in changes of the functional states of the protein and inducing downstream biochemical and biological effects. Therefore, protein deformation, or protein conformational change under mechanical force, is an excellent candidate for the molecular mechanism of mechanotransduction [10, 11, 24].

Similar to well-known physical (e.g., temperature increase) or chemical (e.g., changes in pH) events that alter protein conformation [23, 25], which may transform proteins from a native or biologically active state to a denatured or inactive state, the application of mechanical forces to a protein can lead to protein conformational changes including domain motion (Figures 12.3(A) and 12.3(B)), domain deformation (Figures 12.3(C) and 12.3(D)) and domain unfolding (Figures 12.3(E) and 12.3(F)), all of which can have significant biological implications. In general, protein domain motion includes domain hinge motion and domain sliding motion. In domain hinge motion (Figure 12.3(B)), the individual globular domains have very limited deformation; the

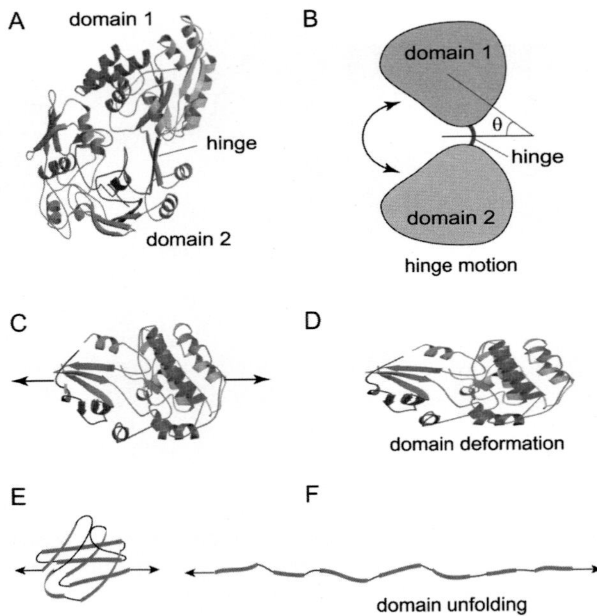

Figure 12.3. Protein deformation modes. Protein conformational changes under force including domain motion (A, B), domain deformation (C, D), and domain unfolding (E, F). These deformation modes occur according to different levels of applied force.

motion largely consists of rotations of domains around a flexible hinge (e.g., loops, α-helices, or β-sheets that join the domains together). In domain sliding motion, two protein globular domains have relative (in-plane or out-of-plane) sliding ("shear") motion. It is also possible to have mixed mode domain motions, such as a combination of rotation about a hinge and twist about an axis of the domain. Protein domain motions can be driven by applied torque as well as Brownian forces or moments ($\sim 1\ K_B T$) [26], which are equivalent to forces on the order of 0.5–10 pN (depending on the degree of motion). With large tensile forces in the range of 50–200 pN, a protein globular domain may unfold [27–30], as illustrated in Figures 12.3(E) and 3(F).

Under applied tensile (compressive) forces, a protein globular domain may elongate (contract), as illustrated in Figures 12.3(C) and 12.3(D). This type of protein conformational change is referred to as domain deformation, which is distinct from pure domain motion (little or no conformational change in the domain itself) and domain unfolding (opening of a globular domain). Domain deformation typically requires forces in the range of 1–100 pN. Although the three protein deformation modes shown in Figure 12.3 have different characteristics and force ranges, there may be coupling between the modes. For example, before domain unfolding occurs, there could be a significant domain deformation. Under mechanical forces, some protein domains may unfold gradually while others may unfold abruptly. It is also true that during domain motion, individual protein domains may have small conformational changes. The amount of coupling between different modes of protein deformation depends on not only the applied forces but also different protein structures, as is expected.

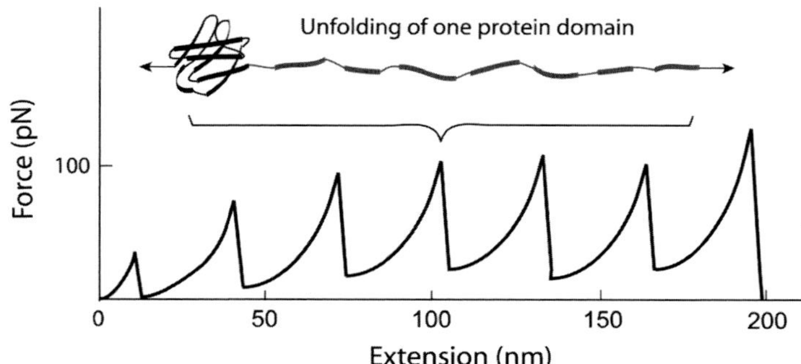

Figure 12.4. An example of protein domain deformation and unfolding. Under applied tensile force, multiple immunoglobulin domains of muscle molecule titin deform and unfold, showing a characteristic sawtooth pattern.

There is ample experimental evidence to suggest that protein domains can sustain deformation or even unfolding under normal physiological forces. It has been revealed that, under stretching with AFM or optical tweezers, the immunoglobulin domains of muscle molecule titin may deform and unfold, with a characteristic sawtooth pattern [27, 28], as shown schematically in Figure 12.4. Clearly, a force of ~ 100 pN is required for domain unfolding of titin. The FN-III domains of the ECM protein tenascin unfold upon stretching in a similar fashion [30]. Single-molecule biomechanics studies, both experimental and theoretical, have been performed for the forced unfolding of ECM molecule fibronectin, which regulates many cellular functions through its binding to integrin [31–33]. Experimental evidence suggests that, during cell adhesion and locomotion, the FN-III domains of fibronectin may unfold and refold due to the tensile and contractile forces the cells apply to the ECM [32]. This unfolding and refolding may have significant implications to cell signaling, since the resulting structural changes of the arginine-glycine-aspantate (RGD) loop seems to regulate the binding between fibronectin and integrin, as revealed by molecular dynamics simulations [33–35].

There are two important issues concerning protein deformation, or protein conformational change under mechanical force. The first is the biochemical and biological consequences of such deformation; the second is how protein deformation is related to protein structures (discussed in the next section). As with protein conformational change due to other chemical or physical events, protein conformational change under mechanical force could result in a wide range of biological consequences. Shown in Figure 12.5 are some examples of the effects of protein deformation in a living cell. Many proteins have specific ligand-binding sites buried initially by a protein domain or a peptide (a "lid"). As illustrated in Figure 12.5(A), upon applying mechanical forces to such a protein, the lid opens, exposing the ligand-binding site. The reverse is also true: Protein deformation can close the "lid" that is initially open, thereby burying the ligand-binding site. Alternatively, a protein globular domain could unfold under mechanical force, exposing the ligand-binding site that is buried inside the globular domain [Figure 12.5(B)]. A specific example is the uncoiling of the glycoprotein von Willebrand Factor (vWF) under blood-flow–induced shear stress.

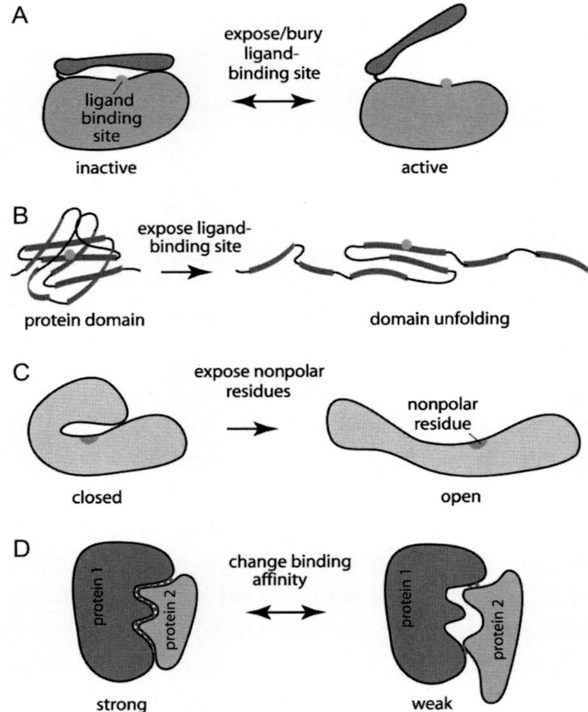

Figure 12 5. Examples of biological consequences of protein deformation. Mechanical forces can (A) switch a "lid" in a protein from "close" to "open" position, or (B) unfold a protein domain, thus exposing the ligand binding site. Protein deformation could also (C) expose the nonpolar residues, causing nonspecific interaction between the protein domain and other biomolecules; or (D) induce a change in binding affinity, altering protein–protein interactions.

It has been suggested that the unfolding of the A1 domain of the vWF is responsible for the increased interaction between vWF and GP1bα on platelets, causing the platelets to adhere to the vessel wall [36]. This has significant implications to platelet adhesion, thrombus formation, and coagulation.

Of the 20 amino acids that form a protein, 10 (including cysteine, glycine, and alanine) are nonpolar, hydrophobic residues (side chains); these residues are usually buried inside a globular or filamentous domain of a protein to avoid direct contact with water [1, 23]. Mechanical forces, however, could unfold the domain and thus expose the nonpolar residues (Figure 12.5(C)), which may cause nonspecific interactions between the protein domains and other biomolecules, and thus alter protein function. It has been demonstrated recently that, when red blood cells were subjected to shear stress, the sterically shielded cysteine residues of the cytoskeletal protein spectrin could be labeled with fluorophores, suggesting that specific domains of spectrin have undergone forced unfolding, exposing the otherwise buried cysteine residues [37]. Other nonpolar residues could be exposed in a similar fashion. It is also possible that domain motion or unfolding of protein domains could expose phosphorylation sites that are initially buried.

It is well known that proteins interact with each other based on conformational matches: A good conformational match leads to high binding specificity and affinity between two proteins, while a poor conformational match does the reverse [1]. As shown schematically in Figure 12.5(D), when proteins 1 and 2 have a good conformational match, they have strong interactions to realize their functions, for example, to activate a signaling cascade or facilitate an enzymatic activity. However, when one of the proteins, say protein 2, sustains a force-induced conformational change, the interaction between proteins 1 and 2 becomes weak due to the poor conformational match, thus altering the function of protein 2. The reverse is also true: Deformation of a protein could increase its affinity to another protein that otherwise would not interact due to the poor conformational match in its native state. This concept is not limited to protein–protein interactions; protein–DNA, protein–RNA, and protein–small molecule interactions can be altered by force-induced protein conformational changes as well.

Force-induced protein conformation, or protein deformation, could have many effects on the biology of a cell. In certain cases, the biological consequence of protein deformation is similar to that of protein conformational change due to biochemical events such as allosteric regulation. In some other cases, the biological consequence of protein deformation may be similar to that of protein misfolding. In fact, protein deformation and protein misfolding share many common features, including: (a) no change in amino acid sequence; (b) resulting in two or more stable conformations; (c) altering protein binding/aggregation; and (d) possibly inducing toxic activity or loss of normal biological function [38]. More molecular-level experimental evidence is needed to confirm that protein deformation in a living cell indeed causes a gain or loss of protein function.

A distinctive feature of protein deformation in living cells is the spatial and temporal regulation of forces that alter protein conformation. When a cell is subjected to mechanical forces (or displacements), the intracellular distribution of the forces varies with location and time, depending on not only the applied forces but also the intracellular structure (such as cell cytoskeleton) and its dynamics, as well as cell–ECM and cell–cell interactions. As such, proteins that are well connected to the force-bearing structures of a cell may be easier to deform, whereas proteins that are not "wired" to the force-bearing structures may rely on viscous forces in order to deform. Such viscous forces may be due to the deformation of cells, inducing shear flow of the highly viscous cytosol and nucleoplasm and transmitting shear stress to proteins. Of course the deformability of a protein is critically dependent on its structure, as will be discussed later. It is possible that cells utilize force-induced domain deformation and unfolding in proteins at specific locations and in a time-dependent manner to transduce a mechanical signal into a biochemical reaction. This adds to the richness (and complexity) of the spatial and temporal regulation of mechanotransduction in a cell.

12.3 Protein Mechanics

Protein deformations may play important roles in biology and medicine in general, and mechanotransduction in particular. However, the understanding of the mechanics of proteins is still in its infancy, and many important questions remain open.

For example, how does a protein deform under different applied forces or displacements? How is the deformation of a protein dependent on its three-dimensional structure? How do protein deformations affect protein function? What is the dynamic behavior of proteins? Even at the very fundamental level, protein mechanics is different from the classical mechanics of deformable bodies in terms of the following aspects: (a) Thermal energy ($\sim 1\ k_BT$) and entropic elasticity may become dominant; (b) local quantities such as stress and strain (or stretch ratio) may not be the best parameters to characterize the mechanical behavior of a protein; and (c) the mechanics and chemistry of a protein are intrinsically coupled. Unlike a typical engineering material of which the mechanical behavior does not depend on the chemistry of its surrounding media, for a protein, it is critical to include the aqueous solution (and its ionic strength) in the modeling and experimental studies of conformational changes under force. Similar to all polymeric materials, protein molecules are viscous in nature, and their deformations are rate- and temperature-dependent.

12.3.1 Protein Constitutive Behavior

To describe how a material deforms under mechanical forces or deformation, one typically develops a constitutive model that links deformation to the applied forces. For example, for a linear elastic material that is homogeneous and isotropic, the one-dimensional relationship between stress σ (force per unit area) and strain ε (relative change in length) is given by

$$\sigma = E\varepsilon, \tag{12.1}$$

where E is the elastic modulus. For the arterial wall as a membrane under uniaxial loading, we have [39]

$$\sigma = (\sigma^* + \beta)e^{\alpha(\lambda-\lambda^*)} - \beta, \tag{12.2}$$

where σ is load divided by the initial cross-sectional area A_0; λ is the stretch ratio (current length L divided by the original length L_0); α, β are constants; and (σ^*, λ^*) represents a reference point on the stress–strain curve. For three-dimensional deformations, more complicated constitutive models are required. Constitutive models have two key components: the mathematical formulation and the set of material parameters.

To describe the mechanical behavior of a protein, that is, how a protein deforms under mechanical force, a constitutive model is also required. Perhaps the simplest model so far is the worm-like chain (WLC) model [40] for a protein under uniaxial stretching [30]:

$$\frac{f\xi}{k_BT} = \frac{x}{L} + \frac{1}{4(1 - x/L)^2} - \frac{1}{4}, \tag{12.3}$$

where f is the applied force, ξ is the flexural persistence length, k_B is Boltzmann's constant, T is temperature, x is the extension, and L is the "contour" length of the protein (length of the polypeptide under stretching). For the deformation and unfolding of individual domains of the protein molecule titin under stretching, it has been found that the force–extension relationship of titin can be described by the WLC model (Equation 12.3), with $\xi = 0.4$ nm; $L = \lambda_0 + n{\cdot}\lambda$ ($n = 0, 1, 2, \ldots$), where $\lambda_0 = 58$ nm; and λ ranges from 28 to 29 nm. A similar expression was found to fit very well the measured force–extension curve for tenascin [30]. The WLC model seems to work well for domain deformations of protein molecules containing multiple, individually folded domains with β-sandwich structures [41].

Although new experimental data are needed to develop constitutive models for a wide class of proteins, it is possible that, for a class of globular proteins with single or multiple domains, their domain deformation and unfolding under uniaxial stretching can be described by

$$\frac{f\xi}{K_B T} = \varphi\left(\frac{x}{L}, \alpha_1, \alpha_2, \ldots, \alpha_n\right), \tag{12.4}$$

where φ is a nondimensional function and $\alpha_1, \alpha_2, \ldots \alpha_n$ are parameters that may reflect the specific structure of the protein, rate effect, and the chemical environment. Both the form of the function φ and its dependence on the parameters α_i can be determined by single-molecule experiments of stretching globular proteins.

There is a lack of understanding of how proteins deform in living cells under physiological conditions. Given that most cells generate contractile forces during locomotion, division, and adhesion, it is reasonable to believe that certain proteins deform under stretching or compression in living cells, assume the native conformation upon unloading (relaxation), and such stretching and relaxation could be one- or multidimensional. However, protein deformation under compression is most likely very different from that under stretching, since the globular domains are much more resistant to compression than to tension. It is also possible that certain proteins in a living cell sustain shear or torsional deformation due to, for example, forces generated by motors and other molecular machines; however, currently there is a lack of experimental data at the single-molecule level to provide a basis for constitutive modeling.

12.3.2 Protein Domain Hinge Motion

Many protein molecules undergo hinge motion in cells, which has many biological implications [42] (see also http://bioinfo.mbb.yale.edu/MolMovDB/). In these proteins, one or more domains rotate about a hinge due to chemical reaction (such as ATP hydrolysis), Brownian force, applied force (torque), or molecular interactions (such as ligand binding). While some motor molecules rely on hinge motion to convert chemical energy into mechanical force, other proteins may utilize hinge motion to transduce signals, regulate interactions, and facilitate enzymatic activities. As an example, the structure of *E. coli* periplasmic dipeptide binding protein is displayed in Figure 12.2(A).

It has two domains linked by a hinge consisting of two β-sheets; its hinge motion has been well documented (http://bioinfo.mbb.yale.edu/MolMovDB/).

Some insight can be gained by performing a Langevin dynamics analysis of the protein hinge motion in which proteins having two domains linked by a hinge are considered (Figure 12.3B). The angle of rotation θ of a domain about the center of the hinge is assumed to obey the stochastic differential equation [26]

$$I_0 \frac{d^2\theta}{dt^2} + \eta_r \frac{d\theta}{dt} + k\theta = R(t) + F(t) \tag{12.5a}$$

$$I_0 \frac{d^2\theta}{dt^2} + \int_0^t \alpha(t-s) \frac{d\theta}{ds} ds + k\theta = R(t) + F(t), \tag{12.5b}$$

where I_0 is the moment of inertial of the domain, η_r is the rotational viscous drag coefficient, α(t) is a dissipation function, *k* is the elastic constant of the hinge, and *F(t)* is the applied torque. The Brownian torque *R(t)* in Equation (12.5) is a random function of time and represents the collective effect of the random collisions of the surrounding molecules in the solvent with the protein. The ensemble average of *R(t)* is zero. Similar to almost all the studies of stochastic processes, Boltzmann's ergodic hypothesis is adopted [43] and the time-average of a parameter $\bar{a}$ is assumed to be identical to its ensemble average $\langle a \rangle$. Although Equation (12.5(a)) is simpler, when the aqueous solution surrounding the protein is very viscous (such as in a living cell), Equation (12.5(b)) should be used according to the fluctuation-dissipation theory [44].

Domain hinge motion has a direct relevance to mechanochemical coupling and mechanotransduction. For example, in pumping protons across an insulating membrane or manufacturing ATP from ADP and phosphate, ATP synthase replies on the hinge motion of the β domains of the F_1-ATPase to generate mechanical torque from the free energy of ATP binding, or produce ATP using torques generated by the protonmotive force [45]. Protein domain hinge motion can also affect receptor–ligand binding. As mentioned before, many proteins have the ligand binding site located near the hinge between two globular domains. Therefore, depending on the magnitude and rate of domain hinge motion, receptor–ligand binding may be altered or even blocked. Specifically, if the rate of hinge motion is much faster than the time required for certain ligands to diffuse into the binding pocket, it is likely that binding of these ligands is prohibited. This may serve as a "selection" mechanism for specific receptor–ligand binding.

12.3.3 Receptor–Ligand Binding

Perhaps one of the most important aspects of protein deformation under force is its effect on receptor–ligand binding, an essential process in cells [46], because specific molecular recognition and interactions rely on conformational matches and charge complementarity between the receptor and the ligand. In general, protein–protein and protein–DNA interactions are determined by many factors, including electrostatic double–layer force, van der Waals force, "steric" repulsion forces, hydrogen-bonding, and hydrophobic contacts. Most of these factors operate within short ranges [47]. Therefore, the three-dimensional geometry local to the binding pocket of the receptor and ligand contributes significantly to the characteristics of their binding.

When mechanical forces are applied to a receptor (and ligand), the receptor may deform, thereby altering the conformational match between the receptor and the ligand. The specific effect of mechanical forces on receptor–ligand binding may depend on the mode and the magnitude of the deformation of a receptor or a ligand (or both), which are in turn determined by the structure of the molecules involved, the solvent surrounding them, and loading history (including rate). It is possible that deformation only alters the kinetic rates [48, 49]. However, in certain cases, only protein deformation can expose the binding site, thus switching between the "on" and "off" states of the receptor. It is also possible that protein deformation can change the specificity of a receptor/ligand pair, that is, binding to ligand B instead of ligand A upon deformation. The latter two effects are both well known for molecules such as integrins that can be activated biochemically [50]. It is likely that force-induced protein deformation may yield similar results.

12.3.4 Toward a Protein Structural Basis for Mechnotransduction

From a mechanics point of view, deformation of a material is critically dependent on its structure, be it atomic or macroscopic. For protein deformation, this dependence could be even more profound: Through evolution, the structure of a specific protein that has mechanical functions (load sensing, bearing, and generating) may have been optimized, and the resulting structural features may provide insight for better designs of engineering materials. In fact, most of the fibrous proteins that have rod- or wirelike tertiary structures, including keratins, collagens, and elastins for constructing connective tissues, tendons, bone matrix, and muscle fiber, are optimized for load-bearing functions. However, it remains an important task to establish the structural features of globular proteins that have force-sensing and mechanotransduction functions.

Proteins can be classified into different classes according to their structural features. For example, in the Structural Classification of Proteins (SCOP) database developed by Alexey G. Murzin and his associates, proteins are divided into seven major classes based on the general "structural architecture" of the protein domains (see http://scop.mrc-lmb.cam.ac.uk/sco/): (a) all alpha proteins; (b) all beta proteins; (c) alpha and beta proteins (a/b); (d) alpha and beta proteins (a+b); (e) multidomain proteins (alpha and beta); (f) membrane and cell surface proteins and peptides; and (g) small proteins. Each class of proteins is further grouped into family (with some sequence similarity), superfamily (with sufficient structural and functional similarity), and fold (with similar arrangements of regular secondary structures). The SCOP database provides a broad survey of all known protein folds, detailed information about the close relatives of any particular protein, and a framework for future research and classification.

Table 12.1. *Structural features of key force-sensing or -bearing proteins.*

Protein	Class	Superfamily	Fold
Fibronectin type III domain, tenascin, Type I titin, integrin beta-4 subunit	All beta proteins	Fibronectin type III	Immunoglobulin-like beta-sandwich: *sandwich; 7 strands in 2 sheets*
VCAM-1, ICAM-1, ICAM-2	All beta proteins	Immunoglobulin	Immunoglobulin-like beta-sandwich: *sandwich; 7 strands in 2 sheets*
von Willebrand factor A1 domain, von Willebrand factor A3 domain	Alpha and beta proteins (a/b)	vWA-like	vWA-like: *Mainly parallel beta sheets (beta-alpha-beta units)*
Integrin alpha1-beta1, integrin alpha2-beta1, integrin alpha-x beta2, integrin beta A domain, integrin alpha M, integrin CD11a/CD18	Alpha and beta proteins (a/b)	vWA-like	vWA-like: *Mainly parallel beta sheets (beta-alpha-beta units)*
Integrin beta-3	Small proteins	Plexin repeat	Trefoil/Plexin domain-like: *disulfide-rich fold; common core is alpha+beta with two conserved disulfides*

Table 12.1 shows the classes and structural features of some key proteins that may have force-sensing, force-bearing, and mechanotransduction functions. Interestingly, most of the proteins listed in Table 12.1 are concentrated in two classes: von Willebrand factor A1 and A3 domains and many integrins and integrin domains (integrin molecules $\alpha_1\beta_1$, $\alpha_2\beta_1$, $\alpha_x\beta_2$, CD11a/CD18; integrin α_M and β_A domains) belong to the class of "alpha and beta proteins (a/b)," with a vWA-like domain consisting of mainly parallel beta sheets (beta-alpha-beta units). On the other hand, adhesion molecules VCAM-1, ICAM-1, ICAM-2, ECM; fibronectin type III domain and tenascin; muscle molecule type I titin; and integrin β_2 subunits belong to the class of "all beta proteins," with immunoglobulin-like beta-sandwich domains consisting of a sandwich of seven β-strands in two sheets. The most striking feature revealed by Table 12.1 is that many of the known mechanically stressful globular proteins such as integrins, adhesion molecules, ECM molecules, vWFs, and titins have domains that are rich in β-sheets. This might be due to the fact that these domains are quite compact, thus resisting compression and shear; at the same time, they could be deformed and open under tension, thus exposing ligand binding sites. Although still very preliminary, the structural features shown in Table 12.1 provide some important insight into the mechanics of proteins.

To establish a deeper understanding of the relationship between protein structure and protein deformation, a more systematic structural analysis of proteins that have mechanical load-sensing, -bearing or -generating functions is needed. This includes the analysis of protein structural domains based on similarities of their amino acid sequences and three-dimensional structures using, for example, SCOP. It is also necessary to perform single-molecule experiments [41] and detailed structural analysis using computational tools such as steered molecular dynamics to understand why load-sensing and -bearing proteins have some common structural features, as revealed by Table 12.1. It is anticipated that, by systematically simulating protein conformational dynamics under force with sufficient details, the structural basis for protein deformation and its roles in mechanotransdution will emerge.

12.4 Modeling and Experimental Challenges in Studying Protein Deformation

Understanding mechanotransduction in living cells involves two basic questions. The first is how mechanical forces are being "sensed" by cells, and the second is how such a mechanical signal is being transduced into biochemical and biological responses. As a candidate for the molecular mechanism of mechanotransduction, protein deformation, or protein conformational change under mechanical force, may provide an answer for both. Cells may "sense" mechanical forces by deforming specific proteins at specific locations; such deformations alter (induce or prohibit) molecular interactions such as receptor–ligand binding, causing changes in the signaling pathways and other cellular processes. For example, the receptor-mediated signaling processes in cells can be regulated by deformation of ECM molecules such as fibronectin, resulting in the exposure of an RGD sequence for binding to integrin [50], whose functions include both adhesion and signal transduction. In addition to the ability of switching between resting and active states of proteins upon ligand occupancy, it is conceivable that cells utilize more continuous conformational changes in response to various forces to produce more gradual exposures of the functional domain or affect the binding rate for the downstream signaling molecules, thereby transducing mechanical forces and deformations into biochemical responses. Taken together, it is essential to study protein deformation in order to understand mechanotransduction in living cells.

Despite the significant progress over the last decade or so in developing tools for single-molecule biomechanics studies, including AFM, optical tweezers, magnetic tweezers, micropipette force spectroscopy, and surface force apparatus, experimental probing of the deformation of single proteins is severely limited by at least three factors. First, it is difficult to image or measure the deformations of protein molecules by force due to the small (0.1–1 nm) geometric changes. Therefore, advances in protein mechanics require further progress in bioimaging technology. Second, it is difficult to hold, position, and manipulate single-protein molecules by conventional means. As a result, attaching a protein to devices that impose force or deformation in a controlled manner without affecting its native conformation is a major experimental challenge. Finally, the characteristic time for motion and relaxation of proteins and nucleic acids, which spans the wide range of nanoseconds to seconds, may render experimental studies of the underlying deformation mechanisms a rather difficult task. Since fluorescence resonance energy transfer (FRET) between donor and acceptor fluorophores is extremely sensitive to their relative distance (<10 nm), it is possible that the conformational dynamics of proteins under force can be studied using FRET. In fact, FRET has been used for imaging fibronectin extension and unfolding in cell culture [51, 52].

Theoretical modeling and numerical simulation of the deformation and dynamics of proteins is a very important area, and the steered molecular dynamics simulations of fibronectin and integrin have provided much insight into how globular proteins unfold under tensile forces [34, 35]. However, currently MD simulations can be performed only for a short time period, up to 100 nss, while most biological processes in cells occur on a time-scale of 1 ms. Consequently, the results of MD simulations are often qualitative rather than quantitative. Further, the force fields used in MD simulations (i.e., the potential functions that characterize the atomic and molecular interactions) are usually obtained based on experimental measurements of the interactions of limited pairs of atoms; they may not be applicable to many biomolecular interactions. Therefore, it is necessary to seek a new numerical approach to simplify the simulations while still preserving the basic features of the molecular interactions, to increase our computing power, and to develop more accurate force fields.

In summary, the development of protein mechanics is the key to definitively establishing a molecular mechanism for mechanotransduction in living cells. This requires a quantitative description of the constitutive behavior of proteins, including how proteins deform under different mechanical loads; what are the modes and time-scales of protein motions and deformations; and how such deformations are related to protein structural features. It also requires a better understanding of how protein deformation affects molecular interactions in cells, including protein–protein, protein–RNA, and protein–DNA interactions. The ultimate goal is to understand the roles of protein mechanics in disease processes, and utilize this understanding in disease detection, treatment, and prevention.

Acknowledgments

This work was supported in part by the National Heart Lung and Blood Institute of the NIH as a Program of Excellence in Nanotechnology (HL80711). The author would like to thank Dr. Wonjong Rhee for generating the results shown in Figure 12.1.

REFERENCES

[1] Alberts, B., Bray, D., Lewis, J., Raff, M., Roberts, K., Watson, J. D. (2002) *Molecular Biology of the Cell.* 4th ed. Garland Publishing, New York.

[2] Wang, N., Butler, J. P., Ingber, D. E. (1993) Mechanotransduction across the cell surface and through the cytoskeleton. *Science*, **260**, 1124–1127.

[3] Vogel, V., Sheetz, M. (2006) Local force and geometry sensing regulate cell functions. *Nat Rev Mol Cell Biol*, **7**, 265–275.

[4] Howard, J. (2001) *Mechanics of Motor Proteins and the Cytoskeleton*. Sinauer Associates, Sunderland, MA.

[5] Block, S. M., Goldstein, L. S., et al. (1990) Bead movement by single kinesin molecules studied with optical tweezers. *Nature*, **348**, 348–352.

[6] Stossel, T. P. (1993) On the crawling of animal cells. *Science*, **260**, 1086–1094.

[7] Pelham, R. J. J., Wang, Y. (1997) Cell locomotion and focal adhesions are regulated by substrate flexibility. *Proc. Natl. Acad. Sci. U.S.A.*, **94**, 13661–13665.

[8] Discher, D. E., Janmey, P., Wang, Y. (2005) Tissue cells feel and respond to the stiffness of their substrate. *Science*, **310**, 1139–1143.

[9] Brownell, W. E., Spector, A. A., Raphael, R. M., Popel, A. S. (2001) Micro- and nanomechanics of the cochlear outer hair cell. *Annu. Rev. Biomed. Eng.*, **3**, 169–194.

[10] Bao, G. (2002) Mechanics of biomolecules. *J. Mech. Phys. Solids*, **50**, 2237–2274.

[11] Zhu, C., Bao, G. Wang, N. (2000) Cell mechanics: Mechanical response, cell adhesion, and molecular deformation. *Annu. Rev. Biomed. Eng.*, **2**, 189–226.

[12] Bershadsky, A. D., Balaban, N. Q., Geiger, B. (2003) Adhesion-dependent cell mechanosensitivity. *Annu. Rev. Cell Dev. Biol.*, **19**, 677–695.

[13] Bershadsky, A., Kozlov, M., Geiger, B. (2006) Adhesion-mediated mechanosensitivity: A time to experiment, and a time to theorize. *Curr. Opin. Cell Biol.*, **18**, 472–481.

[14] Jalali, S., del Pozo, M. A., Chen, K., Miao, H., Li, Y., Schwartz, M. A., Shyy, J. Y., Chien, S. (2001) Integrin-mediated mechanotransduction requires its dynamic interaction with specific extracellular matrix (ECM) ligands. *Proc. Natl. Acad. Sci. U.S.A.*, **98**, 1042–1046.

[15] Fung, Y. C. (1990) *Biomechanics: Motion, Flow, Stress, and Growth*. Springer-Verlag, New York.

[16] Fung, Y. C. (1993) *Biomechanics: Mechanical Properties of Living Tissues*. 2nd ed. Springer-Verlag, New York.

[17] Engler, A. J., Sen, S., Sweeney, H. L., Discher, D. E. (2006) Matrix elasticity directs stem cell lineage specification. *Cell*, **126**, 677–689.

[18] Ali, M. H., Schumacker, P. T. (2002) Endothelial responses to mechanical stress: Where is the mechanosensor?. *Crit. Care Med.*, **30**, S198–206.

[19] Wootton, D. M., Ku, D. N. (1999) Fluid mechanics of vascular systems, diseases, and thrombosis. *Annu. Rev. Biomed. Eng.*, **1**, 299–329.

[20] Fisher, A. B., Chien, S., Barakat, A. I., Nerem, R. M. (2001) Endothelial cellular response to altered shear stress. *Am. J. Physiol. Lung. Cell. Mol. Physiol.*, **281**, L529–533.

[21] Hu, H., Sachs, F. (1997) Stretch-activated ion channels in the heart. *J. Mol. Cell Cardiol.*, **29**, 1511–1523.

[22] Ingber, D. E. (2002) Mechanical signaling and the cellular response to extracellular matrix in angiogenesis and cardiovascular physiology. *Circ. Res.*, **91**, 877–887.

[23] Creighton, T. E. (1993) *Proteins*. W. H. Freeman and Company, New York.

[24] Bao, G., Suresh, S. (2003) Cell and molecular mechanics of biological materials. *Nat. Mat.*, **2**, 715–726.

[25] Voet, D., Voet, J. G. (1995) *Biochemistry*. 2nd ed. John Wiley & Sons, New York.

[26] McCammon, J. A., Gelin, B. R., Karplus, M., Wolynes, P. G. (1976) The hinge-bending mode in lysozyme. *Nature*, **262**, 325–326.
[27] Rief, M., Gautel, M., Oesterhelt, F., Fernandez, J. M., Gaub, H. (1997) Reversible unfolding of individual titin immunoglobulin domains by AFM. *Science*, **276**, 1109–1112.
[28] Kellermayer, M. S. Z., Smith, S. B., Granzier, H. L., Bustamante, C. (1997) Folding-unfolding transitions in single titin molecules characterized with laser tweezers. *Science* **276**, 1112–1116.
[29] Tskhovrebova, L., Trinnick, J., Sleep, J. A., Simmons, R. M. (1997) Elasticity and unfolding of single molecules of the giant muscle protein titin. *Nature*, **387**, 308–312.
[30] Oberhauser, A. F., Marszalek, P. E., Erickson, H. P., Fernandez, J. M. (1998) The molecular elasticity of the extracellular matrix protein tenascin. *Nature*, **393**, 181–185.
[31] Krammer, A., Lu, H., Isralewitz, B., Schulten, K., Vogel, V. (1999) Forced unfolding of the fibronectin type III module reveals a tensile molecular recognition switch. *Proc. Natl. Acad. Sci. U.S.A.*, **96**, 1351–1356.
[32] Ohashi, T., Kiehart, D. P., Ericson, H. (1999) Dynamics and elasticity of the fibronectin matrix in living cell culture visualized by fibronectin-green fluorescent protein. *Proc. Natl. Acad. Sci. U.S.A.*, **96**, 2153–2158.
[33] Craig, D., Krammer, A., Schulten, K., Vogel, V. (2001) Comparison of the early stages of forced unfolding for fibronectin type III modules. *Proc. Natl. Acad. Sci. U.S.A.*, **98**, 5590–5595.
[34] Vogel, V., Thomas, W. E., Craig, D. W., Krammer, A., Baneyx, G. (2001) Structural insights into the mechanical regulation of molecular recognition sites. *Trends. Biotechnol*, **19**, 416–423.
[35] Puklin-Faucher, E., Gao, M., Schulten, K., Vogel, V. (2006) How the headpiece hinge angle is opened: New insights into the dynamics of integrin activation. *J. Cell. Biol.*, **175**, 349–360.
[36] Ruggeri, Z. M. (2003) Von Willebrand factor. *Curr Opin Hematol*, **10**, 142–149.
[37] Johnson, C. P., Tang, H. Y., Carag, C., Speicher, D. W., Discher, D. E. (2007) Forced unfolding of proteins within cells. *Science*, **317**, 663–666.
[38] Soto, C. (2001) Protein misfolding and disease; protein refolding and therapy. *FEBS Letters*, **498**, 204–207.
[39] Fung, Y. C. (1967) Elasticity of soft tissues in simply elaongation. *Am. J. Physiol.*, **28**, 1532–1544.
[40] Marko, J. F., Siggia, E. D. (1995) Streching DNA. *Macromolecules*, **28**, 8759–8770.
[41] Schwaiger, I., Schleicher, M., Noegel, A. A., Rief, M. (2005) The folding pathway of a fast-folding immunoglobulin domain revealed by single-molecule mechanical experiments. *EMBO Rep.*, **6**, 46–51.
[42] Subbiah, S. (1996) *Protein Motions*. Chapman & Hall, Austin, TX.
[43] Coffey, W. (1985) In Evans, M.W. (ed.), *Dynamical Processes in Condensed Matter*, pp. 69–252. John Wiley & Sons, New York.
[44] Uhlenbeck, G. E., Ornstein, L. S. (1930) On the theory of the Brownian motion. *Phys. Rev.*, **36**, 823–841.
[45] Boyer, P. D. (1993) The binding change mechanism for ATP synthase – Some probabilities and possibilities. *Biochem. Biophys. Acta*, **1140**, 215–250.
[46] Lauffenburger, D. A., Linderman, J. J. (1993) *Receptors*. Oxford University Press, New York.
[47] Israelachvili, J. (1992) *Intermolecular and Surface Forces*. Academic Press, San Diego, CA.
[48] Evans, E., Ritchie, K. (1997) Dynamic strength of molecular adhesion bonds. *Biophys. J.*, **72**, 1541–1555.

[49] Merkel, R., Nassoy, P., Leung, A., Ritchie, K., Evans, E. (1999) Energy landscapes of receptor-ligand bonds explored with dynamic force spectroscopy. *Nature*, **397**, 50–53.

[50] Sanchez-Mateos, P., Cabanas, C., Sanchez-Madrid, F. (1996) Regulation of integrin function. *Semin. Cancer Biol.*, **7**, 99–109.

[51] Baneyx, G., Baugh, L., Vogel, V. (2001) Coexisting conformations of fibronectin in cell culture imaged using fluorescence resonance energy transfer. *Proc. Natl. Acad. Sci. U.S.A.*, **98**, 14464–14468.

[52] Baneyx, G., Baugh, L., Vogel, V. (2002) Fibronectin extension and unfolding within cell matrix fibrils controlled by cytoskeletal tension. *Proc. Natl. Acad. Sci. U.S.A.*, **99**, 5139–5143.

13

Translating Mechanical Force into Discrete Biochemical Signal Changes: Multimodularity Imposes Unique Properties to Mechanotransductive Proteins

Vesa P. Hytönen, Michael L. Smith, and Viola Vogel

13.1 Introduction: Mechanical Force Can Regulate Molecular Function

Cells can sense and transduce a broad range of mechanical forces into distinct sets of biochemical signals that ultimately regulate cellular processes, including adhesion, migration, proliferation, differentiation, and apoptosis. But how is force translated at the molecular level into biochemical signal changes that have the potential to alter cellular behavior? Is it just the rigidity of matrices that is sensed by cells, or can force applied to the extracellular matrix switch their functional display? How about other proteins that are part of the force-bearing protein networks that connect the extracellular matrix to the contractile cytoskeleton: Can their molecular recognition sites be altered if mechanically stretched? The advent of nanotech tools, particularly atomic force microscopy and optical tweezers (Fisher et al., 2000; Kellermayer et al., 1997; Rief et al., 1997; Tanase et al., 2007; Tskhovrebova et al., 1997), were a major milestone in recognizing the unique mechanical properties of proteins. After a decade of new insights into single molecule mechanics, the focus now turns to addressing how force-induced mechanical unfolding could potentially change protein functions (for reviews, see Bustamante et al., 2004; Discher et al., 2005; Gao et al., 2006; Giannone and Sheetz, 2006; Orr et al., 2006; Vogel, 2006; Vogel and Sheetz, 2006). Beyond the molecular recognition sites that confer biochemical specificity to proteins, are there common mechanical design criteria by which structural motifs are assembled to confer unique mechanical properties to proteins? If so, is it possible that cell generated tension is sufficient to mechanically unfold proteins that are part of force-bearing protein networks in living tissues? How are proteins stabilized against mechanical unfolding, and do cells switch protein functions by force to regulate or even switch between intracellular signaling networks?

Through careful investigations of the mechanical characteristics of isolated proteins that are mechanically stretched *in vitro* and computational simulations that provide high-resolution structural information of the unfolding pathways of proteins, key design principles are beginning to emerge that describe how intracellular, extracellular, and transmembrane proteins might sense mechanical forces and convert them into biochemical signal changes (for reviews see Clausen-Schaumann et al., 2000; Fredberg and Kamm, 2006; Vogel, 2006; Vogel and Sheetz, 2006). First, these studies have demonstrated that mechanical stability is not only regulated by the type of protein fold assumed under equilibrium, but also by the orientation of key force-bearing backbone hydrogen bonds with respect to the force vector. Second, many proteins contain buried cryptic binding sites in their hydrophobic interior that are only exposed once the protein unfolds (Craig et al., 2001, 2004; Gao et al., 2002a, 2003; Grater et al., 2005; Johnson et al., 2007; Krammer et al., 1999, 2002; Lee et al., 2007; Li et al., 2005; Marszalek et al., 1999; Sawada et al., 2006). Third, a large variety of proteins that are part of force-bearing protein networks linking the intracellular cytoskeleton and extracellular matrix have multimodular structures where each individual module often carries one or more unique binding sites. What are the advantages of linking such modules equipped with different functionalities into macromolecules? What functional aspects are we missing if we functionalize materials with peptides that contain single molecular recognition motifs for biomedical technological applications, as done, for example, with the cell adhesive tripeptide RGD (Pytela et al., 1985, 1986) from the 450-kDa fibronectin molecule, instead of using the full-length protein?

Beyond addressing these questions, we will particularly focus on how and by what rules multimodularity might confer unique properties to mechanotransductive proteins. Do the modules in such multimodular proteins break stochastically or in a well-defined sequence of events? How are active sites in proteins regulated by mechanical force? What might be their physiological relevance?

13.1.1 Significance of Force-Bearing Protein Networks

Mechanical forces regulate cell function. The cellular nanomachinery is thereby subjected to both exogenously applied forces and cell generated forces that the cells apply to their local extracellular matrix (ECM) and to neighboring cells. Forced protein unfolding may be one important mechanism by which mechanical factors are translated into biochemical signal changes in a variety of tissues and cell types. The large number of proteins that are part of force-bearing networks that connect the cell interior with the exterior implicates a similarly large list of candidate proteins potentially acting as mechanosensors (for review see Bao and Suresh, 2003; Bustamante et al., 2004; Chen et al., 2004; Orr et al., 2006; Vogel, 2006; Vogel and Sheetz, 2006). Most every cell in the body is exposed to mechanical forces. In addition to cell–cell and cell–ECM contacts, flowing blood and urine impart shear stress on the endothelial blood vessel lining and epithelial urinary tract, respectively, which actively influences their morphology and function. Lung expansion and contraction imposes stress on lung tissues that triggers the release of surfactants onto the epithelial surface. Mechanical loading of bone leads to remodeling. Furthermore, cells are not passive members of tissue but actively associate, stretch, and remodel their surroundings using a variety of specialized structures. Focal adhesions (FA) and focal complexes (FX; together focal contacts) are structures specific to cell–ECM

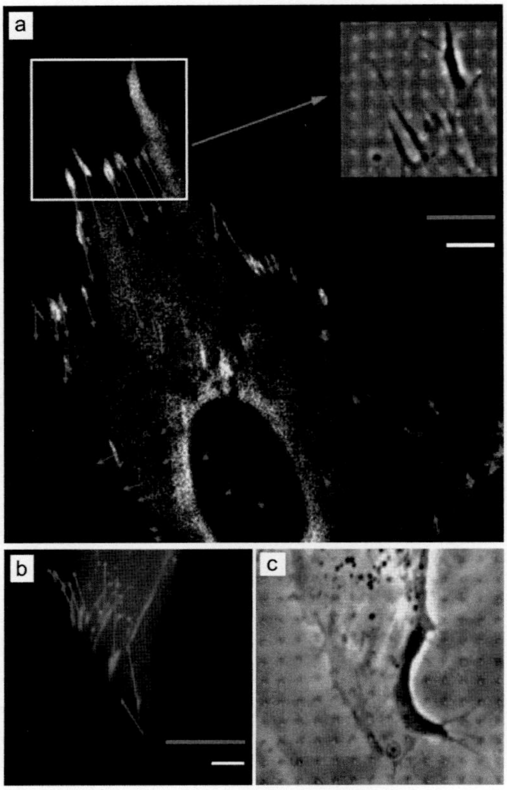

Figure 13.1. Focal adhesions and the spatially resolved force vectors. Using a micropatterned elastomeric substrate and GFP-labeled vinculin, Balaban et al. were able to quantify the mechanical forces that are applied to the substrate via focal adhesions (Balaban et al., 2001).

interactions. These structures mechanically link the cell cytoskeleton and force-generating machinery within the cell to the ECM. Intracellular traction can generate large forces that are easily visualized as strain within the underlying stretchable substrates (Figure 13.1, Plate A; Balaban et al., 2001; Harris et al., 1980; Tan et al., 2003). In addition, focal contacts are not passively resistant to force, but force actively induces focal contact strengthening through the recruitment of additional focal adhesion proteins and the initiation of intracellular signaling events (Chrzanowska-Wodnicka and Burridge, 1996; Galbraith et al., 2002; Helfman et al., 1999; Riveline et al., 2001). Focal contacts grow in size and resistance to stress if they are mechanically loaded (Giannone et al., 2003; Riveline et al., 2001).

Sensation of physiological forces may be equally as diverse. For instance, the vascular endothelium presents a thick layer of proteoglycans and other components on its apical surface that retards fluid flow (Smith et al., 2003; Vink and Duling, 1996), and digestion of this layer inhibits flow-mediated NO release by excised vessels and cultured endothelial cells, suggesting that glycocalyx deformation or glycocalyx-mediated stress application to intracellular proteins may underlie mechanosensation (Florian et al., 2003; Mochizuki et al., 2003). Traction forces at the edge of spreading cells lead to force application and downstream activation of Rap1, which is blocked if p130Cas levels are reduced using siRNA (Sawada et al., 2006). Are these and other observations mediated by force-regulated changes of the structure–function relationship of proteins?

13.1.2 Cell-Generated Forces Can Unfold Proteins *In Vivo*

Conclusive evidence that force unfolds proteins and that the unfolding imparts new functions to the protein rely on the demonstration that mechanical stress unfolds proteins *in vivo*. Until recently, limited tools were available that could be used to study nonequilibrium protein conformations in their native settings. Cellular sensation of mechanical forces and responsiveness to them at the molecular level remains the next frontier in our understanding of mechanotransduction processes, and it is thus necessary to determine the mechanical properties of biomolecules, their force-induced protein unraveling pathways, as well as the related changes in their functional states. Mechanotransductive ion channels have received significant attention (Booth et al., 2007; Kung, 2005; Sotomayor et al., 2005; Vollrath et al., 2007). Ciliary vibrations lead to openings of the ion channels in auditory sensory cells, permitting auditory sensation through the influx of ions, while the exposure of bacteria to a hypo-osmotic environment leads to membrane tension that opens ion channels, thus allowing solutes to escape the cell (for reviews see Booth et al., 2007; Kung, 2005; Vollrath et al., 2007). Inside eukaryotic cells at sites of cell-adhesion to the extracellular environment, cell-derived tension or exogenously applied force can lead to the unfolding of a number of proteins. A recent investigation demonstrated that free cysteines, which are buried in the nonstressed state of proteins, can become exposed within the cell due to externally applied shear stress or internally generated cellular tension. Proteins that expose free cysteines when mechanically stretched within a living cell include nonmuscle myosin IIA, vimentin, and spectrin (Johnson et al., 2007). Furthermore, p130Cas, a key component of the force-bearing network of proteins that link integrins to the cell cytoskeleton, is not only unfolded by force, but unfolding initiates its phosphorylation by a Src family kinase (Figure 13.2; Sawada and Sheetz, 2002; Sawada et al., 2006; Tamada et al., 2004). Also on the extracellular side, cell-generated tensile forces acting on ECM fibers are sufficiently strong to unfold fibronectin, as visualized in the false colors in Figure 13.3 (Plate 14) (Baneyx et al., 2001, 2002; Smith et al., 2007). The assembly of fibronectin into matrix fibers is a tightly regulated process (Geiger et al., 2001; Larsen et al., 2006; Mao and Schwarzbauer, 2005; Zaidel-Bar et al., 2004; Zamir et al., 2000). Force-imposed conformational changes of fibronectin have been hypothesized to mediate a variety of functional alterations, from cell binding to the exposure of cryptic sites (Baneyx et al., 2001, 2002; Smith et al., 2007; Vogel, 2006; Vogel and Sheetz, 2006).

13.2 Unique Protein Features Result from Multimodularity

Multimodular proteins account for most of the eukaryotic proteins. Evaluations of sequenced genomes have revealed that 70–80% of eukaryotic proteins have more than one domain (Apic et al., 2001; Wright et al., 2005). Often multimodular proteins

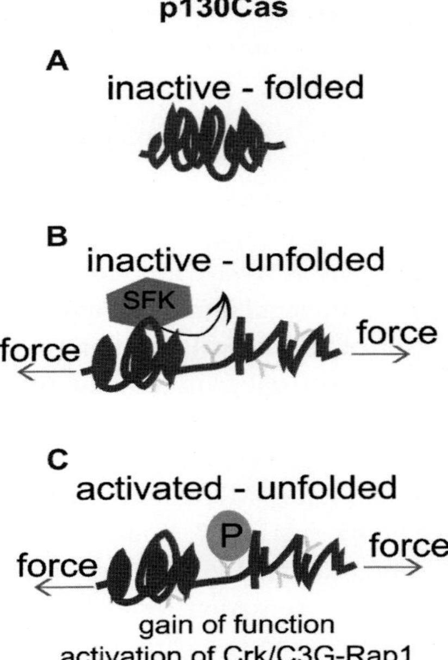

Figure 13.2. The recruitment and stretching of p130Cas to cell adhesion sites regulate cell signaling events. p130Cas unfolding initiates its phosphorylation by a Src family kinase. p130Cas is shown to be phosphorylated when stretched by mechanical force (Sawada et al., 2006). (A) In the relaxed state, p130Cas is not phosphorylated. (B) Mechanical force actively regulates the function of p130Cas by exposing phosphorylation sites, thus making them accessible to SFK. (C) Phosphorylated p130Cas activates Crk/C3G-Rap1.

such as the extracellular proteins Fn and polycystin-1 and the intracellular protein titin are large proteins assembled from numerous modules, each individually folded into an equilibrium structure with particular secondary and tertiary motifs. The focus of this chapter is on multimodular proteins that are often integrated into supramolecular entities within cells or within extracellular networks that are subjected to tensile forces.

While the number of known proteins is rapidly increasing as more and more genomes are sequenced, only a limited number of protein folds (~10,000) seem to exist (Koonin et al., 2002), a few of which are shown in Figure 13.4. β-strands are connected together via polypeptide backbone hydrogen bonds to form β-sheets. Important for the later discussions on mechanical aspects is that β-strands can associate in an anti-parallel or parallel orientation, which impacts their mechanical stability. β-sheets are relatively rigid motifs due to interbackbone hydrogen bonds formed between adjacent β-strands. In contrast, α-helices are maintained by intrabackbone hydrogen bonds within single polypeptides. An α-helix can be stretched and unraveled by sequential breakage of those backbone hydrogen bonds if the force is applied along the helix axis. α-helices packed against each other are only stabilized by polar or nonpolar side-chain interactions. The same is true for the interaction of multiple β-sheets. Table 13.1 gives an overview of the frequency by which the modules discussed here are found inserted into large, multimodular human proteins. For example, most common to all of the proteins analyzed here are immunoglobulin (Ig)-like repeats, which are composed of seven β-strands arranged in two opposing β-sheets. Approximately 0.4% (14,455) of proteins contain Ig-like domains among the 4,134,341 proteins found in databases listed in InterPro, and among them are 3006 human proteins of 68,949 analyzed

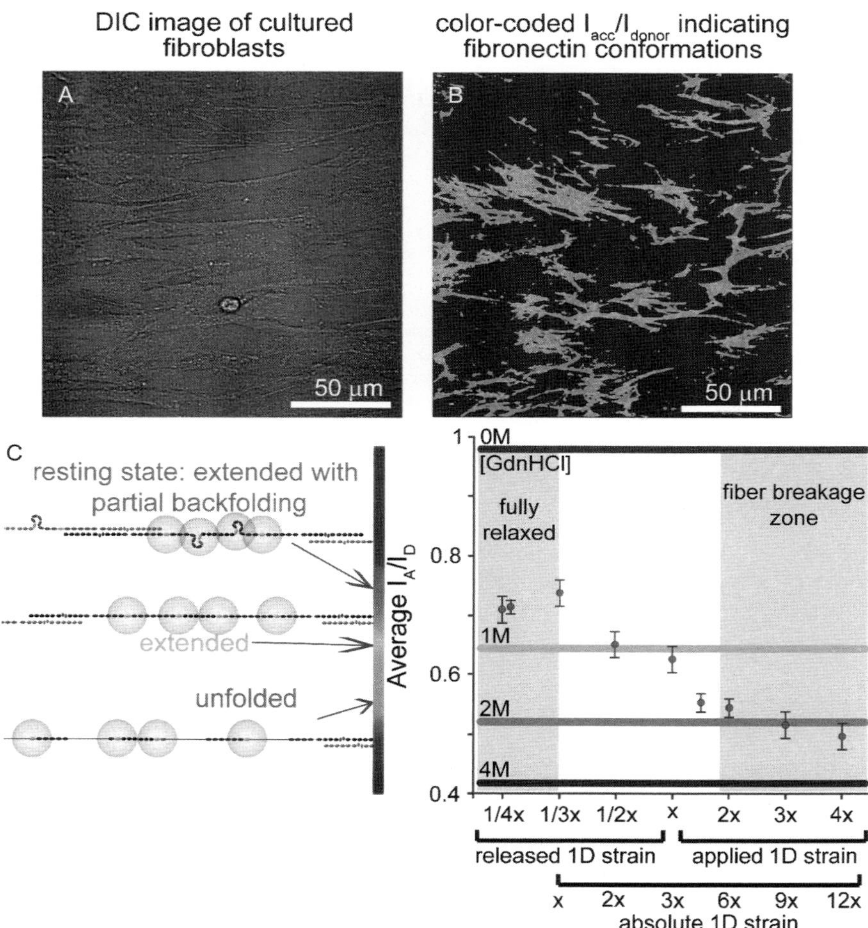

Figure 13.3. Living cells can unfold fibronectin of ECM fibrils. Fibroblasts in the culture (A) apply cell traction forces that lead to dynamic levels of stress within the ECM. Some regions (blue-green fibrils) are significantly unfolded, while other regions (red-yellow fibrils) contain fibronectin with intact secondary/tertiary structure (B). Fluorescence resonance energy transfer (FRET) is employed based on the increase in the distance of covalently attached donor and acceptor pairs during the stretching of fibronectin (C). Increased average separation of the multiple donors and acceptors leads to decreased FRET. In order to calibrate FRET with the degree of strain, a strain device was custom-built to hold a transparent silicone sheet (not shown). The absolute strain scale represents the average I_A/I_D values of fibronectin fibers that are fully relaxed or extended to the point where they begin to break (5–6×). A plot of the average $I_A/I_D \pm$ standard deviation versus applied or released strain to/from manually deposited fibronectin fibers is shown (C). Released strain is acquired by pre-straining the silicone substrate before pulling fibers and then relaxing the strain device. I_A/I_D values are calibrated to chemical denaturing data. The absolute scale of strain is shown as well, assuming that 1/3× is the point of full relaxation. (Figure from Little et al., 2008.)

human proteins (4.4% of listed human proteins contain Ig-like domain(s)). These proteins can range in location and function from ECM proteins to signaling molecules such as vascular endothelial growth factor receptors (VEGFR). Ankyrin repeats, composed of a bundle of α-helices, are also relatively common; 13,195 proteins are

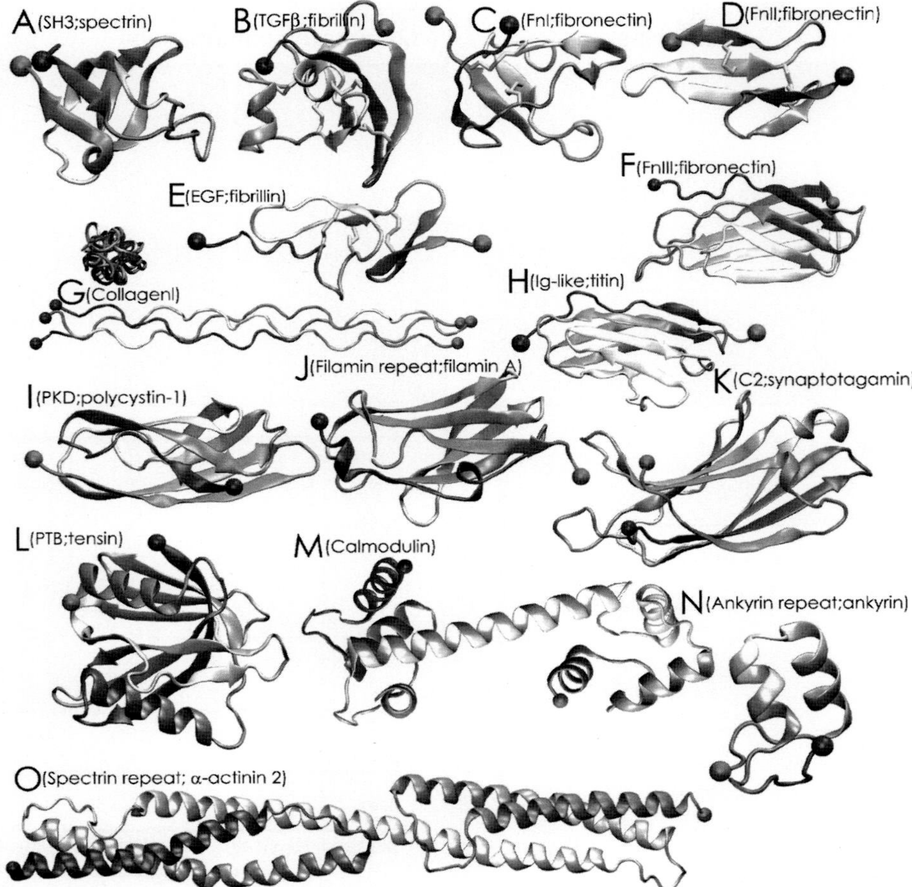

Figure 13.4. The structural richness of protein modules that are often found to be integrated into force-bearing protein networks. Representative modules from proteins that are possibly subjected to mechanical forces *in vivo* are shown. Proteins are represented as cartoon models. Cα-atoms at the termini are shown as spheres. (A) SH3 domain from spectrin (PDB 1AEY). (B) The TGFβ-like domain from fibrillin has four disulfides and is also called the 8-cysteine domain (PDB 1APJ). (C) FnI domain (PDB 1QO6). (D) Human FnII domain (PDB 1QO6). (E) EGF domain from fibrillin with three disulphide bonds (PDB 1EMN). (F) Human FnIII domain (PDB 1TTG). (G) Structure of collagen-like model peptide (PDB 1CAG) and a rotated view along the protein axis. (H) Ig-like domain from human titin (I27; PDB 1TIT). (I) PKD domain from polycystin-1 (PDB 1B4R). (J) Filamin repeat 21 from filamin A (PDB 2BRQ). (K) C2A domain from synaptotagamin (PDB 1RSY). (L) PTB domain from tensin (PDB 1WVH). (M) Calmodulin contains two EF-hand motifs (PDB 1CLL). (N) Ankyrin repeat (PDB 1N11). (O) Tandem spectrin repeats from α-actinin (PDB 1HCI).

known to carry ankyrin repeat(s) even though they perform diverse sets of functions. And finally, src homology-3 (SH3) domains are found in many intracellular or membrane-associated proteins, for instance, neutrophil cytosol factor 2, spectrin, and cortactin.

Table 13.1 *Common structural modules found in multimodular proteins*[a]

Protein domain	InterPro (human proteins)	Pfam	PROSITE pattern/profile
SH3	4306 (609)	2992	—/3308
TGFβ	947 (74)	933	856/—
FnI	109 (32)	105	115/108
FnII	209 (40)	207	190/208
EGF	4476 (589)	—	—
FnIII	4206 (540)	3294	—/3647
Collagen Triple helix repeat	2048 (232)	2048	—
Ig-like	14455 (3006)	—	—/14576
PKD	1272 (33)	731	—/605
Filamin repeat	284 (36)	211	—/279
C2 domain	2946 (380)	2497	—/2148
PTB	527 (128)	369	—/435
EF-hand	8116 (587)	6277	11014/7889
Ankyrin repeat	14014 (669)	7087	6864/7618
Spectrin repeat	647 (147)	503	—

[a] The number of frequently occurring modules found in multimodular proteins that form force-bearing protein networks as shown in Figure 13.4 were found by searching the InterPro database (http://www.ebi.ac.uk/interpro/). InterPro calculates the total number of unique hits in member databases. The number of hits for each protein domain and also the number of human protein hits carrying the particular domain in InterPro are shown. Since different databases have different definitions for protein domains, we have also shown the number of protein domain hits in two InterPro member databases (Pfam and PROSITE). The number of protein sequences searched (March 13th 2007) in the UniProtKB database (release 10) was 4,134,341 and included 68,949 human proteins. The InterPro accession numbers are as follows: SH3, IPR001452; TGFβ, IPR001839; FnI, IPR000083; FnII, IPR000562; EGF, IPR006210; FnIII, IPR003961; Collagen triple helix repeat, IPR008160; Ig-like, IPR007110; PKD, IPR000601; Filamin repeat, IPR001298; C2, IPR000008; PTB, IPR006020; EF-hand, IPR002048; Ankyrin repeat, IPR002110; Spectrin repeat, IPR002017. In the PROSITE database, two definitions for protein domains are used (pattern and profile).

The proteins further discussed here are like pearl necklaces, where individual modules represent pearls on a string. For most multimodular proteins, these necklaces are made of different types of pearls, each with distinct molecular recognition sites. Is there physiological significance associated with the scaffold that is used to present molecular recognition sites? For example, why does fibronectin contain multiple FnIII modules versus the structurally similar Ig domains? VCAM-1 contains mainly Ig-like domains, whereas in titin the Ig-like domains are combined with FnIII modules and a number of other types of domains. While molecular recognition sites could be exposed in the loop regions of both scaffolds, modules made from different structural motifs might differ in their mechanical stabilities. Consequently, one major advantage of linking modules, each of which exposes a different set of functionalities, might be to permit exposure of a range of binding sites from the same protein tuned through mechanical force. In the absence of force, a complex scaffold with well-controlled distances of recognition sites might be presented. If loaded by cell or external forces, however, the geometry of this scaffold might be altered, whereby each of the structural motifs might show a different force-dependent response.

13.2.1 Response of Multimodular Proteins to Force

The molecular architectures and topological designs of proteins are critical determinants of their mechanical stabilities. Since the weakest force-bearing interactions are broken first, the orientation of the force vector relative to force-bearing hydrogen bonds within the protein structure defines the structural stability of motif. If hydrogen bonds subjected to mechanical force need to be broken in parallel, they can withstand significantly higher force compared to an altered geometry, where the bonds would be unzipped one by one. It was thus noticed that the mechanical stability might not correlate with the thermal stability of proteins (Carrion-Vazquez et al., 1999, and further discussed in Bustamante et al., 2004). Thermodynamic stability is defined by all of the interactions in the protein, while mechanical stability is defined by the location of critical energy barriers along the vectorial direction along which the protein is forced to unfold. The mechanical stability of protein domains is thus often defined by a very limited number of key interactions. Calmodulin provides an interesting example as it is thermally very stable (Brzeska et al., 1983) but shows negligible mechanical stability (Carrion-Vazquez et al., 2000), which might not be surprising since it is not an integral part of force-bearing protein networks. Another design feature relates to the relative position of the termini: In cases where the force is applied to the terminal ends of a protein under physiological conditions, it might play a role in whether the termini are positioned at opposing ends of the module or in close proximity to each other. For example, the two terminal ends of the eight anti-parallel stranded β-sandwich C2A (Sutton et al., 1999) point in the same direction, whereas they point in opposite directions for the seven-stranded Ig-like domains. The zipper configuration indeed leads to decreased resistance against unfolding compared to Ig-like domains, where the termini are on the opposite sides of the module. These findings were suggested by steered molecular dynamics (SMD) simulations (Lu and Schulten, 1999) and confirmed by atomic force microscopy (AFM) experiments showing that the unfolding of C2A occurs at forces of ~60 pN (pulling speed of 0.6 nm/ms; Carrion-Vazquez et al., 2000). The C2 domains are positioned at the C-terminus of the membrane-trafficking proteins synaptotagamins that have an N-terminal transmembrane region and a variable linker connecting to the C2 domains (C2A and C2B). However, this rule is not a firm predictor of relative mechanical stabilities since the green fluorescence protein (GFP) whose two termini are also in close proximity seems mechanically at least as stable as $FnIII_{10}$ (Abu-Lail et al., 2006; Perez-Jimenez et al., 2006). The effect of the geometrical arrangement of bonds in regulating the mechanical but not the thermal properties of proteins was further elegantly demonstrated with DNA oligonucleotides: If the hydrogen bonds formed between hybridized strands are broken one by one by "unzipping" (Neuert et al., 2007), the force needed is significantly lower compared to rupturing all of the

Table 13.2. *Peak-forces to mechanically unfold various protein modules.*[a]

Protein module	Conformation	Unfolding force (pN)	Pulling speed (nm/ms)	Reference
Calmodulin	α-Helical	<20	0.6	(Carrion-Vazquez et al., 2000)
Spectrin repeats R13–R18	α-Helical	25–35	0.3	(Rief et al., 1999b)
Spectrin repeats R16, R17	α-Helical	25–35	0.3	(Randles et al., 2007)
Fibronectin III_1	β-Sandwich	220	0.6	(Oberhauser et al., 2002)
Fibronectin III_2	β-Sandwich	220	0.6	(Oberhauser et al., 2002)
Fibronectin III_{10}	β-Sandwich	75	0.6	(Oberhauser et al., 2002)
Fibronectin III_{12}	β-Sandwich	125	0.6	(Oberhauser et al., 2002)
Fibronectin III_{13}	β-Sandwich	89	0.6	(Oberhauser et al., 2002)
PKD domain 1	β-Sandwich	183	0.6	(Forman et al., 2005)
PKD domains 2–4	β-Sandwich	174–248	1.0	(Forman et al., 2005)
PKD domains 6–10	β-Sandwich	~100	0.5–0.7	(Qian et al., 2005)
Titin I27	β-Sandwich	209	0.6	(Oberhauser et al., 2002)
Titin I28	β-Sandwich	264	0.6	(Marszalek et al., 1999)
Yellow fluorescent protein	β-Barrel	69	0.4	(Perez-Jimenez et al., 2006)

[a] The unfolding peak forces measured by AFM for various protein modules are shown together with the pulling speed used in the experiment.

hydrogen bonds at once by "shearing." For instance, 20 pN and 9 pN were reported as sequential unzipping forces for poly(dG-dC) and poly(dA-dT) hybridized in a parallel arrangement, respectively. The force needed for a sequential breakage of multivalent bonds is thus independent of the length of the hybridized strands (Rief et al., 1999a; Strunz et al., 1999). This compares to 250 pN and 35 pN needed to break all bonds at once if poly(dG-dC) and poly(dA-dT) are hybridized in an anti-parallel arrangement, respectively (no exact pulling speeds were reported, but corresponding measurements for λ-DNA were conducted with pulling speeds of 0.15–3 μm/s; Rief et al., 1999a).

While much has been learned about how the mechanical stability of modules made from β-sheets is regulated, the mechanical properties of only a few α-helical proteins have been studied. Calmodulin, for example, forms a dumbbell-shaped structure where the α-helix is capped by two globular regions, each containing two helix-loop-helix EF-hand motifs (Figure 13.4). EF-hand motifs are responsible for Ca^{2+}binding. Interestingly, although calmodulin is highly resistant to thermal denaturation with a melting temperature (T_m) above 100°C in the presence of Ca^{2+} (Tsalkova and Privalov, 1985), it unfolds under very low mechanical force (<20 pN at 0.6 nm/ms; Carrion-Vazquez et al., 2000). While Ca^{2+} plays an important role in determining the thermodynamic stability (Ca^{2+} increases the T_m over 50°C; Tsalkova and Privalov, 1985), it does not increase the mechanical stability (no force peaks detected in the presence or absence of Ca^{2+} in AFM experiments; Carrion-Vazquez et al., 2000).

A brief summary of unfolding forces under similar extension velocities with AFM of various proteins with different protein structures and mechanical stabilities is shown in Table 13.2.

13.2.2 Role of Disulfide Bonds in the Mechano-Regulated Response of Proteins

Disulfide bonds are essential for maintaining the structure and function of many proteins. Disulfide bonds formed between cysteines that are sequentially far apart but in spatial proximity create "clamped" regions inside the proteins, regions that are shielded from stress-induced unfolding (Ainavarapu et al., 2007; Carl et al., 2001). Disulfide (S-S) bonds are far more stable than hydrogen bonds and require forces in the range of one to a few nanoNewtons at a loading rate of 10 nN/sec for rupture (Grandbois et al., 1999). In the case of vascular cell adhesion molecule-1 (VCAM-1), for example, disulfide bonds covalently stabilize about half of the module against forced unfolding (Bhasin et al., 2004). VCAM-1 is a transmembrane cell adhesion molecule important for immune cell trafficking out of the vasculature into inflamed tissues or lymph nodes. One-half of the module can thus be extended easily by force, whereas the other half is protected by the disulfide bond, which blocks further molecular extension (Bhasin et al., 2004; Carl et al., 2001). A second disulfide bond in the first and fourth modules further limits the extent to which these modules can be unfolded by force. Cysteine bonds thus limit the total range of protein extension under force and thereby stabilize intermediate states. In fibronectin, disulfide bonds stabilize FnI and FnII modules. Again, these cysteine bonds are not located close to the module's termini and thus protect only the clamped parts against unfolding, whereas the disulfide-free FnIII modules can be completely unfolded by force.

Cells may also have the capacity to tune locally the mechanical stability of already well-folded proteins through the modulation of the redox state of disulfide bonds, thus further adding to the toolbox of mechanisms by which the mechanical stability of proteins may be tuned by local environmental factors. Despite the oxidizing nature of the extracellular environment, evidence from lymphocytes suggests that cells have the ability to manipulate the local redox state at their extracellular surface (Sahaf et al., 2003), thereby being able to tune the mechanical stability of ECM molecules and of cell surface receptors. Chemical treatments that were known to result in a reduction in intracellular glutathione levels were shown to increase the percentage of reduced thiols on the cell surface. For both the melanoma cell adhesion molecules, Mel-CAM (also known as MCAM, MUC18, and CD146), and VCAM-1, the S-S bond is buried in the core of the folded module, and it becomes solvent when exposed in the early stages of protein elongation (see Figure 13.5; Plate 15). Accordingly, protonation of these disulfides to SH can be catalyzed by their force-induced exposure to solvents (Bhasin et al., 2004; Carl et al., 2001; Wiita et al., 2006). Protonation breaks the force-bearing S-S bond that otherwise protects 50% of these domains from unfolding. The finding that force can regulate the redox state of buried disulfide bonds is also significant because the disulfide bond of the Ig domains of CAMs is almost universally found to span their cores.

Finally, disulfide bonds close to a ligand-binding pocket can significantly stabilize ligand binding under tensile stress conditions as shown for the bacterial adhesin

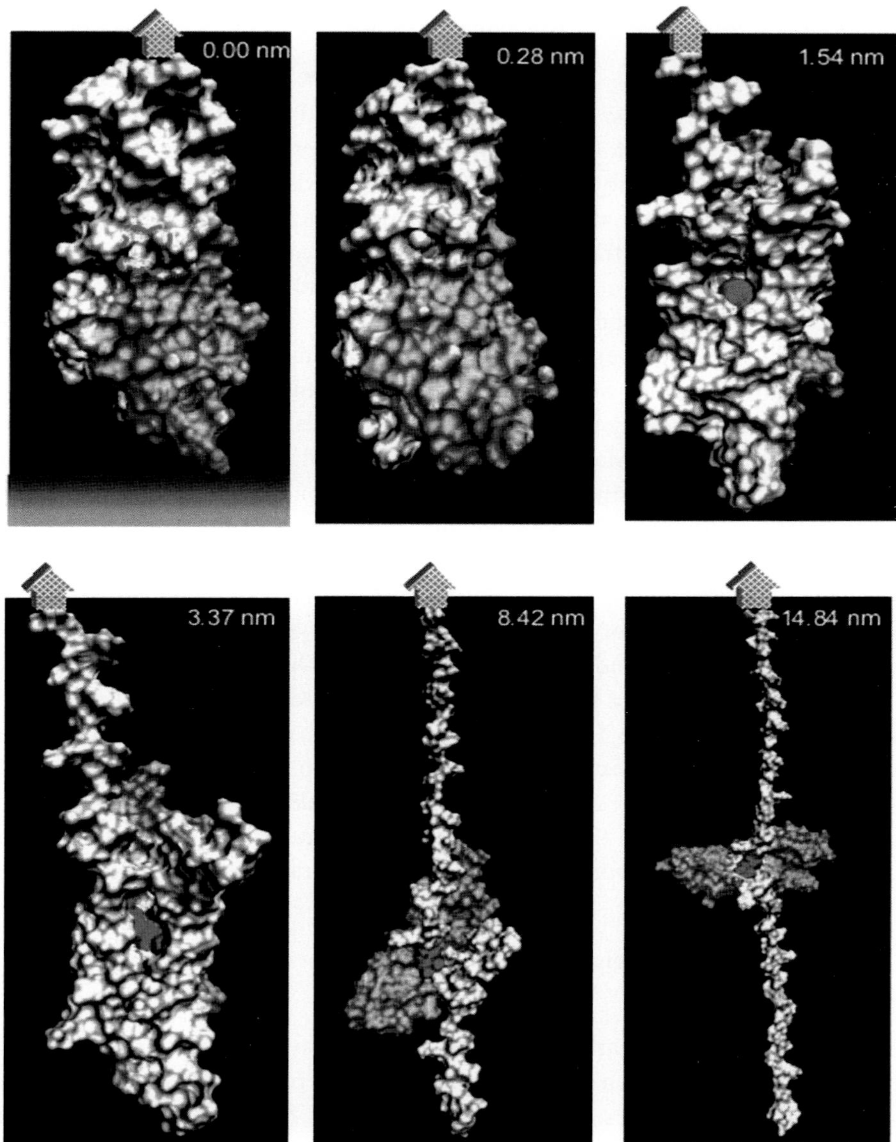

Figure 13.5. Role of a disulphide bond in restricting the unfolding of VCAM-1. VCAM-1 has a disulphide bond (shown in red) that stabilizes the domain against unfolding. An SMD simulation shows the exposure of the disulphide bond to the solvent under external force applied to the protein (Bhasin et al., 2004). The secondary structure element, which is clamped by the disulfide and thus shielded from unfolding, is shown in yellow, while the white regions are not clamped and hence susceptible to forced extension.

FimH. FimH is localized at the tip of *E. coli*'s type I fimbria and mediates specific binding to mannose via catch bond formation (Thomas et al., 2002). Elimination of a disulfide bond underneath the bottom of the binding pocket significantly increases the off-rate under fluid flow even though its removal does not affect adhesion under static conditions (Nilsson et al., 2007).

13.2.3 Module Stability Is Regulated by Key Force-Bearing Backbone Hydrogen Bonds

Steered molecular dynamic simulations have revealed that the passing of major energy barriers along the unfolding pathway of β-sheet proteins is commonly preceded by the breakage of a few key *backbone* hydrogen bonds, and the derived predictions agree with AFM experiments that can probe mechanical stability and determine the existence and positions of transition states (Carrion-Vazquez et al., 2000; Craig et al., 2004; Li et al., 2000; Lu et al., 1998; Sotomayor and Schulten, 2007). Guided by simulations, point mutations of critical residues can provide further insights into critical interactions that regulate the overall mechanical stability of proteins. For example, point mutations were used to confirm that titin's I27 module contains two key hydrogen bonds bridging the A and B β-strands, which stabilize a reversible unfolding intermediate (Marszalek et al., 1999). Mutation of these residues eliminated the unfolding intermediate. It is also widely accepted that the mechanical stabilities of FnIII are regulated by key hydrogen bonds, which are positioned such that they are initially well shielded from attack by water molecules. Strategic point mutations can thus tune the relative mechanical stabilities of the modules and therefore define the sequence of unraveling events (Craig et al., 2001, 2004; Oberhauser et al., 2002; Sotomayor and Schulten, 2007). Variability in the energy barrier that must be overcome to align the β-strands was correlated to very specific amino acid substitutions. For example, $FnIII_7$ and $FnIII_{EDB}$ have a unique hydrogen bond (backbone to side chain) that mechanically connects the two β-sheets and is formed within the hydrophobic core, whereas most other FnIII modules are mechanically weakened by possessing a proline in that particular location, thus lacking the essential amine hydrogen necessary for hydrogen bond formation (Craig et al., 2001).

13.2.4 Physiological Significance of Engineering a Hierarchy of Mechanical Stabilities

If domains of multimodular proteins possessed similar mechanical stabilities, these domains would rupture in a stochastic sequence. The rupture of each domain would change its domain-specific structure–function relation; however, a sequence of force-induced stochastic events could in that case not be translated into a predictable sequence of biochemical changes. Tuning of the relative mechanical stabilities of different modules in multimodular proteins, however, would allow a direct translation of individual mechanoresponses into a well-controlled sequence of biochemical signal changes. If domains of the protein have different molecular recognition sites, those can be functionally switched in response to partial unraveling of the individual domains. Multimodular proteins show variations of their amino acid sequences among their structurally highly homologous repeats. This can lead to significant variations in their relative mechanical stabilities, as was shown for titin's Ig domains (Fowler et al., 2002; Gao et al., 2002b; Kellermayer et al., 1997, 2003; Li and Fernandez, 2003; Lu et al., 1998; Rief et al., 1997, 2000; Tskhovrebova and

Trinick, 2003; Tskhovrebova et al., 1997; Watanabe et al., 2002; Williams et al., 2003) and fibronectin's FnIII modules (Craig et al., 2001, 2004; Ng et al., 2005; Oberhauser et al., 2002; Paci and Karplus, 1999).

FnIII modules serve as good model systems to study the importance of sequence variations in their altered mechanical stabilities since FnIII modules are found in ~1% of all human proteins. Nature not only provides us with a large number of FnIII variants, but an increasing number of FnIII equilibrium structures from various proteins have been resolved in recent years. Despite their low sequence homology, FnIII modules share a remarkable structural homology, and SMD analyses gave the first systematic insights into the relationship between sequence variations and mechanical stability (Craig et al., 2001, 2004; Gao et al., 2003; Ng et al., 2005) which decreases as follows: Ig27 > $FnIII_{12}$ > T-$FnIII_3$ > $FnIII_{13}$ ~ $FnIII_{EDB}$ > ~ $FnIII_{14}$ > $FnIII_{10}$ (Craig et al., 2004). An earlier study established that $FnIII_7$ > $FnIII_9$ ~ $FnIII_8$ > $FnIII_{10}$ (Craig et al., 2001).

Proper protein refolding after forced unfolding may be facilitated by a low sequence homology. The protein folding community found that "low sequence identities could have a crucial and general role in safeguarding proteins against misfolding and aggregation" (Wright et al., 2005). Domains having more than 70% identity are highly prone to aggregation whereas domains having less than 30–40% identity showed no detectable interactions (Wright et al., 2005). This is an important consideration particularly in the context of proteins that are stretched and partially unfolded in fibrillar assemblies under physiological conditions, including titin and Fn. Even in single molecule force spectroscopy studies, module misfolding has been observed in polyproteins that are composed of multiple identical modules (Oberhauser et al., 1999). It is thus noteworthy that the sequence identity of FnIII modules is less than 20%.

While all FnIII domains studied so far are mechanically less stable than titin's Ig domains (Oberhauser et al., 1998; Rief et al., 1998, 2000), the polycystic kidney disease (PKD) modules, which constitute nearly half of the extracellular region of polycystin-1, structurally resemble FnIII modules (see Figure 13.4), yet show an enhanced mechanical strength similar to that of titin's I27 (Table 13.2). It is proposed that PKD domain 1 from human PC1 can form a force-stabilized intermediate where additional hydrogen bonds are formed between the A–B loop and the G strand. However, this force-stabilized intermediate was observed only at low forces (Forman et al., 2005).

13.3 Examples of Multimodular Proteins with Potential Force-Regulated Functions

13.3.1 The Extracellular Matrix Protein Fibronectin

Fibronectin is a dimeric protein of more than 440 kDa that is a pervasive component of the ECM during development and within healing wounds (Hynes, 1990; Pankov and Yamada, 2002). How could stretching impact its functions? Fibronectin displays a number of surface-exposed molecular recognition sites for cells, including integrin binding sites such as the RGD loop, PHSRN synergy site, and LDV sequence, and binding sites for other ECM components, including collagen, heparin, and fibrin (Figure 13.6). A number of cryptic binding sites and surface-exposed binding sites have been proposed to be exposed or deactivated, respectively, as a result of force-dependent conformational changes (as reviewed in Vogel, 2006).

Understanding fibronectin unfolding is broadly significant since ~1% of all mammalian proteins contain domains that adopt a similar structural fold to FnIII domains (see Table 13.1). This motivated a large array of AFM and SMD experiments on native modules of fibronectin as well as fibronectin mutants. Unraveling trajectories for FnIII modules show a complex response to forced unfolding. The β-strands of the opposing β-sheets of the FnIII modules are twisted (twisted state). Overcoming the first major energy barrier requires that the two opposing β-sheets of FnIII are rotated against each other such that all β-strands align parallel to the force vector (aligned state; Craig et al., 2001, 2004). Mutations in the core can alter the unraveling force, as was shown experimentally for FnIII modules of tenascin (Rounsevell and Clarke, 2004) and recently for $FnIII_{10}$ (Ng et al., 2007). SMD reveals that overcoming this first energy barrier requires that one or two conserved backbone hydrogen bonds connecting the A and the B strands must be broken, thus allowing one or two water molecules to enter the periphery of the hydrophobic core. Slipping of water between β-sheets allows the β-strands to align with the external force vector. Transition from the twisted state to the aligned state appears to be the major transition that regulates the relative mechanical stability of FnIII modules, and all further events in the unraveling pathway proceed from the aligned state (Craig et al., 2001, 2004).

After passing this first major energy barrier, which is common to all FnIII modules studied so far, FnIII modules can unravel along several pathways, each of which has distinct intermediates (Gao et al., 2002a). While initially predicted by SMD, AFM studies have confirmed the divergence of forced unraveling pathways for $FnIII_{10}$ (Li et al., 2005). Along one pathway, for example, the A- and then the B-strands break away first, followed by the unraveling of the remaining β-sandwich structure, whereas the G-strand is the first to separate along the second pathway. Most FnIII modules have a highly conserved proline in the G-strand, thus reducing the number of hydrogen bonds formed between the F- and G-strands. Single point mutations can alter the probability of the unfolding pathways that is most frequently observed. For a few FnIII modules, including modules $FnIII_1$ and $FnIII_2$, which play major regulatory roles in fibronectin fibrillogenesis (Ingham et al., 1997; Sechler et al., 2001; Zhong et al., 1998), the G-strand does not contain this otherwise highly conserved proline residue, thus enhancing the interaction of the F- and G-strands. Consequently, the A- and B-strands break away first, leading to a functionally intact intermediate that is mechanically more stable than the completely folded $FnIII_1$ (Gao et al., 2003; Oberhauser et al., 2002). The structure of this intermediate is analogous to the 76-residue protein anastellin, which, if added to the cell culture, induces fibronectin fibrillogenesis (Morla et al., 1994). Furthermore, anastellin was proposed

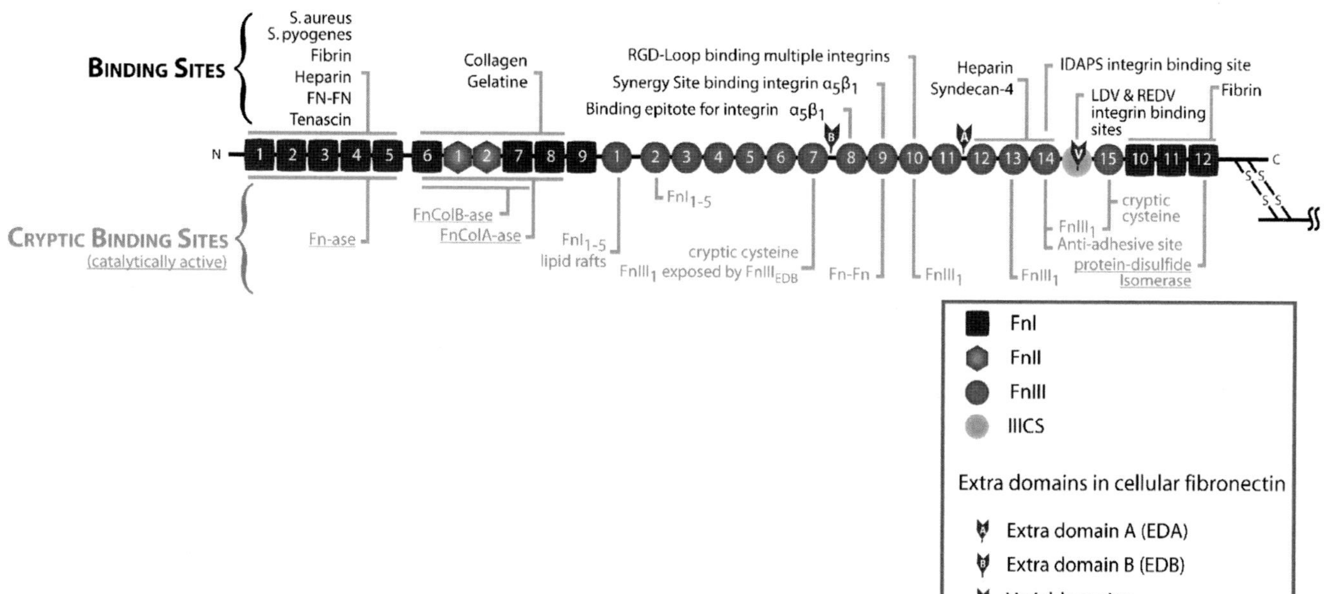

Figure 13.6. Modular structure and binding sites of fibronectin. Fibronectin is a dimeric molecule composed of repeats of type FnI, FnII, and FnIII, which contain alternatively spliced modules as indicated. Dimerization is mediated by two disulphide bonds at the C-terminal end. All the repeats are composed of β-sheet motifs (see Figure 13.4). Fibronectin contains numerous molecular recognition and cryptic sites (lower panel), including the cell binding site RGD (Pierschbacher and Ruoslahti, 1984), the synergy site PHSRN recognized by the $\alpha_5\beta_1$ integrin (Redick et al., 2000), and the IDAPS sequence in the $FnIII_{13-14}$ junction (Sharma et al., 1999), which supports $\alpha_4\beta_1$-dependent cell adhesion (Mould and Humphries, 1991). The cryptic sites include various fibronectin self-assembly sites whose exposure is needed to induce fibronectin fibrillogenesis (Sechler et al., 2001), a cryptic fragment from $FnIII_1$ that localizes to lipid rafts and stimulates cell growth and contractility (Hocking and Kowalski, 2002), and a binding site for tenascin (Ingham et al., 1997). Other cryptic sites with enzymatic activity include a metalloprotease in the collagen-binding domain of plasma Fn capable of digesting gelatin (FnCol-ase); helical type II and type IV collagen; α- and β-casein; insulin β-chain (Schnepel and Tschesche, 2000); a proteinase (Fn-ase) specific to fibronectin, actin, and myosin (Schnepel et al., 2001); and a disulfide isomerase (Langenbach and Sottile, 1999). The figure is partially reproduced from Vogel (2006).

to have anti-tumoral, anti-metastatic, and anti-angiogenic properties *in vivo* (Briknarova et al., 2003).

A variety of approaches have established that fibronectin fibrils are highly extendable within the ECM. A Fn-green fluorescent protein (Fn-GFP) chimeric construct expressed in fibroblast cells was used to show that fibronectin fibrils are elastic and under considerable tension; a fibronectin fibril that was released or broken by cells or severed artificially quickly retracted, indicating prior tensile stress (Ohashi et al., 1999, 2002). Fibrillogenesis, or the production of fibronectin fibrils from cell-secreted or soluble fibronectin, requires mechanical force acting on the protein (Baneyx and Vogel, 1999). Fibrillogenesis is blocked if cell contractility is manipulated through disruption of the actin cytoskeleton (Barry and Mosher, 1988; Wu et al., 1995), inhibition of Rho-mediated contractility (Zhong et al., 1998), or culture of cells in a released collagen gel that limits a cell's ability to apply force to the ECM (Halliday and Tomasek, 1995). Despite the ubiquitous presence of binding sites whose function could be altered by mechanical strain and the conclusive demonstration that fibronectin fibrils are under strain within the ECM, until recently it remained highly controversial whether extensibility involved unfolding of FnIII modules.

Two models were proposed to account for the extensibility of fibronectin fibers. The *quaternary structural model* proposed that fibronectin in unstretched fibers exists in the compact conformation, where its two dimeric arms are crossed over each other (Abu-Lail et al., 2006). Force was predicted to break apart the two arms, followed by a progressive extension and alignment of the molecule along the force vector until a completely linear alignment without loss of secondary/tertiary structure was reached (extended state without loss of secondary structure) (Abu-Lail et al., 2006). In contrast, the *unfolding model* proposed that fully relaxed fibrils are composed of fibronectin in an already extended conformation, and that further stretching of fibrils leads to partial unfolding of fibronectin modules (Baneyx et al., 2001, 2002). The unfolding model was initially proposed based on estimates of the free energy of denaturation and extension of individual FnIII modules in comparison to the force generated by single myosin or kinesin motor proteins (Erickson, 1994; Soteriou et al., 1993).

Since the question of whether fibronectin can be mechanically unfolded by cell-generated forces has major implications on the models put forward to explain the not-yet understood mechanism of fibrillogenesis, and on the question of whether the ECM itself can act as a mechanochemical signal converter, conclusive experiments were needed. Recent work from our group employed two different fibronectin labeling schemes for FRET with sensitivity to either compact-to-extended without loss of secondary structure or compact-to-unfolded conformations (Smith et al., 2007). These studies confirmed previous results that fibrillar fibronectin within a human fibroblast ECM is partially unfolded by cell-generated tensile forces (Figure 13.3; Plate 14). While we see by FRET that fibronectin assumes the compact conformation in solution, the spectroscopic analysis of the ECM of living fibroblasts did not show any indications that a fraction of the fibrillar fibronectin population exists in the compact conformation with crossed-over arms. In addition, a fibroblast matrix was grown on pre-stretched polydimethylsiloxane (PDMS) membranes, which could then be relaxed up to four-fold after decellularization of the matrix, and, importantly, the compact conformation with significant quaternary structure was again not detected. However, relaxed fibrillar fibronectin is not fully extended either. The fibrils show a certain degree of quaternary structure that could originate from some backfolding of the fibronectin arms upon themselves (Figure 13.3; Plate 14). Although the exact structure of the fully relaxed conformation of fibronectin, which retains some quaternary structure, remains to be determined, high-resolution cryo-scanning electron microscopic images of cell-derived fibronectin nanofibrils were either straight or contained nodules 10–15 nm in diameter that bulged from the otherwise smooth surface of the fibronectin nanofibrils (Chen et al., 1997; Peters et al., 1998). Smooth nanofibrils were attached at each end, while nodular structures had one free end and more likely represented fully relaxed fibrils. Taken together, these data suggest that fully relaxed fibronectin may be composed of extended fibronectin molecules where a few FnIII modules aggregate to form protruding nodular structures. Consequently, fibronectin extensibility has two components, straightening of the dimeric arms through elimination of those nodules and partial unfolding of the fibronectin modules.

Fibronectin might thus play a significant role in mechanotransduction processes. In addition, correlation of stretch or relaxation of the underlying PDMS substrate with changes in FRET revealed that fibronectin fibril strain within native cell cultures varied by more than two-fold length change within a typical field of view containing ~15–20 confluent cells. Thus, cell traction force results not only in unfolded regions of fibronectin within the ECM, but the large degree of heterogeneity suggests that cells may mechanically tune their matrix environment with high spatial and temporal precision (Baneyx et al., 2001, 2002; Smith et al., 2007), thus switching the biochemical display of their modules. The latter hypothesis still awaits experimental validation (see Table 13.2).

13.3.2 The Cytoplasmic Protein p130Cas

While the regulatory role of force on intracellular signaling is well established, little information was available about which intracellular molecules might act as mechanochemical signal converters. The first evidence for force-regulated exposure of cryptic sites on intracellular cytoskeleton-associated proteins was derived through the creative use of stretchable substrates and extracted cell cytoskeletons. "Triton cytoskeletons" were first generated by removing the membranes of cells anchored to a stretchable substrate using Triton-X (Sawada and Sheetz, 2002; Tamada et al., 2004). Triton cytoskeletons were next incubated with cytoplasmic protein extracts, and cytoplasmic proteins that bound to stretched Triton cytoskeletons were compared against cytoplasmic protein extracts that bound control or relaxed substrates. Thus, intracellular signaling molecules that bound to cryptic binding sites exposed by mechanical strain could be identified. Paxillin, focal adhesion kinase (FAK), p130Cas, and the protein kinase PKB/Akt (Sawada and Sheetz, 2002), as well as

C3G and the adapter protein CrkII (Tamada et al., 2004), bound preferentially to stretched Triton cytoskeletons. Triton cytoskeletons were able to transduce tensile force into Rap1 activation via tyrosine phosphorylation of p130Cas by Src family kinases (Tamada et al., 2004). Using an *in vitro* protein stretch assay, Sawada et al. (2006) showed that mechanical unraveling of the p130Cas substrate domain remarkably enhanced its tyrosine phosphorylation with constant kinase activity and that unraveling of the p130Cas substrate domain *in vivo* correlates with regions of higher tensile force. The Cas family of scaffolding molecules is known to possess important intracellular signaling functions (Defilippi et al., 2006), although limited structural information is available. p130Cas contains an N-terminal SH3 domain, a proline-rich region, a large substrate binding domain containing 15 repeats of a YxxP sequence, a serine rich region, and a C-terminal domain (Figure 13.7). Tyrosine phosphorylation of p130Cas is regulated by growth factors and integrin-mediated adhesion; however, it has only recently been shown that force can upregulate the phosphorylation of p130Cas through exposure of otherwise cryptic sites (Sawada et al., 2006).

13.3.3 The Muscle Protein Titin

The muscle protein titin spans over half of the length of the vertebrate striated-muscle sarcomere (reviewed in Gao et al., 2006; Tskhovrebova and Trinick, 2003). Titin is formed by a single polypeptide of up to 4 MDa containing up to 300 repeating modules, thus making it the largest protein in the human genome. These modules are predominantly composed of immunoglobulin-like and a smaller fraction of FnIII folds (Figure 13.8). However, they also contain a number of other types of modules such as titin kinase, the only known catalytically active domain of titin. Titin is essential for numerous physiological processes, including structural organization during muscle development and control of the total range of strain through which sarcomeres can be stretched before slippage occurs between the intercalated filaments. While titin spans the entire half-sarcomere, it is firmly anchored at each end of the I-band. Since passive force is developed when a relaxed muscle is stretched, it was originally proposed that titin elasticity may be derived from the reversible unfolding of its Ig and FnIII domains (Erickson, 1994; Soteriou et al., 1993). Indeed, titin's Ig domains were the first to be unfolded in single molecule pulling experiments using both AFM (Rief et al., 1997) and laser tweezers (Kellermayer et al., 1997). Based on the hypothesis that mechanical force unfolds titin's domains, titin kinase has been suggested to serve as a force sensor and SMD data were recently used to propose a structural mechanism of force-induced activation of the kinase domain (Grater et al., 2005). It was postulated from simulations that applied force caused conformational changes in the titin kinase such that the catalytic site was released from an autoinhibited state.

The mechanical properties of the entire titin molecule as well as of numerous recombinant fragments have been extensively studied with AFM, optical tweezers, and SMD simulations (reviewed in Gao et al., 2006). Single molecule pulling experiments revealed that during the forced extension of single titin molecules, the PEVK

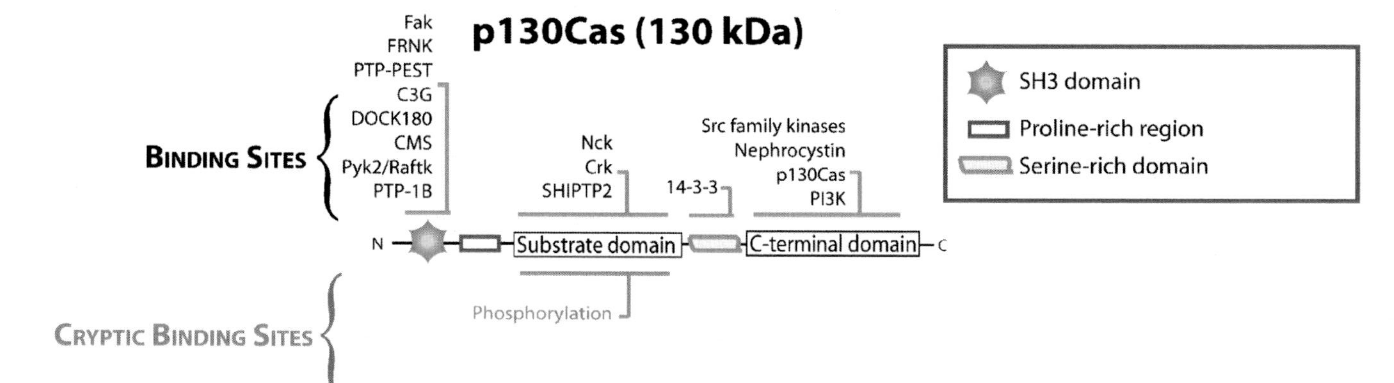

Figure 13.7. Modular structure and binding sites of p130Cas. p130Cas is an intracellular protein that functions as a scaffolding adaptor protein and has a role in cellular signaling events. The interactions of p130Cas are shown according to previous review articles (Alexandropoulos et al., 2003; Bouton et al., 2001; Defilippi et al., 2006; O'Neill et al., 2000). p130Cas contains a SH3 domain in the N-terminus followed by a proline-rich region. The substrate domain of p130Cas contains 15 YxxP-repeats, which are binding sites for SH2 and PTB domains. These repeats are phosphorylated in response to multiple factors, including growth factors and hormones (Bouton et al., 2001). Interestingly, mechanical force applied to p130Cas unfolds the protein, permitting phosphorylation of the substrate domain of p130Cas (Sawada et al., 2006). The serine-rich domain of p130Cas is a bundle of four α-helices (PDB 1Z23) based on NMR spectroscopy.

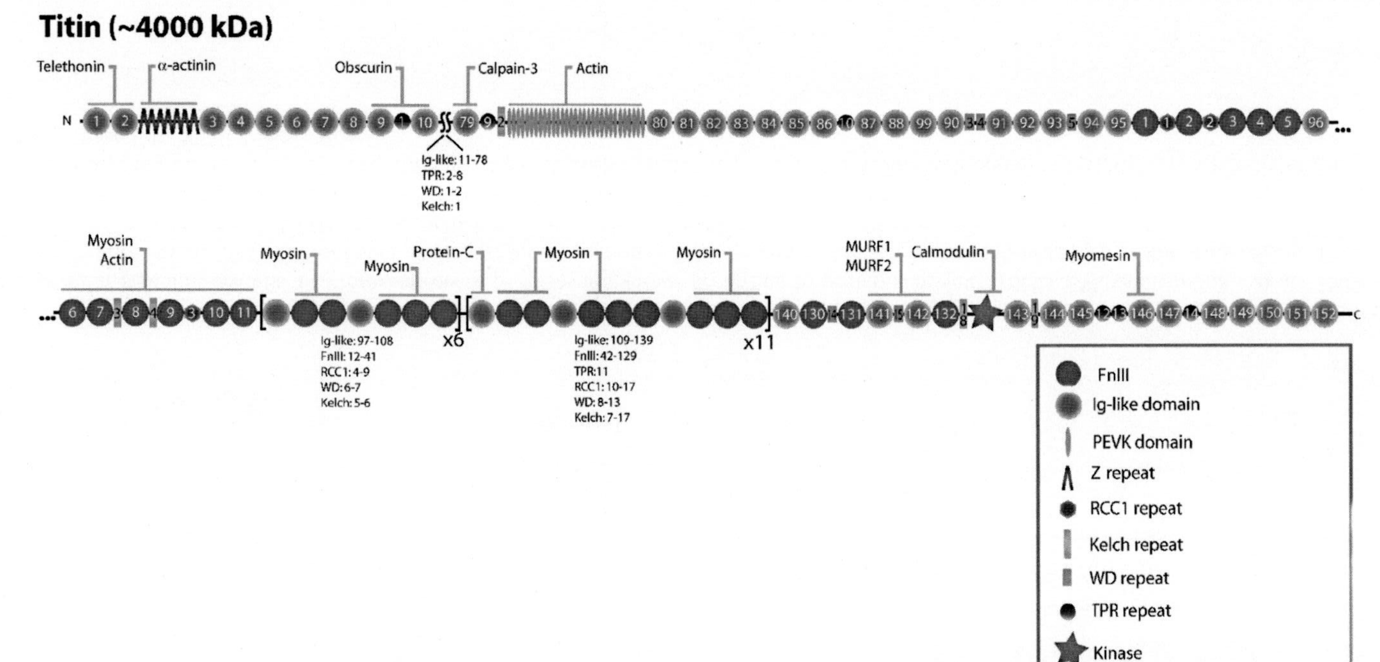

Figure 13.8. Modular structure and binding sites of titin. Titin is the largest known protein with a molecular weight of 4 MDa. There are various isoforms of titin in humans (Tskhovrebova and Trinick, 2003), and the human muscle isoform is shown (Swissprot accession number Q8WZ42). The interaction partners of titin are shown according to Tskhovrebova and Trinick (2004). There are two regions containing domain repeats, and the domains present within the N-terminal region constituted of mainly Ig-like domains are shown in the lower panel.

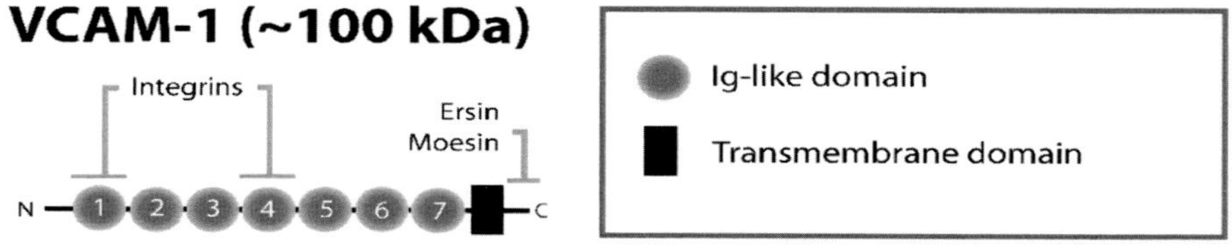

VCAM-1 (~100 kDa)

Integrins
Ersin
Moesin
N 1 2 3 4 5 6 7 C

Ig-like domain
Transmembrane domain

Figure 13.9. Modular structure and binding sites of VCAM-1. VCAM-1 is a transmembrane protein made of seven extracellular Ig domains that each contain at least one solvent-buried disulphide bond. Integrins $\alpha_4\beta_1$, $\alpha_4\beta_7$, $\alpha_D\beta_2$, and $\alpha_M\beta_2$ are known to bind to VCAM-1 (reviewed in Barthel et al., 2006). Integrin binding sites are located in domains 1 and 4 (Vonderheide and Springer, 1992). According to TMPred (http://www.ch.embnet.org/software/TMPRED_form.html) and THMM (http://www.cbs.dtu.dk/services/TMHMM/), VCAM-1 has a 21–23-residue C-terminal transmembrane domain (TMPred, residues 700–720; THMM, residues 699–721). The cytoplasmic tail domain binds ersin and moesin (Barreiro et al., 2002) and is ~20 residues in length.

region unfolds first at the lowest forces. However, the Ig and FnIII modules unfold under the largest extensions found *in vivo*. AFM studies have shown that the Ig modules from human cardiac titin (I27–I34) have a wide range of mechanical stabilities (see Table 13.2) and unfold at forces ranging from 150 pN to 300 pN at pulling rates of 1 nm/ms (Rief et al., 1997). The PEVK region was also found to be elastic, and it unfolds with negligible force (<5 pN; pulling rate not clearly indicated) in AFM experiments (Li et al., 2001). It is now generally believed that reversible unfolding of immunoglobulin domains is physiologically significant (Gao et al., 2006; Li et al., 2002; Minajeva et al., 2001).

13.3.4 The Vascular Cell Adhesion Molecule VCAM-1

Vascular cell adhesion molecule-1 (VCAM-1) is a multimodular transmembrane protein that is variably spliced to contain six or seven extracellular repeating modules. VCAM-1 with seven Ig-like modules is shown in Figure 13.9. These modules are composed of two β-sheets that are structurally similar to immunoglobulin modules. Each of the domains contains one core disulfide bond. The first and fourth modules contain a second disulfide bridge that appears solvent-exposed in the crystal structure. VCAM-1 is a member of a large family of multimodular proteins that are essential for an inflammatory response to pathogens (for review, see Yusuf-Makagiansar et al., 2002). In order to fight infection, leukocytes must first exit the vascular compartment to combat pathogens that are resident in tissue. The endothelium, an organ consisting of a one-cell–thick layer that lines all blood vessels, provides adhesive ligands to passing leukocytes, driven by high shear forces, which mediate adhesion between these two cell types. A critical step in this process is the adhesion of integrins, which are expressed on the leukocyte surface, to various integrin ligands present on the vascular surface of the endothelium, including members of the immunoglobulin superfamily of glycoproteins such as intercellular adhesion molecule-1 (ICAM-1) and VCAM-1. VCAM-1 is a receptor for $\alpha_4\beta_1$, $\alpha_4\beta_7$, and $\alpha_D\beta_2$ integrins and mediates the adhesion of monocytes and lymphocytes to the vasculature. VCAM-1 has two homologous binding sites for $\alpha_4\beta_1$, although the most crucial interaction appears to be in domain 1 of VCAM-1 (Figure 13.9). Thus, it is likely that Ig-like domains of VCAM-1 are subjected to significant tensile force during leukocyte adhesion in the presence of fluid drag.

The possibility of force-dependent unfolding of endothelial-expressed Ig superfamily members is supported by the significant fluid drag forces imposed by the passing blood on adherent leukocytes. A single leukocyte adhered to the vascular endothelium is estimated to resist fluid drag forces that were measured experimentally in the range of 200 to 8000 pN distributed over many adhesins (House and Lipowsky, 1988), a range of forces that is significant in the context of forced protein unfoldings. Single molecule AFM studies revealed partial unfolding of individual domains of VCAM-1 at only 30–50 pN under a constant pulling velocity of 1 nm/ms (Bhasin et al., 2004). Even in the presence of multiple integrin–VCAM-1 bonds, the high forces imparted by fluid flow as well as the significant time-scales involved (leukocytes may adhere for minutes prior to extravasation across the endothelial barrier and hence loss of drag forces) might lead to forced unfoldings of VCAM-1 domains. However, future work is necessary to demonstrate that VCAM-1 is mechanically unfolded during leukocyte adhesion *in vivo* and to elucidate the impact of partial unfolding on the bond lifetime, and hence the adhesive stability, of $\alpha_4\beta_1$-integrin binding to VCAM-1.

13.3.5 The Transmembrane Protein Polycystin

Polycystin-1 (PC1) is another large multimodular protein (~520 kDa) that is thought to function as a cell adhesion receptor (Ibraghimov-Beskrovnaya et al., 2000). The extracellular portion of the protein is composed of leucine-rich repeats (LRR), a cell wall integrity and stress response component (WSC), a C-type lectin domain, a low-density lipoprotein-like domain (LDL), 16 immunoglobulin-like polycystic disease domains (PKD; schematically shown in Figure 13.10), and a region that is homologous to the sea urchin protein–named receptor for jelly egg (REJ; Sandford et al., 1997). Polycystin-1 also interacts with the mechanosensitive ion channel polycystin-2, a member of the mechanosensitive transient receptor potential (TRP) cation channels. Polycystin-1 has been localized to focal adhesion complexes and desmosomes.

PC1 might work as a mechanoreceptor sensing flow rates within renal tubules. Interestingly, PC1 is proteolytically cleaved in association with alterations in mechanical stimulation. Cleavage releases its C-terminal cytosolic tail, which then enters the nucleus and initiates signaling processes (Chauvet et al., 2004). It has been found that cells lacking functional PC1 are not capable of a normal influx of Ca^{2+} in response to fluid flow (Nauli et al., 2003). Therefore, the epithelial response to mechanical stimulation was blocked in the cells expressing mutated PC1, and this may contribute to the cyst formation observed in mouse models (reviewed in Nauli et al., 2003).

To study the mechanical properties of PKD domains forming roughly half of the extracellular domain of PC1 (see Figure 13.10), Forman et al. (2005) constructed polyproteins composed of I27 and PKD domains. Interestingly, although the

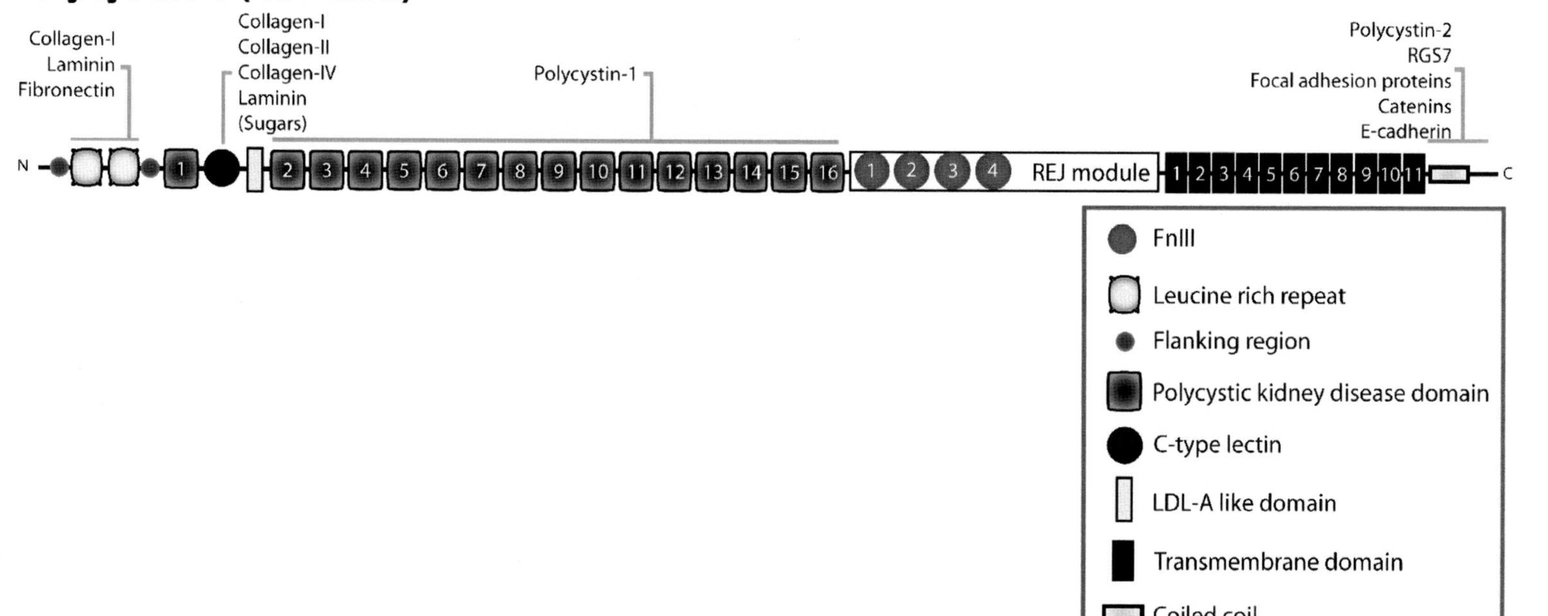

Figure 13.10. Modular structure and binding sites of polycystin-1. Polycystin-1 is a large transmembrane protein with a number of transmembrane domains in the C-terminus. Polycystin-1 makes homophilic interactions via PKD domains in the middle of the protein that most likely lead to both intra- and intercellular protein dimerization (Ibraghimov-Beskrovnaya et al., 2000). The C-type lectin domain of polycystin-1 contains binding sites for various carbohydrates (Weston et al., 2001) and many important proteins of the ECM, for example, collagen-I. The N-terminal leucine-rich repeats have been found to interact with Fn, collagen-I, and laminin (Malhas et al., 2002). Importantly, polycystin-1 can associate with focal adhesion proteins (Geng et al., 2000) and RGS7, a regulator of G-protein signal (Kim et al., 1999). Polycystin-1 complexes with E-cadherin and catenins (Huan and van Adelsberg, 1999).

stability against the denaturants urea and GdnHCl was low, they found that PKD domain 1 possessed mechanical stability that was similar to that of I27 (Forman et al., 2005; Qian et al., 2005). Unfolding occurred with a maximum force around 200 pN at pulling speeds ranging from 0.3 to 2.5 nm/ms (Forman et al., 2005). They also studied the mechanical stability of a construct containing PKD domains 2–4 and found that they are comparable or even mechanically more stable than PKD domain 1. In addition, SMD simulations showed that PKD domain 1 can form a mechanically stabilized unfolding intermediate that is about 8 Å longer than the native state. This intermediate appears to possess a large energy barrier to further unfolding, explaining the high mechanical stabilities observed in AFM experiments. Furthermore, the LRR and WSC domains are reported to have very low mechanical stabilities, while the PKD domains 6–10 have a broad hierarchy of mechanical stabilities: Unfolding forces measured for individual PKD domains ranged between 50 and 250 pN (Qian et al., 2005).

13.3.6 The Microfilament Protein α-Actinin

α-actinin is a member of the spectrin family and includes α- and β-spectrins and dystrophin (Broderick and Winder, 2005). α-actinin is an anti-parallel homodimer that consists of two calponin homology domains (CH1, CH2), four spectrin repeats, and an EF-hand motif (EF2) (Broderick and Winder, 2005) (Figure 13.11). The structure of the full-length protein was determined by cryo-EM (Liu et al., 2004) and consists of four spectrin repeats in the rod domain, which shows some torsional twist (Ylanne et al., 2001). The structure of the rod domain was determined by x-ray crystallography (Ylanne et al., 2001).

α-actinin has an actin binding domain (ABD) in the head domain consisting of CH1 and CH2 modules, and the structure of the ABD was recently resolved by x-ray crystallography (Borrego-Diaz et al., 2006). α-actinin functions as a thin-filament cross-linking protein in the muscle Z-disks (Masaki et al., 1967). Moreover, it is found close to the plasma membrane and mechanically links the actin cytoskeleton to integrins (Otey et al., 1990) and structurally stabilizes the adhesion site (Rajfur et al., 2002). In nonmuscle cells it is an important component of stress fibers (for review, see Otey and Carpen, 2004). Thus, α-actinin is found in force-bearing networks that link the ECM to the intracellular cytoskeleton and is thereby subjected to tensile stress.

Spectrin repeats have important mechanical properties. Spectrin domains in α-actinin are essential for anti-parallel dimerization of the protein, which increases both its stability and rigidity (Djinovic-Carugo et al., 2002). Pasternak et al. (1995) showed with a dystrophin protein knockout mouse that dystrophin increases the mechanical stability of the sarcomere. Similarly, spectrin is thought to be responsible for the remarkable mechanical properties of erythrocytes (reviewed by Bennett and Gilligan, 1993). Spectrin repeats are shown to unfold cooperatively under force (Law et al., 2003) and helical linkers between spectrin repeats are thought to explain the extensibility of the erythrocyte cytoskeleton. Furthermore, tandem repeats are

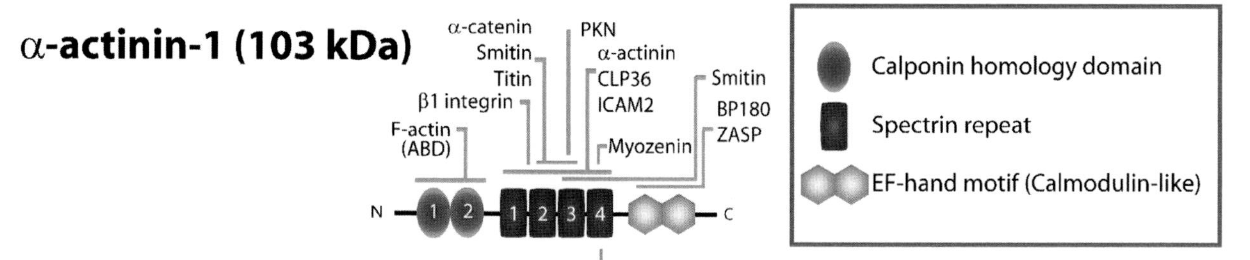

Figure 13.11. Modular structure and binding sites of α-actinin. α-actinin, similar to filamin, has two calponin homology domains in the N-terminus forming the actin binding domain. The central rod is composed of four spectrin repeats, which are responsible for the anti-parallel dimerization of α-actinin. Proteins interacting with α-actinin-1 are shown. CLP36 is a PDZ-LIM family protein (Vallenius et al., 2000). α-catenin (Nieset et al., 1997) is a cytoplasmic protein that connects transmembrane cadherins to the cytoskeleton. PKN is a fatty acid– and Rho-activated serine/threonine protein kinase (Mukai et al., 1997). The binding site of integrin β1A to α-actinin was recently determined by cryoelectron microscopy (Kelly and Taylor, 2005). α-actinin also binds to the muscle protein titin (Young et al., 1998), smitin (a titin-like protein; Chi et al., 2005), and to intercellular adhesion molecule-2 (ICAM-2; Heiska et al., 1996). BP180, also called collagen XVII, is a transmembrane protein that is part of a hemidesmosome with adhesion of the epidermis to dermis (Borradori and Sonnenberg, 1999; Gonzalez et al., 2001).

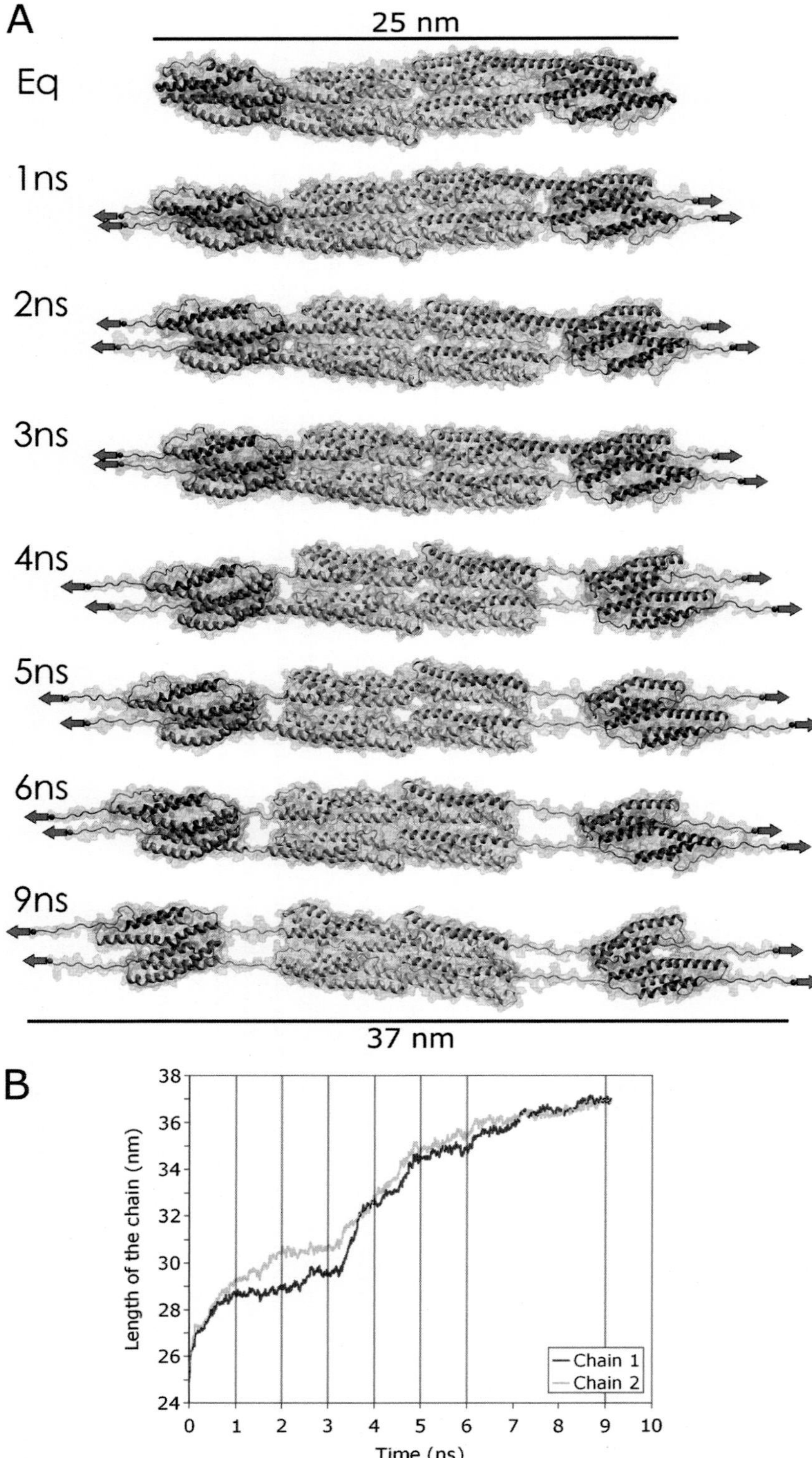
A
25 nm
Eq
1ns
2ns
3ns
4ns
5ns
6ns
9ns
37 nm
B
Length of the chain (nm)
Time (ns)
Chain 1
Chain 2
38
36
34
32
30
28
26
24
0
1
2
3
4
5
6
7
8
9
10

thermodynamically more stable than individual repeats (MacDonald and Pozharski, 2001).

The triple helical bundles of spectrin can form stable unraveling intermediates when subjected to external forces, yet mainly mechanical functions have been proposed (Altmann et al., 2002). It is not known whether any of these intermediates of spectrin-like repeats have regulatory functions or whether the variety of potential unraveling pathways (Altmann et al., 2002) are of regulatory significance. The potential involvement of a cryptic site was implicated in the binding of α-actinin to a titin-like protein, smitin. The smitin binding sites are located in the smooth-muscle α-actinin R2-R3 spectrin-like repeat rod domain and a C-terminal domain formed by cryptic EF-hand structures (Chi et al., 2005). It is not known whether this cryptic site is force-regulated (Vogel, 2006).

The mechanical stability of multimodular proteins may also be regulated by the structure of the linker regions between modules. In contrast to the peptide chains that connect the β-strand repeats of extracellular proteins, the linkers connecting the α-helical bundles of the spectrin repeats are made of α-helices (Figures 13.4 and 13.12(A)). Spectrin domains are stabilized by each other in equilibrium experiments (MacDonald and Pozharski, 2001) and can be observed to unravel cooperatively in AFM experiments (Law et al., 2003). Upon linker unraveling, two proximal loops from each of the repeats that normally sequester and "protect" the linker from the solvent lift away under the applied force (Ortiz et al., 2005). The linker connecting

Figure 13.12. (A) Steered molecular dynamics simulation of human α-actinin 2. An SMD simulation for the human α-actinin 2 rod (residues 272–746, PDB 1HCI) was conducted (Hytönen and Vogel, unpublished) using NAMD (Kale et al., 1999). Initially, the protein was placed in a box filled with explicit TIP3 water molecules and 32 sodium ions to neutralize the system. The complete system contains 237600 atoms. After energy minimization, the temperature of the system was increased to 300 K and the system was equilibrated for 1 ns in constant pressure and temperature (1 bar, 300 K). To study the behavior of α-actinin under mechanical force, the equilibrated structure was subjected to SMD simulation where 300 pN constant force was applied to each terminus (600 pN force per dimer). Two force vectors at the ends of the molecules were fully parallel to each other.

The antiparallel dimer contains eight spectrin repeats. During the 9-ns simulation, four of the six linkers unraveled. Interestingly, linkers connecting repeats 1–2 and 3–4 unraveled first while those connecting repeats 2–3 in the middle of the molecule did not unravel during the experiment. Previous studies showed that spectrin repeats in α-actinin are hotspots for unfolding (Ortiz et al., 2005). The major difference between their study and the data presented here is that we have included a dimeric pair of four spectrin repeats instead of monomeric tandem repeats. Importantly, we see some co-operativity in the unfolding process: Unfolding of the linker is reflected to the neighboring molecule across the dimer interface. Note that the structure is symmetric (antiparallel dimer). The structure snapshots are rendered using VMD (Humphrey et al., 1996) and the secondary structure is calculated for each snapshot using STRIDE (Frishman and Argos, 1995). (B) The end-to-end distances of α-actinin chains during SMD simulation. The time points corresponding to snapshots presented in (A) are shown with light lines. The lengths of the chains during the equilibrium simulation (1 ns) were constant from 24.6–25.4 nm, and only a slight increase in the overall length relative to the starting structure was seen (on average for two chains 0.7 nm).

the spectrin repeats 2–3 shows five- to sixfold greater exposure to water than the linker connecting repeats 1–2. Accordingly, for all computationally derived trajectories at the slower (0.5 nm/ns) pulling rate, the linker between repeats 1–2 never unraveled, whereas the linker between repeats 2–3 was reported to always unravel (Ortiz et al., 2005). The contact between the linker and the sequestering loops is primarily between hydrophobic residues.

In order to determine if quaternary structure and dimerization affects the mechanical properties of α-actinin, we conducted constant-force SMD simulations for the α-actinin dimer starting with the structure determined using x-ray crystallography (Figure 13.12; Hytönen and Vogel, unpublished data). All previous SMD experiments for spectrin repeats were carried out for single spectrin repeats or their combinations. In the constant-force SMD simulation of the α-actinin rod dimer (Figure 13.12), the linkers between repeats 1–2 and 3–4 unraveled before the linker between repeats 2–3. Constant force application of 300 pN to each terminus led to an average pulling velocity of ~1.3 nm/ns over the simulation (Figure 13.12(B)). To compare our data to those from simulations of monomeric tandem repeats (Ortiz et al., 2005), we measured solvent-buried surface areas of the linkers connecting helical repeats during a 1-ns equilibration simulation. The linkers between repeats 1–2 and 2–3 were defined as in Ortiz et al. (2005), and the linker between the repeats 3–4 was defined to contain 20 residues (residues 622 to 642). We found that the linker between repeats 1–2 contains the largest surface area of solvent exclusion (on average for two chains 1617 Å^2), while linker 2–3 had the second largest area of exclusion (1337 Å^2). Interestingly, the linker between repeats 3–4 contained the least buried surface area in the structure (1173 Å^2). It appears that there is some discrepancy between the results obtained with monomeric tandem repeats (Ortiz et al., 2005) and those from our simulations carried out for the dimeric molecule (Figure 13.12) since linker 2–3 seems to be the most resistant to unfolding in the dimeric α-actinin rod. Interestingly, the buried areas of linkers 2–3 and 3–4 are virtually independent of the dimerization of the molecule, whereas linker 1–2 is significantly buried by the dimer pair (the buried area in the monomer is 1222 Å^2 calculated from data from the simulation carried out for the dimer). The observed differences in the unfolding process due to oligomerization are not unique for α-actinin, as discussed in the following chapter.

Many multimodular proteins form homophilic oligomers. It is therefore important to know whether and to what extent interdomain interactions can alter mechanical stability. Experiments suggest that titin molecules may globally associate into oligomers that mechanically behave as independent worm-like chains (WLCs). Although oligomers may form globally via head-to-head associations of titin, the constituent molecules otherwise appear independent from each other along their contour since they appear to unravel independently from each other (Kellermayer et al., 2003). About 500 spectrin repeats that have homologous triple-helical repeat motifs can be found in the human genome, including cytoskeletal actin filament-associated proteins such as spectrins, dystrophin, utrophin, and α-actinins (Altmann et al., 2002). Does lateral association facilitate or oppose forced unraveling? In the case of anti-parallel

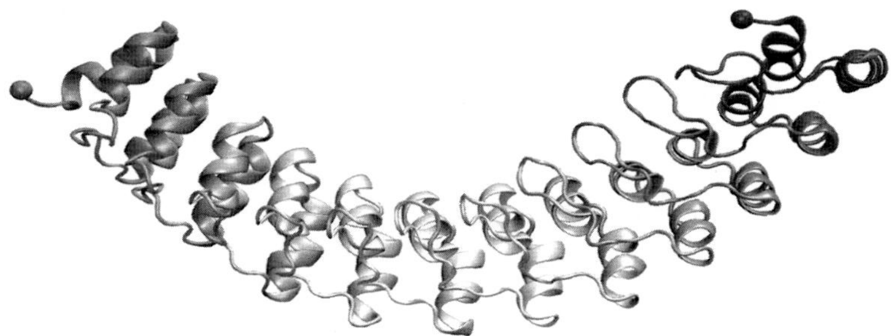

Figure 13.13. Ankyrin. Ankyrins are adaptor proteins linking membrane proteins to the cytoskeleton. Human ankyrin has 24 ankyrin repeats, and the structure of a segment containing 12 repeats (repeats 13–24) that was recently resolved using x-ray crystallography (PDB 1N11) is shown in a cartoon model.

spectrin heterodimers, the associated chains in a dimer can stay together and unravel simultaneously in addition to unraveling independently. Strong lateral interactions lead to coupled unravelings of laterally adjacent repeats, whereas weak lateral interactions are surprisingly neutral in their effect on the force to unravel a repeat (Law et al., 2003). However, the frequency of tandem repeat unraveling events is low relative to single repeat events, suggesting that tandem unraveling does not propagate as well in two laterally associated heterodimer chains compared with a single chain (Law et al., 2003), at least for the cases studied so far. Our simulations for α-actinin suggest a role for dimerization in the regulation of the unfolding hierarchy of linkers connecting spectrin repeats (Figure 13.12).

13.3.7 The Adaptor Protein Ankyrin

Ankyrin repeats are short (33-residue) domains that are among the most frequently observed amino acid motifs in proteins (for review, see Gao et al., 2006; Mosavi et al., 2004). Each domain consists of two anti-parallel α-helices and connecting loops (Figure 13.4). Ankyrin family proteins contain a large number of ankyrin domains where the helices are tilted with respect to the axis of the rod-like quaternary structure (Figure 13.13). SMD simulations carried out for 17 and 24 cytoplasmic ankyrin repeats of the mechanosensitive TRP channels showed that polyankyrin domains respond to larger forces in two phases: first by straightening out the banana-shaped curvature of the helix bundle without unfolding of helices, and, second, by rupture of the stack of repeats and unfolding of its repeats (Sotomayor et al., 2005). The first phase causes an elastic deformation (spring constant ~5 mN/m) with reversible nanospring behavior (Lee et al., 2006; Sotomayor et al., 2005).

13.3.8 The Actin Crosslinking Protein Filamin

Filamins are actin-binding proteins. There are three isoforms of filamin in humans (filamin-a (FLNa), FLNb, and FLNc). The molecular mass of filamin is ~280 kDa, which dimerizes via the last C-terminal filamin repeat. Filamin binds to actin (Stossel and Hartwig, 1975) by the N-terminal actin-binding domain, a region that is composed of two calponin homology domains that are similar to those in α-actinin. The rod domain of α-actinin is made of 24 filamin repeats, which have an anti-parallel β-sheet secondary structure that forms β-sandwiches resembling immunoglobulin domains (Figure 13.4). Those repeats are connected by flexible loops of roughly 30 residues. Interactions of FLNa are shown in the Figure 13.14.

13.4 Putting the Pieces Together

Biochemists initially considered the surface features of a protein's equilibrium fold to be most important for its function. Protein surface chemistry of the equilibrium topography was believed to be the sole determinant of the relationship between structure and function. However, recent findings highlight the importance of the scaffold design of a protein's core, as it determines the protein's resistance to mechanical unfolding. We now know that the structure of a protein might change in response to tensile forces, implying that a single unique structure–function relation might not be intrinsic to a protein, but that this relation could be force-regulated. Mechanical force thus represents an additional dimension of functional regulation at the protein level. In living systems, however, mechanical forces are not applied directly to isolated modules in solution, but they are transmitted through extended protein networks. Either the mechanically weakest physical linkage will break, or, if all are sufficiently stable, the mechanically weakest module will unfold first. An example of how tightly the mechanical properties of a proteinous tether are tuned to the mechanical characteristics of a receptor-ligand interaction is given by the bacterial type I fimbriae. To extend the lifetime of *E. coli* in the surface bound state, the fimbriae uncoil under high shear flow conditions, which would otherwise rapidly break the force-activated fimH–mannose bond (Thomas et al., 2002; Forero et al., 2006). Thus, in order to build a more quantitative picture of the cellular response to force, we need to consider the mechanical responses of all of its components, as well as all the physical linkages by which they are interconnected. This includes determining their mechanical hierarchies.

Various engineering principles by which mechanical stability can be regulated have been derived recently. Mechanical stability is not a universal molecular property, but it depends on the direction of force with respect to the force-bearing hydrogen bonds, as well as on the geometrical arrangement of the stabilizing backbone strands. Finally, the presence of molecules that bind either to the equilibrium or to the mechanically strained conformation might alter the module's mechanical response, which includes receptor–ligand interactions or oligomerization. Future advancements in computational resources and modeling techniques (Stone et al., 2007) may make it possible, for instance, with SMD, to predict the relative stabilities of force-bearing bonds and the physical linkages formed between proteins integrated into force-bearing networks (see Section 13.5). Thus, as the size of systems that can

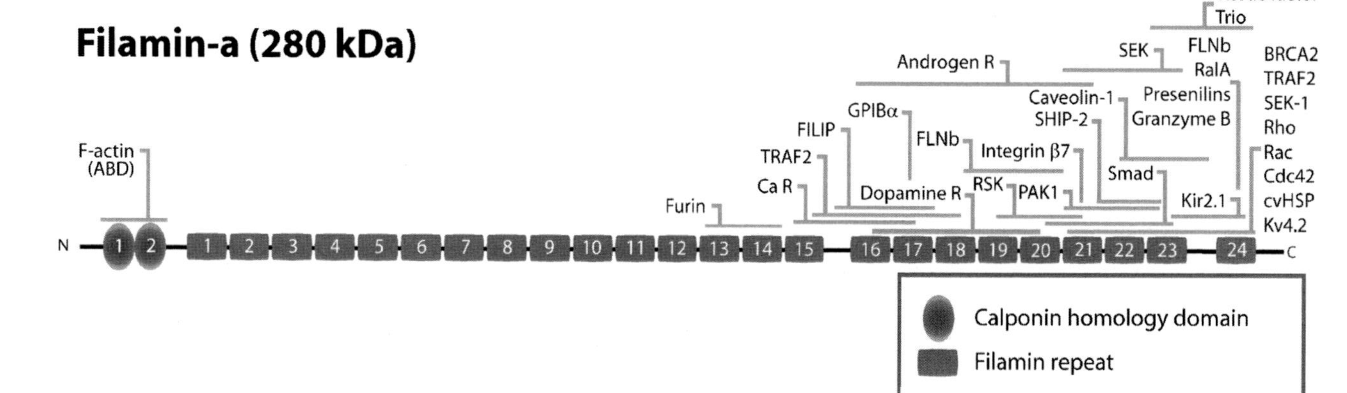

Figure 13.14. Modular structure and binding sites of filamin. Filamins are actin-binding proteins. There are three isoforms of filamin in humans (filamin-a (FLNa), FLNb, and FLNc). The molecular mass of filamin is ~ 280 kDa, which dimerizes via the last C-terminal filamin repeat. Filamin binds to actin (Stossel and Hartwig, 1975) by the N-terminal actin-binding domains, which is composed of two calponin homology domains, as in the case of α-actinin. The rod domain of α-actinin is made of 24 filamin repeats, which have anti-parallel β-sheet secondary structures forming β-sandwiches resembling immunoglobulin domain (Figure 13.4). Those repeats are connected by flexible loops of roughly 30 residues. Interactions of FLNa are shown in the figure essentially according to Stossel et al. (2001). Integrins β_{1A}, β_{1D}, β_2, β_3, and β_7 are known to interact with filamin (Calderwood et al., 2000), but only β_7 is indicated in the figure since it was recently crystallized in complex with filamin repeat 21 (Kiema et al., 2006) (Figure 13.15; Plate B).

be simulated computationally is rapidly increasing, quantitative insights will emerge about how mechanical forces regulate function.

In this chapter we focused on molecular components related to integrin-mediated cell adhesions. To transmit mechanical forces from the outside to the inside, and vice versa, the integrin receptor needs to be mechanically linked to both the cytoskeleton and the extracellular environment. Mature focal adhesions are known to contain more than 150 different proteins (Geiger, 2006; Zaidel-Bar et al., 2007); however, it is critical to differentiate between those components of focal adhesions that are integral components of force-bearing networks and those that only interact with these force-bearing structures. Even if only a few intracellular proteins have so far been shown to be mechanically unfolded (Johnson et al., 2007; Sawada et al., 2006), the remarkable breadth of adhesive structures that link cells to other cells and cells to different ECM compositions makes it very unlikely that only one or a few mechanotransductive proteins exist.

How does the integrin cytoplasmic domain couple to intracellular proteins that are part of the force-bearing network? In the simplest example of a force-bearing linkage formed between fibronectin and the actin cytoskeleton, at least five components are most frequently found: fibronectin or another ECM molecule, integrins, talin, filamin and actin (Figure 13.15; Plate B). Thus far, only a limited number of high-resolution structures are available of at least some fragments of those protein complexes. The structure of filamin repeat 21 complexed with the β_7-integrin cytoplasmic domain, for example, was determined using x-ray crystallography (Kiema et al., 2006). In this complex (PDB 2BRQ), the integrin tail incorporates with the filamin repeat as an additional β-strand (Figure 13.15; Plate B). Phospho-mimicking mutations (T → E) for threonines 783, 784, and 785 caused impaired filamin repeat 21 binding. Therefore, phosphorylation of the integrin β_7 tail may regulate integrin-filamin interactions (Kiema et al., 2006). The complex of talin F2 and F3 (FERM domains) with the cytoplasmic domain of integrin β3 was also determined by x-ray crystallography (Figure 13.15; Plate B). Similar to the integrin–filamin complex, the β_3-integrin tail incorporates into the β-sheet of the talin head. In this case, however, the β-sheet is relatively short, consisting of only about three to five residues. Tyrosine 747 in the integrin tail, which is known to be essential for binding, binds to an acidic pocket within the talin head (Calderwood et al., 2003). Therefore, although the talin head resembles a phosphotyrosine binding (PTB) domain, it does not appear to be phosphorylated in this position due to the structure of the binding site (Calderwood et al., 2003). This is in line with experimental findings: Phosphorylation of integrin tails serves as a negative regulator of complex formation by decreasing the binding affinity to talin (Tapley et al., 1989).

What are potential functions of those proteins that are present in focal contacts yet are not part of the force-bearing network? At least some of the force-bearing components are hypothesized to unfold under force; however, additional sensory components are necessary to detect these nonequilibrium components and initiate downstream responses. Some kinases might act on their substrates in a force-regulated manner, as illustrated for the force-exposed phosphorylation sites of p130Cas

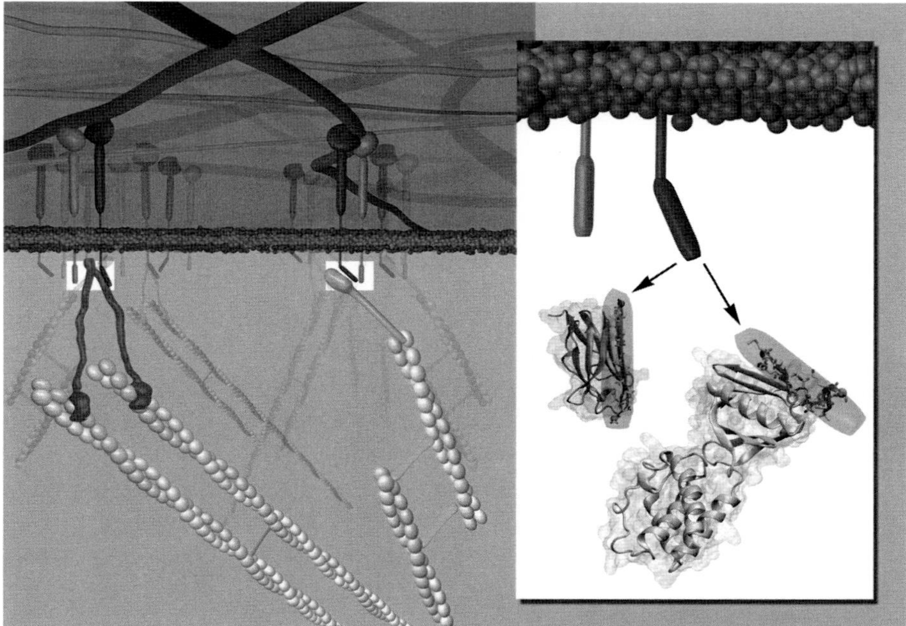

Figure 13.15. Contact between cytoskeleton and integrins. A schematic representation of the connection between ECM and cytoskeleton is shown. The integrin receptors are bound to the ECM proteins (for example, Fn). The cytoplasmic tails of integrins serve as binding sites for few proteins, among them talin and filamin, which are shown here. The binding site of talin head in the β-integrin cytoplasmic domain overlaps with the binding site of filamin so that four integrin residues are shared between the x-ray crystallographically determined complexes (Garcia-Alvarez et al., 2003; Kiema et al., 2006). These particular proteins are also able to bind to actin, thus forming a physical connection between ECM and cytoskeleton. The experimentally determined structure of β_7-integrin cytoplasmic tail (PDB 2BRQ) complexed with filamin repeat 21 (right part of the figure) and the β_3-integrin cytoplasmic domain (PDB 1MK7) complexed with the F2 and F3 talin head domains (left part of the figure) are also shown. The atomic stick model of the bound integrin peptide is given together with the cartoon model whereas talin and filamin are represented as cartoon models with terminal Cα-atoms presented as spheres (N-terminus blue and C-terminus red).

(Sawada and Sheetz, 2002; Sawada et al., 2006; Tamada et al., 2004). Alternatively, protein recruitment to force-bearing proteins might be upregulated if otherwise cryptic binding sites are exposed by force. Talin, for example, is a component of force-bearing networks that has been hypothesized to transduce mechanical force (Calderwood and Ginsberg, 2003; Giannone et al., 2003; Jiang et al., 2003; Lee et al., 2007). The talin rod domain contains up to 11 binding sites for vinculin (Gingras et al., 2005), and at least most of these sites are inactive under equilibrium conditions. Therefore it has been suggested that mechanical stress applied to talin, which physically connects integrin cytoplasmic domains to the actin cytoskeleton, could expose vinculin binding sites in talin (Papagrigoriou et al., 2004). High-resolution structural models of mechanical force–mediated activation of vinculin binding to talin have been derived (Hytönen and Vogel, 2008; Lee et al., 2007). As the search for mechanochemical signal converters continues, it is likely that many other force-regulated mechanisms will be discovered by which protein function can be altered.

Since mechanical forces critically regulate cellular behavior and tissue development, dysfunctional responses to mechanical force are hypothesized to directly initiate the progression or symptoms of a variety of seemingly unrelated diseases. Pathologies as diverse as osteoporosis, congenital deafness, musculodystrophies, and polycystic kidney disease are believed to be due to dysfunctional mechanochemical signal conversion (reviewed in (Ingber, 2003)). Therefore, a new focus is needed not only on the genetic basis of diseases, but also on how the mechanoresponsive machinery of a cell might be altered in response to point mutations or environmental stress. Consequently, understanding the molecular basis of mechanotransduction has a variety of potential medical applications. For instance, adult stem cells can be directed to differentiate into different lineages based on the degree of spreading or stiffness of their substrates (Engler et al., 2006; McBeath et al., 2004). Furthermore, the stiffness of tumor tissue is often elevated relative to the surrounding tissue. Emerging evidence indicates that tumor stiffness promotes the malignant behavior of resident tumor cells (Paszek et al., 2005), thus implicating the physical environment in tumor pathology.

Understanding the underlying design principles of mechanically regulated proteins may also benefit more applied fields such as tissue engineering. Even the rigidity of a matrix can have such profound regulatory roles as directing stem cell lineage specification (Engler et al., 2006). Functional regulation via physical parameters combined with mechanochemical signal converters will be a new frontier in tissue engineering. Some outlines for engineering of novel protein-based materials can be drawn: Stiff and robust protein modules can be used as a scaffold for novel materials, examples of which include collagen (Koide, 2005) and silk (Altman et al., 2003). To include force-sensitive signals, one may consider the inclusion of domains containing cryptic binding sites or binding sites that disappear simultaneously with domain unfolding. Many examples of such domains are found within fibronectin (Figure 13.6). Engineered proteins with mechanically tunable functions may also be envisioned through selection of modules and assembly with traditional tools in molecular biology, or through the synthesis of biosynthetic hybrid materials. Although engineered ECMs are popular tools for wound healing or tissue engineering applications, to our knowledge the use of engineered matrices that consider the mechanical response of their components have not been realized. For example, a cryptic recognition site for a protease, that could be then exposed by mechanical tension, could lead to force-activated degradation of the material. Thus it might be possible to build programmable materials that can be cleaved once a threshold level of mechanical strain is reached.

In conclusion, Linus Pauling wrote in 1948, "To understand all of these great biological phenomena we need to understand atoms, and the molecules that they form by bonding together" (Slade and Pauling, 1948). Since then, molecular biologists have focused largely on the importance of the equilibrium structure of proteins.

In the future, we must focus our efforts not just on how molecules bond together, but how force transmission alters molecular structure.

13.5 Appendix: Experimental Approaches to Study Force-Induced Conformational Changes of Proteins

Our understanding of the mechanically stabilized nonequilibrium conformations of multimodular proteins *in vivo* is rapidly improving due to the development of new tools capable of investigating tension-dependent nonequilibrium protein conformations *in vitro* and *in vivo*. We discuss how otherwise short-lived intermediates (Eaton et al., 2000) might be stabilized by tensile force for experimentally relevant time periods. However, new tools are needed that are capable of verifying that such intermediates states exist in living systems and that they transduce mechanical force into a biochemical responses. Although NMR-spectroscopy and x-ray crystallography provide structural information with atomic resolution, these tools cannot currently be used to derive nonequilibrium structures of strained proteins.

13.5.1 Tools to Stretch Populations of Proteins

In order to investigate potential functional alterations of nonequilibrium conformations of proteins, one must be able not only to stretch the protein but also to stabilize the higher energy intermediates of the protein on a time-scale that permits the investigation of possible functional alterations. While nanoanalytical tools gave considerable insights into the unfolding trajectory of single proteins, biochemical and cellular assays are now needed to establish how their structure–function relations are changed by force. This can best be done not on single molecules but on larger populations of proteins if assays are developed that allow us to uniformly strain protein populations.

13.5.1.1 *Stretching Proteins in Solution*

Creating a population of proteins in well-controlled or tunable conformations in solution was made possible recently by realizing an allosteric spring probe. By covalently linking a protein on each end to a single strand of DNA, stress can be applied to the protein through hybridization of the DNA strand, thus increasing its stiffness, which will straighten the hybridized section of base pairs. Choi and Zocchi (2006) recently described and utilized this tool to directly demonstrate the activation of protein kinase A in the absence of an activating ligand. A protein chimera was created with terminal cysteines covalently linked to ssDNA via a maleimide-activated cross-linker. After addition of the complementary strand and hybridization to the protein chimera, they observed activation of the enzyme, which normally happens only after binding of the ligand cAMP. Importantly, they were able to regulate the applied force by using complementary strands with variably complementary sequences, thereby exposing loop regions in the hybridized molecule. This concept of engineering an allosteric spring probe is unique since it can be used to apply tension to a population of proteins in solution, thus providing a method for studying the role of force-induced conformational changes in enzyme functions. Although the exact conformational changes of the enzyme that are induced by DNA hybridization are not known, enzyme function in the presence of applied force could be easily measured using traditional biochemical approaches.

13.5.1.2 *Proteins Bound to Stretchable Surfaces*

A widely used technique to study strained proteins is to attach them to a stretchable substrate. Protein attachment can be accomplished by using specific attachment tags such as covalent attachments with chemical cross-linkers or by simply adsorbing proteins to the surface. With this approach, the substrate strain can be externally adjusted and tuned; however, the force necessary to induce a conformational change in the bound protein remains unknown. Sawada et al. (2006) recently used this technique for the study of the strain response of biotinylated p130Cas attached to an avidinylated latex surface. Stretching of the substrate by 50% then led to increased phosphorylation of p130Cas (Figure 13.2). This simple technique was used to demonstrate the extension of p130Cas that was necessary to expose otherwise cryptic phosphorylation sites. Zhong et al. (1998) also used a stretchable substrate to demonstrate that strain exposes a buried sequence in fibronectin that was known to play a crucial role during fibronectin fibrillogenesis. By attaching fibronectin molecules to a stretchable substrate and assaying for binding of a monoclonal antibody to this cryptic sequence, Zhong et al. were able to demonstrate that strain was sufficient for epitope exposure in a reductionist system.

In our laboratory, stretchable substrates were recently used to stabilize non-equilibrium conformations of fibronectin in its fibrillar state. Deposition of fibronectin fibrils to stretchable PDMS makes it possible to tune the conformational states of the fibrillar fibronectin (Little et al., 2008). Fibronectin fibers were drawn from a concentrated fibronectin solution (Brown et al., 1994; Ejim et al., 1993; Little et al., 2008) and deposited onto a stretchable substrate that was mounted onto a custom-built, one-dimensional strain device. To optically monitor the changes in fibronectin conformation, a small fraction of FRET-labeled fibronectin was added to the solution, from which the fibers were deposited. FRET imaging showed that manually deposited fibers and fibronectin within ECMs of living fibroblasts had similar molecular conformations; however, the manually pulled fibers presented a much narrower, and hence more defined, range of molecular conformation that could be specifically tuned by straining the substrate.

13.5.2 Probing Cryptic Epitope Exposure by Mechanical Force Using Ligands

Steady-state binding experiments can be used to probe the presence of specific epitopes in cell cultures that result from cell-derived or exogenously applied tension. Ligand or antibody binding to cryptic epitopes only occurs when the protein is stretched by force. Therefore, this technique serves as a qualitative indicator of the presence of strained proteins. Detection of cryptic epitopes or sequences using specific antibodies provides one of the simplest methods to detect conformational changes of multimodular proteins. Monoclonal antibodies have successfully been generated against cryptic protein epitopes. This can be achieved, for example, by digesting proteins with a protease and generating antibodies against peptides that are typically buried in the protein structure. This approach is useful since the conformational change can be localized to a specific region of the protein. For example, the experiments described above by Zhong et al. utilized a monoclonal antibody that was generated in this way (Chernousov et al., 1987; Zhong et al., 1998).

13.5.3 Single Molecule Techniques to Study the Mechanoresponse of Proteins

13.5.3.1 *Atomic Force Microscopy*

Single molecule force spectroscopy using the AFM is a powerful tool that can be used to stretch proteins with sub-nanometer resolution. This approach has been reviewed thoroughly elsewhere; however, here we briefly introduce the advantages and disadvantages of the technique since numerous mechanical properties of multimodular proteins were discovered using the AFM (for review see Bao and Suresh, 2003; Fisher et al., 2000). With this method single proteins are stretched between a sharp cantilever tip and a flat surface using a high-resolution piezoelectric controller. The technique is schematically illustrated in Figure 13.16: during forced unfolding of a multimodular titin construct, a sawtooth pattern is measured (Oberhauser et al., 2001). The peak of the sawtooth represents the force required to unfold each individual module, while the length from peak to peak can be used to calculate the number of amino acid residues in the unfolded module. Although unfolding of modules is a probabilistic event, large data sets can be used to generate histograms of unfolding forces for a specific protein module. AFM has been used to identify or confirm numerous properties of multimodular proteins, including the presence of unfolded intermediate conformations (Li et al., 2002) and mechanical hierarchies of unfolding (Oberhauser et al., 2002). Although this information is invaluable to our understanding of forced protein unfolding, AFM also has some limitations. For instance, forced unfolding of individual proteins might not accurately represent the unfolding behavior of proteins in their native environments. For instance, fibronectin unfolding within ECM fibrils may be different than forced unfolding of single molecules due to steric considerations or the degree of packing between adjacent molecules. In addition, AFM data can most easily be interpreted using recombinantly expressed proteins, for instance, to limit the total diversity or number of domains in a single pull or to engineer multidomain polyproteins that contain modules with known mechanical signatures. Expression of recombinant protein chimeras is labor-intensive and will thus inherently limit the number of proteins that can be studied.

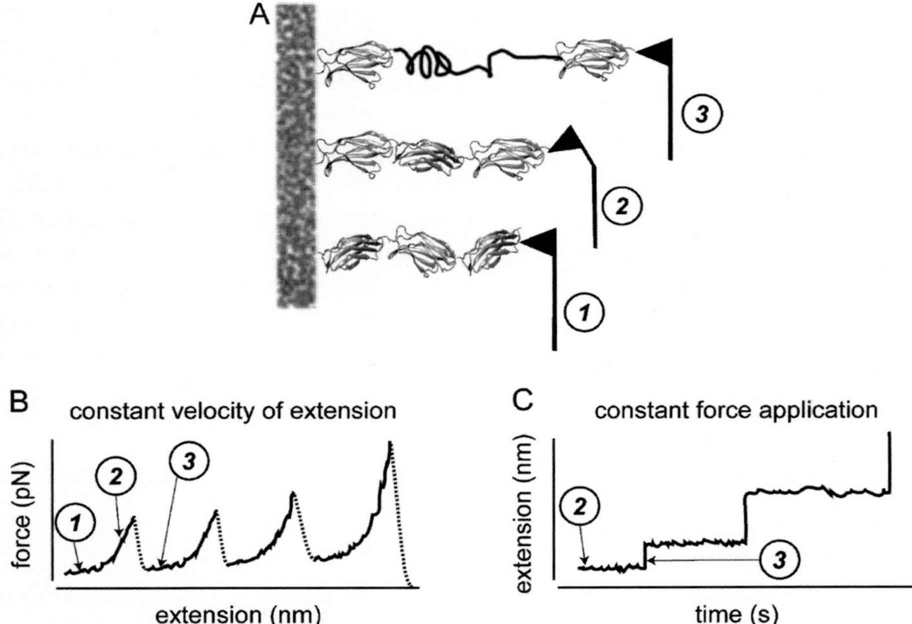

Figure 13.16. Multidomain protein unfolding probed either under constant velocity or constant force. Hypothetical unfolding data are shown for a multimodular protein containing three FnIII modules (A). Protein unfolding occurs first by aligning the modules (1 to 2) along the vector of force and the subsequent unfolding of a module (3). Under a constant velocity of extension (B), the AFM cantilever is moved linearly with a constant velocity, and the pulling force needed to unfold the protein is measured. A sawtooth pattern is obtained in which force peaks correspond to single domain unfolding events. The contour length of each module can thus be determined. The last peak (higher force) corresponds to the detachment of the protein from the cantilever, leading to a drop of force to zero. With application of a constant force (C), the time necessary to unfold each module is measured. Extension of the unfolded polypeptide proceeds rapidly until the constant force is once again achieved (adopted from Oberhauser et al., 2001).

AFM experiments can also be used to study the refolding of proteins under residual force. For tension-dependent unfolding of proteins *in vivo*, it is important to determine whether protein refolding itself provides a restoring force, or whether protein folding only proceeds after abrogation of force application. In these experiments, the AFM tip is used to unfold the protein, and subsequently the tip is moved back to its original position. Upon reapplication of force with the AFM, the presence of sawtooth patterns indicates that the module had refolded during the relaxation phase and was unfolded again during the second pull (Oberhauser et al., 2002). Such experiments can be used to determine the time necessary to refold through the pause time between consecutive pulls (Oberhauser et al., 2002) and were also used to demonstrate that the titin I27 domain refolds under residual force, which is significantly lower than the force necessary for initial protein unfolding (Carrion-Vazquez et al., 1999).

13.5.3.2 *Steered Molecular Dynamics*

Computational simulations can be used to study molecular behavior under force with atomic resolution. Although numerous experimental methods are available, computational methods are sometimes the only way to explore the details of force-induced structural changes, and multiple computational predictions have been experimentally verified (for reviews see Gao et al., 2006; Sotomayor and Schulten, 2007; Vogel, 2006). Simulations can be conducted if high-resolution protein structures are available. The protein module is first equilibrated preferentially in a box filled with explicit water (molecular dynamics, MD), followed by steering atoms of choice apart (mostly the terminal ends of a protein) either with constant force or constant velocity (steered molecular dynamics, SMD; Figure 13.12). SMD is ideally suited to identify critical events and bonds that have to be broken to enable the passage across major energy barriers in the unfolding trajectory. In constant-force simulations, plateau regions in the extension-time plots indicate the existence of intermediate states, whereas the molecule extends rapidly when passing an energy barrier (Figure 13.17). In constant-velocity simulations, the passage of major energy barriers is seen as force peaks in force–time plots. In this review, we discussed how SMD simulations were applied to study the relative mechanical stabilities of modules (Craig et al., 2004). Due to limitations in computational resources, it is typically necessary to apply relatively high forces or pulling velocities to proteins to observe unfolding events in an achievable time-scale. Therefore, unfolding forces measured by SMD on nanosecond time-scales are typically higher compared to those measured by AFM on millisecond time-scales. However, it was possible to correctly predict the relative mechanical stabilities of proteins, which suggests that the same energy barriers were probed in this case by SMD and AFM (Craig et al., 2004; Oberhauser et al., 2002).

13.5.4 Fluorescence Approaches

13.5.4.1 *Green Fluorescence Protein (GFP) and Its Analogues*

GFP and its analogues are widely used to genetically express fluorescently labeled proteins, which are of particular interest for the study of intracellular proteins in living cells. For instance, force applied to the cell by Fn-coated beads was used to show the localization of a chimeric GFP-α-actinin protein to the site of force application (von Wichert et al., 2003). Alternatively, GFP-labeled vinculin was used to demonstrate that it was recruited to cell adhesion sites in a force-dependent manner (Balaban et al., 2001). GFP derivatives are also used to probe force-dependent changes in protein conformations. While the fluorophore of GFP forms spontaneously inside the β-barrel, which is composed of 11 β-sheets, Ghosh et al. (2000) were able to produce nonfluorescent complementary fragments of GFP by cutting one loop region of the β-barrel. These become fluorescent only when they are in physical contact. Later, this system was applied to study conformational changes in maltose binding protein (MBP) in which the nonfluorescent halves of GFP were fused to the

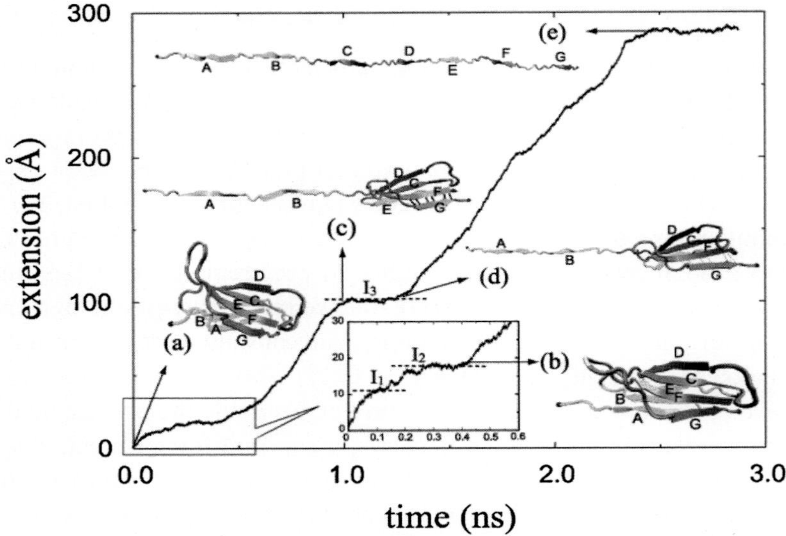

Figure 13.17. Single $FnIII_{10}$ unfolding trajectories obtained from SMD. Constant force (500 pN) SMD simulation was performed for $FNIII_{10}$ (Gao et al., 2002a). Extension-time profile and representative snapshots (a)–(e) from unfolding trajectory are shown: (a) native structure; (b) structure in orientation in which β-strands are aligned; (c) intermediate state at 100 Å in extension; (d) breakage of hydrogen bonds between β-strands G and F; and (e) fully unfolded structure. Hydrogen bonds between β-strands F and G in (c) are shown as thick black lines and ruptured hydrogen bonds in (d) are shown as broken black lines.

termini of MBP and the fluorescence recovery after maltose binding to MBP was observed (Jeong et al., 2006). Alternatively, GFP halves tagged to the termini of a protein of interest could be used to indicate the forced extension of the protein by probing the loss of fluorescence from the separation of the GFP halves. This approach was recently successfully utilized to demonstrate that strain applied to avidin-coated latex membranes also resulted in extension of a biotinylated p130Cas substrate domain that was bound to the substrate (Sawada et al., 2006).

13.5.4.2 *Fluorescence Resonance Energy Transfer*

Limited tools are available to probe the different conformational states of proteins in their native states *in vitro* or *in vivo*. Single molecule FRET techniques are extremely powerful since distances can be calculated between fluorophores that are covalently linked to the protein of interest, and hence nanometer-scale resolution can be obtained if single donor and acceptor pairs are used (Förster, 1948). In the context of mechanotransduction processes, vinculin dynamics within the focal adhesions of cells were studied by introducing the CFP-YFP FRET pair into a chimeric vinculin protein through genetic engineering and following its activation in living cells via changes in the distance of fluorescent proteins (Chen et al., 2005). Intramolecular FRET using multiple donors and acceptors was further applied for the study of a large range of conformational changes of fibronectin within the ECM, as discussed

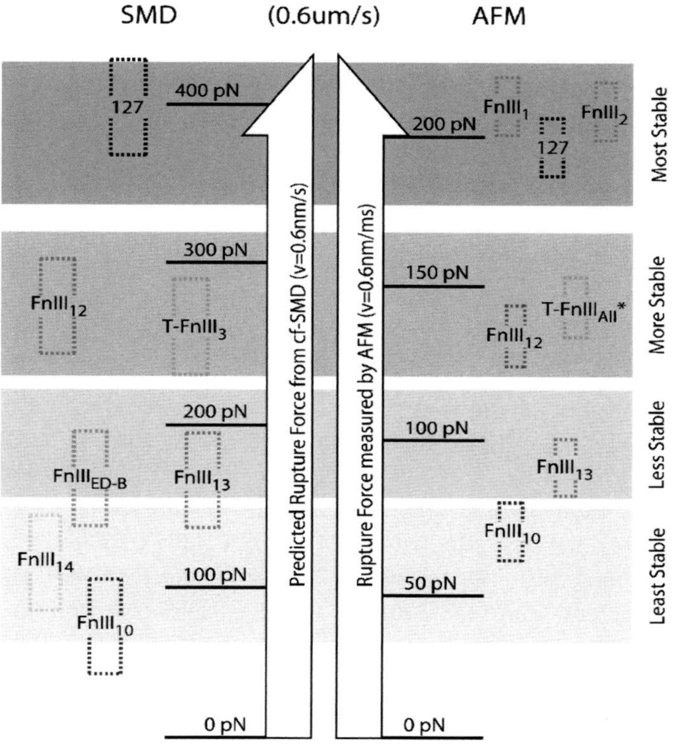

Figure 13.18. Comparison of relative mechanical stabilities of FnIII domains. Comparison of mechanical stability of FNIII modules determined by AFM (right) and by SMD (left) (Craig et al., 2004). AFM results are taken from (Oberhauser et al., 1998) and (Oberhauser et al., 2002).

earlier. Multiple labeling schemes were developed that took advantage of the existence of two free cysteines per fibronectin monomer within modules III_7 and III_{15} (Antia et al., 2006; Baneyx et al., 2001, 2002; Barker et al., 2005; Lai et al., 1993; Wolff and Lai, 1988, 1989, 1990). FRET-labeled fibronectin was produced by labeling fibronectin on all four free cysteines with an equal quantity of both acceptors and donors or on all four sites with Alexa 546 acceptors and on seven random amines with Alexa 488 donor fluorophores (Smith et al., 2007). The Förster radius of this fluorophore pair is ~6 nm; therefore, energy transfer is limited to less than 12 nm of $FnIII_7$ or $FnIII_{15}$ (fading spheres in Figure 13.18 left panel). By labeling the protein on $FnIII_7$ and $FnIII_{15}$ with acceptors and randomly labeling free amines with donors, a labeling scheme results that provides broad conformational sensitivity since energy transfer can occur in both the compact and extended conformations, but energy transfer is lost in the unfolded conformation (Smith et al., 2007). However, using the cysteine-only labeling scheme with two donors and two acceptors randomly attached only to $FnIII_7$ and $FnIII_{15}$, sensitivity is limited to differentiation between the compact (dimeric arms of Fn cross over each other) and extended conformations since fluorophores attached to III_7 of each monomer arm are in close proximity as a result of the folding of the two fibronectin arms across one another, but energy transfer between cysteines on opposing arms of fibronectin cannot occur when the arms are separated since $FnIII_7$ and $FnIII_{15}$ on a single linear monomer arm are separated by ~25 nm and the two $FnIII_{15}$ modules on the disulfide–cross-linked arms would be separated by ~16 nm (Antia et al., 2006; Baugh and Vogel, 2004). With these two labeling schemes, the spatially resolved ratio of acceptor to donor intensities was quantified within the ECM of living fibroblasts cultured on glass using confocal microscopy (Smith et al., 2007).

13.6 Acknowledgments

The authors gratefully acknowledge Kristopher Kubow, Delphine Gourdon, William Little, and John Saeger for their helpful discussions and Sheila Luna for help in the preparation of Figure 13.15. This work was financially supported by the ETH Zurich, the Nanomedicine Development Center (NDC) Nanotechnology Center for Mechanics in Regenerative Medicine (NIH grant PN2 EY016586), that participates in the NIH Nanomedicine Development Center Network (NNDCN), and by postdoctoral fellowships from the Human Frontier Science Program Organization (MLS) and from the Academy of Finland (project no. 109425, VPH).

REFERENCES

Abu-Lail, N. I., Ohashi, T., Clark, R. L., Erickson, H. P., and Zauscher, S. 2006. Understanding the elasticity of fibronectin fibrils: Unfolding strengths of FN-III and GFP domains measured by single molecule force spectroscopy. *Matrix Biol.* **25**: 175–184.

Ainavarapu, S. R., Brujic, J., Huang, H. H., Wiita, A. P., Lu, H., Li, L., Walther, K. A., Carrion-Vazquez, M., Li, H., and Fernandez, J. M. 2007. Contour length and refolding rate of a small protein controlled by engineered disulfide bonds. *Biophys. J.* **92**: 225–233.

Alexandropoulos, K., Donlin, L. T., Xing, L., and Regelmann, A. G. 2003. Sin: good or bad? A T lymphocyte perspective. *Immunol. Rev.* **192**: 181–195.

Altman, G. H., Diaz, F., Jakuba, C., Calabro, T., Horan, R. L., Chen, J., Lu, H., Richmond, J., and Kaplan, D. L. 2003. Silk-based biomaterials. *Biomaterials.* **24**: 401–416.

Altmann, S. M., Grunberg, R. G., Lenne, P. F., Ylanne, J., Raae, A., Herbert, K., Saraste, M., Nilges, M., and Horber, J. K. 2002. Pathways and intermediates in forced unfolding of spectrin repeats. *Structure.* **10**: 1085–1096.

Antia, M., Islas, L. D., Boness, D. A., Baneyx, G., and Vogel, V. 2006. Single molecule fluorescence studies of surface-adsorbed fibronectin. *Biomaterials.* **27**: 679–690.

Apic, G., Gough, J., and Teichmann, S. A. 2001. Domain combinations in archaeal, eubacterial and eukaryotic proteomes. *J. Mol. Biol.* **310**: 311–325.

Balaban, N. Q., Schwarz, U. S., Riveline, D., Goichberg, P., Tzur, G., Sabanay, I., Mahalu, D., Safran, S., Bershadsky, A., Addadi, L., and Geiger, B. 2001. Force and focal adhesion assembly: A close relationship studied using elastic micropatterned substrates. *Nat. Cell Biol.* **3**: 466–472.

Baneyx, G., Baugh, L., and Vogel, V. 2001. Coexisting conformations of fibronectin in cell culture imaged using fluorescence resonance energy transfer. *Proc. Natl. Acad. Sci. USA.* **98**: 14464–14468.

Baneyx, G., Baugh, L., and Vogel, V. 2002. Fibronectin extension and unfolding within cell matrix fibrils controlled by cytoskeletal tension. *Proc. Natl. Acad. Sci. USA.* **99**: 5139–5143.

Baneyx, G., and Vogel, V. 1999. Self-assembly of fibronectin into fibrillar networks underneath dipalmitoyl phosphatidylcholine monolayers: Role of lipid matrix and tensile forces. *Proc. Natl. Acad. Sci. USA*. **96**: 12518–12523.

Bao, G., and Suresh, S. 2003. Cell and molecular mechanics of biological materials. *Nat. Mater.* **2**: 715–725.

Barker, T. H., Baneyx, G., Cardo-Vila, M., Workman, G. A., Weaver, M., Menon, P. M., Dedhar, S., Rempel, S. A., Arap, W., Pasqualini, R., Vogel, V., and Sage, E. H. 2005. SPARC regulates extracellular matrix organization through its modulation of integrin-linked kinase activity. *J. Biol. Chem.* **280**: 36483–36493.

Barreiro, O., Yanez-Mo, M., Serrador, J. M., Montoya, M. C., Vicente-Manzanares, M., Tejedor, R., Furthmayr, H., and Sanchez-Madrid, F. 2002. Dynamic interaction of VCAM-1 and ICAM-1 with moesin and ezrin in a novel endothelial docking structure for adherent leukocytes. *J. Cell Biol.* **157**: 1233–1245.

Barry, E. L., and Mosher, D. F. 1988. Factor XIII cross-linking of fibronectin at cellular matrix assembly sites. *J. Biol. Chem.* **263**: 10464–10469.

Barthel, S. R., Johansson, M. W., Annis, D. S., and Mosher, D. F. 2006. Cleavage of human 7-domain VCAM-1 (CD106) by thrombin. *Thromb. Haemost.* **95**: 873–880.

Baugh, L., and Vogel, V. 2004. Structural changes of fibronectin adsorbed to model surfaces probed by fluorescence resonance energy transfer. *J. Biomed. Mater. Res.* **69A**: 525–534.

Bennett, V., and Gilligan, D. M. 1993. The spectrin-based membrane skeleton and micron-scale organization of the plasma membrane. *Annu. Rev. Cell Biol.* **9**: 27–66.

Bhasin, N., Carl, P., Harper, S., Feng, G., Lu, H., Speicher, D. W., and Discher, D. E. 2004. Chemistry on a single protein, vascular cell adhesion molecule-1, during forced unfolding. *J. Biol. Chem.* **279**: 45865–45874.

Booth, I. R., Edwards, M. D., Black, S., Schumann, U., and Miller, S. 2007. Mechanosensitive channels in bacteria: Signs of closure? *Nat. Rev. Microbiol.* **5**: 431–440.

Borradori, L., and Sonnenberg, A. 1999. Structure and function of hemidesmosomes: More than simple adhesion complexes. *J. Invest. Dermatol.* **112**: 411–418.

Borrego-Diaz, E., Kerff, F., Lee, S. H., Ferron, F., Li, Y., and Dominguez, R. 2006. Crystal structure of the actin-binding domain of alpha-actinin 1: Evaluating two competing actin-binding models. *J. Struct. Biol.* **155**: 230–238.

Bouton, A. H., Riggins, R. B., and Bruce-Staskal, P. J. 2001. Functions of the adapter protein Cas: Signal convergence and the determination of cellular responses. *Oncogene*. **20**: 6448–6458.

Briknarova, K., Akerman, M. E., Hoyt, D. W., Ruoslahti, E., and Ely, K. R. 2003. Anastellin, an FN3 fragment with fibronectin polymerization activity, resembles amyloid fibril precursors. *J. Mol. Biol.* **332**: 205–215.

Broderick, M. J., and Winder, S. J. 2005. Spectrin, alpha-actinin, and dystrophin. *Adv. Protein Chem.* **70**: 203–246.

Brown, R. A., Blunn, G. W., and Ejim, O. S. 1994. Preparation of orientated fibrous mats from fibronectin: Composition and stability. *Biomaterials*. **15**: 457–464.

Brzeska, H., Venyaminov, S., Grabarek, Z., and Drabikowski, W. 1983. Comparative studies on thermostability of calmodulin, skeletal muscle troponin C and their tryptic fragments. *FEBS Lett.* **153**: 169–173.

Bustamante, C., Chemla, Y. R., Forde, N. R., and Izhaky, D. 2004. Mechanical processes in biochemistry. *Annu. Rev. Biochem.* **73**: 705–748.

Calderwood, D. A., Fujioka, Y., de Pereda, J. M., Garcia-Alvarez, B., Nakamoto, T., Margolis, B., McGlade, C. J., Liddington, R. C., and Ginsberg, M. H. 2003. Integrin beta cytoplasmic domain interactions with phosphotyrosine-binding domains: A structural prototype for diversity in integrin signaling. *Proc. Natl. Acad. Sci. USA*. **100**: 2272–2277.

Calderwood, D. A., and Ginsberg, M. H. 2003. Talin forges the links between integrins and actin. *Nat. Cell Biol.* **5**: 694–697.

Calderwood, D. A., Shattil, S. J., and Ginsberg, M. H. 2000. Integrins and actin filaments: Reciprocal regulation of cell adhesion and signaling. *J. Biol. Chem.* **275**: 22607–22610.

Carl, P., Kwok, C. H., Manderson, G., Speicher, D. W., and Discher, D. E. 2001. Forced unfolding modulated by disulfide bonds in the Ig domains of a cell adhesion molecule. *Proc. Natl. Acad. Sci. USA*. **98**: 1565–1570.

Carrion-Vazquez, M., Oberhauser, A. F., Fisher, T. E., Marszalek, P. E., Li, H., and Fernandez, J. M. 2000. Mechanical design of proteins studied by single-molecule force spectroscopy and protein engineering. *Prog. Biophys. Mol. Biol.* **74**: 63–91.

Carrion-Vazquez, M., Oberhauser, A. F., Fowler, S. B., Marszalek, P. E., Broedel, S. E., Clarke, J., and Fernandez, J. M. 1999. Mechanical and chemical unfolding of a single protein: A comparison. *Proc. Natl. Acad. Sci. USA*. **96**: 3694–3649.

Chauvet, V., Tian, X., Husson, H., Grimm, D. H., Wang, T., Hiesberger, T., Igarashi, P., Bennett, A. M., Ibraghimov-Beskrovnaya, O., Somlo, S., and Caplan, M. J. 2004. Mechanical stimuli induce cleavage and nuclear translocation of the polycystin-1 C terminus. *J. Clin. Invest.* **114**: 1433–1443.

Chen, C. S., Tan, J., and Tien, J. 2004. Mechanotransduction at cell-matrix and cell-cell contacts. *Annu. Rev. Biomed. Eng.* **6**: 275–302.

Chen, H., Cohen, D. M., Choudhury, D. M., Kioka, N., and Craig, S. W. 2005. Spatial distribution and functional significance of activated vinculin in living cells. *J. Cell. Biol.* **169**: 459–470.

Chen, Y., Zardi, L., and Peters, D. M. 1997. High-resolution cryo-scanning electron microscopy study of the macromolecular structure of fibronectin fibrils. *Scanning*. **19**: 349–355.

Chernousov, M. A., Faerman, A. I., Frid, M. G., Printseva, O., and Koteliansky, V. E. 1987. Monoclonal antibody to fibronectin which inhibits extracellular matrix assembly. *FEBS Lett.* **217**: 124–128.

Chi, R. J., Olenych, S. G., Kim, K., and Keller, T. C., 3rd. 2005. Smooth muscle alpha-actinin interaction with smitin. *Int. J. Biochem. Cell Biol.* **37**: 1470–1482.

Choi, B., and Zocchi, G. 2006. Mimicking cAMP-dependent allosteric control of protein kinase A through mechanical tension. *J. Am. Chem. Soc.* **128**: 8541–8548.

Chrzanowska-Wodnicka, M., and Burridge, K. 1996. Rho-stimulated contractility drives the formation of stress fibers and focal adhesions. *J. Cell Biol.* **133**: 1403–1415.

Clausen-Schaumann, H., Seitz, M., Krautbauer, R., and Gaub, H. E. 2000. Force spectroscopy with single bio-molecules. *Curr. Opin. Chem. Biol.* **4**: 524–530.

Craig, D., Gao, M., Schulten, K., and Vogel, V. 2004. Tuning the mechanical stability of fibronectin type III modules through sequence variations. *Structure*. **12**: 21–30.

Craig, D., Krammer, A., Schulten, K., and Vogel, V. 2001. Comparison of the early stages of forced unfolding for fibronectin type III modules. *Proc. Natl. Acad. Sci. USA*. **98**: 5590–5595.

Defilippi, P., Di Stefano, P., and Cabodi, S. 2006. p130Cas: A versatile scaffold in signaling networks. *Trends Cell Biol.* **16**: 257–263.

Discher, D. E., Janmey, P., and Wang, Y. L. 2005. Tissue cells feel and respond to the stiffness of their substrate. *Science*. **310**: 1139–1143.

Djinovic-Carugo, K., Gautel, M., Ylanne, J., and Young, P. 2002. The spectrin repeat: A structural platform for cytoskeletal protein assemblies. *FEBS Lett.* **513**: 119–123.

Eaton, W. A., Munoz, V., Hagen, S. J., Jas, G. S., Lapidus, L. J., Henry, E. R., and Hofrichter, J. 2000. Fast kinetics and mechanisms in protein folding. *Annu. Rev. Biophys. Biomol. Struct.* **29**: 327–359.

Ejim, O. S., Blunn, G. W., and Brown, R. A. 1993. Production of artificial-orientated mats and strands from plasma fibronectin: A morphological study. *Biomaterials*. **14**: 743–748.

Engler, A. J., Sen, S., Sweeney, H. L., and Discher, D. E. 2006. Matrix elasticity directs stem cell lineage specification. *Cell*. **126**: 677–689.

Erickson, H. P. 1994. Reversible unfolding of fibronectin type III and immunoglobulin domains provides the structural basis for stretch and elasticity of titin and fibronectin. *Proc. Natl. Acad. Sci. USA*. **91**: 10114–10118.

Fisher, T. E., Marszalek, P. E., and Fernandez, J. M. 2000. Stretching single molecules into novel conformations using the atomic force microscope. *Nat. Struct. Biol.* **7**: 719–724.

Florian, J. A., Kosky, J. R., Ainslie, K., Pang, Z., Dull, R. O., and Tarbell, J. M. 2003. Heparan sulfate proteoglycan is a mechanosensor on endothelial cells. *Circ. Res.* **93**: 136–142.

Forero, M., Thomas, W. E., Sokurenko, E. V., and Vogel, V. 2006. Uncoiling mechanics of *E. coli* type I fimbriae are optimized for catch bonds. *PLoS Biol.* **4**: 1–8.

Forman, J. R., Qamar, S., Paci, E., Sandford, R. N., and Clarke, J. 2005. The remarkable mechanical strength of polycystin-1 supports a direct role in mechanotransduction. *J. Mol. Biol.* **349**: 861–871.

Fowler, S. B., Best, R. B., Toca Herrera, J. L., Rutherford, T. J., Steward, A., Paci, E., Karplus, M., and Clarke, J. 2002. Mechanical unfolding of a titin Ig domain: Structure of unfolding intermediate revealed by combining AFM, molecular dynamics simulations, NMR and protein engineering. *J. Mol. Biol.* **322**: 841–849.

Fredberg, J. J., and Kamm, R. D. 2006. Stress transmission in the lung: Pathways from organ to molecule. *Annu. Rev. Physiol.* **68**: 507–541.

Frishman, D., and Argos, P. 1995. Knowledge-based protein secondary structure assignment. *Proteins.* **23**: 566–579.

Förster, T. 1948. Intermolecular energy transference and fluorescence *Ann. Physik.* **2**: 55.

Galbraith, C. G., Yamada, K. M., and Sheetz, M. P. 2002. The relationship between force and focal complex development. *J. Cell. Biol.* **159**: 695–705.

Gao, M., Craig, D., Lequin, O., Campbell, I. D., Vogel, V., and Schulten, K. 2003. Structure and functional significance of mechanically unfolded fibronectin type III1 intermediates. *Proc. Natl. Acad. Sci. USA.* **100**: 14784–14789.

Gao, M., Craig, D., Vogel, V., and Schulten, K. 2002a. Identifying unfolding intermediates of FN-III(10) by steered molecular dynamics. *J. Mol. Biol.* **323**: 939–950.

Gao, M., Lu, H., and Schulten, K. 2002b. Unfolding of titin domains studied by molecular dynamics simulations. *J. Muscle Res. Cell Motil.* **23**: 513–521.

Gao, M., Sotomayor, M., Villa, E., Lee, E. H., and Schulten, K. 2006. Molecular mechanisms of cellular mechanics. *Phys. Chem. Chem. Phys.* **8**: 3692–3706.

Garcia-Alvarez, B., de Pereda, J. M., Calderwood, D. A., Ulmer, T. S., Critchley, D., Campbell, I. D., Ginsberg, M. H., and Liddington, R. C. 2003. Structural determinants of integrin recognition by talin. *Mol. Cell.* **11**: 49–58.

Geiger, B. 2006. A role for p130Cas in mechanotransduction. *Cell.* **127**: 879–881.

Geiger, B., Bershadsky, A., Pankov, R., and Yamada, K. M. 2001. Transmembrane crosstalk between the extracellular matrix–cytoskeleton crosstalk. *Nat. Rev. Mol. Cell Biol.* **2**: 793–805.

Geng, L., Burrow, C. R., Li, H. P., and Wilson, P. D. 2000. Modification of the composition of polycystin-1 multiprotein complexes by calcium and tyrosine phosphorylation. *Biochim. Biophys. Acta.* **1535**: 21–35.

Ghosh, I., Hamilton, A. D., and Regan, L. 2000. Antiparallel leucine zipper-directed protein reassembly: Application to the green fluorescent protein. *J. Am. Chem. Soc.* **122**: 5658–5659.

Giannone, G., Jiang, G., Sutton, D. H., Critchley, D. R., and Sheetz, M. P. 2003. Talin1 is critical for force-dependent reinforcement of initial integrin-cytoskeleton bonds but not tyrosine kinase activation. *J. Cell. Biol.* **163**: 409–419.

Giannone, G., and Sheetz, M. P. 2006. Substrate rigidity and force define form through tyrosine phosphatase and kinase pathways. *Trends Cell Biol.* **16**: 213–223.

Gingras, A. R., Ziegler, W. H., Frank, R., Barsukov, I. L., Roberts, G. C., Critchley, D. R., and Emsley, J. 2005. Mapping and consensus sequence identification for multiple vinculin binding sites within the talin rod. *J. Biol. Chem.* **280**: 37217–37224.

Gonzalez, A. M., Otey, C., Edlund, M., and Jones, J. C. 2001. Interactions of a hemidesmosome component and actinin family members. *J. Cell Sci.* **114**: 4197–4206.

Grandbois, M., Beyer, M., Rief, M., Clausen-Schaumann, H., and Gaub, H. E. 1999. How strong is a covalent bond? *Science.* **283**: 1727–1730.

Grater, F., Shen, J., Jiang, H., Gautel, M., and Grubmuller, H. 2005. Mechanically induced titin kinase activation studied by force-probe molecular dynamics simulations. *Biophys. J.* **88**: 790–804.

Halliday, N. L., and Tomasek, J. J. 1995. Mechanical properties of the extracellular matrix influence fibronectin fibril assembly in vitro. *Exp. Cell Res.* **217**: 109–117.

Harris, A. K., Wild, P., and Stopak, D. 1980. Silicone rubber substrata: A new wrinkle in the study of cell locomotion. *Science.* **208**: 177–179.

Heiska, L., Kantor, C., Parr, T., Critchley, D. R., Vilja, P., Gahmberg, C. G., and Carpen, O. 1996. Binding of the cytoplasmic domain of intercellular adhesion molecule-2 (ICAM-2) to alpha-actinin. *J. Biol. Chem.* **271**: 26214–26219.

Helfman, D. M., Levy, E. T., Berthier, C., Shtutman, M., Riveline, D., Grosheva, I., Lachish-Zalait, A., Elbaum, M., and Bershadsky, A. D. 1999. Caldesmon inhibits nonmuscle cell contractility and interferes with the formation of focal adhesions. *Mol. Biol. Cell.* **10**: 3097–3112.

Hocking, D. C., and Kowalski, K. 2002. A cryptic fragment from fibronectin's III1 module localizes to lipid rafts and stimulates cell growth and contractility. *J. Cell Biol.* **158**: 175–184.

House, S. D., and Lipowsky, H. H. 1988. In vivo determination of the force of leukocyte-endothelium adhesion in the mesenteric microvasculature of the cat. *Circ. Res.* **63**: 658–668.

Huan, Y., and van Adelsberg, J. 1999. Polycystin-1, the PKD1 gene product, is in a complex containing E-cadherin and the catenins. *J. Clin. Invest.* **104**: 1459–1468.

Humphrey, W., Dalke, A., and Schulten, K. 1996. VMD: Visual molecular dynamics. *J. Mol. Graph.* **14**: 33–38, 27–28.

Hynes, R. O. 1990. *Fibronectins*. Springer-Verlag, New York.

Hytönen, V., and Vogel, V. 2008. How force might activate talin's vinculin binding sites: SMD reveals a structural mechanism. *PLoS Comp. Biol.* 2009 **4**(2): e24.

Ibraghimov-Beskrovnaya, O., Bukanov, N. O., Donohue, L. C., Dackowski, W. R., Klinger, K. W., and Landes, G. M. 2000. Strong homophilic interactions of the Ig-like domains of polycystin-1, the protein product of an autosomal dominant polycystic kidney disease gene, PKD1. *Hum. Mol. Genet.* **9**: 1641–1649.

Ingber, D. E. 2003. Mechanobiology and diseases of mechanotransduction. *Ann. Med.* **35**: 564–577.

Ingham, K. C., Brew, S. A., Huff, S., and Litvinovich, S. V. 1997. Cryptic self-association sites in type III modules of fibronectin. *J. Biol. Chem.* **272**: 1718–1724.

Jeong, J., Kim, S. K., Ahn, J., Park, K., Jeong, E. J., Kim, M., and Chung, B. H. 2006. Monitoring of conformational change in maltose binding protein using split green fluorescent protein. *Biochem. Biophys. Res. Commun.* **339**: 647–651.

Jiang, G., Giannone, G., Critchley, D. R., Fukumoto, E., and Sheetz, M. P. 2003. Two-piconewton slip bond between fibronectin and the cytoskeleton depends on talin. *Nature.* **424**: 334–337.

Johnson, C. P., Tang, H. Y., Carag, C., Speicher, D. W., and Discher, D. E. 2007. Forced unfolding of proteins within cells. *Science.* **317**: 663–666.

Kale, L., Skeel, R., Bhandarkar, M., Brunner, R., Gursoy, A., Krawetz, N., Phillips, J., Shinozaki, A., Varadarajan, K., and Schulten, K. 1999. NAMD2: Greater scalability for parallel molecular dynamics. *J. Comp. Phys.* **151**: 283–312.

Kellermayer, M. S., Bustamante, C., and Granzier, H. L. 2003. Mechanics and structure of titin oligomers explored with atomic force microscopy. *Biochim. Biophys. Acta.* **1604**: 105–114.

Kellermayer, M. S., Smith, S. B., Granzier, H. L., and Bustamante, C. 1997. Folding-unfolding transitions in single titin molecules characterized with laser tweezers. *Science.* **276**: 1112–1116.

Kelly, D. F., and Taylor, K. A. 2005. Identification of the beta1-integrin binding site on alpha-actinin by cryoelectron microscopy. *J. Struct. Biol.* **149**: 290–302.

Kiema, T., Lad, Y., Jiang, P., Oxley, C. L., Baldassarre, M., Wegener, K. L., Campbell, I. D., Ylanne, J., and Calderwood, D. A. 2006. The molecular basis of filamin binding to integrins and competition with talin. *Mol. Cell.* **21**: 337–347.

Kim, E., Arnould, T., Sellin, L., Benzing, T., Comella, N., Kocher, O., Tsiokas, L., Sukhatme, V. P., and Walz, G. 1999. Interaction between RGS7 and polycystin. *Proc. Natl. Acad. Sci. USA*. **96**: 6371–6376.

Koide, T. 2005. Triple helical collagen-like peptides: Engineering and applications in matrix biology. *Connect Tissue Res.* **46**: 131–141.

Koonin, E. V., Wolf, Y. I., and Karev, G. P. 2002. The structure of the protein universe and genome evolution. *Nature*. **420**: 218–223.

Krammer, A., Craig, D., Thomas, W. E., Schulten, K., and Vogel, V. 2002. A structural model for force regulated integrin binding to fibronectin's RGD-synergy site. *Matrix Biol.* **21**: 139–147.

Krammer, A., Lu, H., Isralewitz, B., Schulten, K., and Vogel, V. 1999. Forced unfolding of the fibronectin type III module reveals a tensile molecular recognition switch. *Proc. Natl. Acad. Sci. USA*. **96**: 1351–1356.

Kung, C. 2005. A possible unifying principle for mechanosensation. *Nature*. **436**: 647–654.

Lai, C. S., Wolff, C. E., Novello, D., Griffone, L., Cuniberti, C., Molina, F., and Rocco, M. 1993. Solution structure of human plasma fibronectin under different solvent conditions. Fluorescence energy transfer, circular dichroism and light-scattering studies. *J. Mol. Biol.* **230**: 625–640.

Langenbach, K. J., and Sottile, J. 1999. Identification of protein-disulfide isomerase activity in fibronectin. *J. Biol. Chem.* **274**: 7032–7038.

Larsen, M., Artym, V. V., Green, J. A., and Yamada, K. M. 2006. The matrix reorganized: Extracellular matrix remodeling and integrin signaling. *Curr. Opin. Cell Biol.* **18**: 463–471.

Law, R., Carl, P., Harper, S., Dalhaimer, P., Speicher, D. W., and Discher, D. E. 2003. Cooperativity in forced unfolding of tandem spectrin repeats. *Biophys. J.* **84**: 533–544.

Lee, G., Abdi, K., Jiang, Y., Michaely, P., Bennett, V., and Marszalek, P. E. 2006. Nanospring behaviour of ankyrin repeats. *Nature*. **440**: 246–249.

Lee, S. E., Kamm, R. D., and Mofrad, M. R. 2007. Force-induced activation of talin and its possible role in focal adhesion mechanotransduction. *J. Biomech.* **40**: 2096–2106.

Li, H., Carrion-Vazquez, M., Oberhauser, A. F., Marszalek, P. E., and Fernandez, J. M. 2000. Point mutations alter the mechanical stability of immunoglobulin modules. *Nat. Struct. Biol.* **7**: 1117–1120.

Li, H., and Fernandez, J. M. 2003. Mechanical design of the first proximal Ig domain of human cardiac titin revealed by single molecule force spectroscopy. *J. Mol. Biol.* **334**: 75–86.

Li, H., Linke, W. A., Oberhauser, A. F., Carrion-Vazquez, M., Kerkvliet, J. G., Lu, H., Marszalek, P. E., and Fernandez, J. M. 2002. Reverse engineering of the giant muscle protein titin. *Nature*. **418**: 998–1002.

Li, H., Oberhauser, A. F., Redick, S. D., Carrion-Vazquez, M., Erickson, H. P., and Fernandez, J. M. 2001. Multiple conformations of PEVK proteins detected by single-molecule techniques. *Proc. Natl. Acad. Sci. USA*. **98**: 10682–10686.

Li, L., Huang, H. H., Badilla, C. L., and Fernandez, J. M. 2005. Mechanical unfolding intermediates observed by single-molecule force spectroscopy in a fibronectin type III module. *J. Mol. Biol.* **345**: 817–826.

Little, W. C., Smith, M. L., Ebneter, U., and Vogel, V. 2008. Assay to mechanically tune and optically probe the conformation of fibrillar fibronectin from fully relaxed to breakage. *Matrix Biol.* **27**(5):451–461.

Liu, J., Taylor, D. W., and Taylor, K. A. 2004. A 3-D reconstruction of smooth muscle alpha-actinin by CryoEm reveals two different conformations at the actin-binding region. *J. Mol. Biol.* **338**: 115–125.

Lu, H., Isralewitz, B., Krammer, A., Vogel, V., and Schulten, K. 1998. Unfolding of titin immunoglobulin domains by steered molecular dynamics simulation. *Biophys. J.* **75**: 662–671.

Lu, H., and Schulten, K. 1999. Steered molecular dynamics simulations of force-induced protein domain unfolding. *Proteins*. **35**: 453–463.

MacDonald, R. I., and Pozharski, E. V. 2001. Free energies of urea and of thermal unfolding show that two tandem repeats of spectrin are thermodynamically more stable than a single repeat. *Biochemistry*. **40**: 3974–3984.

Malhas, A. N., Abuknesha, R. A., and Price, R. G. 2002. Interaction of the leucine-rich repeats of polycystin-1 with extracellular matrix proteins: Possible role in cell proliferation. *J. Am. Soc. Nephrol.* **13**: 19–26.

Mao, Y., and Schwarzbauer, J. E. 2005. Fibronectin fibrillogenesis, a cell-mediated matrix assembly process. *Matrix Biol.* **24**: 389–399.

Marszalek, P. E., Lu, H., Li, H., Carrion-Vazquez, M., Oberhauser, A. F., Schulten, K., and Fernandez, J. M. 1999. Mechanical unfolding intermediates in titin modules. *Nature*. **402**: 100–103.

Masaki, T., Endo, M., and Ebashi, S. 1967. Localization of 6S component of a alpha-actinin at Z-band. *J. Biochem. (Tokyo)*. **62**: 630–632.

McBeath, R., Pirone, D. M., Nelson, C. M., Bhadriraju, K., and Chen, C. S. 2004. Cell shape, cytoskeletal tension, and RhoA regulate stem cell lineage commitment. *Dev. Cell.* **6**: 483–495.

Minajeva, A., Kulke, M., Fernandez, J. M., and Linke, W. A. 2001. Unfolding of titin domains explains the viscoelastic behavior of skeletal myofibrils. *Biophys. J.* **80**: 1442–1451.

Mochizuki, S., Vink, H., Hiramatsu, O., Kajita, T., Shigeto, F., Spaan, J. A., and Kajiya, F. 2003. Role of hyaluronic acid glycosaminoglycans in shear-induced endothelium-derived nitric oxide release. *Am. J. Physiol. Heart Circ. Physiol.* **285**: H722–726.

Morla, A., Zhang, Z., and Ruoslahti, E. 1994. Superfibronectin is a functionally distinct form of fibronectin. *Nature*. **367**: 193–196.

Mosavi, L. K., Cammett, T. J., Desrosiers, D. C., and Peng, Z. Y. 2004. The ankyrin repeat as molecular architecture for protein recognition. *Protein Sci.* **13**: 1435–1448.

Mould, A. P., and Humphries, M. J. 1991. Identification of a novel recognition sequence for the integrin alpha 4 beta 1 in the COOH-terminal heparin-binding domain of fibronectin. *Embo. J.* **10**: 4089–4095.

Mukai, H., Toshimori, M., Shibata, H., Takanaga, H., Kitagawa, M., Miyahara, M., Shimakawa, M., and Ono, Y. 1997. Interaction of PKN with alpha-actinin. *J. Biol. Chem.* **272**: 4740–4746.

Nauli, S. M., Alenghat, F. J., Luo, Y., Williams, E., Vassilev, P., Li, X., Elia, A. E., Lu, W., Brown, E. M., Quinn, S. J., Ingber, D. E., and Zhou, J. 2003. Polycystins 1 and 2 mediate mechanosensation in the primary cilium of kidney cells. *Nat. Genet.* **33**: 129–137.

Neuert, G., Albrecht, C. H., and Gaub, H. E. 2007. Predicting the rupture probabilities of molecular bonds in series. *Biophys. J.* **93**: 1215–1223.

Ng, S. P., Billings, K. S., Ohashi, T., Allen, M. D., Best, R. B., Randles, L. G., Erickson, H. P., and Clarke, J. 2007. Designing an extracellular matrix protein with enhanced mechanical stability. *Proc. Natl. Acad. Sci. USA*.

Ng, S. P., Rounsevell, R. W., Steward, A., Geierhaas, C. D., Williams, P. M., Paci, E., and Clarke, J. 2005. Mechanical unfolding of TNfn3: The unfolding pathway of a fnIII domain probed by protein engineering, AFM and MD simulation. *J. Mol. Biol.* **350**: 776–789.

Nieset, J. E., Redfield, A. R., Jin, F., Knudsen, K. A., Johnson, K. R., and Wheelock, M. J. 1997. Characterization of the interactions of alpha-catenin with alpha-actinin and beta-catenin/plakoglobin. *J. Cell Sci.* **110**: 1013–1022.

Nilsson, L. M., Yakovenko, O., Tchesnokova, V., Thomas, W. E., Schembri, M. A., Vogel, V., Klemm, P., and Sokurenko, E. V. 2007. The cysteine bond in the Escherichia coli FimH adhesin is critical for adhesion under flow conditions. *Mol. Microbiol.* **65**: 1158–1169.

O'Neill, G. M., Fashena, S. J., and Golemis, E. A. 2000. Integrin signalling: A new Cas(t) of characters enters the stage. *Trends Cell Biol.* **10**: 111–119.

Oberhauser, A. F., Badilla-Fernandez, C., Carrion-Vazquez, M., and Fernandez, J. M. 2002. The mechanical hierarchies of fibronectin observed with single-molecule AFM. *J. Mol. Biol.* **319**: 433–447.

Oberhauser, A. F., Hansma, P. K., Carrion-Vazquez, M., and Fernandez, J. M. 2001. Stepwise unfolding of titin under force-clamp atomic force microscopy. *Proc. Natl. Acad. Sci. USA*. **98**: 468–472.

Oberhauser, A. F., Marszalek, P. E., Carrion-Vazquez, M., and Fernandez, J. M. 1999. Single protein misfolding events captured by atomic force microscopy. *Nat. Struct. Biol.* **6**: 1025–1028.

Oberhauser, A. F., Marszalek, P. E., Erickson, H. P., and Fernandez, J. M. 1998. The molecular elasticity of the extracellular matrix protein tenascin. *Nature*. **393**: 181–185.

Ohashi, T., Kiehart, D. P., and Erickson, H. P. 1999. Dynamics and elasticity of the fibronectin matrix in living cell culture visualized by fibronectin-green fluorescent protein. *Proc. Natl. Acad. Sci. USA*. **96**: 2153–2158.

Ohashi, T., Kiehart, D. P., and Erickson, H. P. 2002. Dual labeling of the fibronectin matrix and actin cytoskeleton with green fluorescent protein variants. *J. Cell Sci.* **115**: 1221–1229.

Orr, A. W., Helmke, B. P., Blackman, B. R., and Schwartz, M. A. 2006. Mechanisms of mechanotransduction. *Dev. Cell.* **10**: 11–20.

Ortiz, V., Nielsen, S. O., Klein, M. L., and Discher, D. E. 2005. Unfolding a linker between helical repeats. *J. Mol. Biol.* **349**: 638–647.

Otey, C. A., and Carpen, O. 2004. Alpha-actinin revisited: A fresh look at an old player. *Cell Motil. Cytoskeleton*. **58**: 104–111.

Otey, C. A., Pavalko, F. M., and Burridge, K. 1990. An interaction between alpha-actinin and the beta 1 integrin subunit in vitro. *J. Cell Biol.* **111**: 721–729.

Paci, E., and Karplus, M. 1999. Forced unfolding of fibronectin type 3 modules: An analysis by biased molecular dynamics simulations. *J. Mol. Biol.* **288**: 441–459.

Pankov, R., and Yamada, K. M. 2002. Fibronectin at a glance. *J. Cell. Sci.* **115**: 3861–3863.

Papagrigoriou, E., Gingras, A. R., Barsukov, I. L., Bate, N., Fillingham, I. J., Patel, B., Frank, R., Ziegler, W. H., Roberts, G. C., Critchley, D. R., and Emsley, J. 2004. Activation of a vinculin-binding site in the talin rod involves rearrangement of a five-helix bundle. *Embo. J.* **23**: 2942–2951.

Pasternak, C., Wong, S., and Elson, E. L. 1995. Mechanical function of dystrophin in muscle cells. *J. Cell Biol.* **128**: 355–361.

Paszek, M. J., Zahir, N., Johnson, K. R., Lakins, J. N., Rozenberg, G. I., Gefen, A., Reinhart-King, C. A., Margulies, S. S., Dembo, M., Boettiger, D., Hammer, D. A., and Weaver, V. M. 2005. Tensional homeostasis and the malignant phenotype. *Cancer Cell.* **8**: 241–254.

Perez-Jimenez, R., Garcia-Manyes, S., Ainavarapu, S. R., and Fernandez, J. M. 2006. Mechanical unfolding pathways of the enhanced yellow fluorescent protein revealed by single molecule force spectroscopy. *J. Biol. Chem.* **281**: 40010–40014.

Peters, D. M. P., Chen, Y., Zardi, L., and Brummel, S. 1998. Conformation of fibronectin fibrils varies: Discrete globular domains of type III repeats detected. *Microscopy. Microanal.* **4**: 385–396.

Pierschbacher, M. D., and Ruoslahti, E. 1984. Cell attachment activity of fibronectin can be duplicated by small synthetic fragments of the molecule. *Nature*. **309**: 30–33.

Pytela, R., Pierschbacher, M. D., Ginsberg, M. H., Plow, E. F., and Ruoslahti, E. 1986. Platelet membrane glycoprotein IIb/IIIa: Member of a family of Arg-Gly-Asp–specific adhesion receptors. *Science*. **231**: 1559–1562.

Pytela, R., Pierschbacher, M. D., and Ruoslahti, E. 1985. Identification and isolation of a 140 kd cell surface glycoprotein with properties expected of a fibronectin receptor. *Cell*. **40**: 191–198.

Qian, F., Wei, W., Germino, G., and Oberhauser, A. 2005. The nanomechanics of polycystin-1 extracellular region. *J. Biol. Chem.* **280**: 40723–40730.

Rajfur, Z., Roy, P., Otey, C., Romer, L., and Jacobson, K. 2002. Dissecting the link between stress fibres and focal adhesions by CALI with EGFP fusion proteins. *Nat. Cell Biol.* **4**: 286–293.

Randles, L. G., Rounsevell, R. W., and Clarke, J. 2007. Spectrin domains lose cooperativity in forced unfolding. *Biophys. J.* **92**: 571–577.

Redick, S. D., Settles, D. L., Briscoe, G., and Erickson, H. P. 2000. Defining fibronectin's cell adhesion synergy site by site-directed mutagenesis. *J. Cell Biol.* **149**: 521–527.

Rief, M., Clausen-Schaumann, H., and Gaub, H. E. 1999a. Sequence-dependent mechanics of single DNA molecules. *Nat. Struct. Biol.* **6**: 346–349.

Rief, M., Gautel, M., and Gaub, H. E. 2000. Unfolding forces of titin and fibronectin domains directly measured by AFM. *Adv. Exp. Med. Biol.* **481**: 129–136; discussion 137–141.

Rief, M., Gautel, M., Oesterhelt, F., Fernandez, J. M., and Gaub, H. E. 1997. Reversible unfolding of individual titin immunoglobulin domains by AFM. *Science.* **276**: 1109–1112.

Rief, M., Gautel, M., Schemmel, A., and Gaub, H. E. 1998. The mechanical stability of immunoglobulin and fibronectin III domains in the muscle protein titin measured by atomic force microscopy. *Biophys. J.* **75**: 3008–3014.

Rief, M., Pascual, J., Saraste, M., and Gaub, H. E. 1999b. Single molecule force spectroscopy of spectrin repeats: Low unfolding forces in helix bundles. *J. Mol. Biol.* **286**: 553–561.

Riveline, D., Zamir, E., Balaban, N. Q., Schwarz, U. S., Ishizaki, T., Narumiya, S., Kam, Z., Geiger, B., and Bershadsky, A. D. 2001. Focal contacts as mechanosensors: Externally applied local mechanical force induces growth of focal contacts by an mDia1-dependent and ROCK-independent mechanism. *J. Cell Biol.* **153**: 1175–1186.

Rounsevell, R. W., and Clarke, J. 2004. FnIII domains: Predicting mechanical stability. *Structure.* **12**: 4–5.

Sahaf, B., Heydari, K., Herzenberg, L. A., and Herzenberg, L. A. 2003. Lymphocyte surface thiol levels. *Proc. Natl. Acad. Sci. USA.* **100**: 4001–4005.

Sandford, R., Sgotto, B., Aparicio, S., Brenner, S., Vaudin, M., Wilson, R. K., Chissoe, S., Pepin, K., Bateman, A., Chothia, C., Hughes, J., and Harris, P. 1997. Comparative analysis of the polycystic kidney disease 1 (PKD1) gene reveals an integral membrane glycoprotein with multiple evolutionary conserved domains. *Hum. Mol. Genet.* **6**: 1483–1489.

Sawada, Y., and Sheetz, M. P. 2002. Force transduction by Triton cytoskeletons. *J. Cell Biol.* **156**: 609–615.

Sawada, Y., Tamada, M., Dubin-Thaler, B. J., Cherniavskaya, O., Sakai, R., Tanaka, S., and Sheetz, M. P. 2006. Force sensing by mechanical extension of the Src family kinase substrate p130Cas. *Cell.* **127**: 1015–1026.

Schnepel, J., and Tschesche, H. 2000. The proteolytic activity of the recombinant cryptic human fibronectin type IV collagenase from E. coli expression. *J. Protein Chem.* **19**: 685–692.

Schnepel, J., Unger, J., and Tschesche, H. 2001. Recombinant cryptic human fibronectinase cleaves actin and myosin: Substrate specificity and possible role in muscular dystrophy. *Biol. Chem.* **382**: 1707–1714.

Sechler, J. L., Rao, H., Cumiskey, A. M., Vega-Colon, I., Smith, M. S., Murata, T., and Schwarzbauer, J. E. 2001. A novel fibronectin binding site required for fibronectin fibril growth during matrix assembly. *J. Cell Biol.* **154**: 1081–1088.

Sharma, A., Askari, J. A., Humphries, M. J., Jones, E. Y., and Stuart, D. I. 1999. Crystal structure of a heparin- and integrin-binding segment of human fibronectin. *Embo. J.* **18**: 1468–1479.

Slade, R. E., and Pauling, L. 1948. The nature of forces between large molecules of biological interest. *Royal Institution of Great Britain.*

Smith, M. L., Gourdon, D., Little, W. C., Kubow, K. E., Eguiluz, R. A., Luna-Morris, S., and Vogel, V. 2007. Force-induced unfolding of fibronectin in the extracellular matrix of living cells. *PLoS Biol.* **5**: e268.

Smith, M. L., Long, D. S., Damiano, E. R., and Ley, K. 2003. Near-wall micro-PIV reveals a hydrodynamically relevant endothelial surface layer in venules in vivo. *Biophys J.* **85**: 637–645.

Soteriou, A., Clarke, A., Martin, S., and Trinick, J. 1993. Titin folding energy and elasticity. *Proc. Biol. Sci.* **254**: 83–86.

Sotomayor, M., Corey, D. P., and Schulten, K. 2005. In search of the hair-cell gating spring elastic properties of ankyrin and cadherin repeats. *Structure.* **13**: 669–682.

Sotomayor, M., and Schulten, K. 2007. Single-molecule experiments in vitro and in silico. *Science.* **316**: 1144–1148.

Stone, J. E., Phillips, J. C., Freddolino, P. L., Hardy, D. J., Trabuco, L. G., and Schulten, K. 2007. Accelerating molecular modeling applications with graphics processors. *J. Comput. Chem.* **28**: 2618–2640.

Stossel, T. P., Condeelis, J., Cooley, L., Hartwig, J. H., Noegel, A., Schleicher, M., and Shapiro, S. S. 2001. Filamins as integrators of cell mechanics and signalling. *Nat. Rev. Mol. Cell Biol.* **2**: 138–145.

Stossel, T. P., and Hartwig, J. H. 1975. Interactions between actin, myosin, and an actin-binding protein from rabbit alveolar macrophages. Alveolar macrophage myosin Mg-2+-adenosine triphosphatase requires a cofactor for activation by actin. *J. Biol. Chem.* **250**: 5706–5712.

Strunz, T., Oroszlan, K., Schafer, R., and Guntherodt, H. J. 1999. Dynamic force spectroscopy of single DNA molecules. *Proc. Natl. Acad. Sci. USA.* **96**: 11277–11282.

Sutton, R. B., Ernst, J. A., and Brunger, A. T. 1999. Crystal structure of the cytosolic C2A-C2B domains of synaptotagmin III. Implications for Ca(+2)-independent snare complex interaction. *J. Cell Biol.* **147**: 589–598.

Tamada, M., Sheetz, M. P., and Sawada, Y. 2004. Activation of a signaling cascade by cytoskeleton stretch. *Dev. Cell.* **7**: 709–718.

Tan, J. L., Tien, J., Pirone, D. M., Gray, D. S., Bhadriraju, K., and Chen, C. S. 2003. Cells lying on a bed of microneedles: An approach to isolate mechanical force. *Proc. Natl. Acad. Sci. USA.* **100**: 1484–1489.

Tanase, M., Biais, N., and Sheetz, M. 2007. Magnetic tweezers in cell biology. *Methods Cell Biol.* **83**: 473–93.

Tapley, P., Horwitz, A., Buck, C., Duggan, K., and Rohrschneider, L. 1989. Integrins isolated from Rous sarcoma virus-transformed chicken embryo fibroblasts. *Oncogene.* **4**: 325–333.

Thomas, W. E., Trintchina, E., Forero, M., Vogel, V., and Sokurenko, E. V. 2002. Bacterial adhesion to target cells enhanced by shear force. *Cell.* **109**: 913–923.

Tsalkova, T. N., and Privalov, P. L. 1985. Thermodynamic study of domain organization in troponin C and calmodulin. *J. Mol. Biol.* **181**: 533–544.

Tskhovrebova, L., and Trinick, J. 2003. Titin: Properties and family relationships. *Nat. Rev. Mol. Cell. Biol.* **4**: 679–689.

Tskhovrebova, L., and Trinick, J. 2004. Properties of titin immunoglobulin and fibronectin-3 domains. *J. Biol. Chem.* **279**: 46351–46354.

Tskhovrebova, L., Trinick, J., Sleep, J. A., and Simmons, R. M. 1997. Elasticity and unfolding of single molecules of the giant muscle protein titin. *Nature.* **387**: 308–312.

Vallenius, T., Luukko, K., and Makela, T. P. 2000. CLP-36 PDZ-LIM protein associates with nonmuscle alpha-actinin-1 and alpha-actinin-4. *J. Biol. Chem.* **275**: 11100–11105.

Watanabe, K., Muhle-Goll, C., Kellermayer, M. S., Labeit, S., and Granzier, H. 2002. Different molecular mechanics displayed by titin's constitutively and differentially expressed tandem Ig segments. *J. Struct. Biol.* **137**: 248–258.

Weston, B. S., Bagneris, C., Price, R. G., and Stirling, J. L. 2001. The polycystin-1 C-type lectin domain binds carbohydrate in a calcium-dependent manner, and interacts with extracellular matrix proteins in vitro. *Biochim. Biophys. Acta.* **1536**: 161–176.

Wiita, A. P., Ainavarapu, S. R., Huang, H. H., and Fernandez, J. M. 2006. Force-dependent chemical kinetics of disulfide bond reduction observed with single-molecule techniques. *Proc. Natl. Acad. Sci. USA.* **103**: 7222–7227.

Williams, P. M., Fowler, S. B., Best, R. B., Toca-Herrera, J. L., Scott, K. A., Steward, A., and Clarke, J. 2003. Hidden complexity in the mechanical properties of titin. *Nature*. **422**: 446–449.

Vink, H., and Duling, B. R. 1996. Identification of distinct luminal domains for macromolecules, erythrocytes, and leukocytes within mammalian capillaries. *Circ. Res.* **79**: 581–589.

Vogel, V. 2006. Mechanotransduction involving multimodular proteins: Converting force into biochemical signals. *Annu. Rev. Biophys. Biomol. Struct.* **35**: 459–488.

Vogel, V., and Sheetz, M. 2006. Local force and geometry sensing regulate cell functions. *Nat. Rev. Mol. Cell Biol.* **7**: 265–275.

Wolff, C., and Lai, C. S. 1988. Evidence that the two amino termini of plasma fibronectin are in close proximity: A fluorescence energy transfer study. *Biochemistry*. **27**: 3483–3487.

Wolff, C., and Lai, C. S. 1989. Fluorescence energy transfer detects changes in fibronectin structure upon surface binding. *Arch Biochem Biophys*. **268**: 536–545.

Wolff, C. E., and Lai, C. S. 1990. Inter-sulfhydryl distances in plasma fibronectin determined by fluorescence energy transfer: Effect of environmental factors. *Biochemistry*. **29**: 3354–3361.

Vollrath, M. A., Kwan, K. Y., and Corey, D. P. 2007. The micromachinery of mechanotransduction in hair cells. *Annu. Rev. Neurosci.* **30**: 339–365.

von Wichert, G., Haimovich, B., Feng, G. S., and Sheetz, M. P. 2003. Force-dependent integrin-cytoskeleton linkage formation requires downregulation of focal complex dynamics by Shp2. *Embo J.* **22**: 5023–5035.

Vonderheide, R. H., and Springer, T. A. 1992. Lymphocyte adhesion through very late antigen 4: Evidence for a novel binding site in the alternatively spliced domain of vascular cell adhesion molecule 1 and an additional alpha 4 integrin counter-receptor on stimulated endothelium. *J. Exp. Med.* **175**: 1433–1442.

Wright, C. F., Teichmann, S. A., Clarke, J., and Dobson, C. M. 2005. The importance of sequence diversity in the aggregation and evolution of proteins. *Nature*. **438**: 878–881.

Wu, C., Keivens, V. M., O'Toole, T. E., McDonald, J. A., and Ginsberg, M. H. 1995. Integrin activation and cytoskeletal interaction are essential for the assembly of a fibronectin matrix. *Cell.* **83**: 715–724.

Ylanne, J., Scheffzek, K., Young, P., and Saraste, M. 2001. Crystal structure of the alpha-actinin rod reveals an extensive torsional twist. *Structure*. **9**: 597–604.

Young, P., Ferguson, C., Banuelos, S., and Gautel, M. 1998. Molecular structure of the sarcomeric Z-disk: Two types of titin interactions lead to an asymmetrical sorting of alpha-actinin. *Embo J.* **17**: 1614–1624.

Yusuf-Makagiansar, H., Anderson, M. E., Yakovleva, T. V., Murray, J. S., and Siahaan, T. J. 2002. Inhibition of LFA-1/ICAM-1 and VLA-4/VCAM-1 as a therapeutic approach to inflammation and autoimmune diseases. *Med. Res. Rev.* **22**: 146–167.

Zaidel-Bar, R., Cohen, M., Addadi, L., and Geiger, B. 2004. Hierarchical assembly of cell-matrix adhesion complexes. *Biochem Soc Trans.* **32**: 416–420.

Zaidel-Bar, R., Itzkovitz, S., Ma'ayan, A., Iyengar, R., and Geiger, B. 2007. Functional atlas of the integrin adhesome. *Nat. Cell Biol.* **9**: 858–867.

Zamir, E., Katz, M., Posen, Y., Erez, N., Yamada, K. M., Katz, B. Z., Lin, S., Lin, D. C., Bershadsky, A., Kam, Z., and Geiger, B. 2000. Dynamics and segregation of cell-matrix adhesions in cultured fibroblasts. *Nat. Cell Biol.* **2**: 191–196.

Zhong, C., Chrzanowska-Wodnicka, M., Brown, J., Shaub, A., Belkin, A. M., and Burridge, K. 1998. Rho-mediated contractility exposes a cryptic site in fibronectin and induces fibronectin matrix assembly. *J. Cell Biol.* **141**: 539–551.

14

Mechanotransduction through Local Autocrine Signaling

Nikola Kojic and Daniel J. Tschumperlin

14.1 Introduction

While mechanotransduction within the cytoskeleton, cell membrane, and cell-matrix interface are well known, multicellular tissue organization provides an additional opportunity for mechanotransduction in the extracellular space through local autocrine signaling. In many tissues and organs, cells are organized into relatively dense structures separated by narrow interstitial spaces. The interstitial fluid in these spaces facilitates nutrient and waste transport, and provides a conduit for the exchange of autocrine and paracrine signals. When mechanical loads are applied to tissues, the resulting pressure gradients can locally redistribute interstitial fluid, altering the geometry of the interstitial spaces separating cells. These changes in interstitial geometry, in the presence of a constitutively active autocrine signaling environment, provide a means for coupling mechanical loading to changes in local ligand concentration and receptor activation. In this chapter we discuss these behaviors using the example of airway epithelial cells, which exhibit constitutive autocrine signaling localized to an interstitial space that deforms under physiological levels of loading. We highlight evidence demonstrating that changes in autocrine ligand concentration arising from the deformation of the local interstitial geometry are sufficient to drive alterations in receptor activation. We then use a computational modeling approach to explore the unique characteristics of this mechanical signaling system, and discuss its broader implications.

14.2 Mechanical Stress–Induced Signaling in Bronchial Epithelium

The mechanical environment shapes the development and function of the lung's airways from the first stages of airway morphogenesis *in utero* (Tschumperlin and Drazen, 2006). In the adult lung the mechanical environment is defined by a unique balance of surface, tissue, and muscle forces, but the underlying load-bearing structure is similar to many organs of the body: an epithelium-lined structure surrounded by extracellular matrix, vasculature, and mesenchymal cells (e.g., smooth muscle). In diseases such as asthma, activation of the airway smooth muscle wrapped around

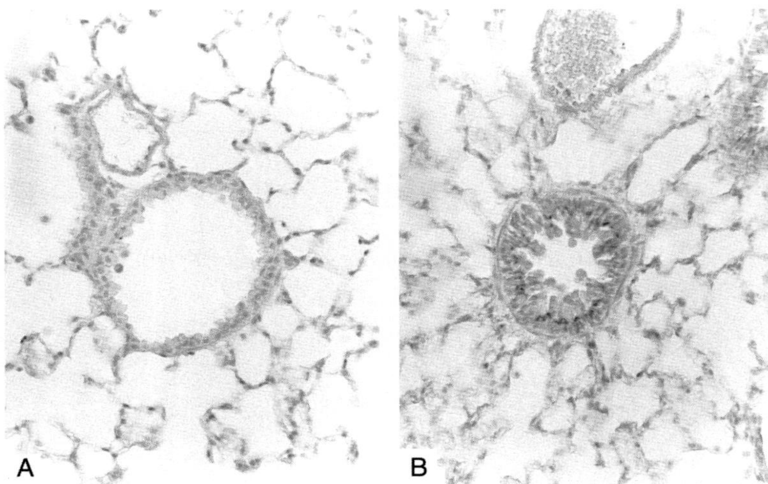

Figure 14.1. Sections of mouse lungs showing similar generation airways fixed by tracheal perfusion of paraformaldehyde after PBS (A) or methacholine (B) were perfused through the airways. Methacholine activates airway smooth muscle shortening and airway narrowing, changing the stress state within the airway wall tissue.

airways abruptly narrows the airway lumen (Figure 14.1; Plate 16) and profoundly changes the stress state within the tissue of the airway wall (Wiggs et al., 1997). Because mechanical stress–induced remodeling is well known in many tissues, and tissue remodeling is frequently observed in asthmatic airways (Bousquet et al., 2000), two related question arise: Are the cells of the airway wall responsive to the mechanical stresses that accompany smooth muscle shortening (bronchoconstriction), and if so, how do the cells sense changes in their mechanical environment?

Because the airways of the lung are embedded deep within a delicate parenchyma, *in situ* investigations of mechanotransduction are inherently challenging. Thus, *in vitro* methods have been developed to study the cells of the airways in specialized systems that mimic the *in situ* environment (Tschumperlin and Drazen, 2006). In one such system primary bronchial epithelial cells are grown on microporous substrates at an air–liquid interface. In the presence of appropriate soluble factors, the cells take on a pseudostratified structure with a mix of mucous-secreting, ciliated, and basal cells similar to that observed *in situ* (Perez-Vilar et al., 2003). In this preparation, traditional mechanical manipulations such as stretch are impossible to impose due to the inelasticity of the culture surface. However, the cells in this culture system can be subjected to an apical to basal pressure gradient (Figure 14.2) that produces a compressive stress similar in magnitude to that generated by maximal airway smooth muscle activation *in vivo* (Ressler et al., 2000; Swartz et al., 2001; Wiggs et al., 1997). Human and rat airway epithelial cells grown in this system respond in a graded fashion to increasing physiological levels of loading (Ressler et al., 2000; Swartz et al., 2001; Tschumperlin et al., 2002). Sustained exposure to stress leads to intracellular signaling through the mitogen-activated protein (MAP) kinase cascade (Tschumperlin et al., 2002), increased expression of pro-fibrotic genes

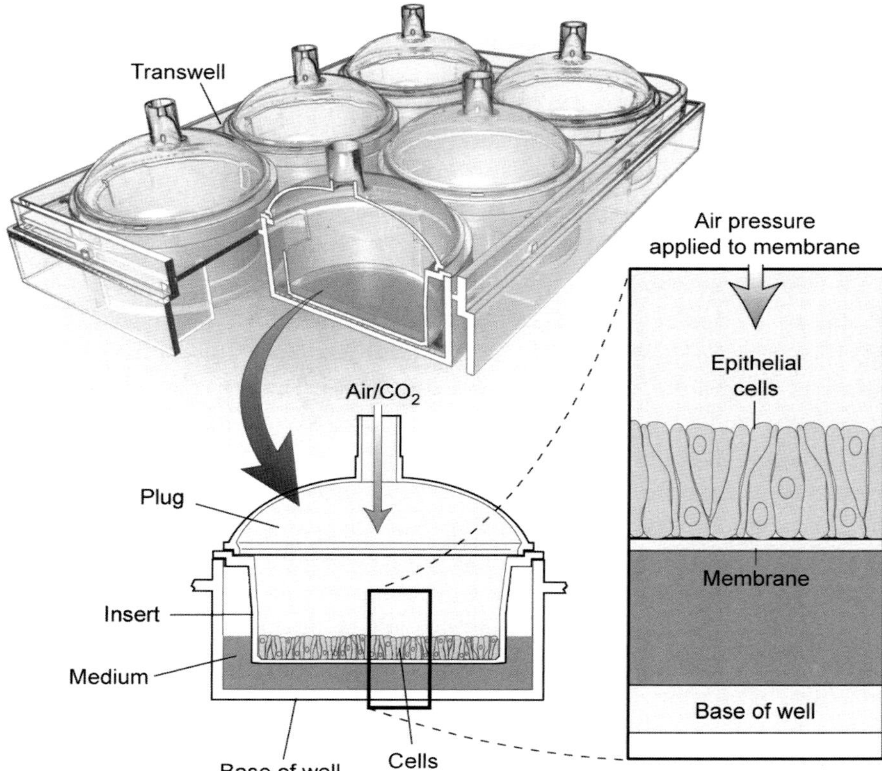

Figure 14.2. Schematic of the cell culture setup for application of apical to basal transcellular pressure. Cells are grown on a microporous membrane at an air–liquid interface. To apply stress, the apical compartment is sealed with a plug, and the apical pressure is increased by application of 5% CO_2 via an inlet port. The basal media remains at atmospheric pressure.

and gene products (Tschumperlin et al., 2003), and communication of a pro-fibrotic signal to co-cultured human lung fibroblasts (Swartz et al., 2001). Taken together these findings implicate the transduction of mechanical stress by airway epithelial cells in the deposition of an extracellular matrix around the airways, a pathological finding associated with the gradual decline in lung function in chronic asthma (Fish and Peters, 1999). These studies highlight the potential physiological role for mechanotransduction in the airway wall but leave open the question of how mechanical forces are coupled to molecular signaling.

14.3 Measuring Changes in Cellular and Extracellular Geometry

While the *in vitro* system for applying a transcellular pressure gradient is simple to implement, the cellular deformations induced by applying such a load across a confluent epithelium grown on a microporous substrate are not trivial to predict, or visualize. What is needed is a method to noninvasively measure cellular-level deformations under loading. One such method ideally suited to this purpose is two-photon microscopy, which utilizes simultaneous excitation of a fluorophore by two photons within a thin imaging plane, allowing reconstruction of three-dimensional fluorescence images with minimal photobleaching (So et al., 2000). Application of this new approach builds on the pioneering work of Ken Spring and colleagues at the NIH, who developed the first methods to visualize the interstitial spaces separating living epithelial cells (Spring and Hope, 1978), and first described the changes in geometry of these spaces under pressure gradients (Spring and Hope, 1978; Timbs and Spring, 1996).

To visualize cell bodies in the two-photon imaging system, cells can be loaded with any of a number of fluorescent, cell-permeant cytoplasmic dyes (e.g., Celltracker, calcein AM). To visualize the extracellular space, fluorophore-conjugated dextrans or other molecular tracers can be used, as long as they are prevented from entering the cells. Two-photon images obtained from bronchial epithelial cells grown to a pseudostratified morphology similar to that observed *in situ* are shown in Figure 14.3(A) (Plate 17). Typical of cuboidal or columnar epithelia, the cells are joined by tight junctions at their apical surface and are separated along their lateral surface by an interstitial space commonly referred to as the lateral intercellular space (LIS). In the cultures shown here, the LIS comprises roughly 15% of the total tissue volume under resting conditions (Tschumperlin et al., 2004).

Repeated imaging of the same region of cells before and during application of apical to basal transcellular pressure reveals that mechanical loading decreases the thickness of the cell layers ($-10.6\pm1.9\%$), but does not significantly change the total cellular volume ($-1.6\pm2.3\%$) (Tschumperlin et al., 2004). The implication of these results is that the cells themselves change shape but not volume, while the overall tissue volume shrinks via fluid leaving the LIS (Figure 14.3(B); Plate 17). This shrinkage of the LIS is strikingly apparent in single plane sections of the same region obtained before loading and $\sim$ 20 min after the onset of tonic loading (Figure 14.3(A); Plate 17). The finding that the interstitial spaces separating cells are susceptible to large percentage volume changes under physiological loading provided a new and unexpected candidate locus for mechanical signaling in deforming tissues.

14.4 Local Signaling Loops and Their Constitutive Activity

Signal transduction from ligand to receptor is a central paradigm of intercellular communication. While such signals can be transmitted over relatively long distances through the convection and diffusion of ligands, it is also common to find ligands and their receptors expressed in the same cells, with receptors often present in great molar excess to ligands (Wiley et al., 2003). This co-localization of ligands and receptors establishes the potential for autocrine or paracrine signaling circuits operating within spatially localized domains. Such is the case for the epidermal growth factor receptor (EGFR) system, which is present in airway epithelial cells and widely expressed in many tissues. In the EGFR system, EGF-family ligands such as EGF, TGF-alpha, and HB-EGF are synthesized as transmembrane precursors;

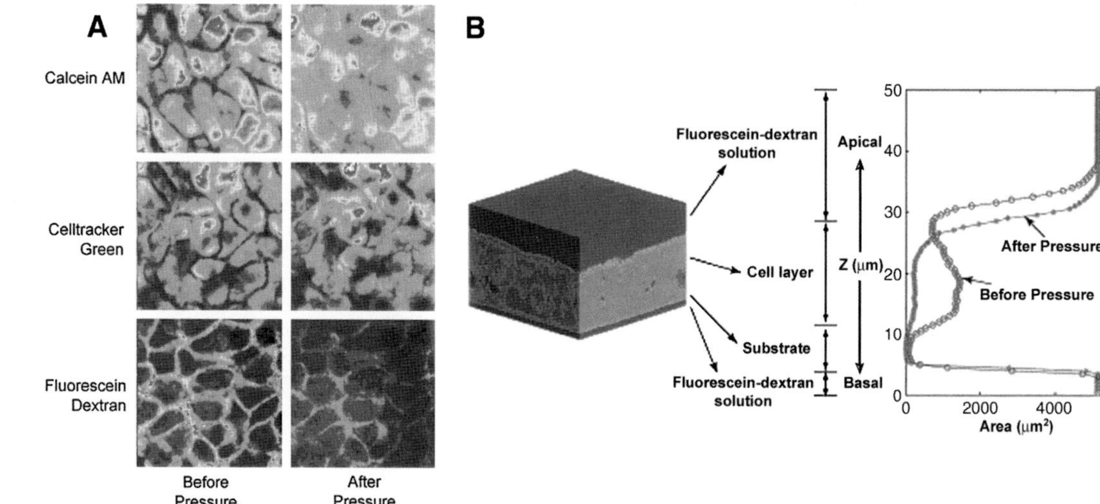

Figure 14.3. (A) Transverse (x-y) two-photon microscopy sections of living bronchial epithelial cells showing the cytoplasm (Calcein AM, Celltracker Green) or lateral intercellular spaces (Fluorescein Dextran) before or during application ($\sim$ 20 min after onset) of apical to basal transcellular pressure. (B) Three-dimensional reconstruction of extracellular spaces (red) above, below, and internal to bronchial epithelial cells (blue-green), and quantitation of the extracellular content in the same tissue volume before and after application of apical to basal transcellular pressure. Adapted from Tschumperlin et al. (2004).

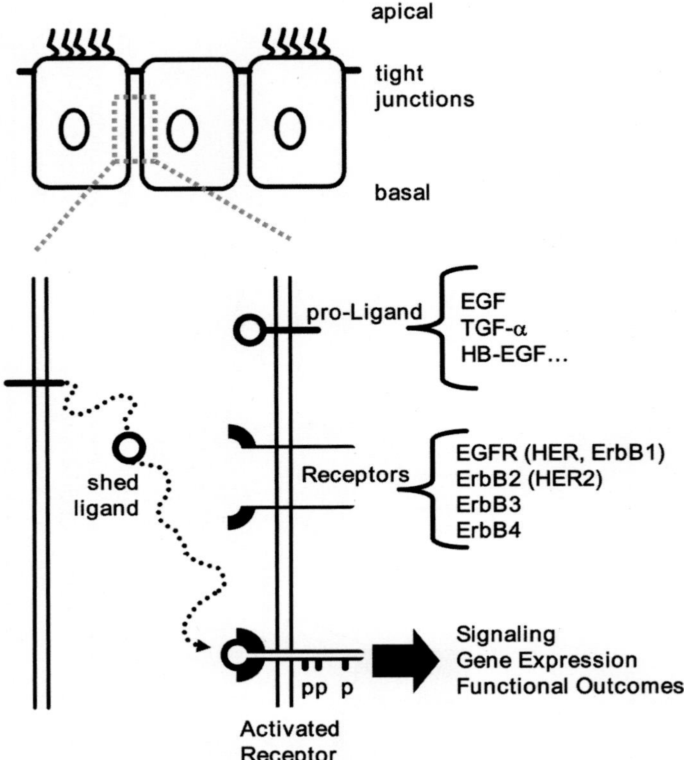

Figure 14.4. Schematic of typical local autocrine loop. In this case, a close-up of the lateral intercellular space of an epithelial tissue is shown to detail the ligands and receptors of the EGFR autocrine system.

upon constitutive or stimulated proteolytic processing, the ectodomains of these ligands are shed from the cell surface, releasing a biologically active peptide that traffics through the extracellular space, free to interact with cognate receptors of the EGFR (ErbB or HER) family (Figure 14.4) (Harris et al., 2003). These spatially localized EGFR loops have been shown, using quantitative systems-level experimental and computational approaches, to play key developmental and migratory roles (Wiley et al., 2003).

In local EGFR signaling loops, cells are highly effective at capturing ligands before they escape into the bulk medium. Thus the presence and functionality of autocrine loops can be masked under typical conditions and the production of autocrine ligands difficult to measure (DeWitt et al., 2001). This complication is amplified in the interstitial geometry of the LIS, where the receptor-bearing cell surface area is large relative to the interstitial volume, and the path length required for ligands to escape into the bulk medium is relatively long. However, the presence and constitutive activity of autocrine loops can be experimentally tested using methodologies employing receptor neutralizing antibodies that prevent ligand–receptor binding (Dempsey and Coffey, 1994; DeWitt et al., 2001). In the presence of saturating anti-EGFR antibodies, bronchial epithelial cells can be shown to shed both EGF and approximately ten times greater amounts of TGF-alpha into their basal medium (Kojic et al., 2006b). Ligand accumulation without the EGFR-blocking antibody is significantly lower, at the detection limit of the assays. Together these observations demonstrate that bronchial epithelial cells constitutively shed ligands of the EGF family and capture >90% of the ligands they produce.

The functionality of constitutive autocrine loops can be further confirmed by examining the effects of ligand or receptor blockades on baseline signaling and gene expression. The EGFR in cultured bronchial epithelial cells exhibits low levels of receptor phosphorylation under normal culture conditions (which includes EGF in the medium). Omission of EGF from the medium does not significantly decrease the phospho-EGFR levels, suggesting that the endogenous ligand production discussed above is sufficient to maintain baseline receptor activation (Kojic et al., 2006b). Incubation of resting cells with a metalloprotease inhibitor (galardin) reduces phospho-EGFR levels, likely through the blockade of endogenous ligand shedding. Incubation of cells with an inhibitor of EGFR kinase activity (AG1478) decreases the baseline expression of several known EGFR target genes (e.g., egr-1, c-fos, HB-EGF) (Chu et al., 2005; Kojic et al., 2006b). Together these results establish that constitutive EGFR autocrine loops are present and functional within the LIS of bronchial epithelial cells. Accumulating evidence indicates that a variety of constitutive autocrine loops are present in a number of cellular systems (Adams, 2002; DeWitt et al., 2001; Heasley, 2001; Hofer et al., 2004; Janowska-Wieczorek et al., 2001; Manabe et al., 2002). While local autocrine loops are known to be important in developmental and migratory contexts (Wiley et al., 2003), the full utility of constitutive autocrine signaling is not fully understood. A role in mechanotransduction, as described in the following, may provide an additional key biological rationale for constitutive local signaling loops.

14.5 Testing the Autocrine Mechanotransduction Hypothesis

The presence of functional autocrine loops within a deformable extracellular space establishes the necessary preconditions for the proposed mechanotransduction mode. The challenge that remains is to implement experimental and computational strategies to rigorously test whether these elements and their convergence are necessary and sufficient to constitute a mechanotransduction response.

An obvious first test of the involvement of the EGFR loop in mechanotransduction is to assess whether the onset of loading increases the phosphorylation and activation of the EGFR. This has been shown to be the case, as apical to basal transcellular pressure induces an increase in the phosphorylation of tyrosine residue 1068 on the EGFR within 5 min (Tschumperlin et al., 2004). When phosphorylated, this tyrosine residue on the EGFR serves as a docking site for the recruitment of Grb2, which couples receptor activation to the downstream ERK1/2 MAP kinase cascade through SOS and Ras. As expected, apical to basal transcellular pressure strongly induces ERK1/2 phosphorylation, with a slight time lag relative to EGFR phosphorylation (Tschumperlin et al., 2004). The magnitude of ERK1/2 phosphorylation response scales with the pressure gradient applied, demonstrating a proportional response to mechanical stimulus (Tschumperlin et al., 2002). Blockade of the tyrosine kinase domain of the EGFR strongly attenuates the mechanical stress–induced phosphorylation of ERK1/2, as does a function-blocking EGFR antibody. Together these findings confirm a role for both the extracellular ligand-binding domain and the intracellular kinase domain of the EGFR in the mechanotransduction response to transcellular pressure. A broad spectrum inhibitor of matrix metalloprotease activity that blocks ectodomain shedding of EGF family precursors (Dong et al., 1999; Prenzel et al., 1999) attenuates the transcellular pressure-induced increase in ERK1/2 phosphorylation. Finally, a function-blocking antibody for HB-EGF, but not function-blocking antibodies for EGF and TGF-alpha, attenuates the ERK1/2 phosphorylation response to stress (Tschumperlin et al., 2004). Taken together these results suggest that mechanical stress elicits signaling through an autocrine pathway involving shedding of the EGF family member HB-EGF.

Clearly, then, the EGFR system is activated by mechanical stress and is necessary for the ERK1/2 phosphorylation response. However, establishing that mechanical stress initiates signaling at the EGFR through geometric effects alone, and that this stimulus is sufficient to account for the signaling response, remains a challenging task. Other modes of signaling can co-opt the EGFR system through transactivation, especially G-protein–coupled receptors (Correa-Meyer et al., 2002; Gudi et al., 1996; Prenzel et al., 1999), making unambiguous identification of the primary mechanotransducer difficult. To date, pertussis toxin sensitive G-protein and ATP signaling have been ruled out as primary mechanotransducers in bronchial epithelial compression experiments (Tschumperlin et al., 2004), but exhaustively ruling out all alternative primary mechanotransducer candidates remains a daunting prospect.

A more fruitful approach is to test whether the changes in ligand concentration induced by mechanical stress and geometric changes are sufficient to account for, and are temporally compatible with, the EGFR signaling response. The steady-state concentration of ligand molecules in the extracellular space can be approximated by solving the mass-balance equation for ligand shedding and diffusion, assuming an idealized, parallel-plate LIS geometry and constant and uniform ligand shedding along the lateral cell walls (Figure 14.5) (Tschumperlin et al., 2004). The steady-state ligand concentration c reduces to $c = c_m + (h^2 - x^2)q/(Dw)$, where the concentration in the LIS exceeds the concentration in the medium (c_m) because of autocrine ligand shedding (rate $= q$). Due to the polarity of the epithelium, which allows free transport through the basal LIS opening but not across the tight junctions at the apical surface of the cells, the concentration depends on the distance x measured from the apex of the LIS (height $= h$). The difference in ligand concentration from that in the basal medium is inversely proportional to the diffusion constant of the ligand D and the LIS width w. Thus, this analysis predicts that the effective ligand concentration to which cells are exposed is strongly dependent on the geometry of the LIS, which is itself a function of the mechanical environment of the tissue.

Based on this steady-state analysis one can directly compare the predicted effects of mechanical stress with the measured effects of exogenous ligand stimulation on

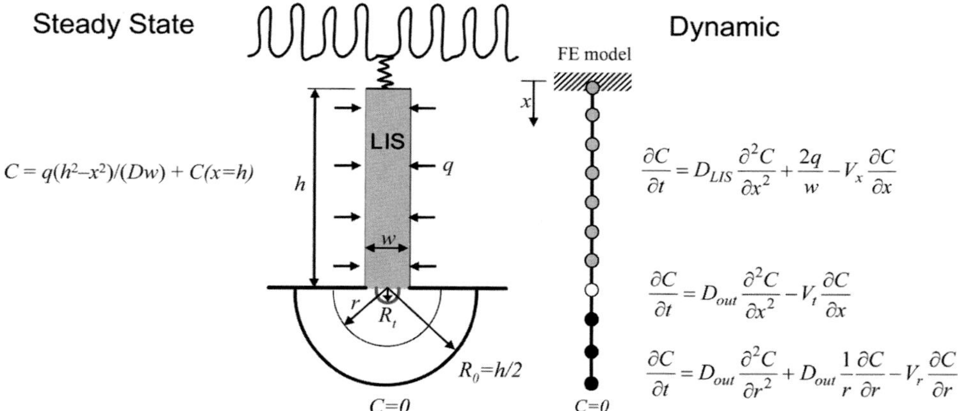

Figure 14.5. Idealized models of ligand concentration in the local environment of the LIS. The governing diffusion-convection equations in three domains are used to fully describe the ligand concentration in the one-dimensional idealized geometry (right). Under steady-state conditions, the concentration in the LIS can be solved analytically to express the ligand concentration as a function of shedding rate (q), position (x), diffusion coefficient (D), and LIS width (w). Adapted from Kojic et al. (2006a).

EGFR signaling. Given the measured 87% decrease in LIS width in imaging studies, the equation above predicts that the concentration of ligand should increase ~ 8-fold in the new LIS configuration under compressive stress (neglecting the low concentration of ligand in the bulk medium). In agreement with this prediction, mechanical stress was shown to produce a change in ERK1/2 phosphorylation equivalent to a roughly 10-fold increase in ligand concentration (Tschumperlin et al., 2004). Taken together, the results from imaging, analytical modeling, and biochemical approaches all strongly support the hypothesis that mechanical stress–induced changes in LIS geometry, and the accompanying changes in extracellular ligand concentration, are sufficient to account for the signaling response observed under loading.

14.6 The Challenge of Dynamics

While the imaging, steady-state modeling, and biochemical experiments discussed in the previous section all support the hypothesis that mechanotransduction can occur through autocrine signaling in a changing local geometry, each of these approaches is limited by the lack of a temporal component. The imaging and biochemical experiments provide isolated snapshots in time for what is doubtless a dynamic process. Similarly, the analytical relationship for ligand concentration in the LIS is valid only under steady-state conditions, and cannot be used to compute the dynamic changes in ligand distribution that accompany changes in LIS geometry. Fortunately, advanced methodologies are emerging that will enable the dynamic measurements and kinetic models needed to more rigorously evaluate extracellular mechanotransduction. In the following section we detail a modeling approach that allows the kinetics of mechanotransduction to be explored; this is followed by an introduction to methodologies that will provide the quantitative measurements and model parameters necessary for a fuller kinetic analysis of extracellular mechanotransduction.

14.7 Solving the Kinetics of Ligand Concentration in a Changing Geometry

To move from an analytical steady-state model to a dynamic assessment of ligand concentration in a changing geometry, a solution to the time-varying ligand diffusion-convection equation is required (Kojic et al., 2006a). As in the previously described analytical approach, the extracellular space separating neighboring cells can be approximated as bounded by idealized parallel plates (cell walls) with an impermeable tight junction at the apical surface. The basal opening of the interstitial space is open to a large reservoir (e.g., the underlying media) such that sufficiently far below the basal cell surface the ligand concentration is assumed to be zero (see Figure 14.5). To account for convective effects, as well as to determine how the concentrations at and below the LIS boundary change during a collapse, three domains are included in the model: a one-dimensional Cartesian LIS domain extending from the tight junction to the basal boundary, a transitional domain for coordinate conversion, and a radial domain external to the LIS, described in cylindrical coordinates. The governing transport equations for each domain are shown in Figure 14.5, where D_{LIS} and D_{out} are the ligand diffusivities in the LIS and outside space, respectively; V_x is the bulk fluid velocity in the LIS caused by changes in LIS dimensions; V_t is the fluid velocity in the transitional domain (assumed to be uniform); and V_r is the radial fluid velocity at a radius r measured from the LIS boundary. Note that in governing equations for the transitional and radial domains there are no q (ligand shedding) terms, because shedding is assumed to occur only from the lateral surfaces of the LIS. By the conservation of mass and fluid incompressibility it can be shown that in the LIS, $V_x = (\dot{w}/w)x$, where $\dot{w} = dw/dt$ is the rate of change of LIS width.

The transitional regime is included to avoid numerical difficulties that can occur when switching coordinate systems; it extends to a distance $R_t = w/\pi$ from the LIS boundary. The velocity field within this domain is approximated as uniform and equal to the bulk velocity at the LIS exit $V_t = V_x(x = h)$. The radial domain encompasses the region between R_t (end of the transitional domain) and $R_0 = h/2$ (where we assume the ligand concentration to be zero). Mathematically, the concentration of ligand will approach zero as the distance from the LIS approaches infinity, but for efficient numerical simulations $R_0 = 7.5$ μm is sufficient such that further increasing R_0 has little effect on the overall concentration profile.

The system of diffusion-convection equations and the accompanying boundary conditions can be solved using the finite element (FE) method (Kojic et al., 2006a). Using the previously measured change in LIS width (Tschumperlin et al., 2004) occurring over an arbitrary 60-duration, the model can be used to compute the time-varying ligand concentration profile in the LIS (Figure 14.6). Another way to visualize the change in ligand concentration is to plot the fold change in the mean

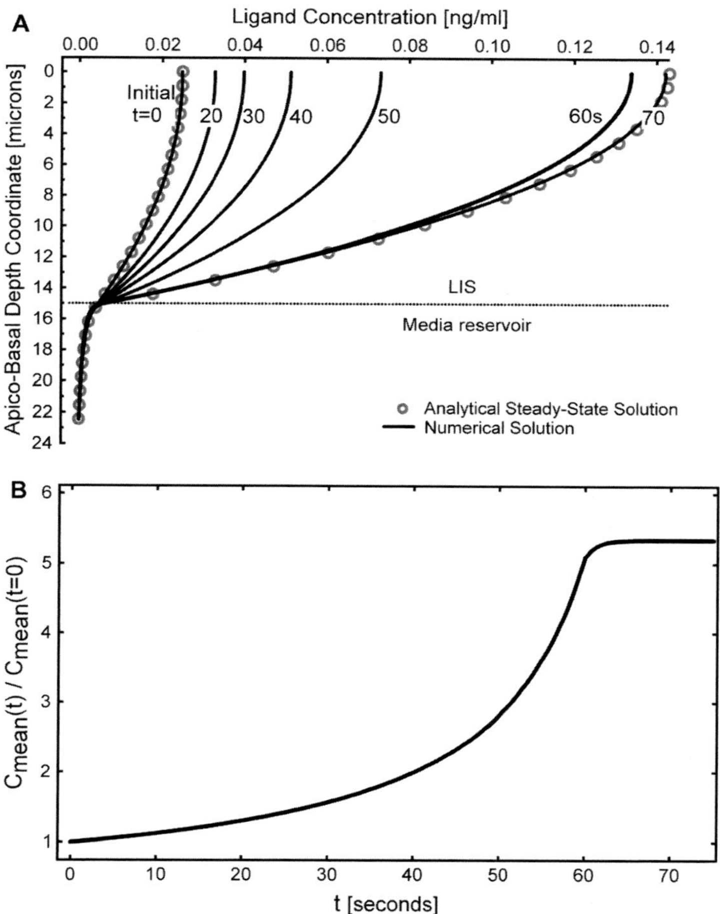

Figure 14.6. FE analysis of the ligand diffusion-convection equations during a change in LIS width (85% decrease) over 60 s provides the ligand concentration profiles at various times (A), and the normalized change in mean ligand concentration in the LIS as a function of time (B) (Kojic et al., 2006a).

ligand concentration ($C_{mean}(t)/C_{mean}(t = 0)$) as a function of time (Figure 14.6), where $C_{mean}(t) = (1/h) \int_0^h C(x,t)\,dx$ and $C_{mean}(t = 0)$ is the mean LIS ligand concentration just prior to the change in LIS width.

14.8 Exploring the Parameters That Govern Autocrine Mechanotransduction

Across cellular systems and different ligand–receptor systems, ligands can be released at a variety of rates. Variation in ligand shedding rate affects only the absolute value of ligand concentration, not the fold change in concentration induced by changes in LIS geometry, minimizing the role of the shedding rate in mechanotransduction. In addition to variations in shedding rates, signaling molecules exhibit a range of diffusivities depending on their molecular size and charge characteristics, including the ligands within the EGF family (Harris et al., 2003; Kovbasnjuk et al.,

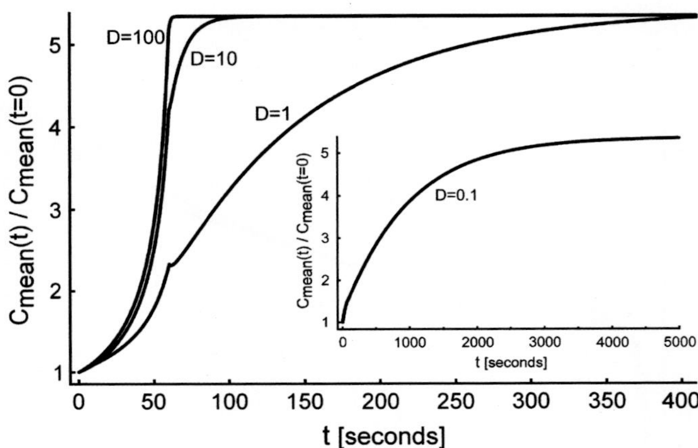

Figure 14.7. The effect of ligand diffusion coefficient *(D)* on the kinetics of ligand concentration change during a 60-s linear decrease in LIS width (85% decrease in width). The inset ($D = 0.1$) uses a longer time-scale (5000 s) to capture the slow kinetics of this scenario (Kojic et al., 2006a).

2000; Xia et al., 1998). The FE analysis of ligand diffusion-convection demonstrates that the smaller the ligand diffusivity (or, conversely, the larger the ligand), the slower the change in the normalized mean concentration during LIS width change (Figure 14.7). The diffusivities used in Figure 14.6(A) range from 100 $\mu m^2/s$, representing free diffusion of small molecules ($\sim$ 0.1–10 kDa), to 0.1– 1$\mu m^2/s$, representing hindered diffusion of large molecules in a crowded space (Kovbasnjuk et al., 2000).

Because the kinetics of ligand concentration in a changing LIS geometry depend so strongly on the diffusivity within the LIS, it is important to estimate ligand diffusivities as accurately as possible. However, diffusion coefficients obtained in free diffusion experiments may be misleading (Aukland and Reed, 1993). In the LIS of MDCK epithelial cells, which share a similar architecture with human bronchial epithelial cells, molecules larger than 3 kDa diffuse in a manner significantly hindered relative to free diffusion. Ligand-specific hindered diffusion could thus factor prominently in the relative roles of ligands in mechanosensing. For instance, the bioactive shed form of HB-EGF has a molecular weight of about 22 kDa (Harris et al., 2003; Raab and Klagsbrun, 1997) and is heavily charged (Raab and Klagsbrun, 1997) relative to TGF-alpha and EGF, which are $\sim$ 5.5 kDa in size (Harris et al., 2003; Raab and Klagsbrun, 1997). Previous studies have also shown that interactions between charged molecules and the extracellular matrix or glycocalyx can hinder diffusion beyond the expected size effect (Dowd et al., 1999). Therefore, HB-EGF diffusion in the LIS is expected to be significantly hindered due both to its size and charge interactions. Based on experimental studies in a similar system (Kovbasnjuk et al., 2000), we have approximated the HB-EGF diffusion coefficient in the LIS as $D_{LIS} = 1.8\ \mu m^2/s$ (Kojic et al., 2006a); in cell culture we assume that outside the LIS, HB-EGF undergoes free diffusion ($D_{out} = 75\ \mu m^2/s$). Because of their smaller size and relative lack of charge, we assume that TGF-alpha and EGF diffuse in a relatively unhindered manner both inside and outside the LIS ($D = 120\ \mu m^2/s$) (Kojic

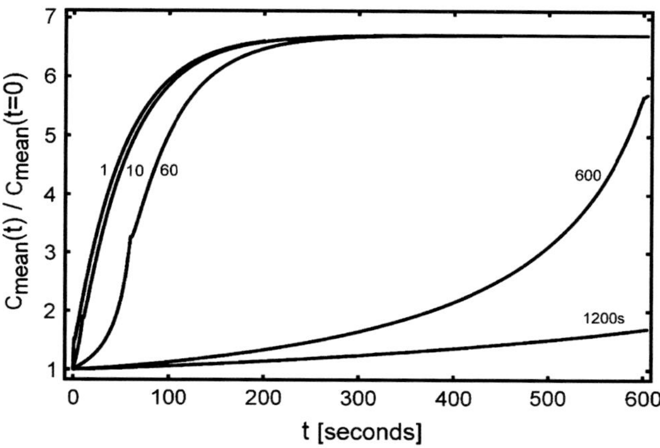

Figure 14.8. The kinetics of ligand concentration change under various loading conditions that lead to a linear decrease in LIS width by 85% over 1, 10, 60, 600, and 1200 s. (Adapted from Kojic et al., 2006a.)

et al., 2006a). As will be discussed later, experimental approaches for measuring ligand diffusion within the LIS would help to clarify these important aspects of ligand behavior.

14.9 Rate Sensitivity of Autocrine Mechanotransduction

Biomechanical loading can develop on a range of time-scales, from milliseconds for traumatic injury to hours or days for cellular proliferation and tissue morphogenesis. In the case of the *in vitro* application of apical to basal transcellular pressure, the kinetics of LIS geometry changes are unknown, but the relevant time-scale appears to be on the order of seconds to minutes. Specifically, by 20 min after the onset of transcellular compressive stress a new steady state in LIS geometry is established (Tschumperlin et al., 2004). Using the simplest approximation of a linear change in LIS dimensions with time, one can employ the FE analysis to explore rate effects by computing the ligand concentration profiles and change in mean concentration during deformations completed over a range of durations (Figure 14.8). Incorporating the estimates for hindered diffusions of HB-EGF inside the LIS and free diffusion outside of the LIS, the model solutions demonstrate how the rate of change in LIS geometry defines the kinetics of the changing ligand concentration.

The rate aspect of ligand-receptor signaling is a potentially important cue because receptor activation and downstream signaling are influenced not only by the magnitude, but also by the rate of change of ligand concentration in the cellular microenvironment (Sasagawa et al., 2005). To explore this kinetic aspect of mechanotransduction, one useful metric is the maximum rate of concentration change, which can be computed by differentiating normalized C_{mean} curves with respect to time. The results shown in Figure 14.9 emphasize that the rate at which mechanical force is converted into changes in LIS geometry can profoundly influence the rate at which the ligand concentration changes in the cellular microenvironment; this finding

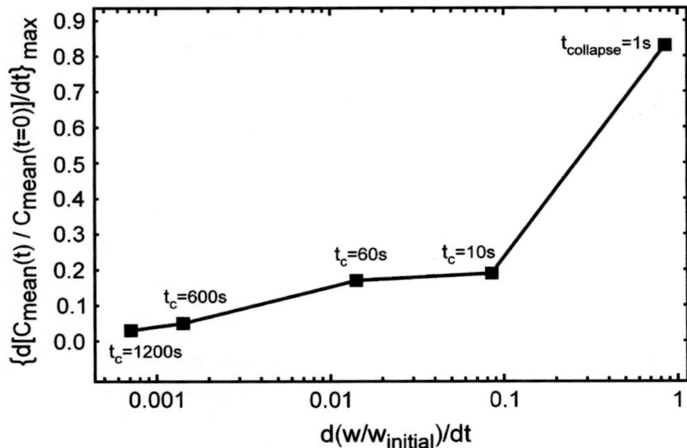

Figure 14.9. Maximum rate of change of mean ligand concentration as a function of the rate of change of LIS width (t_c = time for complete reduction in LIS width by 85%), using the same loading rate scenarios as in Figure 14.8 (Kojic et al., 2006a).

suggests a mechanism by which cells could use rate-sensitive receptor signal processing to discriminate between mechanical processes occurring over a range of rates.

14.10 Ligand-Specific Concentration Dynamics in the LIS

While bronchial epithelial cells shed a number of autocrine ligands, including those of the EGF family, the mechanotransduction response appears to depend predominantly on HB-EGF (Tschumperlin et al., 2004). How could this be explained? If one employs the estimated diffusivities for HB-EGF (D_{LIS} = 1.8 μm^2/s, D_{out} = 75 mm^2/s) and EGF/TGF-alpha ($D_{LIS} = D_{out}$ = 120 μm^2/s), and further assumes all ligands are shed at the same rate (q = 10 molecules/cell/min), the solution of the governing diffusion-convection equations during a 60-s collapse yields the absolute mean concentration curves (not normalized) shown in Figure 14.10(A). For hindered HB-EGF in the LIS, the mean absolute concentrations are an order of magnitude higher than those of TGF-alpha (or EGF). A corollary to this result is that in order to have a similar LIS concentration for all ligands, the cell must shed TGF-alpha (or EGF) at a rate ~ 10 times higher than that of HB-EGF.

The hindered diffusion of HB-EGF in the cellular microenvironment thus provides two potential explanations for the selective role of HB-EGF in extracellular mechanotransduction. In the first case, the different concentrations that arise in the LIS as a consequence of different ligand diffusivities could place HB-EGF and TGF-alpha on different portions of an EGFR–ligand binding curve (Lauffenburger et al., 1998a) (see Figure 14.10(B)), making the cells more or less responsive to mechanical stress–induced changes in particular ligand concentrations. A second potential explanation is that the molecular sieving properties embedded within the LIS (Kovbasnjuk et al., 2000) might be altered by the geometric decrease in LIS dimensions. Specifically, the shrinking of the intercellular space under loading could create a more densely packed barrier to diffusion, especially for large, highly charged

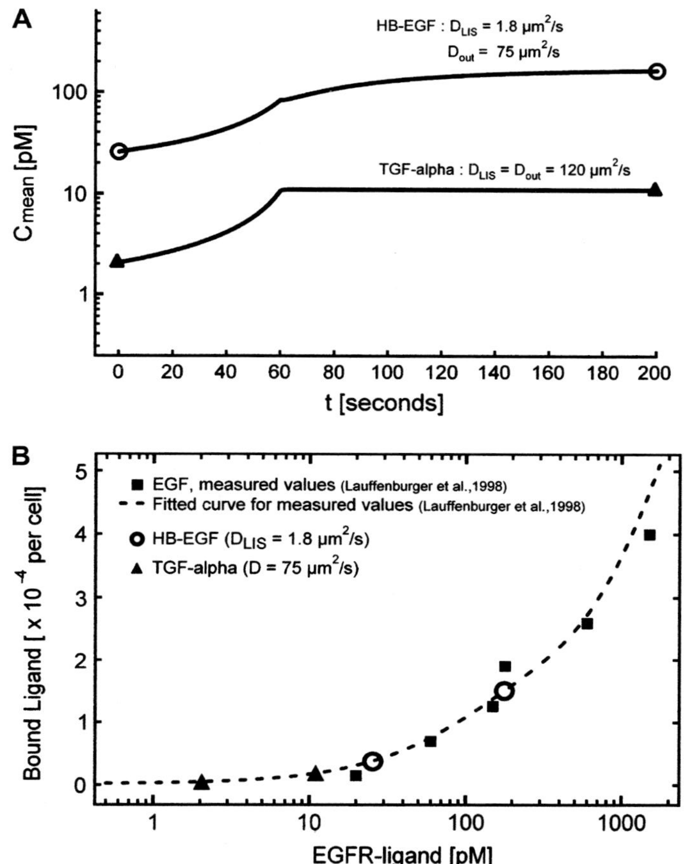

Figure 14.10. (A) Ligand concentrations for HB-EGF and TGF-alpha (or EGF) during a change in LIS width (85% reduction), assuming equivalent shedding rates for each ligand, and using the indicated diffusivities for each ligand. (B) EGFR-ligand binding curve (dashed line) as a function of ligand concentration, fitted from experimental observations (squares) (Lauffenburger et al., 1998b). The mean concentrations of HB-EGF and TGF-alpha pre- and post-change in LIS width (from panel A) are shown to indicate their relative position on the ligand–receptor binding curve (Kojic et al., 2006a).

molecules like HB-EGF, while leaving smaller molecules like EGF and TGF-alpha less affected. A decrease in D_{LIS} during collapse would amplify the increase in local ligand concentration and potentially magnify cellular mechanotransduction effects.

14.11 New Methods to Bridge Modeling and Experiments

With a computational framework in hand to analyze the kinetics of extracellular mechanotransduction, new approaches and technologies are needed to measure those model inputs that until now have been estimated. Moreover, to test the ultimate fidelity of the hypothesized mechanotransduction mechanism, approaches must be found that will allow direct comparison of the model output – the time-varying ligand concentration response to geometry changes – to experimental measures of the same outcome (or a reasonable proxy thereof).

The most important missing information needed to test the mechanotransduction mechanism is the kinetics of the extracellular space deformation under loading. New generations of imaging devices, such as high-speed two-photon microscopy, are now available to dynamically assess changes in geometry. Custom-built high-speed two-photon microscopes can acquire up to 10 frames/s (300 × 400 pixels per frame) using a polygonal mirror scanner rather than a conventional galvanometer-driven scanner, allowing repeated imaging in a single plane at high temporal resolution and faster acquisition of three-dimensional stacks to reconstruct the geometry of the cells and the extracellular space as they change under loading (Kim et al., 1999). The dynamics of the geometric changes obtained with these improved methods can be analyzed to provide the time-varying LIS width for input into the existing FE model. Alternatively, the spatial information in the three-dimensional reconstructions could be used to solve the time-varying ligand diffusion-convection equations in three dimensions; this more computation-intensive approach could provide new insights into the role local geometries play in mechanotransduction.

The ability to measure dynamic changes in extracellular and cellular geometries during mechanical loading will make possible a fuller characterization of the physical responses to loading. Important questions remain as to the dependence of geometry changes on stress magnitude (e.g., threshold effect) and to the reversibility of geometry changes under transient and sustained loadings. Moreover, the general questions of how cellular and extracellular deformations under this type of loading are resisted by various intracellular (e.g., cytoskeleton, tonicity) and extracellular (e.g., glycocalyx, matrix) components, and how the effects of loading vary across different cell types and tissues, remain largely unexplored (Aukland and Reed, 1993). Eventually the two-photon imaging approach should be used in a variety of intact tissues to understand how interstitial spaces vary under loading *in situ*. The ultimate goal of these studies would be to establish micro-scale constitutive laws relating loading to the deformation behavior of cells and their extracellular spaces.

New approaches for measuring receptor activation kinetics will provide the spatiotemporal information needed for comparison to the changes in ligand concentration predicted from dynamic imaging and modeling studies. Enzyme-linked immunosorbent assays and antibody arrays employing phospho-specific antibodies are available that will provide efficient characterizations of receptor phosphorylation under loading; these methods will allow the kinetics of receptor activation to be reconstructed after the fact (Nielsen et al., 2003). Alternatively, to visualize receptor activation in real time, cells expressing genetically encoded fluorescent reporter systems could be utilized (Ting et al., 2001; Wang et al., 2005). This approach could provide both the spatial coordinates and the temporal patterns of receptor activation. In coordination with the dynamic imaging and modeling approaches described earlier, these methods could provide a direct test of the correspondence between local geometry changes and local mechanical signaling.

The modeling results described above emphasize the critical importance of accurately measuring ligand diffusivity and shedding rates. To date there is little information available on the relative diffusion coefficients of ligands within local interstitial microenvironments. What is known comes from measuring the diffusions of various size photoactivated fluorescent probes within the LIS of MDCK cells (Kovbasnjuk et al., 2000; Xia et al., 1998). A similar approach can be applied using two-photon fluorescence correlation microscopy to measure the local diffusion of fluorescent tracers (Alexandrakis et.al., 2004). This method could be used in future studies to measure both the LIS geometry and the diffusivity of various size molecular tracers under resting conditions and during loading-induced changes in LIS geometry.

The ultimate goal of these approaches is to understand how mechanical stress–induced changes in extracellular geometry are transduced into biochemical signaling. A decisive test of the hypothesis will require a direct comparison of modeling and experimental outputs, a task thus far frustrated by the different outcomes provided by experimental (receptor phosphorylation) and modeling (ligand concentration) approaches. To bridge this gap, one approach would be to modify the modeling approach to incorporate ligand–receptor interactions and receptor phosphorylation and dephosphorylation (Resat et al., 2003). However, unambiguous incorporation of this approach will require measuring relevant parameters for the cell type of interest under the appropriate conditions (mechanical loading). Alternatively, one could attempt to modify the experimental approach to provide a measure more directly comparable to the existing model output. For instance, cells engineered to shed fluorescently tagged ligands could be used to monitor changes in fluorescence in the LIS in real time, providing a direct comparison to the current computational model. However, the application of this approach will be challenging given the low shedding rates and ligand concentrations in the LIS, the difficulty in distinguishing transmembrane from free ligands, and the likely effects of fluorescent tags on ligand shedding and diffusivity.

Despite these obstacles, many tools and techniques are emerging that will continue to improve our ability to measure and interrogate aspects of autocrine mechanotransduction. While a fuller understanding of the autocrine mechanotransduction mechanism awaits further developments in these areas, the supportive facts gathered to date warrant a consideration of the potential roles for this mode of mechanotransduction.

14.12 Broader Implications of Autocrine Mechanotransduction

A distinguishing characteristic of mechanotransduction occurring through autocrine signaling in a changing geometry is its relative insensitivity to small or transient perturbations. Cellular viscoelasticity; fluid drag; and the kinetics of ligand shedding, diffusion, and convection combine to create a lag between the onset of mechanical loading and changes in local ligand concentration (see Figures 14.7 and 14.8). Depending on the exact kinetics of the system (and assuming that some or all of the changes in LIS geometry are reversible), it is apparent that mechanical loading would need to be sustained for some duration to evoke meaningful changes in ligand concentration and receptor activation via this mode of mechanotransduction. As a result of these system characteristics, the signaling effects of transient deformations are likely to be substantially attenuated relative to the responses to sustained loading. An example from the lung helps to illustrate the usefulness of this property: Cyclic breathing motions, which are small and transient, would likely elicit little response via this mechanotransduction mechanism, while sustained loading, such as that occurring during bronchoconstriction, would maximize changes in geometry and subsequent alterations in ligand concentration and receptor activation. This capacity to favor the transduction of sustained mechanical loading might be particularly useful for cellular decisions on whether to mount energetically costly responses to mechanical stress.

While the discussion here has focused on autocrine mechanotransduction through the EGFR system, other ligand–receptor systems could be similarly co-opted to participate in mechanotransduction. To a large extent the participation of other ligand–receptor systems likely depends on characteristics unique to each system, and to the local concentration of ligands present under resting conditions. For a typical ligand–receptor system responsive to changes in ligand concentration over two or more orders of magnitude (e.g. Figure 14.10(B)), changes in interstitial spacing of 50%, 90%, and 99%, would be within the dynamic range of the receptor (assuming a resting concentration near the low end of responsiveness) and would yield changes in ligand concentration of 2-, 10- and 100-fold, respectively. How these changes in ligand concentration are processed and interpreted by the cellular circuitry and control systems would play a dominant role in shaping the cellular response, but in theory cells and tissues could use various ligand–receptor systems to respond in unique ways to loading. Further flexibility and complexity in signaling responses would emerge from the ability of cells to mount unique receptor-mediated responses to identical ligand stimuli in cell- and context-specific manners (Freeman and Gurdon, 2002), and likewise to the capacity of cells to respond in unique ways depending on the kinetics and durations of receptor stimulations (Sasagawa et al., 2005). Mechanotransduction through autocrine signaling can thus be envisioned to utilize various ligand–receptor systems and their built-in signal processing capabilities to discriminate between and respond appropriately to a variety of loading conditions in a tissue-specific manner.

In what tissues might these types of mechanical responses be present? As discussed in the introduction, the only absolute prerequisites are the presence of constitutive local signaling loops and an extracellular geometry sensitive to local loading conditions. Local autocrine signaling loops are now recognized from a diverse and growing family of ligand–receptor systems (Adams, 2002; Heasley, 2001; Hofer et al., 2004; Janowska-Wieczorek et al., 2001; Singh and Harris, 2005). With regard to extracellular geometries, one need only consider the distribution of the most abundant material within the human body – water. While $\sim$ 2/3 of the body's water is found within cells, the vast majority ($\sim$ 75%) of the remaining 1/3 is extracellular and interstitial (not in plasma) (Aukland and Reed, 1993). This interstitial fluid is in a continuous slow state of turnover as it flows from the vasculature to the lymphatics. But external or internal forces that create pressure gradients will cause local redistribution of interstitial fluid. In soft tissues with a high cell density, the net effect of such interstitial fluid redistributions will be prominent changes in local intercellular spacings. The evidence collected in this chapter demonstrates that in tissues presenting the appropriate autocrine signaling environment, changes in intercellular spacings can be sufficient to initiate a mechanotransduction response. Recent evidence of local autocrine loops in intact heart (Yoshioka et al., 2005) and modeling that suggests a role for autocrine mechanotransduction in cardiac signaling (Maly et al., 2004) together raise the exciting possibility that the autocrine mechanotransduction mechanism described here is not unique to airway epithelium, and may be applicable in a variety of systems.

In summary, the unique characteristics inherent to extracellular mechanotransduction, including its capacity to filter out transient deformations, its magnitude and rate dependence, and its co-opting of ligand-receptor signal processing all combine to create a unique, modular, potent, and subtly complex mechanical signaling mechanism. While its ultimate utility in explaining various mechanoresponses remains to be evaluated, key processes like tissue morphogenesis and remodeling, both of which require cellular populations to respond in a unified manner to sustained changes in mechanical environment, appear especially likely to employ this mode of mechanotransduction. Future work will enhance our understanding of the molecular and kinetic aspects of this mechanotransduction mechanism and establish its wider contributions in a variety of biological systems.

REFERENCES

Adams GR. 2002. Invited Review: Autocrine/paracrine IGF-I and skeletal muscle adaptation. *J Appl Physiol* **93**(3):1159–67.

Alexandrakis G, Brown EB, Tong RT, McKee TD, Campbell RB, Boucher Y, Jain RK. 2004. Two-photon fluorescence correlation microscopy reveals the two-phase nature of transport in tumors. *Nat Med* **10**(2):203–7.

Aukland K, Reed RK. 1993. Interstitial-lymphatic mechanisms in the control of extracellular fluid volume. *Physiol Rev* **73**(1):1–78.

Bousquet J, Jeffery PK, Busse WW, Johnson M, Vignola AM. 2000. Asthma. From bronchoconstriction to airways inflammation and remodeling. *Am J Respir Crit Care Med* **161**(5):1720–45.

Chu EK, Foley JS, Cheng J, Patel AS, Drazen JM, Tschumperlin DJ. 2005. Bronchial epithelial compression regulates epidermal growth factor receptor family ligand expression in an autocrine manner. *Am J Respir Cell Mol Biol* **32**(5):373–80.

Correa-Meyer E, Pesce L, Guerrero C, Sznajder JI. 2002. Cyclic stretch activates ERK1/2 via G proteins and EGFR in alveolar epithelial cells. *Am J Physiol Lung Cell Mol Physiol* **282**(5):L883–91.

Dempsey PJ, Coffey RJ. 1994. Basolateral targeting and efficient consumption of transforming growth factor-alpha when expressed in Madin-Darby canine kidney cells. *J Biol Chem* **269**(24):16878–89.

DeWitt AE, Dong JY, Wiley HS, Lauffenburger DA. 2001. Quantitative analysis of the EGF receptor autocrine system reveals cryptic regulation of cell response by ligand capture. *J Cell Sci* **114**(Pt 12):2301–13.

Dong J, Opresko LK, Dempsey PJ, Lauffenburger DA, Coffey RJ, Wiley HS. 1999. Metalloprotease-mediated ligand release regulates autocrine signaling through the epidermal growth factor receptor. *Proc Natl Acad Sci USA* **96**(11):6235–40.

Dowd CJ, Cooney CL, Nugent MA. 1999. Heparan sulfate mediates bFGF transport through basement membrane by diffusion with rapid reversible binding. *J Biol Chem* **274**(8): 5236–44.

Fish JE, Peters SP. 1999. Airway remodeling and persistent airway obstruction in asthma. *J Allergy Clin Immunol* **104**(3 Pt 1):509–16.

Freeman M, Gurdon JB. 2002. Regulatory principles of developmental signaling. *Annu Rev Cell Dev Biol* **18**:515–39.

Gudi SR, Clark CB, Frangos JA. 1996. Fluid flow rapidly activates G proteins in human endothelial cells. Involvement of G proteins in mechanochemical signal transduction. *Circ Res* **79**(4):834–9.

Harris RC, Chung E, Coffey RJ. 2003. EGF receptor ligands. *Exp Cell Res* **284**(1):2–13.

Heasley LE. 2001. Autocrine and paracrine signaling through neuropeptide receptors in human cancer. *Oncogene* **20**(13):1563–9.

Hofer AM, Gerbino A, Caroppo R, Curci S. 2004. The extracellular calcium-sensing receptor and cell-cell signaling in epithelia. *Cell Calcium* **35**(3):297–306.

Janowska-Wieczorek A, Majka M, Ratajczak J, Ratajczak MZ. 2001. Autocrine/paracrine mechanisms in human hematopoiesis. *Stem Cells* **19**(2):99–107.

Kim KH, Buehler C, So PTC. 1999. High-speed, two-photon scanning microscope. *Appl Opt* **38**:6004–9.

Kojic N, Kojic M, Tschumperlin DJ. 2006a. Computational modeling of extracellular mechanotransduction. *Biophys J* **90**(11):4261–70.

Kojic N, Tian D, Drazen JM, Tschumperlin DJ. 2006b. Evidence for constitutive activity of an epidermal growth factor receptor (EGFR) autocrine loop in human bronchial epithelial cells (abstract). *Proc Am Thoracic Soc* **3**:A819.

Kovbasnjuk ON, Bungay PM, Spring KR. 2000. Diffusion of small solutes in the lateral intercellular spaces of MDCK cell epithelium grown on permeable supports. *J Membr Biol* **175**(1):9–16.

Lauffenburger DA, Oehrtman GT, Walker L, Wiley HS. 1998. Real-time quantitative measurement of autocrine ligand binding indicates that autocrine loops are spatially localized. *Proc Natl Acad Sci USA* **95**(26):15368–73.

Maly IV, Lee RT, Lauffenburger DA. 2004. A model for mechanotransduction in cardiac muscle: Effects of extracellular matrix deformation on autocrine signaling. *Ann Biomed Eng* **32**(10):1319–35.

Manabe I, Shindo T, Nagai R. 2002. Gene expression in fibroblasts and fibrosis: Involvement in cardiac hypertrophy. *Circ Res* **91**(12):1103–13.

Nielsen UB, Cardone MH, Sinskey AJ, MacBeath G, Sorger PK. 2003. Profiling receptor tyrosine kinase activation by using Ab microarrays. *Proc Natl Acad Sci USA* **100**(16): 9330–5.

Perez-Vilar J, Sheehan JK, Randell SH. 2003. Making More MUCS. *Am J Respir Cell Mol Biol* **28**(3):267–70.

Prenzel N, Zwick E, Daub H, Leserer M, Abraham R, Wallasch C, Ullrich A. 1999. EGF receptor transactivation by G-protein-coupled receptors requires metalloproteinase cleavage of proHB-EGF. *Nature* **402**(6764):884–8.

Raab G, Klagsbrun M. 1997. Heparin-binding EGF-like growth factor. *Biochimica et Biophysica Acta* **1333**:F179–F199.

Resat H, Ewald JA, Dixon DA, Wiley HS. 2003. An integrated model of epidermal growth factor receptor trafficking and signal transduction. *Biophys J* **85**(2):730–43.

Ressler B, Lee RT, Randell SH, Drazen JM, Kamm RD. 2000. Molecular responses of rat tracheal epithelial cells to transmembrane pressure. *Am J Physiol Lung Cell Mol Physiol* **278**(6):L1264–72.

Sasagawa S, Ozaki Y, Fujita K, Kuroda S. 2005. Prediction and validation of the distinct dynamics of transient and sustained ERK activation. *Nat Cell Biol* **7**(4):365–73.

Singh AB, Harris RC. 2005. Autocrine, paracrine and juxtacrine signaling by EGFR ligands. *Cell Signal* **17**(10):1183–93.

So PT, Dong CY, Masters BR, Berland KM. 2000. Two-photon excitation fluorescence microscopy. *Annu Rev Biomed Eng* **2**:399–429.

Spring KR, Hope A. 1978. Size and shape of the lateral intercellular spaces in a living epithelium. *Science* **200**(4337):54–8.

Swartz MA, Tschumperlin DJ, Kamm RD, Drazen JM. 2001. Mechanical stress is communicated between different cell types to elicit matrix remodeling. *Proc Natl Acad Sci USA* **98**(11):6180–5.

Timbs MM, Spring KR. 1996. Hydraulic properties of MDCK cell epithelium. *J Membr Biol* **153**(1):1–11.

Ting AY, Kain KH, Klemke RL, Tsien RY. 2001. Genetically encoded fluorescent reporters of protein tyrosine kinase activities in living cells. *Proc Natl Acad Sci USA* **98**(26): 15003–8.

Tschumperlin DJ, Dai G, Maly IV, Kikuchi T, Laiho LH, McVittie AK, Haley KJ, Lilly CM, So PT, Lauffenburger DA, et al. 2004. Mechanotransduction through growth-factor shedding into the extracellular space. *Nature* **429**(6987):83–6.

Tschumperlin DJ, Drazen JM. 2006. Chronic effects of mechanical force on airways. *Annu Rev Physiol* **68**:563–83.

Tschumperlin DJ, Shively JD, Kikuchi T, Drazen JM. 2003. Mechanical stress triggers selective release of fibrotic mediators from bronchial epithelium. *Am J Respir Cell Mol Biol* **28**(2):142–9.

Tschumperlin DJ, Shively JD, Swartz MA, Silverman ES, Haley KJ, Raab G, Drazen JM. 2002. Bronchial epithelial compression regulates MAP kinase signaling and HB-EGF-like growth factor expression. *Am J Physiol Lung Cell Mol Physiol* **282**(5):L904–11.

Wang Y, Botvinick EL, Zhao Y, Berns MW, Usami S, Tsien RY, Chien S. 2005. Visualizing the mechanical activation of Src. *Nature* **434**(7036):1040–5.

Wiggs BR, Hrousis CA, Drazen JM, Kamm RD. 1997. On the mechanism of mucosal folding in normal and asthmatic airways. *J Appl Physiol* **83**(6):1814–21.

Wiley HS, Shvartsman SY, Lauffenburger DA. 2003. Computational modeling of the EGF-receptor system: A paradigm for systems biology. *Trends Cell Biol* **13**(1):43–50.

Xia P, Bungay PM, Gibson CC, Kovbasnjuk ON, Spring KR. 1998. Diffusion coefficients in the lateral intercellular spaces of Madin-Darby canine kidney cell epithelium determined with caged compounds. *Biophys J* **74**(6):3302–12.

Yoshioka J, Prince RN, Huang H, Perkins SB, Cruz FU, MacGillivray C, Lauffenburger DA, Lee RT. 2005. Cardiomyocyte hypertrophy and degradation of connexin43 through spatially restricted autocrine/paracrine heparin-binding EGF. *Proc Natl Acad Sci USA* **102**(30):10622–7.

15

The Interaction between Fluid-Wall Shear Stress and Solid Circumferential Strain Affects Endothelial Cell Mechanobiology

John M. Tarbell

15.1 Introduction

Endothelial cells (EC) lining blood vessel walls are exposed to both the wall shear stress (WSS) of blood flow and the circumferential strain (CS) and associated circumferential stress driven by the wall motion induced by pulsing pressure. Most *in vitro* studies of EC response to mechanical forces and mechanotransduction have focused on the either the WSS or the CS, but not their interaction. This is in spite of the fact that in the arterial circulation that is most susceptible to disease, the WSS and the CS are imposed concurrently. While there have been relatively few studies of simultaneous WSS and CS, several recent investigations have revealed that the response of endothelial cells to combined stresses is exquisitely sensitive to the temporal phase angle between them, suggesting that when they are applied in a highly out-of-phase manner, a pro-atherogenic response is produced, whereas when they are applied in-phase, the response is more favorable.

In this chapter we first review the physiological background on WSS and CS in the circulation to focus on those regions where their interaction is significant. In the process we uncover a fascinating pattern that suggests that the WSS and the CS are most asynchronous (out-of-phase temporally) in precisely those regions of the circulation where atherosclerosis is localized. This background is followed by a consideration of the *in vitro* experiments in which the WSS and the CS have been applied simultaneously to the EC. There we uncover dramatic influences of the phase angle between the WSS and the CS indicating that out-of-phase forces induce a pro-atherogenic EC phenotype. Animal experiments that are consistent with this view are reviewed and possible countermeasures are described.

Finally, we discuss mechanotransduction mechanisms by which simultaneous WSS and CS and the phase angle between them could induce unique cellular responses not characteristic of either force applied separately.

15.2 Physiological Background

In the arterial circulation of humans, pulsatile blood flow produces oscillatory WSS with mean values on the order of 10 dyn/cm^2 (higher or lower depending on the exact

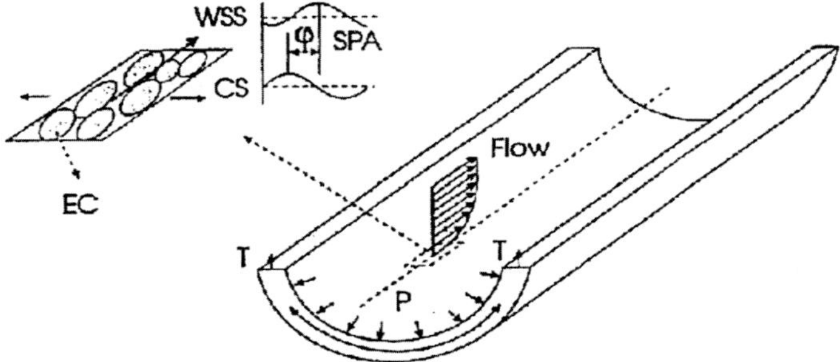

Figure 15.1. Schematic diagram of the stress pattern applied to the EC layer. WSS is the axial wall shear stress; CS is the circumferential strain; SPA is the stress phase angle ≡ φ(CS-WSS); T is the circumferential stress; P is the pressure. (From Qiu and Tarbell, 2000.)

location – more on this later) and peak values that may be an order of magnitude higher in the proximal aorta and lower as one moves distally in the circulation (Lipowsky, 1995; Tarbell and Hollis, 1982; Greve et al., 2006). Pulsatility is greatly diminished in the microcirculation and the venous circulation (Milnor, 1989).

Large arteries expand predominantly in the circumferential direction, as longitudinal expansion is constrained due to blood vessel branching and tethering (Patel and Fry, 1964). During normal blood flow, the arterial wall is constantly subjected to pulsing transmural pressure that drives variations in vessel diameter (D). The circumferential strain is defined as CS $\equiv (D - D_{\text{mean}})/D_{\text{mean}}$, and the difference between the maximum and the minimum CS over the pressure pulse in large arteries such as the thoracic aorta, carotid artery, femoral artery, coronary arteries, and pulmonary artery ranges from 2 to 18% (Dobrin, 1978). On the venous side of the circulation, there is almost no diameter variation due to the low pressure pulse.

In the larger arteries, ECs are subjected simultaneously to oscillatory WSS and CS that act in approximately perpendicular directions in straight vessels (Figure 15.1). Due to the temporal phase angle between the pressure and flow (impedance phase angle, IPA) and the temporal phase angle between the WSS and flow induced by the inertial effects of blood flow, the WSS is not always in-phase with the CS. This out-of-phase interaction of the CS and WSS can generate a complex, time-varying mechanical force pattern on the EC layer that can be characterized by the temporal phase angle between the CS and the WSS that has been referred to in the literature as the stress phase angle (SPA – Qiu and Tarbell, 2000) and is illustrated in Figure 15.1.

As part of the physiological background for this review, it is important to understand the factors that contribute to variations in SPA throughout the circulation. To gain insight, Qiu and Tarbell (2000) expressed the SPA as the phase angle (φ) between the vessel diameter (D) and the WSS (τ) and denoted it $\varphi(D - \tau)$. This definition recognized that the CS is synchronous with the variation in vessel diameter (D). The SPA was decomposed into two parts through the following identity:

$$\varphi(\mathrm{D} - \tau) = \varphi(\mathrm{D} - \mathrm{Q}) - \varphi(\tau - \mathrm{Q}) \sim \varphi(\mathrm{P} - \mathrm{Q}) - \varphi(\tau - \mathrm{Q}), \qquad (15.1)$$

where $\varphi(D - Q)$ is approximately equal to the impedance phase angle (IPA), $\varphi(P - Q)$, since the diameter (D) and pressure (P) are nearly in-phase (elastic vessel), and $\varphi(\tau - Q)$ is the phase angle between the WSS and the flow rate (Q).

$\varphi(P - Q)$ is determined by the resistance, compliance, and wave reflections in the circulation distal to the vessel of interest. $\varphi(P - Q)$ of the first harmonic of a vascular waveform is about $-45°$ in the aorta and larger arteries that feed into a high impedance flow circuit (the coronaries are an exception due to their location on the beating heart), approaches $0°$ in small arteries due to a reduction in distal compliance, and also approaches $0°$ in veins that feed into a low impedance flow circuit (Milnor, 1989; Nichols et al., 1977).

The shear-flow phase angle $\varphi(\tau - Q)$ in straight vessels is dictated by the relative importance of unsteady inertia and viscous forces and depends strongly on the unsteadiness parameter ($\alpha \equiv a\sqrt{\omega/\upsilon}$, where a is the vessel radius, ω is the fundamental radian frequency of the heartbeat, and υ is the kinematic viscosity of blood). For large straight arteries and veins with high α (order 20), $\varphi(\tau - Q)$ approaches $+45°$; for small arteries, veins, and microvessels with low α, $\varphi(\tau - Q)$ approaches $0°$ (Womersley, 1955). Therefore, we have the following approximations for the SPA distribution in straight vessels:

$$\text{Large (straight) artery : SPA} = -45^{\circ} - 45^{\circ} = -90^{\circ}$$

$$\text{Large (straight) vein : SPA} = 0^{\circ} - 45^{\circ} = -45^{\circ}$$

$$\text{Small (straight) artery : SPA} = 0^{\circ} - 0^{\circ} = -0^{\circ}$$

$$\text{Small (straight) vein : SPA} = 0^{\circ} - 0^{\circ} = -0^{\circ}$$

It is apparent from these considerations that the CS and WSS will be more asynchronous (SPA more negative) in large arteries than in veins. Equation (15.1) also indicates more generally that the SPA is dictated by systemic factors that control the impedance phase angle $\varphi(P - Q)$ and local factors that control $\varphi(\tau - Q)$.

15.2.1 Factors Influencing $\varphi(\tau - Q)$

In addition to the unsteady inertia characterized by the unsteadiness parameter that contributes as much as $45°$ to $\varphi(\tau - Q)$, the nonlinear convective acceleration that is induced by vessel geometry can make even greater contributions. In two studies of stretch and shear in geometries that are at risk for the development of intimal hyperplasia and atherosclerosis, large effects were induced in $\varphi(\tau - Q)$. Qiu and Tarbell (1996) numerically simulated flow in an end-to-end, 16% undersized graft anastomosis with a normal IPA of $-45°$. The SPA was found to be $-55°$ in the proximal converging flow region (high shear) while dropping to $-140°$ in the distal diverging flow region (low shear). Thus $\varphi(\tau - Q)$ was strongly dependent on the spatial position in the nonlinear geometry, varying from 10° in the high shear region to 95° in the low shear region. Lee and Tarbell (1997) performed experimental studies of a compliant abdominal aortic bifurcation model with a normal IPA of −30°. The SPA was observed to be −40° on the flow divider (high shear) and −100° on the outer wall (low shear region where atherosclerosis arises – Mark et al., 1989). Again, $\varphi(\tau - Q)$ was strongly dependent on the spatial position, varying from 10° in the high shear region to 70° in the low shear region.

The mechanism that leads to larger $\varphi(\tau - Q)$ in low shear regions was investigated by flow visualization in the aortic bifurcation model by Lee and Tarbell (1997). They observed that reversed (upstream) flow started earlier on the outer wall of the bifurcation (low shear), during mid-systole of the flow cycle, while the flow was still in the forward direction near the inner wall (low shear). This is equivalent to an effective phase lead of wall shear (τ) on the outer wall relative to the inner wall. This mechanism seems to be quite general and has been observed in all vascular geometries to date (anastomosis, bifurcation, curvature).

The most extreme effects of geometry on $\varphi(\tau - Q)$ have been uncovered in the bifurcation of the common carotid artery into the internal and external carotid arteries as reported in simulations by Tada and Tarbell (2005). The model geometry is shown in Figure 15.2 (top). The fluid was assumed Newtonian with a density of 1.05 g/cm^3 and a viscosity of 3.5 cp. The vessel wall was modeled as an incompressible, linear elastic material with Young's modulus of 5.0×10^6 dynes/cm^2. The numerical simulation was performed by a fluid–structure interaction method that accounted for the coupling between the fluid and solid phases.

Physiological flow and pressure waveforms were imposed in the simulation, and the input impedance spectrum computed for the common carotid artery was in accord with measurements reported in the literature. Circumferential strain variations of 3–4% over the pulsatile pressure cycle were predicted, again in accord with observations. WSS waveforms at important locations are displayed in Figure 15.2 (middle). They show the characteristic distinction between the internal carotid artery (ICA) outer wall (low shear) and the ICA inner wall (high shear). The new result that was demonstrated for the first time in this study is the spectrum of the stress phase angle (Figure 15.2 – bottom). At the outer wall of the ICA (low shear region), the SPA of the first harmonic approaches –180°, whereas the SPA is nearly 0° at the inner wall of the ICA (high shear region). The IPA of the first harmonic was −36°, and $\varphi(\tau - Q)$ varied from −36° in the high shear region to 144° in the low shear region – an extreme variation of 180°. It is also clear that the complete spectra of SPA for these two regions are widely divergent, whereas the spectra in the external carotid artery (ECA) are convergent. This simulation shows highly negative SPA in a region well known to be susceptible to atherosclerotic plaque localization.

15.2.2 Factors Influencing $\varphi(P - Q)$

In addition to the well-known differences between the input impedances of arteries and veins [$\varphi(P - Q)$ more negative in arteries than veins], systemic factors including

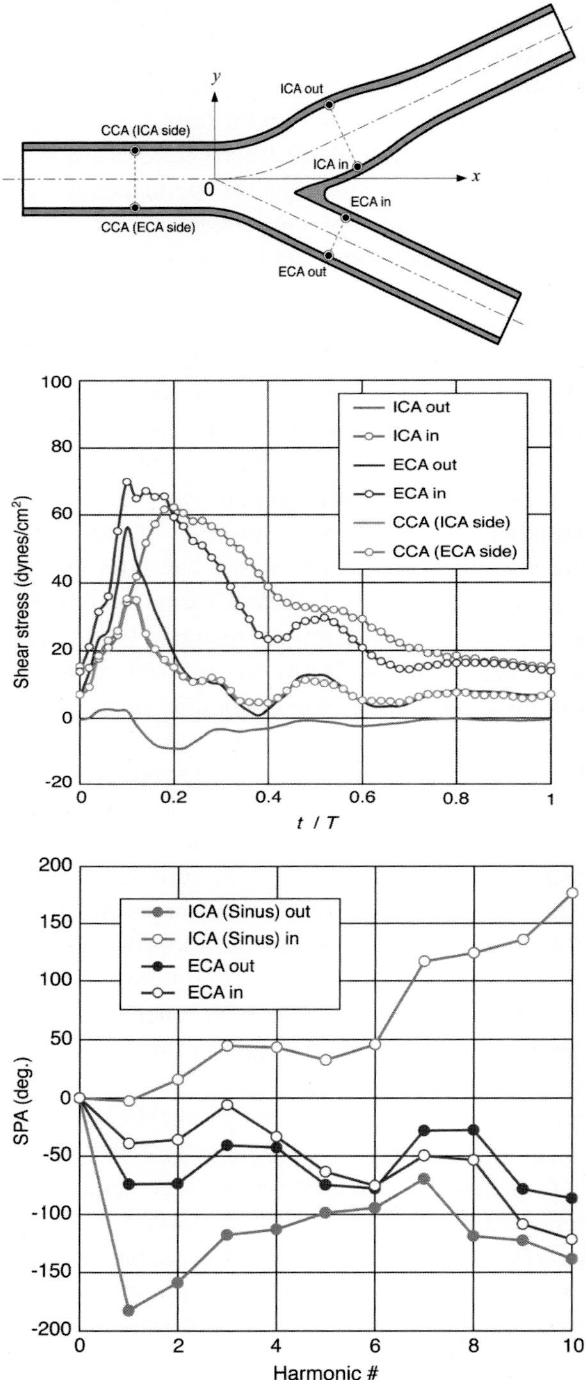

Figure 15.2. Top: Mid-plane of the carotid bifurcation model. Middle: WSS waveforms at various locations. Bottom: SPA spectrum at various locations. (Adapted from Tada and Tarbell, 2005.)

hypertension and vasoactive drugs can affect arterial input impedance. Hypertension, a well-known risk factor for cardiovascular disease, and vasoconstrictor drugs both reduce $\varphi(P - Q)$ to more negative values resulting in more asynchronous CS and WSS, whereas vasodilator drugs have the opposite effect (Ting et al., 1986; O'Rourke and Taylor, 1967; Lee and Tarbell, 1998).

The impedance of the coronary circulation of the left ventricle, where atherosclerosis is prominent, is unique because of the position of the distal blood vessels within the contracting heart muscle. During ventricular systole, when pressure (and in turn CS) is high, flow (and WSS) is low because of the increased hydrodynamic resistance associated with vessels collapsing within the contracting heart muscle. During diastole, when the ventricle relaxes, the flow (WSS) is high and the pressure (CS) is low. This leads to $\varphi(P - Q)$ of the first harmonic approaching $-180°$. Using pressure and flow data measured in the left circumflex coronary artery of a resting dog (Atabek et al., 1975), Qiu and Tarbell (2000) estimated $\varphi(P - Q)$ of the first harmonic to be $-222°$. This unique impedance characteristic of the coronary arteries of the left ventricle ensures that the CS and WSS will be highly asynchronous. Qiu and Tarbell (2000) also computed the SPA of a compliant curved tube model of a coronary artery experiencing the pressure and flow waveforms of Atabek's dog. The tube curvature simulated the contour of the epicardial surface of the heart. The SPA on the inside wall of the curved tube (low shear region) was $-250°$ and on the outside wall (high shear region) it was $-220°$. Thus $\varphi(\tau - Q)$ varied only modestly from $-2°$ (high shear) to $28°$ (low shear).

The relatively small spatial variation of SPA in the coronary artery (30°) is quite distinct from the large spatial variation of SPA in the internal carotid artery (180°). However, both arteries can experience highly asynchronous CS and WSS: SPA $=$ $-180°$ at the outer wall of the ICA and $-220°$ to $-250°$ in the coronary artery. The big difference between the two locations is that in the coronary artery the asynchrony is driven by the impedance of the coronary network whereas in the internal carotid it is the nonlinear inertia induced by the geometry of the carotid sinus that dominates the asynchrony.

15.3 Biological Response of Endothelial Cells to Simultaneous CS and WSS

15.3.1 Methods for Imposing Simultaneous Stretch and Shear *In Vitro*

Experimental studies of simultaneous stretch and shear were first undertaken using tubular systems by Moore et al. (1994), Benbrahim et al. (1994), Ziegler et al. (1998), and Peng et al. (2000). In all of these investigations, cultured endothelial cells were plated on the inside surfaces of straight, compliant, silicone tubes and placed in a flow loop with a pulsatile flow driving the WSS and pulsatile pressure driving the CS. In these initial studies the concept of the SPA was not yet recognized and there was no attempt to generate asynchronous CS and WSS by using the impedance of the flow loop ($\varphi(P - Q)$). As a result, it appears that only modest levels of asynchrony (not measured or reported) were generated, primarily from the Womersley effect on $\varphi(\tau - Q)$ (estimated to be less than 30°). As a result, these were devices that generated nearly synchronous CS and WSS with SPA close to 0°.

A planar device for applying stretch and shear simultaneously was described recently by Owatverot et al. (2005). It incorporated a planar moving boundary to generate Couette flow and defined WSS on a cell-laden silicone membrane that was being stretched uniaxially. The system was operated with an SPA of 0° in the only biological study that was described.

Building on the initial tubular studies and recognizing the potential significance of asynchronous CS and WSS that had been described by Lee and Tarbell (1998), a compliant tubular device was developed, but this time with a variable impedance element in the flow loop distal to the cell-coated test section (Qiu and Tarbell, 2000) to allow control over the SPA. An air compliance chamber in series with an adjustable length of elastic tubing and a variable resistance clamp provided an adjustable impedance network allowing control of $\varphi(P - Q)$. Using this flow loop, it was possible to generate physiological levels of SPA between −14° and −102°.

Recognizing that the complete asynchrony of the coronary arteries and the carotid sinus could not be approached with the conventional impedance network employed by Qiu and Tarbell (2000), a new flow device was developed that was first used by Dancu et al. (2004) and was more thoroughly described by Dancu and Tarbell (2006). This device also employed compliant silicone tubes to support the endothelial cells, but a unique pumping mechanism to allow complete control of the SPA was devised. A pair of pulsatile piston pumps, one proximal and one distal to the test section, that were operated in a "push–pull" mode, allowed the generation of oscillatory flow (WSS) without changing the volume (or CS) of the test section. By enclosing the test section in a rigid chamber connected to an external piston pump, it was possible to vary the volume of the test section (and in turn the CS) independently of the flow (WSS). By adjusting the temporal phase of the external piston pump with respect to the in-line piston pumps, it was possible to generate a complete range of SPA between 0° and −180°.

15.3.2 Studies with SPA Approximately 0°

One of the most characteristic responses of endothelial cells to mechanical forces is morphological adaptation. In response to WSS by itself, endothelial cells become elongated and align in the direction of the applied stress (Levesque and Nerem, 1985), which in the case of a straight, cylindrical vessel is in the direction of the tube axis. Stretch by itself causes endothelial cell elongation and alignment as well, but in a direction perpendicular to the applied stretch. For a cylindrical vessel this is the tube axis as well. Moore et al. (1994) showed that elongation and alignment of ECs was significantly enhanced at 24 h when the WSS and CS were applied concurrently (in-phase) compared to the WSS by itself. Owatverot et al. (2005) showed that when stretch and shear (in-phase) were applied in perpendicular directions, as occurs physiologically in straight vessels, their effects on cell alignment were reinforcing, as observed by Moore et al. (1994). When these forces were applied in parallel, the alignment response was suppressed.

Casey et al. (2001) measured nitric oxide (NO) release from human saphenous vein ECs cultured on tubes under steady and pulsatile shear conditions for both rigid and compliant cases. When cells were exposed to pulsatile flow in a compliant system, the highest levels of NO were produced. Independent decreases in compliance, flow, and pulsatility resulted in significantly lower rates of NO production than those in the group with these conditions intact. This study indicates that combined CS and WSS applied in-phase (SPA = 0°) lead to the most favorable result as NO production is considered a positive sign of endothelial health.

Peng et al. (2003) exposed BAECs on compliant and stiff tubes to pulsatile perfusion with a mean shear stress of 7 dynes/cm^2 without shear reversal for 2 h. In compliant tubes, pulsatile perfusion doubled protein kinase B (Akt) phosphorylation above nonpulsatile flow levels, whereas P-Akt declined to static levels in stiff tubes. Endothelial nitric oxide synthase (eNOS) phophorylation (S-1179) similarly increased with pulsatile perfusion in compliant tubes but was nearly undetectable with increased pulsatile perfusion in stiff tubes. The novel finding of these studies was that stiff tubes profoundly inhibited P-Akt and P-eNOS. In a subsequent study using the same cells and tubes, this group showed that pulse perfusion of cells in compliant tubes had twice the protection against H_2O_2–stimulated apoptosis compared to cells on stiffer tubes (Li et al., 2005). Since reduced arterial compliance is a major risk for cardiovascular disease (Zieman et al., 2005), these studies support the concept of in-phase stretch and shear as vasoprotective.

In further support of the benefits of in-phase cyclic stretch, Kwak et al. (2005), using the same system as Ziegler et al. (1998), exposed cultured endothelial cells to various shear, stretch, and pressure conditions and measured the expression of connexin 43 (C × 43) protein after 4 and 24 h of exposure. Their results showed that cyclic shear stress and circumferential stretch, but not pressure, affected C × 43 levels. Most notably, 4% cyclic stretch up-regulated C × 43 by 7-fold at 4 h and 3.5-fold at 24 h for a reversing oscillatory shear case. For uni-directional shear, C × 43 was also up-regulated by stretch, but less dramatically.

The only study that seemed to contradict the notion that in-phase cyclic stretch is beneficial to endothelial cells was reported by Ziegler et al. (1998), who showed that both eNOS and endothelin-1 (ET-1) mRNA from BAECs grown on silicone tubes were not significantly affected by cyclic stretch added to nonreversing oscillatory flow (6 ± 3 dynes/cm^2:mean ± amplitude) or reversing oscillatory flow (0.3 ± 3 dynes/cm^2).

15.3.3 Studies with Variable SPA

While the association of asynchronous CS and WSS with atherosclerosis-susceptible regions of the circulation had been recognized by the late 1990s (as described in section 15.2), it had not been demonstrated that the SPA could have any influence on the biological response of ECs to simultaneous WSS and CS. The first experimental study to probe the influence of SPA on EC biology was reported by Qiu and Tarbell

(2000). They plated BAECs on the inner wall of elastic tubes and exposed the cells to identical sinusoidal WSS waveforms (10 ± 10 dynes/cm^2) and CS waveforms ($\pm$ 4%) at either SPA = $-14°$ or $-102°$ for 4 h. The more asynchronous case (SPA = $-102°$) induced the production of the vasoconstrictor, ET-1, and suppressed the production of the vasodilators, NO, and prostacyclin (PGI_2), relative to SPA = $-14°$. Since ET-1 is pro-atherogenic (a smooth muscle mitogen) and NO and PGI_2 are athero-protective (they inhibit smooth muscle proliferation), there *in vitro* data provided a link between the asynchronous hemodynamics characteristic of disease-susceptible regions of the circulation and a biological response of ECs that was pro-atherogenic.

The study also showed that when the elastic tubes were surrounded by a rigid sleeve to prevent stretch, and the same shear stress waveform was imposed, the production levels of PGI_2, NO, and ET-1 were nearly identical to the more asynchronous case without the sleeve, suggesting that asynchronous hemodynamics induce a response that is characteristic of a rigid vessel. This work supports the hypothesis that asynchronous hemodynamic conditions in compliant vessels induce pro-atherogenic cellular responses similar to rigid vessels.

A follow-up to the initial study of Qiu and Tarbell (2000) was an investigation of the expression of three associated genes: eNOS, cyclooxygenase-2 (COX-2), and ET-1 at 5- and 12-h time points (Dancu et al., 2004). The study was again carried out with BAECs on silicone tubes, but using the new hemodynamic simulator. The sinusoidal flow cases (SPA = $0°$ and SPA = $-180°$) had the same WSS (10 ± 10 dynes/cm^2) and the same CS (4 ± 4 %) – the two cases differed only by the SPA. The control cases examined the influence of steady shear (WSS = 10 dynes/cm^2) alone (SS) and steady pressure alone (PC) that induced steady strain (CS = 4%). All of these cases were compared to the static control (SC) with WSS = 0 dynes/cm^2 and CS = 0%. Figure 15.3 shows the results for eNOS and ET-1.

At both 5 and 12 h the SPA = $0°$ and $-180°$ cases are significantly different from each other for eNOS, and these differences are greater than the differences between SS and SC or PC and SC, implying that the phase angle is more influential than either steady shear stress or steady stretch. For ET-1, the differences between SPA = $0°$ and $-180°$ are also highly significant at 12 h. Similarly significant effects of SPA have been observed for COX-2 by Dancu et al. (2004), where it is also shown that the respective products measured in the media (NO, PGI_2, and ET-1) follow the patterns of their associated genes, suggesting that their production at these time points is controlled by transcription. These data indicate that hemodynamics with asynchronous CS and WSS (SPA = $-180°$) induces an atherogenic gene expression pattern (low eNOS, high ET-1) compared to synchronous hemodynamics (SPA = $0°$).

Because the coronary arteries are highly susceptible to atherosclerosis and are characterized by highly asynchronous hemodynamics, Dancu and Tarbell (2006) continued the theme of earlier *in vitro* studies by examining *in vitro* gene expression profiles in rabbits for eNOS and ET-1 in the left anterior descending coronary artery (cLAD) and left circumflex coronary artery (cCIRC) relative to the more synchronous aorta. Three sections of the aorta were sampled: the outer wall and the inner wall of the aortic arch and a straight section of the thoracic aorta.

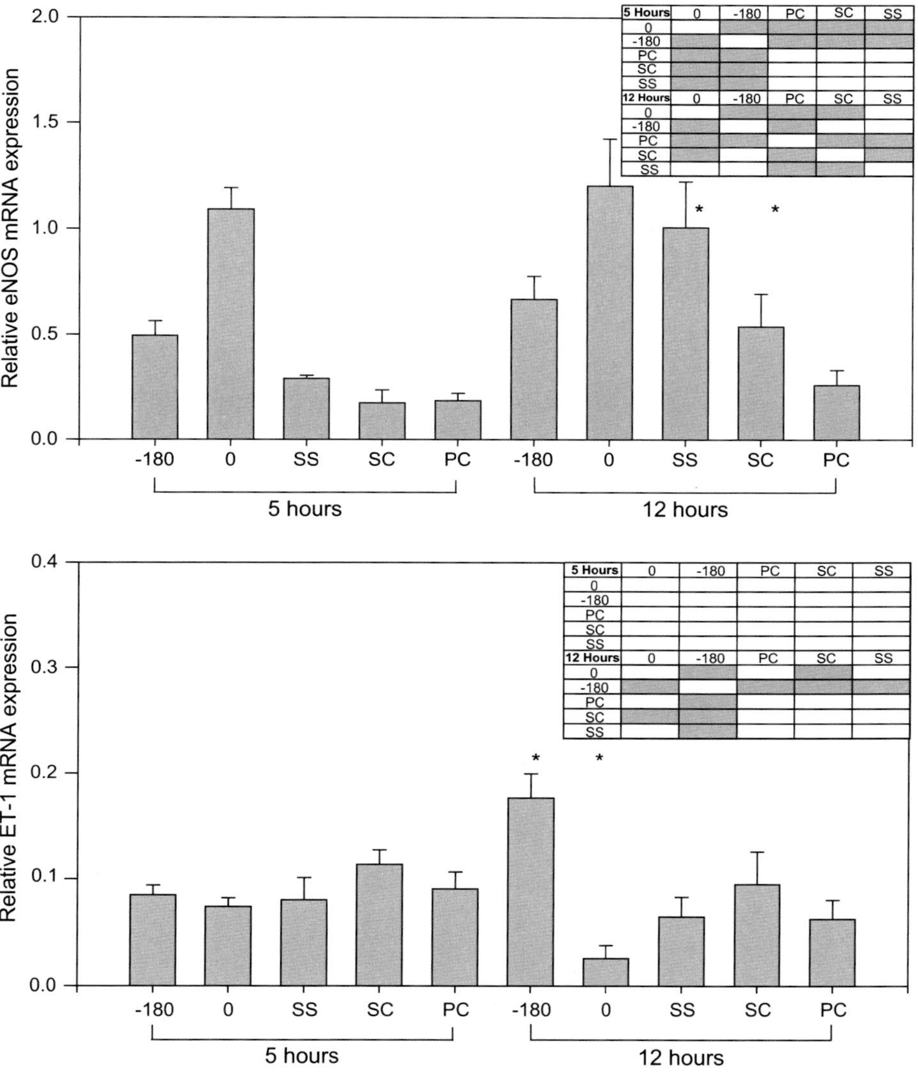

Figure 15.3. eNOS (top) and ET-1 (bottom) gene expressions at 5 and 12 h normalized to the housekeeping gene, HPRT. The inset table indicates pairwise significant differences ($P < 0.05$) by the shaded boxes and the asterisk for 5 versus 12 h. (From Dancu et al., 2004.)

The relative gene expressions for eNOS and ET-1 in the aortic and coronary regions are shown in Figure 15.4. It is striking to note the distinct suppression of eNOS and elevation of ET-1 in the coronary arteries relative to the aorta. When all of the coronary results are combined and then compared to the combined aortic results, the differences are highly significant ($P < 0.01$). These results are consistent with the *in vitro* studies (Dancu et al., 2004) comparing gene expression under synchronous (high eNOS, low ET-1) and asynchronous (low eNOS, high ET-1) hemodynamic conditions (Figure 15.3).

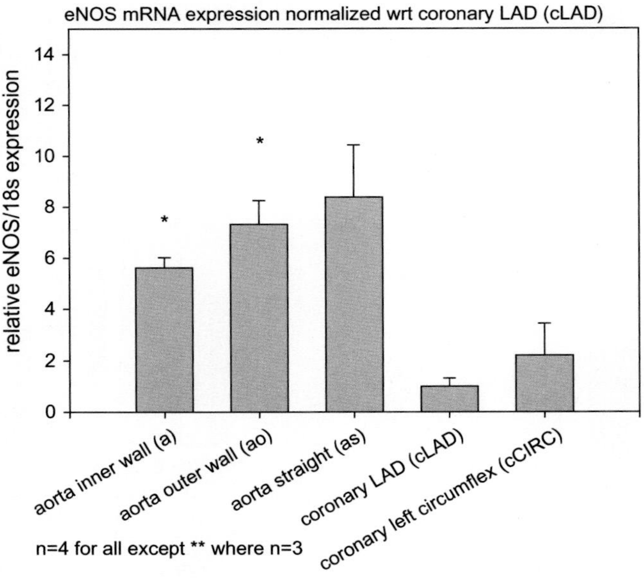

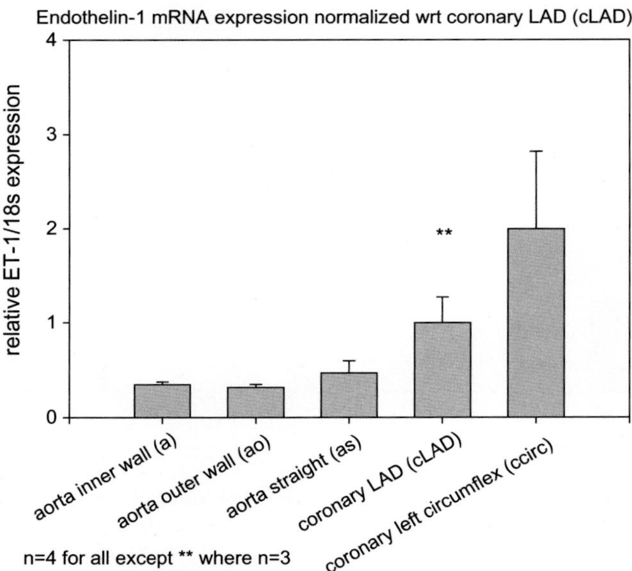

Figure 15.4. eNOS (top) and ET-1 (bottom) gene expressions in the aorta and coronary arteries. In both cases $n = 5$; $*P < 0.05$ compared to cLAD. (From Dancu and Tarbell, 2006.)

To gain insight into the local hemodynamics in the regions where mRNA was sampled, Dancu and Tarbell (2006) used a nuclear staining method that has been used widely in the literature (Flaherty et al., 1972; Staughton et al., 2001) to assess, in a semi-quantitative manner, the local WSS characteristics. They measured the length (major axis) and width (minor axis) of endothelial nuclei and the orientation of the major axis relative to a fixed reference direction (e.g., the vessel axis in a straight section). The results for the major/minor axis ratio are shown in Figure 15.5. The

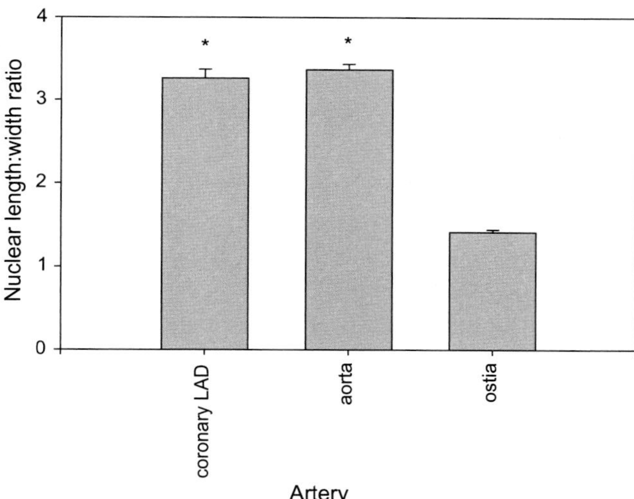

Figure 15.5. Nuclear length:width ratios of endothelial cells from the coronary LAD, straight aorta, and intercostal ostia. Each mean and SEM was based on measurements of approximately 100 cells. $*P < 0.05$ compared to ostia. (From Dancu and Tarbell, 2006.)

coronary LAD and the straight section of the aorta displayed nearly identical ratios that were significantly different from cells near the ostia of an aortic intercostal artery, where flow is well known to be disturbed (Staughton et al., 2001). The standard deviation of the major axis orientation angle (not shown here) was found to be less than 5° for both the cLAD and the straight aorta whereas it was more than 35° near the aortic ostia. These data suggest that WSS characteristics in the cLAD and the straight aorta are similar and are not associated with disturbed flow. It should be recognized that EC and their nuclei will elongate (align) in the direction of WSS and perpendicular to the direction of CS (Wang et al., 2001). For regions such as the cLAD and the straight aorta, these two directions coincide (in the tube axis direction). Therefore, it must be emphasized that EC nuclear shape and orientation do not reflect the complete hemodynamic environment because they do not distinguish synchronous from asynchronous hemodynamics as characterized by the SPA. Dancu and Tarbell (2006) hypothesized that the dramatic differences between cLAD and the straight aorta in gene expression (Figure 15.3) are not associated with disturbed flow conditions in one region versus the other but, rather, with significant differences in SPA.

The results of Dancu and Tarbell (2006) draw attention to a dilemma that has long been recognized in studies of mechanical factors in vascular disease: Mechanical features that are inherent in the "design" of the cardiovascular system, such as the asynchronous hemodynamics of coronary arteries and the unique geometry of the carotid sinus, play a significant role in predisposing these vessels to the development of atherosclerosis. A key question that follows from this dilemma is, even if the most atherogenic features of the mechanical environment are established, how can this knowledge be used to treat vascular disease? We are not likely to contemplate surgical re-plumbing of the coronary circulation to create hemodynamics that are more synchronous! On the other hand, one mechanical approach that has been implemented in the clinic is enhanced external counterpulsation (EECP), a noninvasive, pneumatic technique that provides beneficial effects for patients with chronic, symptomatic angina pectoris. EECP uses the sequential inflation of three sets of pneumatic cuffs wrapped around the lower extremities. The cuffs are inflated sequentially (from calf to thigh to buttock) at the onset of diastole, producing aortic counterpulsation and diastolic augmentation. Several large studies in patients have shown that EECP provides symptomatic relief of angina, improves blood flow to ischemic areas of the myocardium, eliminates or reduces nitrate use, and improves exercise tolerance (Wu et al., 2006). Pressure measurements in patients (Michaels et al., 2002) and calculations using a circulatory model (Ozawa et al., 2001) show significant increases in diastolic coronary artery pressure and flow induced by EECP. Since circumferential stress (CS) is approximately in-phase with pressure, and wall shear stress (WSS) is approximately in-phase with flow in the coronary arteries, it is clear that EECP induces synchronous CS and WSS in the coronaries. Of course, flow augmentation in the coronary arteries during diastole should be beneficial by itself, but the synchronous CS and WSS induced by EECP may be a previously unrecognized mechanism through which the procedure provides benefit.

A more biologic approach to treating the consequences of an adverse mechanical environment might be to ask whether there are pharmaceuticals or nutraceuticals that can normalize the atherogenic gene expression profile induced by hemodynamics. To begin to explore this, Dancu et al. (2007) administered the nutraceutical, conjugated linoleic acid (CLA), a dietary fatty acid found in dairy products and meats of ruminant animals, and a peroxisome proliferator-activated receptor (PPAR) $-\alpha$ and $-\gamma$ activator. BAECs were exposed to synchronous (SPA $= 0°$) and asynchronous (SPA $= -180°$) hemodynamic environments for 5 and 12 h with or without CLA, and gene expression profiles were compared. CLA-treated systems showed enhancement of eNOS that was suppressed by asynchrony when CLA was not present, and inhibition of ET-1 that was induced by asynchrony when CLA was not present. This study demonstrated the potential for pharmacological treatment to normalize pro-atherogenic gene expression profiles induced by a natural hemodynamic state.

15.4 Mechanism of Stretch and Shear

While the experimental observations that have been described are intriguing and strongly support the importance of the interaction between solid strain (CS) and fluid shear (WSS), there has been limited theoretical assessment of the manner by which these forces might interact or the proposal of a mechanism whereby the phase angle between them (SPA) could be influential. This leads to considerations of mechanotransduction mechanisms for CS and WSS on endothelial cells. As several review papers have emphasized, the plasma membrane and its associated glycocalyx, the intercellular junctions (adherens junctions), the basal adhesion plaques, and the cytoskeleton are structures that can mediate mechanotransduction driven by WSS and CS (Tarbell and Pahakis, 2006; Secomb et al., 2001; Kamm and Kaazempur-Mofrad, 2004; Ingber, 1998; Davies, 1995; Banes, 1993; Fry et al. 2005). It seems quite clear from the outset that endothelial cells do not possess a "phase angle sensor." Rather, it is the interaction of the oscillating WSS and CS forces that is modulated by the phase angle (SPA).

The first model to consider the simultaneous effects of CS and WSS on endothelial cells was developed by Tada et al. (2007). This biomechanics model treated the endothelial cell as a membrane (plasma membrane) covering a viscoelastic, solid cell body (cytoplasm). They built upon an earlier theoretical analysis by Fung and Liu (1993) that was extended by Wiesner et al. (1997) to consider how WSS (alone) on the plasma membrane could alter the strain energy density (SED) of the membrane and in turn the opening of ion channels mediating signal transduction. These earlier studies were based on steady-state (time-averaged) equations for the membrane and assumed that the circumferential tension was zero everywhere. Tada et al. (2007) allowed for nonzero circumferential tension (CS) and considered the time-dependent equations for the membrane so that the SPA could be introduced naturally and analyzed. The model shows a strong influence of SPA on the local SED of the membrane, particularly in the upstream and downstream regions where signal transduction is predicted to occur.

The model makes the very interesting prediction that the time-average value of the SED for the deformable model (CS = $\pm$ 2.5%), at both the up- and downstream ends of the plasma membrane, approaches the value for a rigid model (CS = 0%) as the SPA takes on highly negative values (approaching $-180°$). Related to this prediction, it is important to remember that Qiu and Tarbell (2000) investigated the interaction of sinusoidal WSS and CS on the endothelial production of PGI_2, NO, and ET-1 using rigid (CS = 0%) and compliant (CS = $\pm$ 4%) tubes. Bovine aortic endothelial cell production of PGI_2, NO, and ET-1 for the compliant tubes with a highly negative SPA ($-102°$) showed very similar trends to those of rigid tubes – reduced PGI_2 and NO and elevated ET-1, atherogenic characteristics. This suggests that variations of the SED of the plasma membrane near the up- and downstream ends of the cell play a role in the mechanotransduction of simultaneous CS and WSS. Related to this, recall that Li et al. (2005) compared pulsatile flow effects on BAECs plated on rigid or compliant tubes and showed that rigid tubes suppressed Akt-dependent anti-apoptosis signaling (pro-atherogenic) compared with compliant tubes where stretch and shear were approximately in-phase (Li et al., 2005). All of this supports the concept that noncompliant vessels and compliant vessels with highly asynchronous CS and WSS induce an atherogenic phenotype in endothelial cells.

Further elaborations of the initial model of Tada et al. (2007) incorporating additional cellular structures (basal adhesion plaques, intercellular junctions, etc.) should prove useful in understanding mechanotransduction driven by CS and WSS over a physiological range of SPA.

15.5 Concluding Remarks

This review has been brief because there has been relatively little systematic study of the combined forces of solid mechanical strain and fluid mechanical shear on endothelial cells. There appear to be at least two reasons for the lack of emphasis on combined forces in the literature. One is related to the difficulty of applying combined forces experimentally *in vitro*. Further emphasis on the development of convenient systems for application of simultaneous forces *in vitro* will surely advance the field. The second is simply a lack of appreciation of the possibility for unique endothelial cell responses to combined forces, not characteristic of either force alone. Hopefully this review will contribute to the awareness of the range of responses made possible by combined forces.

It is clear, however, that in a healthy circulatory system, both CS and WSS are significant forces in arteries whereas CS is diminished in veins and the microcirculation. In arteries, CS and WSS are highly asynchronous in precisely those regions of the circulation where atherosclerosis most commonly occurs (coronary arteries, carotid sinus, inner curvatures of bends, and outer wall of bifurcations). Compliant arteries experiencing asynchronous CS and WSS display a phenotype that is similar to stiffened arteries, suggesting the existence of common pro-atherogenic mechanical mechanisms. While it may be impractical to alter the endothelial phenotype induced by asynchronous forces using mechanical means, pharmaceutical (nutraceutical) approaches are feasible and should be investigated more thoroughly.

The mechanisms by which strain and shear simultaneously exert their influences on endothelial cells remain almost completely unexplored. Certainly biomolecular pathways that are operative when each force is applied separately are likely to be important when they are applied in concert. Whether new pathways that are uniquely activated by combined forces are significant or it is the modulation of pathways characteristic of separate forces that dominate responses remains a challenging question for the future.

Acknowledgment

This work was supported by NIH Grants HL 35549 and HL 086543.

REFERENCES

Atabek HB, Ling SC, Patel DJ. Analysis of coronary flow fields in thoracotomized dogs. *Circulation Engng* 1975; **107**: 307–315.

Banes AJ. Mechanical strain and the mammalian cell; in *Physical Forces and the Mammalian Cell*, Frangos JA (ed). 1993, 81–123, Academic Press, New York.

Benbrahim A, L'italien GJ, Milinazzo BB. A compliant tubular device to study the influences of wall strain and fluid shear stress on cells of the vascular wall. *J Vasc Surg* 1994; **20**: 184–194.

Casey PJ, Dattilo JB, Dai G, Albert JA, Tsukurov OI, Orkin RW, Gertler JP, Abbott WM. The effect of combined arterial hemodynamics on saphenous venous endothelial nitric oxide production. *J Vasc Surg* 2001; **33**: 1199–1205.

Dancu MB, Tarbell JM. Large negative stress phase angle (SPA) attenuates nitric oxide production in bovine aortic endothelial cells. *J Biomech Eng* 2006; **128**: 329–334.

Dancu MB, Tarbell JM. Coronary endothelium expresses a pathologic gene expression pattern compared to aortic endothelium: Correlation of asynchronous hemodynamics and pathology *in vivo*. *Atherosclerosis* 2007; **192**(1): 9–14.

Dancu M, Berardi D, Vanden Heuvel JP, Tarbell JM. Asynchronous hemodynamic shear stress and circumferential strain characteristics of coronary arteries reduces eNOS and COX-2 but induces ET-1 gene expression in endothelial cells. *Arterioscler Thromb Vasc Biol* 2004; **24**: 2088–2094.

Dancu M, Berardi D, Vanden Heuval JP, Tarbell JM. Atherogenic responses to asynchronous hemodynamics are mitigated by CLA. *Ann Biomed Eng* 2007; **35**(7): 1111–19.

Davies PF. Flow-mediated endothelial mechanotransduction. *Physiol Rev* 1995; **75**: 519–560.

Dobrin PB. Mechanicalproperties of arteries. *Physiol Rev* 1978; **58**: 397–460.

Frye SR, Yee A, Eskin SG, Guerra R, Cong X, McIntire LV. cDNA microarray analysis of endothelial cells subjected to cyclic mechanical strain: importance of motion control. *Physiol Genomics* 2005; **21**: 124–130.

Flaherty JT, Pierce JE, Ferrans VJ, Patel DJ, Tucker WK, Fry, DL. Endothelial nuclear patterns in the canine arterial tree with particular reference to hemodynamic events. *Circ Res* 1972; **30**: 23–33.

FungYC, Liu SQ. Elementary mechanics of the endothelium of blood vessels. *J Biomech Eng* 1993; **115**: 1–12.

Greve JM, Les AS, Tang BT, Draney Blomme MT, Wilson NM, Dalman RL, Pelc NJ, Taylor CA. Allometric scaling of wall shear stress from mice to humans. *Am J Physiol* 2006; **291**: H1700–1708.

Ingber DE. Cellular basis of mechanotransduction. *Biol Bull* 1998; **194**: 323–327.

Kamm RD, Kaazempur-Mofrad MR. On the molecular basis for mechanotransduction. *Mech Chem Biosyst* 2004; **1**: 201–209.

Kwak BR, Silacci P, Stergiopulos N, Hayoz D, Meda P. Shear stress and cyclic circumferential stretch, but not pressure, alter connexin43 expression in endothelial cells. *Cell Commun and Adhesion* 2005; **12**: 261–270.

Lee CS, Tarbell JM. Wall shear rate distribution in an abdominal aortic bifurcation model:effects of vessel compliance and phase angle between pressure and flow waveforms. *J Biomech Eng* 1997; **119**: 333–342.

Lee CS, Tarbell JM. Influence of vasoactive drugs on wall shear stress distribution in the abdominal aortic bifurcation: An in vitro study. *Ann Biomed Eng* 1998; **26**: 200–212.

Levesque MJ, Nerem RM. The elongation and orientation of cultured endothelial cells in response to shear stress. *J Biomech Eng* 1985; **107**: 341–347.

Li M, Chiou K-R, Bugayenko A, Irani KD, Kass DA. Reduced wall compliance suppresses Akt-dependent apoptosis protection stimulated by pulse perfusion. *Circ Res* 2005; **97**: 587–595.

Lipowsky HH. Shear stress in the circulation; in *Flow-Dependent Regulation of Vascular Function*, Bevan JA et al. (eds). 1995, Oxford University Press, New York.

Mark FF, Bargeron CB, Deters OJ, Friedman MH. Variation in geometry and shear rate distribution in cases of human aortic bifurcations. *J Biomecahnics* 1989; **22**: 577–582.

Michaels AD, Accad M, Ports TA, Grossman W. Left ventricular systolic unloading and augmentation of intracoronary pressure and Doppler flow during enhanced extertnal counterpulsation. *Circulation* 2002; **106**: 1237–1242.

Milnor, WR. *Hemodynamics, second edition*, 1989, Williams & Wilkins Co., Baltimore, MD.

Moore JE, Burki E, Suciu A, Zhao S, Burnier M, Brunner HR, Meister JJ. A device for subjecting vascular endothelial cells to both fluid shear stress and circumferential cyclic stretch. *Ann Biomed Eng* 1994; **22**: 416–422.

O'Rourke MF, Taylor MG. Input impedance of the systemic circulation. *Circ Res* 1967; **20**: 365–380.

Owatverot TB, Oswald SJ, Chen Y, Wille JJ, Yin FC-P. Effect of combined cyclic stretch and fluid shear stress on endothelial cell morphological responses. *J Biomech Eng* 2005; **127**: 374–382.

Ozawa ET, Bottom KE, Xiao X, Kamm RD. Numerical simulation of enhanced external counterpulsation. *Ann Biomed Eng* 2001; **29**: 284–297.

Patel DJ, Fry DL. In situ pressure-radius-length measurements in ascending aorta of anesthetized dogs. *J Appl Physiol* 1964; **19**: 413–426.

Peng X, Haldar S, Deshpande S, Irani K, Kass DA. Wall stiffness suppresses Akt/eNOS and cytoprotection in pulse-perfused endothelium. *Hypertension* 2003; **41**: 378–381.

Peng X, Recchia FA, Byrne BJ, Wittstein IS, Ziegelstein RC, Kass DA. In vitro system to study realistic pulsatile flow and stretch signaling in cultured vascular cells. *Am J Physiol* 2000; **279**: C797–805.

Qiu Y, Tarbell JM. Computational simulation of flow in the end-to-end anastomosis of a rigid graft and a complaint artery. *ASAIO J* 1996; **42**: M702–M709.

Qiu Y, Tarbell JM. Numerical simulation of pulsatile flow in a compliant curved tube model of a coronary artery. *J Biomech Eng* 2000; **122**: 77–85.

Qiu Y, Tarbell JM. Interaction between wall shear stress and circumferential strain affects endothelial cell biochemical production. *J Vasc Res* 2000; **37**: 147–157.

Secomb TW, Hsu R, Pries AR. Effects of the endothelial surface layer on transmission of fluid shear stress to endothelial cells. *Biorheology* 2001; **38**: 143–150.

Staughton TJ, Lever MJ, Weinberg PD. Effect of altered flow on the pattern of permeability around rabbit aortic branches. *Am J Physiol* 2001; **281**: H53–59.

Tada S, Dong C, Tarbell JM. Effect of the stress phase angle (SPA) on the strain energy density of the endothelial plasma membrane. *Biophys J* 2007; **93(9)**: 3026–3033.

Tada S, Tarbell JM. Computational study of flow in a complaint carotid bifurcation: a new hemodynamic factor (the stress phase angle) correlates with plaque location. *Ann Biomed Eng* 2005; **33**: 1202–1212.

Tarbell JM, Chang LJ, Hollis TM. A note on wall shear stress in the aorta. *J Biomech Eng* 1982; **104**: 343–345.

Tarbell JM, Pahakis MY. Mechanotransduction and the glycocalyx. *J Internal Med* 2006; **259**: 339–350.

Ting CT, Brin KP, Lin SJ, Wang SP, Chang MS, Chiang BN, Yin FCP. Arterial hemodynamics in human hypertension. *J Clin Invest* 1986; **78**: 1462–1471.

Wang JH, Goldschmidt-Clermont P, Willie J, Yin FC. Specificity of endothelial cell reorientation in response to cyclic mechanical stretching. *J Biomechanics* 2001; **34**: 1563–1572.

Wiesner TF, Berk BC, Nerem RM. A mathematical model of the cytosolic-free calcium response in endothelial cells to fluid shear stress. *Proc Natl Acad Sci* 1997; **94**: 3726–3731.

Womersley JR. Method for the calculation of velocity, rate of flow and viscous drag in arteries when the pressure gradient is known. *J Physiol* 1955; **127**: 553–563.

Ziegler T, Bouzourene K, Harrison VJ, Brunner HR, Hayoz D. Influences of oscillatory and unidirectional flow environments on the expression of endothelin and nitric oxide synthase in cultured endothelial cells. *Arterioscler Thromb Vasc Biol* 1998; **18**: 686–692.

Zieman SJ, Melenovsky V, Kass DA. Mechanisms, pathophysiology, and therapy of arterial stiffness. *Arterioscl Thromb Vasc Biol* 2005; **25**: 932–943.

16

Micro- and Nanoscale Force Techniques for Mechanotransduction

Nathan J. Sniadecki, Wesley R. Legant, and Christopher S. Chen

16.1 Introduction

Mechanical forces can act as insoluble cues that affect cellular events such as migration, differentiation, growth, and apoptosis. The response to mechanical stimuli leads to adaptive and functional changes in tissue that contribute to physiological homeostasis (Hughes-Fulford 2004; Ingber 2006). Since many diseases occur in a setting where cells are exposed to abnormal forces, it is now evident that alterations in the mechanical context of healthy tissue contributes to pathological responses, such as in hypertension, asthma, and cancer (Ingber 2003; Huang and Ingber 2005). Mechanical forces that affect cellular responses also arise from within cells. Cells generate traction forces through myosin motors and cytoskeletal filaments that are essential for their locomotion and contraction (Lauffenburger and Horwitz 1996; Ridley et al. 2003). These traction forces appear to regulate the same cellular events that are observed with external forces, suggesting a common mechanism for transducing forces into biochemical responses (Chen et al. 2004). For these reasons, identifying the underlying principles in mechanotransduction has been an active area of research.

Depending on the tissue system, cells experience different kinds of external forces. Impulsive forces occur in the musculoskeletal system where strains of 3000–4000 $\mu\epsilon$ are common in bone and forces up to 9 kN have been reported in tendons during physical exertion (Lanyon and Smith 1969; Wang 2006). Rhythmic mechanical forces are pervasive in the normal physiology of the vascular or pulmonary systems. Cardiac or ventilatory cycles produce a combination of shear, tensile, and compressive stresses as blood or air flows across the cell surface and pressure levels rise and fall (Davies 1995; Waters et al. 2002). These forces act locally at the site of force but are also dispersed through viscoelastic tissues. These forces propagate along a network of macromolecules that composes the extracellular matrix (ECM), which surrounds the cells, as well as through cell–cell contacts that link adjacent cells. Because these forces are distributed throughout the tissue, the magnitudes of forces acting at the cellular level are not as large as their tissue-level counterparts and range from pico- to nano-Newtons. Yet, even these small forces

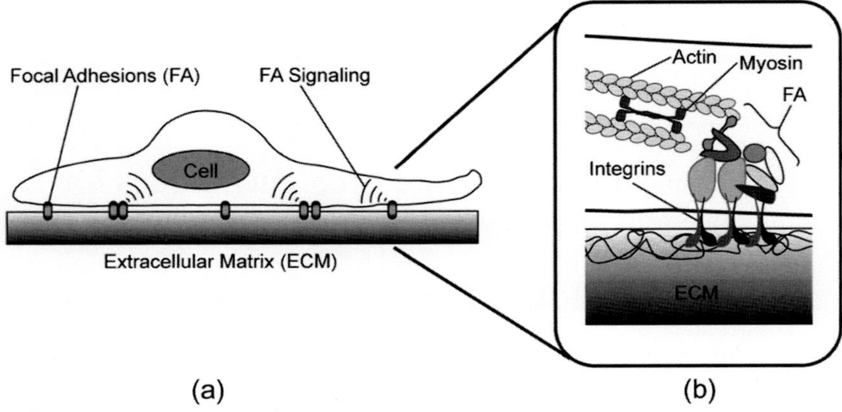

Figure 16.1. Mechanotransduction at focal adhesions. (a) Cells attached to the extracellular matrix at assemblies of proteins known as focal adhesions. Attachment to ECM enables focal adhesion signal transduction that influences cellular growth, proliferation, differentiation, and apoptosis. (b) Integrins cluster together and bind to the extracellular matrix, enabling focal adhesion proteins and actin microfilaments to fully assemble. Intracellular force generated by myosin motors acting on actin produces force at the focal adhesion, which leads to growth and stability of the adhesion site.

are able to elicit mechanotransductive responses from cells. Normal physiological processes expose cells to a variety of mechanical stimuli that differ in magnitude, frequency, and direction, but how cells sense and respond to forces at the molecular level to produce orchestrated responses is currently under investigation.

Studies have indicated that cells primarily sense mechanical forces at their adhesion sites, where integrin receptors connect the cell to the ECM (Alenghat and Ingber 2002). Integrins cluster together and bind extracellularly to ligands in the ECM and intracellularly to adaptor proteins that connect them to the cytoskeleton. These assemblies of proteins, known as focal adhesions (FAs), mediate cellular attachment to the ECM at many punctate locations (1–3 μm^2) across the cell surface (Geiger et al. 2001). Clustered integrins have numerous signaling proteins associated with them, and so FAs serve as signaling hubs to transduce intracellularly generated traction forces and extracellularly generated external forces into biochemical changes (Figure 16.1). In fact, cells subjected *in vitro* to mechanical forces experience dynamic changes in protein accumulation and signaling at FAs (Davies et al. 1994; Sawada and Sheetz 2002). However, these techniques apply forces across the entire cell and therefore limit the ability to scrutinize the exact proteins and organelles involved in mechanotransduction. As a consequence, it is difficult to discern if the observed responses are due to stresses applied to FAs or to other structures like the cell membrane, cytoskeleton, or nuclear envelope. Therefore, applying forces at FAs with precision at the micro- and nanoscale levels can elucidate the role of mechanical stresses in biological responses.

In this chapter, we review the approaches and tools used to investigate the response of cells to mechanical forces applied at their adhesion sites and to measure the strength of traction forces generated from within cells. For each tool or technique, we

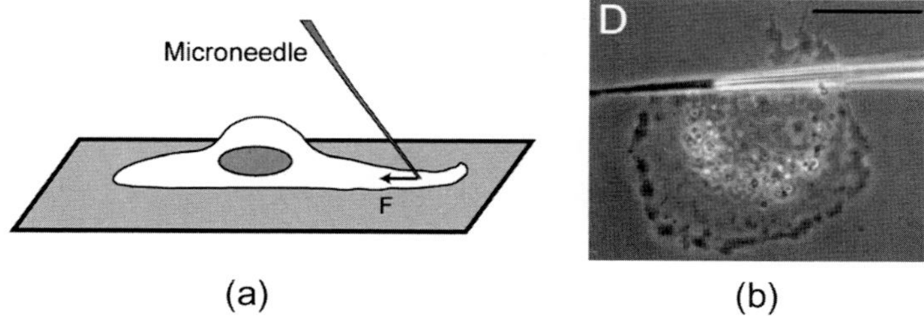

Figure 16.2. Microneedles for studying mechanotransduction. (a) A pulled-glass capillary or pipette is placed into contact and then dragged across the surface of a cell to apply force. (b) Phase contrast micrograph of cell pulled by microneedle. Adapted from Riveline et al. (2001). Scale bar: 20 μm.

give a general description, an explanation of the governing physical principles, and a summary of the findings. We conclude the chapter with a discussion of future directions that could substantially enhance our understanding of mechanotransduction.

16.2 Applying Forces to Cells

Groups or individual cells subjected *in vitro* to mechanical forces by fluid shear flow or mechanical stretch show activated signaling proteins, up-regulated gene expression, and increased protein synthesis (Davies 1995; Orr et al. 2006). These studies mimic closely the physiological forces that a cell experiences *in vivo* and have shown that FAs exhibit dynamic changes in protein accumulation and signaling with applied force (Davies et al. 1994; Sawada and Sheetz 2002; Sawada et al. 2006). However, it is unclear how individual FAs lead to these widespread phenotypic changes within cells. For this reason, several advanced techniques have been developed to apply forces with spatial control at the micro- and nanoscale for a deeper understanding of mechanotransduction that occurs at individual FAs. To achieve this precision, forces have been applied directly with fine-tipped probes like microneedles or atomic force microscopes and indirectly with optical or magnetic microparticles that are manipulated with laser light or magnetic fields.

16.2.1 Microneedles

The use of microneedles to physically probe cells has provided insights into mechanotransduction that have been fundamental to the field. A glass capillary tube with a long, flexible tip is coated with ECM proteins and positioned into contact with an adherent cell in culture to allow integrins to bind to its surface. Once adhesions have fully formed, the tip is translocated to apply force to a small area of contact with the cell surface (Figure 16.2). Changes have been observed in the size of focal adhesions, the arrangement of the cytoskeleton, and the function of organelles.

The tips of the glass capillary tubes are formed by heating a section of the tube until it is molten and semi-liquid and then pulling from both ends to create a tapered tip that has 0.1–5 μm diameter and 100–1000 μm length. The tip segment of the capillary tube acts like a simple cantilever in which the deflection of the tip is proportional to the applied force. In the experiments, the base of the glass capillary is tightly mounted on a micropositioner, and the glass tip is permitted to move freely. From Euler-Bernoulli beam theory (Lardner and Archer 1994), the force F applied at the end of a cantilever is proportional to the displacement δ of the tip and is given by

$$F = \frac{3EI(x)}{L^3}\delta, \tag{16.1}$$

where E is the modulus of elasticity of glass, L is the length of the tip, and $I(x)$ is the second moment of area, which varies in x because of the tapered diameter of the tip. It is difficult to determine $I(x)$ exactly due to the small dimensions and tapered shape of the tip, and so the spring constant of each tip ($K = 3EI/L^3$) is obtained empirically by calibrating each capillary with a small weight placed at the tip. Typically, the tips have stiffness values of $K = 1$–10 pN/μm and, due to the precision of the micromanipulation, the applied force can vary from 0.1 to 100 nN.

Using microneedles to apply force directly to the cell through integrin receptors was observed to cause local growth in the sizes of FAs (Riveline et al. 2001; Kaverina et al. 2002). The response was independent of Rho-kinase, a myosin activation–associated protein, but dependent on mDia1, an actin polymerization–associated protein, suggesting that cytoskeletal remodeling rather than acto-myosin contractility is involved in the process. In addition to FAs, the cytoskeleton may be force-responsive since actin and microtubule polymerization can occur locally at the site of force application and an increase in cytoskeletal contraction against the tip has been observed (Heidemann et al. 1999; Kaverina et al. 2002). The response may not be solely due to FAs or cytoskeletal polymerization, but may involve a coupled process involving both mechanisms because a high degree of connectivity exists between the sites of adhesions and the cellular structure. In fact, applying force at adhesion sites can cause large distortions in the cytoskeleton and at the nuclear envelope, which strongly indicates the existence of intracellular force transmission (Maniotis et al. 1997). Even though the complete mechanism is not well understood, studies involving microneedles have shown that external force can affect cell fate. In neuronal cells, applied force can induce elongation of growth cones and stimulate axonal development (Bray 1984; Lamoureux et al. 2002).

16.2.2 Atomic Force Microscopy

Atomic force microscopes (AFM) are used to map the surface topography of a sample through tracing a microscale stylus across it, but, like microneedles, they can also be used to directly apply forces at a single point on the surface of a living cell (Figure 16.3). AFMs are mechanical microscopes that enable fine resolution of forces in real time through monitoring the deflection of the stylus tip. However, the tip can damage or scrape the sample, so a tapping mode is typically used, where the tip is vibrated above the surface and briefly taps the surface as it rasters across the area

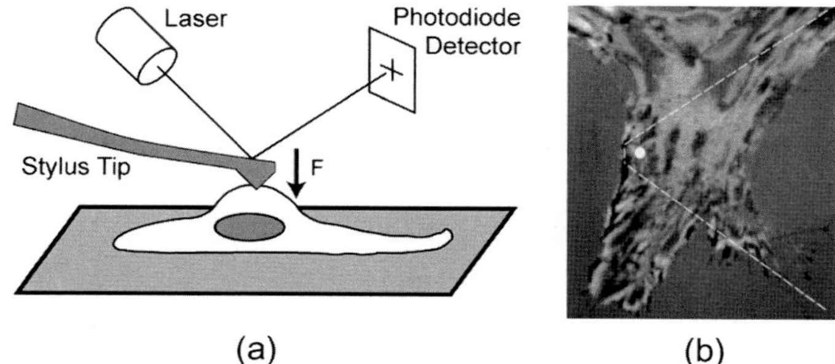

Figure 16.3. Atomic force microscopy for cellular mechanotransduction. (a) Stylus tip applies a vertical force onto the surface of an adherent cell. An optical laser beam reflected off the back of the stylus tip toward a photodiode detector measures the deflection of the tip with subnanometer precision (<1 nm). (b) Micrograph taken with internal reflection microscopy highlights the focal adhesions of a vascular smooth muscle cell as dark patches or spots. The dashed lines represent the position of the AFM tip and the white spot denotes the point of contact with the cell. Adapted from Trache and Meininger (2005).

of interest. This is opposite of the contact mode, where the tip is dragged across the surface to improve surface imaging, which can damage the cell. To enable cellular adhesion to the stylus tip, it is functionalized with ECM proteins. In some cases, the sharp stylus tip (10 nm radius) is replaced with a round 1–30 μm diameter microbead to prevent damage to the cell membrane.

The tip of the AFM acts like a cantilever to impart forces in the vertical direction as opposed to the lateral forces exerted with a microneedle. The deflection of the cantilever is measured from a laser beam that is reflected off the back of the stylus tip toward an array of photodiode detectors. For a rectangular cantilever AFM tip, the applied force is proportional to the measured deflection δ of the stylus tip and is given by

$$F = \frac{Ewt^3}{4L^3}\delta, \tag{16.2}$$

where E is the modulus of elasticity of cantilever material, which is typically silicon; w is the width of the cantilever; t is the thickness; and L is the length. The spring constant ($K = Ewt^3/4L^3$) is given by the manufacturer of the AFM tip, but is often confirmed experimentally. AFM systems can also use V-shaped cantilevers to stabilize against lateral forces. The spring constants for both cantilever types vary from 10 to 30 nN/μm. Due to the optical detection scheme, AFMs have greater control and range than microneedles and can apply forces from 1 pN to 100 nN.

AFMs can measure the mechanical properties of substrates through nanoscale force probing and are used to investigate cell mechanics and cytoskeletal elasticity (Radmacher 1997; Rotsch and Radmacher 2000; Mathur et al. 2001; Alcaraz et al. 2003; Laurent et al. 2005). When used for mechanotransduction studies, cellular adhesion strength was measured to increase with contact time to the AFM tip, ruptured between 20 and 100 pN, and strongly depended on the presence of growth factors in the media (Sun et al. 2005). When compression or tensile forces were

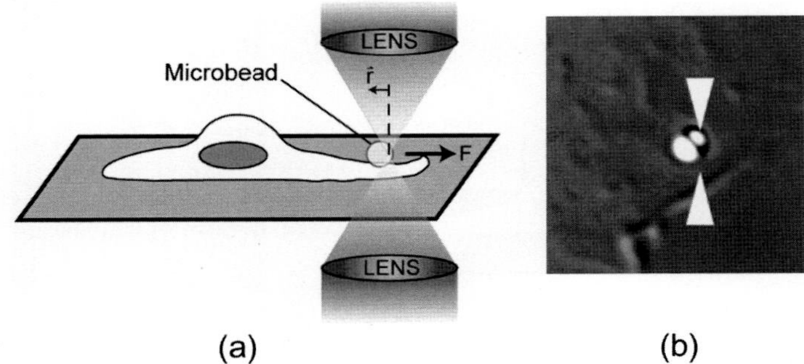

Figure 16.4. Optical tweezers for applying forces to focal adhesions. (a) A transparent microbead is held at the focal point of a focused laser by means of a photonic restoring force. (b) Differential interference contrast (DIC) micrograph of cellular lamellipodia pulled with a polystyrene microbead in an optical tweezer setup. Arrowheads designate the focal point of the laser. Adapted from Galbraith et al. (2002).

applied to the apical surface of cells, the adhesions on the basal surface reorganized and grew in area, again suggesting connectivity within cells (Mathur et al. 2000; Trache and Meininger 2005). Applied force was also observed to open mechanosensitive ion channels that depended on the structural integrity of the cytoskeleton, which suggests that integrins may have other partners in transducing force (Charras and Horton 2002).

16.2.3 Optical Tweezers

Microneedles and AFM tips impart force on the cell surface during the length of the experiment, but these direct methods do not allow the force to be turned off or modulated at high frequency, which falls short in replicating the transient or cyclic forces that are transmitted through the ECM. To overcome this limitation, indirect methods can apply forces to cells when a laser or magnetic field is on and negligible force when the field is off, and can rapidly change the strength of the applied force by modulating the laser power or position of the focal point. Optical tweezers (or optical traps) use a single laser beam to impart a net force on a transparent, dielectric microbead from photons passing through its surface. The laser traps a 1–6 μm diameter microbead in the center of the focal point of the beam (Figure 16.4). Positioning the laser so that the trapped bead is at the apical surface of a cell allows integrins to bind to ligands coated on the surface. Once bound, the optical restoring force at the microbead is transferred to the adhesion site.

Photons generate force due to the change in momentum of light from reflection or refraction. When a single-beam laser is focused on a microbead with a diameter R that is larger than the wavelength of the laser ($\lambda \sim$ 800–850 nm), individual rays of light passing through the bead refract at the interface between the medium and the microbead (Ashkin 1992). The direction of force generated from each refracted ray depends on the angle of incidence θ and the indexes of refraction n_1 and n_2 of the medium and the microbead, respectively. Integrating the contribution of the individual refracted rays yields a net force on the bead $\vec{F}$. To combine all the factors involved in the laser-microbead-medium system, optical tweezers are usually characterized as having a quality factor $\vec{Q}$ that encompasses the material and geometric properties of the system, and so $\vec{F}$ is given by

$$\vec{F} = \frac{n_1 P}{c} \vec{Q}(n_1, n_2, \theta, \vec{r}, R), \tag{16.3}$$

where P is the laser power, c is the speed of light, and $\vec{r}$ is the position of the focal point of the laser relative to the center of the bead. The quality factor can be approximated as having a magnitude that increases linearly with distance from the focal point of the trap $|\vec{r}|$ and acting counter to the direction of displacement $(-\hat{r})$. This relation arises because the angle of incidence increases with distance from the focal point $(\theta \propto |\vec{r}|)$, which causes a change in the direction of the refracted rays. As the distance increases, the restoring force $\vec{F}$ rises and acts to pull the bead back to the center of the trap $(r \approx 0)$. With the linear approximation, optical tweezers are generally described by a simple spring relationship $\vec{F} = -k\vec{r}$, where k is the trap stiffness. Typical optical traps have $k = 20$ pN/μm and can generate forces from 1 to 100 pN, which is dramatically smaller than the forces possible with direct methods. When the edge of the microbead is positioned outside the focal point $(|\vec{r}| > R)$, refraction is lost and the bead is free from the trap. The use of precision optical components enables fine positioning of the focal point of the laser relative to the center of the microbead and subsequently allows for a high degree of control over the location and magnitude of force applied to the cell. Although laser power can cause local heating, optical tweezers in practice are equipped with infrared wavelengths and used at low power so as not to harm cells.

Microbeads held in the optical trap experience a restoring force as described but also a counterforce at the adhesion site from the cell. When microbeads are placed on the lamellipodia region, active force generation through retrograde actin flow can pull the microbead from the laser focal point (Schmidt et al. 1993; Felsenfeld et al. 1996; Choquet et al. 1997). The counterforce of the cell was observed to increase in strength and could eventually overcome the laser strength, allowing the cell to pull the microbead free from the trap. Once outside the trap, the microbeads could not be recaptured with the optical tweezers unless a higher laser power was applied. The need for a higher restoring force to trap the microbead indicates that there exists active reinforcement at the adhesion site, which arises by increased coupling of the adhesion site to the actin cytoskeleton. Inhibition of tyrosine phosphatase activity abrogated the response and indicated that this reinforcement was due to biochemical changes (Choquet et al. 1997). Subsequent studies have shown that reinforcement is mediated by the recruitment of adaptor proteins that link the adhesion at the microbead to the cytoskeleton (Galbraith et al. 2002; Giannone et al. 2003; Jiang et al. 2003) and by activation of signaling proteins (von Wichert et al. 2003a, 2003b) associated with focal adhesions. In fact, these biochemical changes associated with external force application appear to emit from the site of applied force and can propagate rapidly through the cell (Wang et al. 2005).

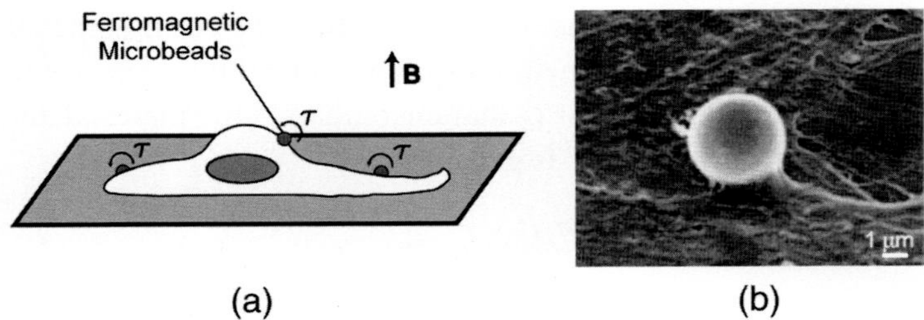

Figure 16.5. Magnetic twisting cytometry. (a) Ferromagnetic microbeads are twisted under a magnetic field (B) and impart torque (τ) on the surface of a cell. (b) Micrograph from a scanning electron microscopy shows a microbead in contact with the apical surface of a cell. Adapted from Fabry et al. (2003).

16.2.4 Magnetic Twisting Cytometry

Ferromagnetic microbeads with 4–6 μm diameters are coated with specific integrin ligands and are allowed to bind the surfaces of adherent cells. Microbeads are magnetized with a horizontally applied uniform field and then magnetically rotated by a uniform field perpendicular to their magnetic orientation, a technique known as magnetic twisting cytometry (MTC; Figure 16.5). Rotation of the microbead is due to magnetic torque and causes a shear stress at each adhesion site. Unlike single-cell experiments with optical tweezers, magnetic beads can be sprinkled onto cells in culture to impart mechanical stress across a group of cells for population response studies.

To create a uniform magnetic torque, a horizontal field is pulsed at high strength ($B = 1000$ Gauss) to magnetize the microbeads and align their dipole moments $\vec{\mu}$ in the same direction. A perpendicular field is applied at a lower strength ($B = 0$–25 Gauss) to avoid remagnetization but strong enough to cause a magnetic torque τ that is given by

$$\vec{\tau} = \vec{\mu} \times \vec{B}. \tag{16.4}$$

The torque imparts rotational strain at the adhesion that can be related to an applied shear stress on the cell. Initial designs applied a constant field and indirectly measured the rotation of the microbeads with an inline magnetometer that detected the bulk orientation of $\vec{\mu}$ of all microbeads. Later implementations recorded the lateral displacements of microbeads with optical microscopy under an oscillating field. Shear stresses from 1 to 90 Pa have been reported, which equates to a range of force from 1 to 1000 pN, depending on the contact area between the microbead and the cell.

The use of MTC for mechanotransduction studies showed that focal adhesions provided resistance to the rotation of the microbeads. Greater rotations were possible with inhibition of integrin binding, reduction in ligand density on the microbead, and disruption of cytoskeletal integrity (Wang et al. 1993; Wang and Ingber 1994). These findings further support that adhesion strength is highly dependent on the degree of coupling to the cytoskeleton via integrins at focal adhesions.

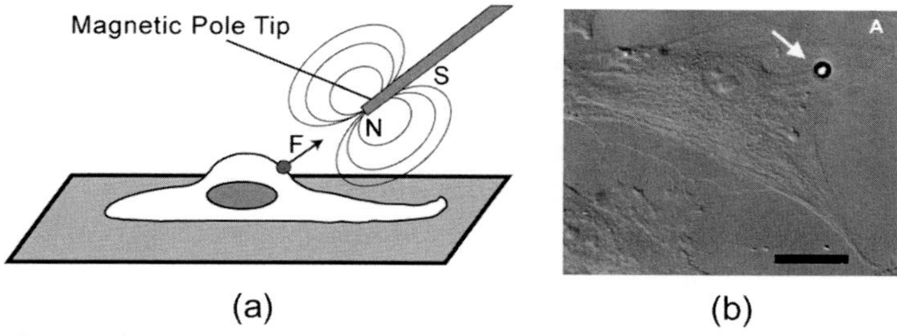

Figure 16.6. Magnetic tweezers for applying forces to cells. (a) Ferromagnetic microbead is pulled toward the gradient field created by a magnetic pole tip and generates a force on the surface of a cell. (b) DIC micrograph of a microbead attached to a cell and pulled with a magnetic tweezer. Arrow indicates the microbead. Adapted from Matthews et al. (2006).

Interestingly, microbeads twisted at the apical surface induced displacements at basal focal adhesions due to force transmission through the cytoplasm that was dependent on the integrity of the cytoskeleton (Hu et al. 2003; Wang and Suo 2005).

A key use of MTC is to characterize the rheological properties of cells by applying an oscillatory force and recording the microbead rotations (Wang and Ingber 1994; Hubmayr et al. 1996; Fabry et al. 2001a, 2001b; Bursac et al. 2005; Hoffman et al. 2006). With this approach, the dynamic mechanical properties of the cytoskeleton – how it deforms, flows, and reorganizes – could be assayed in response to different loading conditions and inhibition of biochemical regulators. Underlying these mechanical adaptations are changes in gene expression associated with MTC stimulation; ribosomes and mRNA translocate to the site of force and cells show up-regulation of gene transcription with twisting (Chicurel et al. 1998; Meyer et al. 2000).

16.2.5 Magnetic Tweezers

Magnetic tweezers use ferromagnetic microbeads in a nonuniform magnetic field to apply linear forces to cells instead of twisting stresses as with MTC. The microbeads are coated with ECM protein prior to seeding onto the cells, and a needle-like magnetic tip is positioned within a few micrometers of the targeted microbead. When the magnetic field is turned on, the nonuniform field emanating from the tip creates a magnetic force that pulls the microbeads toward it (Figure 16.6). With this approach, higher forces are possible than MTC, but, like optical tweezers, only single-cell experiments are currently possible.

The field from the magnetic tweezer is nonuniform and has a gradient field strength that decreases with distance from the tip. A magnetic tweezer consists of an electromagnetic coil with a fine ferromagnetic needle tip ($\sim$ 10 μm tip radius) mounted on a micropositioner. The field from the magnetic tweezers generates a force on the microbead that is proportional to the field gradient $\nabla\vec{B}$ and is given by

$$\vec{F} = \vec{\mu} \bullet \nabla\vec{B}. \tag{16.5}$$

Because the gradient increases with proximity to the tip, the force is strongly dependent on the nanoscale placement of the probe. Through precision positioning of the tip, magnetic tweezers produce forces from 0.1 to 10 nN. Other implementations of this system use a large magnet positioned above the cells to form a vertical gradient, which pulls the microbeads perpendicular to the substrate.

The pulling force from magnetic tweezers caused the FA strength to increase by recruitment of vinculin and actin to the interface between the cell membrane and the microbead (Alenghat et al. 2000; Matthews et al. 2004). As seen with MTC, mechanical forces at apical focal adhesions resulted in the displacement of basal adhesions, and this effect was mitigated with the inhibition of the Src pathway (Mack et al. 2004). Mechanosenstive ion channels may also play a role in the mechanotransduction process because rapid calcium influx is seen at the onset of force application (Glogauer et al. 1995, 1997; Matthews et al. 2006). Gene expression is affected by external force because mechanical stimulation activates the transcription factor Sp1, leading to up-regulation of filamin A, which binds to actin and increases its rigidity (D'Addario et al. 2001, 2002, 2003, 2006). A consequence of force stimulation is greater cytoskeletal stiffness that may act as a protection mechanism against forces that would lead to cellular damage (Glogauer et al. 1998; Matthews et al. 2006).

16.2.6 Summary of Applying External Forces for Mechanotransduction

These techniques use different methods to apply forces to cells at the micro- and nanoscale, but their findings show common biological responses. FA adaptor proteins recruit to the site of force application (Riveline et al. 2001; Galbraith et al. 2002; Kaverina et al. 2002; Matthews et al. 2004) and can cause distant changes at FAs on the basal surface (Mathur et al. 2000; Hu et al. 2003; Mack et al. 2004). Recruitment couples actin microfilaments of the cytoskeleton to the site of applied force and enhances adhesion strength (Wang et al. 1993; Choquet et al. 1997; Jiang et al. 2003; von Wichert et al. 2003a; Sun et al. 2005). Moreover, force stimulation causes changes in the structure and strength of the cytoskeleton (Glogauer et al. 1998; Heidemann et al. 1999; Kaverina et al. 2002; Bursac et al. 2005; Matthews et al. 2006) and nuclear shape (Maniotis et al. 1997). The responses of FAs and the cytoskeleton are in part due to biochemical changes that occur locally at the adhesion site (Glogauer et al. 1997; von Wichert et al. 2003a, 2003b) and may propagate throughout the cell (Wang et al. 2005), or through the gating of mechanosensitive ion channels (Glogauer et al. 1995; Charras and Horton 2002). It is unclear which mechanisms dominate or if both act together in the process of mechanotransduction. Downstream of these biochemical signals, external forces initiate the activation of transcription factors (Meyer et al. 2000; D'Addario et al. 2002, 2006) and up-regulation of protein expression (Chicurel et al. 1998; D'Addario et al. 2001). As a result, the application of force can lead to dramatic changes in the fate and function of cells (Bray 1984; Garcia-Cardena et al. 2001; Lamoureux et al. 2002; Fass and Odde 2003; Dai et al. 2004). Taken together, these techniques have convincingly shown that external forces acting at FAs can elicit biological responses in cells. However, cells generate traction forces that also stimulate the FA signaling pathways that regulate their proliferation, differentiation, and migration. Therefore, methods to better understand the role of these intrinsic forces have been developed and are discussed in the next section.

16.3 Measuring Traction Forces

The fundamental studies into the contractile response of muscle cells laid the groundwork with which to understand traction forces within other cell types (Hill 1922, 1938; Fung 1967). A similar molecular mechanics of myosin motors acting on actin microfilaments that create force in myofibril organelles is present in the stress fibers that generate traction forces. However, unlike the highly organized and parallel myofibrils, stress fibers form a heterogeneous network throughout the cell. As a result, force generation is not unidirectional, as seen in myocyte contraction, but instead traction forces are nonuniformly oriented and distributed across different cellular regions of nonmuscle cells.

Early work into measuring traction forces studied the micro-Newton forces generated in contracting nonmuscle tissue explants (James and Taylor 1969). Even though the force contribution per cell was found by dividing the total traction force by the number of cells in the explants or tissue construct, the values obtained were remarkably close to the measurements from current techniques. However, these methods relied on the assumption that cells throughout the explant exist in a homogeneous contractile state, a condition that, aside from being nearly impossible to obtain experimentally, abrogates any signaling heterogeneity in regions of the cell or tissue. Because the magnitude and area of forces in single cells were below the capabilities of traditional force measurements, new technologies with micro- and nanoscale spatial resolutions were required to measure forces at individual focal adhesions. To achieve this degree of resolution, cells have been grown on flexible substrates that deform in proportion to the traction forces exerted by the cells. Techniques to measure this deformation include observing wrinkles in a thin film, tracking the displacement of fiduciary markers in an elastic substrate, and microfabricated cantilevers that deflect proportionally to the local traction force.

16.3.1 Wrinkling Silicone Membranes

In this method, cells are cultured on a thin film of rubber on top of a lubricating liquid layer. The top film arises from unpolymerized silicone fluid on a glass coverslip that is thermally cross-linked by briefly exposing it to an open flame or hot wire. The silicone at the top surface solidifies into a thin rubber film, which shrinks slightly and acts as a tempered surface or drumhead lying over a lubricating fluid of un-cross-linked silicone. The final surface is hydrophobic, nonbiofouling, and mildly cell adhesive. As cells spread and migrate, they contract and wrinkle the top film

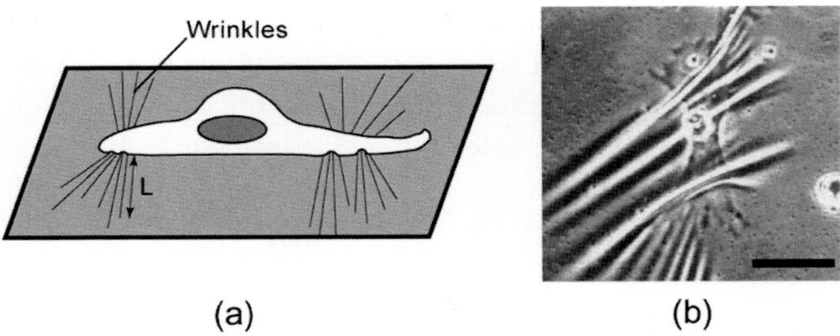

Figure 16.7. Wrinkling silicone membranes for measuring traction forces. (a) Cells plated onto a thin film of silicone exert traction forces that wrinkle the membrane. Measurement of cellular tension is inferred qualitatively by the length (L) and number of wrinkles in the silicone film. (b) Phase contrast micrograph of a cell on a wrinkling silicone membrane. Adapted from Kelley et al. (1987). Scale bar: 50 μm.

(Figure 16.7). Although it is not strictly quantitative, the size and locations of these wrinkles may be used to estimate cell tractions.

The physics of thin film buckling are difficult to describe mechanically for the complex patterns of cell tractions, but they give a qualitative measurement of traction force. In practice, applying a force with a calibrated glass microneedle generates wrinkles that propagate from the site of force application. The longitudinal lengths of these wrinkles can be used to linearly report the local force. Depending on the compliance of the film, this technique has been used to measure forces ranging from 10 to 1000 nN. For more complex traction fields, wrinkles are typically classified into two categories. Compression wrinkles are generated underneath cells and are oriented perpendicular to the applied force. Tension wrinkles radiate away from the cell and are aligned in the direction of traction force (Harris 1984). Using both the linear approximations of wrinkle length and force, counting the number of wrinkles, and classifying them by wrinkle category permitted the first studies into the temporal and spatial generation of traction forces exerted by a single cell.

Wrinkling membranes provided insights that opened and expanded the field for many of the techniques that are discussed later. They confirmed unequivocally that the contraction of tissue is in large part due to cell contraction and not solely to enzymatic activity, dehydration, or biochemical alterations of the ECM. They also provided the resolution necessary to discern the spatial distribution of traction forces for single cells. Nonmuscle cells were shown to produce traction forces underneath the cell, just behind its leading edge, and directed perpendicular to the cell boundary through adhesions between the plasma membrane and the substrate (Harris et al. 1980; Harris 1984). Moreover, different cell types were observed to generate different degrees of contractility, with fibroblasts, smooth muscle cells, and pericytes being the most contractile (Harris et al. 1981; Kelley et al. 1987). By using internal reflectance microscopy (IRM), strong forces were correlated to regions of dense focal contacts (Harris et al. 1980; Harris 1984). Wrinkling membranes provided evidence that muscle cells exert traction-like forces against the ECM through vinculin-rich costameres (Danowski et al. 1992). More compliant films, weakened by exposing the cross-linked layer to UV energy, were used to monitor traction-mediated cleavage of the

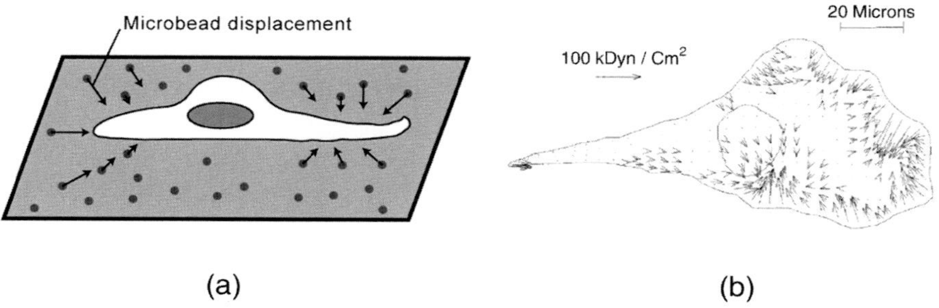

Figure 16.8. Traction force microscopy. (a) Cells exert traction forces that deform the elastic substrate and displace the microbeads that are embedded within it. (b) Example of the spatial traction force vector maps that are inferred from the displacement field. The scale arrow denotes the range of traction forces measured. Adapted from Dembo and Wang (1999).

intracellular bridge after cytokinesis of adherent mammalian cells and to formulate theories about the mechanism behind keratocyte migration (Burton and Taylor 1997; Burton et al. 1999). In addition to cell traction investigations, silicone wrinkling membranes have been adopted to assay the biochemically induced changes in traction force and have convincingly shown that the small GTPase Rho pathways (Rho, Rac, Cdc42) and the Ca^{2+}-calmodulin pathway are direct regulators of cell contractility (Chrzanowska-Wodnicka and Burridge 1996; Helfman et al. 1999; Kiosses et al. 1999).

16.3.2 Elastic Substrata

Due to difficulties in quantifying wrinkle formation, researchers improved the use of deformable elastic substrates by embedding discrete markers in the film and quantifying their displacement (Lee et al. 1994; Oliver et al. 1995; Dembo et al. 1996). Initially, latex microbeads were brushed onto the uncured silicone fluid. When cross-linked, these microbeads became permanently embedded into the film. The drumhead tension or film pre-stress was then increased so as to be high enough to prevent buckling yet still permit deformation by traction stresses. Instead of qualitatively measuring wrinkle formation, tracking the displacement of the microbeads in the deformed gel was used to infer the cellular traction forces. The versatility of this technique was expanded significantly by replacing the thin silicone film with a continuous polyacrylamide (PAA) gel for traction force microscopy (TFM) or microfabricated silicone rubber patterned with arrays of microdots.

In TFM, a suspension of randomly distributed, fluorescent microbeads is mixed into a PAA gel that is spread to a thin film on a glass coverslip. Chemically cross-linking the gel results in a nontoxic film with high compliance. However, the film is also hydrophilic and nonadhesive to cells. Therefore, it is necessary to covalently link ECM proteins to the surface. After ECM incorporation, cells can adhere, spread, and migrate across the gel, displacing the fluorescent microbeads as they deform the surface (Figure 16.8).

To measure the traction forces, an image of the microbead positions in the deformed gel is acquired and then the cell is removed via trypsinization to yield an image of the undeformed gel. Microbead displacements are found from one-to-one correspondence of their positions between the paired images or by pattern correlation within user-defined windows using image analysis software. This analysis yields a map of displacement vectors at discrete points within the gel. The deformable gel can be regarded mechanically as a semi-infinite elastic half-space if one assumes that traction forces act as shear forces on the surface, that the overall traction stress only deforms the gel in its elastic regime, and that displacements are small compared to the gel thickness. Under these assumptions, the relationship between the traction stress field $T(\vec{r})$ and the displacement field $d(\vec{m})$ is given by the integral relationship

$$d_{ij} = \int \int G_{ij}(\vec{m} - \vec{r}) T_i(\vec{r}) dr_1 dr_2, \tag{16.6}$$

where $i,j \leqslant 2$ for the basis vectors in a two-dimensional half-space and $G_{ij}(\vec{m} - \vec{r})$ is Green's function that relates the displacement at position $\vec{m}$ that arises from the force at position $\vec{r}$. Obtaining the traction stress field requires inverting and discretizing Equation (16.6), which is an inverse problem. The inversion can result in an ill-posed mathematical equation, which, in addition to being computationally intensive to solve, often does not guarantee a unique solution. If the coefficient matrix to be inverted is ill-conditioned, the solution becomes very sensitive to small noise levels in the displacement map. Therefore, regularization schemes are required which include assumptions about static equilibrium; constraints on the reasonable locations of traction forces, for example, inside the cell boundary or at FAs; and requirements that the least complex solution be chosen (Dembo and Wang 1999). One option to avoid the computationally intensive numerical solutions to Equation (16.6) is to perform the solution in Fourier space (Butler et al. 2002). This dramatically reduces the computational time but may introduce artifacts at the boundaries of the stress field. Another approach has utilized a three-dimensional finite element method to compute the stress field, eliminating errors from the infinite elastic half-space assumption but still requiring precise measurements of bead displacements (Yang et al. 2006). Through these methods, this technique has produced detailed traction maps with a spatial resolution of a few microns and a stress resolution of approximately 50 pN/um^2 (Dembo and Wang 1999).

Regular arrays of lithographically patterned markers have been created to avoid difficulties in measuring the displacement field of randomly distributed markers in TFM. In this technique, polydimethylsiloxane (PDMS) substrates are cast against a topographically patterned silicon wafer containing orthogonal rows and columns of markers (Figure 16.9). PDMS is a silicone elastomer that, when cured against the surface of a wafer, forms a reverse topographical copy with nanometer accuracy. The resulting surface is hydrophobic, moderately cell adhesive, and contains a regular array of markers visible with phase or fluorescent microscopy (Balaban et al. 2001). Afterwards, ECM protein is physically absorbed to the PDMS surface to further promote cell adhesion.

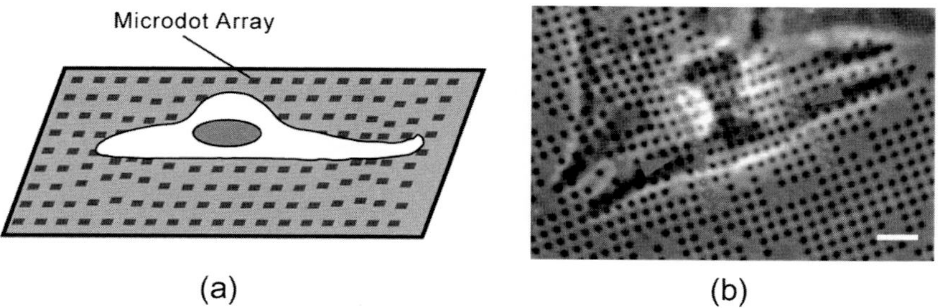

Figure 16.9. Microdot array for measuring traction forces. (a) Regular array of fiduciary markers are micropatterned into an elastic substrate. Cellular tension distorts the rows and columns of markers and reports the local traction forces. (b) Phase contrast micrograph of a cell exterting tension on a microdot array substrate. Adapted from Balaban et al. (2001). Scale bar: 6 μm.

For strong forces, the displacement vectors are easily detected from the observed deviations in the markers. Weaker forces cause smaller displacements, and so image analysis is required for high-resolution traction mappings. Once the displacement field is obtained from the deformed and undeformed images of the array, an identical procedure to solve Equation (16.6) is performed to generate the traction force map; however, the regular array of markers reduces the number of possible solutions as compared to TFM. Regularization schemes are still needed, but this technique yields improved spatial resolution (down to 3 μm) and increased certainty in the presented solution (Schwarz et al. 2002).

Elastic substrata were the first techniques to produce high-resolution traction maps of forces exerted by a single cell. This increase in resolution permitted distributions of forces to be attributed to specific regions of the cell, such as the lamellipodia or the retracting tail during migration, where there may be different biochemical regulators of force generation (Dembo and Wang 1999). The use of an orthogonal array of fluorescent markers further increased the spatial resolution such that traction forces could be attributed to individual FAs (Balaban et al. 2001). With this accuracy, it was possible to convincingly demonstrate that FAs have dynamic and geometric correlations with traction forces. Using green fluorescent protein (GFP) fusions of FA-associated proteins, it was observed that the FA area, protein density, and structural orientation have positive correlations with local traction force magnitude and direction. Moreover, the introduction of a soluble inhibitor of actomyosin activity led to a simultaneous decrease in traction forces and disassembly in FAs. The temporal correlation may not be so straightforward because nascent FAs show an inverse relationship where traction forces decrease as the protein recruitment density increases (Beningo et al. 2001). This discrepancy begs further investigation because external forces, generated from a nearby microneedle to locally deform the PAA gels, could cause a direct increase in FA size (Wang et al. 2001).

These techniques also allowed versatility in controlling the adhesive properties of the microenvironment to assay the response in traction forces. Selectively modulating the amount of ECM proteins covalently bound to the gel surface allowed control over the ligand density presented to cells. In these studies, traction forces were shown to change in magnitude in accordance with the adhesivity of the ECM (Gaudet et al. 2003; Rajagopalan et al. 2004; Reinhart-King et al. 2005). Changing the ligand density also caused an increase in the total spread area of cells, which was found to also positively correlate with total traction forces. Additionally, tailoring the elasticity of the ECM by changing the degree of cross-linker in the PAA gels showed that traction forces are larger on stiffer substrates (Lo et al. 2000; Wang et al. 2001; Paszek et al. 2005). Cells will migrate preferentially from soft regions toward stiffer regions in the ECM, which is a phenomenon called durotaxis (Lo et al. 2000). Interestingly, interruption of focal adhesion kinase (FAK) activity caused insensitivity to ECM stiffness in migration and traction forces (Wang et al. 2001), suggesting that cells actively probe their microenvironment and receive feedback at their focal adhesions. The versatility and ease of implementation of these techniques have led to the rapid incorporation of elastic substrata into laboratories that study mechanotransduction.

16.3.3 Microcantilevers

Elastic substrata have greatly enhanced our understanding of the traction fields generated by cells and have advanced the field of mechanotransduction; however, these techniques are complicated by the fact that displacements due to discrete forces applied at focal adhesions are convoluted and transmitted throughout the continuous surface of the gel. This condition stimulated the creation of new techniques to measure traction forces in a mechanically decoupled manner. Through the use of technology to fabricate the mechanical components of micrometer size, arrays of microcantilevers were developed to provide independent force measurements at discrete locations along a cell.

The first application of this technique utilized an array of horizontal cantilevers created on top of a silicon wafer. This technique is similar to an inverted AFM tip, but instead of vertical deflections, the tip bends in the plane of traction forces. Each cantilever lies underneath the top surface of the silicon wafer and has a small pad at its tip ranging in area from 4 to 25 um^2 at which the cell's traction forces can be measured. A square opening around each pad allows the cells to contact the pad and displace it as they exert force (Figure 16.10). The base of the cantilever is anchored to the rigid substrate, and therefore the pad has a displacement that is independent of the other cantilevers on the device.

As with an AFM cantilever, the deflection of the micromachined cantilever is described by Euler-Bernoulli beam theory. For a simple rectangular cantilever, the relationship between the applied cellular force F and observed pad displacement δ is given by

$$F = \frac{Ew^3 t}{4L^3}\delta, \tag{16.7}$$

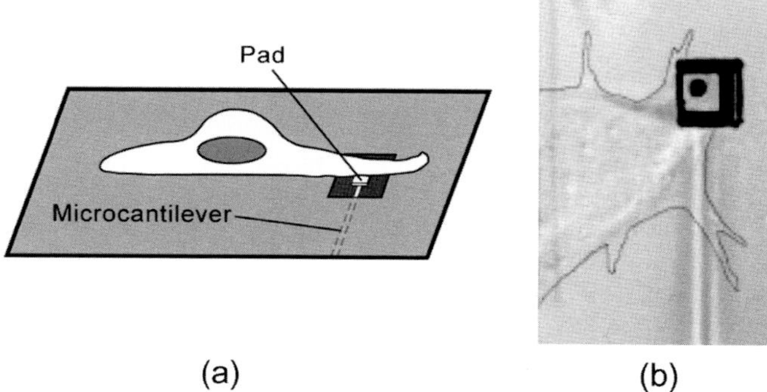

Figure 16.10. Micromachined cantilever for reporting local traction force. (a) A migrating cell attaches to the pad of a microcantilever that is located underneath the top surface of a silicon wafer. Local traction force causes deflection of the cantilever. (b) Phase contrast micrograph of cellular attachment to the pad of a microcantilever. Adapted from Galbraith and Sheetz (1997).

where E is the modulus of elasticity of polycrystalline silicon, w is the width of the cantilever, t is the thickness, and L is the length. In practice, however, the measurement of these parameters is difficult to ascertain because of subtle dimensional variations incurred during microfabrication. Therefore, the relationship between applied force and pad displacements was determined empirically for each substrate using a calibrated micropipette. Once the spring constant ($K = Ew^3t/4L^3$) for each pad was known, this technique permitted real-time observation of the traction forces exerted by cells at a single location. The cantilever can only deflect in one direction and restricts the measurement of force $F_{observed}$ to be along one axis. As a consequence, the actual magnitude of the traction force F_{cell} must be estimated from the assumption that traction forces act in a parallel direction to migration such that $F_{cell} = F_{observed}/\sin(\theta)$, where θ is the angle between the migration direction and the axis of the cantilever length. This poses a stipulation for data analysis as it explicitly confines the direction in which a cell could be exerting forces.

Micropost arrays provide an improvement over micromachined cantilevers because a dense array of vertical cantilevers is used to generate high-resolution maps of traction forces with independent measurements. Arranging cantilever elements along the vertical axis allows a tight packing of force sensors (Figure 16.11). In this manner, the base is attached to the substrate, the length is oriented perpendicular to the substrate, and the tip is permitted to move freely in the two-dimensional plane. The fabrication process is similar to that for the microdot arrays in that PDMS is cast against a topographically patterned microfabricated silicon wafer. The resulting PDMS surface consists of an array of cylindrical microposts that are typically 3 μm in diameter, 11 μm tall, and at 9 μm intervals; however, these dimensions may be adjusted to change the spacing or stiffness of the posts. After fabrication, the tops of the microposts are coated with ECM proteins and the sidewalls and base layer are made nonadhesive with a surfactant coating. In this manner, cells can only adhere to

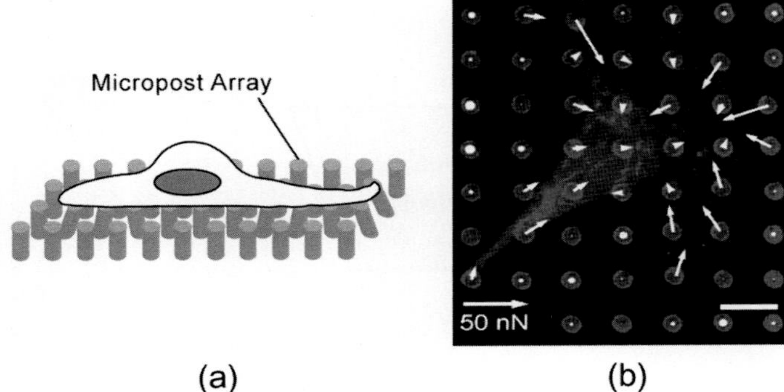

Figure 16.11. Micropost array for force measurements. (a) Microfabricated array of vertical microposts deflects and reports the local traction force of the adherent cell on the tops of the microposts. (b) Representative micrograph of the measured displacements of the micropost tips and the corresponding traction forces. Scale arrow indicates the range of traction forces. Adapted from Tan et al. (2003). Scale bar: 10 μm.

the tops of the microposts and the presentation of adhesive area can be precisely controlled.

A traction force acting at the top of a micropost bends it in proportion to the magnitude and in the direction of the line of action. Micropost deflections may be observed in real time using phase or fluorescence microscopy and referenced to their undeflected positions. Since the microposts deflect like cantilevers, the relationship between the applied load F and the resulting tip displacement δ is given as

$$F = \frac{3\pi E r^4}{4L^3}\delta, \tag{16.8}$$

where E is the modulus of elasticity of PDMS, r is the micropost radius, and L is the height of the microposts. The spring constant for the microposts ($K = 3\pi E r^4/4L^3$) was verified using a calibrated micropipette and was found to agree with Equation (16.8). Further refinements of this technique have reduced the micropost size and array pitch to increase the spatial and force sensitivity to approximately 4 μm and 50 pN, respectively (du Roure et al. 2005).

Horizontal micromachined cantilever devices were first used to observe oscillating traction forces of migrating fibroblasts and keratocytes, which contained both low- and high-frequency components (Galbraith and Sheetz 1997, 1999). These arrays successfully decoupled the force measurement at a single point from the surrounding substrate; however, results were limited because the pad was only able to measure forces along a single axis and at a single point along the cell. Vertical micropost arrays permitted a more complete mapping of the traction field and reported traction force measurements at a comparable resolution to the continuous elastic substrates (Tan et al. 2003). Because the cells only come into contact with the tops of the posts, substrate compliance or available ECM area may be altered while keeping the geometry and chemical composition presented to the cells identical. In this manner, many of the findings on traction forces from TFM were confirmed. Average traction force was observed to increase with spread area (Tan et al. 2003). Controlling the radius and length of the microposts showed that traction forces increase with substrate stiffness (Saez et al. 2005). Analyzing the FA area at individual posts verified the positive correlation between traction force and focal adhesion assembly, but also confirmed the presence of small FAs with large traction forces (Tan et al. 2003). Additionally, microposts have been used as a contractility assay to gauge the biochemical contribution that FAK mutants and isoforms of nonmuscle myosin II have in the generation of traction forces (Cai et al. 2006; Pirone et al. 2006). A significant advantage of the tool is the capability to measure traction forces in multicellular groups. Because each micropost is mechanically decoupled, independent force measurements can be made with high accuracy. From these studies into forces within a monolayer of cells, it was observed that traction forces are largest at the edges of the group and rapidly decay toward the interior (du Roure et al. 2005; Nelson et al. 2005). The larger forces at the edge appear to provide a proliferative signal in the regulation of tissue growth.

16.3.4 Summary of Traction Force Measurements

The contractile phenotype of cells has been shown to be highly dependent on the physical qualities of the microenvironment and on internal contributions from biochemical signals. Traction forces have been shown to modulate in accordance with substrate stiffness (Lo et al. 2000; Wang et al. 2001; Saez et al. 2005), ligand density (Gaudet et al. 2003; Rajagopalan et al. 2004; Reinhart-King et al. 2005), and available area (Reinhart-King et al. 2003; Tan et al. 2003). These effects likely are integrated together during durotaxis, where cells migrate toward regions of higher stiffness, or haptotaxis, where cells crawl toward regions of higher ligand densities (Carter 1967; Lo et al. 2000). Moreover, the role of traction forces and ECM stiffness may play a role in tumor formation and stem cell differentiation (Paszek et al. 2005; Engler et al. 2006). The findings from these measurement techniques indicate that FAs are organelles that perceive the mechanical and adhesive properties of the cellular microenvironment. FA size has been shown to positively correlate with traction forces in a spatial and temporal manner (Balaban et al. 2001; Tan et al. 2003) and inhibition of vital focal adhesion proteins eliminates their mechanosensitivity (Wang et al. 2001; Pirone et al. 2006).

Previous studies have led to the common paradigm that cells use several biochemical pathways to regulate cytoskeletal contractility (Kolodney and Elson 1993, 1995; Chrzanowska-Wodnicka and Burridge 1996; Helfman et al. 1999; Kiosses et al. 1999; Wozniak et al. 2003; Beningo et al. 2006; Cai et al. 2006) to apply forces at FAs in order to probe and adapt to mechanical cues. Given that traction forces vary across the different regions of a cell (Dembo et al. 1996; Galbraith and Sheetz 1997; Burton et al. 1999), these structures have defined roles in providing mechanotransduction signals to a cell. Going further, the difference in traction forces at the edge of a monolayer to those in the interior may act as a mechanical cue or positional marker in the growth or repair of tissue during embryonic development or wound healing (Ingber 2005). It remains to be seen if these edge effects have equivalent *in vivo* manifestations that lead to similar responses.

16.4 Conclusion

The field of mechanotransduction has been vitally guided by the development of tools to study the biological phenomena of interest. Techniques that apply external forces with precision and measure traction forces with accuracy have shown that FAs are involved in sensing the composition and compliance of the ECM; however, the exact interplay between passive and active contributions to FA mechanosensing, cell spreading, and traction force generation is still under investigation. Advancements from the fields of optical sciences, electromagnetism, material research, and mechanical engineering have enabled the measurement of complex cellular tractions and probing of mechanical responses. As current techniques are further refined and more fully utilized, it is important to concurrently develop new and novel methods to assess the role of mechanically mediated cell signaling.

The spatial resolutions of many of the methods in this chapter are close to the optical detection limitations of current light microscopes. Next-generation techniques will have to incorporate novel methods of detecting forces in order to attain nanometer resolution to understand the mechanics involved at the level of individual proteins. Moreover, molecular interactions and focal adhesion dynamics may occur over the course of a few milliseconds to several hours. Improvements to the temporal resolution and long-term feasibility of force application and traction measurements will allow researchers to probe both of these extremes. Furthermore, FAs and the cytoskeleton observed *in vivo* differ from their *in vitro* counterparts in terms of architectural organization and associated signaling proteins (Cukierman et al. 2001; Walpita and Hay 2002). Current techniques that attempt to recreate an *in vivo* microenvironment to measure traction forces can be used to examine the causes behind these different phenotypes (Kolodney and Wysolmerski 1992; Beningo et al. 2004), but improvements in applying force and measuring traction forces in three dimensions will be central to determining why these changes in mechanotransduction occur.

Current techniques have largely focused on measuring responses from single focal adhesions. However, it is unclear if signals from many interconnected adhesions are integrated to produce a cellular response. Therefore, as we further refine single adhesion measurements, it will also be important to develop techniques to examine how changes in a single focal adhesion are transmitted throughout the cell. Combining these advances with new techniques to track the temporal and spatial activation of signaling proteins within cells (Wang et al. 2005; Ballestrem et al. 2006; Pertz et al. 2006) should lead to breakthrough accomplishments in separating and quantifying the effects of mechanical and chemical parameters. Through simultaneous efforts by researchers on all of these fronts, we will attain a better understanding of the relationship between cells and their environment, which will likely provide new approaches for therapeutic design.

REFERENCES

Alcaraz, J., L. Buscemi, M. Grabulosa, X. Trepat, B. Fabry, R. Farre and D. Navajas (2003). "Microrheology of human lung epithelial cells measured by atomic force microscopy." *Biophys J* **84**(3): 2071–9.

Alenghat, F. J., B. Fabry, K. Y. Tsai, W. H. Goldmann and D. E. Ingber (2000). "Analysis of cell mechanics in single vinculin-deficient cells using a magnetic tweezer." *Biochem Biophys Res Commun* **277**(1): 93–9.

Alenghat, F. J. and D. E. Ingber (2002). "Mechanotransduction: all signals point to cytoskeleton, matrix, and integrins." *Sci STKE* **2002**(119): PE6.

Ashkin, A. (1992). "Forces of a single-beam gradient laser trap on a dielectric sphere in the ray optics regime." *Biophys J* **61**(2): 569–82.

Balaban, N. Q., U. S. Schwarz, D. Riveline, P. Goichberg, G. Tzur, I. Sabanay, D. Mahalu, S. Safran, A. Bershadsky, L. Addadi and B. Geiger (2001). "Force and focal adhesion assembly: A close relationship studied using elastic micropatterned substrates." *Nat Cell Biol* **3**(5): 466–72.

Ballestrem, C., N. Erez, J. Kirchner, Z. Kam, A. Bershadsky and B. Geiger (2006). "Molecular mapping of tyrosine-phosphorylated proteins in focal adhesions using fluorescence resonance energy transfer." *J Cell Sci* **119**(Pt 5): 866–75.

Beningo, K. A., M. Dembo, I. Kaverina, J. V. Small and Y. L. Wang (2001). "Nascent focal adhesions are responsible for the generation of strong propulsive forces in migrating fibroblasts." *J Cell Biol* **153**(4): 881–8.

Beningo, K. A., M. Dembo and Y. L. Wang (2004). "Responses of fibroblasts to anchorage of dorsal extracellular matrix receptors." *Proc Natl Acad Sci USA* **101**(52): 18024–9.

Beningo, K. A., K. Hamao, M. Dembo, Y. L. Wang and H. Hosoya (2006). "Traction forces of fibroblasts are regulated by the Rho-dependent kinase but not by the myosin light chain kinase." *Arch Biochem Biophys* **456**(2): 224–31.

Bray, D. (1984). "Axonal growth in response to experimentally applied mechanical tension." *Dev Biol* **102**(2): 379–89.

Bursac, P., G. Lenormand, B. Fabry, M. Oliver, D. A. Weitz, V. Viasnoff, J. P. Butler and J. J. Fredberg (2005). "Cytoskeletal remodelling and slow dynamics in the living cell." *Nat Mater* **4**(7): 557–61.

Burton, K., J. H. Park and D. L. Taylor (1999). "Keratocytes generate traction forces in two phases." *Mol Biol Cell* **10**(11): 3745–69.

Burton, K. and D. L. Taylor (1997). "Traction forces of cytokinesis measured with optically modified elastic substrata." *Nature* **385**(6615): 450–4.

Butler, J. P., I. M. Tolic-Norrelykke, B. Fabry and J. J. Fredberg (2002). "Traction fields, moments, and strain energy that cells exert on their surroundings." *Am J Physiol Cell Physiol* **282**(3): C595–605.

Cai, Y., N. Biais, G. Giannone, M. Tanase, G. Jiang, J. M. Hofman, C. H. Wiggins, P. Silberzan, A. Buguin, B. Ladoux and M. P. Sheetz (2006). "Nonmuscle myosin IIA-dependent force inhibits cell spreading and drives F-actin flow." *Biophys J* **91**(10): 3907–20.

Carter, S. B. (1967). "Haptotaxis and the mechanism of cell motility." *Nature* **213**(73): 256–60.

Charras, G. T. and M. A. Horton (2002). "Single cell mechanotransduction and its modulation analyzed by atomic force microscope indentation." *Biophys J* **82**(6): 2970–81.

Chen, C. S., J. Tan and J. Tien (2004). "Mechanotransduction at cell-matrix and cell-cell contacts." *Annu Rev Biomed Eng* **6**: 275–302.

Chicurel, M. E., R. H. Singer, C. J. Meyer and D. E. Ingber (1998). "Integrin binding and mechanical tension induce movement of mRNA and ribosomes to focal adhesions." *Nature* **392**(6677): 730–3.

Choquet, D., D. P. Felsenfeld and M. P. Sheetz (1997). "Extracellular matrix rigidity causes strengthening of integrin-cytoskeleton linkages." *Cell* **88**(1): 39–48.

Chrzanowska-Wodnicka, M. and K. Burridge (1996). "Rho-stimulated contractility drives the formation of stress fibers and focal adhesions." *J Cell Biol* **133**(6): 1403–15.

Cukierman, E., R. Pankov, D. R. Stevens and K. M. Yamada (2001). "Taking cell-matrix adhesions to the third dimension." *Science* **294**(5547): 1708–12.

D'Addario, M., P. D. Arora, R. P. Ellen and C. A. McCulloch (2002). "Interaction of p38 and Sp1 in a mechanical force-induced, beta 1 integrin-mediated transcriptional circuit that regulates the actin-binding protein filamin-A." *J Biol Chem* **277**(49): 47541–50.

D'Addario, M., P. D. Arora, R. P. Ellen and C. A. McCulloch (2003). "Regulation of tension-induced mechanotranscriptional signals by the microtubule network in fibroblasts." *J Biol Chem* **278**(52): 53090–7.

D'Addario, M., P. D. Arora, J. Fan, B. Ganss, R. P. Ellen and C. A. McCulloch (2001). "Cytoprotection against mechanical forces delivered through beta 1 integrins requires induction of filamin A." *J Biol Chem* **276**(34): 31969–77.

D'Addario, M., P. D. Arora and C. A. McCulloch (2006). "Role of p38 in stress activation of Sp1." *Gene* **379**: 51–61.

Dai, G., M. R. Kaazempur-Mofrad, S. Natarajan, Y. Zhang, S. Vaughn, B. R. Blackman, R. D. Kamm, G. Garcia-Cardena and M. A. Gimbrone, Jr. (2004). "Distinct endothelial phenotypes evoked by arterial waveforms derived from atherosclerosis-susceptible and -resistant regions of human vasculature." *Proc Natl Acad Sci USA* **101**(41): 14871–6.

Danowski, B. A., K. Imanaka-Yoshida, J. M. Sanger and J. W. Sanger (1992). "Costameres are sites of force transmission to the substratum in adult rat cardiomyocytes." *J Cell Biol* **118**(6): 1411–20.

Davies, P. F. (1995). "Flow-mediated endothelial mechanotransduction." *Physiolog Rev* **75**(3): 519–60.

Davies, P. F., A. Robotewskyj and M. L. Griem (1994). "Quantitative studies of endothelial cell adhesion. Directional remodeling of focal adhesion sites in response to flow forces." *J Clin Invest* **93**(5): 2031–8.

Dembo, M., T. Oliver, A. Ishihara and K. Jacobson (1996). "Imaging the traction stresses exerted by locomoting cells with the elastic substratum method." *Biophys J* **70**(4): 2008–22.

Dembo, M. and Y. L. Wang (1999). "Stresses at the cell-to-substrate interface during locomotion of fibroblasts." *Biophys J* **76**(4): 2307–16.

du Roure, O., A. Saez, A. Buguin, R. H. Austin, P. Chavrier, P. Siberzan and B. Ladoux (2005). "Force mapping in epithelial cell migration." *Proc Natl Acad Sci USA* **102**(7): 2390–5.

Engler, A. J., S. Sen, H. L. Sweeney and D. E. Discher (2006). "Matrix elasticity directs stem cell lineage specification." *Cell* **126**(4): 677–89.

Fabry, B., G. N. Maksym, J. P. Butler, M. Glogauer, D. Navajas and J. J. Fredberg (2001). "Scaling the microrheology of living cells." *Phys Rev Lett* **87**(14): 148102.

Fabry, B., G. N. Maksym, J. P. Butler, M. Glogauer, D. Navajas, N. A. Taback, E. J. Millet and J. J. Fredberg (2003). "Time scale and other invariants of integrative mechanical behavior in living cells." *Phys Rev E* **68**(4, Pt. 1): 041914.

Fabry, B., G. N. Maksym, S. A. Shore, P. E. Moore, R. A. Panettieri, Jr., J. P. Butler and J. J. Fredberg (2001). "Selected contribution: time course and heterogeneity of contractile responses in cultured human airway smooth muscle cells." *J Appl Physiol* **91**(2): 986–94.

Fass, J. N. and D. J. Odde (2003). "Tensile force-dependent neurite elicitation via anti-beta1 integrin antibody-coated magnetic beads." *Biophys J* **85**(1): 623–36.

Felsenfeld, D. P., D. Choquet and M. P. Sheetz (1996). "Ligand binding regulates the directed movement of beta1 integrins on fibroblasts." *Nature* **383**(6599): 438–40.

Fung, Y. C. (1967). "Elasticity of soft tissues in simple elongation." *Am J Physiol* **213**(6): 1532–44.

Galbraith, C. G. and M. P. Sheetz (1997). "A micromachined device provides a new bend on fibroblast traction forces." *Proc Natl Acad Sci USA* **94**(17): 9114–8.

Galbraith, C. G. and M. P. Sheetz (1999). "Keratocytes pull with similar forces on their dorsal and ventral surfaces." *J Cell Biol* **147**(6): 1313–24.

Galbraith, C. G., K. M. Yamada and M. P. Sheetz (2002). "The relationship between force and focal complex development." *J Cell Biol* **159**(4): 695–705.

Garcia-Cardena, G., J. Comander, K. R. Anderson, B. R. Blackman and M. A. Gimbrone, Jr. (2001). "Biomechanical activation of vascular endothelium as a determinant of its functional phenotype." *Proc Natl Acad Sci USA* **98**(8): 4478–85.

Gaudet, C., W. A. Marganski, S. Kim, C. T. Brown, V. Gunderia, M. Dembo and J. Y. Wong (2003). "Influence of type I collagen surface density on fibroblast spreading, motility, and contractility." *Biophys J* **85**(5): 3329–35.

Geiger, B., A. Bershadsky, R. Pankov and K. M. Yamada (2001). "Transmembrane extracellular matrix–cytoskeleton crosstalk." *Nat Rev Mol Cell Biol* **2**(11): 793–805.

Giannone, G., G. Jiang, D. H. Sutton, D. R. Critchley and M. P. Sheetz (2003). "Talin1 is critical for force-dependent reinforcement of initial integrin-cytoskeleton bonds but not tyrosine kinase activation." *Journal of Cell Biology* **163**(2): 409–19.

Glogauer, M., P. Arora, D. Chou, P. A. Janmey, G. P. Downey and C. A. McCulloch (1998). "The role of actin-binding protein 280 in integrin-dependent mechanoprotection." *J Biol Chem* **273**(3): 1689–98.

Glogauer, M., P. Arora, G. Yao, I. Sokholov, J. Ferrier and C. A. McCulloch (1997). "Calcium ions and tyrosine phosphorylation interact coordinately with actin to regulate cytoprotective responses to stretching." *J Cell Sci* **110**(Pt 1): 11–21.

Glogauer, M., J. Ferrier and C. A. McCulloch (1995). "Magnetic fields applied to collagen-coated ferric oxide beads induce stretch-activated Ca2+ flux in fibroblasts." *Am J Physiol* **269**(5 Pt 1): C1093–104.

Harris, A. K., Jr. (1984). "Tissue culture cells on deformable substrata: biomechanical implications." *J Biomech Eng* **106**(1): 19–24.

Harris, A. K., D. Stopak and P. Wild (1981). "Fibroblast traction as a mechanism for collagen morphogenesis." *Nature* **290**(5803): 249–51.

Harris, A. K., P. Wild and D. Stopak (1980). "Silicone rubber substrata: a new wrinkle in the study of cell locomotion." *Science* **208**(4440): 177–9.

Heidemann, S. R., S. Kaech, R. E. Buxbaum and A. Matus (1999). "Direct observations of the mechanical behaviors of the cytoskeleton in living fibroblasts." *J Cell Biol* **145**(1): 109–22.

Helfman, D. M., E. T. Levy, C. Berthier, M. Shtutman, D. Riveline, I. Grosheva, A. Lachish-Zalait, M. Elbaum and A. D. Bershadsky (1999). "Caldesmon inhibits nonmuscle cell contractility and interferes with the formation of focal adhesions." *Mol Biol Cell* **10**(10): 3097–112.

Hill, A. V. (1922). "The maximum work and mechanical efficiency of human muscle, and their most economical speed." *J Physiol* **56**: 19–41.

Hill, A. V. (1938). "The heat of shortening and the dynamic constants of muscle." *Proc Roy Soc B* **141**: 104–117.

Hoffman, B. D., G. Massiera, K. M. Van Citters and J. C. Crocker (2006). "The consensus mechanics of cultured mammalian cells." *Proc Natl Acad Sci USA* **103**(27): 10259–64.

Hu, S., J. Chen, B. Fabry, Y. Numaguchi, A. Gouldstone, D. E. Ingber, J. J. Fredberg, J. P. Butler and N. Wang (2003). "Intracellular stress tomography reveals stress focusing and structural anisotropy in cytoskeleton of living cells." *Am J Physiol Cell Physiol* **285**(5): C1082–90.

Huang, S. and D. E. Ingber (2005). "Cell tension, matrix mechanics, and cancer development." *Cancer Cell* **8**(3): 175–6.

Hubmayr, R. D., S. A. Shore, J. J. Fredberg, E. Planus, R. A. Panettieri, Jr., W. Moller, J. Heyder and N. Wang (1996). "Pharmacological activation changes stiffness of cultured human airway smooth muscle cells." *Am J Physiol* **271**(5 Pt 1): C1660–8.

Hughes-Fulford, M. (2004). "Signal transduction and mechanical stress." *Sci STKE* **2004**(249): RE12.

Ingber, D. E. (2003). "Mechanobiology and diseases of mechanotransduction." *Ann Med* **35**(8): 564–77.

Ingber, D. E. (2005). "Mechanical control of tissue growth: Function follows form." *Proc Natl Acad Sci USA* **102**(33): 11571–2.

Ingber, D. E. (2006). "Mechanical control of tissue morphogenesis during embryological development." *Int J Dev Biol* **50**(2–3): 255–66.

James, D. W. and J. F. Taylor (1969). "The stress developed by sheets of chick fibroblasts in vitro." *Exp Cell Res* **54**(1): 107–10.

Jiang, G. Y., G. Giannone, D. R. Critchley, E. Fukumoto and M. P. Sheetz (2003). "Two-piconewton slip bond between fibronectin and the cytoskeleton depends on talin." *Nature* **424**(6946): 334–7.

Kaverina, I., O. Krylyshkina, K. Beningo, K. Anderson, Y. L. Wang and J. V. Small (2002). "Tensile stress stimulates microtubule outgrowth in living cells." *J Cell Sci* **115**(Pt 11): 2283–91.

Kelley, C., P. D'Amore, H. B. Hechtman and D. Shepro (1987). "Microvascular pericyte contractility in vitro: comparison with other cells of the vascular wall." *J Cell Biol* **104**(3): 483–90.

Kiosses, W. B., R. H. Daniels, C. Otey, G. M. Bokoch and M. A. Schwartz (1999). "A role for p21-activated kinase in endothelial cell migration." *J Cell Biol* **147**(4): 831–44.

Kolodney, M. S. and E. L. Elson (1993). "Correlation of myosin light chain phosphorylation with isometric contraction of fibroblasts." *J Biol Chem* **268**(32): 23850–5.

Kolodney, M. S. and E. L. Elson (1995). "Contraction due to microtubule disruption is associated with increased phosphorylation of myosin regulatory light chain." *Proc Natl Acad Sci USA* **92**(22): 10252–6.

Kolodney, M. S. and R. B. Wysolmerski (1992). "Isometric contraction by fibroblasts and endothelial cells in tissue culture: a quantitative study." *J Cell Biol* **117**(1): 73–82.

Lamoureux, P., G. Ruthel, R. E. Buxbaum and S. R. Heidemann (2002). "Mechanical tension can specify axonal fate in hippocampal neurons." *J Cell Biol* **159**(3): 499–508.

Lanyon, L. E. and R. N. Smith (1969). "Measurements of bone strain in walking animal." *Res Vet Sci* **10**(1): 93.

Lardner, T. J. and R. R. Archer (1994). *Mechanics of Solids: an Introduction*. New York, McGraw-Hill.

Lauffenburger, D. A. and A. F. Horwitz (1996). "Cell migration: a physically integrated molecular process." *Cell* **84**(3): 359–69.

Laurent, V. M., S. Kasas, A. Yersin, T. E. Schaffer, S. Catsicas, G. Dietler, A. B. Verkhovsky and J. J. Meister (2005). "Gradient of rigidity in the lamellipodia of migrating cells revealed by atomic force microscopy." *Biophys J* **89**(1): 667–75.

Lee, J., M. Leonard, T. Oliver, A. Ishihara and K. Jacobson (1994). "Traction forces generated by locomoting keratocytes." *J Cell Biol* **127**(6 Pt 2): 1957–64.

Lo, C. M., H. B. Wang, M. Dembo and Y. L. Wang (2000). "Cell movement is guided by the rigidity of the substrate." *Biophys J* **79**(1): 144–52.

Mack, P. J., M. R. Kaazempur-Mofrad, H. Karcher, R. T. Lee and R. D. Kamm (2004). "Force-induced focal adhesion translocation: effects of force amplitude and frequency." *Am J Physiol Cell Physiol* **287**(4): C954–62.

Maniotis, A. J., C. S. Chen and D. E. Ingber (1997). "Demonstration of mechanical connections between integrins, cytoskeletal filaments, and nucleoplasm that stabilize nuclear structure." *Proc Natl Acad Sci USA* **94**(3): 849–54.

Mathur, A. B., A. M. Collinsworth, W. M. Reichert, W. E. Kraus and G. A. Truskey (2001). "Endothelial, cardiac muscle and skeletal muscle exhibit different viscous and elastic properties as determined by atomic force microscopy." *Biomech* **34**(12): 1545–53.

Mathur, A. B., G. A. Truskey and W. M. Reichert (2000). "Atomic force and total internal reflection fluorescence microscopy for the study of force transmission in endothelial cells." *Biophys J* **78**(4): 1725–35.

Matthews, B. D., D. R. Overby, F. J. Alenghat, J. Karavitis, Y. Numaguchi, P. G. Allen and D. E. Ingber (2004). "Mechanical properties of individual focal adhesions probed with a magnetic microneedle." *Biochem Biophys Res Commun* **313**(3): 758–64.

Matthews, B. D., D. R. Overby, R. Mannix and D. E. Ingber (2006). "Cellular adaptation to mechanical stress: Role of integrins, Rho, cytoskeletal tension and mechanosensitive ion channels." *J Cell Sci* **119**(Pt 3): 508–18.

Meyer, C. J., F. J. Alenghat, P. Rim, J. H. Fong, B. Fabry and D. E. Ingber (2000). "Mechanical control of cyclic AMP signalling and gene transcription through integrins." *Nat Cell Biol* **2**(9): 666–8.

Nelson, C. M., R. P. Jean, J. L. Tan, W. F. Liu, N. J. Sniadecki, A. A. Spector and C. S. Chen (2005). "Emergent patterns of growth controlled by multicellular form and mechanics." *Proc Natl Acad Sci USA* **102**(33): 11594–9.

Oliver, T., M. Dembo and K. Jacobson (1995). "Traction forces in locomoting cells." *Cell Motil Cytoskeleton* **31**(3): 225–40.

Orr, A. W., B. P. Helmke, B. R. Blackman and M. A. Schwartz (2006). "Mechanisms of mechanotransduction." *Dev Cell* **10**(1): 11–20.

Paszek, M. J., N. Zahir, K. R. Johnson, J. N. Lakins, G. I. Rozenberg, A. Gefen, C. A. Reinhart-King, S. S. Margulies, M. Dembo, D. Boettiger, D. A. Hammer and V. M. Weaver (2005). "Tensional homeostasis and the malignant phenotype." *Cancer Cell* **8**(3): 241–54.

Pertz, O., L. Hodgson, R. L. Klemke and K. M. Hahn (2006). "Spatiotemporal dynamics of RhoA activity in migrating cells." *Nature* **440**(7087): 1069–72.

Pirone, D. M., W. F. Liu, S. A. Ruiz, L. Gao, S. Raghavan, C. A. Lemmon, L. H. Romer and C. S. Chen (2006). "An inhibitory role for FAK in regulating proliferation: A link between limited adhesion and RhoA-ROCK signaling." *J Cell Biol* **174**(2): 277–88.

Radmacher, M. (1997). "Measuring the elastic properties of biological samples with the AFM." *IEEE Engin Med Biol Mag* **16**(2): 47–57.

Rajagopalan, P., W. A. Marganski, X. Q. Brown and J. Y. Wong (2004). "Direct comparison of the spread area, contractility, and migration of balb/c 3T3 fibroblasts adhered to fibronectin- and RGD-modified substrata." *Biophys J* **87**(4): 2818–27.

Reinhart-King, C. A., M. Dembo and D. A. Hammer (2003). "Endothelial cell traction forces on RGD-derivatized polyacrylamide substrata." *Langmuir* **19**(5): 1573–9.

Reinhart-King, C. A., M. Dembo and D. A. Hammer (2005). "The dynamics and mechanics of endothelial cell spreading." *Biophys J* **89**(1): 676–89.

Ridley, A. J., M. A. Schwartz, K. Burridge, R. A. Firtel, M. H. Ginsberg, G. Borisy, J. T. Parsons and A. R. Horwitz (2003). "Cell migration: integrating signals from front to back." *Science* **302**(5651): 1704–9.

Riveline, D., E. Zamir, N. Q. Balaban, U. S. Schwarz, T. Ishizaki, S. Narumiya, Z. Kam, B. Geiger and A. D. Bershadsky (2001). "Focal contacts as mechanosensors: externally applied local mechanical force induces growth of focal contacts by an mDia1-dependent and ROCK-independent mechanism." *J Cell Biol* **153**(6): 1175–86.

Rotsch, C. and M. Radmacher (2000). "Drug-induced changes of cytoskeletal structure and mechanics in fibroblasts: An atomic force microscopy study." *Biophy J* **78**(1): 520–35.

Saez, A., A. Buguin, P. Silberzan and B. Ladoux (2005). "Is the mechanical activity of epithelial cells controlled by deformations or forces?" *Biophys J* **89**(6): L52–4.

Sawada, Y. and M. P. Sheetz (2002). "Force transduction by Triton cytoskeletons." *J Cell Biol* **156**(4): 609–15.

Sawada, Y., M. Tamada, B. J. Dubin-Thaler, O. Cherniavskaya, R. Sakai, S. Tanaka and M. P. Sheetz (2006). "Force sensing by mechanical extension of the Src family kinase substrate p130Cas." *Cell* **127**(5): 1015–26.

Schmidt, C. E., A. F. Horwitz, D. A. Lauffenburger and M. P. Sheetz (1993). "Integrin-cytoskeletal interactions in migrating fibroblasts are dynamic, asymmetric, and regulated." *J Cell Biol* **123**(4): 977–91.

Schwarz, U. S., N. Q. Balaban, D. Riveline, A. Bershadsky, B. Geiger and S. A. Safran (2002). "Calculation of forces at focal adhesions from elastic substrate data: The effect of localized force and the need for regularization." *Biophys J* **83**: 1380–94.

Sun, Z., L. A. Martinez-Lemus, A. Trache, J. P. Trzeciakowski, G. E. Davis, U. Pohl and G. A. Meininger (2005). "Mechanical properties of the interaction between fibronectin and alpha(5)beta(1)-integrin on vascular smooth muscle cells studied using atomic force microscopy." *Am J Physiol Heart Circul Physiol* **289**(6): H2526–35.

Tan, J. L., J. Tien, D. M. Pirone, D. S. Gray, K. Bhadriraju and C. S. Chen (2003). "Cells lying on a bed of microneedles: An approach to isolate mechanical force." *Proc Natl Acad Sci USA* **100**(4): 1484–9.

Trache, A. and G. A. Meininger (2005). "Atomic force-multi-optical imaging integrated microscope for monitoring molecular dynamics in live cells." *J Biomed Opt* **10**(6): 064023.

von Wichert, G., B. Haimovich, G. S. Feng and M. P. Sheetz (2003a). "Force-dependent integrin-cytoskeleton linkage formation requires downregulation of focal complex dynamics by Shp2." *Embo J* **22**(19): 5023–35.

von Wichert, G., G. Jiang, A. Kostic, K. De Vos, J. Sap and M. P. Sheetz (2003b). "RPTP-alpha acts as a transducer of mechanical force on alphav/beta3-integrin-cytoskeleton linkages." *J Cell Biol* **161**(1): 143–53.

Walpita, D. and E. Hay (2002). "Studying actin-dependent processes in tissue culture." *Nat Rev Mol Cell Biol* **3**(2): 137–41.

Wang, H. B., M. Dembo, S. K. Hanks and Y. Wang (2001). "Focal adhesion kinase is involved in mechanosensing during fibroblast migration." *Proc Natl Acad Sci USA* **98**(20): 11295–300.

Wang, J. H. (2006). "Mechanobiology of tendon." *J Biomech* **39**(9): 1563–82.

Wang, N., J. P. Butler and D. E. Ingber (1993). "Mechanotransduction across the cell surface and through the cytoskeleton." *Science* **260**(5111): 1124–7.

Wang, N. and D. E. Ingber (1994). "Control of cytoskeletal mechanics by extracellular matrix, cell shape, and mechanical tension." *Biophys J* **66**(6): 2181–9.

Wang, N. and Z. Suo (2005). "Long-distance propagation of forces in a cell." *Biochem Biophys Res Commun* **328**(4): 1133–8.

Wang, Y., E. L. Botvinick, Y. Zhao, M. W. Berns, S. Usami, R. Y. Tsien and S. Chien (2005). "Visualizing the mechanical activation of Src." *Nature* **434**(7036): 1040–5.

Waters, C. M., P. H. Sporn, M. Liu and J. J. Fredberg (2002). "Cellular biomechanics in the lung." *Am J Physiol Lung Cell Mol Physiol* **283**(3): L503–9.

Wozniak, M. A., R. Desai, P. A. Solski, C. J. Der and P. J. Keely (2003). "ROCK-generated contractility regulates breast epithelial cell differentiation in response to the physical properties of a three-dimensional collagen matrix." *J Cell Biol* **163**(3): 583–95.

Yang, Z., J. S. Lin, J. Chen and J. H. Wang (2006). "Determining substrate displacement and cell traction fields – a new approach." *J Theor Biol* **242**(3): 607–16.

17

Mechanical Regulation of Stem Cells: Implications in Tissue Remodeling

Kyle Kurpinski, Randall R. R. Janairo, Shu Chien, and Song Li

17.1 Introduction

Stem cells, which can self-renew and differentiate into cells with specialized functions, are usually classified as embryonic stem cells (ESCs) and adult stem cells. ESCs are derived from the inner cell mass of a blastocyst, can self-renew indefinitely, and can give rise to cell types of all somatic lineages from the three embryonic germ layers. Adult stem cells have been found in many types of tissues and organs such as bone marrow, blood, muscle, skin, intestine, fat, and brain. Bone marrow is one of the most abundant sources of adult stem cells and progenitor cells. Mesenchymal stem cells (MSCs), hematopoietic stem cells (HSCs), and endothelial progenitor cells (EPCs) can be isolated from bone marrow. These bone marrow MSCs are pluripotent stromal cells. MSCs can be expanded into billions of folds in culture, and can be stimulated to differentiate into a variety of cell types. HSCs give rise to blood cells. These cells, along with EPCs, are mobilized in response to growth factors and cytokines released upon injury in tissues and organs, and therefore can be isolated from peripheral blood in addition to the bone marrow. Both embryonic stem cells and adult stem cells have tremendous potential for cell therapy and tissue repair.

The function and fate of stem cells are regulated by various microenvironmental factors. In addition to chemical factors, mechanical factors can also modulate stem cell survival, organization, migration, proliferation, and differentiation. It is widely accepted that mechanical forces are critical not only in growth and development, but also in the maintenance and function of tissues such as the remodeling of cardiovascular tissues, the production of bone, and the contraction of muscle. Recent research has been extended to the mechanobiology of stem cells and progenitor cells. These studies can shed light on the role of mechanical factors in the development and remodeling of tissues and organs at the cellular level. Furthermore, stem cells and progenitor cells can be used for cell therapies and the construction of functional tissues. Dissecting the mechanical regulation of stem cells will not only help us to understand how transplanted cells respond to local mechanical stimuli, but also provide a rational basis for controlling the differentiation of stem cells into desired cell

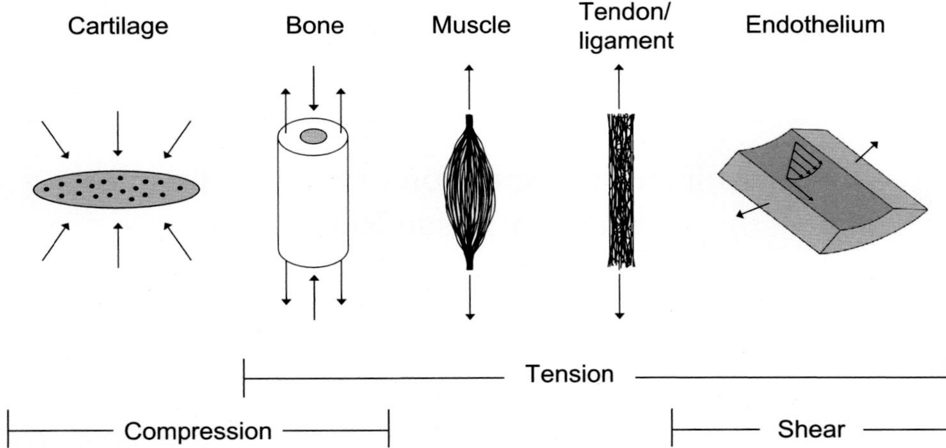

Figure 17.1. Examples of tissues subjected to mechanical loads *in vivo*. The three basic mechanical forces (tension, compression, and shear) provide varying degrees of stimulation to tissues throughout the body. Consequently, these tissues are structured to withstand these forces. Common examples include compressive strain of cartilage and bone; tensile strain of muscle, tendon, and ligament; and shear strain of endothelial cells lining blood vessels. Bone and blood vessels are examples in which multiple forces act on a single tissue.

types. In this chapter, we will focus on the regulation of tissue remodeling and stem cells by tension and compression—the mechanical loading relevant to the microenvironment in many tissues such as cardiovascular tissues, cartilage, bone, and muscle.

17.2 Mechanical Stimulation in the Body

Since as early as the nineteenth century, it has been suggested that mechanical forces play a role in the development and maintenance of human tissues. This idea is certainly reasonable, considering the numerous physical forces that are exerted on our bodies on a daily basis. Without even moving, natural external forces such as gravity and pressure act constantly on the cells in the body. The tissues and cells in the body also generate their own internal forces, including muscle contraction (both voluntary and involuntary), fluid shearing caused by digestion and blood flow, and distension of the lungs during breathing. Physical movements including walking, stretching, exercising, and any number of other activities result in a complicated balance of tension, compression, and shear, or even more intricate modes of strain such as torsion and bending. In most cases, however, the forces acting on these tissues can be reasonably modeled using just one or two primary modes of force. Some basic examples of *in vivo* mechanical stimulation are depicted in Figure 17.1. The field of mechanobiology has been developed to investigate how cells, tissues, and organs respond to these mechanical forces.

Bone remodeling is one of the most widely known and well-understood phenomena in mechanobiology, and it is readily observable as the gain or loss of bone mass due to varying physiological loads [1, 2]. It has been noted that during periods of inactivity or reduced physical load, bone tissue has a tendency to degenerate. For example, patients can suffer from atrophy of bone tissue after an extended period in bed. Astronauts often experience similar bone loss after prolonged durations of weightlessness, which also lead to reduced mechanical loads on the skeleton. Conversely, heavy exercise and increased loads can enhance the generation of bone mass.

These and many other examples make it evident that such remodeling is in response to mechanical loading, and it can also be achieved through cellular activity. This response is predominantly achieved through cellular activity. Bone tissue consists of both a densely mineralized extracellular matrix and bone cells, including osteoblasts, osteoclasts, and osteocytes. Osteoblasts are responsible for laying down a new matrix and for differentiating into mature osteocytes that populate and maintain the established bone tissue. Conversely, osteoclasts are responsible for the resorption/degradation of bone. Maintenance and remodeling of bone tissue is a continual process that requires a precise balance between the resorption of old bone and the formation of new bone, which is regulated by a balance of forces. Physical loads are transmitted to the body, distributed to the tissues such as bone, and reacted to by the residing cells.

However, this type of "mechanically responsive" cell behavior is not restricted to bone, and it is relatively common among many (if not all) connective tissues. This trend makes sense when one considers that most load-bearing and force-generating tissues are in fact part of the mesenchyme: the connective tissues of the body, including bone, cartilage, muscles, tendons, and ligaments. In addition to bone remodeling, load-induced responses have been noted in other cellular phenomena. For example, endochondral ossification, the transformation of hyaline cartilage into densely mineralized bone, is regulated by mechanical loading in areas of bone development [3–7]. In vascular tissue, the phenotype of smooth muscle cells (SMCs) is modulated in part by mechanical factors, due to the cyclic distension of the vascular wall during blood flow [5–9]. Excessive mechanical loading of skeletal muscle causes microdamage to organelles and results in amplified cell growth [10, 11]. This phenomenon is most readily observed in muscular strength training [12, 13].

These mechanically induced responses are a recurring theme throughout the various cells and tissues of the body, particularly within load-bearing tissues. It is also clear that these examples demonstrate the mechanical plasticity of various cell types. In other words, cell fate is often adaptable and may be influenced by mechanical factors. With this in mind, it is important to note that the most plastic cell type in the human body is the stem cell. The pluripotency of stem cells has important implications for how these cells may react to mechanical stimuli, particularly during early development with regard to embryonic stem cells and during tissue repair and regeneration with regard to adult stem cells. How stem cells react to the physical forces generated during these processes is a fundamental question underlying both developmental biology and stem cell biology.

17.3 Theories of Mechanical Regulation of Tissue Remodeling

Utilizing the knowledge of the forces depicted in Figure 17.1, many researchers have conceived intuitive models to describe the responses of cells and tissues to mechanical forces. With these theories and hypotheses, they have laid the groundwork for much of the mechanobiological experimentation done to date.

One of the earliest theories on the mechanical regulation of tissue remodeling was formulated by the German surgeon Julius Wolff (c. 1892). Wolff's Law, as it became known, states that the internal structure of bone is constantly being remodeled to optimally adapt to mechanical loading. That is, the microstructure of bone corresponds to the distribution of the principal mechanical stresses it experiences. In 1895, the German anatomist and experimental embryologist Wilhelm Roux hypothesized that during development, differentiation of connective tissues can be attributed to mechanical stresses. In 1924, Alfred Benninghoff specifically theorized that mechanical shear stress stimulates cartilage formation. By analyzing the works of these and other scientists and by conducting his own research, Friedrich Pauwels helped create the theoretical basis for the mechanical regulation of tissue remodeling. Pauwels hypothesized the roles of mechanical stimuli in mesenchymal tissue differentiation and healing. Specifically, he asserted that elongation and hydrostatic pressure are specific to mesenchymal tissue differentiation and as such could also play a major role in its healing. Pauwels hypothesized that elongation (tension) could contribute to connective tissue healing, while hydrostatic pressure (compression) could play a significant role in cartilage and bone healing [14, 15].

Building upon the tissue healing theories of Pauwels, Stephan M. Perren formulated his interfragmentary strain theory in 1979, in which he stated, "A tissue which ruptures or fails at a certain strain level can not be formed in a region of precursor tissue which is experiencing strains greater than this level" [16]. This critical strain level is called the "interfragmentary strain." Perren calculated theoretical values for "interfragmentary strains" for bone (2%), cartilage (10%), and granulation tissue (100%). In 1987, Dr. Harold M. Frost developed his "mechanostat" theory, which stated that bone remodeling occurs similarly to the action of a thermostat. Just as a thermostat response is triggered by a temperature change above or below a set temperature, Frost predicted that bone is remodeled analogously when the mechanical loads on bone tissue differ from a conceptual effective strain. Specifically, Frost's model claimed that overuse of bone leads to increase in bone mass, while disuse leads to its decrease [17].

In the late 1980s, Dennis R. Carter and his colleagues Patricia R. Blenman and Gary S. Beaurpre, generated new theories about the effects of mechanical loading on tissue differentiation [18–20]. Carter's hypothesis stated that under healthy vascular conditions during initial fracture healing, tension and lower levels of cyclic dilatational compression tend to transform connective tissue into bone, while higher levels of compression encourage transformation into cartilage. On the other hand, poor vascular conditions promote transformation into cartilage. Carter also asserted that the history of both tensile strain and compression determines the final differentiated tissue type. For example, high tensile strain leads to a high net production of fibrous tissue.

A decade later, Patrick J. Prendergast proposed the significance of fluid in the mechanoregulation of mesenchymal tissue differentiation and remodeling [21].

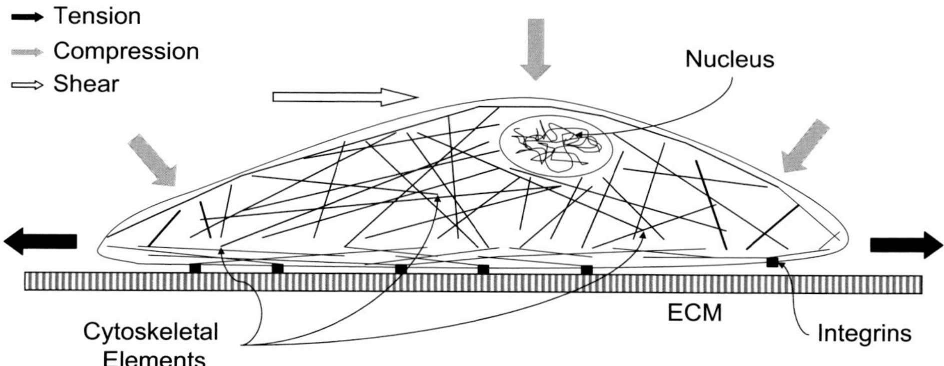

Figure 17.2. Forces experienced by a single cell. A cell experiences mechanical stimuli in the form of tensile, compressive, and/or shear force(s). These forces may ultimately be converted to changes in cellular activity via activation of specific signal transduction pathways and changes of gene expression. A cell may also experience mechanical stimuli transduced via the underlying ECM. Integrins play a key role in the sensing of such changes by a cell. The ultimate cellular reaction to mechanical stimuli, whether direct or indirect, may vary greatly with the mechanical forces the cells may experience.

Prendergast asserted that fluid flow over time acts to increase stresses felt by cells. In accordance with Pauwels' hypotheses and Perren's interfragmentary strain theory, high strains influence granulation tissue formation, intermediate strains allow cartilage formation, and bone forms at low strains. If high fluid flow over time is taken into account in addition to these strains, each tissue phenotype will exhibit a decrease in differentiation potential. If the strain/fluid flow becomes low enough, resorption can occur due to a lack of mechanical stimulation.

It is evident that the majority of these mechanobiological theories originated from experimental studies on the behavior and mechanics of tissues, most notably bone, while the remainder are based on modeling. Verification of these hypotheses often proves difficult, since it is not easy to accurately reproduce complex combinations of mechanical forces in a controlled environment. Indeed, to address this difficulty, finite element analysis has been utilized in measurable situations such as fracture healing, distraction, and osteogenesis, showing the formation of cartilage, bone, and fibrous tissue as predicted.

As alluded to previously, in order to further verify and understand these theories in mechanobiology, one must eventually examine them at the scale of the fundamental unit of living tissue, that is, the cellular level. Just like living tissues, individual live cells experience three main types of mechanical stimuli – tension, compression, and shear (Figure 17.2). Cellular responses to each type of mechanical stimulus, or a combination of them, are ultimately translated to the response observed at the tissue level.

Stem cells are potentially one of the main players in the phenotype determination of a tissue in response to mechanical loading. However, it was not until recently that experimental evidence has been obtained to determine the effects of mechanical stimuli on stem cells, especially their differentiation. This was mainly because stem cells have only recently been characterized and made readily accessible for experimentation. In the next section, we will discuss recent findings regarding stem cell response to tension and compression.

17.4 Effects of Tension and Compression on Stem Cells

As mentioned in the previous section and depicted in Figure 17.1, there are three primary mechanical forces that act on tissues in the body: tension, compression, and shear. Here we will focus on the effects of tensile and compressive strains on stem cells. Tensile loading applies a positive strain, which results in the elongation of cells. In contrast, compressive loading applies a negative strain, which results in an overall reduction or compaction of cell volume and shape. The effects of these mechanical stimuli on cellular differentiation are varied and can be related to the direct effects of strain on the cell itself. Referring back to Figure 17.2, it is readily apparent that tensile and compressive strains can have direct effects on cell morphology and structure, including changes in cell membrane, shape, and volume and in cytoskeletal structure and organization. These physical changes can be converted into changes in cell signaling and transcriptional activities in the nucleus to cause alterations in cellular differentiation, proliferation, and migration. For example, cell surface receptors such as integrins and ion channel–linked receptors can be activated. Specific cytoplasmic elements including organelles and signaling molecules may change proximity with each other during mechanical loading, thereby changing their activities. In the tissue, mechanical strain usually acts through the substrate, that is, the extracellular matrix (ECM), to which cells are attached. These changes are transduced through the focal adhesion complexes of the cells and dispersed through the interconnected cytoskeleton network.

The effects of mechanical loading are dependent on the various parameters of the strain, including magnitude (% elongation), mode (cyclic loading vs. a single "static" step), frequency (if cyclic), duration, directionality (equiaxial, uniaxial, anisotropic, etc.), and dimensionality (cell cultures in two dimensions vs. three dimensions. *In vivo*, these variables change dramatically depending upon the specific location in the body. For example, SMCs in the vascular walls receive relatively constant cyclic tensile strains at moderate magnitudes. Conversely, bone cells receive tensile strains of extremely low magnitudes and at sporadic intervals. To accurately simulate the effects of these forces on stem cells, researchers utilize a variety of devices to deliver tensile and compressive strains to cells cultured *in vitro*. Examples of these devices are depicted in Figures 17.3 and 17.4, respectively.

The devices in Figure 17.3 can be used to reasonably model many of the most common modes of tensile strain found in the body. By seeding cells on a flexible substrate such as an elastic membrane or gel, tensile strain can be applied to these materials, which is then transduced to the attached cells. For each experiment, the user can dictate the magnitude and duration of the tensile strain, and for cyclic loading, the frequency of the strain may be adjusted as well. While most tensile strain devices focus on cells cultured on a two-dimensional substrate, some

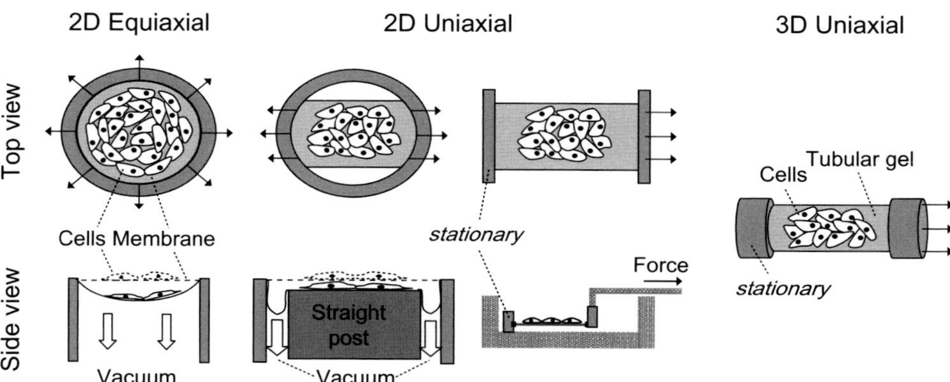

Figure 17.3. Examples of mechanical devices for applying tensile strain to stem cells *in vitro*. Various methods can be employed to apply tensile strain to cells, including the use of vacuum pressure or pulling force applied to a flexible substrate to which the cells are attached. The specific geometry of the substrate and the direction of the force determine whether the mode of tensile strain is equiaxial, uniaxial, or more complex. Cells can be seeded as a monolayer on a flat two-dimensional substrate or within a three-dimensional construct such as a gel. There are many variations of mechanical stretch devices.

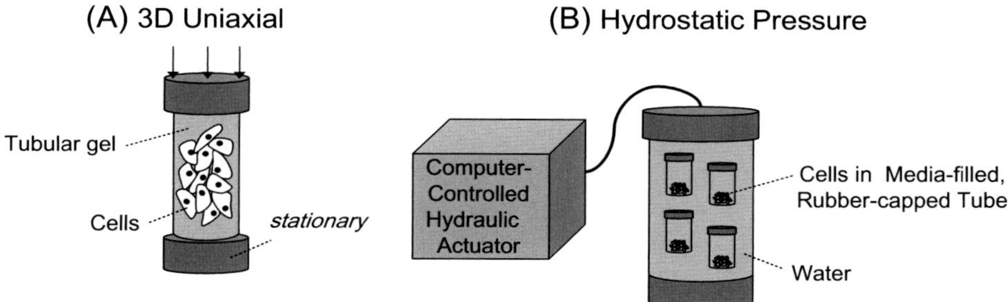

Figure 17.4. Examples of mechanical devices for applying compressive strain to cells *in vitro*. Various devices have been used to study the effects of compressive strain on cells. In general, however, studies have been performed by applying compressive stimuli to three-dimensional, cell-seeded gel constructs (uniaxially) or stem cell aggregates in fluid-filled containers (equiaxially). Although three-dimensional constructs are predominately used in compressive strain studies, the use of two-dimensional substrates is also feasible.

custom-built apparatuses allow cells to be cultured and tested in a three-dimensional matrix for a more accurate representation of *in vivo* mechanical stimulation.

The application of compressive strain on cells has mainly been achieved in three-dimensional culture, as seen in Figure 17.4. Cells seeded in gels or placed in media as aggregates "feel" compressive strain via the gel or pressure change in the fluid. The forces applied to create the strain in these constructs may be varied between constant "pressure" and repeated, cyclic compressive strain. Using a cellular setup like those depicted in Figure 17.4 allows further mimicry of *in vivo* conditions by creating a three-dimensional environment similar to those in the tissue. Although two-dimensional compressive strain experimentation is uncommon, it is still feasible. For example, a pre-stretched flexible substrate can be used as the platform for seeding cells, which will experience uniaxial compression when the flexible substrate is relaxed.

Numerous variations of the devices in Figures 17.3 and 17.4 are in use today, and they have provided us with a reasonable understanding of how stem cells react to compression and tension *in vitro*. In the following sections, we focus mainly on the effects of tensile and compressive strains on MSCs and ESCs.

Tensile strain has been shown to have different effects on stem cell differentiation when compared to compressive strain. In particular, tensile strain causes stem cells to differentiate toward various mesenchymal lineages including bone, muscle, and ligament. Furthermore, these tensile strain–induced responses are dependent on the various mechanical parameters mentioned previously. Current scientific findings suggest that the magnitude and frequency of tensile strain are particularly important in determining the type of mechanically induced differentiation that stem cells will undergo.

At low magnitudes of tensile strain (3% or lower), MSCs cultured on flexible two-dimensional substrates have been shown to differentiate toward an osteogenic phenotype. This mechanically induced bone differentiation has been noted in both human and rat MSCs, with an up-regulation of osteoblastic genes and matrix mineralization, and a decrease in proliferation upon stimulation with low magnitudes of cyclic equiaxial tensile strain [22, 23]. Surprisingly, this phenomenon is not restricted to low frequencies of cyclic strain as might be expected for bone differentiation. At high frequencies, osteogenic differentiation was observed, although at a much lower magnitude of strain (see Figure 17.5 for details).

At larger magnitudes of tensile strain (5% or higher), stem cells differentiate more toward a cardiovascular or muscular phenotype. Several independent studies have confirmed this phenomenon with various mechanical parameters. In particular, two separate studies have suggested that two-dimensional equiaxial strain may induce cardiovascular differentiation of stem cells [24, 25]. Under conditions of two-dimensional equiaxial tensile strain, ESCs have been shown to differentiate toward either a capillary or a cardiomyocyte phenotype. Conversely, under similar conditions, MSCs have been shown to differentiate toward an aortic valve phenotype. These findings suggest that two-dimensional equiaxial tensile strain may induce cardiovascular differentiation, but the final phenotype may be dependent on the type of stem cells being stimulated (Figure 17.5).

However, when the mode of tensile strain is changed to two-dimensional uniaxial instead of equiaxial, MSCs tend to differentiate toward a vascular SMC phenotype including up-regulation of smooth muscle markers such as SM α-actin, SM22-α, and h1 calponin [26, 27]. However, uniaxial cyclic strain in two-dimensions also causes adherent stem cells to realign perpendicularly to the axis of strain – a phenomenon first observed in SMCs [28]. It has been suggested that the effects of two-dimensional uniaxial cyclic strain are transient and begin to diminish once the cells have realigned because overall tensile strain on the cytoskeleton is reduced in this orientation. A newer technique utilizes micropatterned grooves on the two-dimensional elastic substrate that are oriented parallel to the axis of strain [29]. The "microgrooves" provide contact guidance that keeps the stem cells aligned with the axis of strain instead of reorienting perpendicular to the strain. MSCs stimulated

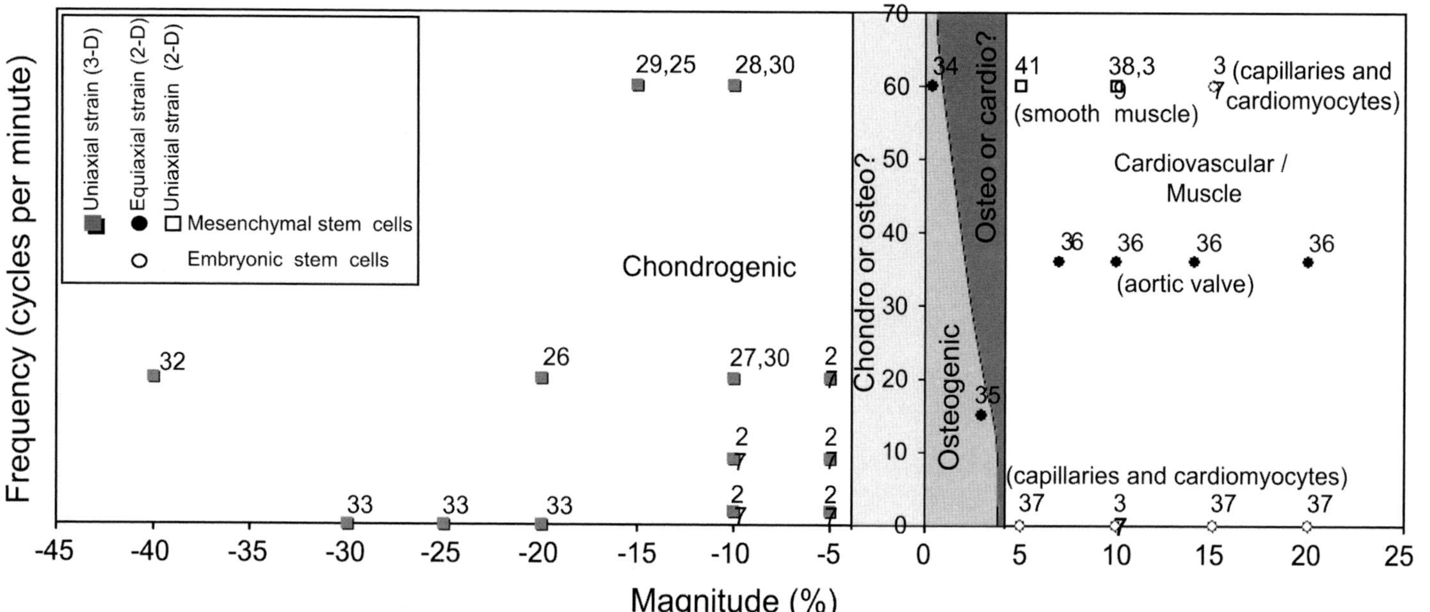

Figure 17.5. Effects of tensile and compressive strains on stem cells. At low magnitudes of cyclic tensile strain, stem cells may differentiate toward an osteogenic lineage, while at larger magnitudes, stem cells tend to differentiate toward a vascular or muscle phenotype. Equiaxial tensile strain tends to induce a cardiovascular phenotype (such as cardiomyocytes or aortic valve cells), while uniaxial tensile strain tends to induce a vascular SMC phenotype. Compressive strain predominately causes stem cell differentiation down the chondrogenic lineage. When investigating stem cell potential to differentiate under uniaxial compressive strain, three-dimensional culture systems were primarily utilized.

with uniaxial cyclic tensile strain on these membranes remain oriented with the strain axis, which leads to decreases in cartilage and bone markers and at least partial differentiation toward a vascular SMC phenotype over the course of several days.

Earlier *in vivo* studies suggested that the compression of mesenchymal progenitor cells is an important stimulus for cartilage formation and repair [30–32]. These findings have been reinforced by subsequent investigations on the effects of controlled compressive strain on MSCs *in vitro*, which showed that the application of compressive strain modes of varying magnitudes and frequencies to MSCs *in vitro* tend to cause them to differentiate down the chondrogenic lineage.

At moderately low magnitudes of compressive strain (−5% to −15%), chondrogenic differentiation of MSCs has been exhibited over a range of frequencies of strain as low as 1.8 cycles per minute (cpm) to as high as 60 cpm. Under loading regimens within these ranges of magnitude and frequency, chick, rabbit, bovine, and human MSCs have all been shown to increase the expression of cartilage-specific markers. These markers include cartilage-specific ECM components such as collagen types II and X, aggrecan, glucosaminoglycans (GAGs), and Sox9 (a transcription factor shown to increase cellular expression of type II collagen and promote chondrogenesis) [33–38].

At higher magnitudes of compressive strain (−20% to −40%), chondrogenic differentiation of MSCs has been observed, but only at frequencies of strain lower than 20 cpm, including "static" loading [39–41] (Figure 17.5). To date, no data have been published regarding compressive strain effects on embryonic stem cells.

The effects of tensile and compressive strain on stem cell differentiation are not limited to a single lineage. Specific mechanical parameters such as magnitude, frequency, and mode of strain all play a role in determining the course of mechanically induced differentiation. In addition, the type of stem cells being stimulated has an effect on the overall outcome. Thus, the mechanical stimulation of adult stem cells and ESCs does not necessarily have the same effects. The various effects of tensile and compressive strains on MSCs and ESCs are summarized in Figure 17.5. The current scientific evidence summarized in this figure matches well with the original hypotheses at the tissue level as mentioned in Section 17.3, and provides an explanation at the cellular and molecular levels as to how mechanical forces result in tissue remodeling. The map created in Figure 17.5 allows us to predict the specific effects of different mechanical regimes on stem cells *in vitro*. Based on the published experimental data, we have created hypothetical zones of mechanically induced differentiation for specific cellular lineages. As expected, compressive loading of stem cells results in chondrogenic differentiation of stem cells at all magnitudes and frequencies. What is somewhat surprising is that there are no reports of osteogenic differentiation under compressive loading. Instead, osteogenic differentiation of stem cells occurs at very low magnitudes of tensile strain. At higher magnitudes of tensile loading, stem cells tend to differentiate toward a cardiovascular or muscular phenotype.

The trends depicted in Figure 17.5 support what is already known about the mechanophysiology of the human body. Compressive loads in the body are absorbed mostly by cartilage and bone. However, bone is also subjected to low levels of tensile strain as well as torsion and bending, whereas cartilage is not. The trends in Figure 17.5 match this idea well. It seems that compression alone drives chondrogenic but not osteogenic differentiation, and that osteogenic differentiation requires small amounts of tensile loading. Interestingly, the deformation of tissue may result in interstitial fluid flow and shear stress on the cell surface, which has been shown to regulate bone and cartilage remodeling. This will add a third dimension to the map in Figure 17.5. In addition, higher magnitudes of tensile strain lead to cardiovascular or muscular stem cell differentiation, which also matches well with known *in vivo* loading. The exact cellular lineage produced from larger magnitudes of tensile strain appears to be dependent on the cell type and mode of strain.

This chapter has focused mainly on compressive and tensile strains alone, but more intricate modes of mechanical stimulation, as well as mechanical stimulation supplemented with growth factors, can result in more complex responses. For example, one study showed that concurrent cyclic axial and torsional strains applied to MSCs in a three-dimensional matrix resulted in differentiation toward a ligament phenotype [42]. Another study examined the differentiation of mouse ESCs within the three-dimensional mechanical environment of a microporous polymer tube [43]. When the ESCs were treated with a combination of vascular endothelial growth factor (VEGF) and pulsatile flow, including fluid shear stress on the luminal surface of the tube and circumferential tensile strain within the tube wall, the luminal layer differentiated into an endothelial cell (EC) phenotype while the cells in the deeper layers of the tube wall displayed an SMC phenotype, similar to the natural organization of ECs and SMCs in a blood vessel. Yet another study showed that the combination of mechanical compression and growth factors of the TGF-β family enhanced the chondrogenic differentiation of MSCs with a more than just additive effect [33]. These findings suggest that the combination of multiple mechanical and chemical factors may involve more complicated signaling mechanisms, and assessment of the relative importance of each factor needs further investigations.

A recent study has shown that the elasticity of the ECM also has profound effects on MSC differentiation. MSCs differentiate toward a bone, muscle, or neuronal phenotype when grown on hard, medium, or soft surfaces, respectively, that are akin to the stiffnesses of their respective tissues [44]. The expression of these differentiation markers is more pronounced when cells are grown on their "native" rigidities in addition to being fed chemical factors known to promote the particular phenotype, compared to cells grown in chemical factors alone. There is also evidence that the differentiation of MSCs into smooth muscle and cartilage cells is modulated by ECM elasticity (unpublished observation). How ECM elasticity and the dynamic mechanical loading work together to regulate stem cell differentiation will be an important research area for the future.

17.5 Conclusion

Recent studies have demonstrated the vast possibilities for the mechanical regulation of stem cells, as well as the need for better control of the parameters involved.

First, with the advancement of various technologies, new tools and systems can be developed to manipulate cells with mechanical and chemical factors. For example, micro- and nanoscale patterning can be used to direct cell orientation and organization to mimic that *in vivo*, new matrix materials and the structure can be fabricated to better mimic natural ECM, and mechanical devices can be built to apply multiple mechanical loads. Second, more studies are needed to define the effects of multiple mechanical and chemical factors on stem cells. Third, the molecular mechanisms of mechanical regulation of stem cells need to be elucidated. Finally, bioreactors can be designed to guide stem cell differentiation and fabricate functional tissues under mechanical loading regimes.

The phenomena summarized in this chapter have considerable implications for the use of stem cells in tissue engineering as well as for our understanding of developmental biology, tissue regeneration and repair, and the mechanobiology of stem cells in general. As more data become available, our understanding of the effects of mechanical stimulation on stem cells will continue to grow.

Based on the current literature, we would like to propose a general hypothesis on the mechanical regulation of stem cells: Stem cells can sense the mechanical forces in a tissue, and these tissue-specific mechanical forces can promote stem cell differentiation toward the phenotype of the cells residing within this tissue. This hypothesis needs the support of more experimental evidence from the research in the near future.

17.6 Acknowledgment

This work was supported in part by grants from the National Institutes of Health (NHLBI grant HL078934 and NIGMS grant R25 GM56847).

REFERENCES

[1] Duncan, R. L. and C. H. Turner, Mechanotransduction and the functional response of bone to mechanical strain. *Calcif Tissue Int*, 1995. **57**(5): 344–58.

[2] Ehrlich, P. J. and L. E. Lanyon, Mechanical strain and bone cell function: A review. *Osteoporos Int*, 2002. **13**(9): 688–700.

[3] Carter, D. R. and M. Wong, Mechanical stresses and endochondral ossification in the chondroepiphysis. *J Orthop Res*, 1988. **6**(1): 148–54.

[4] Wong, M. and D. R. Carter, A theoretical model of endochondral ossification and bone architectural construction in long bone ontogeny. *Anat Embryol (Berl)*, 1990. **181**(6): 523–32.

[5] Reusch, P., et al., Mechanical strain increases smooth muscle and decreases nonmuscle myosin expression in rat vascular smooth muscle cells. *Circ Res*, 1996. **79**(5): 1046–53.

[6] Li, Q., et al., Stretch-induced proliferation of cultured vascular smooth muscle cells and a possible involvement of local renin-angiotensin system and platelet-derived growth factor (PDGF). *Hypertens Res*, 1997. **20**(3): 217–23.

[7] Kanda, K., T. Matsuda, and T. Oka, Mechanical stress induced cellular orientation and phenotypic modulation of three dimensional cultured smooth muscle cells. *Asaio J*, 1993. **39**(3): M686–90.

[8] Birukov, K. G., et al., Stretch affects phenotype and proliferation of vascular smooth muscle cells. *Mol Cell Biochem*, 1995. **144**(2): 131–9.
[9] Tock, J., et al., Induction of SM-alpha-actin expression by mechanical strain in adult vascular smooth muscle cells is mediated through activation of JNK and p38 MAP kinase. *Biochem Biophys Res Commun*, 2003. **301**(4): 1116–21.
[10] Czerwinski, S. M., J. M. Martin, and P. J. Bechtel, Modulation of IGF mRNA abundance during stretch-induced skeletal muscle hypertrophy and regression. *J Appl Physiol*, 1994. **76**(5): 2026–30.
[11] Yang, H., et al., Changes in muscle fibre type, muscle mass and IGF-I gene expression in rabbit skeletal muscle subjected to stretch. *J Anat*, 1997. **190**(Pt 4): 613–22.
[12] Komi, P. V., et al., Effect of isometric strength training of mechanical, electrical, and metabolic aspects of muscle function. *Eur J Appl Physiol Occup Physiol*, 1978. **40**(1): 45–55.
[13] Hakkinen, K. and P. V. Komi, Alterations of mechanical characteristics of human skeletal muscle during strength training. *Eur J Appl Physiol Occup Physiol*, 1983. **50**(2): 161–72.
[14] Pauwels, F., *Biomechanics of the Normal and Diseased Hip*. 1976, Berlin: Springer-Verlag.
[15] Pauwels, F., *Biomechanics of the Locomotor Apparatus*. 1980, Berlin: Springer-Verlag.
[16] Perren, S. M., Physical and biological aspects of fracture healing with special reference to internal fixation. *Clin Orthop Relat Res*, 1979. **138**: 175–96.
[17] Frost, H. M., The mechanostat: A proposed pathogenic mechanism of osteoporoses and the bone mass effects of mechanical and nonmechanical agents. *Bone Miner*, 1987. **2**(2): 73–85.
[18] Blenman, P. R., D. R. Carter, and G. S. Beaupre, Role of mechanical loading in the progressive ossification of a fracture callus. *J Orthop Res*, 1989. **7**(3): 398–407.
[19] Carter, D. R., Mechanical loading history and skeletal biology. *J Biomech*, 1987. **20**(11–12): 1095–109.
[20] Carter, D. R., P. R. Blenman, and G. S. Beaupre, Correlations between mechanical stress history and tissue differentiation in initial fracture healing. *J Orthop Res*, 1988. **6**(5): 736–48.
[21] Prendergast, P. J., R. Huiskes, and K. Soballe, ESB Research Award 1996. Biophysical stimuli on cells during tissue differentiation at implant interfaces. *J Biomech*, 1997. **30**(6): 539–48.
[22] Jing, Y., et al., The effect of mechanical strain on proliferation and osteogenic differentiation of bone marrow mesenchymal stem cells from rats. *Sheng Wu Yi Xue Gong Cheng Xue Za Zhi*, 2006. **23**(3): 542–5.
[23] Simmons, C. A., et al., Cyclic strain enhances matrix mineralization by adult human mesenchymal stem cells via the extracellular signal-regulated kinase (ERK1/2) signaling pathway. *J Biomech*, 2003. **36**(8): 1087–96.
[24] Ku, C. H., et al., Collagen synthesis by mesenchymal stem cells and aortic valve interstitial cells in response to mechanical stretch. *Cardiovasc Res*, 2006. **71**(3): 548–56.
[25] Schmelter, M., et al., Embryonic stem cells utilize reactive oxygen species as transducers of mechanical strain-induced cardiovascular differentiation. *Faseb J*, 2006. **20**(8): 1182–4.
[26] Park, J. S., et al., Differential effects of equiaxial and uniaxial strain on mesenchymal stem cells. *Biotechnol Bioeng*, 2004. **88**(3): 359–68.
[27] Hamilton, D. W., T. M. Maul, and D. A. Vorp, Characterization of the response of bone marrow-derived progenitor cells to cyclic strain: Implications for vascular tissue-engineering applications. *Tissue Eng*, 2004. **10**(3–4): 361–9.

[28] Kanda, K., T. Matsuda, and T. Oka, Two-dimensional orientational response of smooth muscle cells to cyclic stretching. *Asaio J*, 1992. **38**(3): M382–5.
[29] Kurpinski, K., et al., Anisotropic mechanosensing by mesenchymal stem cells. *Proc Natl Acad Sci USA*, 2006. **103**(44): 16095–100.
[30] O'Driscoll, S. W., F. W. Keeley, and R. B. Salter, Durability of regenerated articular cartilage produced by free autogenous periosteal grafts in major full-thickness defects in joint surfaces under the influence of continuous passive motion. A follow-up report at one year. *J Bone Joint Surg Am*, 1988. **70**(4): 595–606.
[31] Wakitani, S., et al., Repair of large full-thickness articular cartilage defects with allograft articular chondrocytes embedded in a collagen gel. *Tissue Eng*, 1998. **4**(4): 429–44.
[32] Tagil, M. and P. Aspenberg, Cartilage induction by controlled mechanical stimulation in vivo. *J Orthop Res*, 1999. **17**(2): 200–4.
[33] Campbell, J. J., D. A. Lee, and D. L. Bader, Dynamic compressive strain influences chondrogenic gene expression in human mesenchymal stem cells. *Biorheology*, 2006. **43**(3–4): 455–70.
[34] Elder, S. H., et al., Effect of compressive loading on chondrocyte differentiation in agarose cultures of chick limb-bud cells. *J Orthop Res*, 2000. **18**(1): 78–86.
[35] Elder, S. H., et al., Chondrocyte differentiation is modulated by frequency and duration of cyclic compressive loading. *Ann Biomed Eng*, 2001. **29**(6): 476–82.
[36] Huang, C. Y., et al., Effects of cyclic compressive loading on chondrogenesis of rabbit bone-marrow derived mesenchymal stem cells. *Stem Cells*, 2004. **22**(3): 313–23.
[37] Huang, C. Y., P. M. Reuben, and H. S. Cheung, Temporal expression patterns and corresponding protein inductions of early responsive genes in rabbit bone marrow-derived mesenchymal stem cells under cyclic compressive loading. *Stem Cells*, 2005. **23**(8): 1113–21.
[38] Mauck, R. L., et al., Regulation of cartilaginous ECM gene transcription by chondrocytes and MSCs in 3D culture in response to dynamic loading. *Biomech Model Mechanobiol*, 2007. **6**(1–2): 113–25.
[39] Angele, P., et al., Cyclic hydrostatic pressure enhances the chondrogenic phenotype of human mesenchymal progenitor cells differentiated in vitro. *J Orthop Res*, 2003. **21**(3): 451–7.
[40] Angele, P., et al., Cyclic, mechanical compression enhances chondrogenesis of mesenchymal progenitor cells in tissue engineering scaffolds. *Biorheology*, 2004. **41**(3–4): 335–46.
[41] Takahashi, I., et al., Compressive force promotes sox9, type II collagen and aggrecan and inhibits IL-1beta expression resulting in chondrogenesis in mouse embryonic limb bud mesenchymal cells. *J Cell Sci*, 1998. **111**(Pt 14): 2067–76.
[42] Altman, G. H., et al., Cell differentiation by mechanical stress. *Faseb J*, 2002. **16**(2): 270–2.
[43] Huang, H., et al., Differentiation from embryonic stem cells to vascular wall cells under in vitro pulsatile flow loading. *J Artif Organs*, 2005. **8**(2): 110–8.
[44] Engler, A. J., et al., Matrix elasticity directs stem cell lineage specification. *Cell*, 2006. **126**(4): 677–89.

18

Mechanotransduction: Role of Nuclear Pore Mechanics and Nucleocytoplasmic Transport

Christopher B. Wolf and Mohammad R. K. Mofrad

18.1 Introduction

Cells, especially in multicellular organisms during development, are subject to a wide range of forces that help shape their overall fate and response to a variety of physiologically important stimuli. During such environmental interactions, it is important that the cell maintains its own "sensory" system, not only with biochemical receptors but also with regard to mechanical signals as well. Mechanotransduction is common in a wide variety of biological and physiological phenomena, from developmental biology to the development of pathophysiological conditions that have wide-ranging health and medical implications. In the case of atherosclerosis, for example, it manifests as a potential progenitor to thrown clots that can cause strokes or myocardial infarctions. In the case of developmental pathways, constructions on the proportions of oocyte and/or egg growth will have adverse consequences for the entire developing organism, should it manage to even continue growing. Several hypotheses have been proposed, as articulated in the individual chapters of this book, for describing the mechanism by which the cell senses mechanical forces and converts them into a cascade of biochemical signals that affect the phenotype of the cell in health and disease. It has also been shown that the deformation of the nucleus is observed in a large variety of important events within the cell such as replication and response to mechanical forces. In this chapter, we focus on the nuclear pore complex (NPC) and examine the role of the NPC and nucleocytoplasmic transport in the regulation and mediation of mechanotransduction.

18.2 The Nuclear Pore Complex

The NPC is a large pore complex that spans the nuclear envelope of eukaryotic cells and is 90 nm in height and 120 nm in outermost diameter, with a channel in the core lumen that forms anywhere from 10 to 50 nm bounded by phenylalanine-glycine (FG) repeats. It is composed of a multitude of protein subunits, each of which has a biomechanical and biological function in performing the overarching goal of the NPC, facilitating nucleocytoplasmic transport. Both the cytoplasmic and nucleoplasmic faces of the NPC serve as anchors to longer fibrils, mostly comprised of FG repeats, that extend outward and serve to increase the probability of binding cargo complexes that have karyophilic affinities for the repeats prior to the initiation of the main transport event. The NPC's superstructure is subdivided into eight symmetric, large spokes that span the nuclear envelope (see Figure 18.1, Plate 18, and Figure 18.2). The most striking feature of the NPC is its eightfold symmetry, each section of which is comprised of macromolecular constructs (gp210 proteins). This construction is also very highly conserved in higher organisms, with practically no change to the overall superstructure from yeast NPCs to human and mammalian NPCs observed. With regard to other transport channels and pores, this type of construction is unique, especially considering the large cargo complexes that traverse the NPC at rapid rates. The one and only gateway for nucleocytoplasmic transport, the NPC serves as everything from the final regulatory checkpoint before mRNA export and transcription to the import of material required for preparing the nucleus for the events leading up to cellular replication. The aforementioned processes are dramatically demanding since they occur at varying levels of intensity across the cell, nearly constantly, and require the transport of cargoes that are relatively large in a very short time frame. These requirements are only increased during special events, like replication or extensive growth of the cell or tissue of which the cell type comprises. Due to these requirements, the NPC has evolved to be capable of transporting 125 MDa/s, in either direction; in the time span of 1 s this is approximately the molecular weight of one whole NPC that is being transported (Kiseleva et al., 1998). In addition, many of the cargo complexes that the NPC has to transport are virtually the size of the core lumen of the NPC itself (Kiseleva et al., 1998; Panté and Kann, 2002). Yet the NPC has rarely been observed to natively clog at body temperature, but does so at very low temperatures, despite the high rates of continuous nucleocytoplasmic transport (Tomoyuki et al., 2004). Another interesting characteristic of the NPC is the threshold it presents for transport; if a molecule is under 40 kDa in weight, it is passively transported through the NPC without requiring an NLS (nuclear localization signal) and the subsequent binding to the karyopherins Importin-β and Importin-α (see Figure 18.3). It is not yet clear if these sub-40-kDa cargo complexes use the core channel or possibly one of the potential side channels that are formed by the protrusions from the gp210 spokes, as they appear to be fully unimpeded by the FG (phenylalanine-glycine) repeats, which comprise large polymer brushes around the cytoplasmic face of the NPC and extend into the core lumen and the nucleoplasmic face of the NPC (see Figure 18.1; Plate 18).

Given that the NPC is such a large complex (see Figure 18.2) with a high nucleocytoplasmic transport throughput, it is reasonable to assume that multiscale, biomechanical responses play a role in the ability of the NPC to perform its role in the cell. The first evidence for this came from two direct observations. The first had to do with the reconformations and dilated states the NPC assumed during nucleocytoplasmic transport, and the second observation illustrated that the number of spokes that made up the NPC was not arbitrary and was, in fact, optimized for eight spokes

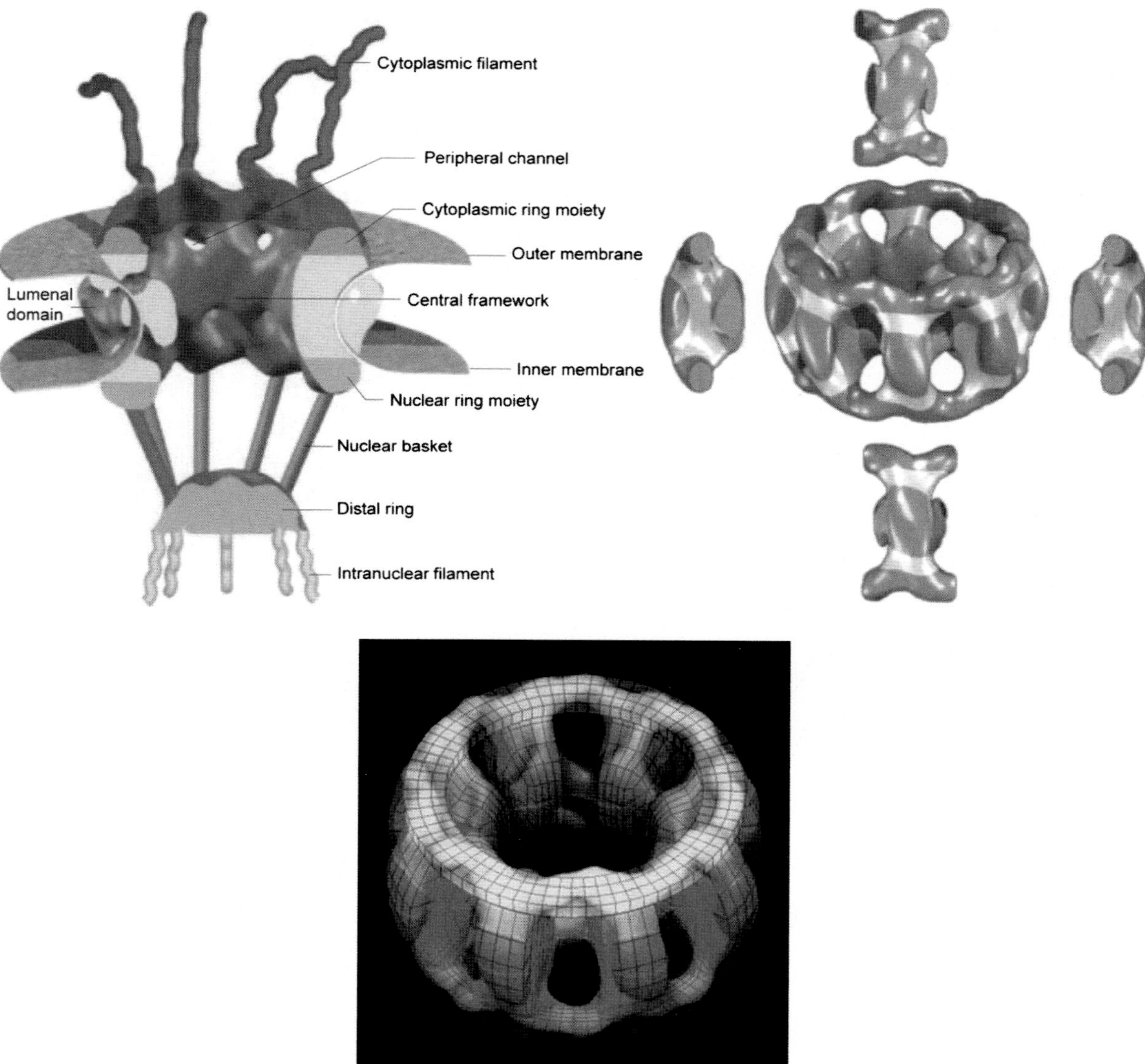

Figure 18.1. Top Panels: An anatomical view of the NPC; easily seen are the eight spokes that are connected to form the pore superstructure. The NPC is approximately 150 nm from the cytoplasmic ring to the nucleoplasmic ring in depth and 120 nm in diameter. Left illustration prepared using a visual programming environment developed at The Scripps Research Institute (Hinshaw et al., 1992; Stoffler et al., 1999), top right illustration from Panté and Aebi (1993). Bottom Panel: A biomechanical finite element analysis (FEA) model of the NPC (Wolf and Mofrad, 2008)

as the ninefold and tenfold spoke mutants were a significantly small statistical minority in a large survey of the NPCs in living organisms (Hinshaw et al., 1992; see Figure 18.4). Intrigued by this and other observations about the NPC *in vivo*, we investigated the mechanical properties of the NPC, especially the eightfold

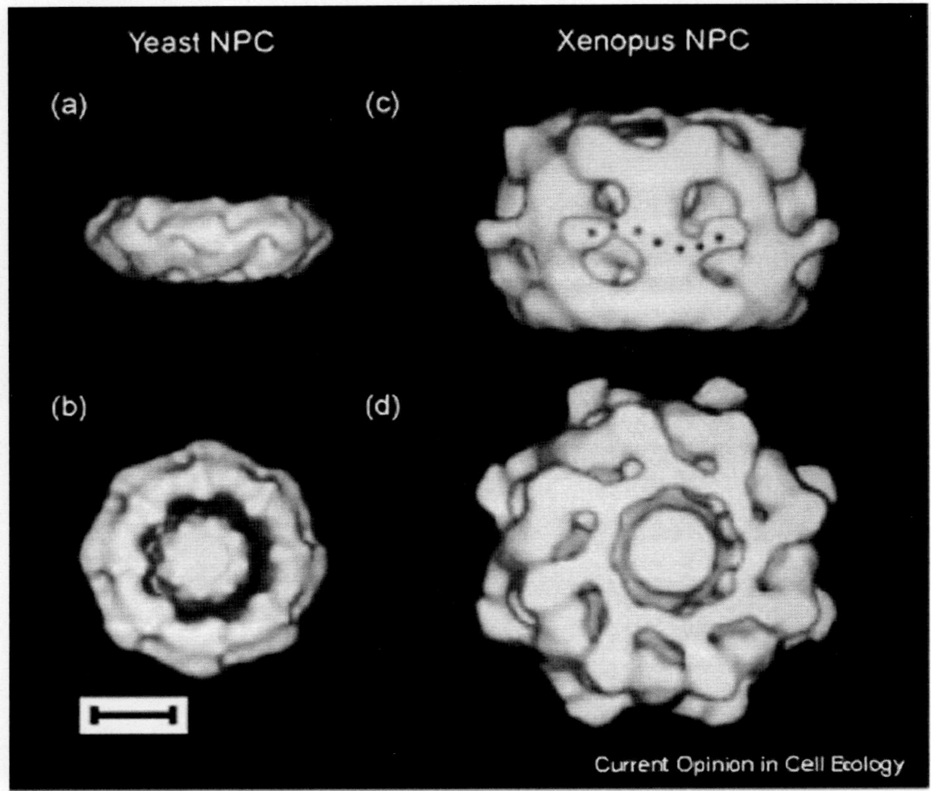

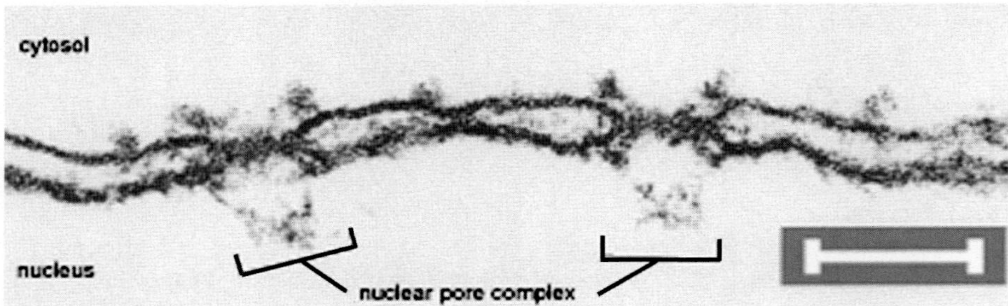

Figure 18.2. Top: Yeast and vertebrate nuclear pore complexes. Front (a, b) and top (c, d) of yeast and vertebrate (Xenopus) NPCs illustrating the relatively larger size of the vertebrate NPC in the front view and the ubiquitous octagonal shape of both in the top view (Yang et al., 1998; Stoffler et al., 1999). The striking octagonal shape (eightfold symmetry) of the NPC is evident. The scale bar is 30 nm. Bottom: An electron micrograph of a nuclear membrane with two NPCs present (Franke et al., 1970). The scale bar is 100 nm.

symmetry that is such a prominent feature (Wolf and Mofrad, 2008). The octagonal symmetry of the NPC is an explicit showcase of the role biomechanics plays in the operation of this complex. The gains become evident when one realizes that the NPC is not a static construct traversing the nuclear envelope. As the electron micrographs have shown, there is a significant amount of conformational changes that the NPC undergoes during the nucleocytoplasmic transport of cargo complexes,

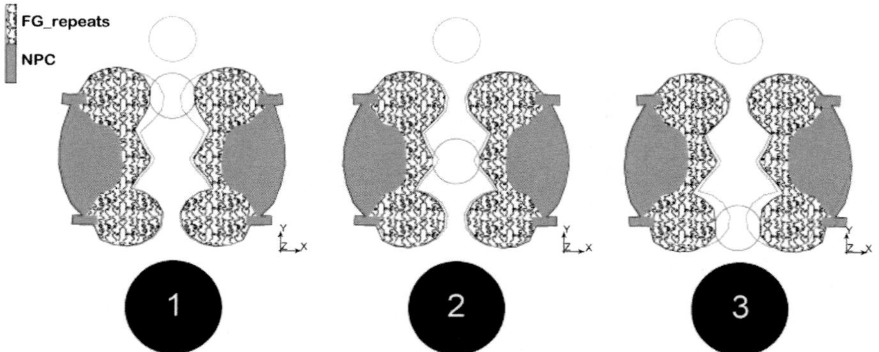

Figure 18.3. A showcase of the deformations encountered for a cargo complex as it traverses the NPC in the various stages of nucleocytoplasmic transport and illustrates the deformations of the FG brushes in conjunction with the re-alignment of the spokes; the stage numbers at the bottom correspond to the stage numbers in Figure 18.11, for reference.

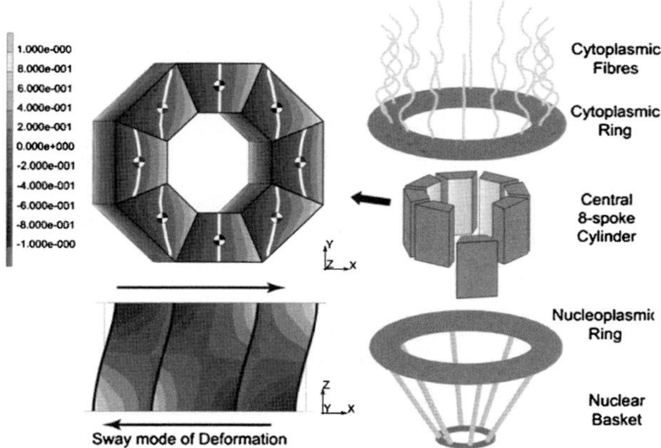

Figure 18.4. Normalized bending stresses plotted on deformed finite element model (left) of unconnected octagonal segments from NPC central eight-spoke cylinder (right) caused by sway deformation. The bending stress is normalized by dividing all values by the maximum bending stress. The maximum stress is proportional to the amount of sway. Instead of an integral cross section, the segmentation of the NPC leads to each spoke ectending, twisting, or bending about its own axis that collectively contribute to the structural stiffness of the NPC. In this case, each segment bends about its own neutral axis passing through each segment's centroid where the bending stress is zero (white line).

especially large ones. We explored the question of the eightfold symmetry of the NPC and why it appears to be such an important component defining the biological transport functionality as well as the overall shape. Looking at the large spokes that comprise the octagonal superstructure of the NPC (Figure 18.2), we see that these subunits can be visualized as trapezoidal cross sections (Figure 18.5). Viewing the octagonal subunits of the NPC in this manner, we can calculate a multitude of biomechanical properties of the entire NPC. In an initial study, we speculated that the reason behind the existence of the spoke subunits was to maximize the stiffness of each spoke by distributing the area into an octagonal symmetry to increase the minimum principal moment of inertia and therefore the flexural bending stiffness of the spoke. Simply stated, this means that the spoke stiffness is maximized, and, therefore, any observed deformations of the NPC overall will most likely result from the spokes bending about their own axes to form several of the conformational modes shown in Figure 18.6 that are observed both in simulations and *in vivo*, experimentally. Further detail about the biomechanics of the NPC can be found in Wolf and Mofrad (2008).

18.3 Molecular Biology of Nucleocytoplasmic Transport

Nucleocytoplasmic transport plays a key role in the biology of practically every eukaryotic organism. It is a required gateway and checkpoint to the classical biological dogma of the progression from DNA to mRNA and finally protein transcription. For the mRNA strands to successfully describe the synthesis of a new protein, they must be allowed export into the cytoplasm and undergo the protein transcription process with an associated ribosome. Subsequently, the resulting proteins are the basic "machine" units of the cell that perform the vast majority of all the vital functions required for sustained cellular vitality. Because of this, nucleocytoplasmic transport can be viewed as a component in a control feedback system representation of the cell, including mechanotransduction, as depicted in Figure 18.7. Given the biological importance of nucleocytoplasmic transport and its eventual effect on the mechanotransduction of necessary signals through the cell, it is important to understand the underlying biophysics behind the transport process and the changes that will have the largest impact on nucleocytoplasmic transport, and hence the remainder of the cell. Considering the amount of traffic that traverses the nuclear envelope going in both directions, nucleocytoplasmic transport must be a robust and durable process along with a certain amount of regulation; otherwise, there would be little to no reason for the sequestration of the nuclear contents by the nuclear envelope. There are yet more stringent requirements of nucleocytoplasmic transport, in that it must be able to handle very large cargo complexes, and, furthermore, these larger cargo complexes must not require excessive time or resources to be transported relative to the smaller cargo complexes. Following from these requirements, we see that nucleocytoplasmic transport does indeed have a threshold for what cargo complexes may transit the NPC of their own volition and which ones require the addition of special nucleocytoplasmic transport factors, called karyopherins. For almost all cargo complexes under 40 kDa, there is no requirement for karyophilic binding to undergo nucleocytoplasmic transport. However, above this limit, two karyophilic factors are required for successful native nucleocytoplasmic transport, namely, Importin-β and Importin-α, as seen in Figure 18.8 (Plate 19). These two Importins, as the karyopherins with nuclear import directionality are named, bind together whereupon Importin-α then binds to a nuclear localization signal, or NLS, on the cargo. This comprises the cargo complex that is to undergo nucleocytoplasmic transport. After transport, this cargo complex binds with

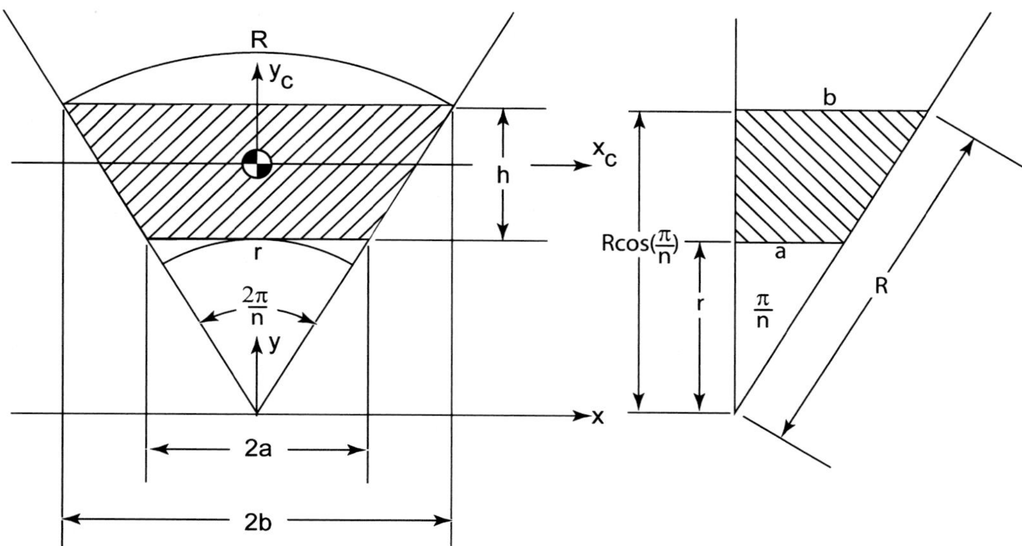

Figure 18.5. One segment of an n-sided annular polygon contained by the NPC's concentric rings (see Wolf and Mofrad, 2008) forms an isosceles trapezoid whose section properties can be determined from radil of the concentric rings and the number of segments, n. The subscript "C" refers to those axes that pass through the centroid of the trapezoid. The minimum principal moment of inertia, lxc, of the segment becomes a key factor in determining the octagonal shape of the central NPC framework.

RanGTP and the cargo is released, sans the Importins, into the nucleoplasm. So far, we have focused on the NPC without the cargo complex. This has been a reasonable assumption up until this point, where we considered the cargo complex as simply playing a passive role in nucleocytoplasmic transport through the NPC without any further biomechanical effects on the pore or the transport itself and since we modeled the inclusion of the cargo simply through changes in the boundary values in the previously referenced analyses. This, however, is not the case as, put simply, there must be a difference between the macromolecular cargo complexes that require the karyopherins Importin-β and Importin-α for transport and the sub-40-kDa cargo that passes through unimpeded. It has already been shown that the Balbiani rings partially disassemble before nucleocytoplasmic transport (Kiseleva et al., 1998) and that certain viral capsids are required to be hard and dense to undergo successful nuclear import localization (Ward et al., 1996; see Figure 18.9; Plate 20). Given that certain xenogenic nucleocytoplasmic transport, such as the nuclear import and localization of oversized viral capsids, may occur in the absence of an NLS on the xenogenic cargo or binding to the karyophilic Importins, it is crucial to understand the abilities of such viral capsids (e.g., Hepatitis B viral capsids) and the roles they play in opening the gate of the NPC. Understanding the techniques that these viral capsids use will help in the interpretation of the process of nucleocytoplasmic transport and the factors, excepting the karyopherins, that uniquely affect it.

Other obvious components of the NPC are the FG repeat brushes that crowd both faces of it and part of the core lumen. These FG repeats are hydrophobic,

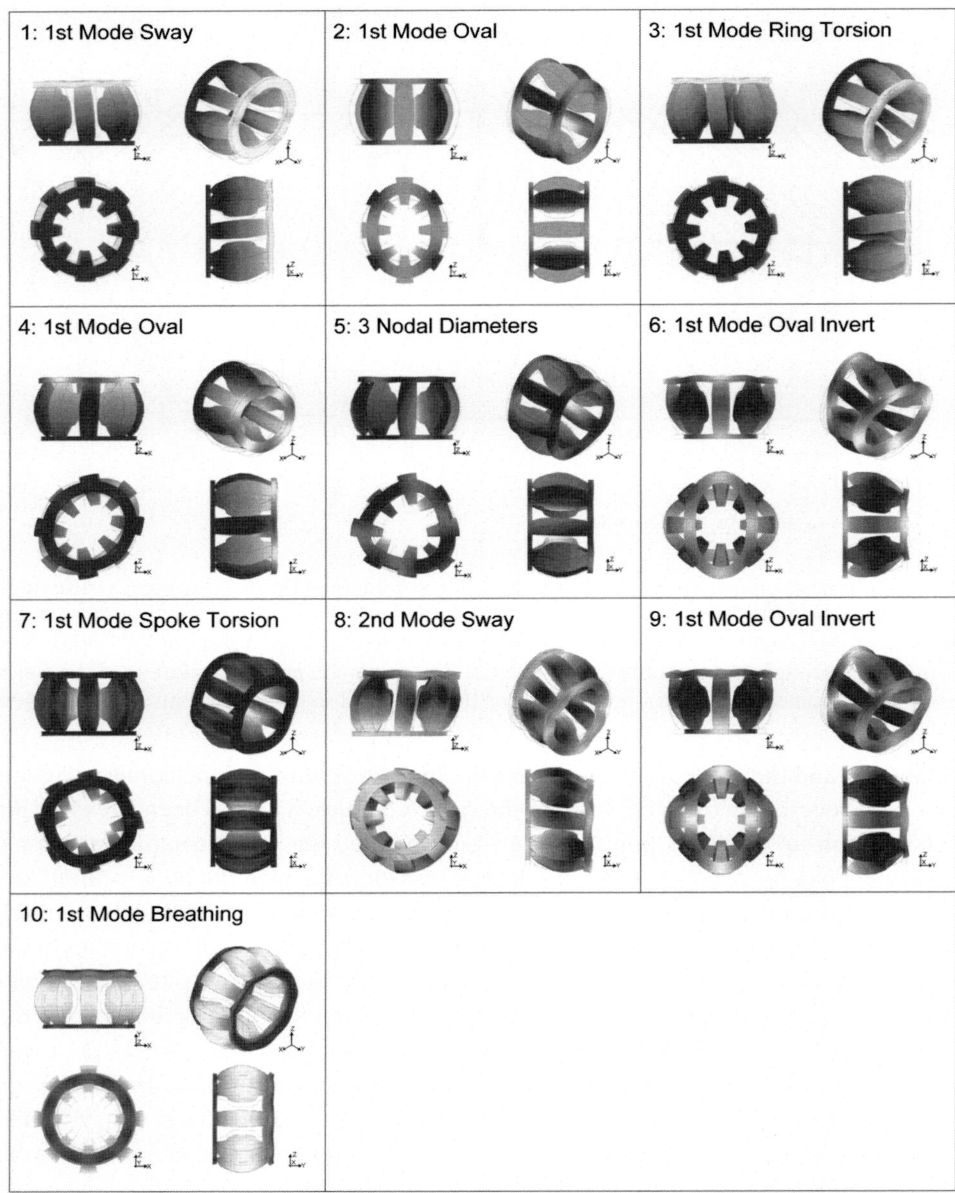

Figure 18.6. Mode shapes that the NPC model assumes during a modal analysis; the three modes that are most obviously observed experimentally, at steady state, are the three modes that are the lowest in energy. Notice that the dilation of the NPC is predicted, along with other deformative modes that were predicted, namely, inversion and eversion of the NPC. This is also in agreement with experimental observations regarding electron micrographs of the NPC undergoing NCT of a large cargo (Wolf and Mofrad, 2008).

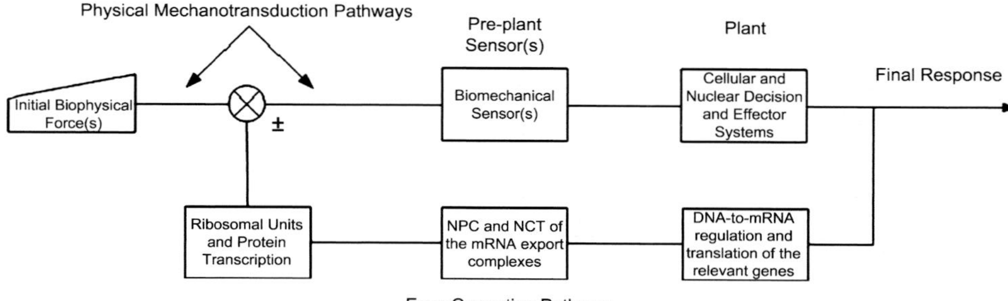

Figure 18.7. A hypothetical control feedback system layout of the relations between mechanotransduction, the NPC, and NCT. Here we can see how small changes affecting the NPC and/or NCT can lead to larger overall changes affecting the result of this system.

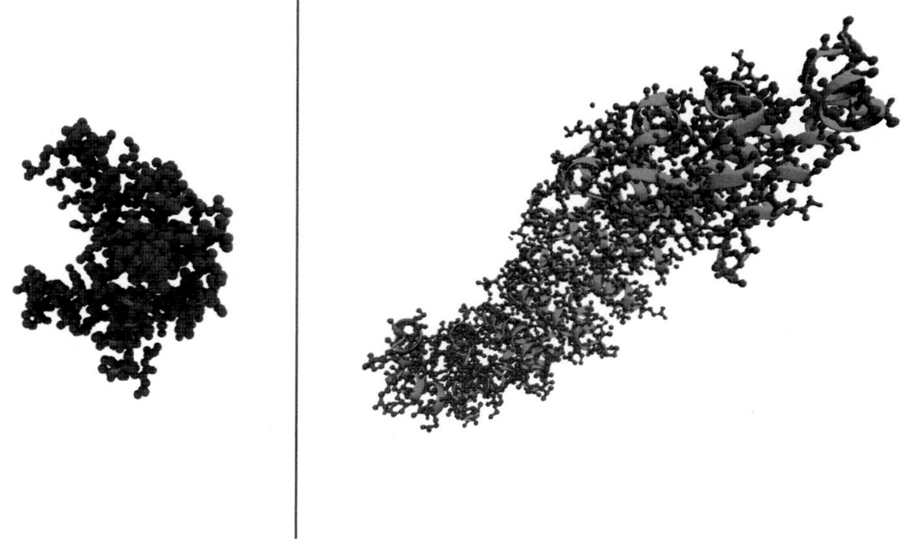

Figure 18.8. A hybrid rendering of both Importin-β (left) and Importin-α (right); the binding site for the FG repeats on Importin-β are visible in the posterior and bottom portion in the illustration. The superhelical structure of Importin-α is easily seen in the illustration; both Importins are all-α-helix motif proteins.

long-chain polymers that act like lipids but pack significantly better (see Figure 18.10; Plate 21). Due to these properties, hydrogels that have been made of the FG repeats have the expected properties of being viscous and "greasy" with a large amount of surface tension. Indeed, the issues encountered at the border of any viscous, almost fluid-like, medium on these scales are enough to significantly perturb transport and foul mechanisms of understanding the biophysical operation of such constructs. Certainly, surface tension can be dealt with by the usage of surfactants on the topology of the cargo complex, and this is actually done to some extent. However, the biochemical properties of the karyopherins Importin-β and Importin-α, albeit important for initiating and completing nucleocytoplasmic transport, are not the sole characteristics that endow the cargo complex with the ability to traverse the nuclear pore complex. If they were, then viral capsids that have little to no karyophilic motifs on their outside coating would not be nearly enough to overcome the attractive forces at the cytoplasmic interface of the FG core for long enough to let the capsid through. Looking solely at this interface, such a requirement increases greatly in proportion to the size and molecular weight of the cargo until a point is reached where the entire surface topology and structure must be covered by karyophilic regions, or at least significantly hydrophilic regions, and remain at the interface for a short period of time to even part the FG repeats, assuming they were a plug that filled the entire space of the core lumen. The aforementioned inconsistencies, some calculations regarding the radius of gyration of the FG repeats (Lim et al., 2007), and high-resolution experimental observations of analogs to the NPC structure with the inclusion of FG repeats (Beck et al., 2007) elucidate the existence of a central channel, devoid of FG repeat units, in the core lumen of the NPC instead of a simply solid plug. But what is the exact purpose of this central channel, and how does it operate to aid transport?

The first stage of nucleocytoplasmic transport, after the cargo complex has attached and localized near the NPC by means of the free-floating FG fibrils that are attached to the cytoplasmic ring, begins when Importin-β first binds to the closest FG repeat chain that is within a large FG polymer brush. This is mediated, from the perspective of Importin-β, by a binding region on the molecule that accommodates 14 binding sites for the phenylalanine residues on a mostly linear topology to act as a "zipper" of sorts. Once this binding takes place, Importin-β is now guided along a long FG repeat into the local of the inlet of the core channel. The motive forces proposed to inch Importin-β along the FG repeat strand are a combination of Brownian diffusive motion and thermokinetic vibrations where the 14-site binding topology on the molecule acts like a one-way catch where, once it has shifted an increment in the allowed direction, it is very difficult or even impossible to move back in the other direction while still attached to the same FG repeat strand. During this, the cargo is tethered to Importin-β by another Importin, Importin-α, which has a superhelical structure and, like importin-β, is another all-α-helix motif protein that is relatively rigid (see Figure 18.8; Plate 19). The cargo itself is drawn closer to the inlet of the core channel formed by the FG repeat brushes and begins to become a mechanical inclusion. At this stage, the cargo begins to compress and slightly pack the FG repeats at the topmost FG brush and, since there is little to no friction between the FG repeats and the topology of the cargo complex, the cargo can easily slip into position within the core channel. At this stage, due to the mechanical inclusion of the cargo complex, the NPC has responded by attaining the first transport mode whereby it inverts, to a certain extent, to further accommodate the cargo complex and minimize the increase in orthogonal pressure applied to the cargo complex via the FG brushes (see Figure 18.3). Due to being constrained and localized by both the channel and the NPC, the mean free path of diffusion (λ) is minimized in the horizontal direction, parallel to the nuclear envelope, and is only large in the direction of transit while the diameter of the core channel, which is comprised of flexible FG repeats in biomechanically compliant FG polymer brushes, is adjusted to nearly the same diameter of

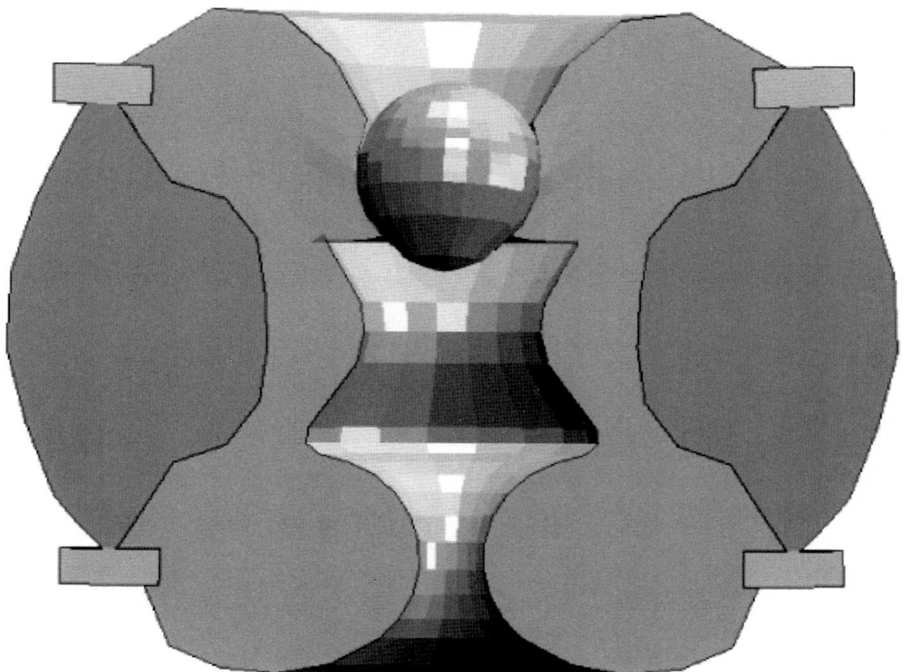

Figure 18.9. This is an FEA simulation of the soft, mature capsid attempting to traverse the NPC, much like the hard, immature capsid did. However, due to additional energy taken up to deform not only the FG repeat brushes but also the very soft capsid itself along with energy losses due to increased internal viscosities of the mature capsid, it is unable to progress past the attempt to initiate nucleocytoplasmic transport; furthermore, this is the best case simulation, out of a number of them performed, as most other mature capsids this soft did not even make it far enough in to present as a viable cargo complex inclusion to the NPC.

the cargo complex itself. Let's take the analogy of forcing a large metal sphere through rubber tubing versus forcing it through an octagonally segmented mechanical pore that is more compliant through all stages of transport due to the high bending stiffness of the spokes; this is akin to the first stage of transport where the cargo complex begins to fit into the core channel of the NPC.

18.4 Effect of NPC Mechanics on Nucleocytoplasmic Transport

The octagonal construction of the NPC, as opposed to a monostructural cylindric representation of the NPC, can be rationalized in that it accommodates the cargo complex during nucleocytoplasmic transport far better, with more compliance and less change in energy required, than would a design consisting of simply two large subunits or just one large cylinder, forming the NPC as a cylinder (see Wolf and Mofrad, 2008). This can be conceptually abstracted as a force insertion problem of a large sphere through two different pore structures, a cylindrical section of rubber tubing versus a pore constructed of eight rigid spokes that can pivot around their midpoints, which are connected by elastic elements at the inlet, middle cross section,

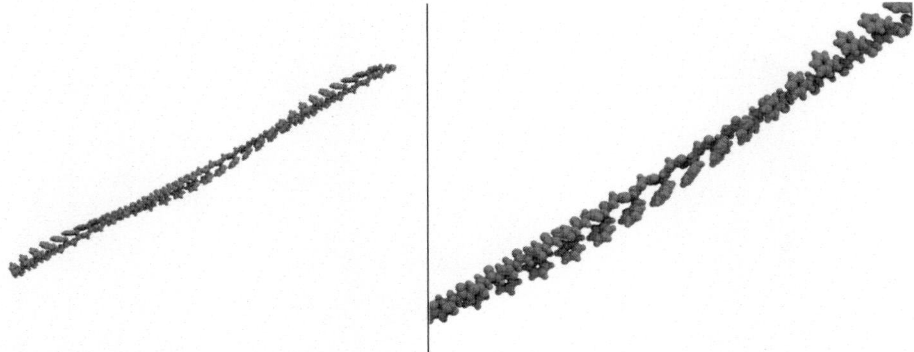

Figure 18.10. A rendering of a segment of a long FG repeat chain; left image, whole FG chain segment; right image, close-up of FG repeat chain segment. The torqued helical structure of these FG repeat chains, a helix with its own axial backbone twisted slightly, aids in tighter packing when they are pushed aside by cargo complexes undergoing nucleocytoplasmic transport. At high rates of transport, the manifest shear strain induces crystalline lattice packing forms of the constitutive FG repeat polymers of the FG brushes due to their long length, hydropathic similarity in that they are all hydrophobic, and helical structure, giving them the ability to be packed into a tighter space when compressed; this is a similar effect to that which occurs to the constitutive proteins of silk, Fibroin and Sericin, when exposed to the shear strain resulting from being forced through a narrow nozzle with a high shear strain rate. This type of additional packing lends extra stability to the formation of the channel that the cargo complex moves through during the transport event.

and outlet (see Figure 18.1; Plate 18). The energy required to force the sphere through the rubber tubing, in this analogy, is greater than that required to move it through the latter construct, whose spokes allow it to conform more easily around the sphere as it is traversing the "lumen" of this mechanical analog of the pore. An example of this is given in Figure 18.11, where the three main modes of the NPC, in the process of transporting a cargo complex, adapt better to the mechanical inclusion of the cargo due to the higher bending stiffness of the spokes versus the degrees of freedom they have. In addition, contact with the cargo will be maintained along only one contour throughout the duration of nucleocytoplasmic transport, which is beneficial since this minimizes impedances to transport that arise due to excessive interactions and pressure against large surfaces. In this manner, the NPC minimizes the energy that is required for the transit of cargo through itself since the compliance is maximized with the spokes bending as required, instead of a continuous plane deforming, as in the case of a single large cylinder, generating excess high stresses in the material that would comprise a single monocylindrical pore construct. Due to the segmentation, it is easier for the NPC to flex, invert, and accommodate the cargo complex throughout the nucleocytoplasmic transit whereas, if it was just one large cylinder that behaved like rubber tubing, it would have to stretch and generate excessive stresses at the topology between the stretched portion of the tube and the relaxed portion. Further, the entire "rubber cylinder" would also apply pressure wherever it came into contact with the cargo complex versus the segmentation of the NPC, where, through mechanical interaction with the cargo complex slipping

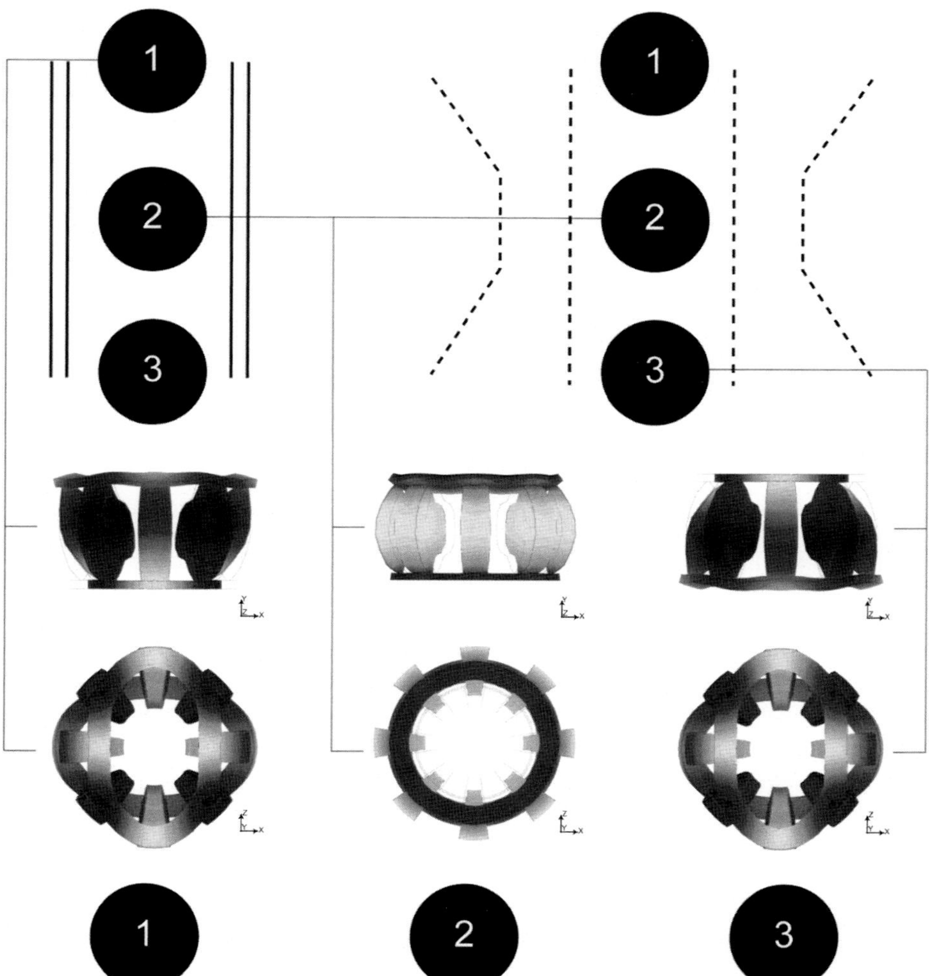

Figure 18.11. The three main modes of the NPC, in the process of transporting a cargo complex, adapt better to the mechanical inclusion of the cargo.

through the core channel during nucleocytoplasmic transport, the NPC is able to readjust and only contact, and hence apply a small amount of pressure, on the cargo complex only where necessary (always at the maximum diameter of the cargo, usually this is also the midpoint of the cargo complex). After this point, the cargo complex proceeds to travel into the core channel farther, all the way to the midpoint of both the channel and the NPC, where there is a predicted lower density of FG repeats as compared to the denser FG brushes at both the cytoplasmic face and the nucleoplasmic face; this is where the superstructure that forms the skeleton the FG core channel is formed upon narrows, it is reasonable to assume there are fewer FG repeats here, to keep it less crowded, than at either the inlet or the outlet of the NPC. The NPC then dilates slightly, due to the inclusion of larger cargo complexes, as the cargo complex traverses the remainder of the core channel. The final stage is the

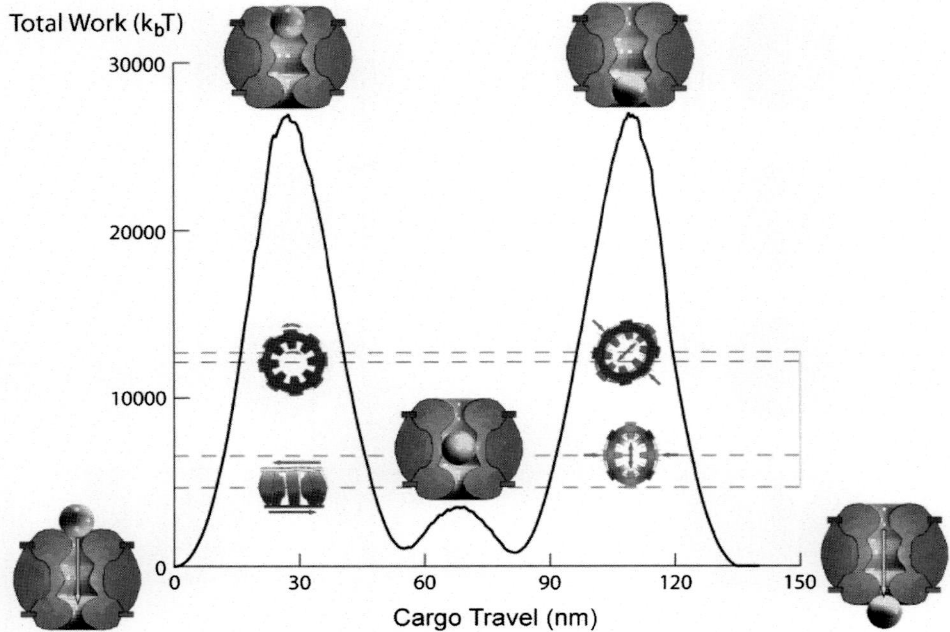

Figure 18.12. The total work as the cargo transverses the NPC through all of its conformations (Wolf and Mofrad, 2008). The FG brushes are denser near each face of the NPC and are sparser toward the core lumen channel. The cargo in this case is a 39 nm by 40 nm hard sphere, emulating the cargo complex used in the work of Panté and Kann (2002), which in turn was similar to immature Hepatitis B Viral capsids, which are hard as well before maturation (Ward et al., 1996).

eversion step, basically the inversion step at the beginning, just reflected about the horizontal axis, where the cargo complex is released from the NPC and binds to the nuclear basket, awaiting the RanGTP molecule to unlock it from the basket and release the cargo freely into the nucleoplasm. However, this phosphorylation step provides only about 25 k_bT of energy, which is not enough to account for the larger, on the order of 10K–13K k_bT of energy that results from the entire nucleocytoplasmic transport process as well as calculations (Wolf and Mofrad, 2008; Photos et al., 2007); see Figure 18.12. In addition to this effect, the reconfiguration of the NPC also helps the FG brushes conform around the cargo complex better as opposed to their ability to surround the cargo complex if the NPC were simply static. Most of the modes of deformation that have been illustrated tend to have the pivot point of the flexion in contact with the maximum diameter of the cargo complex and be roughly parallel to a tangent at that interface. This in turn further encourages the formation of a slip flow channel since the maximum packing also occurs at the point of maximum cargo complex diameter and is always kept at an optimum interface with the FG repeat brushes. A proposed reason as to why the channel exists as it does in the NPC is that it enhances the transport rates of a given cargo complex by the mechanism of molecular slip flow, where, instead of a traditional flow against a frictive surface interface, much like water flowing in a pipe, that stipulates the no-slip boundary condition, we have the condition of near-constant slippage along the walls of the "pipe" or channel flow guide. This effect has, traditionally, been noted in carbon nanotubes with diameters on the order of 10 to 2 nm transporting, or diffusing in the case of membranes comprised of these nanotubes, particles that are relatively large in comparison to the channel diameter (Holt et al., 2006). In addition, this effect also increases the transport rate greatly, past that which would be expected should one apply the boundaries of traditional flow to such a transport. One of the metrics that is used to determine when slip flow begins to occur is the Knudsen number, $K_{nu} = \lambda/D$, where λ is the mean free path between molecules and D is the diameter of the NPC. This dimensionless quantity allows us to ascribe a threshold for when slip flow begins to occur between the cargo complex and the almost completely smooth walls of the core channel. The channel walls that the cargo complex encounters are, due to how the FG repeats pack themselves, comprised mostly of the exposed phenyl rings of the phenylalanine residues of the FG repeats themselves. These rings present a stable surface and are especially comparable to the rings that compose the carbon nanotubes where this effect is particularly noticeable.

18.5 Cellular Mechanotransduction Regulated by NPC Mechanics and Nucleocytoplasmic Transport

During the process of transporting a cargo complex, the NPC is not simply a static pore that just facilitates the transport. Rather, as described earlier in this chapter, the NPC participates actively in the process of nucleocytoplasmic transport. Two points still remain: How does the NPC "know" when to undergo the conformational change to prepare for transport during the initial stages of the cargo complex binding to the cytoplasmic-facing FG brushes, and where exactly does a large amount of the energy for this transport event come from, if not from the phosphorylative event itself? We propose that answers to these questions lie within the big picture of the mechanics of mechanotransduction. To understand how the energy can be, literally, mechanically transduced and then stored in the NPC, we must first look at the anatomy of the cell, nucleus, and the associated fibers of the cytoskeleton and nucleoskeleton. The cytoskeleton is one of the major, if not the primary, load-bearing structure; along with the nucleus, it is one of the largest organelles in the cell. Both cytoskeletal and nucleoskeletal networks take the majority of forces that are transduced into the cell, especially due to the fact that they are comprised of a network of filaments, ranging in material properties but usually relatively stiffer than most other parts of the cell, including the cell membranes. These filamental networks are viable, dynamically remodeled pathways of force, and hence energy, transduction. Taking these observations into consideration, it is reasonable to assume that the majority of energy that is transduced into the NPC is derived from the lamin that composes the majority of the nucleoskeleton.

The lamin composes most of the nucleoskeleton, especially that which is located within the proximity of the underside of the nuclear envelope. This layer of lamin comprises a sublayer right below the nuclear envelope that is called the lamina and is

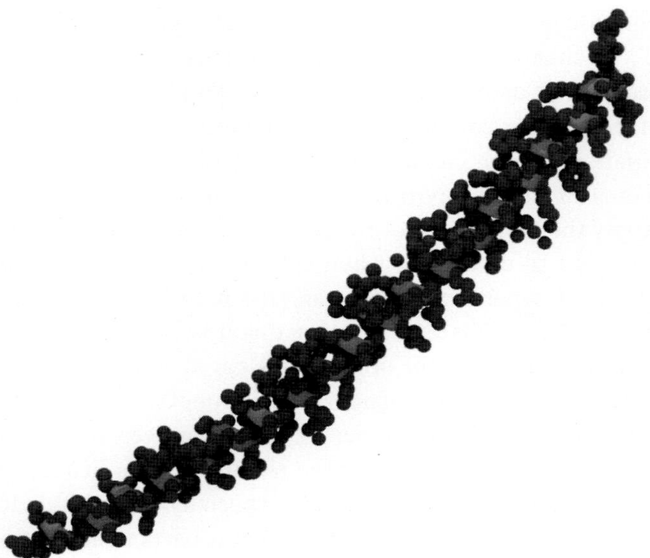

Figure 18.13. A hybrid render of the lamin molecule; in the nucleoskeleton, the lamin fibers usually form in dimerized pairs that comprise the nuclear lamina. Data structure obtained from Strelkov et al. (2004).

connected to practically all transmembrane proteins that are embedded within the nuclear envelope; see Figure 18.13 (Plate 22). The nucleoskeleton is connected to the cytoskeleton via LINC complexes, which are similar in mechanical response and material properties to the NPC relative to the nuclear envelope. The LINC complexes, serve as the anchorage points on the nuclear envelope to the cytoskeleton. This provides an excellent mechanotransductive pathway into the nucleus and has been shown in various studies (see Chapter 9) to result in a definite biophysical and physiological response in both the overall properties of the cell as well as those of the nucleus itself. These observations tend to suggest a potential pathway for mechanotransduction. It was mentioned previously that there are energy states associated with the various modes of deformation that the NPC assumes as it acts as a gateway for facilitating nucleocytoplasmic transport (see Figure 18.12). This leads to the question brought up previously pertaining to the origin of the energy. A viable pathway for the transduction of these amounts of energy lies in the mechanical response of the nuclear lamina and how it transduces vibrations, particularly larger scale thermokinetic vibrations, and the attachment of the lamin fibers directly to large transmembrane complexes such as the nuclear pore complex. A large amount of the energy, in the form of vibrations and impulse deformations ranging from large to small, that the cell receives is transduced through the cytoskeletal elements, encountered by the nucleoskeletal network, and finally arrives at a given NPC, whereupon it results in a period of excitation for the complex. The energy transduced through these means, and eventually attained by a given NPC, must not be underestimated. The energy required for the steady state and partial amounts of transport-stage NPC conformations fall well within the energies obtained from the constant thermokinetic vibrations that are encountered in practically the entire cell, from membranes to large complexes like the NPC. The stiffness of the NPC also plays a large role in that it is able to quickly derive energy and not only transmit it, but "store" the energy via steady-state thermokinetic vibrations with very little energy lost to damping effects. Currently planned are theoretical modeling and experiments where we can further examine the transmission of energy into the cell and nucleus, and finally to the NPC itself, and the subsequent effect it has on nucleocytoplasmic transport.

18.6 Role of NPC Mechanics and Nucleocytoplasmic Transport in Mechanotransduction

Not only do mechanical effects influence the NPC, but, we propose, the mechanical processes are mediated by the NPC. For example, take an endothelial cell that is being exposed to shear stresses from blood flow; it has been shown that this type of flow can eventually cause the nucleus to harden as the nucleus itself is a large load-bearing organelle in the cell (Deguchi et al., 2005; see Chapter 9). However, this type of response is common as a transient response, initially, but it is further maintained in endothelial cells exposed to cyclical flow patterns. This suggests that there is a component that factors in gene expression, for example, coding for stress proteins and cytoskeletal remodeling proteins, which are activated due to the biophysical forces that are experienced. In addition to responding to the initial forces, the NPC now plays the role of the final gateway when it comes to exporting the mRNA that will code, for example, a protein that is key in the major remodeling of the cytoskeleton. In this sense, the NPC is not only directly affected by the response of biophysical forces that are mechanically transduced upon the cell, but also operates in a feedback pathway to modulate the cell's response to such forces by being the final checkpoint for nucleocytoplasmic transport. Depending on what mobile nucleoporins and karyopherins are available, different mRNA export complexes are transported into the cytoplasm. In mutant cells where the NPCs are missing certain nucleoporins, they are unable to export particular mRNA sequences to the cytoplasm. Furthermore, this effect also occurs in reverse with import, as nucleoporins such as nup98 appear to play a key role in the successful import and integration of HIV-1V and other similar retroviruses. A similar checkpoint gating role is seen in cellular development as well, where the NPC can effectively be used to modulate the end protein products that are transcribed in the heightened activity that is observed during the developmental stages. It has been speculated in research literature on developmental biology that mechanotransduction has a critical role in the formative stages of development of everything from drosophila oocytes to the differentiation of stem cell progenitors. Yet, in such a manner, the final response of the cell to the mechanically transduced force is dependent on the synthesized proteins that result as expression products from the appropriate genes. This final response is dependent on the mRNA export complexes that are translated from these now up-regulated genes, a process that is the final checkpoint before the NPC. This process can be visualized as a hypothetical control feedback system, as shown in Figure 18.7. Viewing this process as a control feedback system allows us to more easily understand the role the NPC and nucleocytoplasmic transport might have in the effect on the forces "felt," or mechanically transduced initially, on the cell itself. Whereas the overall physiology and morphology of the cell control its initial transient response to the first stages of the biophysical perturbation, the final equilibrated response of the cell and the long-term effects are dictated by the changes in biological structure and, further, in the gene expression that was initially triggered by the biophysical forces. In this manner, we can also see the potential use the NPC and nucleocytoplasmic transport can be in both an applied clinical/medical aspect, with new treatments and therapeutic methodologies, as well as new laboratory techniques that can either enhance transfection or inhibit a large amount of potential protein products, that would otherwise be expressed, in a manner that is far easier to control and more precise than current techniques. In detail, one of the main advantages that is imparted from this understanding is that the system response can be altered in the control feedback system without the entire system attempting to compensate for a change that is desired by a clinician or investigator to be artificially maintained for an extended period of time. For example, take the shear-based hardening of vasculature due to the angle of incident flow shearing it and the flow rate. If the same control feedback methodology is valid in this case, we can begin to understand how we can control for the stiffness or compliance of the endothelial vasculature cells in two ways; in practical scenarios, it is sometimes desirable to stiffen the endothelial vasculature to prevent potential aneurysm sites from forming when the shear that is encountered by the endothelial cells is either not large enough or, in certain cases, too diffuse and not directed in a uniform direction as well. In this case, the cell is encountering biophysical forces from the environment and extracellular matrix, being part of both the vasculature ECM and the blood itself, and the normal morphological response would form elongated cell shapes that have the ability to pack together tightly and provide a uniform surface for the blood flow. To correct abnormalities that form aberrant bulges and distensions, resulting from improper flow or divergent flow eddying near the endothelial vasculature surface, it is possible to use pharmaceuticals to affect the entire vasculature biomechanical response. Say we had a pharmaceutical that let us control how much is sensed by the cell, the pre-plant biophysical sensors, or the plant effector directly (see Figure 18.7). A change in the direct pathway, due to the pharmaceutical agents targeting the mechanotransductive pathway and the rest of the direct flow, will engender a response from the error correction pathway that will attempt to return the system to a steady state or a set point. Like a physiological adaption response to addictive drugs, the system simply readjusts to the point where more of the pharmaceutical agent needs to be applied to maintain the clinically desirable levels of cell stiffening or compliance; an approach in this manner will reach the extrema of the system inevitably and will cause excessive biological loading due to the required amounts of the pharmaceutical that can have toxic side effects and/or consequences.

Now let us apply a pharmaceutical that affects not the direct pathway, but an effector in the error correction pathway, particularly the NPC and the associated nucleocytoplasmic transport with cargo complex–specific capability. This would enable us to affect the system without engendering a continual response from the system that attempts to correct for the perturbation by ramping up the corrective effects. There is now no more need to apply large amounts of the therapeutic agent for extended periods of time due to the response, or lack thereof, in this case, of the system to counter the perturbation. Furthermore, on closer inspection of the error correction pathway and treating the NPC and nucleocytoplasmic transport as a selective gating element, it is far easier to target and use the NPC and the transport it promotes as a sort of selective control valve for most of the cell, since it governs the pathways of transit that are crucial for the continued operations and ever-evolving response of cells *in vivo*. This is in comparison to similar techniques that affect this pathway via, for instance, proteins. Although proteins have a wide variety of effects as operational units within the cell, it requires a large amount of work to specifically target one, and only one, new protein and modulate its activity versus the molecule(s) it operates on without adversely affecting any other protein enzymes in the cell. Thus, the NPC can be used in the manner of a master control valve for certain manipulations of the cell to achieve desirable results with regard to clinical/medical usage as well as laboratory techniques where other, very highly specific, methods of response control by changing far smaller parts one by one would prove to be too costly, too complicated, or impractical to apply with the given resources and situation.

18.7 Conclusion

Throughout the course of its lifetime, the cell has to respond to the surrounding environment in an appropriate, quick, and proportional manner. It is because of many of these biophysical forces that we see a significant portion of orchestrated behaviors within the cell that occur whenever a particular stimuli is encountered, such as shear stress in the case of endothelial cells in the vasculature, or stress buildup within bone contributing to differential remodeling to compensate. Just like humans, cells must interact and respond to their environment through sensory modalities that transduce the various biophysical forces that are applied to the cell via changes in the environment. This characteristic is intrinsic to all cells and is evident in both classical signaling pathways and mechanotransductive pathways, the latter of which still has new, large-impact discoveries being made concerning individual pathways, their interactions, and how they affect the cell *in vivo*. In many ways, mechanotransduction can be likened to the sibling of cellular biochemical signaling in that they operate on the same organism(s), they can result in similar effects, and they have the ability to be the final checkpoints for a variety of crucial cellular processes such as division, intracellular and intercellular transport, morphological/physiological changes, and developmental dynamics. The primary difference between the two methods of transmitting the information to effector plants in the cell, which act as the elements that receive the signal and directly perform an action, is their mediums; classical cellular signaling is conducted through a mostly biochemical medium whereas mechanotransduction relies on biophysically, usually biomechanically, reactive transductive pathways. Both play crucial roles in defining the boundaries and dynamics of cellular life on a day-to-day basis, the understanding of which will prove to be a very valuable addition with ramifications ranging from new insights into developmental processes, to new laboratory techniques, to new clinical and medical therapeutics that use both pathways to minimize adverse side effects to the patient while still maintaining very high efficacy in their designated treatment goals. We believe that further understanding in all the roles that biomechanics plays in cellular activity, in everything from biomechanical responses to stresses and strains to, as discussed, nucleocytoplasmic transport via the NPC, will result in major gains to the whole state of understanding of biological life engineering and sciences, especially in new anti-viral therapies.

REFERENCES

Alber, F., S. Dokudovskaya, L. M. Veenhoff, W. Zhang, J. Kipper, D. Devos, A. Suprapto, O. Karni-Schmidt, R. Williams, B. T. Chait, A. Sali, M. P. Rout. 2007. The molecular architecture of the nuclear pore complex. *Nature* **450**: 695–701.

Beck, M., V. Lucic, F. Förster, W. Baumeister, O. Medalia. 2007. Snapshots of nuclear pore complexes in action captured by cryo-electron tomography. *Nature* **449**(7162): 611–615.

Bodoor K, S. Shaikh, D. Salina, W. H. Raharjo, R, Bastos, M. Lohka, B. Burke. 1999. Sequential recruitment of NPC proteins to the nuclear periphery at the end of mitosis. *J Cell Sci* **112**(13): 2253–2264.

Cingolani G., C. Petosa, K. Weis, C. W. Müller. 1999, Structure of importin-beta bound to the IBB domain of importin-alpha. *Nature* **399**(6733): 221–229.

Conti E., J. Kuriyan. 2000, Crystallographic analysis of the specific yet versatile recognition of distinct nuclear localization signals by karyopherin alpha. *Cell Press (Structure)* **8**(3): 329–338.

Deguchi, S., K. Maeda, T. Ohashi, M. Sato. 2005. Flow-induced hardening of endothelial nucleus as an intracellular stress-bearing organelle. *J Biomech* **38**: 1751–1759.

Fahrenkrog, B., U. Aebi. 2003. The nuclear pore complex: Nucleocytoplasmic transport and beyond. *Nat Rev Mol Cell Biol* **4**: 757–766.

Franke, W. W., U. Scheer. 1970. The ultrastructure of the nuclear envelope of amphibian occytes: A reinvestigation. I. The mature oocyte. *J Ultrastruct Res* **30**: 288–316.

Frey, S., R. P. Richter, D. Görlich. 2006. FG-rich repeats of nuclear pore proteins form a three-dimensional meshwork with hydrogel-like properties. *Science* **314**(5800): 815–817.

Hearps, A. C., K. M. Wagstaff, S. C. Piller, D. A. Jans. 2008. The N-terminal basic domain of the HIV-1 matrix protein does not contain a conventional nuclear localization sequence but is required for DNA binding and protein self-association. *Biochemistry* **47**(7): 2199–2210.

Hinshaw, J. E., B. O. Carragher, R. A. Milligan. 1992. Architecture and design of the nuclear pore complex. *Cell* **69**: 1133–1114.

Hinshaw, J. E., R. A. Milligan. 2003. Nuclear pore complexes exceeding eightfold rotational symmetry. *J Struct Biol* **141**: 259–268.

Holt, J. K., H. G. Park, Y. Wang, M. Stadermann, A. B. Artyukhin, C. P. Grigoropoulos, A. Noy, O. Bakajin. 2006. Fast mass transport through sub-2-nanometer carbon nanotubes. *Science* **312**: 1034–1037.

Garbitt, R. A., K. R. Bone, L. J. Parent. 2004. Insertion of a classical nuclear import signal into the matrix domain of the Rous sarcoma virus Gag protein interferes with virus replication. *J Virol* **78**(24): 13534–13542.

Kiseleva E, M. W. Goldberg, T. D. Allen, C. W. Akey. 1998. Active nuclear pore complexes in Chironomus: Visualization of transporter configurations related to mRNP export. *J Cell Sci* **111**: 223–236.

Lim. R. Y., B. Fahrenkrog, J. Köser, K. Schwarz-Herion, J. Deng, U. Aebi. 2007. Nanomechanical basis of selective gating by the nuclear pore complex. *Science* **318**(5850): 640–643.
Panté, N., M. Kann. 2002. Nuclear pore complex is able to transport macromolecules with diameters of about 39 nm. *Mol Biol Cell* **13**(2): 425–434.
Panté, N., U. Aebi. 1993. The nuclear pore complex. *J Cell Biol* **122**(5): 977–984.
Photos, P. J., H. Bermudez, H. Aranda-Espinoza, J. Shillcock, D. E. Discher. 2007. Nuclear pores and membrane holes: Generic models for confined chains and entropic barriers in pore stabilization. *Soft Matter* **3**: 364–371.
Stoffler, D., B. Fahrenkrog, U. Aebi. 1999. The nuclear pore complex: From molecular architecture to functional dynamics. *Curr Opin Cell Biol* **11**: 391–401.
Strelkov SV, Schumacher J., P. Burkhard, U. Aebi, H. Herrmann. 2004. Crystal structure of the human lamin A coil 2B dimer: Implications for the head-to-tail association of nuclear lamins. *J Molec Biol* **343**(4): 1067–1080.
Tomoyuki, O. E. C. Schirmer, T. Nishimoto, L. Gerace. 2004. Energy- and temperature-dependent transport of integral proteins to the inner nuclear membrane via the nuclear pore. *J Cell Biol* **167**(6): 1051–1062.
Ward, P. L., W. O. Ogle, B. Roizman. 1996. Assemblons: Nuclear structures defined by aggregation of immature capsids and some tegument proteins of herpes simplex virus 1. *J Virol* **70**(7): 4623–4631.
Wolf, C. B., M. R. K. Mofrad. 2008. On the octagonal structure of the nuclear pore complex: Insights from coarse-grained models. *Biophys J* **95**: 2073–2085.
Yang, Q., M. P. Rout, C. W. Akey. 1998. Three-dimensional architecture of the isolated yeast nuclear pore complex: Functional and evolutionary implications. *Mol Cell* **1**: 223–234.

19

Summary and Outlook

Mohammad R. K. Mofrad and Roger D. Kamm

19.1 Introduction

The primary objective of this book was to bring together various points of view on cellular mechanotransduction. This final, closing chapter attempts to summarize the various viewpoints discussed in previous chapters and establish a horizon for future research directions toward understanding the underlying processes involved in mechanotransduction.

Mechanotransduction is an essential function of the cell, controlling its growth, proliferation, protein synthesis, and gene expression. Extensive data exist documenting the cellular responses to external forces, but less is known about how force affects biological signaling. More generally, the question of how the mechanical and biochemical pathways interact remains largely unanswered. As articulated in the various chapters of this book, many studies during the past two decades have been carried out to shed light on a wide range of cellular responses to mechanical stimulation. It is well known that living cells can sense mechanical stimuli, and that forces applied to a cell or physical cues from the extracellular environment can elicit a wide range of biochemical responses that affect the cell's phenotype in health and disease. It is now widely accepted that stresses experienced *in vivo* are instrumental in a wide spectrum of pathologies. One of the first diseases found to be linked to cellular stress was atherosclerosis, where it was demonstrated that hemodynamic shear influences endothelial function and that conditions of low or oscillatory shear stress are conducive to the formation and growth of atherosclerotic lesions. Similarly, the disease process of calcification in heart valves can be understood as a mechanotransduction phenomenon as valvular endothelial and interstitial cells tend to respond to pathophysiological stresses from disturbed blood flow patterns directly linked with abnormal deformations in the valve tissue. The role of mechanical stress on bone growth and healing was probably the first to be widely recognized, and since then many other stress-influenced cell functions have been identified.

Many researchers have investigated the signaling cascades that become activated as a consequence of mechanical stress, and these are generally well characterized. The initiating process, however, by which cells convert the applied force into a biochemical signal, termed mechanotransduction, is much more poorly understood, and only recently have researchers begun to unravel some of these fundamental mechanisms. Several theories have been proposed to explain the process of mechanotransduction, but most are still in their infancy. Various mechanisms have been proposed to explain different aspects and forms of this phenomenon (see details in Chapters 1–17). Some studies have suggested that a change in membrane fluidity acts to increase receptor mobility, leading to enhanced receptor clustering and signal initiation (Chapter 3). Stretch-activated ion channels (Chapter 6) or strain-induced activation of G-proteins represent other means of mechanotransduction (Chapter 4). Others have focused on the role of mechanosensing surface glycocalyx in endothelial mechanotransduction (Chapters 1 and 2). Mechanical disruption of microtubules (Chapter 10) or forced deformations within the nucleus have also been proposed (Chapters 8 and 9). Constrained autocrine signaling is yet another mechanism whereby the strength of autocrine signaling is regulated by changes in the volume of extracellular compartments into which the receptor ligands are shed (Chapter 14). Changing this volume by mechanical deformation of the tissues can increase the level of autocrine signaling. Finally, others have proposed conformational changes in intracellular proteins in the force transmission pathway connecting the extracellular matrix with the cytoskeleton through FAs as the main mechanotransduction mechanism (Chapters 11–13). While all or a subset of these theories may contribute to mechanotranduction, little direct evidence has been presented in their support.

In the following, we will outline some of the challenges and opportunities facing the general field of cellular mechanotransduction. Deficiencies of the current tools/methods will be discussed and the necessary new tools/capabilities needed to be developed are briefly described.

19.2 Mechanotransduction: Tools and Measurements, Challenges and Opportunities

The study of cellular mechanotransduction has accelerated recently with the advent of more precise experimental tools and techniques probing and peeking into the cell. While these experimental techniques have largely improved in dealing with biochemical and biomechanical investigations on molecular "populations," they have been lacking atomic level resolution. Computational techniques have matured to complement the experimental single molecule investigations and provide insight into atomic resolution mechanisms underlying cellular mechanotransduction. Molecular dynamics, Monte Carlo methods, energy perturbations, and Brownian dynamics calculations have all become the computational tools of molecular probes into cellular mechanics and mechanotransduction.

Experimental investigations have clearly shown that mechanotransduction is a phenomenon not just of biochemical signal transduction and transcriptional changes but also of mechanical changes in molecules resulting in altered activities.

When external forces are applied to cellular surfaces the cytoskeletal filaments and associated molecules rearrange to form focal adhesions at the site of the external force. These focal adhesions act to strongly adhere cytoskeletal actin filaments to membrane-bound integrin molecules. The process of focal adhesion formation is a biochemomechanical process, and it has been hypothesized that the mechanism of mechanical activation of focal adhesion formation is due to the existence of mechanosensory molecules. Computational studies have shown mechanisms of activation via mechanical stimulation in potential mechanosensory molecules. Molecular dynamics simulations of locally applied forces to talin have shown activation of its cryptic vinculin binding site (Chapters 11 and 13). Activation of the talin–vinculin interaction results directly in the linkage of actin filaments to integrin molecules. Molecular mechanosensation is believed to be an atomic resolution phenomenon and could have only been demonstrated by computational simulation.

Computational tools have been used to study the phenomena of cellular mechanosensation and more generally the mechanical properties of cellular interaction with the extracellular matrix (ECM). Experimental studies have demonstrated a rigid binding interaction between integrin and ECM molecules, and computational simulations have shown that the rigid binding results from the strong interaction between the membrane-bound molecule integrin with the arginine-glycine-aspartate (RGD) peptide in ECM filaments. This interaction forms a multitude of stabilizing hydrogen bonds and accounts for the strength of the interaction between the cell and the substrate (Chapters 11 and 13).

Investigation of these biological systems using computational tools has matured and several investigative techniques have emerged. Molecular dynamics has emerged as a technique for simulating conformational changes along a molecule's free energy landscape by overcoming energy barriers with externally applied forces (Chapters 11–13). Simulated annealing and umbrella sampling techniques have emerged as methods of determining the local and global minimum structures of molecules. Normal mode analysis has advanced as a technique for determining the natural vibrational states of molecules (Chapter 18). Together these computational techniques are used to determine the mechanical and physical properties of molecules that account for their functional roles and that can demonstrate the mechanisms of biomechanical phenomena.

The computational techniques have demonstrated wide applicability to the study of cellular mechanotransduction. The atomic resolution of computational simulations reveals fascinating molecular properties and mechanisms of biological functionality. However, there are limitations to these studies. The greatest limitation to atomic resolution investigations is the availability of protein structures solved by x-ray crystallography and other experimental techniques. Also, the force fields used in simulations are always maturing and the accuracy of the simulations is continually limited by the accuracy of these force fields. The treatment of solvent has always been a major computational challenge; implicit solvent representations are more computationally feasible, while explicit solvent representations *may* better account for first-layer solvent electrostatic effects. The computational techniques are unable to simulate covalent modification and phosphorylation and are limited in their applicability to several important physiological phenomena. In efforts to mimic some force-induced phenomena in biology, several recent molecular dynamics studies have been performed using steered molecular dynamics techniques that involve the application of forces on single protein molecules pulling at a constant velocity or with constant forces (Chapters 11–13). A common challenge in these simulations is how to mimic *in vivo* mechanics, for example, how does the cell regulate the force application and what is the exact direction of force? Further challenges exist in determining the influence of steric effects of protein complexes and what boundary conditions are most appropriate in the isolation of protein molecular or multimolecular complexes. Perhaps the most challenging aspect centers around how to integrate the molecular simulations into the cell level to prove relevance in cellular mechanotransduction. Furthermore, these studies are also dependent on multiscale simulations to prove relevance. Given that biochemical and mechanical phenomena involved in cellular mechanotransduction span a large range of time-scales (from femtoseconds to days and years) and length scales (from nanometers to meters), one increasingly emerging challenge is to develop novel algorithms and modeling approaches that can seamlessly link these disparate scales. Finally, regardless of the advancement of the theoretical computational models, the technological limits of computing power prevent computational simulations with statistical sampling, or simulations with time-scales larger than several nanoseconds.

Advances in biochemomechanical techniques and technologies have benefited both experimental and computational investigations of cellular mechanotransduction. The computational tools are complemented by experimental studies for validation, and the experimental studies are complemented by computational studies for atomic resolution analysis of mechanisms. There have been many successful implementations of computational molecular mechanics investigations, but there remain many limitations and challenges.

Similar to computational challenges, numerous experimental milestones can be sighted for molecular biomechanics. Protein mechanics experiments, both *in vivo* and *in vitro*, are essential to confirm the molecular dynamics (in silico) experiments. The existing methods (e.g., optical techniques) are useful but lack sufficient details of conformational change. We need to be able to label proteins at specific locations using small tags to facilitate *in vivo* and *in vitro* experiments. Unlike cellular scale manipulations, where it is also possible to apply forces to the outsides of cells using beads that are manipulated with a magnet or optical tweezers, or by applying fluidic shear stress, or stretching the substratum (see Chapter 16), we currently lack methods to apply forces to molecular complexes inside of cells; applying forces to a bead that is nonspecifically localized inside of cells cannot localize onto specific molecular complexes. Therefore, there is also a need for a method to attach a force applicator – such as a magnetic bead – to a particular protein domain inside a live cell.

Currently there is a need for methods to measure or sense forces quantitatively on target molecules inside living cells, so that we can determine how forces are applied to molecules of interest *in vivo*. This will allow researchers to apply and measure forces on complicated molecular complexes to understand the force response of systems that bridge the length scale between single molecules and cells.

Numerous open questions and opportunities exist on the road to understanding cellular mechanotransduction. One most relevant question to understanding the molecular basis of mechanotransduction is to specifically identify the mechanosensing proteins, that is, what proteins are likely to be deformed? What are the extent and importance of mechanosensing characteristics of different proteins in the molecular machinery of the cell? How can we control molecular conformational changes and proteins' modes of deformation to ascertain the biomechanical phenomena under question. To demonstrate experimentally that protein deformations actually occur *in vivo* and link them to certain biological functions, we need to distinguish with specificity the distinct molecular conformations. We also need to be able to alter the molecular conformation of mechanosensing proteins or nucleic acids with threshold specificity. Given that protein domain motions can often work as switches between distinct biological functions, experimental techniques are necessary to characterize and quantify how domain deformation and unfolding alter interactions. In order to ascertain the relevance of these localized molecular events to overall biological phenomena, it is essential to identify how conformational changes at one location may potentially change binding affinities at another location. It is also important to characterize allosteric effects. Optical fluorescent techniques involving green fluorescent proteins (GFP), which can be genetically incorporated in a robust manner, have recently made major contributions to many biomechanical and biological discoveries, yet one must note that the GFP is itself approximately 5 nm in size, that is, comparable to most proteins that are to be tagged. This caveat points out the need for less invasive approaches for marking proteins *in vivo*. New experimental techniques are also required for the mechanical regulation of proteins and other biomolecules. Similar to computational approaches, multiscale experimental settings are also essential to the study of the molecular basis of cellular mechanotransduction. Taking into account the spatial organization of proteins and cellular complexes, and how forces are distributed in molecular complexes, proves critical in this context. Many of these phenomena require breakthrough thinking as in many cases one needs to identify, devise, and build the appropriate building blocks for such complexes and incorporate the suitable interactions and force fields, similar to the computational approaches. It is vitally important to devise surrogate cell models *in vitro* where we can add one component at a time and study that by applying mechanical stimuli.

In summary, many challenges and milestones loom ahead, both experimental and computational/theoretical, for the advancement in our understanding of cellular mechanotransduction. Experimental techniques are required to capture intracellular forces and molecular conformations. More sophisticated techniques are essential to dissect changes in molecular conformations. New computational modeling techniques are needed to be able to model protein clusters and apply forces on them to mimic biologically meaningful settings. How do we find alternative protein states and the effects of force on them? Experimental methods are then essential to objectively test the central hypothesis of the structural basis of cellular mechanotransduction. Mutation experiments are valuable tools to ascertain the effect of localized changes in the molecular constructs of the protein machinery on the overall mechanical behavior of such systems. Perhaps the most subtle step is to couple these computational/theoretical and experimental aspects. With the development of new techniques and theories and algorithms, it will become increasingly feasible to understand the multiscale biological phenomena over a range of time and length scales. This represents both a challenge and an opportunity to understand a disease or biological process in its entirety. Many disease examples remain to be explored – atherosclerosis, cancer, and infectious diseases, to name a few – in all of which mechanics may play an important role. In atherosclerosis, for example, after more than 30 years of multifaceted investigations, we still do not know the molecular mechanisms involved in the initiation and development of this disease. Or in cancer, what molecular factors play a role in metastasis? What steps are required for a multiscale study on metastasis incorporating the key mechanical signaling pathways involved?

These computational/theoretical and experimental aspects must be integrated to help understand the complex processes underlying cellular mechanotransduction over a range of time and length scales. This represents both a challenge and an opportunity and calls for close coordination between efforts in molecular biomechanics and the investigations focused on cell, tissue, and organ biomechanics, necessitating the developments of theoretical and computational approaches to bridge the gap between these disparate scales.

19.3 Closure

To understand the phenomena related to mechanotransduction in living cells, the mechanics and chemistry of single molecules, and of molecular complexes, that form the biological signaling pathways acting in concert with the mechanics must be examined. To understand the mechanobiology of the cell requires a multiscale/multiphysics view of how externally and internally applied stresses or traction forces are transmitted through the complex machinery consisting of focal adhesion receptors, integrins, and cytoskeletal filaments, and how they are distributed throughout the cell and elicit biological responses ultimately in the nucleus of the cell and thereby affect the phenotype of the cell and the overarching tissue and organ. This presents both a challenge and an opportunity for further research into understanding the intrinsically coupled mechanobiological phenomena that eventually determine the macroscopic behavior and function of the cell. We hope that this monograph has been able to intrigue new researchers with fresh ideas stimulated toward understanding how cells sense and respond to mechanical signals.

Index

ABD. *See* actin binding domain
actin(s)
  collapsing, in lamellipodia, 189
  lamellipodia and, 190
  myosin proteins and, 190
actin binding domain (ABD), 310
actin filaments, 28
  in microtubular mechanotransduction, 241
α-actinin, 121, 310, 312, 315
  ABD and, 310
  conformational states of, 123f
  modular structure/binding sites of, 311f
  SMD for, 312–13
  spectrin repeats in, 310, 315
β-actinin, 121
  conformational states of, 123f
  RGD peptides and, 122
adenomatous polyposis coli (APC), 239
adhering cells, 225
  in ECs, 228
adhesion-complex proteins, 127, 280
affixin, 189
agonist-independent constitutive activity, 102
amino acids, 274
anchorage-dependent cells, 212
angiotensins, 104–5
  vasoactive peptides, 104
anisotropic cell spreading, 182
ankyrin(s), 171
ankyrin proteins, 315–16
ankyrin repeats, 291–92
antibodies
  ligand density and, 135
  nonperturbing, ligand density and, 137
aortic valve leaflets, 50
APC. *See* adenomatous polyposis coli
aPKC. *See* atypical protein kinase C
apoptotic blebbing, 187
arterial walls, thickness of, 1
arthritis, 2
associated cytoskeleton (CSK), 90
  cell membranes and, 90
atherogenesis, 22
  development of, 254–55
  glycocalyx and, 35
atherogenic phenotypes, in ECs, 62
atheroprotective phenotypes, in endothelium mechanotransduction, 47
atherosclerosis, 22, 35–36, 269
  calcific aortic valve sclerosis, 48–51
  G-proteins and, 89
  risk factors for, 89
atomic force microscopy (AFM), 286
  for ECs, 63
  for endothelial mechanotransduction, 26
  for focal adhesion complexes, 254
  mechanical forces with, 380–82, 381f
  protein conformational change with, 323–25
  in vitro tests with, 5
atypical protein kinase C (aPKC), 239
autocrine signaling, 339–57
  bridge modeling for, 353–56
  in bronchial epithelium, 340–41
  cell culture setup for, 341f
  cellular/extracellular geometry changes, 341, 344

autocrine signaling *(cont.)*
cellular viscoelasticity and, 356
constrained, 11–12
in EGFR system, 344–5
FE method for, 349
fluid drag and, 356
governing parameters for, 349–51
hypothesis testing for, 346–8
implications of, 356–57
interstitial geometry and, 339
ligand concentrations and, 344, 346f, 347–9
LIS and, 341, 344, 346
local loops, 344–45, 345f
in lungs, 340f, 341
through MAPK, 341
MD for, 347
rate sensitivity of, 351–52
two-photon microscopy and, 343f, 353

B_2. *See* bradykinin
BAECs. *See* bovine aortic endothelial cells
basal cells, 161
basal membrane, 61
BDM. *See* butadiene monoxime; 2,3-butanedione monoxime; 2,3-butanedione monoxime (BDM)
bead forcing, *in vitro* tests with, 5
benzyl alcohol, 93
blebbing motility, 187–88
apoptotic, 187
cell spreading and, 187
mitosis and, 187
MRLC phosphorylation and, 187
myosin proteins and, 187
PLC and, 188
tumor cell migration and, 187
blebbistatin, 139, 144
BMP4. *See* bone morphogenic protein 4
bone cells, 269
endochondrial ossification in, 405
osteoblasts, 405
osteoclasts, 405
remodeling, 404–5
bone morphogenic protein 4 (BMP4), 51
bone remodeling, 404–5
bovine aortic endothelial cells (BAECs), 69, 70f
in Cl^- ion channels, 164, 166
in flow-sensitive ion channels, 163, 174
G-proteins and, 89
shear stress and, 89
bradykinin (B_2), 104–5
bronchial epithelium, autocrine signaling in, 340–42
MAP and, 342
butadiene monoxime (BDM), 133

calcific aortic valve sclerosis, 48–51
BMP4 and, 51
causes of, 49
features of, 48–50, 49f
leaflets in, 50
calcium, 126
in intracellular oscillations, 162
calmodulin, 294–95
cancers, 15
carotid bifurcation model, 364f
cartilage, tensegrity in, 209
cause-and-effect hypothesis, for endothelial mechanotransduction, 48
caveolae
cholesterol in, 101
coat proteins in, 42, 101
in ECs, 68
sphingolipids in, 101
cell(s)
adhering, 225
anchorage-dependent, 212
bone, 269
compression struts in, 207
cytoplasmic region of, 223–24, 230
eukaryotic, 169, 236
force levels in, 12–15, 13t
gated ion channels in, 234–35
gene transcription in, 212
GFP-LamA in, 224f
localized distortion in, 212
mechanical force and, molecular function, 286–89
mechanical forces on, 269–70, 407
mechanical forces on, responses by, 379
mechanosensing in, 181–91, 442
as mechanotransducer, 212–13
metabolism of, 197
muscle, 209, 269
nucleated, 203f
nuclei measurements for, 222

organelles in, 230
permeabilized, 204
properties of, 184–91
signal transduction in, 212
stem, 143, 269
stiffness in, 204
as tensegrity structures, 202–6
towed growth in, 234
cell–cell junctional proteins, 63
CellMAP program, 187–88
cell membrane fluctuations (CMF), 99–100
free volume theory and, 100
in red blood cells, 100
Saffman hydrodynamic theory and, 100
cell membranes,
altered gene expression in, 64
in animals, 65
basal, 61
bilayer modulation in, 64–65
CMF and, 99–100
CSK and, 90
diffusive dynamics of, 100
flow-sensitive ion channels and, 171–72
fluidity of, 65–67, 172
ion channels and, 161, 166
lipid molecules in, 65
as mechanosensor, 64–75
phospholipids in, 65
pressure profiles in, 91, 91f
structure of, 65
cell metabolism, tensegrity and, 194
cell migration, 135, 138
body translocation in, 138
computational modeling for, 135
debris after, 138
lamellipodium protrusion in, 138
limited neural crest, 135
regulation of, 136
substratum adhesions in, 138
three-dimensionality of ECM, 146, 149f
trailing edge retraction in, 138
cell signaling, endothelial, mechanotransduction in, 21
cell spreading, 182
anisotropic, 182
blebbing motility and, 187
isotropic, 182
cell surface integrins, 161
cell-to-cell forces, in ECM, 129
cellular cytoskeletal network, 161
cellular mechanotransduction. *See* mechanotransduction, cellular
cholera-toxin-B (CT-B), 69
cholesterol, in ECs, 66
in caveolae, 101
integrin-mediated mechanotransduction and, 72–73
in lipid bilayers, 93
in lipid domains, 69
chromatins, in nuclear mechanics, 226
chromosomes, nuclear mechanics and, 227
ciliary vibrations, 289
circumferential stress (CS), 360–74
carotid bifurcation model for, 364f
in ECs, 359, 365–72
influencing factors for, 362–63, 364
intercellular junctions and, 373
physiological background of, 360–65
SED and, 373
simultaneous stretch/shear methods in, 365–66
SPA and, 361–74
stretch/shear mechanism for, 372–73
tension and, 373
CLA. *See* conjugated linoleic acid
cLAD. *See* left descending anterior coronary artery
Cl^- ion channels, 164–66
BAECs in, 164, 166
desensitization in, 165
ECs in, 165
fluid flow in, 165
in HAECs, 166
VRAC in, 166
CMF. *See* cell membrane fluctuations
coat proteins, 42, 101
collagen, 140
in microtubular mechanotransduction, 234
complex modulus, 29–30
compression strain, on stem cells, 408–10, 412–13
devices for, 409f
compression struts, 207
conjugated linoleic acid (CLA), 372
constitutive models, of proteins, 276–77

continuous protrusion, in cell mechanosensing, 188–89
  affixin and, 189
  WAVE1 and, 189
contractile forces, 129
contractile waves, 189
control feedback system, NPCs as, 425f, 434
CPC. *See* phospholipase C
CREEP program, 187
CS. *See* circumferential stress
CSK. *See* associated cytoskeleton
CT-B. *See* cholera-toxin-B
cysteines, 289
  disulfide bonds and, 295–96
cytoplasm
  nuclear mechanics in, 223–24, 230
  organelles in, 230
cytoskeletal proteins
  cellular network of, 161
  ECM v., 120–21
  integrins and, 319f
  mechanical signals in, 182
  mechanosensing for, 181–83
  prestresses in, 205f, 206
  substrate priming of, 182
  tensegrity and, 197, 209–10
  traction forces in, 129, 130–31
cytoskeletal transduction, 10
  in endothelium, 10–11, 36–40
  mechanosensors in, 10

DCSA network. *See* distributed cytoplasmic structural actin network
diabetes mellitus, 89
Dictyostelium chemotaxis, 140
Dicyostelium discoideum, 138
direct activation, of ion channels, 169–71, 170f
diseases
  arthritis, 2
  atherosclerosis, 22, 35–6, 269
  cancer, 15
  diabetes mellitus, 89
  disturbed flow in mechanotransduction and, 2
  Emery-Dreifuss muscular dystrophy, 227
  focal adhesion complexes and, 250
  Hutchinson-Gifford progeria syndrome, 227
  hyperlipidemia, 89
  hypertension, 89
  KLF2, 269
  malaria, 15
  monocytes in, 2
  nuclear mechanics and, 227–28
  PKD, 2
  pulmonary, 2
  thrombosis, 269
distributed cytoplasmic structural actin (DCSA) network, 31
disulfide bonds, 295–97
  cysteine bonds and, 295–96
  ligand binding in, 296–97
  Mel-Cam and, 296
  VCAM-1 and, 295, 297f
domain deformation, 271–72
domain motion, 271
  elongation in, 272
  hinge, 271–72, 277–78
  sliding, 272
domain unfolding, 271
double-membrane envelope, 223
durotaxis theory, 140, 141
  haptotaxis and, 141
dynamic instability, of microtubules, 238–40
  deformation effects on, 242
  epithelial cells and, 238
  fibroblast cells and, 238
  kinetochores and, 238
  signaling and, 238–40
  stochastic switching and, 238

ECM. *See* extracellular matrix
ECs. *See* endothelial cells
EECP. *See* enhanced external counter-pulsation (EECP)
EGF receptor (EGFR), in ECM, 150
  ligand density in, 353f
EGFR system. *See* epidermal growth factors receptor system
elastic modulus, 29
elastic substrata, 389–92
  TFM for, 389
ELCs. *See* essential light-chains
electron microscopy, 121
Emery-Dreifuss muscular dystrophy, 227
endochondrial ossification, 405

endothelial cells (ECs)
adhering cells in, 228
AFM for, 63
arterial expansion in, 361
atherogenic phenotypes in, 62
ATP release in, 169
BAECs in, 69, 70f
biological responses of, 365–72
biomechanics of, 28–33
blood flow in, 63f
caveoli in, 68
cholesterol content in, 66
in Cl^- ion channels, 165
conformation in, 162
CS/WSS in, 361f
dysfunction in, 62
ECM and, 121
EPCs, 403
fatty acid chain saturation in, 66
FCS for, 76–78, 80f
finite element analysis for, 73–74
fluid shear stress on, 3f
glycocalyx in, 62, 70, 171
hereditary integral for, 28
heterogeneity in, 51–52
homeoviscous adaptation for, 67
HUVECs, 94, 95f, 269
hydrophobic mismatch in, 67–68
ion channels in, 161, 164, 172–74
K^+ ion channels, 164
linear viscoelastic models for, 28, 29t
lipid diffusion within, 74–75
lipid-mediated models for, 79, 80f, 81
mechanosensitive functions for, 72–73
membrane fluidity in, 65–67
membrane tension effects on, 74–75
micrograph of, 229f
in microtubular mechanotransduction, 234
molecular organization of, 64f
nuclear mechanics and, 220
phospholipids in, 65
in plasma membrane mechanotransduction, 61
protein conformational change and, 269
protein content in, 66
SMCs and, 171
steady flow mediation of, 167–68
stress–strain tissue relationships in, 28
traction force generation by, 133
vascular flow in, 221f
wound closure under, 72
endothelial nitric oxide synthase (eNOS), 90
SPA with, 369
endothelial progenitor stem cells (EPCs), 403
endothelium
atherogenesis and, 22
as blood "gatekeeper," 22
cardiovascular development and, 22
eNOS in, 90
functions of, 22
mechanosensing in, 161
NO productivity in, 90
shear stress on, 105
vascular permeability in, 22
endothelium, mechanotransduction in, 20–52, 62–63
actin deformation in, 27
AFM for, 26
aging and, 32
arterial hemodynamics in, 46f
arterial heterogeneity in, 47–48
atheroprotective phenotypes in, 47
calcific aortic valve sclerosis and, 48–51
cause-and-effect hypothesis for, 48
cell signaling in, 21
complex modulus for, 29–30
through cytoskeleton, 36–40, 39f
DCSA network in, 31
as decentralized mechanism, 23–26, 25f
deformation in, 22
descriptors for, 22–23
disturbed flow for, 47
downstream/immediate signaling responses, 27
elastic modulus for, 29
filament bundling in, 31
filament displacement in, 27
as flow-mediated mechanism, 20
fluid shear stress in, 39–40
focal adhesion in, 72
FRET in, 37
gene expression in, 47
GFPs in, 38
glycocalyx in, 26, 34–36
G-protein activation in, 21, 33, 34
hemodynamics and, 21, 51–52
heterogeneity in, 51–52

endothelium, mechanotransduction in (cont.)
immunocytochemistry in, 45
integrin-mediated, 40–44, 72
intracellular prestress in, 38
ion channel activation in, 21, 33
KLF-2 in, 48
in lamellipodia, 37–38
magnetic bead twisting cytometry, 32
mechanical force conversion in, 27
mechanosensors in, 25
mechanotransducers in, 21
mechanotransmission in, 23, 26–27
membrane fluidity in, 21
microdomains in, 68
microtubule networks in, 27, 30
NO in, 35
PAK in, 44–45
PECAM-1 in, 24, 44
physical deformation in, 26
PKC in, 47–48
plasma transport characteristics in, 23
primary cilium in, 22, 26, 36
scale for, 51–52
shear stress on, 21f, 33–6, 44–45
at single cell level, 51
sinusoidal input functions, 29
soft glassy material models in, 32
spatial gradients in, 62
spatiotemporal flow responses in, 26–27
strained focusing for, 38–39
substratum adhesion sites in, 24
TCSPC microscopy for, 75
temporal gradients in, 62
tension field theory for, 33
viscoelastic models for, 30, 31t, 32
viscous modulus for, 30
in vitro phenotypes in, 45–47
in vivo phenotypes in, 45–47
VRAC in, 33
enhanced external counterpulsation (EECP), 373
eNOS. *See* endothelial nitric oxide synthase
EPCs. *See* endothelial progenitor stem cells
epidermal growth factors receptor (EGFR) system, 344–45
hypothesis testing for, 346–48
epithelial cells
bronchial, autocrine signaling in, 340–41
dynamic instability and, 238
ECM and, 121
tensegrity in, 209
ERK. *See* extracellular-signal-regulated kinase
essential light-chains (ELCs), 130
MLCK, 130
eukaryotic cells
ion channels in, 171
microtubular mechanotransduction in, 236
as multimodular, 289
protein unfolding in, 289
Euler-Bernoulli beam theory, 392
Euler Bucking model, 237
extracellular matrix (ECM), 41f, 120–51
calcium in, 126
cell homeostasis by, 120
cell-to-cell forces in, 129
collagen and, 140
computational methods for, 440
contractile forces in, 129
cytoskeletal proteins v., 120–21
Dictyostelium chemotaxis and, 141
durotaxis theory and, 140–41
ECs and, 121
EGFR in, 150
elasticity of, 133
epithelial cells and, 121
FAK in, 143
fibrinogen and, 134, 137
fibroblasts in, 134, 138
fibronectins in, 120, 124, 134, 150
focal adhesion complexes in, 251
force-bearing protein networks and, 287
gels in, 207
Hooke's Law and, 140
hydrogels in, 140
integrin-mediated mechanotransduction and, 43, 253–54
integrins in, 120–1
lamellar ruffling in, 140
laminin and, 121
ligand density and, 133–44
magnesium in, 126
manganese in, 126–27
mechanical forces on, 270, 377
mechanical strain on, 408
mechanosensing compliance and, 139–41, 143–44

melanoma cells in, 127
microfibrillar network in, 130
MMPs and, 150
myosin proteins in, 143, 251
Poisson's ratio and, 140
protein conformational change and, 269
protrusive forces in, 129
RGD peptides in, 440
rheological properties of, 120, 139
rigidity of, 133–34
sensing in, 120
shear modulus in, 140
spring models for, 141, 143
stem cells in, 143
stiffness in, 141
tensegrity and, 197, 203, 208–9
three-dimensionality of, 144–51
traction forces in, 129–33, 139
vitronectin and, 134
wound healing and, 134
extracellular-signal–regulated kinase (ERK), 72
integrins and, 127

FAK. *See* focal adhesion kinase
fatty acid chains, in ECs, 66
FCS. *See* fluorescence correlation spectroscopy
FCs. *See* focal complexes
FE method. *See* finite element method
FERM domains, 126
in multimodular proteins, 318
PTB domains and, 126
FG repeats. *See* phenylalanine-glycine repeats
fibrillar adhesions (FX), 129
fibrillogenesis, 302
fibrinogen, 134, 137
fibroblasts
dynamic instability and, 238
ECM and, 134, 138
in focal adhesion complexes, in cellular mechanotransduction, 253
three-dimensionality of ECM and, 146
fibronectins, 297–300, 302–3
cell recognition sites on, 300
in ECM, 120, 124, 134, 150
fibrillogenesis and, 302
in focal adhesion complexes, 252
FRET for, 302
GFP and, 302
lamellipodia and, 189
in multimodular proteins, 291f, 299
PDMs and, 302
as proteins, 280
quaternary structural model for, 302, 303
structure of, 301f
unfolding of, 300, 302
filaments
actin deformation of, 27
bundling of, 31
in endothelial mechanotransduction, 27
filamin, 316
cytoskeleton/integrin contact in, 319f
modular structure/binding sites of, 317f
filopodia, 186
filopodial motility, 186–87
capping proteins and, 186
CellMAP program for, 186–87
CREEP program for, 187
lamellar protrusion during, 187
paxillin molecules and, 186
protein families and, 186
finite element analysis, for stress, in ECs, 73–74
finite element (FE) method,
for autocrine signaling, 349
for ligand density, 349
FLIM. *See* fluorescence lifetime imaging microscopy
flow-sensitive ion channels, 161, 163–64
BAECs in, 163, 174
cell membranes and, 171–72
HAECs in, 163
HUVECs in, 163
MscL, 171
TRP in, 163
fluid drag, 356
fluid shear stress
on EC, 3f
in endothelial mechanotransduction, 39–40
fluorescence correlation spectroscopy (FCS), 75–76
confocal volume for, 76
for ECs, 76–78, 80f
fluorescence lifetime in, 78–79
for lipid bilayers, 94, 95f, 96
for unilamellar vesicles, 76–78

fluorescence lifetime, 78–79
IRF and, 78
fluorescence lifetime imaging microscopy (FLIM), 79, 80f
fluorescence recovery after photobleaching (FRAP), 71
lipid bilayers and, 94, 96
tensegrity and, 207
fluorescence resonance energy transfer (FRET), 81
for fibronectins, 302
for multimodular proteins, 291f, 326, 328
for protein conformational change, 282, 323
focal adhesion complexes,
cellular mechanotransduction in, 7f, 9–10, 250–64, 378
in endothelium, mechanotransduction in, 72
FAK, 40
FC in, 128
IAEDANS in, 10
in integrins, 128f
maturation of, 127–28
mechanical forces and, 287
mechanotransduction in, 7f, 9–10
shear stress and, 74f
tensegrity and, 203
Triton X and, 9
focal adhesion complexes, mechanotransduction in
AFM for, 254
BDM in, 252–53
biochemical signaling pathways in, 250
diseases and, 250
in ECM, 251
experimental studies on, 252–54
FAK in, 251
fibroblasts in, 253
fibronectin microbeads in, 252
forced-induced assembly of, 252–54
force regulation of, 261–62
force sensing proteins in, 254–59, 261
forces exerted by, 253–54
initial, 251
integrin-mediated ECM in, 253–54
maturation stages of, 251–52
MD for, 254
numerical studies on, 254–59, 261
Rho proteins in, 251–52
role of, 251
sustained force in, 251
talin in, 254–59, 255f
vinculin in, 254–55
focal adhesion kinase (FAK), 40
in cellular mechanotransduction, 251
in ECM, 143
integrins and, 121
three-dimensionality of ECM, 146–47
traction forces and, 392
focal complexes (FCs), 128
FA conversions to, 128
paxillin molecules in, 128
force-bearing protein networks, 287
ECM and, 287
significance of, 287–89
force levels, in cells, 12–15, 13t
force sensing proteins, 254–59, 261
formyl peptide receptors (FPRs), 104–5
Förster resonance energy transfer (FRET), 37
FPRs. *See* formyl peptide receptors
FRAP. *See* fluorescence recovery after photobleaching
free volume, 97–98
free volume theory, 100
FX. *See* fibrillar adhesions

GAG polymer. *See* glycosaminoglycan polymer
gated ion channels, 234–35
GDP release, 107
GEF. *See* guanine exchange factor
gene expression, in cellular mechanotransduction, 10–11
in cell membranes, 64
in endothelium, 47
in proteins, 270
generalized fluorescence polarization (GP), 99
gene transcription, 212
GFP. *See* green fluorescent protein
GFP-LamA, 224f
giant unilamellar vesicles (GUVs), 97, 98f
glycocalyx
arterial role of, 35
atherogenesis and, 35

atherosclerosis and, 35–36
in cellular mechanotransduction, 10
in ECs, 62, 70, 171
in endothelial mechanotransduction, 26, 34–36, 70
endothelial mechanotransduction and, 26
GAG polymer in, 34
GPI-anchored proteins in, 34
in ion channels, 171
in shear stress, 5
structure of, 34–35
in vascular physiology, 69–70
glycosaminoglycan (GAG) polymer, 34
glycosylphoshatidylinositol (GPI)-anchored proteins, 34
GP. *See* generalized fluorescence polarization
GPCR. *See* G-protein-coupled receptors
GPI proteins. *See* glycosylphoshatidylinositol-anchored proteins
G-protein–coupled receptors (GPCR), 101–5
activation of, 103
agonist-independent constitutive activity, 102
angiotensins and, 104–5
bradykinin and, 104–5
conformational changes in, 102
direct stretching in, 102
dynamics and structure of, 101–3
FPRs and, 104–5
GDP release from, 107
GEF activity in, 105
in ion channels, 161
membrane protein activation in, 107
as "plastic," 102
rhodopsin in, 101
shear stress in, 102
signal amplification cascades in, 102
transduction pathways and, 105–6
G-proteins, in cellular mechanotransduction, 8–9, 33–34, 89–109
activation of, 21
activation pathways, 105–8
atherosclerosis and, 89
in BAECs, 89
CMF in, 99–100
CSK and, 90
GPCR, 101–5
heterotrimeric, 90–1
hydrolysis of, 67–68
ion channels and, 161
lipid bilayers and, 91–101
NO production in, 107–8
green fluorescent protein (GFP), 38, 294
conformational change for, 326–27, 328
fibronectins and, 302
tensegrity in, 207
for traction forces, 391
guanine exchange factor (GEF), 42
in GPCR, 105
microtubular mechanotransduction and, 244
GUVs. *See* giant unilamellar vesicles

HAECs. *See* human aortic endothelial cells
haloalthane, 68
haptotaxis, 134f
durotaxis theory and, 141
hearing, cellular mechanotransduction in, 2
heart valve tissue, calcification of, 1
hematopoietic stem cells (HSCs), 403
hemodynamic shear stress, 1
in endothelial mechanotransduction, 21
Hooke's Law, 28–29
ECM and, 140
HSCs. *See* hematopoietic stem cells
human aortic endothelial cells (HAECs), 163
Cl^- ion channels in, 166
human umbilical endothelial cells (HUVECs), 94, 95f, 269
in flow-sensitive ion channels, 163
Hutchinson-Gifford progeria syndrome, 227
HUVECs. *See* human umbilical endothelial cells
hydrophobic matching, in lipid bilayers, 92
hydrophobic mismatch, 67–68
hydrostatic pressure, 6
hypercholesterolemia, 51
hyperlipidemia, 89
hypertension, 89

IAEDANS (fluorescent dye), 10
ICAM-1. *See* intracellular adhesion molecule-1
Ig-like domains, 291
in VCAM-1 proteins, 293

immunocytochemistry, 45
Importin-α, 418, 423, 425f, 426
Importin-β, 418, 423, 425f, 426
importins, in nucleocytoplasmic transport, 423, 425–26
indirect activation, of ion channels, 169, 170f
"inside-out signaling," 126
instrument response function (IRF), 78
integrin(s), 120–21
  β-actinin, 121
  activation of, 124–27
  adhesion-complex proteins in, 127
  adhesion receptors in, 121
  antibodies for, 124
  cell surface, in ion channels, 161
  in ECM, 120–21
  electron microscopy for, 121
  ERK and, 127
  FAK and, 121, 127
  FERM domains and, 126
  filamin and, 319f
  focal adhesion complexes in, 128f
  FX in, 129
  "inside-out signaling" for, 126
  interference reflection microscopy for, 121
  interleukin-2 receptors and, 124
  lateral diffusion coefficient of, 136
  ligand binding to, 122, 136
  molecular structural basis for, 124
  "outside-in signaling" for, 126–27, 139
  as proteins, 280
  PTB domains for, 126
  RGD peptides and, 122
  tensegrity and, 203
  three-D adhesions in, 129
integrin-mediated cell adhesions, 318
integrin-mediated mechanotransduction, 40–4, 41f, 72
  α-actinin, 121
  ECM composition, 43
  FAK in, 40
  flow conditioning for, 42, 43–44
  in focal adhesion complexes, 253–54
  GEF in, 42
  ligands and, 122f
  MAPK in, 42
  MCL in, 42
  PAK activation in, 44–45
  reduced adhesion sites in, 40
  RGD peptides and, 122
  spatiotemporal displacement patterns in, 40
interference reflection microscopy, 121
interfragmentary strain theory, 406–7
interleukin-2 receptors, 124
intracellular adhesion molecule-1 (ICAM-1), 307
intracellular contraction, 12
intracellular loading, 230
intracellular traction, 288
in vitro tests, of mechanotransduction
  with AFM, 5
  arterial hemodynamics in, 46f
  with bead forcing, 5
  in endothelium, 45–47
  hydrostatic pressure in, 6
  of mechanosensation, 3–6, 4f
  with shear stress, 3–5
  with substrate stretch, 5–6
in vivo phenotypes, of mechanotransduction,
  arterial hemodynamics in, 46f
  in endothelium, 45–47
  immunocytochemistry and, 45
ion channels, 161–74
  activation of, 21, 33, 166–68
  ankyrin in, 171
  ATP release in, 169
  basal cell surfaces in, 161
  cell membranes and, 161, 166
  cell surface integrins in, 161
  cellular cytoskeletal network in, 161
  Cl^-, 164–66
  direct activation of, 169–71, 170f
  in ECs, 161, 164, 172–74
  in eukaryotic cells, 171
  flow mechanisms of, 168–74, 170f
  flow-mediated vasoregulation of, 161
  flow-sensitive, 161, 163–64
  gated, 234–35
  glycocalyx in, 171
  GPCR in, 161
  G-proteins and, 161
  indirect activation of, 169, 170f
  inhibition data for, 166–67
  intracellular calcium oscillations in, 162
  intracellular space and, 162
  K^+, 164
  MAPK induction in, 162

mechanosensitive, 162–66
MscL, 171
oscillatory flow in, 168, 172–74
PECAM-1 and, 161
SA, 198
in SMCs, 174
steady flow in, 168, 173f
stretch-sensitive, 163
vascular wall remodeling in, 161
IRF. *See* instrument response function
isotropic cell spreading, 182

karyopherins, 418
in nucleocytoplasmic transport, 422, 426
katanin, 243
key force-sensing/bearing proteins, structural features of, 280t
kinesin, 237
kinetochores, 238
K^+ ion channels, 164
KLF2. *See* Kruppel-like Factor-2
Kramers reaction rate theory, 96
Kruppel-like Factor-2 (KLF2), 48, 269
shear stress–induced changes in, 270f

lamellar ruffling, 140
lamellipodia, 37–38
actins and, 190
cell spreading in, 182
extension of, 190
fibronectin and, 189
microtubular mechanotransduction in, 239
periodic contractions in, 189–90
laminin, 121
lamins
in cell nuclei, 226
in NPCs, 432
in nucleocytoplasmic transport, 432
tensegrity in, 210
large unilamellar vesicles (LUVs), 99
lateral intercellular space (LIS), 342, 344, 346
ligand density and, 348–49, 352–54
LDL. *See* low-density lipoprotein
left descending anterior coronary artery (cLAD)
EC length in, 371f
SPA and, 370–71
leucine-rich repeats (LRR), 308
leukocytes, VCAM-1 proteins and, 307
ligand(s), 214
ligand binding
in disulfide bonds, 296–97
protein conformational change and, 273, 323
ligand density
antibodies and, 135
autocrine signaling and, 344f, 346, 347–49
cell migration and, 135
cellular mechanotransduction regulation by, 134–39
convection with, 356
diffusion and, 356
diffusion of, 351f
ECM and, 133–44
in EGF receptor, 353f
FE method for, 349
during haptotaxis, 134f
integrin binding to, 122, 136
kinetics of, 351f
LIS and, 347–49, 352–53
maximum rate of change for, 352f
motility control by, 135, 138
nonperturbing antibodies and, 137
regulation of, 137f
RGD peptides and, 122
three-dimensionality of ECM and, 148
lipid bilayers, 91–101
caveolae in, 101
cholesterol in, 93
CMF in, 99–100
composition of, 97
diffusion rates for, 93
FCS for, 94, 95f, 96
FRAP and, 94, 96
free volume with, 97–98
GP in, 99
GUVs in, 97, 98f
HUVECs and, 94, 95f
hydrophilic headgroups in, 98
hydrophobic matching in, 92
Kramers reaction rate theory, 96
lateral fluidity of, 93–96
lateral stretching in, 92
LUVs in, 99
mechanical properties of, 91–93
membrane fluidity and, 97f, 281–82
membrane heterogeneity in, 100–1
membrane protein function in, 97

lipid bilayers *(cont.)*
  membrane tension between, 92–93, 99–100
  molecular rotors in, 94
  nonlamellar prone lipids in, 101
  perturbation in, 93
  polarity in, 98–99
  raft domains in, 100
  shear stress in, 92
  structure of, 91–93
  SUVs in, 98
  thickness of, 92
  transmembrane pressure profile in, 91, 91f
lipid diffusion, within ECs, 74–75
lipid domains, 68–69
  BAECs in, 69, 70f
  cholesterol in, 69
  CT-B in, 69
  membrane perturbation of, 71–72
  membrane tension effects on, 74–75
  rafts of, 68
  shear stress in, 71f
  wound closure and, 72
lipid molecules, 65
  fluidity of, 65, 67
  G-proteins and, 67–68
  hydrophilic portion of, 66
  prestresses in, 207
lipid perturbation, measurement tools for, 75–79
  FCS, 75–76
  FLIM, 79, 80f
  fluorescence resonance energy transfer, 81
  TCSPC, 75
lipid rafts, 68
LIS. *See* lateral intercellular space
low-density lipoprotein (LDL), 308
LRR. *See* leucine-rich repeats
lungs
  autocrine signaling in, 340f, 341
  mechanical forces and, 287, 377
  tensegrity in, 209
LUVs. *See* large unilamellar vesicles

magnesium, 126
magnetic beads, tensegrity and, 204
magnetic bead twisting cytometry, 32
magnetic tweezers, 385–86
magnetic twisting cytometry, 384–85
malaria, 15
manganese, 126, 127
MAPK. *See* mitogen-activated protein kinase
matrix metalloproteinases (MMPs), 150
MD. *See* molecular dynamics
mechanical forces, 286–89
  with AFM, 380–82, 381f
  blood flow and, 287
  bone loading as, 287
  on cells, 269–70, 407f
  on ECM, 270, 377
  examples of, in tissues, 404f
  focal adhesion complexes and, 287, 288f
  focal contracts and, 288
  intracellular traction as, 288
  lung motion as, 287, 377
  with magnetic tweezers, 385–86
  with magnetic twisting cytometry, 384–85
  with microneedles, 379–80, 379f
  with optical tweezers, 382–83
  protein unfolding from, 289
  traction as, 288
  urine flow and, 287
mechanobiology, 404–5
  bone remodeling in, 404–5
  definition of, 61
  tissue remodeling in, 405–8
mechanome, 15
mechanosensation,
  design criterion for, 12
  in endothelium, 161
  in G-protein mechanotransduction, 101–5
  in vitro tests of, 3–6, 4f
mechanosensing, in cells, 181–91, 442
  cell spreading in, 182
  continuous protrusion in, 188–89
  cytoskeletal organization in, 181–84
  hierarchy of, 183
  motility modules in, 181–82
  motility phenotypes in, 181
  nonlinearity in, 183
  periodic lamellipodial contractions in, 189–90
  properties of, 184–91
mechanosensitive channels of large conductance (MscL), 8, 14
  as flow-sensitive, 171
mechanosensors, 7
  CSK, 90

in cytoskeletal transduction, 10
definition of, 61
in endothelial mechanotransduction, 25
membranes as, 61, 64–75
mechanostat theory, 406
mechanotransducers, 21
mechanotransduction, cellular. *See also* *in vitro* tests, of mechanotransduction; microtubules, cellular mechanotransduction in
advances in study of, 439–43
through autocrine signaling in, 339–56
breaking in, 234–44
cell–cell junctional proteins in, 63
cell force levels in, 12–15, 13t
cell mechanosensing, 181–91
computational tools for, 439–41
cytoskeletal transduction in, 10
definition of, 1
design criterion for, 12
in disease, 2–3
disturbed flow in, 2
ECM and, 120–51
in endothelium, 20–52, 62
external forces for, 386–87
in focal adhesion complexes, 7f, 9–10, 247–61, 378
force-bearing protein networks and, 287
gene expression in, 10–11
glycocalyx in, 10
G-protein activation in, 89–109
G-proteins in, 8–9, 33–34
hearing and, 2
hemodynamic shear stress and, 1
historical development of, 1–2
implications of, 211–13
intracellular contraction in, 12
ion channels in, 161–74
ligand density regulation of, 134–39
MD for, 439
mechanisms of, 62–63
mechanomes and, 15
mechanosensors in, 7
in membranes, 8–9
membrane tensions during, 13
microscale force techniques for, 377–96
Monte Carlo methods for, 439
multimodularity in, 286–328
nanoscale force techniques for, 377–96
NPC regulation of, 431–35
nuclear mechanics and, 220–30
in nuclear pore mechanics, 417–36
nucleocytoplasmic transport regulation of, 417–36
plasma membrane in, 61–81
protein conformational change and, 269–82
protein domain hinge motion and, 278
protein structural basis for, 279–81
protein unfolding in, 11
proteoglycans in, 63
reversible expansion in, 212
at single molecule scale, 14
soft tissue remodeling and, 1
solvent treatment in, 440–41
stress generation in, 2
stretch-activated channels in, 7–8
tensegrity in, 31–32, 196–214
TFM in, 12
tissue formation and, 213
vascular physiology and, 62
mechanotransmission, 23, 26–27
melanoma cell adhesion molecules (Mel-CAM), 296
melanoma cells, 127
Mel-CAM. *See* melanoma cell adhesion molecules
membrane heterogeneity, in lipid bilayers, 100–1
phases of, 100
membrane mechanotransduction, 8–9
G-proteins in, 8–9
in MAPK activation, 9
stress in, 9
membrane perturbation, of lipid domains, 71–72
ERK and, 72
FRAP and, 71
membrane proteins, 107
rhodopsin, 101, 107
transducin, 107
mesenchymal stem cells (MSCs), 403
differentiation in, 413
tension strain on, 411, 413
microbeads
fibronectin, 252
magnetic, tensegrity and, 204
in optical tweezers, 383

microcantilevers, 392–95
  Euler-Bernoulli beam theory and, 392
  horizontal, 394
  with micropost arrays, 393–94
microfibrillar network, 130
microfilament polymers, prestresses in, 210
microneedles, 379–80, 379f
micro-Newton forces, 387
microscale force techniques, for cellular mechanotransduction, 377–96
  micro-Newton forces, 387
microtubule networks, 27, 30
  nuclear mechanics in, 228
  tensegrity in, 207
microtubules, cellular mechanotransduction in, 234–44
  actin filaments in, 241
  APC and, 239
  aPKC and, 239
  bending of, 242–43
  breaking of, 243–44
  as cell transport, 237–38
  collagen and, 234
  deformation forces in, 240–44
  depolymerization of, 243–44
  direct mechanical stimulus in, 234
  disruption of, 236
  dynamic instability in, 238–40
  in ECs, 234
  in eukaryotic cells, 236
  Euler Bucking model for, 237
  externally applied force on, 241–42
  GEF and, 244
  katanin and, 243
  kinesin in, 237
  kinetochores in, 238
  lamellipodium protrusion in, 239
  molecular motors in, 241
  myosin motors in, 241
  nerve growth cone advance in, 238
  PDGF and, 244
  plus-end binding proteins in, 239
  polarity, 237
  polymerization forces in, 240, 242
  roles of, 235–40
  signaling cascades in, 239
  signaling in, 238–40
  stochastic switching in, 238
  structure of, 235–37
  thermal fluctuations in, 240
  "whipped" structure in, 236
mitogen-activated protein kinase (MAPK)
  autocrine signaling through, 342
  in integrin-mediated mechanotransduction, 42
  in ion channels, 162
  in membrane mechanotransduction, 9
mitosis
  blebbing motility and, 187
  kinetochores during, 238
  nuclear mechanics and, 226–27
MLC. *See* myosin light chain
MLCK. *See* myosin light-chain kinase
MMPs. *See* matrix metalloproteinases
molecular dynamics (MD), 254
  for autocrine signaling, 348
  for cellular mechanotransduction, 439–40
  for protein conformational change, 282
  SMD, 294
molecular motors, 241
molecular rotors, 94
molecule force spectroscopy, 263
monocytes, 2
Monte Carlo methods, 439
motility modules, 181–82
  blebbing, 187–88
  elements of, 190–91
  filopodial, 186–87
  global coordination, 183
  hierarchy of, 183
  myosin proteins and, 182
  protein coding sequences in, 183
  schematic for, 184f
  types of, 184–85, 185f
motor proteins, 121
  myosin superfamily in, 130
MRLC. *See* myosin regulatory light-chains
MscL. *See* mechanosensitive channels of large conductance
MSCs. *See* mesenchymal stem cells
multimodularity, in cellular mechanotransduction, 286–328
  in eukaryotic cells, 289
  module stability and, 298
  protein features as result of, 289–94
multimodular proteins, 38, 289–94
  α-actinin, 121, 310, 311f, 312–14, 315
  ankyrin, 315–16

ankyrin repeats in, 291–92
calmodulin, 294–95
common structural modules in, 293t
cytoplasmic domain in, 318
FERM domains in, 318
fibronectins in, 291f, 308, 309f, 310
filamin, 316
force-sensitive signals in, 321
FRET for, 291f, 326, 328
GFP, 38, 294
Ig-like domains in, 291
integrin-mediated cell adhesions and, 318
"low sequence identities" in, 299
mechanical stability of, 298–99, 317, 327f
modular structure/binding sites of, 307f
PC1, 308, 309f, 310
p130Cas, 290f, 303–4
PTB domains in, 318
response to force, 294–95
shearing of, 295
SMD for, 294, 325
β-strands, 290–91, 312
thermodynamic stability of, 294
titin, 280, 304, 307
unfolding of, 325f
unzipping of, 294–95
VCAM-1, 293, 307–8
WLCs and, 315
Murray's Law, 1, 20
Murzin, Alexey, G., 279
muscle cells, 269
SMCs, ECs and, 174
tensegrity in, 209
titin proteins in, 280
musculoskeletal system, impulsive forces on, 377
myosin light chain (MLC), 42
MRLC, 187
myosin light-chain kinase (MLCK), 130
myosin motors, in microtubular mechanotransduction, 241
myosin proteins, 130
actins and, 190
blebbing motility and, 187
blebbistatin for, 139, 144
contractions for, 143
in ECM, 251
ELCs, 130
filopodial motility and, 186
MLCK, 130
motility modules and, 182
protein domain hinge motion and, 277
regulation of, 130
myosin regulatory light-chains (MRLC), 187

nanoscale force techniques, for cellular mechanotransduction, 377–96
nerve growth cone advance, 238
neutrophiles, shear stress and, 105
nitric oxide (NO), 35
in endothelium, 90
in G-protein mechanotransduction, 107–8
with SPA, 367
NO. *See* nitric oxide
nonlamellar prone lipids, 101
NPCs. *See* nuclear pore complexes in
nuclear envelope, 223f
deformability of, 224f
lamina in, 226
nuclear lamins, 210
nuclear mechanics, 220–30
actin filaments and, 228
adhering cells in, 225, 228
in biochemistry, 220
cell measurements in, 222
chromatins in, 225
at chromosome level, 227
in cytoplasmic cell region, 223–24, 230
deformations in, 226
disease and, 227–28
double-membrane envelope in, 223
ECs and, 220
force–deformation relationship in, 221
future directions for, 228, 230
GFP-LamA in, 224f
intracellular loading in, 230
intrinsic strain in, 225
lamina in, 226
macroscopic properties in, 222–24
in microtubules, 228
mitosis and, 226–27
morphological change in, 225
nuclear envelope in, 223f
organelles in, 230
properties in, 221
signaling pathways in, 220
strain state in, 225
subnuclear components in, 225–27

nuclear mechanics *(cont.)*
tensegrity model in, 225
viral infections and, 226
nuclear pore complexes (NPCs), 417–20, 422
anatomy of, 419f
cellular mechanotransduction regulation by, 431–35
as control feedback system, 425f, 434
FG repeats in, 417–18
importins in, 418
karyopherins in, 418
lamins and, 432
mode shapes for, 424f
nucleocytoplasmic transport and, 418, 427–31
plane deforming in, 421f
stress generation within, 421f
tensegrity in, 210
in yeast/vertebrates, 420f
nuclear pore mechanics, 417–36
nucleated cells, 203f
nucleocytoplasmic transport, 417–36
cellular mechanotransduction regulation by, 417–36
control feedback system for, 425f
FG repeats in, 425–26, 430–31
importins in, 423, 425–26
karyopherins in, 422, 426
modes of, 429f
molecular biology of, 422–23, 425–27
NPCs and, 418, 427–31
simulation of, 427f
stages of, 426–27
structure of, 423f
nucleus, in cells,
adhering cells in, 225
intrinsic strain in, 225
isolation of, 222–23
lamina in, 226
measurements for, 222
morphological change in, 225
strain state of, 225

optical tweezers, 286, 382–83
microbeads in, 383
organelles, 230
oscillatory flow
in ion channels, 168, 172–74
in WSS, 366
osteoblasts, 405
osteoclasts, 405
"outside-in signaling," 127–29, 139
tyrosine phosphorylation and, 127

PAK. *See* p-21 activated kinase
paxillin molecules, 128
filopodial motility and, 186
three-dimensionality of ECM and, 146–47
PC1 protein. *See* polycystin-1 protein
PDGF. *See* platelet-derived growth factor
PDMS. *See* pre-stretched polydimethysiloxane
PECAM-1. *See* platelet-endothelial cell adhesion molecule-1
periodic contractions, in lamellipodia, 189–90
collapsing, 189
leading edge extension during, 189
transient extraction during, 189
permeabilized cells, 204
phenotypes, in endothelium mechanotransduction, 45–47
atherogenic, 62
atheroprotective, 47
phenylalanine-glycine (FG) repeats, 417–18
chains for, 428f
in nucleocytoplasmic transport, 425–26, 430–31
phospholipase C (PLC), 188
phospholipids, in cell membranes, 65
phosphotyrosine binding (PTB) domains, 126
FERM domains and, 126
in multimodular proteins, 318
PHSRN peptides. *See* pro-his-ser-arg-asn peptides
PKC. *See* protein kinase C
PKD. *See* polycystic kidney disease
plasma membrane, in cellular mechanotransduction, 61–81
ECs in, 61
homeostasis in, 61
mechanosensors in, 61
platelet-derived growth factor (PDGF), 244
platelet-endothelial cell adhesion molecule-1 (PECAM-1), 24, 44
ion channels and, 161
PLC. *See* phospholipase C
Plexiglas, 236
plus-end binding proteins, 239

Poisson's ratio, 140
polarity, in lipid bilayers, 98–99
polycystic kidney disease (PKD), 2
  modules for, 299
  PC1 and, 308, 310
polycystin-1 (PC1) protein, 308, 310
  LDL and, 308
  LRR and, 308
  modular structure/binding sites for, 309f
  PKD and, 308, 310
  TRP channels, 308
  WSC and, 308
polycystins, 213
  PC1, 308, 309f, 310
p130Cas proteins, 290f, 303–4
  scaffolding molecules in, 303
  structure/binding sites of, 305f
  Triton X and, 303
prestresses, tensegrity and, 197, 199, 205f, 206, 210, 213
  computer models for, 201f
  in cytoskeletal proteins, 205f, 206
  in lipids, 210
  in microfilament polymers, 210
  in viruses, 210
pre-stretched polydimethysiloxane (PDMS), 302
primary cilium, 22, 26, 36
  TRP channels in, 36
pro-his-ser-arg-asn (PHSRN) peptides, 124
protein(s). *See also* G-proteins, in cellular mechanotransduction; membrane proteins; motor proteins; multimodular proteins; myosin proteins; unfolding, of proteins
  adhesion-complex, 279–80
  amino acids, 274
  ankyrin, 316–17
  biomolecule transport in, 270
  catalytic functions of, 270
  cell-cell junctional, 63
  "clamped" regions in, 295
  coat, 42
  conformational change of, 269–82
  constitutive models of, 276–77
  cytoskeletal, ECM v., 120–21
  deformation modes in, 271–72, 272f
  disulfide bonds and, 295–97
  domains, 271–72
  features of, from cellular mechanotransduction, 289–94
  fibronectin, 280, 299–300, 302–3
  filamin, 316
  filopodial motility and, 186
  force-bearing networks of, 287
  force sensing, 254–59, 261
  genetic information transmission in, 270
  GPI-anchored, 34
  integrins, 280
  interactions with other proteins, 275
  key force-sensing/bearing, structural features of, 280t
  lipid fluidity in, 67
  mechanical stability of, 294
  mechanics of, 275–81
  metabolic functions of, 270
  in motility modules, coding sequences for, 183
  multimodular, 289–94
  PC1, 308, 309f, 310
  plus-end binding, 239
  p130Cas, 290f, 303–4
  Rho, 251–52
  SCOP for, 279
  signal transduction in, 270
  single molecule mechanoresponse techniques for, 323–24
  structural features of, 271f, 292f
  structural support within, 270
  titin, 280, 304, 307
  unfolding, 289
  VCAM-1, 293, 307–8
protein conformational change, 269–82
  with AFM, 303–24
  amino acids in, 274
  biological consequences of, 270–75
  domain deformation as, 271–72, 272f
  domain motion as, 271
  domain unfolding as, 271
  ECM and, 269
  ECs and, 269
  experimental approaches to, 320–26, 328
  fluorescence approaches to, 325–26, 328
  force-induced, 275
  FRET for, 282, 323
  GFP and, 326–27, 328
  ligand binding and, 273, 323
  MD simulations for, 282

protein conformational change *(cont.)*
mechanical forces responses by, 269–70
modeling for, 281–82
population stretching tools for, 321–22
single molecule biomechanics studies for, 281–82
SMD and, 324–25
spatial/temporal regulation in, 275
vWF as, 273, 280
protein domain elongation, 272
protein domain hinge motion, 271–72, 277–78
mechanochemical coupling and, 278
mechanotransduction and, 278
myosin proteins and, 277
receptor–ligand binding and, 278–79
protein domain slide motion, 272
protein inclusions, membrane tension effects on, 74–75
protein kinase C (PKC), 47–48
atypical, 239
protein unfolding, 11
computational methods for, 11
proteoglycans, in cellular mechanotransduction, 63
protrusive forces, in ECM, 129
PTB domains. *See* phosphotyrosine binding domains
p-21 activated kinase (PAK), 44–45
pulmonary disease, 2

qTIRFM. *See* quantitative total Internet reflection fluorescence microscopy
quantitative total Internet reflection fluorescence microscopy (qTIRFM), 73

raft domains, 100–1
receptor–ligand binding, 278–79
red blood cells, CMF in, 100
reversible expansion, in mechanotransduction, 212
RGD peptides,
β-actinin and, 122
in ECM, 438
integrin binding and, 122
PHSRN, 124
rhodopsin, 101
Rho proteins, 251–52

Saffman hydrodynamic theory, 100
SA ion channels. *See* stress-activated ion channels
SCOP. *See* Structural Classification of Proteins
SED. *See* strain energy density
shear modulus, 140
shear stress
BAECs and, 89
EC wound closure and, 72
on endothelial mechanotransduction, 21f, 33–36, 44–45
on endothelium, 89, 105
finite element analysis for, 73–74
focal adhesion and, 74f
glycocalyx in, 5
in GPCR, 102
in lipid bilayers, 92
in lipid domains, 71f
on membranes, 73–74
neutrophiles and, 105
quantification of, in membranes, 73–74
in vitro tests with, 3–5
VRAC and, 33
signal amplification cascades, 102
signaling pathways
dynamic instability and, 238–40
in microtubular mechanotransduction, 238–40
in nuclear mechanics, 220
sinusoidal input functions, 29
small unilamellar vesicles (SUVs), 98
SMCs. *See* smooth muscle cells
SMD. *See* steered molecular dynamics
smoking, 89
smooth muscle cells (SMCs), 174
phenotype of, 405
tension strain on, 410
soft glassy material models, 32
soft tissue remodeling, 1
solvents, 440–41
SPA. *See* stress phase angle
spatiotemporal displacement patterns, 40
spectrin repeats, 310, 315
sphingolipids, 101
steady flow, in ion channels, 168, 173f
steered molecular dynamics (SMD), 294
for α-actinin, 313, 315f
for multimodular proteins, 294, 324

protein conformational change and, 324–25
unfolding proteins and, 326f
stem cells, 269
compression effects on, 408–10, 411f, 412–13
compression strain devices for, 409f
in ECM, 143
EPCs, 403
HSCs, 403
mechanical regulation of, 403–14
MSCs, 403
tensile strain devices for, 409f
tension effects on, 408–10, 411f, 412–13
stereocilia of hair cells, in stretch-activated channels, 8
"stick and string" tensegrity model, 208
stochastic switching, 238
strain energy density (SED), 373
β-strands, 290–91, 312
stress-activated (SA) ion channels, 198
mechanical properties of, 211
stresses
in membrane mechanotransduction, 9
within NPCs, 421f
"stress fibers", 206
stress phase angle (SPA)
cLAD and, 370–71
CS/WSS and, 361–74
EECP and, 372
eNOS with, 369
influence factors for, 362–63, 365
NO release with, 367
studies with, 366–67
variable, 367–72
stretch-activated channels, 7–8
conductance changes in, 8
MscL, 8, 14
in stereocilia of hair cells, 8
stretch-sensitive ion channels, 163
Structural Classification of Proteins (SCOP), 279
struts, compression, 207
substrate stretch tests, 5–6
biaxial, 5
uniaxial, 5
SUVs. *See* small unilamellar vesicles
talin
activation of, 262
binding mechanism for, 257, 259, 261, 264f, 320
in focal adhesion complexes, 254–56, 255f
force-induced activation of, 256–59
hydrophobic residues in, 261
under molecule force spectroscopy, 263
solvent effects for, 258–59
structure of, 256, 257f, 259, 260f
VBS in, 256–59, 261, 262f
wild-type, 263
TCSPC microscopy. *See* time-correlated single photon counting microscopy
tensegrity, in cellular mechanotransduction, 31–32, 196–214
architecture of, 199–202
artistic representations of, 200f
in blood cells, 209
in cartilage, 209
cell metabolism and, 197
cells and, 202–6
cell stiffness and, 204
cytoskeletal proteins and, 197, 209–10
definition of, 199
in ECM, 197, 203, 208–9
in epithelial tissues, 209
exogenous mechanical loads and, 196
focal adhesion complexes and, 203
FRAP and, 206
in GFP, 207
hierarchy in, 201f, 208–11
integrins and, 203
ligands and, 204
in lungs, 209
magnetic beads and, 204
in microtubules, 207
at molecular level, 206–8
in musculoskeletal system, 209
in nuclear lamins, 210
in nuclear mechanics, 225
in nuclear pore complexes, 210
in nucleated cells, 203f
permeabilized cells and, 204
prestresses and, 197, 199, 205f, 206, 210, 213
rheological measurements and, 205
SA ion channels and, 198

tensegrity, in cellular mechanotransduction *(cont.)*
at single cell level, 201, 203
spherical, 202f
"stick and string" model for, 208
"stress fibers" and, 206
structural rigidity and, 200
in subcellular structures, 210
three-dimensional mapping technique for, 204
transduction molecules and, 198
tension field theory, 33
tension strain, on stem cells, 408–10, 412–13
devices for, 409f
for ESCs, 410
MSCs, 410, 413
SMCs, 410
TFM. *See* traction force microscopy
thin film buckling, 388
three-dimensionality of ECM, 144–50
cell migration and, 146, 149f
cell shape and, 144–46
cellular mechanosensing and, 150
computational models of, 148–49
FAK and, 146–48
fibers in, 148
fibroblasts and, 147
ligand density and, 148
matrix density and, 150
mechanical tension in, 148
paxillin molecules and, 146–48
three-dimensional adhesions, 129
thrombosis, 269
time-correlated single photon counting (TCSPC) microscopy, 75
optical set-up for, 77f
tissue remodeling, 405–8
titin proteins, 280, 304, 307
modular structure/binding sites of, 306f
towed growth, in cells, 234
traction force microscopy (TFM), 12, 389f
for elastic substrata, 389
traction forces
BDM and, 133
cell production of, 131, 132f
in cytoskeleton, 129–31
EC generation of, 133
in ECM, 129–33, 139
elasticity and, 133
elastic substrata in, 389–92
FAK and, 392
GFP for, 391
intracellular, 288
localized, 132
measurement of, 387
as mechanical force, 288
with microcantilevers, 392–95
microdot arrays for, 391f
pinching force, 131
TFM, 12, 389f
thin film buckling and, 388
wrinkling silicone membranes and, 387–89
transducin, 107
transduction molecules, 198
transient receptor potential (TRP) channels, 36
in flow-sensitive ion channels, 163
HAECs in, 163
PC1 proteins and, 308
Triton, X, 9
p130Cas protein and, 303
TRP channels. *See* transient receptor potential channels
tumor cells, migration of, 187
tweezers. *See* magnetic tweezers; optical tweezers
two-photon microscopy, 343f, 354
2,3-butanedione monoxime (BDM), 252, 253
tyrosine phosphorylation, 127

unfolding, of proteins, 289
ciliary vibrations and, 289
cysteines and, 289
in eukaryotic cells, 289
in fibronectin, 300, 302
mechanical stability in, 296
in multimodular proteins, 325f
peak-forces on, 296t
SMD and, 326f
unilamellar vesicles, FCS for, 76–78

vascular cell adhesion molecule-1 (VCAM-1) proteins, 293, 307–8
disulfide bonds and, 295, 297f
ICAM-1, 307
leukocytes and, 307
modular structure/binding sites of, 307

vascular physiology
  cellular mechanotransduction and, 62
  glycocalyx in, 69–70
  membranes in, as mechanosensors, 64–75
  tensegrity in, 209
vasoactive peptides, 104
VBS. *See* vinculin binding sites
VCAM-1. *See* vascular cell adhesion molecule-1 proteins
vinculin
  activation of, 258f, 259, 262
  binding mechanism of, 259, 261, 264f
  in focal adhesion complexes, 254–55
  hydrophobic residues in, 261
  under molecule force spectroscopy, 263
  structure of, 259, 263
  VBS and, 256–59
vinculin binding sites (VBS), 256–59, 261
  activation of, 258f, 259
  binding simulation of, 263
viral infections, 226
viruses, prestresses in, 210
viscoelastic models, 30, 31t, 32
viscous modulus, 30
vitronectin, 134
volume-regulated anion current (VRAC), 33
  in Cl^- ion channels, 166
von Willebrand Factor (vWF), 273, 280
VRAC. *See* volume-regulated anion current
vWF. *See* von Willebrand Factor
wall integrity and stress component (WSC), 308
wall shear stress (WSS), 360–74
  carotid bifurcation model for, 364f
  in ECs, 361f, 365–72
  influencing factors for, 362–63, 365
  intercellular junctions and, 373
  oscillatory flow in, 366
  physiological background of, 360–65
  SED and, 373
  simultaneous stretch/shear methods in, 365–66
  SPA and, 361–74
  stretch/shear mechanism for, 372–73
  tension in, 373
WAVE1, 189
wild-type talin, 263
WLCs. *See* worm-like chains
Wolff's Law, 1, 406
worm-like chains (WLCs), 315
wound healing
  ECM and, 134
  lipid fluidity and, 72
wrinkling silicone membranes, 387–89
WSC. *See* wall integrity and stress component
WSS. *See* wall shear stress

Zamir's Law, 20

Printed in the United States
By Bookmasters